Edited by Marvalee H. Wake

# Hyman's Comparative Vertebrate Anatomy

*Third Edition*

The University of Chicago Press
*Chicago and London*

The University of Chicago Press, Chicago 60637
The University of Chicago Press, Ltd., London

© 1922, 1942, 1979 by The University of Chicago
All rights reserved. Published 1979
Paperback edition 1992
Printed in the United States of America
07 06 05 04 03 02 01 00 99 98      4 5 6 7 8

Library of Congress Cataloging in Publication Data
Hyman, Libbie Henrietta, 1888–1969.
    [Comparative vertebrate anatomy]
    Hyman's comparative vertebrate anatomy. — 3rd ed. / edited by
Marvalee H. Wake.
       p.    cm.
    Rev. ed. of: Comparative vertebrate anatomy. 2nd ed. 1942.
    Third ed. originally published: 1979.
    Includes bibliographical references and index.
    ISBN 0-226-87013-8
    1. Anatomy, Comparative.  2. Vertebrates—Anatomy.  I. Wake,
Marvalee H.  II. Title.  III. Title: Comparative vertebrate anatomy.
QL805.H99  1992
596´.04—dc20                                           92-18372
                                                            CIP

⊚ The paper used in this publication meets the minimum requirements of the
American National Standard for Information Sciences—Permanence of Paper
for Printed Library Materials, ANSI Z39.48–1992.

# Contents

# Preface to the Third Edition

Several generations of zoology students have used Libbie Hyman's *Comparative Vertebrate Anatomy* as their guide to the study of vertebrate structure and evolution. Some of those students have been stimulated by that first exposure to pursue further studies of vertebrate biology, and several of the contributors to this edition are such former students. When the University of Chicago Press first approached some of us in 1968 about revising *Comparative Vertebrate Anatomy,* Libbie Hyman was living quietly in New York, retired from the American Museum of Natural History. She recognized the need to revise the book but was unable to do so herself. At her suggestion, the Press asked workers in several areas of vertebrate morphology to revise chapters in the book, and they accepted such a revision, especially one incorporating new research and concepts in functional morphology and evolution, as a challenging project. Libbie Hyman has since died, but we have continued our work in the belief that "Hyman" is a unique study guide that should be perpetuated because of its usefulness to both students and teachers.

Each of us has written chapters that fall broadly within our special research interests and expertise. We are all "evolutionary morphologists"; some of us are paleontologists, developmentalists, physiologists, and ecologists as well. As a result these chapters present concepts of structure and function in a broad evolutionary context. Further, each chapter shows an expert's interest in and enthusiasm for the material he or she discusses. Editing has been done so as to retain that individual flavor while assuring that format, terminology, and technique are reasonably consistent from chapter to chapter. Much of our work has been in the text of each chapter, and these sections have been rewritten so that current concepts, including much new information and analysis, are clearly presented. The laboratory directions have also been revised by checking actual dissections and updating terminology and information wherever necessary. One of the great strengths of "Hyman" is its incorporation of extensive dissection instructions with the text of the book. We have continued this format, though the dissection directions form a block at the ends of the "wet" chapters—those on the muscular, visceral, and nervous systems. I hope this allows even more practical treatment of the book—grease and formalin should be somewhat less likely to obscure the text. Literature lists for each chapter are updated and include references to the original literature on specific topics as well as to review papers and books. Many new figures are included in this edition, ranging from drawings from fresh dissections that enhance the laboratory directions to transmission and scanning electron microscope photos chosen to illustrate the text. Different margins have been used to distinguish text and laboratory directions and to improve readability.

All of us saw as our major challenge the presentation of modern, dynamic concepts of vertebrate morphology in a functional and evolutionary framework, within the useful "Hyman" format. We want this book to guide the student's consideration of vertebrate morphology, as we present current ideas and problems with an emphasis on the student's use of his eyes and hands on a variety of real specimens. We hope that students will thereby be stimulated to understand patterns of vertebrate evolution rather than simply to memorize lists of terms or names of structures. Each of us is excited and challenged by the dynamism of vertebrate biology; we hope we have conveyed the essence of this dynamism to students. In the preface to the 1942 edition, Libbie Hyman decried the fact that "all young biologists want to be experimentalists, and hardly anyone can be found interested in the fields of descriptive embryology and anatomy." Today, thirty-odd years later, there appears to be a resurgence of interest in vertebrate morphology because of new approaches, tools, and techniques and a new appreciation for morphology as an integrative, rather than a purely descriptive, science. We hope our efforts to present these ideas will further encourage students to investigate the problems of vertebrate evolutionary morphology.

If students and teachers compare the prefaces to the 1921 and 1942 editions with this one they will sense the evolution of the science. Especially evident is the utility of Libbie Hyman's approach, which we accept as our model for this book. To paraphrase her 1942 preface, our task might be described as making molehills out of mountains, culling from the immense body of material and literature on vertebrate anatomy a few of the more important facts and concepts. We hope that we have not done too badly by these ideas—or by Libbie Hyman.

I wish to thank many people for their efforts toward the completion of this edition. First and foremost—the contributors. This work was a long time a-borning, and some contributors required more cajoling than did others. I thank the authors for staying with the task and for willingly working on material at many months' intervals. All the authors had similar attitudes of responsibility toward the accuracy, evolutionary and functional approach, coverage, and level of information in their chapters and their work reflects their integrity toward their material. I believe that these qualities should augur well for the book. The usefulness of a comparative anatomy text is much enhanced by its illustrations. Each contributor prepared his own rough drafts; in fact, the final illustrations for chapter 5 were done by its author. Steven Gilbert did some of the illustrations in chapters 10, 11, and 12, and Emily Reid executed all the remaining illustrations—most of those in the volume—with her unusual skill, patience, imagination, and enthusiasm. The reviewers of the chapters provided many useful comments. The staff at the University of Chicago Press was supportive and sympathetic during the many vicissitudes of this venture. All have been patient and helpful during the prolonged gestation of this book. We hope it survives its birth pangs to have a life of

useful service. Each of the authors recognizes that we teach very differently and would use this book in various ways. We hope we have provided material within a context that allows for creative teaching, and we
welcome your comments and criticisms.

Marvalee H. Wake

# 1 General Considerations of Animal Form
## Marvalee H. Wake

**A. Introduction**

The purpose of this book is to introduce the morphology of vertebrates in a context that emphasizes a comparison of structure and of the function of structural units. The comparative method involves analysis of the history of structure in both developmental and evolutionary frameworks. The nature of adaptation is the key to this analysis. Adaptation of a species to its environment, as revealed by its structure, function, and reproductive success, is the product of mutation and natural selection—the process of evolution. The evolution of structure and function, then, is the theme of this book. Until the early 1950s, comparative morphology emphasized the evolution of structure alone. Since that time evolutionary morphologists have emphasized the history of change of function (i.e., evolution) as well as of structure, for the concept of adaptation stresses that structures change to provide functional advantage. This book, then, presents, system by system, the evolution of structure and function of vertebrates. Each chapter presents the major evolutionary trends of an organ system, with instructions for laboratory exploration of these trends included so the student can integrate concept with example.

**B. Major Considerations in Morphology**

There are several underlying problems or themes in morphology that can be analyzed in an evolutionary and functional context. These are alluded to in each chapter, though not always explicitly. The student should keep them in mind as he looks at each morphological unit. These themes were analyzed by Darwin and by his predecessors, and they continue to be discussed by his successors even now. They seem to have no real "answers" or solutions, but the student should be aware of them, for evolutionary morphologists are constantly gaining new insights into their meaning. The concepts of symmetry, segmentation, cephalization, homology and analogy, and primitive/generalized vs. advanced/specialized are examples of such recurring themes in morphology.

**1. Symmetry**

The form of symmetrical animals depends upon the arrangement of parts with regard to axes and planes (see part 7). There are four fundamental types of animal symmetry—*spherical, radial, biradial,* and *bilateral.* Since all vertebrates are bilaterally symmetrical, the other types of symmetry will not be considered here.

The parts of a bilaterally symmetrical animal are arranged symmetrically with reference to three axes—longitudinal, transverse, and sagittal;

the two ends of the sagittal axis in any given cross section are unlike. There is but one plane of symmetry in such an animal—that plane which passes through the longitudinal and sagittal axes—namely, the median sagittal plane. It divides the animal into approximately identical right and left halves that are mirror images of each other. The structures of vertebrates are either cut in half by the median sagittal plane, in which case they are spoken of as unpaired structures, or placed symmetrically on each side of this plane, equidistant from it, in which case they are called paired structures. The digestive tract is the only system that does not exhibit a symmetrical relation to the median plane in the adult. It too is bilaterally symmetrical in early embryonic stages, but adult structures in a variety of vertebrates may no longer be symmetrical because of functional constraints. Another example of secondary asymmetry is the developmental change from a symmetrical larva to an asymmetrical adult in the flatfish. The eyes, bilateral in larvae, shift so that both lie on the "dorsal" side in the adult.

## 2. Homology and Analogy

The relation between form and function has occupied the attention of comparative anatomists since the dawn of modern zoology at the end of the eighteenth century. This question involves the concepts of homology and analogy, which are to be understood in terms of the principle of evolution. Homology means intrinsic similarity that indicates a common evolutionary origin. Though homologous structures may seem unlike superficially, they can be proved to be equivalent by the following criteria: similarity of anatomical construction, similar topographical relations to the animal body, and similar courses of embryonic development. A familiar example of homology is the wing of the bird, the flipper of the seal, and the foreleg of the cat; investigation shows that they have a similar arrangement of bones and muscles, have the same position relative to the body, develop in the same way from a similar primordium, and work by the same physiological mechanism. The necessity of including specific physiological function or mechanism as a criterion will become apparent when one considers such structures as the endocrine glands. Although these have a similar anatomical (histological) structure throughout the vertebrates, they frequently differ in position and in the details of their embryonic origin; but their specific function, the ultimate test of their homology, remains the same throughout.

Analogy means similarity of general function or of superficial appearance not necessarily associated with similarity of intrinsic anatomical construction or embryonic origin and development. Thus fish and snakes are both covered with scales for protection (similar general function), but investigation shows that the two types of scales are histologically dissimilar and differ in their embryonic origin. Analogous structures may also differ in precise functional mechanism; thus an insect leg and a cat leg serve the same broad general function—walking—but the mechanism is quite different. When analogous structures look strikingly similar, this is termed *convergence* or *parallelism*. Such correspondences are usually associated with a common environment, that is, they are "adaptations." Whole animals may also come to resemble each other markedly through living in the same environment, as with porpoises and fish. On the other hand, animals closely related by descent may differ greatly in general appearance after long sojourn in different environments, as with seals and cats. This phenomenon is termed *divergence.*

Other workers consider that homology is restricted to structural similarity resulting from common ancestry. The term *homoplasy* is applied to structural similarity that is not the direct result of common ancestry or developmental site. Parallelism is *separate* evolution of structures of common ancestry; convergence is separate evolution of similar structures that do *not* share common ancestry. Both parallelism and convergence are homoplastic conditions. *Anaplasy* is the *functional* similarity of structures as a result of homoplasy in this system of usage.

## 3. Metamerism or Segmentation

*Metamerism,* or *segmentation,* is the regular repetition of body parts along the anteroposterior axis. The body of a segmented animal is composed of a longitudinal series of divisions in each of which all or most of the body systems are represented, either by entire paired organs, or parts, or by a portion of the median unpaired structures. Each such division of the body is termed as *metamere, segment,* or *somite.* The anterior and posterior boundaries of each segment may or may not be marked externally by a constriction of the body wall. In the former case the animal is said to exhibit both *external* and *internal* segmentation; in the latter case only internal metamerism is present.

An ideal segmented animal would consist of a series of identical segments; but no such animal exists, since the head and terminal segments always differ, if only slightly, from the other segments because of their special functions. However, the more primitive segmented worms, such as *Nereis,* closely approach the ideal. The segmentation of the animal body into nearly identical segments is spoken of as *homonomy.* Most segmented animals have *heteronomous* segmentation, in which the various segments differ to a greater or lesser extent. In the evolution of segmented animals there has been a continuous progression from the homonomous to the extreme heteronomous condition. Homonomous segmentation is a primitive and generalized state in which the various segments are more or less independent and capable of performing all necessary functions. As heteronomy progresses the segments become unlike, and different body regions specialize in different functions. Each segment is then no longer capable of carrying on all functions but becomes dependent upon the other segments, with a resulting unification and harmony of performance lacking in homonomously segmented forms. The heteronomous condition is derived from the homonomous through a variety of different processes, such as loss of segments, loss of organs or parts from some segments while they are retained in others, structural changes among the repeated organs or parts so that those of different segments become unlike, and so forth. Homonomy is, in fact, considered a special kind of homology by many workers—*serial homology.*

The groups of segmented animals are the annelids, the arthropods, and the chordates. Whereas relationship between annelids and arthropods is generally acknowledged, most zoologists are now of the opinion that the chordates stem from a quite different line of evolution and hence that in them segmentation has arisen independently of the metamerism of the annelid-arthropod line. Metamerism in the chordate line apparently began in the hemichordates, where some structures, particularly the gonads, show evidence of serial repetition, a condition termed *pseudometamerism.* Segmentation then progressed among the lower chordates, reaching its climax in *Branchiostoma,* and retrogressed among the vertebrates. It seems likely that the primitive chordates never had as fully developed segmentation as do annelids and that no chordate was ever externally segmented. Segmentation in chordates primarily concerns the musculature; that of the endoskeleton and nervous system seems to be secondary. Among vertebrates there is a rapid loss of segmentation, which can be followed to some extent during embryology. Embryonic stages are much more obviously segmented and more homonomous than adults, and heteronomy progresses during the development of a given vertebrate. Adult vertebrates are thus internally and heteronomously segmented.

## 4. Cephalization

In the evolution of animals there is a pronounced tendency for the anterior end of the body to become more and more distinctly separated and differentiated from the rest of the body as a *head.* This differentiation consists chiefly of the localization within the head of the main part of the nervous system—the brain—and of several sense organs. Since the brain and the sense organs control to a very large degree the activities and responses of the rest of the body, the head thus becomes the dominant part of the organism. This centralization or localization of nervous structures and functions, with accompanying dominance of the head, is called cephalization. Cephalization is more and more marked as one ascends in the animal kingdom and is particularly prominent as a structural and functional feature of the vertebrates. Associated with centralization of neural function is the elaboration of the feeding apparatus, especially mouth parts such as jaws, musculature, and sometimes teeth, beaks, tongues, and glands.

In segmented animals the advance in cephalization is correlated with the progression of the heteronomous condition. Heteronomy, in fact, appears first in the head region and gradually progresses posteriorly. The anterior end thus retains the least, and the posterior end the most, resemblance to the original homonomous condition. This produces the illusion of a retreat of certain systems toward the posterior regions of the body, whereas the situation actually arises because these systems have disappeared from the anterior segments and are retained in the posterior segments. In the case of certain vertebrate organs, such as the heart, a real posterior descent occurs as the vertebrates evolve. In the vertebrates, as in other heteronomously segmented animals, the head is produced when a certain number of the most anterior segments fuse, some segments or parts of segments are lost, and nearly all systems except the nervous system disappear from the head segments. As cephalization progresses, the head incorporates more and more of the adjacent segments into its structure, so that in general it may be said that the higher the degree of cephalization the greater is the number of segments composing the head. In advanced cephalization, such as in vertebrates, it is not easy to decipher the number and boundaries of the segments that originally went into the composition of the head; in fact, the problem of the segmentation of the vertebrate head has not been completely solved, although it has received the attention of the foremost vertebrate anatomists.

The vertebrates then, are characterized by bilateral symmetry, internal and markedly heteronomous segmentation, and a high degree of cephalization. The details of their structure are understandable only in the light of these three broad anatomical conditions.

5. Primitive and Generalized vs. Derived and Specialized

Primitive and derived are not synonyms for generalized and specialized. All these forms describe the character states seen in evolutionary trends in morphological structure or function or both. A character in a modern form is primitive if it is similar to that of the ancestors of the form or if it is shared by all living groups presumably related to the one under analysis. For example, the pentadactyl, or five-fingered state, is shared by fossil vertebrates and by the majority of all living vertebrates. Change in that five-fingered condition is considered a derivation, whether it results in reduction to the single digit of the horse or in the elongation of units to provide the support for the bat wing. The terms generalized and specialized refer to function, and to concomitant modification of structure. The five-fingered hand of man is generalized—it retains the ability to perform a variety of functions. The five-toed foot of man, however, is specialized—it is structurally modified to perform restricted functions, especially in bipedal locomotion. Therefore, primitive structures may be generalized or specialized according to their functions and resultant modifications, and so may advanced characters. Conversely, generalized characters may be either primitive or derived, depending on the degree to which they are shared with ancestors or extant groups.

These several concepts pervade analysis of the evolutionary and functional morphology of vertebrates. The student would be wise to keep them in mind as he pursues the details of vertebrate form and function.

6. Descriptive Terms

In most vertebrates the body is carried in the horizontal position, and the various surfaces are designated as follows with reference to this position:

Dorsal—the back or upper side (posterior in human anatomy).
Ventral—the underside (anterior in human anatomy).
Lateral—the sides, right and left.
Anterior, cephalic, or cranial—the head end of the animal (superior in human anatomy).
Posterior or caudal—the tail end of the animal (inferior in human anatomy).
Median—the middle.

Adverbs made by substituting d for the last letter of these words mean "in the direction of," as craniad, toward the head, and caudad, toward the tail.

Other descriptive terms are:

Central—the part of a system nearest the middle of the animal.
Peripheral—the part nearest the surface.
Proximal—near the main mass of the body, as the thigh.

*Distal*—away from the main mass of the body, as the toes.
*Superficial*—on or near the surface.
*Deep*—some distance below the surface.
*Superior*—above.
*Inferior*—below.

**7. Planes and Axes**

The structures of most animals are arranged symmetrically with reference to certain imaginary planes and axes.

1. The *median* plane is a vertical longitudinal plane passing from head to tail through the center of the body from dorsal to ventral surfaces. It divides the body into two nearly identical halves, right and left.
2. The *sagittal* plane or section is any vertical longitudinal plane through the body—that is, the median plane or any plane parallel to it. Sagittal planes other than the median plane are sometimes designated as *parasagittal* to avoid misunderstanding.
3. The *horizontal* or *frontal* plane or section is any horizontal longitudinal section through the body—that is, all planes at right angles to the median plane and parallel to the dorsal and ventral surfaces.
4. The *transverse* or *cross* plane or section cuts vertically across the body at right angles to the sagittal and horizontal planes.
5. The *longitudinal* or *anteroposterior* axis is a line in the median sagittal plane extending from head to tail; a *sagittal* or *dorsoventral* axis is any line in the median sagittal plane extending from dorsal to ventral surfaces; a transverse or mediolateral axis is any line in the transverse plane running from side to side.

**References**

Beer, G. de. 1962. *Embryos and ancestors*. 3d ed. Oxford: At the University Press, xii + 197 pp.

Bock, W. J. 1969. The concept of homology. *Ann. N.Y. Acad. Sci.* 167: 71–73.

Bock, W. J., and Wahlert, G. von. 1965. Adaptation and the form-function complex. *Evolution* 19(3):269–99.

Cain, A. J., ed. 1959. *Function and taxonomic importance*. Systematics Assoc. (London), publ. no. 3, 140 pp.

Gans, C. 1969. Some questions and problems in morphological comparison. *Ann. N.Y. Acad. Sci.* 167:506–13.

Grant, V. 1963. *The origin of adaptations*. New York: Columbia University Press, x + 606 pp.

Gregory, W. K. 1949. The humerus from fish to man. *Amer. Mus. Novit.,* no. 1400, pp. 1–54.

Inglis, W. G. 1966. The observational basis of homology. *Syst. Zool.* 15(3):219–28.

Mayer, E. 1963. *Introduction to dynamic morphology*. New York: Academic Press, x + 545 pp.

Russell, E. S. 1916. *Form and function: A contribution to the history of animal morphology*. London: J. Murray, ix + 383 pp.

Weibel, E. R., and Elias, H. 1967. *Quantitative methods in morphology*. New York: Springer-Verlag, vii + 278 pp.

# 2  The Phylum Chordata: Classification, External Anatomy, and Adaptive Radiation
## *Dennis R. Paulson*

**A. Classification**  Because of the bewildering array of living organisms, it would be very difficult to communicate about them without some system of categories. If each species were known only as an individual entity, not related to other species, we could not make such statements as "all birds have feathers." We could only recite the names of all the organisms we knew had feathers. But because of our classification system, we can categorize a robin first as a thrush, then as a perching bird, then as a modern toothless bird, then as a bird, then as a vertebrate, and so on. Classifying organisms consists of three basic operations. The first is recognizing and describing groups according to principles of population biology (including genetics, ecology, and behavior and physiology as well as morphology) and of phylogeny (the development of evolutionarily related groups). The second operation involves fitting these groups into a formal classification hierarchy, and the third is providing names for the groups described. Both the hierarchical classification system now used and the method of naming groups are modified from the system of Linnaeus, introduced in 1735. The hierarchical taxonomy grouped organisms according to the kinds and numbers of morphological characteristics they shared; the fundamental characteristic of the naming system was that it used two latinized words, for genus and species (*binomial nomenclature*), to designate each "type" of organism. The system used today is much like that of Linnaeus, but its meaning and use are considerably different. Linnaeus and other taxonomists believed that species were fixed, unchanging units, each of which was a *type*. Neither variation within a species nor the possibility of change in a species was considered. With the acceptance of Darwinian concepts of evolution, classification systems used the same form, but with very different operational principles. The concepts of absolute types and divine patterns were rejected; the relatedness of species and their ability to change were recognized; and a population concept of systematic units replaced the typological concept. Certain of the basic principles of morphological evolution discussed in chapter 1, particularly homology and analogy, primitive versus advanced state, and convergence and divergence are considered in assessing the relationships of species and groups of species.

Communication is made easier by categorizing organisms into ever more inclusive groups, the only disadvantage being a rather extensive series of names for biologists to learn. One could argue that "sharks and rays" is a more convenient term than "elasmobranchs." But the terms

Elasmobranchii and Squamata can be recognized as the names for sharks and rays and for lizards and snakes, respectively, by biologists speaking any language, and this is the great advantage of using only Latin and Greek for the roots forming these names. This is appropriate down to the level of species, since many organisms, described by scientists and given scientific names, have no specific common name in any language. Thus *Turdus migratorius* should be recognizable by any ornithologist even if the word "robin" is not part of his or her vocabulary.

The classification system is an attempt to express phylogeny, or the actual relationships of the organisms, and its realism reflects what we know about the organisms, from the standpoint not only of morphology but also of physiology, behavior, ecology, and of course the fossil record connecting them with more distant organisms. In many vertebrates, for which series of fossils are available, the classification probably closely approximates the phylogeny.

**B. Adaptive Radiation**

It may seem bewildering to the student that there are so many kinds of vertebrates (see figures for numbers of living species in various classes and orders). Much of the diversity is accounted for by the large size of the earth and the partitioning of the land masses into continents, so that similar species replace one another on different continents and islands. For example, there is a whole array of small mice in North America, a different array in Europe, another in Africa, and so forth; none of the species occurs naturally on more than one of these continents. There is little doubt that one can go to a distant land and find some small mouse that is more or less a counterpart to our abundant deer mouse of North America. In the West Indies, each major island has one or more species of warblers, different from those on the other islands. Geographic isolation has accounted for a great deal of this diversity, with populations reaching new areas and then becoming genetically distinct from their parent populations. On the other hand, in most parts of North America there are more than one species of mouse and more than one species of warbler resident in a given area. In some cases these different species are specialized for different habitats, so even though eight species of mice occur in an area, any given habitat (deciduous forest, grassland, marsh, etc.) may support only one. Where two species occur in the same habitat, they are often specialized to take different foods or to forage in different ways. These differences are adaptive, and they represent derivation from ancestral types: this is *adaptive radiation*. The pattern of adaptive radiation during evolution is consistent in different places and at different times; only the cast of characters changes. Within a phylogenetic group, members of one species may enter a new *adaptive zone;* that is, they begin through gradual changes brought about by natural selection to do something completely different from the usual pattern in their group. For example, the group might be aquatic, and our particular species becomes terrestrial. This is a major change in life-style, and it removes this species from competition with (or predation by) its relatives. The species should become very successful indeed at this point (assuming it has not run into more severe competition or predation by its change), and over a long time it may evolve into a variety of species in this new adaptive zone, with geographic isolation a necessity to preclude the exchange of genetic material between populations. This then is a major time of adaptive radiation. For example, one can picture a herbivorous species coming from the water onto the land, where there are very few other animals of its size but many smaller species. Not only are plants available for food, but these smaller animals constitute potential prey. Any evolutionary changes toward the ability to eat other animals would be favored, and there would be strong selection for carnivores. If there were many different kinds of plants (or animals), there would ultimately be specialization for different types of food in both the herbivores and the carnivores. This phenomenon can be seen well on islands and archipelagos, for example, in Australia, where the marsupials were the successful mammalian group, and, in the absence of competing placental mammals, evolved many diverse *adaptive types*. This basic process is adequate to explain the thousands of species of ray-finned fishes or perching birds that now exist.

It is interesting to see which groups are successful at present and to speculate on what has fit them so well for life in this time period. The aquatic and the terrestrial media seem equally favorable for vertebrates, as the number of species of aquatic vertebrates (at least the present estimate for teleost fishes) is subequal to the number of terrestrial ones. Among the terrestrial vertebrates, the birds (8,700 species) and squamate reptiles (5,700 species) are the most successful groups, followed closely by the placental mammals (4,300 species). As you learn about them, try to think of aspects of the morphology of these animals that might have made any of them particularly successful.

**C. What Is a Chordate?**

The phylum Chordata comprises primarily the familiar vertebrates, but two other types of organisms ("protochordates") share basic characteristics with the vertebrates. One of these groups includes the lancelets (subphylum Cephalochordata). These common marine animals look much like fish larvae, and one can easily see they are related to vertebrates. These relationships are not so apparent in the other group, which includes the tunicates (subphylum Urochordata). Most of these barrel-shaped animals are sessile, attached by one end to the substrate, although two of the three classes contain only planktonic species. It is in the larval tunicates that one can see the basic characteristics that relate the vertebrates, the cephalochordates, and the urochordates to one another. These characteristics are:

1. *Gill slits* are present in the wall of the pharynx, at least in the embryo or larva, if not throughout life.
2. A skeletal supporting rod, the *notochord,* is present at least in the early embryo of vertebrates, in larval tunicates, and throughout life in cephalochordates.
3. A *dorsal, hollow nerve cord* is found in all groups, though it is lost in adult urochordates.
4. A *tail* is present behind the posterior opening of the digestive tract, again in all embryos, even in those groups in which it is lost in the adult.

These characteristics fundamentally define the phylum Chordata. The acorn worms of the phylum Hemichordata possess pharyngeal gill slits and a poorly developed and in some cases hollow dorsal nerve cord, and they must be considered near the chordates in phylogeny. But they lack a notochord and are in other ways rather different from chordates.

**D. Character-istics of the Vertebrates**

Within the phylum Chordata the most prominent characteristic of the vertebrates (subphylum Vertebrata; also called Craniata) is the evolution of the primitive axial skeleton (notochord) into a skull and vertebral column. In addition, most organ systems of vertebrates are more highly developed and in many cases more efficient than those of the protochordates or most invertebrates.

The following descriptions characterize the entire subphylum Vertebrata:

1. bilaterally symmetrical;
2. usually two pairs of paired, jointed locomotor appendages;
3. covered with protective cellular skin;
4. internally segmented, the skeletal, muscular, and nervous systems highly metameric;
5. coelom well developed, divided in adult into two to four compartments;
6. internal skeleton well developed, of cartilage or bone, consisting of axial skeleton—skull, vertebrae, ribs, and sternum when present—and appendicular skeleton—girdles and appendages;
7. highly developed brain enclosed by skull and nerve cord enclosed by vertebrae;
8. well-developed sense organs—eyes, ears, and nostrils—on head;
9. respiratory system, either gills or lungs, intimately connected with pharynx;
10. closed circulatory system with ventral heart and median dorsal artery;
11. genital and excretory systems closely related, the excretory ducts often serving as genital ducts; and
12. digestive tract with two major digestive glands, the liver and the pancreas, secreting into it.

You will be learning much more about all these characteristics as you proceed through this book.

**E. The Vertebrate Groups**

**1. The Fossil Record**

The fossil record of the vertebrates is excellent, and paleontologists have been able to reconstruct in some detail the phylogeny of many major groups, including transitional forms in the evolution of amphibians from fishes, reptiles from amphibians, and birds and mammals from reptiles. The skeleton of vertebrates is easily fossilized under a variety of environmental conditions, and only in those groups with cartilaginous skeletons or very small or delicate bones are there significant gaps in understanding relationships between the major groups. Thus interpretation of cyclostome and bird phylogeny has suffered from these gaps in the fossil record, whereas the evolutionary history of many groups of large mammals is very well understood.

In the classification presented below, the living groups of vertebrates are emphasized. Names of extinct groups are followed by a dagger (†). Only extinct groups of the rank of subclass or higher are listed. Certain small groups of uncertain affinities are omitted, since they are not enlightening to the study of comparative anatomy.

## 2. Fish-like Vertebrates

The organisms usually thought of as "fishes" actually represent three distinct classes of vertebrates, sometimes placed in a superclass, Pisces. The vertebrates have been split in other ways as well, for example, into the Agnatha and Gnathostomata, the former superclass including only the class Agnatha and the latter superclass including all jawed vertebrates. Another division into superclass places the fishes and amphibians in the Anamniota and the reptiles, birds, and mammals in the Amniota. The amnionic membrane around the embryo was essential in the evolution of an egg that could be laid on land and thus characterizes the Amniota.

All the fishes are aquatic and ectothermal (poikilothermic, "cold-blooded"), with fins as locomotory organs. The respiratory organs are gills, or lungs in some groups, the heart is two-chambered (one atrium and one ventricle), and the yolk sac is the only extraembryonic membrane.

### a. Class Agnatha

Cyclostomes. 45 species. Jawless vertebrates with no fins or poorly developed fins. Living groups with cartilaginous skeleton, one (hagfish) or two (lamprey) pairs of semicircular canals, branchial basket (external to gills) attached to cranium, 6–14 external or concealed gill slits, single nostril, and no paired fins. Extinct (Ordovician to Devonian) heavily armored forms are collectively called ostracoderms. The anatomy and other features of living cyclostomes are covered in more detail in chapter 3.

Order Myxiniformes. Hagfishes. 15 species. Nostril opening externally at tip of snout, internally into pharynx; branchial basket vestigial. Temperate, marine, deepwater. May be parasitic on fishes, but most feed on detritus and carrion.

Order Petromyzontiformes. Lampreys. 30 species. External nostril opening on top of snout, leading into a blind sac; branchial basket well developed. Temperate, anadromous, and freshwater. Parasitic on fishes or nonfeeding as adults, detritus feeders as larvae.

### b. Class Acanthodii†

Early jawed fishes with cartilaginous and bony skeleton and bony scales. Fins well developed, with single large anterior spine on all but caudal fin; paired nostrils. Called "spiny sharks," but closer to bony fishes, to which they seem related by the large operculum, method of articulation of the hyomandibular, and other cranial features. Silurian to Permian periods.

### c. Class Placodermi†

Head and trunk with bony armor, the two jointed in the neck region.

More closely related to the Chondrichthyes. A great radiation in the Devonian period, with six distinct orders. The antiarchs had bony, jointed pectoral fins and small mouths. The later orders had reduced armor, and one genus of the Ptyctodontidae clearly seems transitional between other placoderms and the chimaeras.

### d. Class Chondrichthyes

Cartilaginous fishes. 575 species. With jaws and paired fins; skeleton wholly cartilaginous in living forms; three pairs of semicircular canals; gill arches internal to gills, not united to cranium (although joined to it by connective tissue); nostrils paired; scales (when present) dermal, placoid (base plate bearing tooth-like structures). Males with pelvic clasping organs.

Subclass Holocephali. Chimaeras. 25 species. Gills in four pairs, concealed under single gill opening on each side; spiracle absent; scales lacking; cloaca absent; jaw suspension holostylic; flat bony plates in place of teeth. Temperate, marine, deepwater.

Subclass Elasmobranchii. Sharks and rays. 550 species. Gills and gill openings in five (most) to seven pairs; spiracle present; scales placoid; cloaca present; jaw suspension amphistylic (primitive forms) or hyostylic; teeth derived from placoid scales, deciduous. Worldwide, marine (few freshwater).

Order Squaliformes. Sharks. Spiracle small, lateral; gills lateral; pectoral fins normal in size.

Order Rajiformes. Rays, skates, sawfishes. Spiracle enlarged, dorsal; gills ventral; pectoral fins enormously expanded for locomotion.

1) *External Anatomy and Adaptive Radiation of Cartilaginous Fishes.*

a) *Chimaeras.* The chimaeras are peculiar-looking fishes (fig. 2.1) with large heads and tapering bodies, their caudal fins replaced by long tapering tails, the terminal part filamentous. The shape of the tail doubtless prompted the common name, "ratfish." The dorsal fin is large, and the very large pectoral fins are the main organs of locomotion. The mouth is small and furnished with grinding plates for crushing hard-shelled invertebrates. Sensory pores are arranged along a lateral-line system on the head and body. Most chimaeras seem to have similar habits, and the anatomical radiation in this small group involves the shape of the head; but the functions of the stiletto-like snout of the members of one family (Rhinochimaeridae) and the plow-shaped snout of those of another (Callorhynchidae) are still unknown. A peculiar structure, the frontal clasper, projects from the head in males, but its function is likewise unknown.

b) *Sharks.* Although the remaining members of the Chondrichthyes are separated into two orders (sharks and rays) and are indeed distinct groups, there has been amazing convergence between them, with two families of shark-like rays (Pristidae, Rhinobatidae) and one family of ray-like sharks (Squatinidae). Generally the two orders are easily recognizable, the sharks being elongate with caudal fins well developed for thrust, the rays broad and flat with wing-like pectoral fins and poorly developed tails.

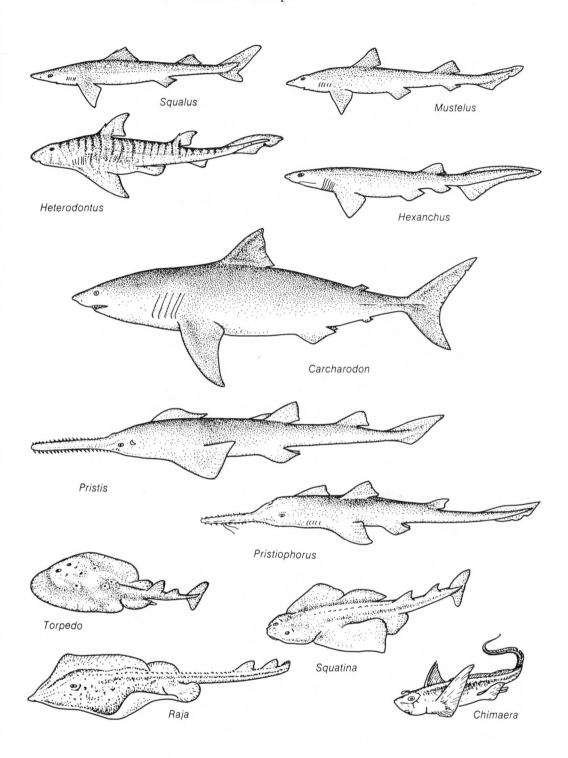

Fig. 2.1.  Representative cartilaginous fishes. After Young 1950.

The sharks are diverse in general appearance and adaptive radiation (fig. 2.1). The so-called requiem sharks (Carcharhinidae) or the smooth dogfish (*Mustelus*) satisfy the layman's concept of a shark and are usually among the common sharks of any coast. However, the species usually available in anatomy laboratories is the spiny dogfish (*Squalus*) of the Squalidae. The long, spindle-shaped body seems designed for quick passage through the medium (water), and the large fins are adapted for propulsion and steering. In most sharks the head is flattened, the rostrum or snout pointed for least resistance during locomotion. The eyes are rather small and lateral, without lids, and the nostrils are on the ventral side of the rostrum. Behind each eye, best perceived by feeling with a finger, is a slight prominence that contains the internal ear. There is no external ear opening, but the ears are connected with the surface by two canals, the endolymphatic ducts, that open just back of the level of the eyes near the dorsal midline of the head. These openings may be difficult to locate. Just behind the internal ear on each side is the first gill opening, the spiracle, and somewhat behind that the other five gill slits. Most sharks have five of these openings, but several families include six-gilled (as in *Hexanchus*) and seven-gilled species. Water enters through the mouth, as the shark moves forward, and exits through the gill slits, having passed over the internal gills where respiration takes place. The mouth is ventral and provided with rows of teeth that are replaced from the rear as they wear or break off.

The skin is composed of minute scales, each with a tiny spine. These dermal denticles are unique to elasmobranchs and give them their characteristic rough surface. The fins are of two types: unpaired or median, and paired or lateral. Of the median fins, the anterior and posterior dorsal fins furnish stability to the moving shark, and the caudal fin is used for propulsion. Most sharks have an anal fin on the ventral surface anterior to the tail, but this is lacking in *Squalus*. Under what conditions might it be advantageous to lack this fin? The caudal fin is asymmetrical (heterocercal); its upper lobe provides more thrust than its lower one, and the shark would tend to somersault were it not for the stabilizing action of the lateral fins and flattened head. The two pairs of lateral fins, the pectoral and pelvic, correspond to the familiar appendages of land vertebrates. In males the pelvic fins are modified into claspers, used to hold the female during mating; thus the sex of any mature individual of this class, including the chimaeras and rays, can easily be determined, unlike the situation in many higher animals. In *Squalus*, *Heterodontus* (horn shark), and others stout spines precede both the dorsal fins; the spines are lacking in most sharks. What function might these spines serve? On the ventral side, where the tail joins the trunk, is the vent. This is the opening from the cloaca, a common chamber receiving urogenital products and digestive wastes.

From the basic type can be derived the mackerel sharks (Isuridae), with slender bodies and tails almost homocercal for high-speed predation (this group includes the man-eating white shark [*Carcharodon carcharias*]); the thresher sharks (Alopiidae), with the upper lobe of the caudal fin tremendously elongated for stunning fish (usually used in schools, as are the saws of sawfishes and saw sharks); and the hammerheads (Sphyrnidae), whose bizarre transversely elongate snout still has not been satisfactorily explained. Do you have any ideas about the significance of these adaptations? Unseen by all but deep-sea fishermen, commercial and scientific, are some of the very elongate and peculiar deepwater groups—the frill sharks (Chlamydoselachidae), goblin sharks (Scapanorhynchidae), and cat sharks (Scyliorhinidae). Some of these are very small, although the smallest shark (*Squaliolus laticaudus,* maturing at 125 mm) is a member of the spiny dogfish family. The largest sharks are open-ocean plankton feeders—the basking shark (*Cetorhinus maximus,* a derivative of the mackeral sharks) and the whale shark (*Rhincodon typus,* the sole member of its family); the latter reaches 16 m in length and is thus the largest living fish. The basking shark strains plankton from the water with long, erectile gill rakers, which are shed during the winter at high latitudes; the shark may also become dormant during that season, when plankton is scarce. The whale shark's straining apparatus is more complex, with lateral processes producing the effect of a fine sieve.

c) *Rays.* The rays are highly modified for bottom life, although many groups have secondarily taken to middle and surface waters. Both the body shape and the fin size and shape are much modified (fig. 2.1) from the basic elasmobranch type, as illustrated by the spiny dogfish.

---

The skates (Rajidae) can be considered typical rays and are usually available for examination. The head and trunk are much flattened and indistinguishable, the tail slender and not developed for locomotion. The scales are much larger than in sharks and found only on certain parts of the body; locate these areas. The spines on some scales may be particularly well developed in males. Note that the dorsal side is colored and the ventral side is not—why is this?

Behind each eye is the spiracle, far separated from the five gill slits that are on the ventral side. The mouth is near enough to the substrate in these bottom-dwellers so that water taken in through it might be mixed with sand or mud. Thus respiratory water enters through the spiracle, which is large and of special importance in this group. There are two nostrils in front of the mouth, each with a flap for closure. Between the angle of the mouth and the nostril on each side is the nasofrontal process, a flap that covers the oronasal groove. This arrangement is the precursor to the closed passage from the nostrils into the mouth that is present in higher vertebrates. Most of the rays possess crushing teeth, adapted as in the chimaeras for feeding on hard-shelled invertebrates.

There are median and paired fins, but the former are much reduced, consisting of two small dorsal fins on the end of the tail. The enormously enlarged pectoral fins furnish locomotion, with ripples of muscular contraction along these fins moving the skate forward. The much smaller pelvic fins are immediately posterior to the pectorals, and in some rays they are continuous with them. The vent is visible as a large opening between the bases of the pelvic fins.

---

The eagle rays (Myliobatidae) have much larger, more pointed pectoral fins and move rapidly through midwater, although they are still tied to the bottom because of their diet of invertebrates. Both these species and the stingray (Dasyatidae) possess venomous spines at the tail base, and the electric rays (Torpedinidae) can equally well defend themselves by discharging an electric current with a strength of more than 100 volts. An evolutionary step beyond the eagle rays is exemplified by the devil rays, or mantas (Mobulidae), huge surface-dwelling plankton-feeders with large mouths and palp-like structures to direct food into the mouth. As in the sharks, the largest members of the group are the plankton-feeders. What might be the significance of this? The most aberrant rays are certainly the sawfishes (Pristidae), looking much like sharks but with the typical ray characteristics of a dorsal spiracle and ventral gill slits. Sawfish, and the sharks with which they are convergent (saw sharks, Pristiophoridae), have long, flattened snouts edged with sharp teeth, with which they disable fish and root for burrowing invertebrates.

### e. Class Osteichthyes

Bony fishes. 17,000+ species. With jaws and paired fins; skeleton partly or mostly of bone; three pairs of semicircular canals; gill arches internal to gills, not united to cranium; gills covered by bony operculum; paired nostrils; dermal scales, not placoid; with a gas bladder (secondarily lost by many). A phylogenetic tree for this class is presented in figure 2.2.

Subclass Dipnoi. Lungfishes. 5 species. Paired fins with a pronounced fleshy, lobed base containing strong skeletal elements; excurrent nostril internal; no typical branchiostegal rays; gas bladder modified as a lung; jaw suspension autostylic; with spiral valve and conus arteriosus. Fresh water; Africa, South America, Australia.

Subclass Crossopterygii. Lobe-fins. 1 species. Paired fins with a pronounced fleshy, lobed base containing strong skeletal elements; true internal nostrils present (although lacking in living species); no typical branchiostegal rays; jaw suspension hyostylic; spiracle present; with spiral valve and conus arteriosus. Living species marine, Indian Ocean.

Subclass Brachiopterygii. Bichirs. 13 species. Pectoral fins with basal lobes; skeleton largely cartilaginous; spiracle present; with spiral valve and conus arteriosus; scales ganoid; nostrils external; no typical branchiostegal rays; ventral, paired lungs opening into ventral wall of esophagus; jaw suspension hyostylic. Freshwater, Africa.

Subclass Actinopterygii. Ray-finned fishes. About 17,000 named species. Paired fins not lobed, fin rays attached directly to girdles; nostrils external; jaw suspension hyostylic; branchiostegal rays usually present. Gas bladder single, dorsal, opening into esophagus or ductless. Ubiquitous from shallow (even temporary) ponds and streams to the oceans' depths.

Superorder Chondrostei. Primitive in every way—skeleton largely cartilaginous; spiracle present; with spiral valve and conus arteriosus;

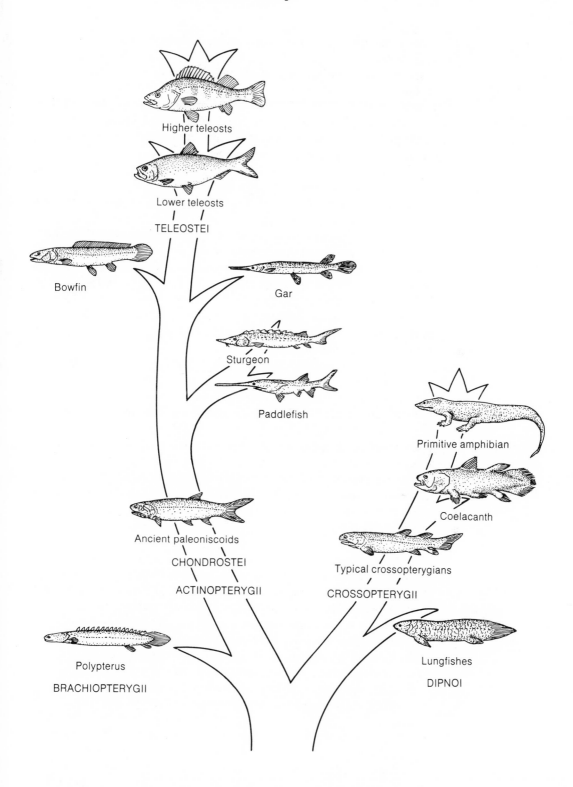

Fig. 2.2. Phylogeny of bony fishes. After Romer 1970.

notochord large, persistent; tail heterocercal; scales ganoid, degenerate in living forms.

Order Acipenseriformes. Sturgeons, paddlefish. Two families, 22 species.

Superorder Holostei. More advanced group, dominant during Mesozoic. Skeleton moderately ossified, vertebral centra usually fairly well developed; tail heterocercal; scales ganoid to cycloid; spiracle lacking; conus arteriosus present; spiral valve vestigial.

Order Amiiformes. Bowfin. 1 species. Scales thin, cycloid; tail tending toward homocercal type.

Order Lepisosteiformes. Gars. 1 family, 6 species. Scales thick, ganoid; tail shortened heterocercal type.

Superorder Teleostei. Most advanced and dominant bony fishes. Skeleton largely of bone; tail homocercal: scales thin, cycloid or ctenoid; no spiracle or spiral valve; conus arteriosus vestigial. It is difficult to characterize the orders of this huge superorder. Different authors recognize from 23 to 30 orders, and there is not yet complete agreement on the content of these orders. The most complete classification of the teleosts to date is that of Greenwood, et al. (1966). They define 30 orders containing 413 families of teleosts.

1) *External Anatomy and Adaptive Radiation of Bony Fishes.* The Osteichthyes represents the largest group of vertebrates. A few species (for example, the mud skipper *Periophthalmus*) forage on land for brief periods, and others are able to glide for rather long distances above the surface of the ocean, but otherwise the group is restricted to water by its mode of respiration, with all possible adaptive zones entered by the modern teleosts. All other aquatic vertebrates together represent a diversity of only about a tenth that of the teleosts, of which it has been estimated there may be over 30,000 species. The size variation in this group is great, from a tiny 12-mm goby to marlins of 5 m and over 600 kg. Some of the diversity in shape and structure is illustrated in figure 2.3.

---

The typical bony fish is streamlined and fusiform for rapid locomotion in its medium. The head and body may be partially or entirely covered by dermal scales or may be scaleless. The scales overlap and are highly variable in shape and size. Many of them have pores, and pored lateral-line scales are a prominent feature of many types. The mouth may be terminal, superior or inferior, the latter two positions adapted for surface- or bottom-feeding. There are two pairs of nostrils, a pair to each olfactory sac, permitting water to circulate through the sacs. The eyes tend to be large and are without lids; they are relatively fixed in the skull but can be moved. The ears are not visible externally. The most prominent features of the head are the opercula, or gill covers, supported by opercular bones. These structures cover the gill openings, within which are the gills. Lift an operculum to observe the gills underneath. The bony gill arches support the gills, each of which consists of a double row of fleshy filaments for respiration. On the anterior surfaces of the gill arches lie the gill rakers, important in keeping food particles out of the gills. In many fishes these rakers have become elongate and fine, and they function in straining plankton from the water for food. The expansible branchiostegal membrane, supported by the bony branchiostegal rays, extends from the ventral edge of the operculum, allowing the gill opening to be kept closed while the operculum opens, creating a vacuum inside the mouth, into which water is drawn. Then the mouth is closed, the branchiostegal membrane folds, and the operculum is drawn inward, forcing water out through the gill slits during the respiratory movements.

The typical fish propels itself by powerful strokes of the caudal fin, the dorsal and anal fins furnishing stability and the pectoral and pelvic fins providing lift and steering. Any moving body tends to revolve around its three axes unless stabilized. Movement around the longitudinal axis is called *roll,* and this is prevented by all the fins (but especially the paired ones), which parallel this axis and project outward from it at right angles to the body. The compressed bodies of many fishes also provide greater stability against rolling. Movement around the vertical axis is called *yaw,* and this is damped by the unpaired fins and by body shape as well. The normal swimming motion of a fish involves lateral undulation of the posterior part of the body, and this in itself is a stabilizing action. Movement around the transverse axis is called *pitch,* and the paired fins compensate for this instability. The fins are clearly important for stability as well as for propulsion and turning, although the body functions similarly. Fish missing part or all of certain fins (a not-uncommon consequence of attempted predation) can still swim, but their efficiency is impaired. It can be seen also that each fin may function in more than one way. Imagine amputating different fins and calculate the consequences to the fish's locomotion. The fins are supported by flexible, jointed rays (lepidotrichia), some of which are replaced by stiff spines in the most advanced teleosts. These spines help deter predators as well as supporting the fins. In certain families (for example, the Salmonidae), a rayless adipose fin of unknown function is present behind the dorsal fin. See chapter 7 for more information about fin structure and function.

In the bony fishes, the urogenital system opens through a urogenital papilla just anterior to the anal fin, and the digestive system opens through the anus just anterior to the urogenital papilla. Thus these two systems have separate openings in this group, as in mammals, but unlike the situation in the cartilaginous fishes, amphibians, reptiles, and birds. Can you think of advantages or disadvantages for each mode? Fertilization is external in most fishes, but in a few groups (e.g., the Poeciliidae, including guppies and mosquitofish) the anal fin is modified into a copulatory organ.

Specimens of some of the more primitive bony fishes (fig. 2.2) should be available for study. The lungfishes have a symmetrical, pointed (diphycercal) caudal fin; peculiar slender paired fins (at least in the South American and African genera); a nonbony operculum over the small gill opening; and internal nares in the roof of the oral cavity, unlike all other living fishes. These open olfactory passages and the well-developed lung (air bladder) are clear signs of relationship to primitive land vertebrates. Sturgeons (Acipenseridae) are superficially shark-

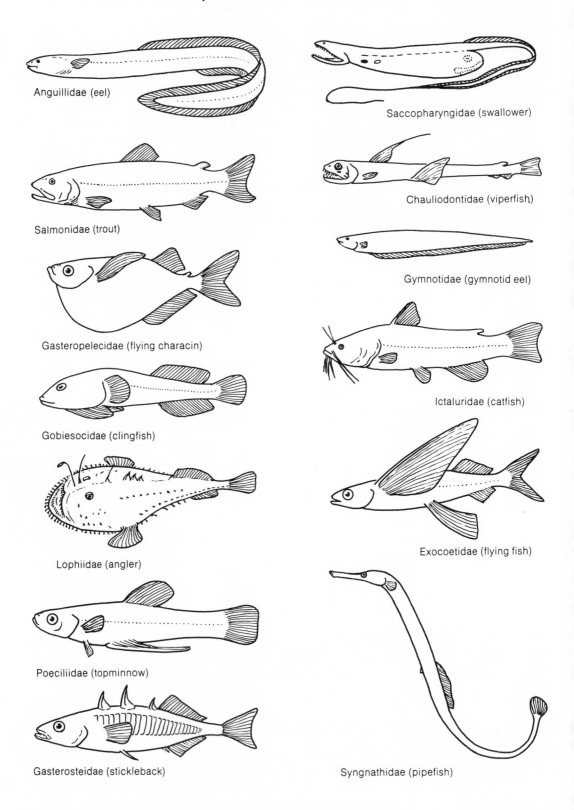

Fig. 2.3.   Representative teleosts. After Greenwood et al. 1966.

Syngnathidae (seahorse)

Scorpaenidae (scorpionfish)

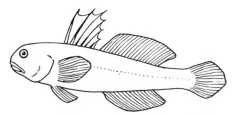

Scaridae (parrotfish)

Echeneidae (remora)

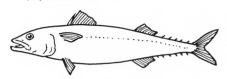

Sphyraenidae (barracuda)

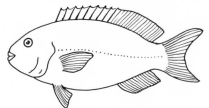

Gobiidae (goby)

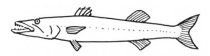

Scombridae (mackerel)

Xiphiidae (swordfish)

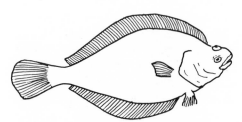

Pleuronectidae (flounder)

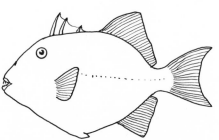

Balistidae (triggerfish)

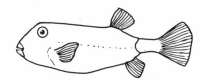

Ostraciontidae (trunkfish)

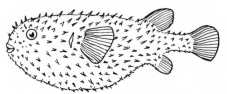

Diodontidae (porcupinefish)

like, with a long rostrum, a ventral mouth, a markedly heterocercal tail, and a spiracle, and they have large ganoid scales in rows on the trunk separated by areas of minute scales. Note the sensory barbels under the rostrum and the lack of teeth. The paddlefish (*Polyodon*) can be called bizarre, with its huge spatulate snout, abundantly provided with sense organs for detecting food. Like the sturgeons, to which it is related, it has a spiracle and heterocercal tail, but it is singularly lacking in external hard parts. The operculum is large, the mouth is large and provided with a branchiostegal membrane without rays, the teeth minute, the gill rakers long and fine. What does this tell you about its food habits? Gars (*Lepisosteus*) are heavily armored, with large ganoid scales of rhombic shape set in diagonal rows. Their jaws are long, with many sharply pointed teeth, and they capture small fish with a quick sideward thrust of the flattened head. Finally, the bowfin (*Amia*) is even more like the modern teleosts, lacking spiracles and with scales thin and imbricated, tail nearly homocercal, and operculum bony with branchiostegal membrane. In the branchiostegal membrane there is a medium gular plate in front and a series of lateral gular plates on each side, transitional between the gular plates of some primitive bony fishes and the more advanced branchiostegal rays of the teleosts.

---

A great degree of change can be seen in the progression from primitive to advanced teleosts, in the shape of the head, the type of scales, the placement and use of the fins, and overall way of life. The herring (*Clupea*) has small pectoral and pelvic fins, similar to each other, the latter well behind the former. In the perch (*Perca*), typifying the advanced groups, the pelvic fin lies under the pectorals, which have moved higher up on the sides. The soft and segmented fin-rays have often been replaced by spines, and the smooth (cycloid) scales have been replaced by spiny (ctenoid) ones. The gas bladder, still functional as a respiratory organ (physostomous) in some primitive teleosts, became closed (physoclistous) in all the advanced ones and was lost entirely in many groups. Thus the primitive teleost is typically a schooling, fast-swimming fish of open water, the advanced one a solitary, often stationary fish of vegetation and bottom. There are many exceptions.

Within the confines of the medium, the locomotory radiation in the Teleostei has been extreme (fig. 2.3). The most powerful swimmers, the tunas and mackerels (Scombridae), have pointed heads, smooth, small-scaled trunks, reduced fins, very narrow caudal peduncles, and stiff, crescent-shaped tails that drive them through the water of the open ocean at speeds of more than 50 kph. From this basic torpedo shape all variations in shape have evolved, with progressive compression or depression to ribbon- or disk-shaped. Many high-bodied fish, adapted for passing through dense vegetation or into crevices, have reduced caudal fins, and the enlarged dorsal and anal fins, together with the caudal peduncle, furnish thrust. The dorsal or anal fins, or both, are important in propulsion in many groups in which the body is held rigid, for example, in fishes with exoskeletons (trunkfishes, Ostraciontidae) or very tough skins (triggerfishes, Balistidae) or those that generate electrical fields (Mormyridae and Gymnotidae). Why do you think electric fish must remain rigid? In these groups either or both fins are undulated in such a way that a sine wave progresses from front to back (or back to front for backward swimming) along the fin. Still other fishes (eels and eel-like families and cyclostomes) undulate the entire body in the same way.

The pectoral fins are important in locomotion in some groups, for example the parrotfishes (Scaridae), in which they are flapped much like bird wings, while the body is held straight. In two groups (flying fishes, Exocoetidae; flying gurnards, Dactylopteridae) the tail propels the fish along the surface until a speed is reached at which the fish becomes airborne by spreading the much-enlarged pectoral fins. The fish may glide for some distance before reentering the water or "taxiing" again with the tail. The flying characins (Gasteropelecidae) actually flap the pectoral fins while out of the water and thus "fly" briefly. This is presumably an adaptation against predation in these species, and fish show many such adaptations. Many are extremely spiny, not only the fin rays but projections from the head and opercula (for example, sticklebacks, Gasterosteidae; and scorpionfishes, Scorpaenidae). There is abundant evidence that these spines confer protection from some predators, as many observers have reported seeing spiny fish spit out by their captors—or, in a few cases, responsible for the death of the predator. Some (porcupinefishes, Diodontidae) are spiny all over, and in addition as soon as they are seized they swallow great quantities of water and become spherical. The porcupinefishes must have few predators, for they are among the few fishes a swimmer can capture by hand.

Some fishes associate closely with large predatory fish and thus gain both protection from other predators and scraps of food. One group, the remoras (Echeneidae), have progressed beyond the stage of association to one of almost obligatory commensalism. The first dorsal fin has become modified into a sucking disk on top of the head, and with this structure the remoras attach to and are carried by sharks, rays, whales, sea turtles, and large bony fishes. Other structures adapted to clinging have formed from either the pelvic fins or skin folds in several groups, for example, the gobies (Gobiidae) and clingfishes (Gobiesocidae). Both are marine groups, but both have representatives in torrential coastal streams, where they use the ventral disk to cling to rocks—a case of a preadaptation allowing entry into a new adaptive zone.

The flatfishes (Pleuronectiformes) represent another specialized way of existence. These fishes are bottom-dwellers with (at first glance) a very depressed shape. However, they are actually highly compressed and swim on one side. From a symmetrical larva, the flatfish gradually shifts over to side-swimming (one side or the other, depending upon the family); only one side is pigmented, and the eye on the off side migrates around so that both are on the upper side.

Sensory adaptations in bony fishes vary from extreme tactile specialization (large barbels on the head as in the catfishes) to great development of visual acuity (large eyes and distance vision in many open-ocean species). Blind, unpigmented cave species have evolved in a number of families, in one case (*Astyanax*) as local populations of an eyed species living just outside the caves.

Feeding adaptations include all those possible in aquatic vertebrates.

There has been a tendency toward enlargement of the head and shortening of the mouth, and in many advanced teleosts the prey is "inhaled" by sudden opening of the mouth and enlargement of the branchial chamber, reducing the interior pressure and causing water and prey to rush in. Many groups are more highly specialized for rapid pursuit of their prey, with large mouths and sharp teeth. Virtually all species swallow the prey whole, but barracudas (Sphyraenidae), with their razor-sharp teeth, can easily slice a fish in pieces, and many coral reef species with heavy teeth can crush hard-shelled invertebrates and eat them piece by piece. Members of many families have become specialized for herbivory, including parrotfish (Scaridae) that graze algae from rocks and characins (Characidae) that take fallen fruit from the surface of rivers. Pipefishes and seahorses (Syngnathidae) are elongate and have tubular snouts and tiny mouths, an adaptation for taking individual planktonic organisms. Billfishes (Istiophoridae) and swordfishes (Xiphiidae) have long, pointed snouts with which they stun fish in a quick rush, then return to swallow them. Anglerfishes (Lophiiformes) develop fishing organs from the highly modified first ray of the dorsal fin, with a "rod," "line," and "lure" hanging just in front of the mouth. The "lure" is moved about, attracting smaller fish that then disappear inside the angler with a barely perceptible gulp. Deep-sea anglers have luminous baits, and, indeed, many groups of deepwater fishes have luminous organs scattered around the body. These may serve for species- and sex-recognition in this otherwise lightless world. Picture interactions within and between species under these conditions. Some deep-sea species have incredibly large mouths and stomachs (for example, the swallower, *Saccopharynx*), probably an adaptation to allow capture of a wide variety of prey sizes in an environment in which encounters may be relatively rare. Similarly, others have an impressive mouthful of very long teeth (viperfish, *Chauliodus*).

3. Tetrapods

The tetrapods include the four classes of "land" animals, although all have aquatic members: amphibians, reptiles, birds, and mammals. These classes fall conveniently into two pairs, the "cold-blooded" (ectothermal) amphibians and reptiles and the "warm-blooded" (endothermal) birds and mammals. The endotherms, able to regulate their body temperature independent of the environment, have had much success colonizing high latitudes where amphibians and reptiles are relatively unsuccessful. This thermoregulatory ability evolved independently in the birds and mammals, both of which are more closely related to particular reptilian ancestors than they are to each other.

### a. Class Amphibia

Amphibians. 2,400 species. First vertebrates with limbs for terrestrial locomotion; lungs (may be lost) and skin as adult respiratory organs, gills variously developed in larvae but retained in a few salamanders; heart with two atria, one ventricle. Aquatic or terrestrial, skin naked (most living forms) or with bony dermal scales.

Order Apoda (Gymnophiona). Caecilians. 158 species. Some species retain the scales (embedded in the skin) of the primitive amphibians.

Worm-like, annulate; lacking limbs and girdles; with compact skull and minute eyes without lids; notochord persistent.

Order Caudata (Urodela). Salamanders. 325 species. Tail retained throughout life; limbs normal.

Order Salientia (Anura). Frogs and toads. 1,900 species. Tailless as adults; hind limbs long, specialized for saltatory locomotion; caudal vertebrae fused into long urostyle.

1) *External Anatomy and Adaptive Radiation of Amphibians.* As would be expected in an ancient group, the living orders of amphibians are very different from one another in body plan, although they have much in common in their general way of life.

a) *Caecilians.* The caecilians are all worm-like (9–140 cm), either burrowing in soil or (a few) aquatic with compressed tail. Some retain the scales characteristic of early amphibians, although they are small and embedded in the annular rings; the latter contribute to the annelid-like appearance. The eyes are degenerate and there are no limbs or girdles and virtually no tail. They have a chemosensory tentacle, unique among vertebrates. All caecilians practice internal fertilization, and many species are viviparous. The group is tropical, but preserved specimens can be seen in most museums.

b) *Salamanders.* The salamanders are constructed on a fairly uniform plan, with an elongate trunk, four legs, and a long tail (fig. 2.4); they vary in length from 5 to 150 cm.

---

*Necturus,* the mudpuppy, is a large salamander that is commonly studied in anatomy courses, but it is not typical of the group in that it is neotenic, of permanent larval form. Most species of salamanders undergo metamorphosis into a terrestrial adult without gills. *Necturus* has a large mouth, and the head is provided with three sets of external sensory receptors. The external nares are widely separated openings just back of the upper lip that communicate with the mouth cavity through the internal nares. Note that this allows external olfactory stimuli, as well as air, to enter the mouth cavity; in the fishes the nostrils were blind sacs. The eyes are small and without eyelids (eyelids are present in terrestrial salamanders), and the ears, as in fishes, are internal. This animal is largely nocturnal and finds its food by smell and touch, although the eyes are functional and allow it to evade predators. Try to picture the aquatic world from the viewpoint of *Necturus;* consider the stimuli that would be received by each of the senses and their relative importance in its life.

The three external gills are conspicuous, and between them the two gill slits open into the pharynx. However, the gills in amphibians are not homologous with those within the gill cavity in fishes; respiration takes place on the surface of these external gills and over the entire body surface. In addition there are lungs, which must be functional, since the *Necturus* may gulp air at the surface of the water. The legs are well developed, with hands and feet both bearing four toes. Five is the typical vertebrate number, but the first digit is missing in this species. The position of the limbs with reference to the body is primitive. The hind limb projects at right angles to the body, all of its parts on a

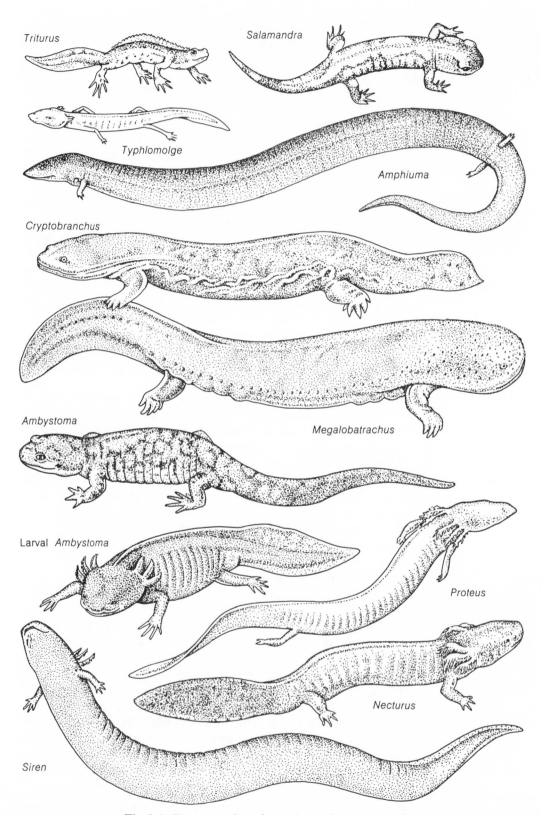

Fig. 2.4.  Representative salamanders. After Young 1950.

plane parallel to the substrate. The anterior and posterior borders and dorsal and ventral surfaces of the limb are obvious. In the forelimb the upper arm is rotated, the forearm is bent downward, and the hand is directed slightly forward (see chap. 7). This allows the salamander to lift itself slightly above the substrate and is an intermediate stage toward the situation in higher vertebrates, in which the limbs support the animal. The flattened tail is bordered by a tail fin that differs from those of the bony fishes in lacking the support of fin-rays. The large vent at the junction of the trunk and tail is the external opening of the cloaca.

Terrestrial salamanders should be examined in addition to *Necturus*. Note the various modifications for life on land: the absence of gills, expanded digits for climbing, larger eyes with eyelids, size and shape of feet, slender body for living in crevices or burrows. In some of these animals (especially the largest family, Plethodontidae) even the lungs have been lost, and respiration is entirely through the skin. In the completely terrestrial species, the larval stage is passed within eggs that are laid on land. Even more aberrant are the large eel-like salamanders, *Amphiuma* (all four legs extremely reduced) and *Siren* (hind limbs absent). These permanently aquatic creatures are common in swamps in the southeastern United States, a region that supports by far the greatest variety of freshwater vertebrates on the North American continent.

Larvae of other species of salamanders should be available for study. These larvae are rather similar to the adult *Necturus* in external appearance, but the number of gill slits may vary from one to three pairs. In very young larvae of some species that live in ponds, there are conspicuous balancers, slender organs extending from the sides of the head and probably both analogous and homologous to the adhesive organs of tadpoles. The pond larvae have very large gills and a high tail fin. Larvae of stream-dwelling salamanders have adequate oxygen in solution for respiration, so their external gills are reduced, and they must avoid being swept away by the current, so the tail fin is reduced and they swim less than the pond species.

c) *Frogs*. Frogs, again, are basically similar to one another in their external anatomy, with variation involving general shape, relative length of legs, and structure of the skin (fig. 2.5). The body is shaped quite differently from that of a salamander, being much shorter and broader, and adults range in length from 1.5 to 25 cm. Some frogs are permanently aquatic (*Pipa, Xenopus*), with extensive webbing between the toes; others are truly amphibious and swim or jump about on land equally well (for example, *Rana*); and still others are entirely terrestrial, with short limbs and rough skin. The toads (*Bufo*) represent one extreme of the variation, their spherical shape adapted to conserving water in the terrestrial environment and their enlarged skin glands containing toxins to protect these otherwise relatively defenseless animals from predators. Many frogs (*Hyla, Polypedates*) are arboreal, with enlarged toe pads for climbing. Most members of this order lay their eggs in water; their larval stage, with round body and high-finned tail, is the

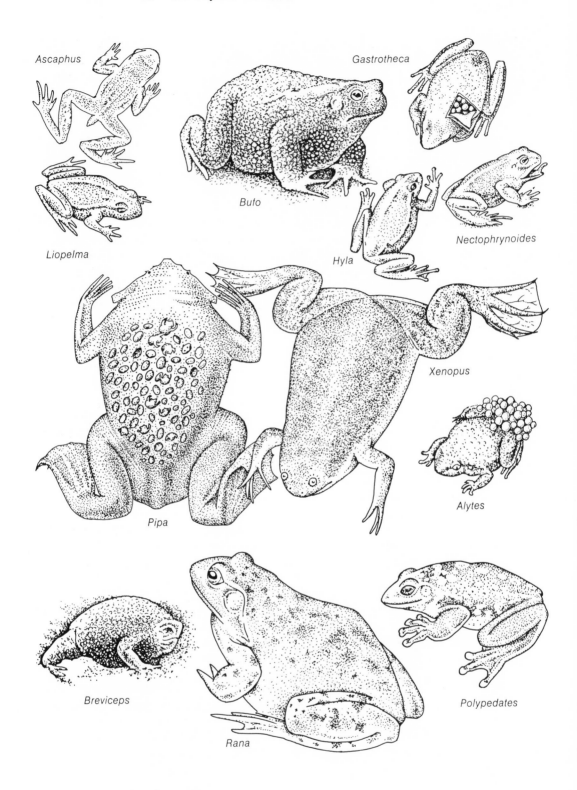

Fig. 2.5.  Representative frogs. After Young 1950.

familiar tadpole that transforms into a tiny frog—a more substantial metamorphosis than in the salamanders. Do you have any thoughts, based on this evidence, about which of the two—frogs or salamanders—is more primitive? Often tadpoles are common enough so that an observer can find individuals in all stages of metamorphosis, and such a series should be available for laboratory study. Tadpoles differ in a number of ways from salamander larvae. Their mouth is adapted for grazing, rather than for eating animals, with horny "jaws" and rows of "teeth." In some stream species the mouth is much enlarged and used as a sucking organ to cling to rocks in the current. Shortly after hatching, two pairs of external gills are present, and there are prominent adhesive organs extending from the sides of the head, to attach the larva to some substrate before it begins to feed actively. With further development the gills become covered by an operculum, and the resultant single opening (spiracle) may be on different parts of the body, depending on the group. During metamorphosis, the horny oral structures are resorbed, along with the tail, gills, and other larval structures, and limbs begin to develop. As in the salamanders, still other frogs (especially tropical-forest species) lay their eggs on land, and these hatch into miniature versions of the adults. In some cases the eggs are carried around on or in the body until they hatch. Thus at least several large families of amphibians have become independent of aquatic situations, although they still need moist air or a moist substrate on which to lay their eggs. Fertilization is external in aquatic species, but in one (*Ascaphus*) the cloaca is elongated into a copulatory organ, an adaptation to facilitate fertilization in swift streams; the mode of fertilization is unknown in the terrestrial species.

### b. Class Reptilia

Reptiles. 6,000 species. First vertebrates well adapted for land, with cleidoic (amniote) egg (embryo with extraembryonic membranes, relatively impermeable shell); lungs for respiration; heart with ventricle partially or completely divided; skin with epidermal scales or bony plates; one occipital condyle in living forms. See figure 2.6 for a phylogenetic tree summarizing the radiation of reptiles.

Subclass Anapsida. Skull without temporal openings, turtles with various degrees of temporal emarginations.

Order Testudinata (Chelonia). Turtles. 320 species. Body relatively short and broad, enclosed in bony armor and epidermal plates; girdles inside the shell; jaws with horny shields in place of teeth.

Subclass Lepidosauria. Skull with two pairs of temporal openings.

Order Rhynchocephalia. Tuatara. One species. Lizard-like *Sphenodon* of New Zealand. Teeth fused to jaw; vertebrae amphicoelous; abdominal ribs present.

Order Squamata. Lizards and snakes. 5,700 species. Skull has lost one or both temporal openings; covered with horny scales; quadrate movable; teeth in sockets; vertebrae usually procoelous; abdominal ribs absent or rudimentary.

Subclass Archosauria. Skull with two pairs of temporal openings. Includes the dinosaurs and the ancestors of birds.

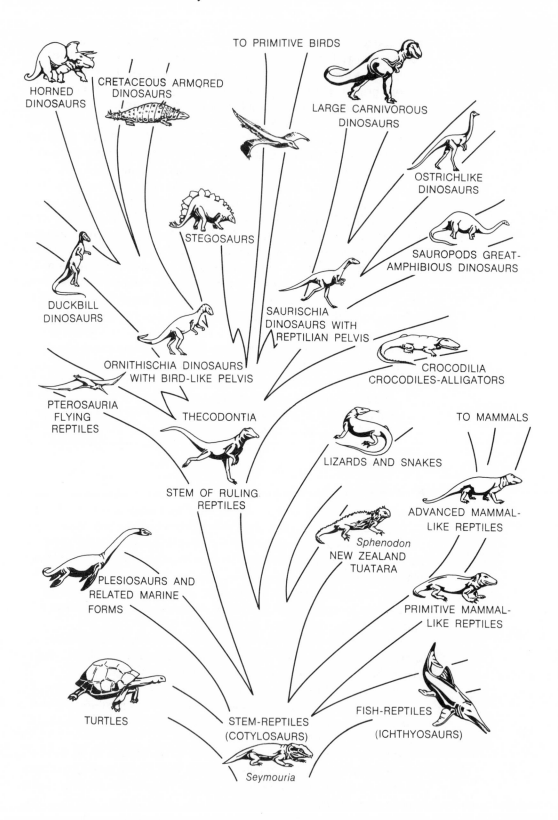

Fig. 2.6. Phylogeny of reptiles. After Romer 1959.

Order Crocodilia. Crocodilians. 21 species. Large, aquatic, with bony plates embedded under epidermis; quadrate fixed; teeth in sockets; abdominal ribs present; heart with two ventricles.

1) *External Anatomy and Adaptive Radiation of Reptiles.* If this book were being written in the Mesozoic, a full chapter could be devoted to radiation in the reptiles. From the fossil record alone, a picture of diversity equaling or exceeding that of any other vertebrate class has emerged, and doubtless the details of the everyday life of these reptiles and their external appearance would be even more fascinating than the skeletal anatomy suggests. Of at least sixteen orders present during the peak of reptilian evolution, only four have survived to the present, and three of these are surely or probably declining.

a) *Turtles.* The living reptiles come in three basic anatomical forms —turtle, lizard, and snake. The turtle form is characterized by a shell and by extreme elongation of the transverse axis (fig. 2.7).

---

The typical turtle, as most often used in laboratories, is the painted turtle (*Chrysemys*). The head is hard and bony at the anterior end, the jaws covered by the horny beak. The external nares are close together at the tip of the snout, allowing the animal to breathe air at the surface while exposing itself minimally. The eyes are well developed, with eyelids and a nictitating membrane in the anterior corner. Just behind the angle of the jaws is the external eardrum. The neck is long and flexible and folds dorsoventrally in most turtles (Cryptodira) to allow the head to be brought within the protective shell. In a smaller group of tropical turtles (Pleurodira) the neck bends around to the side and is thus exposed when the head is retracted; the bizarre matamata (*Chelys*) is in this group.

The trunk is remarkably broad and flat and encased in the shell, a dorsal carapace and a ventral plastron. Horny shields cover the shell, the epidermal scales of all reptiles, and if these are peeled off, one can see the bones that underlie them. Live turtles should be available to show how well the head, neck, limbs and tail can be folded under the margins of the shell. Note the very great protection provided against predation, perhaps important in the survival of this group. In the very best-protected turtles (box turtles, *Terrapene;* mud turtles, *Kinosternon*) the plastron is hinged and the whole shell can be closed tightly with nothing exposed. However, the young and eggs (buried just under the surface of the ground) are very much subject to predation by many animals, and even large adults are taken and crunched up by crocodilians. In *Chrysemys* the male's claws, used for tactile courtship, are considerably longer than the female's. In many turtles the male is smaller than the female and has a depression in the plastron for fitting on top of the female during copulation.

---

The turtles have radiated into all the major environments, with species specialized for marine, freshwater, and terrestrial existence. The marine turtles (*Chelonia, Dermochelys*) are very large (to 2 m in length and 450 kg in weight), and their limbs have been highly modified into flippers like those of marine mammals. They flap the forelimbs in unison to

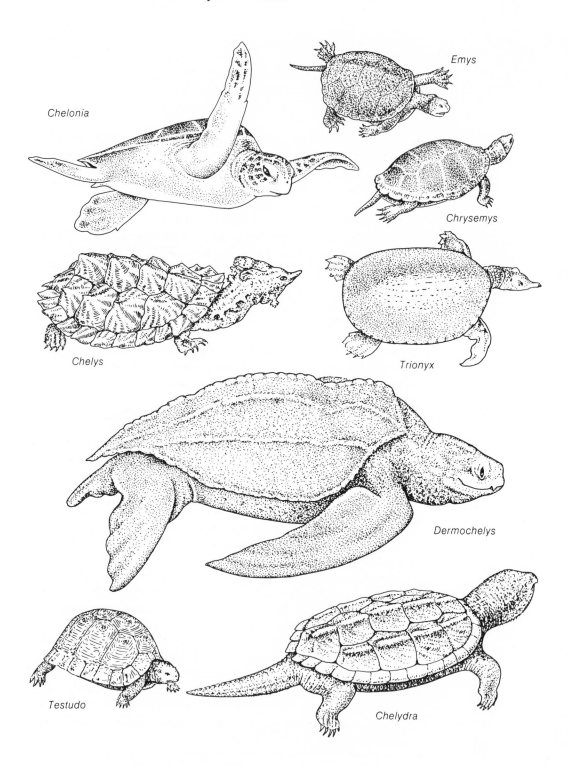

Fig. 2.7. Representative turtles. After Young 1950.

swim, whereas freshwater turtles swim with alternate strokes of diagonal limbs. How can you explain these marked locomotory differences in related organisms? Like all turtles, the marine species must come ashore to lay their eggs, but their powers of locomotion on land are poor. Still other turtles are specialized for a wholly terrestrial existence. These species, the tortoises (*Testudo, Gopherus*), have elephant-like hind feet, adapted to carrying a heavy load, and a high shell to conserve body water. The extreme adaptations to freshwater life are found in the soft-shelled turtles (*Trionyx*), which have reduced armor and are much flattened, with very long necks. They, and some other turtles, can actually respire under water to some extent.

b) *Tuatara.* The lizard form, slender with well-developed limbs and tail, is predominant among the suborder Lacertilia, the crocodilians, and the peculiar tuatara (*Sphenodon*) of New Zealand (fig. 2.8). This sole survivor of the Rhynchocephalia is the most primitive living reptile, but its primitive characters are chiefly skeletal. The young have a very well developed parietal eye, and there is no external copulatory organ, but otherwise the tuatara cannot be distinguished externally from a large lizard (adults are 50–80 cm in length). It is also peculiar in being active at night at lower temperatures (10°C) than any other reptile. If *Sphenodon* can do this, why can't other reptiles, which would make them more successful at higher latitudes?

c) *Crocodilians.* The crocodilians are very large (1–10 m) semi-aquatic reptiles primarily found in the tropics. Compare a crocodilian (alligator or crocodile) with a lizard and note the differences, especially those that signify the aquatic existence of the former. Crocodilians have large nostrils on the tip of the snout, as do turtles, prominent eyes, and external ear openings covered by skin folds. The external covering consists of horny thickenings, underlain by bony plates (osteoderms) on the back and arranged in a regular fashion as in lizards. The webbed feet and compressed tail are typical of aquatic vertebrates. Like turtles, these animals are vulnerable to predators in the egg or young stage, and in many species the female guards the nest and young. The large size of these carnivores probably has allowed them to persist through time in the face of potential predation by and competition with the more advanced vertebrates. But then why did the dinosaurs become extinct?

d) *Lizards.* Although lizards (suborder Lacertilia) have much the same body form as salamanders, they differ in many ways, most of them adaptations for a permanently terrestrial life. The neck is well defined, unlike that of fishes and amphibians, and the body is completely covered with horny scales. Unlike the scales of fishes, these are not discrete, and large parts of the epidermis are shed at once, at regular intervals. In many lizards the scales of the head are very regular and are named for the regions they overlie—for example, nasal, frontal, supraocular. The same condition prevails in snakes. As in all the higher vertebrates, air can be taken in either through the mouth or through the passageway from the external to the internal nares. In most lizards, the eyes are provided with a nictitating membrane and eyelids and there is usually an external ear, with the eardrum sunk into the bottom of a depression. The parietal eye can be seen under a translucent scale on the midline of the

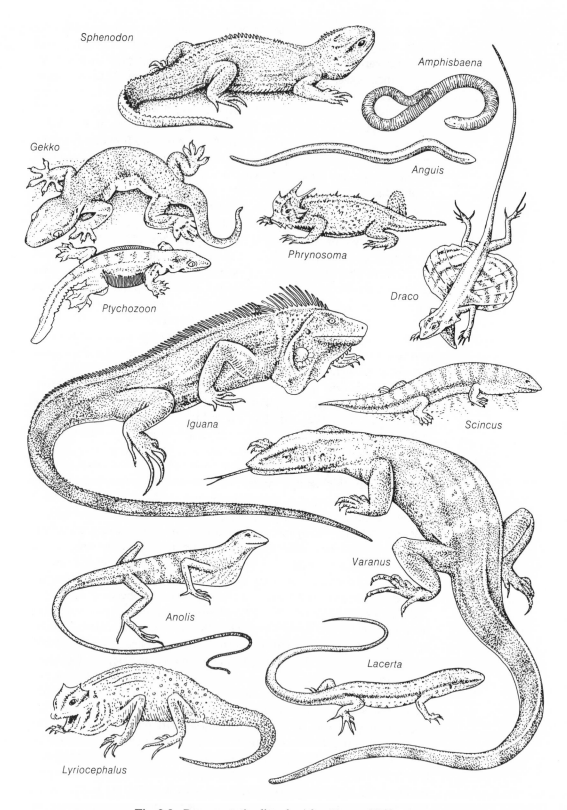

Fig. 2.8.   Representative lizards. After Young 1950.

head just behind the eyes in a variety of lizards. Its function remains a mystery. To what use do you think another visual receptor might be put?

The body structure is highly variable in lizards, from compressed to cylindrical to depressed (fig. 2.8), and the size ranges from 3 to 300 cm. Scalation varies from very coarse, with horns and tubercles, to glass-smooth. This correlates well with habitat; the burrowing forms are smooth for easy transition through the substrate, and the species that spend their time above ground and are thus more commonly exposed to predators (for example, horned lizards, *Phrynosoma*) need the ir-regularities of body surface for camouflage and defense. Many species possess skin folds that can be used in courtship (the gular fold or dewlap in *Anolis,* for example) or defense (the incredible display of the Aus-tralian frilled lizard, *Chlamydosaurus*). In the flying lizards (*Draco*) of southeast Asia, the skin folds along the side are used both in courtship and to support the lizard during its long glides from tree to tree. These folds are actually supported by extensions of the ribs and thus are more anatomically complex than the similar folds of flying squirrels.

The limbs and tail vary similarly. In typical terrestrial lizards the limbs are more effective in holding the animal up off the substrate than they are in salamanders. They typically bear five digits terminating in horny claws. In the hind limb the thigh still extends out at right angles to the body axis but is rotated slightly forward, so that the anatomically dorsal surface is tilted toward the anterior side. The shank is similarly rotated, but the foot retains the primitive position. In the forelimb, the upper arm is rotated in the opposite direction from the thigh, so that the ana-tomically preaxial surface is now dorsal. The upper arm is directed posteriorly, the forearm and hand downward and forward, and so ro-tated that their preaxial borders are facing toward the body axis rather than forward. Climbing lizards are modified in a number of ways, espe-cially in their digits. Chameleons (Chamaeleontidae) are compressed for easy passage between twigs and foliage, some being so highly com-pressed they are effective leaf-mimics. Both their feet and tails are pre-hensile, the former with two pairs of partially fused, opposing digits. Many geckos (Gekkonidae), similarly arboreal, are adapted to larger branches, tree trunks and rocks, and have expanded digits with adhesive pads, as do members of several other families. These pads are actually made up of a great number of very tiny hooks. In burrowing groups, especially the skinks (Scincidae), the limbs are variously reduced and the digits lost, until, in this, the Anguidae, and other families, there have evolved completely legless species, superficially like snakes and as well adapted as the latter group for burrowing. Some of these lizards lack ex-ternal ear openings and have even replaced the eyelids with a transpar-ent scale (spectacle) over the eyes as is the case also in snakes. Can you picture the sequence of evolutionary events in this transition?

In lizards of some families, prominent femoral pores on the under-side of the hind limbs secrete a waxy substance. The function of these pores is not certainly known, but they seem to have something to do with sexual activity. Preanal pores, probably of similar function, are found in fewer species. The vent is a transverse slit, an orientation characteristic of lizards and snakes. In males, either live or preserved, one or both of the copulatory organs (hemipenes) may be extended from the vent. In many lizards sexual dimorphism is highly developed, the males being larger and more brightly colored, but this is not so in other lizards or in snakes.

e) *Snakes.* In general, snakes (suborder Serpentes) can be recognized by their lack of limbs and eyelids, but, as we have seen, there are lizards with the same modifications. In fact, it may be necessary to examine the head skeleton to categorize some small burrowing squamates into one or the other of these suborders. It is apparent that snakes and lizards actually are closely related, although probably more distantly so than some birds that are considered in different orders (for example, kingfishers and woodpeckers). There can be relatively little variation on the legless snake theme although there is great size variation (20–1,000 cm). In most snakes, the transversely expanded ventral scales are important in locomotion, which typically is serpentine, accomplished as a series of sine waves travels down the body, pushing against fixed objects or the substrate itself. Some of the larger, heavy-bodied snakes move like caterpillars, with each ventral scale in turn pushing the body forward. Another mode is like the motion of a concertina, with one end of the snake fixed and the other end pushed or pulled forward by flexion of the body. Finally, some desert species, traveling in loose sand, move by sidewinding, in which the body touches the ground at two separate points and moves laterally. Aquatic snakes are little modified, with the exception of sea snakes (Hydrophidae), many of which have highly compressed tails. A number of modifications have evolved for an arboreal way of life—prehensile tails, I-beam construction (with both ventral and middorsal scales enlarged for structural rigidity), and extreme attenuation both for crossing from branch to branch and for mimicking branches.

Because they have no limbs, snakes have special problems in taking prey. The prey cannot be easily held or torn apart by the combined action of the mouth and legs, as in a turtle or mammal. So snakes must swallow their prey whole. Many species are able to suffocate captured prey by constriction, and of course, the development of venom for rapid killing reaches its most complex level of evolution among snakes. In some groups (Elapidae, Viperidae) enlarged hollow teeth are used to inject venom during a bite. In pit vipers (Crotalinae) the pit, just behind and below the nostrils, is a special heat-detecting apparatus that allows them to strike a warm-blooded prey accurately in darkness. Another special adaptation of this group, in particular the rattlesnakes, is the rattle, produced by a series of tail-tip scales that are shed but remain connected to one another. Do you think the rattle functions in offense or defense? Snake heads and jaws are highly modified for swallowing prey several times the diameter of the predator, with loose articulation, reduced bones, and expansible skin.

Snakes and lizards have been about equally successful in their radiation, with 2,700 species of snakes and 2,900 of lizards. Previously included with the lizards, the 130 species of amphisbaenians (fig. 2.8) are now known to be distinct from both that suborder and the snakes, and these legless, annulate, burrowing reptiles are placed in their own suborder (Amphisbaenia).

### c. Class Aves

Birds. 8,700 species. Feathers only unique characteristic; endothermal; oviparous; forelimbs modified into wings: occipital condyle single; quadrate free; teeth lacking in modern forms, replaced by horny rhamphotheca over bill; heart with two ventricles, no sinus venosus; no urinary bladder; many modifications associated with flight, such as hollow bones, air sacs for increased respiratory capacity, and huge eyes and cerebellum. Most characters used in classification (bill, feet, feather arrangement) are external, but there are minor differences in skeletal, muscular, and visceral anatomy as well.

Subclass Archaeornithes†. Ancestral, very reptile-like birds, with teeth, unfused metacarpals, and long tails. Represented only by the well-preserved *Archaeopteryx* of the Jurassic.

Subclass Neornithes. All modern birds. Teeth lacking (except in one Cretaceous group, the Hesperornithiformes); metacarpals fused; caudal vertebrae shortened into pygostyle. Members of most of the orders of birds are illustrated in figure 2.9.

Order Sphenisciformes. Penguins. 17 species. Flightless diving birds; wings used as paddles; dense scale-like plumage. Cold marine waters of the southern hemisphere.

Order Rheiformes. Rheas. 2 species. Large, flightless walking birds ("ratites") with three toes; plumage lax, wings and tail degenerate; without keel on sternum. South America.

Order Struthioniformes. Ostrich. One species. Largest living bird; like rhea, but with two toes. Africa.

Order Casuariiformes. Emus and cassowaries. 5 species. Ratites with three toes; feathers with long aftershafts, Australia, New Guinea.

Order Dinornithiformes. Kiwis. 3 species. Ratites with four toes; nostrils at tip of bill; feathers hair-like. New Zealand.

Order Tinamiformes. Tinamous. 42 species. Chicken-like but with primitive (ratite) palatal structure; keeled sternum and functional wings and tail. Tropical America.

Order Podicipediformes. Grebes. 18 species. Diving birds with pointed bill; legs far back on body; tarsus and toes compressed, even nails flattened; feet lobed; tail rudimentary.

Order Procellariiformes. Albatrosses, shearwaters, petrels. 95 species. Seabirds with tubular nostrils, through which oil is secreted; bill hooked; wings long and narrow, humerus very long; feet webbed.

Order Pelecaniformes. Pelicans, boobies, cormorants, darters, frigate birds, tropicbirds. 57 species. Marine (some freshwater) birds; bill pointed or hooked, external nostrils rudimentary or absent; with gular pouch (except tropicbirds); all four toes connected by webs (totipalmate).

Order Ciconiiformes. Herons, ibises, storks, flamingos. 123 species. Wading birds with long bill, neck, and legs. Great variation in bill shape for different feeding modes.

Loon
GAVIIFORMES

Grebe
PODICIPEDIFORMES

Albatross
PROCELLARIIFORMES

Penguin
SPHENISCIFORMES

Pelican
PELECANIFORMES

Stork
CICONIIFORMES

Goose
ANSERIFORMES

Hawk
FALCONIFORMES

Pheasant
GALLIFORMES

Rail
GRUIFORMES

Gull
CHARADRIIFORMES

Dove
COLUMBIFORMES

Parrot
PSITTACIFORMES

Cuckoo
CUCULIFORMES

Owl
STRIGIFORMES

Goatsucker
CAPRIMULGIFORMES

Hummingbird
APODIFORMES

Hornbill
CORACIIFORMES

Woodpecker
PICIFORMES

Swallow
PASSERIFORMES

Fig. 2.9.  Representative birds. After Romer 1970.

Order Anseriformes. Waterfowl (ducks, geese, swans) and screamers. 153 species. Waterfowl, swimming birds with broad lamellate bills, webbed feet; screamers, ground birds of South America, bill pointed, feet not webbed, internal anatomy as in waterfowl.

Order Falconiformes. Hawks, vultures. 287 species. Birds of prey; bill hooked, most with naked cere at base; wings long in most; feet raptorial, with long, sharp claws.

Order Galliformes. Gallinaceous birds. 274 species. Chicken-like birds; feathers with prominent aftershaft; wings short and rounded; bill short and stout; feet heavy, with short, broad claws; hind toe (hallux) elevated.

Order Gruiformes. Cranes, rails, and relatives. 204 species. Very diverse assemblage of mostly small, relict families with perforated nostrils. Ground or water birds.

Order Charadriiformes. Shorebirds, gulls, and others. 330 species. Also very diverse, most water birds, either swimmers or waders. Plumage compact; most with long wings. Alcidae (auks, murres, puffins) penguinlike.

Order Gaviiformes. Loons. 4 species. Diving birds with pointed bill; legs far back on body; tarsus compressed; feet webbed. North temperate.

Order Columbiformes. Pigeons. 306 species. Feathers very loosely attached; bill slender with operculum at base.

Order Psittaciformes. Parrots. 339 species. Bill hooked, with prehensile maxilla; tongue fleshy; feet zygodactyl, fourth toe reversible. Tropical.

Order Cuculiformes. Turacos and cuckoos. 150 species. Thin skin; long tail; feet zygodactyl, fourth toe reversible.

Order Strigiformes. Owls. 145 species. Nocturnal birds of prey with soft plumage; eyes huge, fixed forward-looking; ear openings large and asymmetrically located; bill hooked; feet raptorial.

Order Caprimulgiformes. Nightjars. 95 species. Nocturnal; cryptic mottled plumage; very thin skin; bill tiny but gape very large; rictal bristles large; feet small, ill suited for locomotion.

Order Apodiformes. Swift, hummingbirds. 410 species. Wings very long with very short humerus; feet small, ill suited for locomotion, covered with naked skin rather than scales as in other birds.

Order Coliiformes. Mousebirds. 6 species. Plumage hair-like; tail long and slender; first and fourth toes reversible, all can be directed forward. Mouse-like in shape, with weak flight. African.

Order Trogoniformes. Trogons. 34 species. Plumage iridescent; skin very thin, bill stout and flattened; feet weak. Tropical.

Order Coraciiformes. Kingfishers, bee-eaters, hornbills, and others. 195 species. Plumage bright; head large; bill long; feet syndactyl. Mostly tropical.

Order Piciformes. Woodpeckers, barbets, toucans, and others. 391 species. Rather like Coraciiformes but feet zygodactyl. Most families tropical.

Order Passeriformes. Perching birds. 5,048 species. This group is uniform in basic anatomy but very diverse in size, color, shape, and habits. All possess a well-developed incumbent hind toe for perching.

1)  *External Anatomy and Adaptive Radiation of Birds.* The rock dove, or domestic pigeon (*Columba livia*), is a convenient example of this class, and both intact and featherless specimens should be examined. The proportions of the body, when seen without feathers, are somewhat like those of a turtle, with compact trunk, long neck, and short tail (uropygium). The neck is highly flexible because of the number of vertebrae and type of vertebral connections, and the bird can easily reach every part of its body with its bill, which is the equivalent of the grasping organ formed by our own opposable thumb and fingers. The bird with all its feathers presents a rather different appearance, as the wings and tail are then prominent features and the neck is much less obvious in many species. All the small feathers of the body are called contour feathers, and they are arranged in clearly defined tracts, the pterylae (sing., pteryla); the spaces between them are called apteria (sing., apterium). These details can be seen on the plucked specimen. Note that some feathers are hairlike (filoplumes). The feathers important for flight are large and stiff. The wing and tail coverts cover these feathers and grade into the body contour feathers. Note that the skin is thin, one of the many adaptations to lower the weight for flying. It will be seen again and again that the anatomy of birds is highly modified in many ways as an adaptation to flight.

The head is large, with a prominent toothless beak. The earliest birds (*Archaeopteryx* and others) had teeth, but these relatively heavy structures were abandoned in favor of a horny sheath that encases the upper and lower jaws (maxilla and mandible). This sheath becomes worn and is continuously replaced, just as the epidermal scales of reptiles become worn, but the latter are shed and replaced all at once at intervals. The external nares vary from absent to prominent, and they may be surrounded by a soft cere or covered by an operculum (as in the pigeon) or feathers. The eyes are very prominent, those of a large owl, for example, being about the same size as human eyes. Visual perception is extremely well developed in this class, and it is obviously advantageous for a fast-moving aerial animal to have good distance vision and depth perception. The external auditory meatus (ear opening) is prominent, although covered by feathers in the intact bird. Hearing is also well developed in birds, but the olfactory sense is poor compared with that of other vertebrates.

The trunk is compact, with much fusion of the underlying skeleton for structural rigidity. The keel of the sternum, to which the large flight muscles are attached, is obvious on the ventral midline. The anterior limbs are modified into wings and are much altered from the primitive vertebrate condition that is characteristic of their reptilian ancestors. The orientation of the upper arm and forearm is much like that in reptiles, but the wrist and hand are fused, elongated, and directed caudad. There are only three digits, the first, second, and third. The first projects from the base of the hand and supports the alula, an important structure that prevents stalling during low-speed flight. The second digit forms the tip of the hand, and the third is reduced and cannot be seen externally. The outermost remiges (quill feathers; sing., remex), supported by the hand, are called primaries, the next series in, on the forearm, are called the secondaries, and those supported by the upper arm are the tertiaries. During flight the primaries travel the greatest distance and are very important in propulsion; the feathers closer to the base of the wing become increasingly instrumental in providing lift. You should watch some fairly large birds in flight for an understanding of the mechanics of this adaptation.

The hind limbs, then, are the only ones available for contacting the substrate, and all birds are bipedal. This pair of limbs has rotated 90° from the primitive condition, the original dorsal surface now facing anteriad. The more distal parts of the leg—the lower leg, ankle, and foot—are covered with scales homologous to those of reptiles. The tarsometatarsus is the prominent part of the leg proximal to the foot and is made up of components from the tarsals and metatarsals. The fifth digit is lacking from the foot, and the first is directed backwards as an adaptation for perching.

The vent lies beneath the uropygium, and on the top of the uropygium is a prominent papilla, the uropygial gland, which contains the oil the bird uses for preening its feathers. Unlike the mammals, which are abundantly supplied with skin glands, birds have only this one.

---

Most cities have museums with displays of mounted birds, which should be examined. Note especially body form and proportions, relative length and shape of wings and tail, and types of bills and feet. Be cognizant of the great amount of convergence of structures in different groups for the same function and how these structures, originally different, have been modified toward some common plan. Birds vary in size from tiny (50 mm, 3 gm) hummingbirds to the flightless ostrich (2.5 m, 135 kg). Through superficial characters such as the size and shape of feathers and especially the great array of color patterns, a great range of variation is evident in the group. If all the feathers were stripped from a variety of species of similar size within a bird order, they would look virtually identical, and only the relative size and shape of the bill would furnish a clue to the identity of each specimen. It is in bill shape that much of the adaptive modification in birds has been expressed (fig. 2.10). The bills of birds are specialized for their particular feeding modes, although there have been some further modifications for species recognition or display, for example the puffin bill, which becomes much larger and bizarrely colored in the breeding season, after which part of its outer covering is molted. The great range of bill sizes within taxonomic groups is indicative of and correlated with the sizes of food items taken by each species. Comparing the bills of a duck, a hawk, and a hummingbird will acquaint a student with the degree to which this particular structure has been modified for adaptive purposes. In the Anseriformes (ducks) the bill is large and flattened, with lamellate edges (not shown in the illustration) that strain water out while the duck feeds with its head submerged. The extreme is reached in the shovelers, which have huge, spatulate bills that strain plankton from the surface of the water. At the other extreme in this group are the geese, with much smaller, narrower bills and poorly developed lamellae; these birds graze on ter-

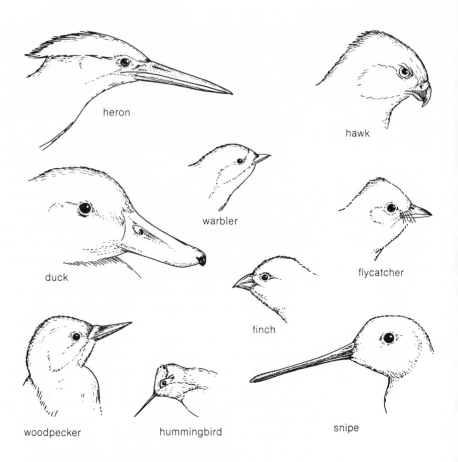

Fig. 2.10. Bird bill types. After Orr 1971.

restrial vegetation. Ducks, such as wigeons, that do much grazing have bills more like those of geese. The mergansers (*Mergus*) have become specialized as fish-eating ducks; their lamellae have become modified into tooth-like structures. The hawk's bill, with its sharp cutting edges and hooked tip, is highly modified for dissecting prey. It is larger and stronger in species that eat large vertebrates and more delicate in the insectivorous species. The falcons (*Falco*), which take very large prey, have an additional "tooth" on the bill for quick killing. Some species are highly specialized, as for example the snail kite (*Rostrhamus*), with its extremely long maxilla for extracting snails from their shells. Finally, the hummingbird's bill has become adapted for probing into flowers to feed on nectar. The tongue in this and other nectarivorous groups is modified into a tubular, brush-tipped structure. The coevolution of hummingbirds and the flowers from which they feed, and which they thus pollinate, has resulted in a great variety of both flower and bill shapes. For feeding in extremely long tubular flowers (thus avoiding competition with any other pollinators) the swordbill (*Ensifera*) has evolved a bill as long as the rest of the bird. Other hummers have deeply curved bills and feed on similarly curved flowers.

From the other adaptive types a number can be singled out for special comment. The very long terete bill of the snipe and other members of the family, although superficially similar to that of the humming-bird, has evolved as a tool for probing for buried invertebrates. Different bill lengths in this group are apparently adaptations for taking prey at different depths in the substrate, and any observer at the beach can watch the shortest-billed shorebirds picking food from the surface while the longer-billed species probe into sand or mud. In the deep-probing species, the bill is much softer than in most birds, and the birds probably can feel their prey with the sensitive tip. The bills of fish-catching birds are usually long and sharply pointed, used either as spears (herons) or forceps (loons). In some groups the tip is hooked (cormorants, gulls) for a better grip on slippery prey; in most of these birds the palate and tongue are provided with backward-pointing projections for additional grasping ability. Some fish-eaters dive for their prey, having sharp bills (terns, kingfishers) for seizing or huge pouches (pelicans) for engulfing the fish. The most common bill type is that of a warbler, a slender, pointed bill that serves as a pair of forceps for taking insect prey. This can easily be modified into an even more slender curved bill for probing in bark, as in creepers and woodcreepers. The woodpeckers are specialized for digging into or flaking off bark, and their chisel-like bill is very effective for this mode of life, as is their barbed tongue with which they extract insects. The heavy, conical bills of the finches are adapted to crack heavy seeds, and finch bills vary in shape and size according to the size of preferred seeds and the proportion of insects in the diet. An extreme modification of this bill type has resulted in that of the crossbills (*Loxia*), in which the tips of the mandibles are narrow and crossed for extracting seeds from under the cone scales of conifers.

Bird feet have become modified for several modes of nonaerial locomotion, with the length and arrangement of the toes greatly variable (fig-2.11). In cursorial (running) birds, there is a strong tendency for elevation and loss of the hallux. The extreme example of this condition is the ostrich, which has only two toes like the large artiodactyl mammals. No other bird has fewer than three toes. In wading birds such as herons and sandpipers the toes are elongate, offering better support on soft substrates, and the legs are usually long. Some birds, such as the jacanas, have even longer toes, allowing them to walk on floating vegetation. The neck length in such birds is usually correlated with the leg length, so that the bird's bill can reach the ground or the surface of water. Some of the wading birds also have partial webs between some of the toes to further increase the surface area of the feet. Complete webs between the front toes are characteristic of swimming birds and have evolved in many groups of swimming and diving species (ducks, gulls, loons, auks, penguins, petrels). All four toes are joined by webs in the order Pelecaniformes, although these birds are no better swimmers or divers than loons and auks. Another type of swimming foot has been developed in the coots, grebes, and phalaropes. In these groups each toe bears a series of lateral lobes, and in the grebes the nails are flattened as well. Note how far back on the body the feet are placed in swimming, and especially diving, birds, and how the tail is reduced in many of them. Why do you think the tail has remained long in some diving birds (cormorants, for example)?

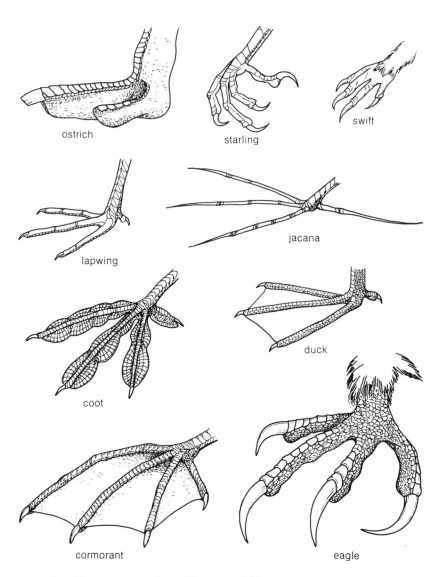

Fig. 2.11.  Bird foot types. After Thomson 1964.

The feet of birds of prey (hawks and owls) are specialized for holding and killing large prey, with deeply cleft toes and strong, sharp, curved nails. Fish-eating hawks like the osprey (*Pandion*) have especially long claws and very rough soles. The largest group of birds, the order Passeriformes, has become specialized for perching on twigs and branches. The hallux is long and at a level with the other toes, and the claws are sharp and curved. When the bird's weight is resting on the feet, tendons that are tightened by the flexion of the heel joint keep the toes bent, thus tightly grasping the perch even while the bird is asleep. Members of many other orders, even with differently constructed feet, can also perch effectively. The perching feet can easily be modified for walking, as in larks and pipits (straight nails, the hind one very long), or for climbing, as in creepers (sharper and more deeply curved nails). The woodpeckers represent a higher degree of specialization for climbing, with the outer toe reversed, thus providing firm support and even sideways motion on a vertical surface. Some birds that spend much time on the wing (swifts, swallows, hummingbirds) have relatively weak feet.

The wings of birds are also highly modified for particular kinds of flying. Some are flightless—the huge cursorial ostrich, rheas, emu, and cassowaries, the penguins, and a variety of birds that live on islands where there are no predators. Why do you think lacking wings would be advantageous to an island bird? The wings of flying birds vary from short, rounded, and almost useless to very long and pointed, the latter (as in swifts and albatrosses) perfect airfoils that permit gliding for hours with virtually no energy output. Certain birds, such as grouse and quails, generally walk but can escape predators by quick takeoff and rapid but relatively short flights, for which their short, very concave wings and rapid wingbeat are well suited. Finally, some birds (auks, penguins) use their wing as flippers for underwater swimming. The penguins are the most highly modified for aquatic life and are superficially much like marine mammals in their adaptations.

The tails of birds are also functionally modified as stabilizers and rudders in flight, as brakes in landing, and for balance while perched. Forked tails are adaptations for rapid directional change in flight by those species that take flying prey. In species that cling to vertical surfaces, such as woodpeckers, woodcreepers, and swifts, the tail feathers may be stiffened and pointed and the tail used as a prop. Tails have undergone much more radiation as display organs than have wings, and some are incredibly long and appear burdensome in flight.

### d. Class Mammalia

Mammals. 4,500 species. Hair and mammary glands only unique characteristics; endothermal; viviparous (oviparous in one order); two occipital condyles; zygomatic arch and secondary palate in skull; teeth in sockets; lower jaw a single bone, the dentary, articulating with the squamosal; quadrate reduced to a middle-ear ossicle; heart with two ventricles, no sinus venosus; with muscular diaphragm. Figure 2.12 illustrates mammalian radiation.

Subclass Prototheria. Oviparous; mammary glands without nipples; pectoral girdle with separate precoracoid, coracoid and interclavicle; oviducts separate; cloaca present; toothless.

Order Monotremata. Monotremes. 6 species. The relict duckbill platypus and echidnas of Australia.

Subclass Metatheria. Viviparous; abdominal skin pouch (marsupium), supported by two epipubic bones, into which embryos crawl and into which mammary glands open; lacking typical placenta; precoracoid and interclavicle absent; vagina double; no cloaca.

Order Marsupialia. Marsupials. 242 species. Australia (most) and New World tropics.

Subclass Eutheria. Viviparous; with allantoic placenta; precoracoid and interclavicle absent; vagina single; no cloaca; mammary glands with external nipples.

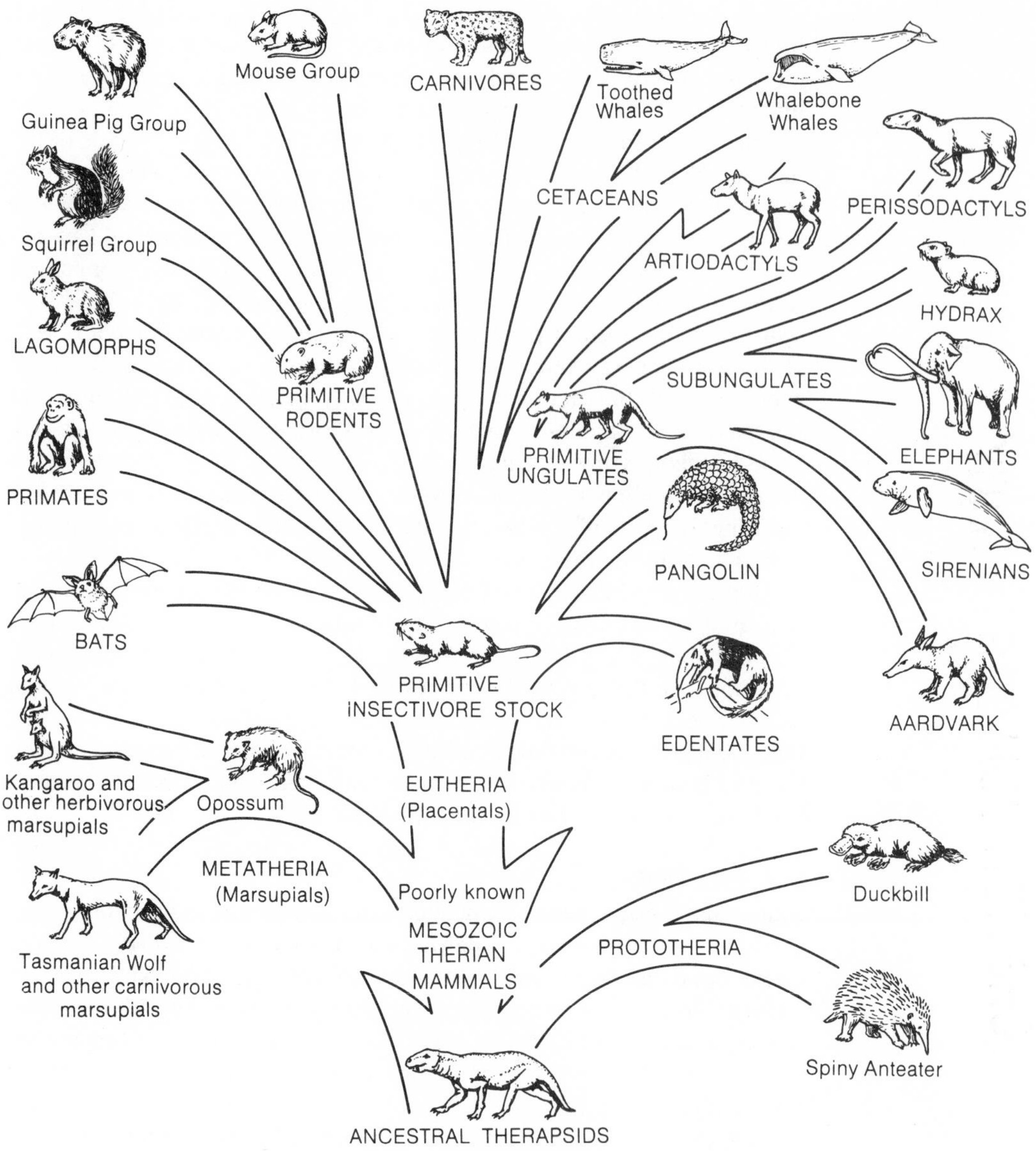

Fig. 2.12.  Phylogeny of mammals. After Romer 1970.

Order Insectivora. Shrews, moles, and others. 409 species. Generally small; snout elongate; feet plantigrade, toes with claws; dentition primitive, the many similar teeth with sharp cusps; no auditory bulla; uterus bicornuate.

Order Chiroptera. Bats. 915 species. Only flying mammals; metacarpals and phalanges greatly elongated to support wing membranes; thumb and hind feet with claws; dentition as in insectivores.

Order Primates. Primates. 176 species. Generalized omnivores; fingers and toes long, nails flat, thumb opposable and gait plantigrade in most; orbital and temporal depressions separated by bony ridge; brain highly convoluted, with large cerebrum. Tropics; humans ubiquitous.

Order Edentata. Anteaters, sloths, armadillos. 31 species. Feet strongly clawed; teeth reduced or lacking. New World tropics.

Order Pholidota. Pangolins. 8 species. Covered (except underside) with horny overlapping scales; snout and tongue long; teeth, zygomatic arch, and clavicles lacking. Old World tropics.

Order Lagomorpha. Rabbits, pikas. 63 species. Ears large; tail short; feet clawed, gait digitigrade; dentition as in rodents, but with a small pair of incisors immediately behind the large upper pair; caecum large.

Order Rodentia. Rodents. 1,685 species. Feet clawed, gait plantigrade; a pair of upper and lower incisors chisel-like, growing throughout life, with enamel on only their front surfaces; diastema between incisors and cheek teeth; caecum large.

Order Cetacea. Whales, dolphins. 69 species. Aquatic (most marine); hairs only on muzzle; shape streamlined, with no apparent neck; forelimbs modified into paddles; a broad horizontal tail fin (fluke); pelvic girdle and hind limbs vestigial.

Order Carnivora. Dogs, bears, weasels, civets, cats, seals, and others. 284 species. Dentition generally for carnivory—canine teeth prominent, anterior cheek teeth sharp for cutting; auditory region well developed, usually with expanded bulla; clavicles reduced or absent. Suborder Pinnipedia (seals) with feet modified into flippers, nails reduced; other groups are in suborder Fissipedia, not highly modified for aquatic life.

Order Tubulidentata. Aardvark. 1 species. Little hair; snout and tongue elongate; teeth reduced, with perforated dentin and no enamel. Africa.

Order Proboscidea. Elephants. 2 species. Little hair; nose and upper lip drawn out into a trunk; two or four incisors enlarged into long tusks; legs without bend at knee and elbow; feet heavy, with hoof-like nails; no clavicles. African and Asian tropics.

Order Hyracoidea. Conies. 11 species. With rodent-type incisors, ungulate-type molars; plantigrade gait; toes four on forelimbs and three on hind limbs, nails flat; no clavicles. Africa and Near East.

Order Sirenia. Dugongs, manatees. 5 species. Little hair; superficially like cetaceans (aquatic with fore flippers and tail fin) but with ungulate-like dentition. Tropical coasts and rivers.

Order Perissodactyla. Horses, tapirs, rhinoceroses. 16 species. Cheek teeth broad, with grinding ridges; hoofed, third digit forming limb axis and others reduced or absent; unguligrade gait; no clavicles; no gallbladder.

Order Artiodactyla. Pigs, camels, deer, giraffes, sheep, antelopes, and others. 171 species. Cheek teeth broad, with round or crescentic cusps; hoofed, third and fourth digits equal in size, limb axis passing between them, others reduced or absent; unguligrade gait; no clavicles; no gallbladder.

1) *External Anatomy and Adaptive Radiation of Mammals*. In this class, the head, neck, trunk, and tail are well developed, but the head is larger and the tail smaller than in the typical tetrapod. There is such variation in shape and relative limb and tail proportions in mammals that this description will refer primarily to the cat (*Felis catus*).

---

The body is clothed with closely set hairs, forming a covering of fur.

The hairs are developed into vibrissae, or whiskers, around the nose or mouth, these long, sensitive structures serving as tactile organs. The head is prolonged into a facial region in front of the eyes, with the cranial region behind the eyes much expanded for the large brain. The lips are well developed and when they are retracted the prominent teeth can be seen. The external nares are large and contained in a distinct nose. This is the only class in which the nares may be moved independently of the head ("sniffing"), indicating the great importance of the olfactory sense in mammals. The eyes have upper and lower lids and a well-developed nictitating membrane. The external ears are composed of the external auditory meatus and prominent pinna, the latter even more mobile than the nose and easily oriented toward sound from any direction.

The trunk is divided into the anterior thorax, supported by the ribs, and the posterior abdomen. On the ventral surface of females are four or five pairs of nipples, the openings of the mammary glands. Toward the posterior end of the ventral surface is the vulva, consisting of the urogenital opening and the labia majora surrounding it. Behind this is the anus, just under the base of the tail.

In the male the scrotum, containing the paired testes, lies in front of the anus, and just anterior to the scrotum is the copulatory organ or penis. The penis is hidden within the prepuce (foreskin) and is extended only for copulation. Note the separation of the external openings of the digestive and urogenital systems; the area around them both is the perineum.

The limbs have undergone a marked change from the primitive position. Instead of extending laterally, they project ventrally and are elongated, so that the body is carried high above the ground. This change has involved a 90° rotation of each limb. The forelimb has rotated posteriorly, so that the postaxial side is inside and the preaxial side is outside. If this were done simply, the toes would point backward, so that further compensation has been necessary, in this case the rotation of the preaxial bone (radius) over the postaxial bone (ulna). Thus in the forearm the preaxial part of the limb is inside. In primates, but not in most other mammals, the radius can be rotated over the ulna to lie on either side, thus allowing the hand to rotate 180°, giving these animals greater manual dexterity. Try this with your own hand: extended in front of you, it is in the prone position when the palm points downward, in the supine position when the palm points upward. You can feel the radius rotating over the ulna. In the hind limb the rotation is forward, so that the preaxial side is inside. Note that with this arrangement the limbs lie in about the same vertical plane as the body, and locomotion can be very efficient in speed, balance, and agility. You should reason this out yourself before proceeding. In the cat there are five digits on each front foot and four on each hind foot (the first is missing). The claws are retractile, as anyone who has had a pet cat knows. Why are cats different from other carnivores in this regard? Finally, examine the tail, which is well developed in this species. It is highly mobile, with many vertebrae and individual muscles, and it serves as an organ of balance during running and jumping. What other functions does it have in the cat?

---

Although it is one of the two most recently evolved classes of vertebrates, the Mammalia encompasses a great array of body forms and modes of locomotion, certainly surpassing the birds, which are limited by their primary means of locomotion, flight; and even this bird specialty is shared with one group of mammals.

Mammals vary in size more than any other vertebrates, from tiny shrews weighing a few grams to the blue whale (*Balaenoptera musculus*), the largest animal that has ever lived, which weighs more than 100 tons and can be up to 30 meters long. On the other hand, the largest land mammals, the elephants and the even larger extinct rhinoceros *Baluchitherium,* never reached the sizes attained by the largest dinosaurs.

The basic mammalian body shape, as exemplified by primitive mammals like the opossum (*Didelphis marsupialis*), has been much modified during adaptation to different locomotory modes. The adaptive radiation in modes of locomotion in some ways parallels the evolution of mammalian orders, like the bats, or suborders, like the pinniped carnivores, but there has also been much convergence of members of different orders toward common body plans best adapted for particular ways of life. This is especially obvious in, for example, aquatic mammals and burrowing mammals.

Familiar mammals such as the opossum or raccoon (*Procyon lotor*) serve as prototypes for the ambulatory mode of locomotion, a mode that may be at the base of the locomotory radiation within the class. These animals walk on the soles of the feet (plantigrade locomotion) and can run, swim or climb with some facility. Note that with progressive specialization for one of these locomotory modes, ability in one or more of the others will be reduced accordingly. Adaptations for running, especially common in mammals of open country, include lengthening of the legs, reduction of foot bones, and a tendency to stand on the tips of the digits (digitigrade). Many carnivores (dogs, cats) illustrate this type, and ungulates as a group represent the extreme modification (unguligrade), in the horses (Equidae) only the hoof (claw) of a single digit touching the ground. The lengthening of the limbs allows longer strides, the reduction of bones effects streamlining, and running on the tips of the digits reduces friction with the substrate. With selective advantages like these, why are there no antelopes, for example, with legs even longer and more slender?

Some of the heavier mammals (rhinoceros, elephant) have much-thickened limb bones and a completely vertical alignment of the girdles and limb bones to support their immense weight. The smaller, more agile running mammals stand with the upper limb bones at an angle to the lower ones and to the girdles and are very loosely put together, with the scapula riding free among the dorsal muscles (as in all mammals) to absorb the shock produced in running and jumping. One line of locomotory evolution, found independently in the kangaroos (Macropodidae), the hares and rabbits, and a variety of rodents, from tiny mice to the large springhaas (*Pedetes*), has led to modification of the hind limbs for jumping. The hind feet and legs are elongate, and locomotion may be bipedal, the front limbs not touching the ground. By this mode of locomotion, jumping mice (Zapodidae) can leap many times their own length, land, change direction, and leap again, a valuable antipredator strategy. The long tails of many of these jumping mammals are important for balance, and a jumping mouse tends to turn somersaults in midair if its tail is missing. From what you know of jumping animals, is a long tail an absolute necessity?

Many terrestrial mammals can climb, especially those that are not extremely modified for other modes of locomotion. Many groups possess members that are specially modified for climbing, and sharp claws for this purpose characterize many rodents and their carnivore predators. Sloths (Bradypodidae), virtually permanently arboreal, hang upside down from very large, curved claws—two or three per foot. Among the rodents, the squirrels (Sciuridae) are excellent climbers, and many species have enlarged, fluffy tails that are used as balancers during long jumps from branch to branch. A variety of marsupials and mice possess the same modifications, but the primates as a group show the greatest adaptive radiation in climbing abilities, from squirrel-like species to the specialized gibbons (*Hylobates*) and others that move rapidly through the forest hanging by their forelimbs (brachiation). The rotatable forearm and opposable thumb are important in climbing as well as in manipulating objects. The evolution of a prehensile (grasping) tail characterizes many mammals of tropical forests, especially in the New World. Gliding certainly arose as an offshoot of climbing, with selection for the ability to make longer and longer jumps. Mammals like flying squirrels (*Glaucomys*) and marsupial gliders have loose folds of skin along the sides of the body and flattened tails, all of which are spread wide during a jump, providing considerable lift. Some of these animals can glide for hundreds of meters.

Even the open air has been conquered in the radiation of this class. Presumably from insectivore ancestors, bats long ago developed the power of flight, and this abundant group takes the place of the primarily diurnal birds as darkness falls, feeding on insects, fruits, nectar, fish, other bats, and even blood. Locomotory modifications include greatly elongated forelimb bones with skin stretched over them; skin similarly placed between the hind limbs and tail; sharp claws, both on the feet and on the thumbs, for hanging from a variety of substrates; and the echolocation apparatus, involving the often-modified nose and ears, an important correlate of night flying.

Mammals of several orders inhabit the soil, either digging burrows for protection and for raising young or permanently burrowing through it. To the latter group belong the moles (Talpidae) and members of a variety of rodent families. This mode of locomotion involves the evolution of a thick, cylindrical body, short and powerful limbs for digging, reduced eyes and external ears, and fur with no definite direction, to allow backing up in the burrow (rub the fur of a mole if possible), or virtually no hair at all in one genus (*Heterocephalus*).

Finally, a series of examples of mammals illustrates the progression toward a fully aquatic life. Most mammals can swim, and many are variously modified to do so. Water shrews (*Sorex*) have fringes of hair on their hind feet, otters (*Lutra*) and muskrats (*Ondatra*) have webbed feet and propulsive tails. All these animals can travel very effectively on land as well, but many of the pinnipeds move only awkwardly on the islands where they go ashore to breed. The true seals (Phocidae) are more highly modified than are the sea lions (Otariidae); the former swim like fishes, with lateral undulations of the vertically oriented hind limbs, and the latter use the forelimbs as paddles. The cetaceans are the most highly modified, never leaving the water and convergent in many ways with the fishes. They are genuine mammals, however, with sparse hair and mammary glands, breathing air and regulating their body temperature within narrow confines. Blubber (fat) under the skin has replaced the fur of other mammals as an insulating material.

As a class, the mammals primarily utilize their senses of smell and hearing to find food, escape predators, and interact socially. The eyes, although well developed in most species, are less important than they are in birds. Some diurnal open-country mammals, however, have excellent eyesight, and in the primates, particularly in humans, vision comes close to being as effective as it is in the birds. The stereoscopically oriented eyes of primates are adapted to judging distance during arboreal locomotion, whereas the eyes of rodents, rabbits, and so forth, are placed laterally to allow nearly full 360° vision for locomotion and predator-detection at the same time. To watch a rabbit is very instructive; the large ears swivel toward the slightest sound; the nose twitches, sifting the incredible arrray of molecules in the air to detect those that signal food, potential predators, or other rabbits; and the large eyes, with retinal rods for well-developed night vision, can be directed at the least signal from the other receptors. Can you explain why most animals are not provided with all three: excellent vision, hearing, and smell? Because most mammals do not have color vision and interact with one another more through their olfactory and tactile senses, few are brightly pigmented. Only some primates and squirrels approach the beautiful colors of the birds, lizards, and fishes.

Mammals exhibit a wide variety of defenses against predators, especially the strategy of remaining quiet and inconspicuous but running away quickly when detected. Some of them possess teeth or claws or both that are effective defenses in themselves, and others depend on body armament (turtle-like scales in armadillos and pangolins, spines in porcupines and hedgehogs).

Convergence among mammals is well illustrated by those species that specialize in eating ants or termites. Members of five different orders, including the anteaters (Edentata), pangolins (Pholidota), aardvark (Tubulidentata), aardwolf (Carnivora), and echidna (Monotremata), have become specialized for taking this particular type of food. Adaptations include thick skin, frequently the lack of teeth, an elongate snout with a long protrusible, sticky tongue, and heavy, sharp claws for tearing open nests. Anyone who has been stung or bitten at an ant nest cannot fail to be impressed by the sight of one of these animals going about its meal with ants swarming over its head.

**F. Patterns in Vertebrate Evolution**

Assuming that the first vertebrates were *Branchiostoma*-like creatures, substantial changes have taken place during the adaptive radiation of this group. Details of these morphological changes are considered for each system in the following chapters of this book. The effects of such changes are summarized here. The course of evolution of fish-like vertebrates especially has involved the production of more and more efficient swimmers. Paired fins, when they appeared in the early filter-feeders, allowed entry to middle and surface waters, opening up new adaptive zones where different kinds of food resources were present. The large and similar pectoral and pelvic fins and heterocercal tail must be effective swimming organs, since they have persisted in recent sharks and primitive bony fishes, which are quite able to find their prey and each other and to avoid their predators. But with progression through time we can see the two pairs of fins becoming more dissimilar, each pair differently specialized. The pectorals now can be used as brakes or sculling organs as well as for stability, and the pelvics have moved forward and underneath the pectorals, where they can be depressed for streamlining or suddenly extended for braking or turning. In many species they are further modified as "walking" organs or suction cups. These advanced fishes can hang in one place in the water column or move slowly through dense vegetation, their gas bladders allowing them to regulate their buoyancy and their homocercal tails providing more efficient forward thrust than their heterocercal antecedents.

Early in the course of this evolution toward more efficient swimming, the fishes changed from jawless filter-feeders to jawed predators. The primitive types have large mouths with abundant teeth, arising from the ancestral anterior gill arch that was already modified to keep larger items that entered the mouth from passing through the gills. The more primitive fishes rush toward their prey and seize them in these toothy mouths, but with teleost evolution a more subtle predatory mode appeared. In this type the head became shorter and the oral cavity larger, so the prey could be engulfed as described previously. With this type of predator lurking in the vegetation, it became harder for smaller fish to detect and escape predation by rapid swimming, and the switch from soft to spiny fin rays, especially in the dorsal and anal fins, probably came about as an adaptation against predation.

While all this was happening, some primitive fishes found themselves in environments that were not conducive to the survival of the fish adaptive type, and there was strong selection for those types that could withstand increasing drought and decreasing oxygen levels in the water. The abilities to move on land and to respire aerially were the keys to survival. The primitive bony fishes must have lived in rather stagnant environments, since many of them can breathe air and presumably could do so millions of years ago when they underwent their radiation. The lobed fins of the crossopterygians provided the base on which the tetrapod limb was built, and at some time during this radiation the air-breathing, land-crawling amphibians appeared. Again, note that this provided the vertebrates with entry into a totally new adaptive zone, which by that time was populated with invertebrates that provided a new food resource.

In a world becoming ever drier, as ours did in the late Paleozoic era, those animals that could adapt to the absence of surface water were favored in many areas, and the development of desiccation-resistant epidermal scales was an integral factor in the success of the reptiles from that time on. In addition, the hard-shelled and relatively impermeable (cleidoic) egg became a feature of the reptiles that allowed them even greater independence from water. As so often happens in evolution, once the reptiles were an evolutionary success many of them moved into adaptive zones in which the amphibians had been dominant and largely replaced them.

Finally, the latitudinal gradient in temperature on the earth precluded the colonization of high-latitude regions by organisms whose body temperature was largely controlled by the environment. Any animal that could regulate its temperature independently and keep its metabolism and therefore its activity levels up was more efficient at foraging and avoiding being eaten than existing terrestrial vertebrates, and thus animals that ultimately were to become birds and mammals began to diverge from their reptilian ancestors. Both groups developed more efficient circulatory systems and epidermal insulation and are now able to thermoregulate within narrow and constant limits over a great range of ambient temperatures. Parental care, much better developed in birds and mammals than in other vertebrates, probably was also important in the move to high latitudes.

There were changes in many aspects of the morphology of vertebrates during this adaptive radiation. The typical fusiform vertebrate body has been modified into almost all imaginable shapes, from spheres to disks to ribbons to very elongate cylinders, with surface irregularities of all sorts. The limbs vary from absent to very well developed, and in some cases only the anterior pair is present. Both fish fins and tetrapod limbs vary as much in shape as do the bodies of their owners, and they also vary greatly both in absolute size and relative to body size. The loss of digits, especially the outer ones, is common. The primitive tetrapod condition consists of the limbs extending laterally from the body, hardly supporting it, and basically fish-like locomotory movements. In the more advanced groups the limbs increasingly support the body above the substrate, and the improved efficiency of locomotion parallels the increased efficiency in swimming brought about by changes in the fins of fishes. Three groups of vertebrates independently evolved the power of flight —the pterosaurs among the reptiles, the birds, and the bats among the mammals. Look at wings or diagrams of wings of members of these three groups and decide which wing type is most efficient.

The outer covering of the body varies from soft, with abundant glands and with or without protective dermal scales, to relatively hard with epidermal structures (scales, feathers, hair) that furnish even more protection from the environment. The mammals are somewhat convergent with the amphibians, both groups having glandular skin and well-developed olfactory senses.

The head tends to increase and the tail to decrease in size relative to the trunk with the evolution of more advanced vertebrates. The former change is associated with the increase in size of the brain, the latter with increased speed and agility. The three pairs of sense organs on the head vary greatly in some groups and relatively little in others. The eyes, as would be expected, are relatively larger in those groups in which vision is important, especially the birds, and they degenerate or disappear in burrowing or cave-dwelling species. The nostrils open into a blind sac in fishes, but in land animals they connect with the pharynx, allowing the animals to breathe without opening the mouth, so that the mouth cavity does not dry out. In the fishes and salamanders the ear is entirely internal, but in frogs there is a middle ear, closed externally by the tympanic membrane. In some reptiles this membrane has sunk beneath the surface, and there is an opening into it. In mammals the external ear is even more complex, with the sound-catching and amplifying pinna around the opening.

The gill slits and gills present in fishes and larval amphibians are not present in most adult amphibians or any of the other tetrapod vertebrates, all of which breathe air directly, through either the skin or the lungs or both.

In most bony fishes, cyclostomes, amphibians, and birds, there are no external genitalia, fertilization being external (fishes and frogs) or internal by cloacal apposition (birds) or the female's picking up a spermatophore from the substrate (salamanders). In elasmobranchs and certain families of bony fishes the pelvic fins and anal fins, respectively, of the males are used as intromittent organs, and in most reptiles and mammals there are single or paired (one used at a time) copulatory organs. The placental mammals, teleost fishes, chimaeras and cyclostomes have separate openings for the urogenital and digestive tracts, but in all other vertebrates the two tracts empty into a cloaca, which then opens to the exterior.

Like the ocean's surface or the edge of a forest, the outside of a vertebrate is only a tantalizing suggestion of the complexity within; but it is surely worth an extended look and full understanding before further penetration.

**References**

Austin, O. L., Jr., and Singer, A. 1961. *Birds of the world.* New York: Golden Press.

Bellairs, A. 1969. *The life of reptiles.* 2 vols. London: Weidenfeld and Nicolson.

Budker, P., and Whitehead, P. 1971. *The life of sharks.* New York: Columbia University Press.

Carter, G. S. 1967. *Structure and habit in vertebrate evolution.* Seattle: University of Washington Press.

Cochran, D. M. 1961. *Living amphibians of the world.* Garden City, N.Y.: Doubleday.

Cockrum, E. L. 1962. *Introduction to Mammalogy.* New York: Ronald Press.

Colbert, E. H. 1955. *Evolution of the vertebrates.* New York: John Wiley. (Paperback by Science Editions, 1961.)

Darlington, P. J., Jr. 1957. *Zoogeography.* New York: John Wiley.

Farner, D. S., and King, J. R. eds. 1971. *Avian biology.* Vol. 1. New York: Academic Press.

Fisher, J., and Peterson, R. T. 1964. *The world of birds*. Garden City, N.Y.: Doubleday.

Goin, C. G., and Goin, O. B. 1971. *Introduction to herpetology*. San Francisco: W. H. Freeman.

Greenwood, P. H.; Miles, R. S.; and Patterson, C., eds. 1973. Interrelationships of fishes. *Zool. J. Linnean Soc. Lond.*, vol. 1, suppl. no. 1, 536 pp.

Greenwood, P. H.; Rosen, D. E.; Weitzman, S. H.; and Myers, G. S. 1966. Phyletic studies of teleostean fishes, with a provisional classification of living forms. *Bull. Amer. Mus. Nat. Hist.* 131:339–456.

Halstead, L. B. 1969. *The pattern of vertebrate evolution*. Edinburgh: Oliver and Boyd.

Herald, E. S. 1961. *Living fishes of the world*. Garden City, N.Y.: Doubleday.

Lagler, K. F.; Bardach, J. E.; and Miller, R. R. 1962. *Ichthyology*. New York: John Wiley.

Lineaweaver, T. H., III, and Backus, R. H. 1969. *The natural history of sharks*. Philadelphia: Lippincott.

Marshall, N. B. 1965. *The life of fishes*. London: Weidenfeld and Nicolson.

Matthews, L. H. 1969, 1971. *The life of mammals*. Vols. 1 and 2. New York: Universe Books.

Noble, G. K. 1931. *The biology of the amphibia*. New York: McGraw-Hill (Dover reprint).

Olson, E. C. 1971. *Vertebrate paleozoology*. New York: Wiley-Interscience.

Orr, R. T. 1971. *Vertebrate biology*. Philadelphia: W. B. Saunders.

Pettingill, O. S., Jr. 1971. *A laboratory and field manual of ornithology*. Minneapolis: Burgess.

Porter, K. R. 1972. *Herpetology*. Philadelphia: W. B. Saunders.

Romer, A. S. 1959. *The vertebrate story*. Chicago: University of Chicago Press.

————. 1966. *Vertebrate paleontology*. Chicago: University of Chicago Press.

————. 1968. *Notes and comments on vertebrate paleontology*. Chicago: University of Chicago Press.

————. 1970. *The vertebrate body*. 4th ed. Philadelphia: W. B. Saunders.

Schmidt, K. P., and Inger, R. F. 1957. *Living reptiles of the world*. Garden City, N.Y.: Doubleday.

Simpson, G. G. 1945. The principles of classification and a classification of mammals. *Bull. Amer. Mus. Nat. Hist.* 85:1–350.

Stahl, B. J. 1974. *Vertebrate history: Problems in evolution*. New York: McGraw-Hill.

Thomson, A. L., ed. 1964. *A new dictionary of birds*. London: Thomas Nelson & Sons.

Van Tyne, J., and Berger, A. J. 1959. *Fundamentals of ornithology*. New York: John Wiley.

Vaughan, T. A. 1972. *Mammalogy*. Philadelphia: W. B. Saunders.

Walker, E. P. 1964. *Mammals of the world*. 3 vols. Baltimore: Johns Hopkins Press.

Welty, J. C. 1962. *The life of birds*. Philadelphia: W. B. Saunders.

Young, J. Z. 1950. *The life of vertebrates*. New York: Oxford University Press.

# 3 Essential Features of Lower Types
## E. J. W. Barrington

The anatomy of members of the phylum Hemichordata, a prechordate group, the chordate subphyla Urochordata and Cephalochordata, and the class Cyclostomata of the subphylum Vertebrata are discussed in this chapter. Special attention is paid to the chordate features and relationships of these groups and to their functional biology.

## A. Phylum Hemichordata: A Prechordate Group

Although the Hemichordata are here separated from the Chordata as a distinct phylum, some authorities prefer to regard them as a subphylum of the phylum Chordata. Reasons for this difference of view will be given later. The hemichordates are divided into two classes: the colonial deep-sea Pterobranchia and the worm-like Enteropneusta. Only the latter group is considered here, the details referring largely to *Saccoglossus kowalevskyi* from the New England coast. (*Saccoglossus* Schimkewitsch 1892 has priority over *Dolichoglossus* Spengel 1893.)

## 1. External Features

Examine the enteropneust in a dish of water. It is an elongated worm-like animal five or six inches long, with a ciliated epidermis that also contains a variety of glandular cells. The epidermal secretion forms a lining for the burrow in which the animal lives and contributes to feeding. The body can be divided into three regions: *proboscis, collar,* and *trunk* (figs. 3.1 and 3.6). Each of these contains a separate division of the coelom, which is thus tripartite, like that of echinoderms. The proboscis is the elongated conical structure at the anterior end, used for burrowing into the sand; it is much longer in *Saccoglossus* than in other genera. The collar is the band encircling the body just behind the proboscis. The proboscis is attached to the inner surface of the dorsal side of the collar by the slender proboscis stalk, below which the collar rim encloses the large, permanently open mouth leading into the buccal cavity inside the collar.

At the base of the proboscis is the preoral ciliary organ, a U-shaped groove bordered by strong cilia.

The proboscis coelom opens to the outside at its base by a *proboscis pore,* and the collar coelom similarly opens by a *collar pore* (or in some species by a pair of pores). The turgor of these coelomic cavities is maintained during movement by water passing in and out through the pores.

The greater part of the animal consists of the trunk, more or less ruffled, with a midventral and middorsal longitudinal ridge. The trunk

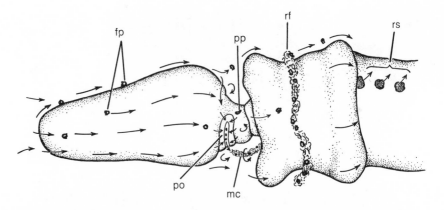

Fig. 3.1. Lateral view of the anterior end of *Protoglossus kohleri;* the direction of movement of food and mucus is indicated by arrows: *fp*, food particles; *mc*, mucous cord, with fine particle leaving the preoral ciliary organ; *po*, preoral ciliary organ; *pp*, proboscis pore; *rf*, rejected material collecting in a band around the collar; *rs*, respiratory stream leaving the gill pores. From Burdon-Jones 1956 (*Kukenthal-Krumbach's Handbuch der Zoologie* 3:57).

can also be divided into three regions—*branchiogenital, hepatic,* and *abdominal* regions—much better marked in some species than in *S. kowalevskyi*. The branchiogenital or thoracic region forms the first part of the trunk and contains the sac-like gonads, which cause bulges on each side known as the genital ridges. In some genera these project as conspicuous, winglike expansions. The sexes are separate. Between the anterior portions of the genital ridges on the dorsal side will be found two longitudinal rows of small parallel slits, the *gill pores,* which are not the true gill slits (see below). The hepatic region, usually a little shorter than the genital region, is so named because here the intestine commonly bears a paired series of lateral pouches, the so-called *hepatic caeca*. These show externally as bulges in only a few enteropneusts (not in *S. kowalevskyi,* which even lacks definite out-pouchings). Although these caeca apparently have some digestive function, there is no evidence that they in any way represent the vertebrate liver. The remainder of the trunk contains a simple tubular intestine opening by a terminal anus.

## 2. Salient Points of the Internal Anatomy

Besides studying the whole animal, you should examine prepared microscope sections. The mouth and buccal cavity lead into the pharynx, which, with its *gill slits* in its dorsolateral walls, is the most chordate-like feature of enteropneusts. The gill slits, a series of vertically oriented openings, are oval in early stages but become U-shaped by the downgrowth from the dorsal end of each slit of a process called the *tongue bar* (fig. 3.2). These tongue bars and the septa between successive gill slits are supported by M-shaped skeletal rods. The gill slits become further subdivided by the development of horizontal bars called *synapticulae*. The tongue bars closely resemble the similarly named structures in *Branchiostoma* and are the main feature held to support the inclusion of hemichordates in the phylum Chordata. The

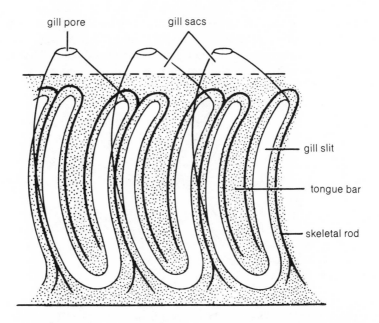

gill pore          gill sacs

gill slit

tongue bar

skeletal rod

Fig. 3.2. Piece of the pharyngeal wall of a balanoglossid, showing three U-shaped gill slits, three tongue bars, gill pores, gill sacs, and skeletal rods.

gill slits do not open directly to the outside in most enteropneusts; instead, each leads into a pouch, the *gill sac,* that opens externally by the gill pore (figs. 3.1, 3.2). The pharynx leads into the esophagus; this is followed by the intestine, which constitutes the main length of the alimentary tract.

Enteropneusts are ciliary feeders, ingesting particles suspended in the seawater or collected from the substratum. This food material is caught up in the mucus secreted by the surface of the proboscis and is carried to the mouth by ciliary movement (fig. 3.1). Food can be rejected by movements of the collar that deflect material from the mouth. This reaction may depend in part upon monitoring by the preoral ciliary organ, which is closely associated with a thickening of the underlying nervous system. Some of the food is driven into the groove of this organ, so that some sampling and sensory function seems likely, although this has not actually been proved to occur. The passage of food and mucus into the mouth is aided by the beat of the strong lateral cilia of the gill bars, which maintains the flow of water which is needed for respiration and also draws in the food material. In contrast to the situation in urochordates and *Branchiostoma,* however, the pharynx plays only a small part in food collection. The finest particles may be filtered by the gill slits, but most material is directed ventrally and thus passes along the nonrespiratory region of the pharynx into the esophagus, and from there into the intestine. This further passage of the food is effected mainly by cilia, although there is some muscular manipulation in the esophagus. Digestion and absorption take place in the intestine, which secretes a full complement of enzymes; amyloclastic activity is also present in the mucous secretion of the proboscis.

From the anterior wall of the buccal cavity a hollow diverticulum, the stomochord, projects forward into the proboscis (fig. 3.3). This diverticulum, which is underlain by a plate of skeletal tissue, the nuchal skeleton, has been thought to represent the notochord, although it does not

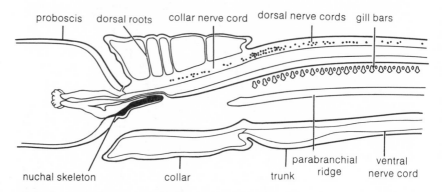

Fig. 3.3. Diagrammatic paramedian section of *Balanoglossus* sp., upon which has been projected all the giant nerve cells of one side (shown as dots). From Bullock 1944 (*J. Comp. Neurol.* 80:355–67).

look like one. Its cells are often vacuolated and thus bear a superficial resemblance to those of a notochord, but the stomochord has no sheath and is neither rigid nor elastic. In some enteropneusts it is said to arise embryologically in the same manner as a notochord, by evagination from the roof of the gut, but in others this is not the case. Further, it differs from a true notochord in that the main dorsal blood vessel lies above it. The homology of the stomochord with the notochord must thus be regarded as dubious. Its functional relationships are certainly different from those of a notochord. The latter serves in vertebrate locomotion as an elastic axial compression strut, in association with the rhythmically contracting musculature of the body wall. The mode of locomotion of enteropneusts, however, is essentially invertebrate; these animals are primarily adapted for burrowing, and for this they rely mainly upon the proboscis and collar (fig. 3.4). Peristaltic movements of the proboscis enable it to extend forward and also to swell, so that, in conjunction with the collar, it can act as an anchor. The remainder of the body is then pulled along after it. This action of the proboscis depends upon its rich blood supply of loosely arranged circular, longitudinal, and radial muscles, the action of the musculature probably being antagonized to some extent by the turgor of the proboscis coelom. This is the principle of the hydrostatic skeleton, which is well developed in annelid worms, for example; but it seems not to be so elaborately differentiated in enteropneusts, for the proboscis coelom is very small. The collar is also well provided with muscles, and it is possible that the stomochord and nuchal skeleton contribute to locomotion by strengthening the neck region during burrowing movements. In general, however, enteropneusts are adapted for a relatively inactive life in their burrows.

The blood vascular system (fig. 3.5) is chiefly invertebrate in nature. There is a main ventral blood vessel in which the blood flows poster-

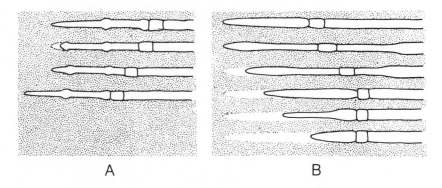

Fig. 3.4. Diagrams showing the proboscis, collar, and trunk of *Saccoglossus* in successive phases of peristaltic locomotion: *A,* burrowing, showing the tip of the proboscis alternately thrust out and withdrawn, forming bulges that travel steadily backward and, being more or less anchored in the sand, tend to move the proboscis forward, while the trunk follows passively; *B,* retreat, showing first the worm at rest; then a contraction appearing posteriorly, while the anterior part of the trunk relaxes and is pulled backward, and the posterior end of the proboscis elongates actively; then the wave of contraction traveling forward until the whole body has contracted. From Knight-Jones 1952.

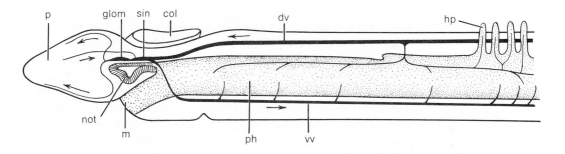

Fig. 3.5. Diagram of the blood system of *Balanoglossus: col,* collar; *dv,* dorsal vessel; *p,* proboscis; *ph,* pharynx; *vv,* ventral vessel. After Bronn, from Young 1950 (*The life of vertebrates.* Oxford: Clarendon Press).

iorly and a main dorsal vessel in which it flows anteriorly, entering the rear end of the heart, situated in the posterior dorsal part of the proboscis above the stomochord. The heart resembles a half-opened vertebrate heart, consisting of a pericardial sac with a muscular ventral wall acting as propulsatory organ, and a blood channel, the *central blood sinus,* representing the heart lumen, situated between the heart sac and the roof of the diverticulum. From the central blood sinus the blood continues forward and immediately enters the presumed excretory organ, called the *glomerulus,* which contains a dense network of blood vessels. It is supposed that waste material passes from the glomerulus into the proboscis coelom. From the glomerulus, branches supply the proboscis, and a pair of vessels run downward and backward, embracing the buccal cavity and uniting to form the ventral vessel.

The nervous system (fig. 3.6) resembles that of echinoderms in some respects but is at a lower level of organization. It consists mainly of a nerve net that still retains its primitive position within the epidermis. A receptor system of scattered epidermal sensory cells connects

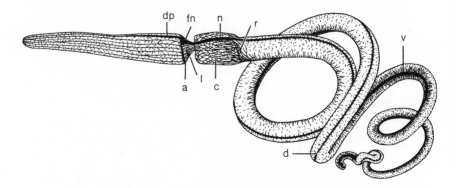

Fig. 3.6. Plan of the nervous system of *Saccoglossus: a,* anterior nerve ring; *c,* nerve fibers of the collar epidermis; *d,* dorsal nerve cord of the trunk; *dp,* dorsal nerve cord of the proboscis; *fn,* fan-shaped thickening; *l,* nerve loop of the ciliary organ; *n,* neurocord seen by transparency; *r,* prebranchial nerve ring; *v,* ventral nerve cord. From Knight-Jones 1952.

with this net, from which fibers pass to innervate the motor organs. There is no anatomically separate motor nervous system. Prominent thickenings of the nerve net form a dorsal nerve cord in the proboscis and trunk and a ventral one in the trunk only. In the collar the nervous tissue invaginates middorsally to form the neurocord (fig. 3.3), with a small central lumen. This structure, which connects the dorsal nerve cords of the proboscis and trunk, recalls in its development the chordate dorsal nerve cord, and, like the pharynx, has been held to justify regarding hemichordates as true chordates. However, its so-called dorsal roots are only epidermal connections, not true nerve roots in the vertebrate sense. But there is no indication that the neurocord has any integrative function, nor is it needed to initiate activity, for the isolated proboscis or collar, or pieces of the trunk, are capable of vigorous movement. The nerve cords of enteropneusts are through-conduction tracts, transmitting motor impulses that pass out from them into the fibers of the nerve net. This can be shown by simple experiments, as, for example, by removing part of the dorsal nerve cord in the proboscis (fig. 3.7). Peristaltic waves stop at this point, even though the rest of the nerve net is undamaged. The main nerve cords transmit impulses in both directions and provide for both slow and rapid conduction. Giant nerve cells, present in the neurocord and sometimes in the dorsal nerve cord of the trunk (fig. 3.3), probably assist in effecting rapid responses, as in many invertebrates, but their contribution has not been precisely defined.

Some enteropneusts have a direct development; others have a *tornaria* larva that closely resembles the bipinnaria larva of starfish. The mode of

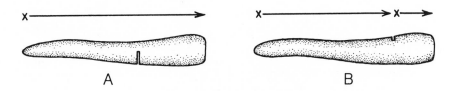

Fig. 3.7. The propagation of burrowing waves along the proboscis of *Sacco-glossus*. Bulges are initiated opposite the points marked X and travel posteri-orly as indicated by the arrows: *A*, peristalsis is uninterrupted by a cut through the ventral half of the proboscis; *B*, peristalsis is interrupted by a lesion of the dorsal nerve cord; an independent series of waves starts behind the lesion. From Knight-Jones 1952.

development of the coelom, its tripartite structure, and its connections with the outside, are other features indicating the close relationship of echinoderms with enteropneusts and, through these, with the Chordata.

**B. The Lower Chordates, Subphylum Urochordata**

The adult urochordates or tunicates bear no resemblance to vertebrates, but the tadpole larva exhibits pronounced chordate characteristics.

**1. External Anatomy**

Place a preserved tunicate in a dish of water. It is an oval sac-like crea-ture, scarcely recognizable as an animal (fig. 3.8). In life the animal is permanently attached to a rock or other object. The end that was attached can be recognized by its rough, irregular form and by the fragments of wood or other material that adhere to it. The opposite end, which in the living state extends free into the water, bears two openings, the *siphons*. When the animal is gently squeezed, jets of water squirt from the siphons; hence the name sea squirt is popularly applied to these animals. One of the two siphons lies more or less terminally; this is the *oral* or *incurrent* siphon. The other, which is shorter, is the *atrial* or *excurrent* one. The region of the body between the two siphons is anterodorsal, since the mouth, already dorsal in the larva, is carried back by the growth changes of metamorphosis, which bend the original anteroposterior axis into a U-shape. The remaining (and much larger) part of the surface is ventral. The outer covering of the animal is a tough membrane, the *tunic* or *test,* secreted by the der-mal epithelium and wandering cells. The tunic consists of a ground substance, *tunicin,* which is remarkable for an animal tissue in being composed of cellulose and protein, the two components varying in their relative proportions from species to species. The protein may be toughened by quinone tanning, like the surface secretions of many in-vertebrates. In some tunicates, however, the test is so transparent that almost all the internal organs are visible through it.

**2. Internal Features**

Remove the tunic by making a slit in its base and then peeling it off, carefully severing its attachment to the siphons. Beneath the test is the soft body wall, also called the *mantle,* consisting of the dermal epi-thelium and connective tissue. It encloses the viscera, to which it closely adheres. The siphons are operated by circular and longitudinal muscles, which can be seen in the mantle (fig. 3.8). The longitudinal

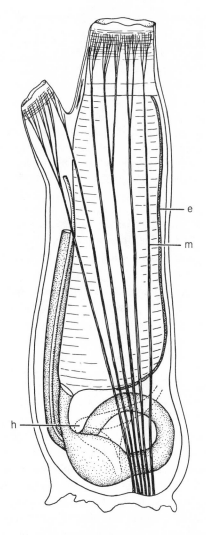

Fig. 3.8. Adult *Ciona* seen from right side. The oral siphon is directed upward, and the atrial siphon toward the left: *e,* endostyle; *h,* heart; *m,* muscle bands. From Berrill 1936 (*Phil. Trans. Roy. Soc.* Lond., ser. B, 226:43–70).

muscles run the length of the body from its base and are inserted on the siphons. Circular muscles are found throughout the body wall, internal to the longitudinal ones, but are more regularly arranged in the siphons.

Fasten the animal in a wax-bottomed dissecting pan by pins through the extreme base and the rims of the siphons; cover with water.

The viscera can usually be seen through the mantle; if it is necessary to remove the mantle, it must be pulled off gently in small strips. The digestive tract is the most noticeable of the internal features. The oral siphon leads into a large thin-walled bag, the pharynx, extending for most or all of the body length and looking to the naked eye like a fine net. The cavity outside the pharynx, between it and the mantle, is the *atrium* (peribranchial sac), formed as an ectodermal ingrowth and opening to the outside by way of the atrial siphon. The lower end of the pharynx narrows abruptly into a short *esophagus,* which opens into the wider *stomach,* situated below or to one side of the base of the pharynx. The stomach leads into the *intestine,* which immediately doubles back to form a loop parallel to the stomach and then extends straight upward toward the atrial siphon, terminating by an *anus* opening into the atrium.

Make a longitudinal slit in the side of the pharynx forward through the oral siphon and spread out the pharyngeal walls. In the midventral line of the pharynx you will notice a conspicuous white cord, the *endostyle* or *hypobranchial groove,* composed of glandular and ciliated cells. Directly opposite the endostyle in the middorsal line of the pharynx is found the *dorsal lamina,* which is either a delicate membranous fold or a fringe-like row of projections called *languets.* At the junction of pharynx and oral siphon there generally is a circle of irregular processes called *tentacles;* posterior to this there is a grooved ridge, the *peripharyngeal band,* in which the anterior ends of the endostyle and dorsal lamina terminate.

The wall of the pharynx has a short prebranchial region; the remainder (branchial region) is perforated by many transverse rows of slit-like openings, the *stigmata,* lined by a ciliated epithelium. They are derived by subdivision of a few *protostigmata* (*primary gill slits*) that are visible in the larva by the end of metamorphosis, and they continue to increase by subdivision throughout the growth of the adult. The structure of the pharyngeal wall differs in detail from genus to genus. In principle, the rows of stigmata are separated by thickenings of the pharyngeal wall, the *interstigmatic transverse bars,* which run between the rows of stigmata. Another series of thickenings, the *longitudinal bars,* runs the length of the pharynx, internal to the transverse bars and supported by them. These bars bear secretory and ciliated cells. Additional transverse (*parastigmatic*) bars, slung from these longitudinal *bars,* runs the length of the pharynx, internal to the transverse bars and into the pharyngeal cavity from the junctions of the transverse and longitudinal bars. Cut out a small piece of the pharyngeal wall, mount in water, spread out flat and examine under low power; or use a prepared slide. Identify as many of the above-mentioned features as you can.

---

The pharynx serves for respiration and feeding, both processes being dependent upon a current of water created mainly by the cilia of the stigmata. This current enters by the oral siphon, passes through the stigmata into the atrial cavity, and leaves by the atrial siphon. Respiratory exchanges take place between the water and the rich blood supply of the pharyngeal wall. An important element of the feeding mechanism is the endostyle, along which extend three pairs of glandular tracts, separated by ciliated cells. These tracts produce a protein-rich secretion that is swept out of the endostyle and carried along the side walls of the pharynx by the cilia of the bars. During its passage it acts as a filtering membrane. Particles in the current of water are caught in it, and these are carried up into the dorsal lamina, whose cilia drive the mixture of secretion and food back into the esophagus, along which it is passed by ciliary action into the stomach. Digestion and absorption take place in the stomach and intestine, the alimentary tract secreting a full complement of digestive enzymes. Undigested feces are passed out of the anus into the atrial cavity and leave the body of the atrial siphon in the excurrent stream of water. There are no differentiated excretory organs; nitrogenous waste is removed by diffusion or accumulated within specialized cells.

---

The central feature of the blood system is the pericardium and heart (fig. 3.8). The pericardium is a membranous sac lying close to the stomach and enclosing the pericardial cavity; this cavity represents the coelom, which is otherwise largely lost in tunicates. The heart is a muscular tube formed by invagination of the pericardial wall. The main vessel leaving it is the ventral vessel, which runs close alongside the endostyle. Blood is returned to the opposite end of the heart after passing through a well-developed blood system. Little can usually be seen of this system of vessels, although it may be possible to inject colored latex into the ventral vessel in suitable specimens. A remarkable feature of the heart is its periodic reversal of direction of beat. Various explanations of this have been advanced. One (due to Anderson) is that there are several pacemaker units in the heart, varying in their frequency as a result of their intrinsic properties. The end of the heart with the greatest frequency of beat initiation at any given moment will be the dominant driver of the circulation.

---

Urochordates are typically hermaphroditic. The ovary may be conspicuous in the intestinal loop, while the testes form a diffuse structure over the intestine. Genital ducts run from these organs alongside the intestine to open into the atrium close to the anus. Germ cells may be released at the same time, but self-fertilization is usually impossible or ineffective.

---

In the mantle, between the two siphons, is a nerve ganglion that constitutes the central nervous system. Nerves may be seen extending from it. These nerves supply the body wall, the siphons, and a visceral plexus overlying the internal organs. Receptors in the dermal epithelium, and particularly in the siphons, are sensitive to various stimuli, including chemical and mechanical ones. For example, touching the outside of a siphon causes it to close (*direct reflex*), and touching the inside causes the other siphon to close (*crossed reflex*). If either of these reactions is combined with contraction of the body water is expelled through one or the other siphon. The crossed reflex (but not the direct one) depends upon the presence of the ganglion, which thus has some of the integrative action expected of a central nervous system. Beneath the ganglion is the subneural gland, best seen in whole mounts or prepared sections of this region. It is probably concerned mainly with phagocytosis rather than with secretion. Many phagocytes, already containing ingested material, enter it from the bloodstream and leave by a duct that opens into the prebranchial region of the pharynx on a ciliated funnel (dorsal tubercle). The association of a gland-like organ with the ganglion has led to suggestions that this neural complex may be homologous with the adenohypophysis/neurohypophysis complex that forms the pituitary gland of vertebrates, but neither embryological nor pharmacological evidence is in accord with this view. It is possible, however, that the dorsal ciliated organ of enteropneusts, the neural complex of ascidians, and Hatschek's pit (p. 69) of *Branchiostoma* are three varied expressions of a tendency of the chordate line to produce anterior glandular and sensory structures, with associated neural developments. This tendency, perhaps correlated with ciliary feeding and the existence of an incurrent stream of water, might conceivably have contributed to the origin of the pituitary.

3. Chordate Features of Tunicate Development

Tunicates pass through a larval stage termed the *ascidian tadpole* (fig. 3.9), from a resemblance of shape to the frog tadpole. The larva has only a brief independent life, its function being to locate a habitat suitable for the adult. Whole mounts of it may be studied if available. The larva is sharply marked off into an oval body and a long slender tail used for locomotion. At the anterior end of the body are three papillae by which the larva attaches before metamorphosis. Behind them is the cerebral vesicle, containing a unicellular statocyst, sensitive to gravity, and an ocellus, sensitive to light. The vesicle is continued into the neural tube,

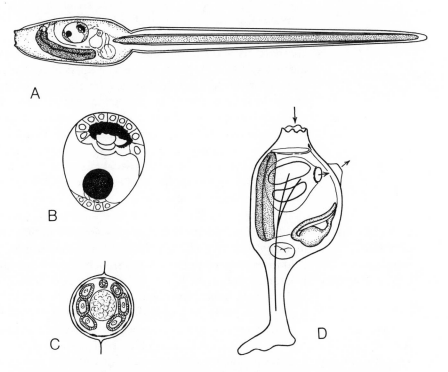

Fig. 3.9. *A*, tadpole larva of *Ciona*. *B*, sensory vesicle of larva, with unicellular otolith and ocellus with three lens cells. *C*, transverse section of tail showing central notochord, dorsal neural tube, and three muscle cells on each side. *D*, first ascidian stage of metamorphosing animal, with functional systems, from left side. From Berrill 1947 (*J. Mar. Biol. Ass. U.K.* 26:616–25.)

differentiated anteriorly into a visceral (larval) ganglion. In favorable specimens the neural tube may be traced from the cerebral vesicle into the tail. Below it will be seen, by careful focusing, the notochord, which is developed in the tail as a stiffening rod. Note that it terminates anteriorly just behind the cerebral vesicle. Notochord and neural tube are typical chordate features and are formed embryologically much as in vertebrates. Below the notochord is the endoderm of the future intestine, enlarged anteriorly into the future pharynx. There is no trace of segmentation at any stage of urochordate development.

The larva usually swims for only a few hours, after which it begins to respond positively to gravity and negatively to light. It then attaches to some shaded object by the three papillae and undergoes a remarkable metamorphosis (fig. 3.9). The tail with its chordate features is resorbed, and the cerebral vesicle with its statocyst and ocellus also disappears. The larval nervous system has previously given rise to the rudiment of the ciliated duct of the subneural gland; this rudiment now differentiates into both the gland and the adult ganglion. The larval pharynx enlarges and differentiates into the parts noted above, while the atrium arises as a pair of pouches that grow in from the outside and eventually fuse into a single cavity. This strange developmental history is difficult to interpret. According to one view it indicates that the urochordates are degenerate remnants of a former advanced chordate group; according to another, the urochordate ancestor was a free-swimming chordate without segmentation, in which the notochord originated in the tail as a locomotor adaptation. It is at least as likely, however, that the sessile life in urochordates is a retention of the primitive microphagous habit of early members of the chordate line, exemplified also in pterobranchs and in certain fossil echinoderms. The larva could have undergone independent free-swimming evolution in connection with its distributive and exploratory functions, a process that would then have necessitated regressive metamorphosis. It has further been suggested that the cephalochordates and the vertebrates could have arisen from such larval types through the neotenous association of sexual maturity with larval organization, although this by no means implies a direct origin of these groups from the urochordates themselves.

**C. The Lower Chordates, Subphylum Cephalochordata**

The little animal known as *Branchiostoma*[1] is of great interest because of its many primitive and generalized chordate features, which also throw light on vertebrate organization. *Branchiostoma* lives along ocean shores in coarse sand or shelly bottom, with the body buried in the substratum and the oral hood protruding.

**1. External Anatomy of *Branchiostoma***

Place a specimen in a dish of water. The body is slender, fish-like, pointed at each end, and compressed laterally. The blunter end is anterior, and the more pointed end posterior; the dorsal surface is sharp, the ventral surface, for the greater part of its length, flattened.

1. *Branchiostoma* Costa 1834 has priority over *Amphioxus* Yarrel 1836, which can now be used only as a trivial name.

At the anterior end is an expanded membrane, the *oral hood,* that encloses a cavity, the *vestibule,* at the bottom of which is the mouth. The borders of the oral hood are extended into a series of stiff *buccal cirri.*

Turn the animal ventral side up and observe that the flattened portion of the ventral surface is bounded laterally by two membranous folds, the *metapleural folds,* extending posteriorly from the oral hood. These folds meet at a point nearly three-fourths of the distance from anterior to posterior end, behind a median opening, the *atriopore.* From this point a median membranous fold, the *fin,* passes to the posterior end of the body, around to the dorsal side, and forward along the dorsal side to the anterior end. The slightly wider portion of this fin that surrounds the pointed posterior end is the *caudal* fin; that along the dorsal side is the *dorsal* fin. The *anus* is situated to the left of the midventral line near the posterior end, just behind the point where the fin widens.

Along the sides of the body, clearly visible through the thin, transparent epidermis, is a longitudinal series of V-shaped *muscle segments* or *myotomes,* separated from each other by connective tissue partitions termed *myosepta.* Note that the myotomes extend nearly to the anterior tip, diminishing in size above the oral hood. The number of myotomes is about sixty. Immediately below the ventral ends of the myotomes will be seen, in some individuals at least, a row of square white masses, the *gonads.* As shown by the metameric arrangement of myotomes and gonads, *Branchiostoma* is segmented. However, because of certain peculiar features of the embryology, the segments of the two sides do not correspond.

## 2. Internal Anatomy

The internal anatomy is most easily studied on small, stained, mounted specimens (figs. 3.10, 3.13). Examine one under low power and identify: the various parts of the fin, containing rectangular bodies, the *fin rays,* serving as the skeletal support of the fins; the myotomes; and the digestive tract, occupying the ventral half of the body. Study the parts of the digestive tract. Note again the oral hood (fig. 3.10) with its cirri, each supported by an internal skeletal rod. In the posterior part of the oral hood on its inner surface is seen the deeply stained *wheel organ,* consisting of patches of ciliated epithelium, arranged as several finger-like projections from a horseshoe-shaped base. The middorsal and longest finger has a central groove, *Hatschek's groove,* seen to terminate anteriorly in a hollow swelling, *Hatschek's pit.* Just behind the base of the wheel organ is seen a vertical membrane, the *velum,* bearing several short projecting *velar tentacles.* The small mouth opening in the velum (impractical to see) leads at once into the *pharynx,* a wide tube extending nearly half the body length (fig. 3.11). At the anterior end of the pharynx, just behind and nearly parallel to the velum, note the deeply stained *peripharyngeal band.* The side walls of the pharynx are composed of slender parallel oblique bars, the gill *bars,* or *branchial bars,* enclosing between them elongated openings, the *gill slits* or pharyngeal clefts. Each gill bar is supported by an in-

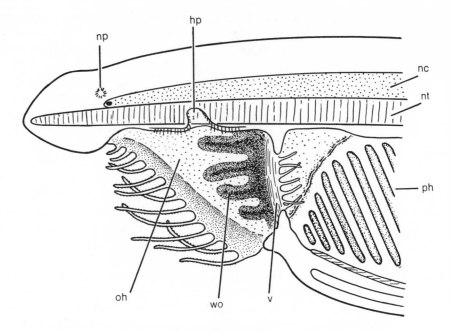

Fig. 3.10. Side view of the head of a young *Branchiostoma*. The oral hood, body wall, and wall of the pharynx have been removed on one side: *hp*, Hatschek's pit; *nc*, nerve cord; *np*, olfactory pit; *nt*, notochord; *oh*, right oral hood; *ph*, pharynx; *v*, velum; *wo*, wheel organ. After Goodrich 1917 (*Quart. J. Mic. Sci.* 62:539–53).

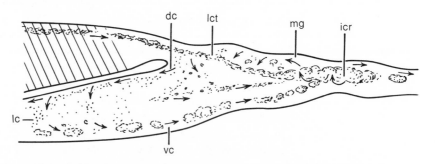

Fig. 3.11. Diagram to show the currents and movements of material in the alimentary tract of *Branchiostoma: dc*, dorsal current carrying small particles into the caecum; *icr*, iliocolon ring, containing a rotating mass of food and secretion; *lc*, small particles deposited and ingested on the lateral wall of the caecum; *lct*, lateral ciliated tract; *mg*, midgut, in which material is thrown forward; *vc*, ventral current carrying secretion from the caecum into the midgut. From Barrington 1965.

ternal skeletal branchial rod (difficult to see), and these rods are united into an arcade at the dorsal ends of the gill bars. These skeletal rods are composed of fibrous connective tissue.

The arrangement of gill bars and skeletal rods differs from that in enteropneusts only in that the tongue bars complete their downward growth (see fig. 3.2) and fuse with the ventral pharynx wall, so that each of the original larval gill slits of *Branchiostoma* becomes divided into two slits. The adult *Branchiostoma* thus has two kinds of gill bars, the *primary* and the *secondary* or *tongue* bars. These are anatomically different in various ways, notably in that the skeletal rod forks at the ventral end of the primary bars, as it does in enteropneusts (fig. 3.2). Note also the fusion of the skeletal rods in the secondary bars in *Branchiostoma*. In older specimens numerous cross-unions, called *synapticulae,* develop between adjacent gill bars; these also occur in enteropneusts.

Surrounding the pharynx is a large cavity, the *atrium,* as in urochordates; its ventral boundary is visible below the pharynx as a line that may be traced to the atriopore. Water, propelled by the cilia on the gill bars, enters the pharynx by way of the mouth, passes through the gill slits into the atrium, and leaves by the atriopore. The postero-dorsal end of the pharynx continues backward as the digestive tube, in which various regions may be recognized (fig. 3.11). The short *esophagus* behind the pharynx soon widens into the *midgut,* from which an outgrowth, the *midgut caecum,* projects anteriorly along the right side of the posterior part of the pharynx. The midgut narrows posteriorly into a deeper-staining region, the *iliocolon ring,* beyond which the narrowed hindgut continues to the anus (fig. 3.13).

---

*Branchiostoma* feeds on diatoms, desmids, and other microscopic organisms. In feeding, the cirri are folded over the entrance to the oral hood, excluding large particles. Sufficiently small particles are carried by the main ciliary current into the pharynx, where they become trapped in a filtering membrane secreted by the endostyle, much as in urochordates. The secretion and particles are passed dorsally by the cilia on the inner faces of the gill bars to the epipharyngeal groove. Any particles in the oral hood that escape the main current are caught and concentrated by the wheel organ with the aid of mucus from Hatschek's groove and pit; and this material, together with particles in the anterior end of the pharynx, is carried dorsally to the epipharyngeal groove along the peripharyngeal band. During these processes the food and secretion are rolled into a food cord that is carried by ciliary action along the epipharyngeal groove and out of the pharynx into the midgut (fig. 3.11). Upon arriving at the iliocolon ring the cord is thrown into a spiral coil that is rotated by the action of the strong cilia of the ring and joined by digestive secretions swept out ventrally from the caecum. Some extracellular digestion probably occurs at this stage, but pieces break off from the rotating mass, and the smallest of these are carried forward by dorsal ciliary currents into the caecum. Here the cells can take up solid particles, and digestion is completed within them. Pieces also break off from the rear end of the rotating mass and are passed back along the hindgut, where some further intracellular digestion occurs. Much of this movement can be observed under a binocular microscope in small specimens that are ingesting carmine particles.

---

Immediately dorsal to the digestive tube and about as wide as the hindgut, is seen a yellowish rod, the *notochord,* extending the body length and running, just above the oral hood, nearly to the extreme anterior tip. Directly above the notochord is situated the much smaller *nerve tube* or *spinal cord* (fig. 3.10), best recognized by the row of black spots it bears. These spots are very simple *eyes* or *ocelli,* each consisting of one ganglion cell and one curved pigment cell, and have been shown to respond to light. The reduced anterior end of the nerve tube seems to represent a brain and may be termed the *cerebral vesicle;* this terminates somewhat behind the anterior tip of the notochord and there bears a black *pigment spot,* shown to have no response to light. Just above the pigment spot on the left side there may be seen a depression, the *flagellated pit,* formerly called the olfactory pit or funnel. Although this connects with the cavity of the cerebral vesicle in the larva, there is no such connection in the adult. The flagellated pit appears to be a sense organ. In the floor of the cerebral vesicle is a group of secretory and ciliated cells called the infundibular organ because it was at one time homologized with the infundibulum of the vertebrate brain. This homology is unjustified. The organ actually secretes a viscous fiber that extends down the spinal cord. The fiber closely resembles Reissner's fiber that arises dorsally in the brain of vertebrates and also extends along the spinal cord.

Two pairs of nerves arise from the brain vesicle; the remainder of the spinal cord gives off a vertebrate-like metameric succession of *spinal nerves.* These consist of the dorsal nerves, chiefly sensory, passing out by way of the myosepta to supply the skin, and the ventral nerves, chiefly motor, passing into the myotomes. Unlike conditions in gnathostome vertebrates, the dorsal and ventral nerves do not unite outside the cord. There is also a well-developed visceral nervous system, composed of multipolar neurons forming nerve nets that are connected with the central nervous system through the spinal nerves. Unlike the nerve nets of the vertebrate alimentary tract, with which the cephalochordate nets are only doubtfully homologous, this system contains both motor and sensory divisions. It is particularly well formed over the atrium, the midgut, and the midgut caecum and is probably functionally associated with the feeding mechanism. For example, tactile or chemical stimulation of the buccal cirri or velar tentacles may evoke expulsion of water through the atriopore by contraction of the atrial floor. This is a cleansing reaction, which serves to get rid of clogging particles. It can be abolished by cutting the nervous connection of the spinal cord with the atrial plexus.

---

Segmentation of musculature and nerves, in conjunction with the presence of the stiff but elastic notochord, makes possible, as in fish, the undulatory movements by which the animal swims and burrows in the substratum. Burrowing is particularly evoked by tactile stimulation of the oral hood. Receptor structures, including free nerve endings, are abundantly developed in the body wall and presumably confer the tactile and chemical sensitivity that is known to influence the animal's selection of a substratum.

**3. Cross Section through the Pharyngeal Region**

Examine the cross section with the lower power and identify as much as possible of the following (fig. 3.12): (*a*) The epidermis, the outer covering of the body composed of a single layer of columnar epithelial cells. (*b*) The dorsal median projection, the dorsal fin, containing the

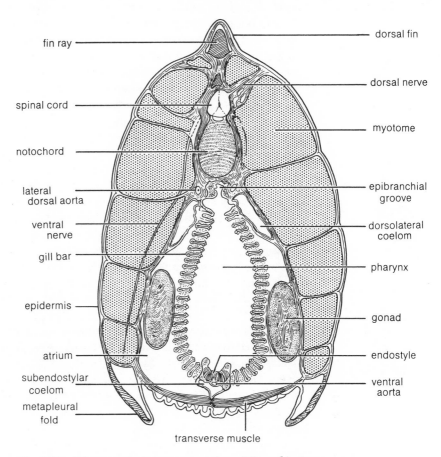

Fig. 3.12. Diagram of a transverse section of *Branchiostoma*.

fin ray, which supports it. (*c*) The two ventrolateral projections, the metapleural folds. There are a number of smaller folds in the ventral wall between the two metapleural folds. (*d*) The myotomes, a series of circular masses filling the dorsal and lateral portions of the body wall and separated from each other by connective tissue partitions. The myotomes are thick dorsally and thin out ventrally. Transverse muscles are present in the ventral body wall, just above the small folds of the epidermis. (*e*) The spinal cord lying between the dorsal portions of the myotomes, below the fin ray. Observe that it contains a central canal, the *neurocoel*. The section may strike one of the eyes described above. (*f*) The notochord, an oval mass much larger than the spinal cord and directly ventral to it. It is provided with muscle fibers, whose contraction, coordinated with that of the myotomes, enables it to act as a hydrostatic skeleton. (*g*) The atrium, the large cavity occupying the ventral half of the section and enclosing the pharynx; it is formed in an advanced larval stage by the invagination of the epidermis. Previous to this occurrence, the larval gill slits open directly to the exterior as in vertebrate embryos. (*h*) The pharynx, occupying the center of the atrium, elongate or heart-shaped. Since the gill bars run obliquely, a cross section cuts a number of them, so that the pharynx appears to be composed of separate pieces, the gill bars, separated by spaces, the gill slits. The gill bar consists chiefly of a tall ciliated epithelium enclosing at the outer side the skeletal rod and blood vessels and coelomic spaces. In the middorsal line of the pharynx is seen the deep *epibranchial groove;* and in its midventral line, the similar *hypobranchial groove* or *endostyle*. The parts and relations of the pharynx and atrium are the same as in tunicates. (*i*) The midgut caecum, an oval, hollow structure lined by a tall epithelium, present to the right side of the pharynx. (*j*) The gonads, if present, are the very large, deeply stained masses at either side, projecting into and nearly filling the atrial cavity; the ovaries consist of cells with large nuclei; the sperm in the testes present a streaky appearance. (*k*) The nephridia, fragments of which may be visible. They are short irregular tubes perched upon the outer surface of the dorsal parts of the secondary gill bars, lying in the dorsal coelomic spaces and opening into the atrium. Each nephridium bears several clusters of *solenocytes,* flagellated cells drawn out into a long tube in which the flagellum plays. Nephridia with solenocytes classify as protonephridia. They are not homologous to the tubules of vertebrate kidneys. (The only other animals known to have comparable protonephridia are certain invertebrates, notably some polychaete annelids, not related to the chordate line.) A vascular glomerulus, in the right wall of the buccal cavity, has also been thought to be excretory. *Branchiostoma* can withstand considerable changes of salinity in the external medium, but the mechanism of adjustment is unknown, nor has it been established whether the protonephridia and glomerulus are involved.

4. Circulatory
System

The circulatory system lacks a heart and is composed of vessels and tissue channels forming a continuous circuit. The contractile vessels have muscular walls, but the others consist only of a thin membrane; and there is no histological distinction among the smaller vessels between arteries, veins, and capillaries. The blood is colorless. From a junction point, often called *sinus venosus,* situated at the posterior end of the pharynx, a contractile vessel variously called the *endostylar artery* or *ventral aorta* runs forward beneath the endostyle and branches to both sides into the primary gill bars (fig. 3.13). Each such branch bears a contractile swelling (*bulbillus*) and then ascends the gill bar as an *aortic arch* that joins the dorsal aorta of that side. (There also spring from each bulbillus two secondary vessels that ascend the primary gill bar; and there are two similar vessels, connected only indirectly with the ventral aorta, in each secondary gill bar.) Before reaching the dorsal aorta, each aortic arch contributes to a plexus of blood vessels in the corresponding nephridium (which also receives branches from the outer of the two vessels in the sec-

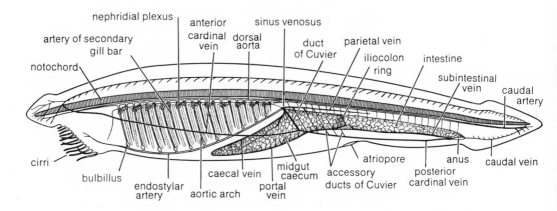

Fig. 3.13. Diagram of the circulatory system of *Branchiostoma*, showing also the digestive tract. After Mozejko 1913.

ondary gill bar). Thus, before reaching the dorsal aortas, the blood has passed through the gills and the nephridia in part. The noncontractile paired *dorsal aortas* begin as fine twigs at the anterior end of the animal and run backward above the pharynx, just below the notochord, receiving the aortic arches and other gill vessels of their respective sides. Shortly behind the rear end of the pharynx the two aortas unite into a single dorsal aorta, which continues backward, becoming the caudal artery of the tail. From the tail the *caudal vein* runs forward and at the anus forks into the *posterior cardinal vein* and the *subintestinal vein* (fig. 3.13). The posterior cardinal vein runs forward along the middle of the body wall just inside the myotomes (fig. 3.13), and opposite the sinus venosus it joins the similar anterior cardinal vein running back from the anterior half of the body. The common *duct of Cuvier,* formed by the union of the two cardinal veins, crosses the atrium and enters the sinus venosus. (Usually there are several accessory ducts of Cuvier parallel to the main one.) The subintestinal vein courses along the ventral side of the intestine and breaks up into a network of small vessels in the intestinal wall, from which network a vein again forms and runs forward along the ventral surface of the midgut caecum, as a portal vein. This breaks up into a network of vessels in the wall of the caecum, and from this network a caecal vein forms and, running along the dorsal surface of the caecum, enters the sinus venosus. There is a pair of *parietal veins* in the dorsal wall beneath the myotomes (fig. 3.13), and all the main vessels give off or receive metamerically arranged branches from the body wall and digestive tube. It will be noted that the portal system of the caecum resembles the hepatic portal system of vertebrates. At one time this was thought to justify homologizing the caecum with the liver, but, as was explained above, the caecum is actually concerned with enzyme secretion and intracellular digestion. The portal system, misleadingly suggestive of hepatic function, is probably related functionally to the removal of the digestive products.

The contractions in the circulatory system are slow and irregular. First the hepatic vein beats toward the sinus; then the sinus contracts,

followed by the ventral aorta, away from the sinus, and by the bul-
billi. The circulation can reverse but rarely does so. The contractile
vessels beat only after they have become distended with blood.

5. Relationships    *Branchiostoma* shows affinities with enteropneusts, tunicates, and vertebrates. The mode of formation of the tongue bars and the arrangement
of the skeletal rods of the gill bars are identical with those of enteropneusts, and the preoral ciliary organ may be the homologue of the
wheel organ. Hatschek's pit and groove are considered by many to be a
possible homologue of the vertebrate hypophysis. Several structures of
the *Branchiostoma* pharynx (endostyle, peripharyngeal band, epipharyngeal groove) are identical with those of tunicates. But much of the
developmental and adult anatomy of *Branchiostoma* is typically vertebrate, such as the mode of formation and relations of the myotomes, notochord, neural tube, digestive tract, and main blood vessels. Notable
vertebrate features of the latter are the ventral and dorsal vessels of
the gill region connected by vessels running through the gill bars, the
characteristic body-wall veins (anterior and posterior cardinals) connected to the contractile center by a duct of Cuvier crossing body
spaces, and the relations of caudal and subintestinal veins.

Â Â Â Â *Branchiostoma* has a larval stage, remarkable for its asymmetry, with
the mouth on the left side and the developing endostyle on the right, and
with the developing gill slits correspondingly displaced. Metamorphosis,
with symmetrization, takes place at varying times, ranging from 75 to
140 days. A giant amphioxides larva may be formed as a result of further
prolongation of larval life, sometimes with gonads appearing before
metamorphosis. This is another example of neoteny, whose possible importance in chordate evolution has already been mentioned (p. 68).

**D. Class
Cyclostomata**    The cyclostomes, which are the only surviving members of the primitively jawless Agnatha, present an interesting blend of primitive and
specialized features, the latter being associated with their semiparasitic
life. They are divided into two subclasses, the Petromyzontia, or lampreys, and the Myxinoidea, or hagfishes. The former occur in streams,
lakes, and the ocean, but the marine forms always ascend freshwater
streams to breed; the myxinoids are exclusively marine.

1. External
Anatomy of a
Lamprey
Â Â Â Â Place the specimen in a dissecting tray. Note the general eel-like
appearance and tough, slimy, naked skin. The body consists of a stout
cylindrical head and trunk and a laterally flattened tail. The posterior
body half bears two *dorsal fins* and a *caudal fin* bordering the tail and
continuous with the second dorsal fin. The fins are supported by
numerous *fin rays,* slender parallel cartilages usually visible through
the skin. Note the total absence of paired fins, which are found in true
(gnathostome) fish and correspond to the fore- and hind limbs of
tetrapods. Their absence in cyclostomes may be a primitive feature,
although paired pectoral fins are present in some fossil agnathans.

Â Â Â Â The anterior end of the head presents a peculiar appearance because
of the lack of jaws (a very primitive character) and the presence of a
large bowl-shaped depression, the *buccal funnel,* directed ventrally.

The edges of the buccal funnel are provided with soft papillae or *lip tentacles,* and its interior is studded with brown horny *teeth,* definitely arranged. At the bottom of the funnel lies the *tongue,* a projection covered with teeth; the mouth is just dorsal to the tongue. Lampreys attach to fishes by using the buccal funnel as a suction cup and rasp off the flesh of their prey by filing movements of the tongue.

On the dorsal surface of the head there is a median *nasohypophyseal opening,* from which a passage leads to the paired olfactory sacs. This median opening also exists in extinct agnathans and contrasts strongly with the paired lateral or ventral nostrils of typical vertebrates. In the lamprey larva the nasal sacs arise as separate epithelial invaginations that are continuous with the hypophyseal invagination (see below). As a result, a common pore is formed, and this is then pushed dorsally by the great development of the buccal funnel. On each side of the head is an eye, without eyelids; and behind each eye is a row of seven oval openings, the gill slits. An area of reduced pigmentation behind the nasohypophyseal opening marks the position of the dorsal pineal and parapineal eyes. These structures, although less differentiated than the lateral eyes, are sensitive to light, and their sensitivity contributes to the diurnal color change of lampreys.

The body-wall musculature is segmented into myotomes separated by myosepta, as in *Branchiostoma,* although here they are W-shaped instead of V-shaped. They are generally detectable through the skin. At the junction of trunk and tail in the midventral line there is a cloacal pit containing the anal opening in front and the urogenital papilla behind.

2. Internal Anatomy

Sagittal section of the anterior end. Make a median sagittal section of your specimen, or study a section so prepared (fig. 3.14). Examine the cut surface and identify as much as possible of the following.

a. Digestive tract: Observe again the buccal funnel with its teeth and tongue. Note the large muscle masses extending posteriorly from the tongue, which effect the rasping movements of the tongue. Find the mouth opening above the tongue and follow it into a passage, the *buccal cavity,* that slopes ventrally. The buccal cavity opens at its posterior end into two tubes, an upper smaller one, the *esophagus,* and a larger ventral one, the *pharynx,* whose wall is pierced by seven oval openings. A fold, the *velum,* is present at the entrance of the buccal cavity into the pharynx. The pharynx, which in the larva constituted the anterior part of the digestive tract, ends blindly in the adult as an adaptation to the peculiar respiratory mechanism. The esophagus, which develops at metamorphosis as a dorsal proliferation of the pharynx, leads into the remainder of the digestive tract.

A liver is present, holding considerable fat stores in the adult; the gallbladder and bile duct are lost at metamorphosis. Pancreatic zymogen tissue is represented by secretory cells that retain their primitive position in the intestinal epithelium. Pancreatic islet tissue is present in a nodule, lying dorsally at the extreme anterior end in association with a small blind caecum of the intestine. Islet tissue is also present as a mass of cells lying between the intestine and liver,

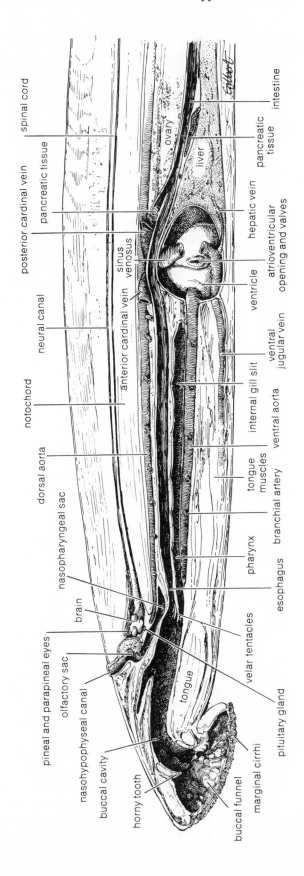

Fig. 3.14. Diagram of a median longitudinal section through the anterior end of an adult lamprey.

partially embedded in the latter (fig. 3.15A, B). A true pancreas appears first in the gnathostomes, but the close association of zymogen and islet tissue in the wall of the lamprey's intestine foreshadows the later differentiation of the organ.

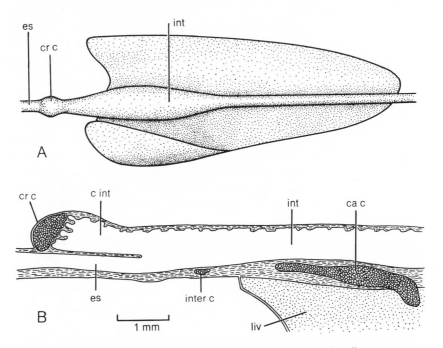

Fig. 3.15. A, dorsal view of the liver and associated region of the alimentary canal of *Petromyzon fluviatilis*. B, vertical longitudinal section of the same region; *c int*, caecum of the intestine; *ca c*, caudal pancreatic cords; *cr c*, cranial pancreatic cords; *int*, intestine; *inter c*, intermediate pancreatic cords; *liv*, liver; *es*, esophagus. After Barrington 1945 (*Quart. J. Micr. Sci.* 85:391–417).

b. Respiratory system: The seven openings in the wall of the pharynx are the internal *gill slits*. They open into much-enlarged *gill pouches* that communicate with the outside through the external gill slits. Water is pumped in and out of the gill pouches through the gill slits, this mode of ventilation being an adaptation to suctorial feeding, which prevents ventilation through the mouth. When not feeding, the animal probably takes in water through the mouth as well. It is thought that much gaseous exchange takes place through the skin.

c. Skeleton: This is entirely cartilaginous. There is a skull enclosing the brain, and associated with it is a complicated lattice termed the *branchial basket*. This is an elastic structure supporting the pharynx and providing insertions for the respiratory muscles. Expiration is effected by muscular contraction, with consequent distortion of the lattice. Inspiration is less well understood, but is probably assisted by elastic recoil of the lattice. The *notochord* is the broad brown rod just dorsal to the esophagus. It is the chief axial skeleton, functioning through the turgor maintained by its large vacuolated cells. The vertebral column is rudimentary, consisting only of small paired cartilages (*arches*) lying dorsolaterally on both sides of the notochord.

d. Vascular system: Posterior to the last gill pouch is a somewhat conical cavity, the pericardial cavity, that contains the heart. This consists of a sinus venosus, auricle (atrium), and ventricle. The blood system is typically vertebrate. In particular, note the ventral aorta passing forward beneath the pharynx to give off paired afferent branchial arteries to the gills. A pair of dorsal aortas lie above the pharynx, joining posteriorly to form the median dorsal aorta of the trunk. A single vessel on the right side conveys blood from the cardinal veins of the body wall into the sinus venosus. Groups of cells lying along the cardinal veins are probably homologues of the cortical and medullary tissue of the adrenal gland of higher vertebrates.

e. Kidneys and gonads: The paired *mesonephric kidneys* extend as dorsal ridges down the length of the coelom and discharge by ducts into the urinogenital papilla. The gonad is a median organ lying between the kidneys and filling much of the coelom in the sexually mature animal. The germ cells are shed into the coelom and pass into the urinogenital papilla by a pair of openings in each kidney duct. A mass of lymphoid tissue in the pericardial cavity is the remains of the *pronephric kidney,* which is functional in the larva.

f. Nervous system and pituitary gland: Above the notochord is a narrow canal, the neural canal, containing the slender spinal cord, which passes anteriorly into the brain. Paired segmental nerves leave the spinal cord; as in *Branchiostoma,* and in contrast to gnathostomes, these do not join to form mixed nerves.

From the nasohypophyseal opening the nasohypophyseal canal slants downward and backward to open into the olfactory sacs, one of which should be visible. Note the much-folded wall of the sac, bearing the olfactory epithelium. Below the sac the canal continues ventrally and posteriorly to end blindly. The adenohypophysis of the pituitary gland lies between the canal and the floor of the brain; in the larva it arises from the canal, but is separate from it in the adult. The neurohypophysis is represented by the floor of the diencephalon (infundibulum), to which the adenohypophysis is closely apposed; it does not, as in higher vertebrates, grow out to form a distinct neural component of the pituitary.

## 3. External Features of a Myxinoid

The relationship of myxinoids to lampreys is uncertain. According to one view they separated only recently; according to another they may have separated early in cyclostome history and have undergone a long period of independent evolution. Favorable to the latter view is the fact that myxinoids are the only vertebrates whose body fluids have an osmotic pressure similar to that of seawater; it appears certain that they have never had a freshwater phase in their history. The common genera are *Myxine* and *Polystotrema* (*Bdellostoma*); whichever is available may be examined. The body is even more slender and eel-like than in lampreys and ends in a short tail bordered by a continuous caudal fin. Enormous quantities of mucus can be discharged from slime glands that open by pores on the sides of the body. Note that the anterior end lacks the buccal funnel of the lampreys and instead has a ventral mouth opening bordered by two pairs of tentacles. At the anterior tip of the head is the *nasopharyngeal canal,* equivalent to the nasohypophyseal canal of lampreys, but opening into the pharynx. The continuous passage so formed is the respiratory passage by which water for the gill pouches enters, to be expelled through the external gill openings. In *Polystotrema* there is a row of from six to fourteen round gill slits on each side, some distance behind the head. In *Myxine* external gill slits are lacking; instead, canals run backward from the gill pouches under the skin and unite on each side to form the *branchial opening.* The two branchial openings lie on the ventral surface about one-third of the body length from the anterior end. The left branchial opening is larger because it receives the *esophageocutaneous duct,* a tube from the pharynx to the exterior that is homologous with a gill slit. This tube also exists in *Polystotrema,* where it opens in common with the last gill slit of the left side. *Myxine* burrows into the substratum and also burrows into its prey, and some of the above-mentioned features are adaptively related to this and to the consequent absence of the suctorial feeding of lampreys. Water is pumped through the pharynx by a muscular velum at its anterior end, with perhaps some contribution from the muscles of the pharynx itself. The esophageocutaneous duct serves to clear large particles from the pharynx; some ejection also takes place through the nasopharyngeal canal.

---

**4. The Ammocoetes Larva**

The lampreys have a larval stage originally called *Ammocoetes* when it was (as so often with larvae) thought to be a distinct genus. It lives, probably for five to six years, in burrows in the bottoms of streams, feeding on the organic content of the mud and on minute organisms. Light-sensitive receptors in the tail ensure that it remains buried in the substratum. The ammocoetes larva is of great phylogenetic importance in illustrating primitive vertebrate structure. The following account includes only those details the student can readily observe; other features will be discussed in later chapters.

---

a. Whole mount: Note general *Branchiostoma*-like appearance. The anterior end forms an expanded *oral hood,* in the back part of which are seen a number of projections, the *oral papillae;* these serve as strainers in feeding. Behind the oral hood will be noticed the *brain,* above; the *gill* or *branchial* region, below. The brain is an elongated, lobed body showing the characteristic vertebrate divisions and closely related to the three sense organs typical of vertebrates—nose, eye, and ear. The most anterior division of the brain, or *telencephalon,* shows two obvious lobes, the *olfactory bulb* in front, and the *olfactory lobe* behind. Directly in front of the olfactory bulb is seen the nasohypophyseal canal ascending to the nasohypophyseal opening, which has a raised rim. (A third part of the telencephalon, the *cerebral hemispheres,* is not readily visible from surface view.) Behind the olfactory lobe is a larger division of the brain, the *diencephalon,* which bears an eye on each side; the eye lies in the body wall and is not functional. The ventral part of the diencephalon projects forward below the olfactory lobe as the infundibulum, most of which represents the neurohypophysis of the pituitary. The anterior part of the roof of the diencephalon bears the pineal and parapineal eyes. The slightly elevated region of the brain above the main mass of the diencephalon and appearing as its dorsal part is really the *mesencephalon* or *midbrain* (*optic lobes*). This is separated by a cleft from the last and largest division of the brain, the hindbrain or *rhombencephalon,* consisting chiefly of the medulla oblongata. Along the anterior ventral part of the medulla, close behind the eye, is seen the large oval *auditory vesicle.* The medulla oblongata tapers abruptly into the *spinal cord,* which extends as a rod nearly to the posterior tip. Immediately below the spinal cord is the much wider notochord, which tapers to an anterior end close behind the auditory vesicle.

Below the brain and notochord are the parts of the digestive tract and a few other organs. The oral hood leads, by way of the mouth, into the buccal cavity, bounded posteriorly by a large flap, the velum, situated in front of the gill slits. The velum really consists of a pair of muscular flaps attached dorsally just in front of the auditory vesicle and to the sides of the buccal cavity. The flaps curve first forward and then diagonally backward. Rhythmic movements of the velar flaps cause a current of water to pass from the mouth into the pharynx, for feeding and respiration. The ammocoetes larva is thus microphagous, like urochordates and *Branchiostoma,* but with muscular action replacing ciliary action and more effectively maintaining the incurrent stream of water. The pharynx is the large tube behind the velum, with seven gill pouches in its walls. The gill lamellae in the walls of the pouches can readily be seen. Each gill pouch opens to the outside by a small gill slit.

Ventrally below the first five gill pouches is a conspicuous elongated body, the endostyle (fig. 3.16), homologous with that of the proto-

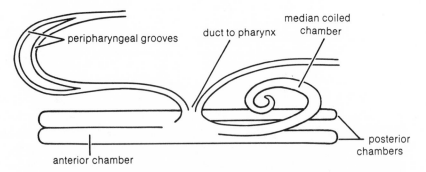

Fig. 3.16. Endostyle of the larval lamprey, seen in diagrammatic lateral view.

chordates. It arises as a trough in the pharynx floor but soon constricts off, remaining connected with the pharynx only by a main median duct that opens between the third and fourth gill pouches, and by two smaller lateral ducts associated with it. In other respects also it is of more complex structure than the protochordate endostyle, consisting of paired anterior and posterior chambers and a median posterior coiled chamber, all with tracts of glandular cells generally resembling those of protochordates. It is uncertain whether this organ plays any part in the feeding mechanism of the larva; there is some evidence that food particles can be trapped in mucus secreted by the pharynx lining and gill lamellae and carried back into the esophagus without participation of the endostylar secretion. Its special interest lies in the fact that at metamorphosis part of its epithelium gives rise to the thyroid gland. Moreover, in the larva the gland can already bind iodine to tyrosine and can synthesize the two thyroid hormones, thyroxine and triiodothyronine. Iodine binding is also demonstrable in the endostyle of protochordates. Thyroxine has been demonstrated in ascidians, and thyroxine and triiodothyronine in *Branchiostoma,* and it is probable that their endostyles synthesize these molecules. It follows that the thyroid gland of vertebrates evolved from the endostyle, and that this organ developed thyroid-like activity before it was transformed into the thyroid gland. It has been suggested that the appearance of iodine-binding in the endostyle may have been determined by the protein metabolism associated with the secretion of the protein-rich filtering membrane used in feeding. Behind the last gill pouch the digestive tract narrows to a slender tube, the *esophagus,* which extends backward for a short distance and then widens into the *intestine.* Below the esophagus are seen two rounded organs; anteriorly and just behind the last gill pouch is the *heart;* and behind this is the larger *liver,* in which there is a conspicuous vesicle, the *gallbladder.* The ill-defined mass above the heart is the pronephros or larval kidney. The intestine continues posteriorly to the anus, which marks the boundary between trunk and tail. Immediately behind the anus the tail fin begins and continues posteriorly around the tail and forward in the middorsal line to about the level of the liver. Lateral fin folds are lacking in the ammocoetes larva.

b. Older larvae: Older larvae may be examined with a hand lens or under a binocular. Note oral hood; nasohypophyseal opening with a thick rim; slanting row of seven gill slits opening into a groove; and myotomes visible through the skin, anus, and caudal and dorsal fins. On the top and sides of the head will be seen short lines of dots, the organs of the *lateral-line system,* a system of skin sense organs peculiar to aquatic vertebrates. Note that the eyes are not visible; as already mentioned, they are deeply sunk and not functional.

c. Section through the pharyngeal region: The section is clothed by a thick epidermis underlain by a thin layer of connective tissue enclosing the layer of myotomes, thicker dorsally than ventrally. Along the sides are the midlateral grooves into which the gill slits open. Middorsally between the myotomes note the large neural canal containing the relatively small, flattened spinal cord, and directly below this the large rounded notochord composed of vacuolated cells and enclosed in a thick sheath. To each side of the notochord is seen the cross section of an anterior cardinal vein, and below the notochord lies the dorsal aorta. The central part of the section is occupied by the large pharynx, whose walls bear the gill lamellae. The appearance of the pharynx depends on whether the section cuts through the anterior or the posterior lamellae of a gill pouch. The septa that bear the gill lamellae slant so that the lamellae on their anterior faces are turned to the outside and those on their posterior faces to the inside. The gill lamellae are little plates projecting horizontally from the septa, so that a whole set of them from dorsal to ventral side is cut crosswise in a section through the pharynx. Each lamella has lateral ridges and is liberally provided with blood vessels. In the middorsal and midventral line of the pharynx there is a projecting *ciliated ridge* of unknown function. The midventral ridge has a single groove or a pair of grooves, depending on the level of section; connected with the endostyle openings, these are continuous around the anterior end of the pharynx as the peripharyngeal grooves and connect with the dorsal ciliated ridge. Below the ventral ridge is the single or paired ventral aorta, and below this the endostyle, the appearance of which depends on the level of section. Anterior to its opening into the pharynx floor the gland consists of two chambers separated by a thin partition. In each chamber, attached to its floor, is seen the cross section of a longitudinal fold of ciliated epithelium, which at four places is infolded to make four conspicuous columns of tall, wedge-shaped cells, termed *cuneiform cells;* these are the glandular tracts. Posterior to the duct the central part of the gland makes an upward spiral coil, as already explained, so that the section cuts through the gland twice and has a complicated appearance.

d. Section through the trunk: Identify dorsal and ventral fin folds, neural canal, spinal cord, notochord, dorsal aorta below the notochord, posterior cardinal veins to each side of the aorta, and coelomic cavity enclosed by the layer of myotomes. The structures in the coelomic cavity will depend on the level of the section. Close behind the pharynx the section shows the esophagus as a small circle of tall epithelium, the paired pronephroi (composed of sections of tubules, situated below the posterior cardinal veins), and the various parts of the heart filled with blood. More posteriorly the liver is encountered, and beneath the cardinal veins are folds of fatty tissue bearing a pronephric duct on each side. Behind the liver the section shows the intestine, with a conspicuous infolding, the spiral fold, filled with tissue enclosing the intestinal artery. Beneath the posterior cardinal veins the fatty tissue supporting the pronephric duct continues, or at certain levels the mesonephric tubules may be present. The single unpaired gonad is seen as a small body to the medial side of the kidney tissue.

e. Metamorphosis: Transformation, which is prolonged, begins during the spring or summer. The oral hood expands to form the buccal funnel, the eyes move to the surface and become functional, the gallbladder disappears, and the bile duct breaks down to form the posterior mass of pancreatic islet tissue. The larval islet tissue, present in the submucosa at the extreme end of the intestine, is carried forward to become the anterior mass of islet tissue of the adult.

Lampreys spawn only once, then die. Brook lampreys (e.g., *Lampetra planeri* in Europe; *L. richardsoni* in western North America) stay in fresh water after metamorphosing. They spawn the following spring without further feeding, living as adults for only about 4–6 months. River lampreys (e.g., *Lampetra fluviatilis* in Europe; *L. ayresi* in western North America) and sea lampreys (*Petromyzon* in the Atlantic, *Entosphenus* in the Pacific) pass down to the sea, where they feed and grow, eventually undertaking an anadromous spawning migration, during which feeding ceases. The esophagus, which develops out of the roof of the larval pharynx, remains closed in *L. planeri,* or opens only briefly, but in the migratory species it is necessarily open during the growth period. It is used not only for feeding but also for marine osmoregulation; lampreys in their marine phase, like marine teleosts, have an osmotic pressure below that of the sea and swallow seawater to regulate the composition of their body fluids—a remarkable example of parallel evolution in these two widely separated groups, contrasting sharply with the situation in the myxinoids (see above).

**References**

Anderson, M. 1968. Electrophysiological studies on initiation and reversal of the heart beat in *Ciona intestinalis. J. Exp. Biol.* 49:363–85.

Barrington, E. J. W. 1937. The digestive system of *Amphioxus. Phil. Trans. Roy. Soc. Lond.* ser. B, 228:269–312.

————. 1965. *The biology of Hemichordata and Protochordata.* Edinburgh and London: Oliver and Boyd.

————. 1968. Metamorphosis in lower chordates. In *Metamorphosis: A problem in developmental biology,* ed. W. Etkin and L. I. Gilbert. New York: Appleton-Century-Crofts.

Barrington, E. J. W., and Jefferies, R. P. S., eds. 1975. *Protochordates.* Symposium of the Zoological Society of London, no. 36. London and New York: Academic Press.

Berrill, N. J. 1950. *The Tunicata.* London: Ray Society.

————. 1955. *The origin of vertebrates.* Oxford: Clarendon Press.

Bone, Q. 1960. The origin of the chordates. *J. Linn. Soc. Lond.* 44:252–69.

————. 1961. The organization of the atrial nervous system of amphioxus *Branchiostoma lanceolatum* (Pallas). *Phil. Trans. Roy. Soc. Lond.* ser. B, 243:241–69.

Brambell, F. W. R., and Cole, H. A. 1939. The preoral ciliary organ of the Enteropneusta: Its occurrence, structure, and possible phylogenetic significance. *Proc. Zool. Soc. Lond.* ser. B, 109:181–93.

Brodal, A., and Fänge, R., eds. 1963. *The biology of Myxine.* Oslo: Universitetsforlaget.

Bullock, T. H. 1946. The anatomical organization of the nervous system of Enteropneusta. *Quart. J. Micr. Sci.* 86:55–111.

Burdon-Jones, C. 1962. The feeding mechanism of *Balanoglossus gigas. Bol. Fac. Fil., Cien. Letr. Univ. S. Paulo,* no. 261. *Zoologia,* no. 24, pp. 255–80.

Elwyn, A. 1937. Some stages in the development of the neural complex in *Ecteinascidia. Bull. Neurol. Inst. N.Y.:* 6:163–77.

Grassé, P. P., ed. 1948. *Traité de zoologie.* 11. *Echinodermes, stomocordés, procordés.* Paris: Masson.

Hardisty, M. W., and Potter, I., eds. 1971. *The biology of lampreys.* London: Academic Press.

Horst, C. H. van der. 1927–36. *Hemichordata. Bronn's Klassen und Ordnungen des Tierreiches,* 4, Div. 4, bk. 2, pt. 2.

————. 1932. Enteropneusta. In *Kükenthal-Krumbach's Handbuch der Zoologie,* vol. 3, part 2. Berlin: de Gruyter.

Hyman, L. H. 1959. *The invertebrates. 5. Smaller coelomate groups.* New York: McGraw-Hill.

Knight-Jones, E. W. 1952. On the nervous system of *Saccoglossus cambrensis* (Enteropneusta). *Phil. Trans. Roy. Soc. Lond.* ser. B, 236: 315–54.

————. 1953. Feeding in *Saccoglossus* (Enteropneusta). *Proc. Zool. Soc. Lond.* 123:637–54.

Lohmann, H.; Huus, J.; and Ihle, I. E. W. 1933. Tunicata. In *Kükenthal-Krumbach's Handbuch der Zoologie,* vol. 5, part 2. Berlin: de Gruyter.

Mozejko, B. 1913. Über das oberflächliche subkutane Gefäs-system von *Amphioxus. Mitt. Zool. Staz. Neapel,* 21:65–103.

Orton, J. H. 1913. The ciliary mechanisms on the gill and the mode of feeding in amphioxus, ascidians, and *Solenomya togata. J. Mar. Biol. Ass. U.K.* 10:19–49.

Pietschmann, V. 1929a. Akrania. In *Kükenthal-Krumbach's Handbuch der Zoologie,* vol. 6, part 1. Berlin: de Gruyter.

————. 1929b. Cyclostomen. In *Kükenthal-Krumbach's Handbuch der Zoologie,* vol. 6, part 1. Berlin: de Gruyter.

Webb, J. E. 1958. The ecology of Lagos Lagoon. III. The life-history of *Branchiostoma nigeriense* Webb. *Phil. Trans. Roy. Soc. Lond.* ser. B, 241:335–53.

Webb, J. E., and Hill, M. B. 1958. The ecology of Lagos Lagoon. IV. On the reactions of *Branchiostoma nigeriense* to its environment. *Phil. Trans. Roy. Soc. Lond.* ser. B, 241:355–91.

Weel, P. B. van. 1937. Die Ernährungsbiologie von *Amphioxus lanceolatus. Pubbl. Staz. Zool. Napoli* 16:221–72.

Werner, E., and Werner, B. 1954. Über den Mechanismus des Nährungserwerbs der Tunicaten, speziell der Ascidien. *Helg. Wiss. Meeresuntersuch.* 5:57–92.

# 4 General Features of Chordate Development
## Marvalee H. Wake

The study of morphology includes an understanding of basic patterns of development. Determinations of evolution of structure and function must be made in terms of similarities and differences of genetic capacity, environmental constraints, and selective factors that influence development. This chapter presents selected examples of vertebrates to illustrate developmental patterns of major groups and some ideas about the way development is initiated, maintained, and controlled.

**A. The Egg**

Development is initiated when a sperm fertilizes an ovum. An ovum is a specialized cell that is at or near maturation. It has a nucleus whose chromosome complement is reduced, or is in the process of being reduced, to the haploid state (a single set of chromosomes); various organelles, such as mitochondria and ribosomes; a full complement of enzymes, including some specialized for fertilization; and a complex cortical chemical organization. It also contains yolk, which is a complex of proteins, phospholipids, and neutral fats. The development of eggs in the ovary is discussed in chapter 12.

The amount of yolk in an egg varies. On this basis chordate eggs are classified as follows: (1) *isolecithal* eggs, with little yolk, evenly distributed, as in *Branchiostoma* and mammals; (2) *telolecithal* eggs with *total* or *holoblastic* cleavage; these have a moderate amount of yolk, which accumulates in part of the egg and retards its development, as in amphibians, cyclostomes, and ganoid fishes; (3) *telolecithal* eggs with *meroblastic* cleavage, in which there is an enormous amount of yolk and the protoplasm is mostly concentrated in a small area, the *germinal disk,* which floats on the surface of the yolk. Meroblastic eggs are characteristic of teleosts, reptiles, and birds, and the mammalian egg has evolved from the reptilian type by loss of the yolk mass. These three types of eggs are illustrated diagrammatically in figure 4.1.

**B. Cleavage and Blastulation**

Sperm develop in the testes, from which they are released after meiosis (cell division that results in haploid gametes) is complete. Further development continues, so that the spherical spermatid becomes a motile sperm. The mature sperm has a head, containing the nucleus, that is capped by the *acrosome,* which facilitates fertilization; a midregion, containing numerous mitochondria (an energy source), and the basal body of the flagellum; a tail, composed of the flagellum, which is equipped with a membrane in some species (see figure 4.2). The sperm is capable

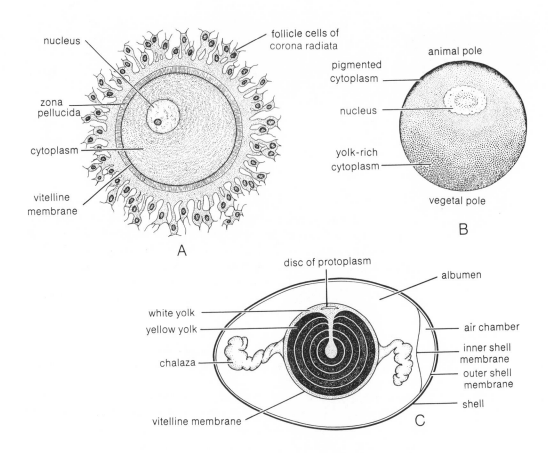

Fig. 4.1.  Types of vertebrate eggs. *A*, isolecithal human egg; *B*, moderately telolecithal frog egg; *C*, highly telolecithal chicken egg. After Arey 1965.

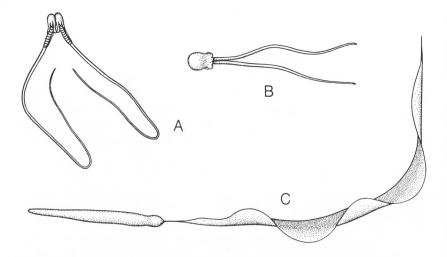

Fig. 4.2.  Types of vertebrate sperm. *A*, opossum; *B*, toadfish; *C*, toad. Note the variation in shape of the heads of the sperm and of the flagella. After Ebert and Sussex 1970.

of short-term, fast movement to reach the egg. It must move in a fluid medium, provided either by the external environment or by cellular secretion.

Fertilization begins when a sperm penetrates the egg; it is complete when egg and sperm nuclei are fused. The chromosomes rotate on a spindle, and the composition of the egg cortex changes, including the formation and elevation of the fertilization membrane and the completion of the nuclear membrane. Division of the nucleus and cytoplasm then occurs in a sequence that initially is carefully timed.

Division of the egg into two, four, etc., cells continues until a large number of cells has been produced, a process called *cleavage*. The timing of division and the shape of cells produced depends on the amount of yolk the egg contains. It should be understood that the yolk is in large part inert material and that the process of development is carried out only by the living protoplasmic portions of the egg; yolk is partitioned into the cells as they are formed by cleavage.

**1. Holoblastic Equal Cleavage**

In the case of isolecithal eggs the entire egg divides and produces a number of approximately equal cells. Such cleavage is said to be *holoblastic* and *equal*. The cells, as they increase in number, gradually withdraw from the center and arrange themselves in a single layer on the surface, thus producing a fluid-filled ball of cells. This ball is called the blastula; its cavity is known as the *segmentation cavity* or *blastocoel*. Such a blastula is produced in the development of *Branchiostoma*. Cleavage and formation of the blastula in *Branchiostoma* are illustrated in figure 4.3*A*.

Study also the models of the cleavage of *Branchiostoma* provided in the laboratory.

**2. Holoblastic Unequal Cleavage**

Holoblastic unequal cleavage occurs in those telolecithal eggs that contain a moderate amount of yolk. The half of the egg that contains more of the yolk is called the *vegetal hemisphere;* that which contains more of the protoplasm is the *animal hemisphere*. Cleavage proceeds faster in the animal hemisphere, since cleavage processes are delayed in the vegetal hemisphere because of the inert yolk. The cells produced in the animal hemisphere are smaller and more numerous than those of the vegetal hemisphere, although the entire egg cleaves. Such cleavage is holoblastic but *unequal*. The cells withdraw from the center, producing a blastula with a somewhat reduced *blastocoel* and a wall several layers of cells thick. The cells of the blastula are of unequal sizes, from the smaller at the animal pole to the larger at the vegetal pole.

This type of development is characteristic of amphibians. It is illustrated in figure 4.3*B*; study also the models of the cleavage of the amphibian egg provided in the laboratory. Then obtain a section through an amphibian embryo in the blastula stage and examine under the low power of the microscope. The blastula is a hollow sphere whose wall is composed of two or three layers of cells. The wall of the animal hemisphere is thin and consists of small cells; it is the future

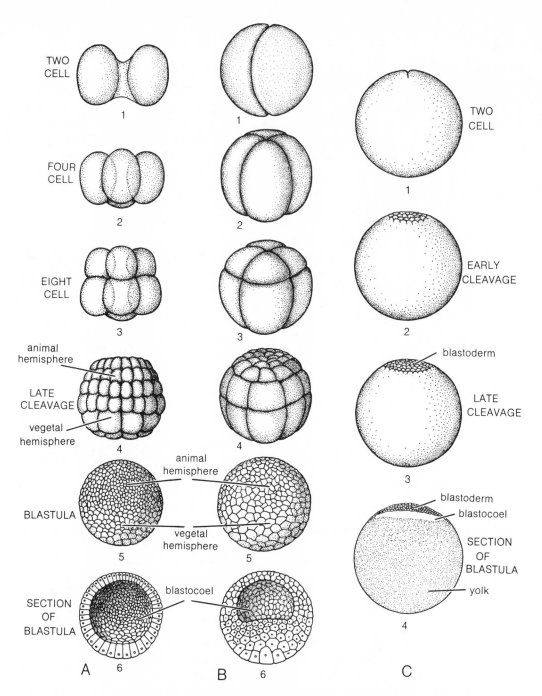

Fig. 4.3. Cleavage of the three types of chordate eggs and formation of the blastula. *A, Branchiostoma:* 1–4, cleavage; 5, external view of the blastula; 6, section of the blastula. Note that the cells of the vegetal hemisphere are only slightly larger than those of the animal hemisphere, the wall of the blastula is one cell layer in thickness, and the blastocoel is large. *B,* frog: 1–4, cleavage; 5, blastula; 6, section of the blastula. The cells of the vegetal hemisphere are considerably larger than those of the animal hemisphere, the wall of the blastula is at least two cell layers in thickness, and the blastocoel is smaller and displaced dorsally. *C,* reptile or bird egg with meroblastic cleavage: 1–3, cleavage; 4, median sagittal section of the blastula. Only the germinal disk cleaves, forming a disk of cells—the blastoderm—resting on the yolk; a slight slit between this and the yolk represents the blastocoel. From Hyman 1942.

dorsal side of the embryo. The wall of the vegetal hemisphere is much thicker and is composed of large cells, laden with yolk and with indistinct boundaries; it is the future ventral side. The blastocoel is smaller than in the blastula of *Branchiostoma* and is displaced dorsally because of the thickness of the ventral wall.

---

3. Meroblastic Cleavage

In eggs containing large quantities of yolk, only the small germinal disk undergoes cleavage. This kind of cleavage is called *meroblastic*. As a result, a minute disk of cells is produced on the surface of the relatively enormous yolk. A slight split appears between the disk and the yolk, and this may correspond to the segmentation cavity of other developing eggs; this state is consequently the blastula stage. Meroblastic cleavage is illustrated in figure 4.3C. The disk of cells produced by meroblastic cleavage soon begins to expand over the surface of the yolk, then being termed the *blastoderm*.

**C. Gastrulation**

Many experiments have been done to determine what cells at various developmental stages would become. Spemann and other workers found that cells from early cleavage stages, when isolated, could continue division and give rise to a complete embryo. Cells at such early stages are considered *totipotent*. Cells from later stages, especially the blastula on, do not have this capacity. Such cells, when marked or transplanted, result in development of a particular cell type. *Differentiation* has occurred, and the "fates" of cells are fixed. *Fate maps* have been constructed that show the cell areas of blastulas that will subsequently give rise to particular kinds of cells. See figure 4.4 for examples, and correlate these fate maps with the diagrams of gastrulas and neurulas that follow.

1. In Eggs of the *Branchiostoma* Type

In such eggs the vegetal hemisphere begins to bend inward as cells at the dorsal lip of the blastopore change shape, and it continues this process of *invagination* until its wall comes in contact with the wall of the animal hemisphere. An embryo with a wall two cell layers thick is thus produced, called a *gastrula*. The outer layer is named the *ectoderm* and the inner layer the *endoderm*. Because of their roles in the subsequent development, these layers are referred to as the first two germ layers. The hollow tube of endoderm is called the *archenteron* or *primitive intestine;* the cavity of the gastrula is the cavity of the archenteron or *gastrocoel;* and the opening of the archenteron to the exterior is the *blastopore*. Note that the blastocoel is eliminated in the production of this type of gastrula. Remember that this is a dynamic, ever-changing process. The formation of the *Branchiostoma* gastrula is illustrated in figure 4.5*A*; study also the models exhibited in the laboratory.

2. In Eggs of the Amphibian Type

In these eggs gastrulation is somewhat modified by the presence of inert yolk in the vegetal hemisphere. It is accomplished partly by the invagination of the endoderm, particularly at the dorsal lip of the blastopore, and partly by the expansion of the ectoderm ventrally, pushing the endoderm into the interior. The result is the same as the foregoing; a gastrula is formed. A small portion of the enclosed yolk-bearing cells commonly

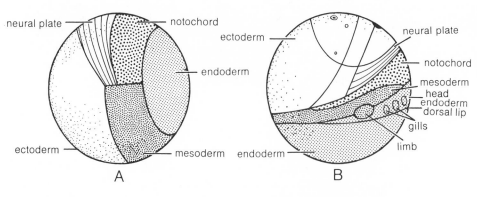

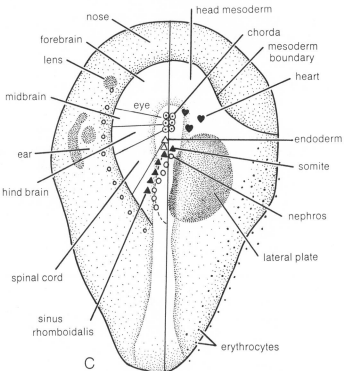

Fig. 4.4. Fate maps of representative chordates. *A*, the blastula of *Branchiostoma* with prospective regions marked; *B*, the early gastrula of a frog with prospective regions mapped; *C*, late gastrula of the chick. *A*, after Arey 1965; *B*, after Vogt 1929 (*Arch. Entw. Org.* 120:384–700); *C*, after Rudnick 1944 (*Quart. Rev. Biol.* 19:187–212).

remains for some time protruding through the blastopore; this portion is called the *yolk plug*.

The formation of the amphibian gastrula is illustrated in figure 4.5*B*; study further the models of amphibian development illustrating this stage, noting especially the sagittal section of the gastrula. Then obtain a slide bearing a sagittal section of the gastrula and study it with the low power of the microscope. The gastrula is slightly elongated in anteroposteriad. The side with the thinner wall is the dorsal side; that

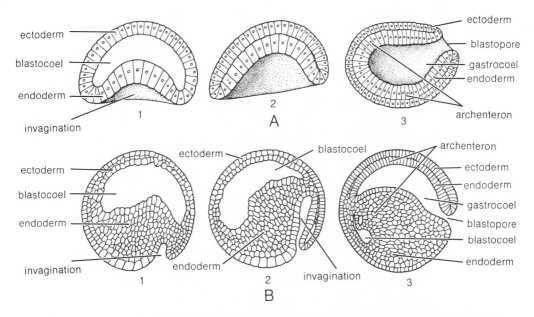

Fig. 4.5. Gastrulation and formation of the endoderm in holoblastic chordate eggs. All shown in median sagittal section. *A, Branchiostoma:* 1, beginning of the invagination; 2, invagination completed; 3, completed gastrula having a wall of two layers, ectoderm and endoderm, and an internal cavity, the gastrocoel. *B,* amphibian: 1, beginning of the invagination; 2, progress of the invagination accompanied by downward growth of the ectoderm; 3, completed gastrula, with very thick endoderm ventrally. In *A*3 and *B*3, the anterior end of the embryo is to the left, posterior end to the right, dorsal surface above, ventral below. After Hyman 1942.

with the thick wall the ventral side; the end with an opening is the posterior end, the opposite end is anterior. The wall consists of two layers, each composed of more than one sheet of cells. The outer and thinner layer is the ectoderm, uniform in width over the whole embryo. The inner layer is the endoderm, separated from the ectoderm by a slight space, and very thick ventrally, where its cells are laden with yolk. The cavity enclosed by the endoderm is the gastrocoel. The opening of the archenteron to the exterior at the posterior end is the blastopore. Ectoderm and endoderm are continuous at the rim of the blastopore. A portion of the endoderm, the yolk plug, protrudes through the blastopore and nearly occludes the opening.

3. In Meroblastic Eggs

The meroblastic type of development of reptiles and birds is an adaptation for providing large stores of food for the embryo, which then completes its development inside the eggshell and hatches as a small replica of the adult. A larval stage is absent. The change from holoblastic to meroblastic development is a drastic one and naturally involves many alterations, especially of the early embryonic processes. According to current theory the dominant process is *delamination*—that is, the downward migration of cells from the underside of the blastoderm to form a new layer, the endoderm, next to the yolk (see fig 4.6C). In reptiles there is also some persistence of endoderm formation by invagination, with the formation of a blastopore and a short archenteron. The endoderm so formed unites with the delaminated endoderm to become the final endoderm. Archenteron formation of the amphibian sort is seen in turtles but among other reptiles is evidently disappearing, and in birds it has been completely lost. In birds the endoderm is formed entirely by delamination, and no traces remain of archenteron or blastopore.

After endoderm formation the blastoderm consists of two strata, the outer ectoderm and the inner endoderm; it lies on the surface of the yolk and by proliferation at its margins gradually spreads over the yolk, eventually enclosing it.

**D. Formation of Mesoderm**

**1. In *Branchiostoma***

After the embryo has attained the gastrula stage, it elongates and presents a flattened dorsal surface, a rounded ventral surface, and recognizable anterior and posterior ends (see fig. 4.5*A*). From the dorsolateral regions of the endoderm, which, you will remember, forms the "inner tube" of the gastrula, hollow pouches begin to grown out in pairs. These pouches are called the *coelomic sacs* or *mesodermal pouches*. The walls of the pouches constitute the *mesoderm,* or *third germ layer,* which, unlike the ectoderm and endoderm, consists of two walls. The pouches grow laterad and ventrad, filling the space between ectoderm and endoderm. The outer wall of the pouches, in contact with the ectoderm, is called the *somatic* or *parietal* mesoderm; the inner wall, in contact with the endoderm, is the *splanchnic* mesoderm. The cavity of the pouches is the *body cavity* or *coelom*. Eventually the anterior and posterior walls of the pouches break down, so that those of each side unite to form a tube. Thus, the coelom, originally segmented, comes to consist of a pair of continuous cavities, one on each side of the embryo.

**2. In Vertebrates**

The outpouching of the archenteron roof to form notochord and mesodermal sacs is usually considered the primitive chordate method, but it does not occur in any vertebrate. In vertebrates with holoblastic unequal cleavage such as cyclostomes and amphibians, a median strand, the notochord, and lateral sheets, the mesoderm, split off from the archenteron roof (from the blastopore forward). The mesoderm sheets are never segmented at first, nor do they contain a coelom. They spread laterally and ventrally, and a central split, the coelom, appears in them, dividing them into somatic and splanchnic walls.

In meroblastic development the yolk must be covered quickly with mesoderm, since the blood vessels are of mesodermal origin and are necessary for carrying the yolk to the embryo as food. Reptiles failed to solve this problem efficiently: a notochord and mesoderm arise much as in amphibians, from a middorsal strand that grows forward from the short invaginated archenteron and therefore represents the archenteron roof. In birds the problem has been satisfactorily solved. As already noted, birds have no archenteron or endodermal invagination. Along the axis of the future embryo the ectoderm begins to migrate and proliferate into the interior. These processes produce a thickening of the ectoderm, and this thickening is so pronounced that it becomes visible to the naked eye as an opaque streak, the *primitive streak*. The ectodermal material thus sent into the interior forms the notochord medially

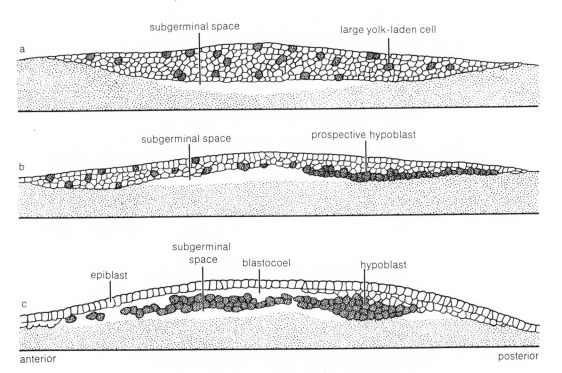

Fig. 4.6. Longitudinal sections through chick blastoderms to show origin of hypoblast: larger yolk-laden cells intermixed with smaller cells (*a*); accumulation of larger cells in subsurface position (*b*); organization of larger cells to form hypoblastic layer, with enlarged germinal of blastocoelic spaces and thinning of epiblast with development (*c*). From Torrey 1967 (*Morphogenesis of the Vertebrates*. New York: John Wiley).

and the mesoderm laterally. The mesoderm spreads rapidly over the yolk as sheets, in which a coelomic split arises as in amphibians. There has long been promulgated in vertebrate embryology the theory that the primitive streak represents an elongated blastopore whose sides have fused together (concrescence theory). This theory now seems definitely erroneous. The primitive streak is a rapid, short-cut method of mesoderm formation. The middorsal ectoderm of the embryo, instead of going through the tedious process of getting into the interior by invagination at the blastopore (not very practical, anyway, in meroblastic eggs), passes directly into the interior by migration and proliferation. The primitive streak therefore represents phylogenetically the formation of the roof of the archenteron, since, like that roof, it is the source of the notochord and the mesodermal sheets.

**E. Formation of the Neural Tube and Notochord**

The ectoderm proliferates on each side of the middorsal line, forming a pair of folds or ridges. These two folds meet and fuse, forming a tube, the *neural tube,* that is the primordium of the brain and spinal cord. From the middorsal wall of the archenteron a solid rod of cells is elevated and separated off; this is the *notochord* or primitive axial skeleton. These processes are illustrated in figure 4.7; see also the models of *Branchiostoma* development.

1. In *Branchiostoma*

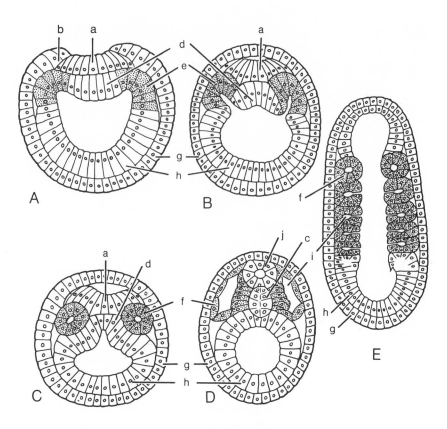

Fig. 4.7. Formation of the neural tube, notochord, mesoderm, and coelom in *Branchiostoma. A–D*, cross sections; *E*, frontal section. *A*, differentiation of the medullary plate *a*, the notochordal plate *d*, the neural folds *b*, and the mesodermal pouches *e. B*, the neural folds have closed across above the medullary plate; the mesodermal pouches are further evaginated. *C*, the medullary and notochordal plates are beginning to close; the mesodermal pouches *f* are completely separated from the endoderm. *D*, the neural tube *j* and the notochord *c* are completed; the mesodermal pouches are increasing in size. *E*, frontal section to show the mesodermal pouches *f* originating from the endoderm segmentally; *a*, medullary plate; *b*, neural fold; *c*, notochord; *d*, notochordal plate; *e*, mesoderm; *f*, mesodermal pouches; *g*, ectoderm; *h*, endoderm or archenteron; *i*, coelom. In all figures the mesoderm is stippled. From Hyman 1942.

2. In Vertebrates   The neural tube or future central nervous system is formed throughout vertebrates in the same manner as in *Branchiostoma*. A pair of longitudinal ectodermal folds, the *neural* or *medullary* folds, rise up in mid-dorsal region of the embryo and fuse together, producing a tube (fig. 4.9).

These processes will be more clearly understood by reference to the accompanying figures (figs. 4.8 and 4.9). Obtain a mounted cross section through an amphibian embryo at the stage of the formation of the mesoderm and examine under the low power. The section is oval in form; it may be still surrounded by the delicate *egg* or *fertiliza-*

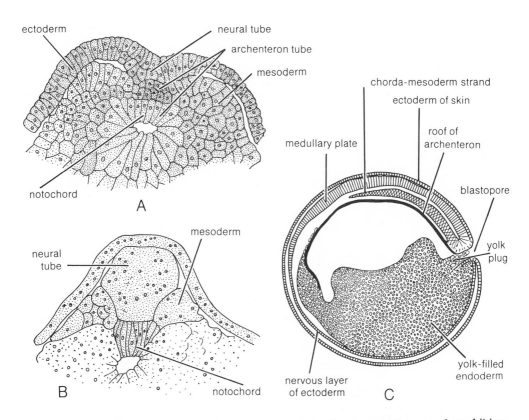

Fig. 4.8. Notochord and mesoderm formation in cyclostomes and amphibians. *A* and *B*, cross sections of the *Petromyzon* embryo, showing separation of notochord and mesoderm from the archenteron roof; neural tube forms as solid invagination. *C*, median sagittal section through the frog embryo, showing chordamesoderm strand (*diagonal hatching*) separating from archenteron roof (*black*); the ectoderm has separated into two layers, an outer one to become the skin epidermis and an inner nervous layer; in the latter the thickening of the medullary plate is seen. From Hyman 1942.

*tion membrane.* The outer layer of the embryo is the ectoderm, relatively thin and of the same depth over the whole surface. In the median dorsal line the ectoderm is producing, or has already produced, the neural tube. In the former case the ectoderm exhibits a pair of *neural folds,* enclosing a thick plate of ectoderm. In the latter case the folds have fused across in the median line, forming a tube, the *neural tube,* that is the oval hollow mass in the median dorsal line, just beneath the ectoderm. The greater part of the section is occupied by the *archenteron* or *primitive intestine,* composed of endoderm. The archenteron has a thin dorsal wall, a thick ventral wall, whose cells contain yolk, and encloses the relatively small *gastrocoel,* which occupies its dorsal part. In the median dorsal region of the archenteron a mass of cells will be seen protruding dorsally, or in some slides this mass of cells may have separated from the archenteron so that it lies between the latter and the neural tube. This mass is the *notochord.* Between the ectoderm and the archenteron on each side is a narrow sheet of cells extending ventrally from the sides of the neural tube. In some slides

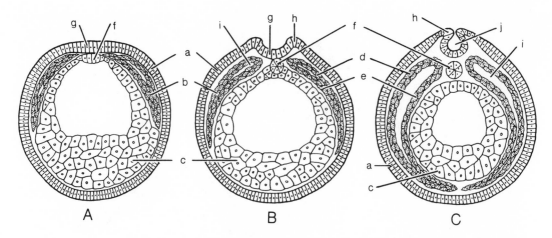

Fig. 4.9. The formation of the neural tube, notochord, mesoderm, and coelom in vertebrates, based on the frog. Cross sections. *A*, differentiation of the notochord (*f*) in the roof of the endoderm; mesodermal plates (*b*) spreading ventrally. *B*, neural folds (*h*) rising at the sides of the medullary plate (*g*); notochord (*f*) separated from the endoderm; mesodermal plates (*b*) extended farther ventrally and developing a central cavity, the coelom (*i*). *C*, neural folds (*h*) nearly closed to form the neural tube (*j*); mesodermal plates have reached the midventral line; coelomic split (*i*) has extended ventrally; *a*, ectoderm; *b*, mesoderm; *c*, endoderm or archenteron; *d*, somatic mesoderm; *e*, splanchnic mesoderm; *f*, notochord; *g*, medullary plate; *h*, neural fold; *i*, coelom; *j*, neural tube. From Hyman 1942.

these sheets will extend only a short distance, while in others they reach nearly to the median ventral line. These sheets are the *mesoderm*.

**F. Cells and Tissue Interactions**

The movements of cells that effect gastrulation are due to (1) changes of shape of individual cells and their active movement, and (2) different rates of division of different cell types. These changes are sequential in space and in time, so that cell movements proceed in careful order. Movements begin at the dorsal lip of the blastopore in several forms mentioned above; those movements seem to trigger other developmental changes. During the 1930s, a series of experiments demonstrated that dorsal lip tissue, when transplanted, caused the development of (1) structures from tissue areas that did not normally give rise to such structures, and (2) duplicate structures. The dorsal lip was called the *primary organizer* because it could elicit a complete secondary embryo. It is now known that the cell movements of gastrulation cause specific tissue interactions. Such interactions are called *inductive* and are necessary for further normal development. For example, movement of the dorsal lip induces development of the neural axis. Such interactions are responsible throughout development for consequent structure. Later in development, for example, both a ureteric bud and nephrogenic tissue must be present to induce formation of ureter and kidney tubules. The processes that give rise to mesoderm, neural tube, and notochord as described above are part of the mechanism of induction. The way induction actually works is not yet fully known. Experiments have demonstrated that "the inducer" is a chemical substance, probably a protein, that is given off by cells, for cells can be separated by a membrane with tiny pores, and normal development will continue. The inducer agent may be slightly different in different cell types.

But induction is not the entire story of development. *Differentiation* can be a product of induction or independent of it. Differentiation is the appearance of *new* properties in cells of a given lineage. Often associated with differentiation is the phenomenon of restriction of the cell's capacity, or *determination*. For example, we have seen that cells from early cleavage can develop into any sort of tissue when transplanted, depending upon the cell's associations. This is not true of "older" cells; for example, a notochord cell cannot develop into kidney mesoderm if transplanted to that presumptive area; its fate has been determined, or fixed. Determination is progressively more specifically fixed as differentiation proceeds.

**G. Differentia- tion and Derivation**

The next sections of this chapter describe some of the stages of differentiation and determination of the primary kinds of embryonic tissues. The major adult structures that are derived from these tissues and their interactions are listed for each of the primary tissue types.

**1. History of the Mesoderm**

The history of the mesoderm is of the utmost importance for understanding vertebrate structure. We have already noted that the mesoderm splits into two layers, an outer or somatic layer and an inner or splanchnic layer, and that the space between the two layers is the body cavity or coelom. The mesoderm grows from each side of the embryonic axis ventrally to the median ventral line, or in meroblastic eggs grows out over the yolk, pushing out between ectoderm and endoderm.

The mesoderm next becomes differentiated into three regions: a dorsal region, called the *epimere,* which lies to each side of the neural tube; a middle region called the *mesomere* or *nephrotome,* situated lateral and ventral to the epimere; and a large ventral region on each side of the archenteron, called the *hypomere* or *lateral plate.* Each of these regions has, of course, both somatic and splanchnic walls (see fig. 4.10*A*). The epimere immediately becomes *segmented;* that is to say, dorsoventral clefts appear in it at regular intervals, the process beginning at the anterior end of the embryo and proceeding posteriorly. Consequently, the epimere becomes divided up into a longitudinal row of blocks, a row on each side of the neural tube. These blocks are epimeres, or *mesoblastic somites.* At first the epimeres are continuous with the mesomere ventrally and laterally, but eventually they are completely cut off from the rest of the mesoderm (see fig. 4.10*B*). The mesomere and the hypomere do not become segmented and remain permanently in close relation to each other. Within the mesomere small tubules develop, which open into the cavity of the hypomere; they are the primary tubules of the kidney (see fig. 4.10*B*). The hypomeres of each side fold around the archenteron, their inner walls coming in contact above and below the archenteron to form double-walled membranes, the *dorsal* and *ventral mesenteries* (see fig. 4.10*B*). The cavities of the two hypomeres become the coelom of the adult; the cavity in each epimere

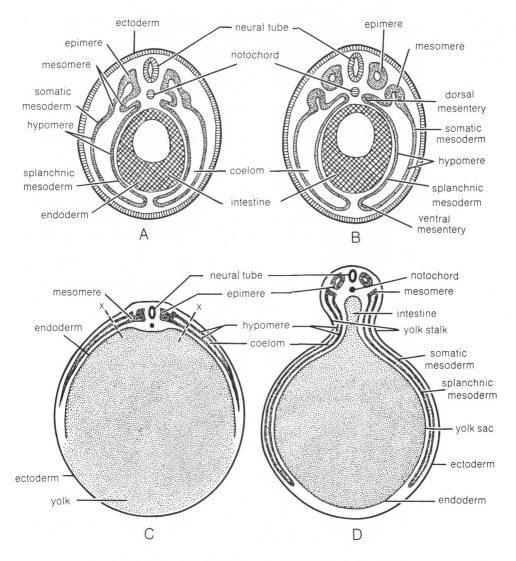

Fig. 4.10. The differentiation of the mesoderm in holoblastic and meroblastic types of development. *A* and *B*, holoblastic type; *C* and *D*, meroblastic type. *A*, differentiation of the mesoderm into epimere, mesomere, and hypomere. *B*, separation of the epimere from the mesomere, appearance of kidney tubules in the mesomere, and closure of hypomere around the intestine to form dorsal and ventral mesenteries. *C*, similar to *A* but in the meroblastic type, showing endoderm and mesoderm growing around the yolk; note how the embryo is spread out on the surface of the yolk; *x*, indicates lines where embryo is cut off from yolk in making sections for microscopic study. *D*, similar to *B* but in the meroblastic type; the endoderm has completely surrounded the yolk; the mesoderm has nearly done so; the yolk sac is seen to be a part of the intestine; the embryo is partly constricted from the yolk sac, the constriction being the yolk stalk. From Hyman 1942.

disappears; and that of the mesomere remains as the cavities of the tubules of the kidney.

In embryos of the amphibian type the archenteron is a closed tube, and the two hypomeres are closed cavities that meet below the archenteron. In embryos resulting from meroblastic eggs, however, the archenteron is open below and spread out on the yolk, and the hypomeres extend out over the yolk. The differences between the two types of embryos are illustrated in figure 4.10A,C. In meroblastic development the embryo is later constricted from the yolk by the formation of deep grooves on all sides. The yolk then hangs from the ventral surface of the embryo, inclosed in a sac of blastoderm, the *yolk sac,* which is connected with the embryo by the yolk stalk, as shown in figure 4.12. At the time of hatching, the yolk has been practically used up, the remnant of the sac is withdrawn into the embryo, and the opening in the body wall where the yolk stalk arises is finally closed over.

---

Obtain a slide bearing a cross section through the trunk of a chick embryo of two days' incubation. As explained above, the intestine of the chick embryo is open below on the yolk. In making such sections the embryo is cut off from the yolk. After understanding the relation of embryo and yolk, examine the section using low magnification. The dorsal boundary of the section is a thin layer, the ectoderm, which is slightly elevated in the median dorsal line; the ventral boundary is another thin layer, the endoderm, which makes a slight upward bend in the median central line, indicating the future intestine. In the median dorsal line just beneath the ectoderm is the oval hollow section of the *neural tube.* Immediately ventral to this is a small circular mass of cells, the *notochord.* On each side of the neural tube is a squarish mass, its cells radiating from the center. This is the *epimere* or *mesoblastic somite.* Lateral to the epimere and continuous with it is a smaller mass, the mesomere or nephrotome, in which one or more tubules with central holes are distinguishable. Beyond the mesomere the mesoderm is observed to split into two layers. This region of the mesoderm is the *hypomere* or *lateral plate.* The outer dorsal layer of the hypomere is the *somatic mesoderm.* It ascends and comes in contact with the ectoderm, the two together constituting the *somatopleure* or body wall. The lower or ventral wall of the hypomere is the *splanchnic mesoderm;* it descends and comes in contact with the endoderm, and the double layer thus formed is the *splanchnopleure* or intestinal wall. The cavity between the somatic and splanchnic walls of the hypomere is the coelom. As already explained, the hypomere in such embryos extends far out over the yolk. Observe that the splanchnopleure contains many holes; these are the cross sections of blood vessels, which convey the food from the yolk sac to the embryo. There is also a large artery in the embryo below each epimere.

---

2. The Fate of the Ectoderm

The ectoderm gives rise to the neural tube, from which develop the brain, spinal cord, and nerves. The ectoderm also forms the external layer of the skin and all its derivatives, such as hair, nails, and feathers.

It also gives rise to the sensory part of all the sense organs, the lining membrane of the nasal cavities, the mouth and anus, the glands and other outgrowths of the nasal and mouth cavities, the glands of the skin, the enamel of the teeth, and the lens of the eye.

**3. The Fate of the Endoderm**

The endoderm is the primitive intestine. It forms the epithelial lining of the adult intestine and the epithelial lining and epithelial cells of all of the outgrowths of the intestine, which include the gill pouches and gills, the larynx, trachea, and lungs, the tonsils, the thyroid and thymus glands, the liver, the gallbladder and bile duct, the pancreas, and the urinary bladder and adjacent parts of the urogenital system. Note that only the epithelial cells of these structures arise from the endoderm.

**4. The Fate of the Mesoderm and the Formation of Mesenchyme**

*a. Mesenchyme*

In the further development of the mesoderm, *mesenchyme* plays an important role. Mesenchyme is not a germ layer but a particular type of tissue. It is a primitive kind of connective tissue, consisting of branched cells, whose branches are more or less united to form a network. Nearly all of the mesenchyme comes from mesoderm, but it may arise from the other germ layers also. Hence tissues and structures that arise from mesenchyme may owe their origin to more than one germ layer. Consequently, to avoid inaccuracy it is usually merely stated that they arise from mesenchyme, without specifying the particular germ layer or layers involved. When a germ layer is about to produce mesenchyme, its cells become loose, separating from their fellows; they lose their epithelial form, and, taking on a branched irregular shape, wander away by amoeboid movements to more or less definite regions, where they give rise to particular tissues. Those parts of the mesoderm that do not become mesenchyme but retain their epithelial characteristics are called *mesothelium.*

*b. The Fate of the Epimeres*

The medial wall of each epimere transforms into a mass of mesenchyme cells that migrate to a position around the notochord and there give rise to the vertebral column. This mass of mesenchyme is known as the *sclerotome* (see fig. 4.11). The outer wall of each epimere transforms into mesenchyme cells that migrate to the underside of the ectoderm and there give rise to the inner layer (dermis) of the skin. This part of the epimere is called the *dermatome* (fig. 4.11). The remainder of the epimere persists in place as mesothelium and is known as a *myotome* or *muscle segment.* Each myotome becomes separated from the adjacent ones by a connective tissue partition, the *myocomma* or *myoseptum.* The myotomes give rise to the voluntary muscles of the body (with certain exceptions). Each grows from its original dorsal position ventrally between the ectoderm and the hypomere to the median ventral line, where it meets its fellow from the opposite side. There is thus produced a complete muscular coat for the body.

*c. The Fate of the Mesomere*

The mesomere gives rise to the kidneys, the reproductive organs, and their

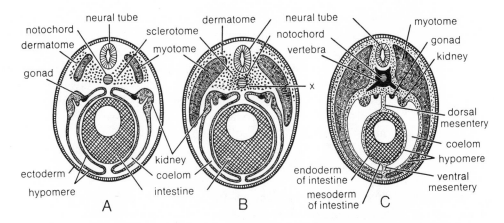

Fig. 4.11. Cross sections of vertebrate embryos to show the differentiation of the epimere into dermatome (skin-producer), myotome (muscle-producer), and sclerotome (skeleton-producer). In B and C the dermatome is seen spreading beneath the ectoderm to form the dermis of the skin; the myotomes are growing ventrally to form the muscle layer of the body wall; the sclerotome is accumulating around the notochord at x to form the vertebrae; and the hypomere encloses the intestine, producing the dorsal and ventral mesenteries and the mesoderm of the intestine. From Hyman 1942.

ducts (the terminal portions of the urogenital ducts may have ectodermal or endodermal linings).

### d. The Fate of the Hypomere

The cavity of the hypomere is the coelom of the adult. The splanchnic walls of the hypomeres of the two sides fold around the archenteron and give rise to mesenchyme, from which are produced the smooth muscle and connective tissue coats of the digestive tract, and also the smooth muscle, connective tissue, or cartilage, as the case may be, of all the derivatives of the digestive tract mentioned above. The hypomere also gives rise to the linings of all the coelomic cavities, the serosa of the viscera, and all of the mesenteries. The splanchnic mesoderm of the hypomere produces the heart. In the region of the gill slits the hypomere produces voluntary muscles.

### e. The Products of the Mesenchyme

The mesenchyme gives rise to all the connective tissue of the body, including cartilage and bone; to all the involuntary or smooth muscles; to the blood cells, the blood vessels, the lymph vessels, and lymph glands; and to the voluntary muscles of the appendages. It has already been stated that the vast majority of the mesenchyme is of mesodermal origin, but a small part arises from the other germ layers.

**H. Sources of Energy During Development**

Yolk has been mentioned as a major component of the eggs of many species of vertebrates. It is a nutrient reservoir of proteins, lipids, and phosopholipids. The ovum also has a complement of mitochondria and ribosomes, so the machinery for metabolism, including biosynthesis, is present. One of the results of fertilization is activation of metabolism.

1. Within
the Egg

The components of yolk are metabolized to provide both energy-yielding substances for use in biosynthesis and chemical units that are available for synthesis into new units, especially proteins and carbohydrates. Further, since all eggs develop in an aqueous medium, water is available to be absorbed to maintain solvent properties in the dividing cells. As a result of metabolism, the biochemical composition of the cell is constantly changing and providing new metabolic substrates. Many organisms have yolk as the sole energy-yielding substance for their entire period of development. They have no other source of nutrients until they hatch or are born and can seek and ingest food.

2. Outside of
the Egg

In contrast to organisms mentioned above that have only yolk for nutrition during development, a number of species in each vertebrate class except Cyclostomata and Aves have developed means of obtaining nutrients directly from females that retain eggs in their bodies for all or part of their development. The evolution of parental care and viviparity is discussed in chapter 12, which cites many different examples among vertebrates. Here let us consider the nature of the nutrients obtained from the female and the structures that facilitate nutrient uptake. Among many lower vertebrates with some parental nutrition, the eggs have a substantial amount of yolk, and fairly advanced embryos use gills, tails, or the yolk sac to effect a "placenta" for gaseous exchange so that oxygen is the primary nutrient obtained (see below). In other groups, such as some fish and the caecilian amphibians, advanced embryos ingest a secretion from glands of the ovarian or oviducal epithelium. The secretion is rich in lipids and lipoproteins. Some of these embryos also eat the epithelium itself, digest it, and metabolize it.

Mammals have solved the problem of nutrient acquisition for retained developing eggs in a different manner. Their eggs have only a slight amount of yolk, so a nutrient reservoir does not exist. Fertilization of the egg occurs in the oviduct; division ensues as the egg moves toward the uterus. Upon reaching the uterus, the developing structure is composed of a number of cells that have formed an inner cell mass that will become the embryo, and an outer cell layer that contributes to the extraembryonic membranes. The outer cell layer comes into contact with the uterine epithelium and lays down an intermediate cell layer, the *trophoblast*. *Implantation* on the uterine wall then takes place. The pole of the cell mass that attaches is species-specific, and the site of attachment is highly determined. Implantation occurs on open wall and directly over a maternal blood vessel. The trophoblast then proliferates. As the need of the embryo for oxygen and nutrients increases, the trophoblast extends villi to other maternal blood vessels, creating an extensive system of maternal blood vessels surrounded by trophoblast tissue. With the establishment of the fetal circulation into the villi, the placenta is established. Oxygen and a variety of nutrients carried in the maternal bloodstream can cross the cell-wall barriers into the fetal circulatory system. In no species is there a "mixing" of maternal and fetal blood, though the walls of maternal and fetal capillaries may be the only barriers. "Small" molecules to those of some size do cross the placenta. Sugars, lipids, proteins of many sizes and structures, enzymes, antibodies, and hormones all cross the placenta in both directions. In effect, the fetus can receive highly refined nutrients, as well as metabolize its own, and can eliminate metabolic waste products into the maternal circulation for disposal.

3. Structure and Function of the Embryonic Membranes

Four embryonic membranes occur in the amniote vertebrates (reptiles, birds, and mammals): the *yolk sac,* the *allantois,* the *amnion,* and the *chorion* (fig. 4.12). These membranes allow the embryo to develop in an aqueous medium within the egg or its components, thus freeing the animals from the "requirement" of placing the eggs in a body of water to facilitate development.

The yolk sac is simply a sac-like expansion of the ventral wall of the intestine, narrowed into a *yolk stalk* near the body of the embryo. It

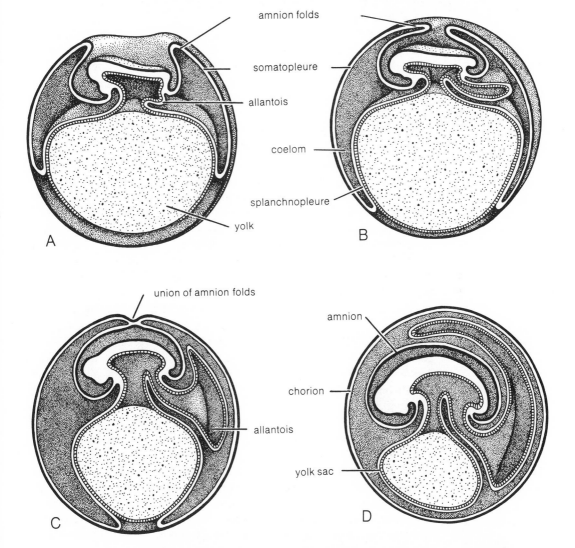

Fig. 4.12. Development of the fetal membranes of the chick. *A* and *B,* the embryo is being enveloped by the growing amniotic folds; *C,* the folds have met and joined; *D,* the chorionic and amniotic membranes result. After Arey 1965.

occurs in the embryos of all vertebrates with meroblastic development, in which it is filled with yolk utilized as food by the embryo. A yolk sac also occurs in mammals as an inheritance from their meroblastic ancestors, and in monotremes it is full of yolk and functions as in reptiles. In other mammals it is empty but may be large and vascular, especially in marsupials, where it functions as a placenta constituting the yolk-sac type of placenta (see below). With the appearance of intrauterine development the mammalian embryo obtains nutrition from the maternal blood and requires no stored food.

The allantois is a large sacciform evagination from the floor of the cloaca and may be regarded as an embryonic urinary bladder. Besides serving to hold embryonic excretory material, it has a more important respiratory function (see below). The adult bladder of amniotes is derived from the allantoic stalk.

The amnion and the chorion are formed simultaneously in the amniote embryo by paired folds of the body wall (somatopleure) that rise up around the embryo, meet above it, and fuse (fig. 4.12). The outer limb of the folds becomes the chorion, the inner limb the amnion. The amnion forms a sac enclosing the embryo except below. The chorion is the outermost membrane of the embryo and is in contact with the eggshell or the uterus, as the case may be. The yolk sac and allantois are between chorion and amnion on the ventral side of the embryo (fig. 4.12).

The yolk sac and the allantois are lined by endoderm, outside which there is a highly vascular layer of mesoderm. The blood vessels of the yolk sac are termed the *vitelline vessels*. The blood vessels of the allantois are termed the *umbilical* or allantoic vessels, and the umbilical veins of mammals are homologous to the ventral or lateral abdominal veins of lower vertebrates. The amnion is lined by ectoderm on the side facing the embryo and has an outer mesodermal layer, and the chorion consists of an outer ectoderm and an inner mesoderm. The amnion and the chorion lack blood vessels.

A yolk sac may occur in any group of vertebrates; but the allantois, amnion, and chorion are found only in reptiles, birds, and mammals, whence these three classes are termed amniotes (Greek, *amnion,* a veil, referring to the amnion covering the embryo like a veil). In birds and reptiles the allantois expands enormously between amnion and chorion, and its outer wall comes in contact and fuses with the chorion to form the *chorioallantoic membrane.* This lies against the eggshell, through which respiratory gases diffuse, so that the blood vessels of the chorioallantoic membrane become the respiratory mechanism of the embryo, carrying gases to and from the embryo.

It was mentioned above, and will be seen in detail in chapter 12, that viviparity may occur in various groups of lower vertebrates; that is, the eggs are retained in the oviducts (or ovaries in some teleosts) and develop there into young miniatures of adults, ready for free existence at birth. Leaving out of account some very bizarre nutritive relations that may develop in certain families of viviparous teleost fishes and some amphibians (discussed in chapter 12), the nutritive arrangement typically consists of projections or folds of one or the other of the embryonic membranes that are in contact with, or dovetail with, similar projections or folds of the oviducal or uterine wall. Among viviparous selachians and reptiles, and also in marsupials, the yolk sac is the embryonic membrane involved, and the placenta is then termed a *yolk-sac placenta* or *omphaloplacenta*. In mammals above marsupials, the chorioallantoic membrane is the embryonic membrane involved, and hence their placenta is called an *allantoic* or *true* placenta. See the preceding section on the development of the mammalian trophoblast and placenta. For the same reason these mammals are termed the *placental* mammals. The name placenta is derived from the round cake-like shape of the human placenta (Latin *placenta,* a cake).

The young of all mammals except monotremes develop in the uterus, a highly modified region of the oviducts, with the aid of a placenta. In marsupials this is a yolk-sac placenta, except in one case, *Parameles,* which has an incipient allantoic placenta. In all other mammals there is an allantoic placenta which, however, shows varying degrees of development. The mammalian placenta consists, in general, of vascular finger-like projections (*villi*) of the chorioallantoic membrane that fit into depressions of the uterine wall or are imbedded in the uterine wall. When these villi occur all over the chorioallantoic membrane, the placentation is termed *diffuse* (many ungulates, whales, some primates). When they occur in separated bunches, called *cotyledons,* the placentation is *cotyledonary* (ruminants). A *zonary* or ring-shaped placenta is found in carnivores, elephants, and *Hyrax;* and a *discoid* placenta is found in insectivores, bats, rodents, and primates, including man. See figure 4.13 for examples of these types. When at birth the embryonic part of the placenta is ejected without taking with it any of the uterine tissue, the placenta is spoken of as *nondeciduate;* when the uterine part of the placenta is also shed, this is termed *deciduate.* The diffuse and cotyledonary placentas are nondeciduate, the others are deciduate.

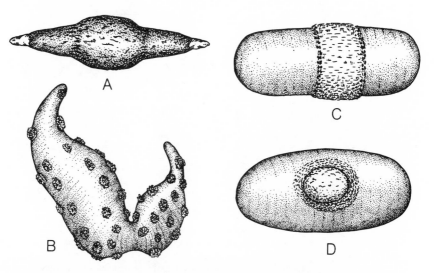

A

C

B

D

Fig. 4.13. Chorionic sacs of representative mammals showing gross forms of the placenta. *A,* diffuse placenta of the pig; *B,* cotyledonary placenta of the sheep; *C,* zonary placenta of the dog; *D,* discoidal placenta of the bear. From Torrey 1967 (*Morphogenesis of the vertebrates.* New York: John Wiley).

The intimacy of the relation between the chorioallantoic membrane and the uterine lining evolved gradually among the placental mammals. In the simplest cases, mostly forms with diffuse placentas, the embryonic and uterine villi merely interdigitate, and all the cell layers of both remain intact, forming the *epitheliochorial* type of placenta (fig. 4.14*A*). In the next, or *syndesmochorial,* type (fig. 4.14*B*), seen in ruminants, the cotyledons fit into depressions of the uterine wall, and the epithelium and

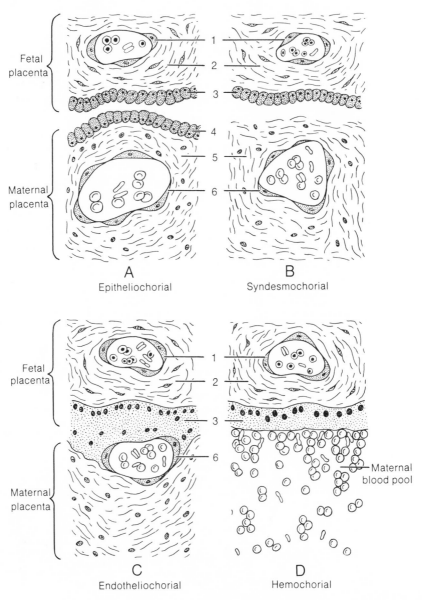

Fig. 4.14.  Histological types of mammalian placentas. 1, endothelium of fetal blood vessel; 2, chorionic connective tissue; 3, chorionic epithelium; 4, uterine epithelium; 5, endometrial connective tissue (mucosa); 6, endothelium of maternal blood vessel. From Torrey 1967 (*Morphogenesis of the vertebrates.* New York: John Wiley).

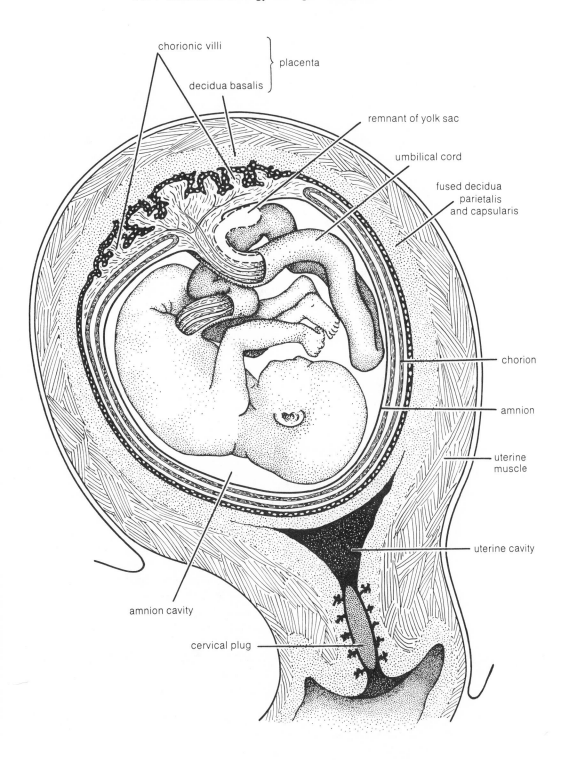

Fig. 4.15. Relationships of the eight-month human fetus and its membranes to the uterus. Note the components of the placenta. From Torrey 1967 (*Morphogenesis of the vertebrates*. New York: John Wiley).

part of the connective tissue of the latter disappear; but at birth the cotyledons are withdrawn and do not carry uterine tissue with them. In the
*endotheliochorial* placenta (fig. 4.14C), found in the zonary type, embryonic and uterine tissues are inseparably intergrown, and the embryonic villi are in contact with the maternal blood vessels. Finally, in
the *hemochorial* placenta (fig. 4.14D), which is of the discoid type, there
is an extensive breakdown of maternal tissues with destruction of blood
vessels, so that the embryonic villi lie in pools of maternal blood (fig.
4.15). Below follow dissection instructions for your examination of typical anamniote and amniote embryos.

### a. Anamniote Embryo of the Dogfish

Cut open the pregnant uterus of a dogfish and remove an embryo, or
examine embryos provided. Note that the embryo is naked or anamniote. From the middle of its ventral wall hangs the large yolk sac, filled
with yolk and attached to the embryo by the narrowed yolk stalk. The
yolk sac is covered externally by a layer of the body wall and internally
consists of the intestinal wall enclosing the yolk. As the yolk is used up,
the yolk sac is gradually withdrawn into the body.

### b. Amniote Embryo of the Cat

If pregnant females are available, open one of the enlargements in the
horns of the uterus. The enlargement contains an embryo. Note that
the embryo is enclosed in a thin membrane, the amnion, which forms a
sac around it. The placenta of the cat is of the zonary type, and on the
inner surface of the uterine wall at the enlargement a thickened vascular ring of tissue will be found, which is the placenta. It will probably
peel off, especially in late stages of pregnancy. Open the amnion and
note the umbilical cord extending from the belly of the embryo to the
inner surface of the amnion, where the latter is applied to the placenta.
The umbilical cord conveys the umbilical blood vessels to and from
the embryonic part of the placenta and is a connection between the
embryo and its own membranes, not a connection between the embryo
and its mother. There is no direct connection between embryo and
mother, but substances can diffuse from the mother's blood into the
embryonic blood by way of the placenta.

**References**

Arey, L. B. 1974. *Developmental anatomy.* 7th ed., rev. Philadelphia:
W. B. Saunders.

Austin, C. R. 1968. *Ultrastructure of fertilization.* New York: Holt,
Rinehart and Winston.

Balinsky, B. I. 1975. *An introduction to embryology.* 4th ed. Philadelphia: W. B. Saunders.

Bodemer, C. W. 1968. *Modern embryology.* New York: Holt, Rinehart
and Winston.

Burnside, B. 1973. Microtubules and microfilaments in amphibian neu-
rulation. *Amer. Zool.* 13:989–1006.

Ebert, J. D., and Sussex, I. M. 1970. *Interacting systems in development.*
2d ed. New York: Holt, Rinehart and Winston.

Hamburger, V. 1960. *A manual of experimental embryology*. Chicago: University of Chicago Press.

Hyman, L. H. 1942. *Comparative vertebrate anatomy*. 2d ed. Chicago: University of Chicago Press.

McMahon, D. 1974. Chemical messengers in development: A hypothesis. *Science* 185:1012–21.

Maderson, P. F. A. 1975. Embryonic induction and evolution. *Amer. Zool.* 15:315–27.

Monroy, A. 1965. *Chemistry and physiology of fertilization*. New York: Holt, Rinehart and Winston.

Nieuwkoop, P. D. 1973. The "organization center" of the amphibian embryo: Its origin, spatial organization, and morphogenetic action. *Adv. Morph.* 10:1–40.

Saunders, J. W. 1966. *Animal morphogenesis*. New York: Macmillan.

Twitty, V. C. 1966. *Of scientists and salamanders*. San Francisco: Freeman.

Waddington, C. H. 1966. *Principles of development and differentiation*. New York: Macmillan.

Willier, B. H., and Oppenheimer, J. M., eds. 1964. *Foundations of experimental embryology*. Englewood Cliffs, N.J.: Prentice-Hall.

Willier, B. H.; Weiss, P.; and Hamburger, V., eds. 1955. *Analysis of Development*. Philadelphia: W. B. Saunders.

# 5 The Comparative Anatomy of the Integumental Skeleton
### Richard J. Krejsa

# Part 1: The Underlying Homology

**A. Introduction**

**1. The Scope of This Chapter**

Just as a single cell or a unicellular organism has a plasma membrane, all metazoans have a body covering, the integument, that forms the interface between the animal and its environment. On, in, or near its surface and derived from it are all the receptors (sense organs) that, in association with the peripheral nervous system, make the animal aware of its external environment. The integumentary morphology often vividly reflects the animal's ecological niche and its behavioral habits. Furthermore, it embodies many of the animal's responses to the environment as it differentiates, grows, matures, reproduces, and dies. The integument is not simply an inert, covering tissue that forms only a waterproof barrier or, occasionally, a mechanically protective exoskeleton. It is, rather, a wide spectrum of cell types and tissues that form, individually or in composite, a truly dynamic, adaptive, and multifunctional organ system.

The literature on skin is generally biased in favor of the tetrapods, especially the homeotherms. Fascinating color patterns as well as protective and thermal qualities provided early investigators with sufficient stimulus to study both mammalian and avian skins. Despite the quality of many recent works, the overall emphasis has remained on the epidermis and its derivatives rather than on the skin as a functional whole. Feathers, for example, have been especially well studied: their presence is the one single character that can be used to classify an entire class of vertebrates. Broadly comparative histological and developmental studies on the entire skin of animals other than man are still wanting. Only recently has the reptilian skin begun to yield its secrets to disciplined research. Amphibian skin, for the most part, has been of more interest to osmoregulatory physiologists than to the comparative developmental biologist. Paleoichthyologists know more about the skin and scales of fossil fishes than is currently understood about their morphogenesis in modern fishes. From a comparative point of view, then, the tetrapods have no exclusive claim to integumentary evolution. Our lack of basic knowledge and, perhaps, too uncritical an acceptance of and dependence upon certain classical interpretations has only made it seem so.

This chapter is divided into two parts: Part 1 introduces the student to the integument and forms a modern review of fundamental similarities in structure, function, and development of the integumental skeleton.

It is hoped that these considerations will make the student's observations in the second part more meaningful and exciting. Part 2 depicts the wide variety of integuments and their derivatives in the various classes of vertebrates.

## 2. Functions of the Skin

The vertebrate skin is only a specialized kind of animal integument. It is a complex composite of many cell types, but functionally and structurally it is no more of an organ system than is the even less well studied epidermal integument of the invertebrates. If they are available, the student should examine prepared slides or photographs of the microanatomy of an insect cuticle, a mollusk shell, or an earthworm epidermis. Shell formation in a snail is no less complicated than tooth formation in a snake.

In addition to its primary skeletal role in supporting and protecting its functionally related soft tissues (see p.169), the skin is also involved in reception and transduction of external stimuli (heat detection, chemoreception, mechanoreception); transport of materials (excretion, secretion, resorption, dehydration, and rehydration); heat regulation; respiration; digestion; nutrition and nutrient storage; locomotion; coloration; behavior (sexual selection, aggression, identification, rearing of young); sound production, and perhaps several other as yet undiscovered functions.

## 3. Basic Structure of the Skin

The basic skin is a *functional unit* composed of an upper *epidermis* and an underlying *dermis* (corium) separated by a microscopic *basement membrane complex,* a very important morphogenetic boundary (figs. 5.1, 5.4). Early anatomists included the "basement membrane" as part of the corium, a view that is only partly correct (see below). Embryologically, the integument and its derivatives originate in the ectoderm, the neural crest (ectomesenchyme), and various mesodermal components. In all animals, the epidermis is derived from the ectoderm. In vertebrates, it usually begins as a single-layered sheath, the *epidermal tube* (see fig. 5.2A), but soon it is composed of a mitotically active inner layer, one or two cells thick, the *stratum germinativum,* and an outer layer of single cells, the *periderm.*

Some investigators believe that any mesenchyme that comes into contact with skin ectoderm will form the *dermis.* The mature dermis therefore is derived from mesenchyme cells originating in the somite mesoderm (dermatome), from the neural crest (ectomesenchyme) and from the somatopleure (see chap. 4). In the embryo, the dermis is little more than part of the basement membrane complex that is the sole means of support for the thin epidermis. In animals in which a thickened epidermal layer is not present, protection against mechanical trauma is probably best provided for by the connective tissues of the dermis, especially the *stratum compactum* layer.

The "basement membrane" of traditional light microscopy is in reality a complex composite structure. Transmission electron microscopy reveals the presence of at least two primary layers: an ultrathin, juxtaepidermal *basal lamina (lamina terminalis,* adepidermal membrane, or dermal membrane of some authors); and, immediately subjacent to this, the *basal lamella* (dermal lamella, *stratum reticulare,* or *stratum compactum*

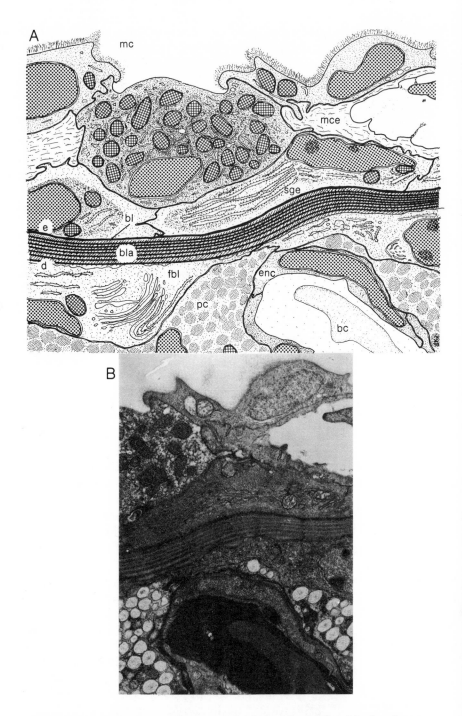

Fig. 5.1. Skin cells. *A*, semidiagrammatic view of developing skin showing epidermis (*e*), basement membrane complex (*bl* and *bla*), dermis (*d*), stratum germinativum cell (*sge*), mucous cell (*mce*), mucous cuticle (*mc*) on surface epithelial cells, fibroblast (*fbl*), pigment cell (*pc*), endothelial cell (*enc*) of capillary with enclosed blood cell (*bc*). Note: heavy dot pattern = nucleus; light dot pattern = melanosome; cross-hatched pattern = mitochondrion. *B*, right half of the electron micrograph (original × 20,000) of developing fish skin on which fig. 5.1*A* is based.

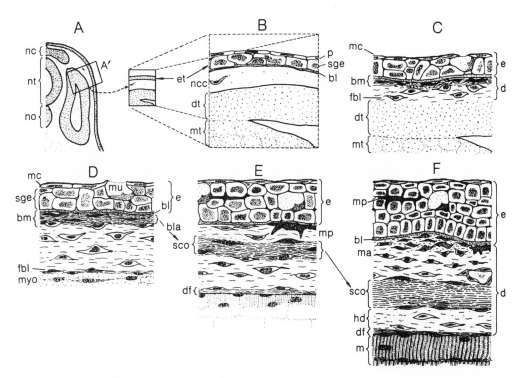

Fig. 5.2. Embryonic development of the generalized skin. These panels give a detailed view of the process generalized in fig. 5.3*A*, 1–3. *A*, section removed from right side of developing embryo showing neural crest (*nc*), neural tube (*nt*), and notochord (*no*). *B*, blow-up of section A′ showing epidermal tube (*et*), dermatome (*dt*), and myotome (*mt*). Note: neural crest cell (*ncc*) in space below epidermal tube, periderm (*p*), stratum germinativum (*sge*), and basal lamina (*bl*). *C*, development of the basement membrane complex (*bm*) and dermis (*d*) by formation of the collagenous basement lamella and invading fibroblasts (*fbl*). *D*, formation of deep fascia layer (*df*) of dermis by fibroblasts adjacent to underlying myoblasts (*myo*). Basement membrane complex is about to delaminate the basement lamella (*bla*). *E*, delamination of basement lamella to form the stratum compactum (*sco*). Note melanophores (*mp*) in epidermis and dermis. *F*, developing skin section at onset of the epithelial-mesenchymal interaction. Note slight pallisading of stratum germinativum in area immediately above aggregated mesenchyme cells (*ma*) and the endothelial cells enclosing a capillary nearby. The so-called hypodermis (*hd*) lies between the stratum compactum and the deep fascia, which covers the muscle tissue (*m*).

*corii* of some authors), a stratified, reticular layer of collagen fibrils seemingly arranged in orthogonal strands (fig. 5.1 and see below).

The basal lamina is found at the basal surface of all epithelia, regardless of the embryonic germ layer from which they were derived. It may vary in thickness from 500 to 1,500 Å. The collagenous basal lamella, although generally considered of dermal origin, is apparently of mixed origin. Some investigators believe part of it is secreted from the basal epidermal cells, while others believe that it is only "influenced" by the presence of overlying epithelia. In the phylum Chordata, the basal lamella represents the simplest and probably the most primitive form of the dermis (the student should examine a histological section of *Branch-iostoma* integument under high power). During subsequent development, both ontogenetic and phylogenetic, the basal lamella gives rise to distinct structures of diverse function such as, for example, the actinotrichia and ceratotrichia of fish fin rays, and the lens and cornea of the eye. Some derivative structures form before, and others after, the basal lamella is invaded by subjacent mesenchyme cells. The stratum compactum of mature skin is a derivative of the embryonic basal lamella.

There are at least two theories that attempt to explain the formation of the basal lamella into layers of collagen fibrils as seen in electron microscopic thin sections (as in figs 5.1 and 5.4). The arguments for each are dependent upon how the collagen fibrils are formed below the basal lamina. The "plywood theory" suggests that the collagen fibrils are deposited in layers (plies) directly parallel to and below the basal lamina. Subsequent parallel layers originate at angles of 90° to the original layer that is, orthogonally, thus giving a plywood appearance to the basal lamella. The "scinduline theory" suggests that the collagen fibrils are budded off the basal lamina in a shallow-angled, flattened spiral that only in perfect cross sections gives the appearance of orthogonal layering. According to some investigators, the scinduline theory more readily explains the dynamics of growth between the basal lamella and the growing, ever expanding epidermis. From the wider perspective of comparative anatomy, however, either theory can be seen as a fundamental part of a widespread developmental phenomenon, the "principle" of *delamination* (see p. 119, and fig. 5.4). Continuing research on this intriguing interface area, the *basement membrane complex,* should yield many of the yet unknown secrets of morphogenesis of the skin.

## 4. Transitional Stages of Development

A "typical" vertebrate skin does not exist, and the student should not be misled into believing that "if you've seen one, you've seen them all." Developing skin is characterized by some variation, and mature skin by much variation, within and between classes of vertebrates. In all vertebrates, there are marked differences in the rates of development of the skin. These occur regionally on the body and are often expressed in anteropostero-, dorsoventro-, or lateral gradients in apical, ventral, or lateral ectodermal ridges. The nature of these gradients is not well understood, but they are critical developmental phenomena. The thickness of any given vertebrate epidermis is apparently dependent upon the rate of aging of the epidermal cells. This primarily involves the amount of cell division and the life span of individual cells but also somehow involves the overall function the epidermis is to serve (see fig. 5.13 and p. 113).

Notwithstanding the admonition above—and because in the early stages, at least, its relative simplicity is uncomplicated by either keratinization or ossification—the fish skin (fig. 5.2) has been chosen to illustrate some of the transitional stages of development that precede the formation of definitive structures that distinguish the various vertebrate classes. In the fish, once the basic embryonic skin stage has been achieved (fig. 5.2*B*) the various layers begin to increase in thickness. Mucous cells and sensory cells are the first to differentiate in the epidermal layer of fishes and amphibians, whereas in reptiles the cells of the presumptive *stratum corneum* are the first to differentiate. Concurrently, great changes are occurring in the dermis (fig. 5.2C). Dermal pigment cells of neural crest origin appear below the basal lamella. Both the number of plies and the diameter of each of the older collagen fibrils in the basal lamella begin to increase. As the basal lamella grows thicker, it is invaded by underlying fibroblasts (fig. 5.2D). These fibroblasts eventually insinuate themselves between the basal lamina and the basal lamella, creating the impression that the basal lamella has split off and has begun to sink inward (figs. 5.2E, 5.4B and see discussion of delamination, p. 119). As might be expected, this "sinking in" of the basal lamella is not synchronized for all parts of the body; the differentiation is regional. For example, in fishes it occurs during fin-ray formation in the embryonic fin fold and also during neuromast formation along the lateral line. The intermediate supra- and infralateral trunk integument of the fish does not so differentiate until scale formation occurs slightly later. In amphibians the delamination of the basal lamella occurs during larval metamorphosis, at which time a new adult basal lamella is also formed at the basal lamina. The regional differentiation of the reptilian and avian integument at this stage is not well documented.

By this time, the dermis appears stratified into distinct layers (fig. 5.2E, F). The most superficial is the basal lamellar portion of the basement membrane complex. Subjacent to this is a layer of loose connective tissue, the *stratum laxum* (also occasionally called "superficial dermis"), in which blood vessels, nerves, and pigment cells are present. Cellular aggregations (dermal papillae) appearing in this layer will soon begin to participate in the formation of most of the derivative structures of the mature skin (fig. 5.3; and text below). Below the stratum laxum, that portion of the basal lamella that earlier had sunk inward is now seen as a dense lamellar layer of wavy collagen fibers and fibroblasts, called the *stratum compactum*. Below the stratum compactum, and separating it from the body musculature, is another connective tissue layer of varying thickness, the *hypodermis* or subcutaneous layer (*tela subcutana*). This layer also contains pigment cells of various types, and adipose tissue frequently is found here. It is separated from the muscle layer by *fascia*.

Concurrently, the number of cell layers in the epidermis has increased by mitotic division in the basal layer of the stratum germinativum, thus separating it from the superficial periderm layer (fig. 5.2E). Epidermal pigment cells (epidermal melanin units) have already made their appearance, and two or three types of mucous cells may be present (fig. 5.2E, F). There is more than just a superficial relationship between the transitional skin and the histology of the vertebrate gut tube. The student would do well at this point to compare a slide of the mouse esophagus, for example, with figure 5.2F.

Beyond these stages of development, the fish skin and vertebrate skin in general are characterized by marked cellular aggregations, invaginations, and evaginations of both the dermis and epidermis. The mature derivative structures, once formed, often lie wholly or partly outside the layer in which they had their origin. Because of this dynamic rearrangement of components in the mature skin, it is often difficult for the student

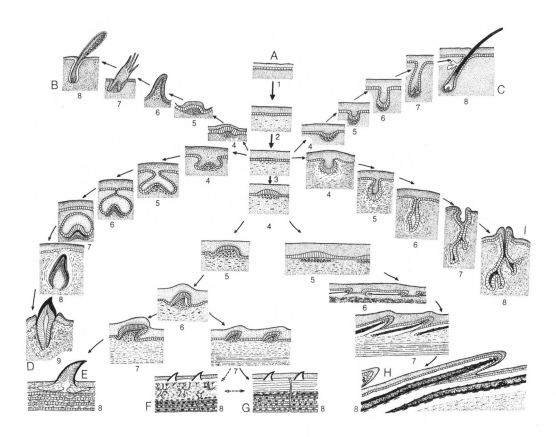

Fig. 5.3.   Early developmental homology and epithelial-mesenchymal interactions depicted as they occur in a variety of vertebrate integuments. *A*, generalized development (1–3) is microscopically similar through first three
stages (see fig. 5.2 for details). Beyond stage 3, epithelial-mesenchymal interactions may result in either invaginations of epidermal cells into dermal
follicles, evaginations of dermal papillae into epidermal cones, or both. Note
that the formation of mesenchymal aggregations is accompanied by recognizable changes in epithelial cell morphology of the adjacent stratum germinativum. *B*, subsequent stages (4–8) of differentiation in a bird feather; *C*, in a
mammalian hair; *D*, in a vertebrate tooth; *E*, in an elasmobranch (placoid)
scale; *F*, in a cosmoid scale; *G*, in a ganoid scale; *H*, in a cycloid or ctenoid
scale; *I*, in a mammary gland. Note that the arrows in panels *F* and *G*, stages
7–8, reflect the uncertainty in the phylogenetic relationship of ganoid and
cosmoid scales.

to establish or determine the homologies so evident in earlier stages of development. A consideration of some of the developmental principles involved should therefore help the student to better comprehend and *enjoy* the variations seen in the end products of these processes.

**B. Some Developmental Considerations**

The precise phylogeny of integumentary derivative structures is not always so easily conceived as is, for example, the concept of a simple mucous cell evolving into a complex glandular organ. In attempting to understand and to establish a morphological basis for reasonable phylogenies, we must first separate ontogenetic fact from phylogenetic speculation. It is necessary, then, to consider several important developmental phenomena.

**1. Epithelial-Mesenchymal Interactions**

In the formation of the integument and its derivatives, all the developmental problems faced by an embryo are paralleled. The activity of any given integumentary cell is the result of its competence, its input of information from neighboring cells, and its response to hormonal and nutritional factors in the surrounding fluids. During various stages of embryonic skin development in vertebrates, parts of two developing layers, the epidermis (epithelium) and the underlying dermis (mesenchyme), interact with each other across the basement membrane complex, and the further differentiation of each layer is thereby influenced. The first microscopically visible consequence of such an interaction between different cell layers is an *aggregation* of mesenchyme cells that in the skin is termed a *dermal papilla*. A series of such mutually inductive interactions forms a common denominator to such diverse adult structures as teeth, fish scales, hair, feathers, scutes, mammary glands, and so forth. The overall phenomenon is illustrated in figure 5.3. That it should occur in skin is but a specific example of the general nature of epithelial-mesenchymal interactions that occur in the development of many vertebrate organ systems.

**2. Delamination**

Another important developmental phenomenon that occurs in, but is not limited to, the vertebrate skin is the *principle of delamination*. Its general applicability to both fossil and recent vertebrates and, in particular, to the integumental skeleton, was elaborated by the Swedish paleoichthyologist Jarvik.

The delamination principle holds that during various stages of development, and partially under the influence of overlying epidermis, potentially skeletogenous ectomesenchymal cell laminae of the corium (dermis) are split off from their position in close association with the basement membrane complex. These cell laminae sink inward to give rise to several superimposed endoskeletal, exoskeletal, or connective tissue generations not necessarily identical in nature. Those formed earliest are most deeply embedded and are often considered to be *endoskeletal,* for example, dermal bones, finrays. Those generations still in contact with or near the basement membrane complex are considered *exoskeletal,* for example, all teeth; scales of all fossil and modern fishes. The principle is illustrated by our own observations with the transmission electron microscope as shown in figure 5.4. Again, as with the interactions be-

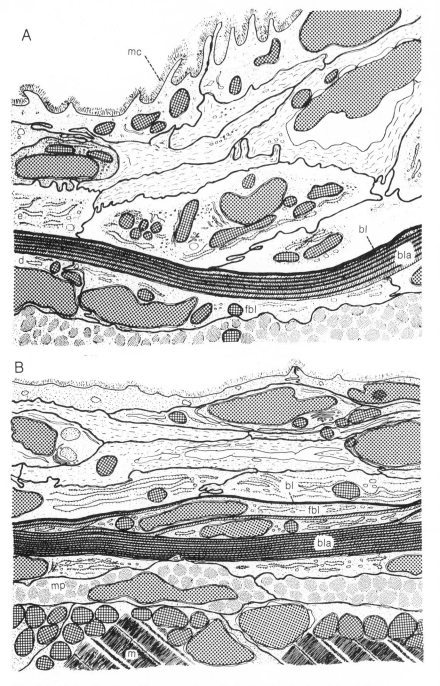

Fig. 5.4. Delamination of basement lamella in the developing skin of the guppy, *Poecilia reticulata*. Semidiagrammatic drawings based on original electron micrographs (×20,000). *A*, epidermis (*e*) separated from dermis (*d*) by basement membrane complex consisting of basal lamina (*bl*) and basement lamella (*bla*). A mucous cuticle (*mc*) is secreted by the surface epithelial cells. Note position of the fibroblast (*fbl*). *B*, basement lamella delaminates from basal lamina as fibroblasts insinuate between plies of collagen originating at the basal lamina. Note muscle tissue (*m*) and melanophores (*mp*) in the dermis. *C*, scale (*sca*) formation occurs in cellular layers between basal lamina and the basement lamella (*bla*), which is now delaminated and referred to as the stratum compactum (*sco*). Lower portion of scale is the fibrillary plate (*fp*). Note: heavy dot pattern = nuclei; light dot pattern = melanosomes; cross-hatched pattern = mitochondria.

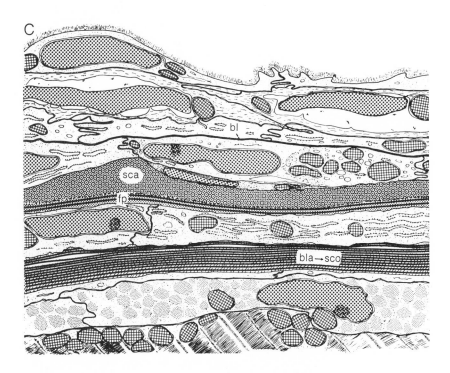

tween epithelial and mesenchymal tissues, the basement membrane complex plays an important role in delamination phenomena.

3. Functional Epithelial Extinction

A spectrum of similar processes vaguely defined and variously termed *desquamation, ecdysis, exfoliation, molting, shedding,* and *sloughing* occurs in the integuments of animals. In all these phenomena, the mature, outermost layer of the integument is periodically or continuously lost in a variety of ways ranging from loss of the entire layer to loss of minute flakes (see p. 146). Among the invertebrates, the loss and replacement of the chitinous cuticle of arthropods is a well-known example (the student should think of and look at others). But among vertebrates, the comparative aspects of such phenomena as sloughing or shedding generally have been inextricably bound to the examination of such keratinized structures and processes of epidermal loss as were known to occur chiefly in amniotes, but also in a few amphibians. Virtually nothing is known about epidermal loss in fishes except that it occurs. That little interest has been engendered for investigating these phenomena in fishes reflects the hitherto prevailing view (see p. 145) that keratinization is mostly a terrestrial phenomenon. Ample evidence now exists in aquatic vertebrates to demonstrate the incompleteness of this older view.

More pertinent, however, is the fact that in some vertebrates loss of epidermal cells is *not* dependent upon their prior keratinization. Keratinization just seems to confuse the picture in higher vertebrates. Recent investigations with electron microscopy suggest that loss of individual nonkeratinized epidermal cells does occur in fish skin. Several fishes have been reported to slough either all or portions of their skin or epidermal derivatives of it (see table 5.1 and p. 156). Similar processes are known to occur in the skin of persistently larval amphibian species as, for example, in *Necturus maculosus,* where the nonkeratinized epidermal cells are periodically sloughed.

However widespread it may eventually prove to be in the poorly studied epidermis of lower vertebrates, replacement and loss of cells without keratinization is characteristic of nonepidermal epithelial cells. The phenomenon of epithelial cell maturation, involving progressive movement of cells away from the basement membrane complex and their eventual shedding into the digestive tract lumen, has been thoroughly documented by medical and dental researchers interested in the normal function and pathology of the digestive and oral mucosa. From the wide vantage point of comparative anatomy, however, periodic or continuous development, loss, and renewal of cells seems to be a primitive characteristic of all epithelia whether or not they are keratinized, and whether they compose the basal layer of the digestive tract mucosa or the stratum germinativum of the epidermis. In order to distinguish it from the process of keratinization, this more primitive ontogenetic phenomenon has been termed "*functional epithelial extinction*" (Krejsa 1970 in list of references).

In the process of functional epithelial extinction, it is hypothesized that, once their attachment to the basement membrane complex is lost, regardless of the species in which they are formed, epithelial cells seemingly commit themselves or are committed to specific functions, all of which eventually lead to separation and eventual loss (extinction) from their subjacent counterparts. More specifically, commitment to a specialized function seems to depend upon separation from the basement membrane complex. Further, once the function is performed for a period of time, the cells seem to "age," and as they are shed or sloughed they become "extinct" and are replaced by new cells that have become newly separated from the basement membrane complex. Precisely which part of this "commitment" is programmed by the genetic mechanism of the separated cell and which part might be due to lack of nourishment or hormonal influences caused by separation from proximate vascularization underlying the basement membrane complex is not known. It seems likely, however, that the commitment to leave the basement membrane complex is controlled by the cell's genetic mechanism, whereas a nutritional effect, if any, may occur later.

The phenomenon of functional epithelial extinction involves both cell division and the life span of the epithelial cell once it leaves the basement membrane complex. For example, it is known that, at least in mammalian epidermis, mitotic activity is hormonally controlled. Recently, however, a series of independent researchers have discovered and isolated chemical substances called *chalones* that are produced by epidermal cells. These are tissue-specific but not species-specific substances that inhibit the mitotic activity in adjacent cells. It is believed that there must be a balance between hormonal stimulation and chalonal inhibition of mitotic activity. Some investigators have even suggested the existence of a feedback mechanism between surface epidermal cells about to be lost and the cells in the stratum germinativum where mitoses occur. While very little is currently known about these phenomena, their relationship to the control of aging of all epithelial cells is of critical importance.

**4. Epidermal Coparticipation**

An attempt to clarify explicitly the interrelationships of the processes of delamination and epithelial-mesenchymal interactions, and implicitly the process of functional epithelial extinction, can be found in the "epidermal coparticipation" hypothesis of Moss (1968), who suggests that the activity of the basal layer of the stratum germination plays a significant unifying role in the formation of all vertebrate integumental skeletal structures. Accordingly, the *coparticipation* of the stratum germinativum with the underlying ectomesenchymal cells results in a variety of morphological end products which vary both in form and composition. These may be classified into three categories, according to the extent of the contribution of each coparticipating layer, as follows:

*a. Structured Epidermal Derivatives*

Structured epidermal derivatives are those products formed exclusively by epidermal cells after an initial inductive interaction with ectomesenchyme, for example, hair, feathers, whale baleen, claws and nails, horns, beaks, reptile scales, keratinized dental ridges of turtles, jaws and teeth of larval amphibians, teeth of cyclostomes. There is no apparent structural contribution of the dermis to any epidermal component.

*b. Structured Epidermal-Ectomesenchymal Derivatives*

Structured epidermal-ectomesenchymal derivatives are those in which, after the initial inductive interaction, both tissues coparticipate in the formation of the structured end product, for example, vertebrate teeth (both oral and pharyngeal), dermal denticles (placoid scales), cosmoid, ganoid, and teleost scales, the dermal "scutes" of armored catfishes, sticklebacks, sea horses, and pipefishes, elasmobranch ceratotrichia, teleost lepido- and actinotrichia.

*c. Structured Mesodermal Derivatives*

Structured mesodermal derivatives are those in which there is as yet no visible or known structured contribution of the epidermis to the matrix of the mesodermal product, for example, calcified plates in *Bufo* spp., calcified tendon, "intermediate" skeletal tissue in the branchial arches of fishes, dermal bones of vertebrates in general, and the osteoderms of reptiles.

**5. Comments on the Developmental Homology**

The essence of the developmental hypotheses described above is to suggest that the unique complexity of the vertebrate integument is closely tied to epithelial-mesenchymal interactions associated with the basement membrane complex interface. These interactions may occur in chronological series and delamination or functional epithelial extinction often follow. For example, the delamination process gives rise to generations of cells or cell products "sinking in" from the basal lamella. Functional epithelial extinction phenomena give rise to generations of cells moving "up and away" from the basal lamina. When these two processes are considered along with the mutually inductive epithelial-mesenchymal tissue interactions occurring across the basement membrane complex, we have the basis of a *unifying developmental theme* within which vertebrate skins, and other animal integuments, may be understood and further studied as functional organ systems (fig. 5.5).

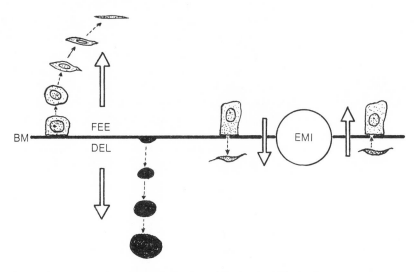

Fig. 5.5. A schema incorporating three developmental phenomena into a unifying theme of integumental development: epithelial-mesenchymal interactions (*EMI*) occur in either or both directions across the basement membrane complex (*BM*); the delamination process (*DEL*) gives rise to one or more generations of potentially skeletogenous cells, or cell products, which "sink" into the dermis from the basement membrane; in functional epithelial extinction (*FEE*), epithelial cells originating in the stratum germinativum, or any basal epithelial layer, give rise to cells or cell generations that move "up and away" from the basement membrane to serve specific functions in the epidermis (or mucosa). Eventually, the mature cells performing these latter specific functions become separated from maturing, underlying cells which replace them and continue their functions.

In the context of the unifying developmental theme, *structured epidermal derivatives* can be envisioned as those in which, after the initial epithelial-mesenchymal interaction(s), the formation of the structured product occurs hand-in-hand with, or is superimposed on, the functional epithelial extinction phenomenon. *Structured epidermal-ectomesenchymal* or *structured mesodermal* derivatives are those in which the process of delamination seems to play a major role after the initial interaction(s) occur.

The main criticism of the interpretations above is that there is contradictory evidence on whether epithelial-mesenchymal interactions actually occur in the ontogeny of *all* structured epidermal or all mesodermal derivatives. Furthermore, the nature of the inductive mechanisms involved in these interactions is not fully understood. The primacy of either epidermis or dermis is still a matter of dispute in many developmental sequences. For example, the development of epidermal derivatives such as feathers and leg scales in birds is generally thought to be determined by the regional origin of the dermis associated with it. In experimental recombinations of leg bud mesoderm with wing bud ectoderm it was found, as expected, that the limb developed the characteristic skeletal morphology, that is, it became a foot. However, the feet bore feathers as well as abnormal scales. Since leg bud ectoderm forms scales but never feathers on the feet, it is apparent that wing bud ectoderm contains feather-forming information and is able to express it in spite of the leg type induction imposed on it by the leg bud mesoderm. Indeed, as in the past, such intriguing problems serve as the focal point of much research in developmental biology today.

6. Some Comments on the Integumental Skeleton

The primary function of any skeleton is the mechanical and structural support or protection of the functionally related soft tissues of the body. This functional definition applies no less to the *integumental skeleton* than to the *endoskeleton*. The *integumental skeleton* is here defined to include all hardened and nonhardened components of the vertebrate integument that subserve the primary function of a skeleton.

Whereas the cartilaginous or calcified *endoskeleton* may be derived from epimeric, neural crest, and endodermal contributions, the *integumental skeleton* is derived from various types of mesenchyme and ectoderm. As was discussed above, many skeletal derivatives of the integument remain on, in, or near the skin surface. Others, however, do not. Often the first generation of integumentary derivatives delaminate (see p. 119) and become fused or otherwise closely associated with the so-called endoskeletal components, for example, dermal bones of the skull, dermal fin rays. Without an intimate knowledge of their precise embryological development, such mature integumentary tissues often are histologically and macroscopically indistinguishable from the endoskeletal tissues with which they are associated.

The classical term *exoskeleton* has long been used to define skeletal derivatives of both the vertebrate and the invertebrate integument. But from a functional, if not always morphological, point of view, the distinctions between the vertebrate exoskeleton and endoskeleton are somewhat blurred. Furthermore, the invertebrate exoskeleton is derived exclusively from the ectoderm whereas the endoskeleton, where present, for example, as in echinoderms, is derived from the mesoderm. The term exoskeleton is well entrenched in comparative vertebrate anatomy but its use might, perhaps, be better limited to the invertebrates, where its definition remains more developmentally precise. On the other hand, cartilage, a tissue formerly thought to be uniquely vertebrate, is now known to occur in invertebrates though as yet there are no invertebrate skeletal tissues known to be formed as *composites* of different embryonic origin. The composite skeletal tissues of vertebrates appear to be unique. The term *integumental skeleton* can accommodate these tissues. However the terminology is finally resolved, at least one conclusion seems warranted at this time:

*Many derivatives of the integumental skeleton, traditionally considered either wholly ectodermal or mesodermal in origin, are in reality composite structures resulting directly or indirectly from mutually in-*

*ductive interactions. The site of such composite skeletal tissue formation is on, in, or near the basement-membrane complex.*

**7. Pattern Formation and Symmetry**

Patterns, whether structural or pigmentary, are universally present in the integument of animals. Perhaps because patterns are so commonplace and esthetically pleasing, most zoologists (with the exception of some taxonomists) tend to overlook them as material subject to comparative analysis. Yet consciously or not it is very often the pattern of a particular animal that calls our attention to it. Color patterns in birds, the squamation of fishes or reptiles, and fingerprints in mammals are most obvious examples, but all vertebrates show similar, if somewhat more cryptic, patterns. Can you think of others?

Many patterns are most obvious during the developmental sequence, but with a little creative probing they can readily be seen and appreciated in the adult. For example, the overall appearance of adult plumage in most birds gives little indication of the underlying *pterylae* (feather tracts) and *apterylae* (featherless zones). (The ratite birds—ostriches, rheas, emus, and cassowaries—and penguins have uniform distribution of feathers over the body.) If not feasible in the comparative anatomy laboratory, student confirmation of feather-tract patterns or, for that matter, hair patterns, is as close as the poultry or slab-bacon section of the local food market.

Patterns occur at all levels of organization from molecular to organismal as, for example, in the repetitive 640 Å spacing within a collagen fiber; the striations in skeletal and cardiac muscle; the incremental lines, or growth rings, present in mineralized tissues such as teeth, bones, and fish scales; the tile-like arrangement of simple squamous epithelial cells; the microoornamentation or sculpturing of the surface of nonkeratinized and keratinized epidermal cells; and the pigmentation and barring patterns of feathers and hair. A few of the many patterns intrinsic to the vertebrate skin and its derivatives are illustrated in figure 5.6. In comparing the external anatomy of specimens present in the laboratory, the student should observe these and also attempt to find and make a list of other, less obvious, examples.

One of the most intriguing, and as yet unsolved, morphogenetic problems is how patterns are formed. It is known, for example, that the primordia of teeth, fish scales, feathers, and hair occur as mesenchymal aggregations (dermal papillae) below the basal lamella of the basement-membrane complex (fig. 5.3). Why, however, do the mesenchyme cells aggregate in such regular patterns? What controls the uniform spacing? What information is being transferred? Does the controlling information reside in the mesenchyme cells or does it originate in the overlying epidermis or, perhaps, in both?

The collagen fibrils of the basal lamella, for example, are arranged in an array of plies that seemingly tend to cross each other at right angles (see p. 128, and fig. 5.6). The epidermis has been implicated as the source of this morphological order, since the first collagen fibrils occurring in the embryonic lamella are randomly oriented, whereas later in development the first ply always forms nearest the epidermis. There is

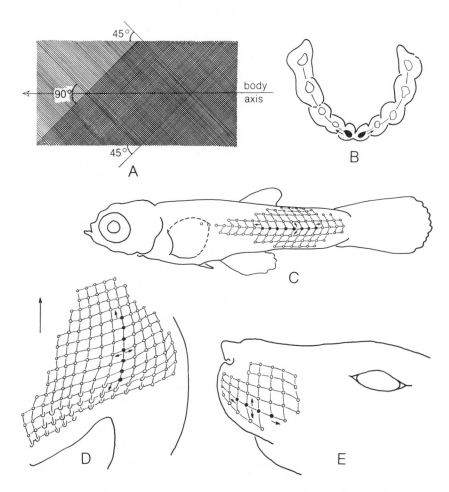

Fig. 5.6. A sampling of intrinsic integumental patterns in vertebrates. *A*, diagram showing a fundamental pattern found in the embryonic basement lamella and the mature stratum compactum: In surface view, collagen fibers are basically arranged in layers intersecting each other at an approximate angle of 90° thereby forming an orthogonal grid pattern. The individual plies of each grid intersect the longitudinal body axis at an angle of approximately 45°. In each of the following (*B–E*) the first primordia (row of origin) of the pattern is indicated by solid circles; arrows indicate the direction, and open circles the location, of subsequent differentiation. *B*, developing deciduous tooth germs in fetal human maxilla (after Applebaum and Cuttita, 1946). *C*, scale primordia in newborn guppy. *D*, humeral feather tract in chick limb bud (after Saunders et al. 1958 [*J. Exp. Zool.* 137:39]). *E*, vibrissal (whisker) follicle pattern in the upper lip of the hooded rat (after Oliver, 1966 [*J. Emb. Exp. Morph.* 6:231]).

little direct experimental evidence existing on the exact role of the epidermis in ordering the collagen fibrils, although it is known that no plies form in cultures of wounded larval amphibian skin that has been stripped of its epidermis. The epidermis adds very little protein to the basal lamella but apparently does contribute polysaccharide to the matrix. In amphibian wound healing, the continuous presence of epidermis seems necessary for basal lamella formation and growth. This has also been shown in fishes.

The orthogonal array of collagen fibrils in the basal lamella and its relationship to pattern formation in developing epidermal and dermal derivative structures is most interesting. In aquatic vertebrates, including cetaceans, and in the squamate reptiles, each ply in the trunk integument describes an angle of about 45° with the body axis (fig. 5.6$A$, $C$). These angles may change and vary somewhat, depending upon the function of the skin during locomotory movements. In terrestrial tetrapodal vertebrates, where the primary locomotor activity is centered more in the limbs than in the trunk, the underlying pattern of fibers in the integumentary basal lamella and stratum compactum is no longer obvious. The regular pattern of intersecting fibers is limited to certain more or less flattened areas of the body, for example, bird limbs, the nail bed of a human finger, the lamina propria of the cornea. Curved surfaces complicate the pattern, even in the aquatic vertebrates. A detailed analysis of the arrangement of fibers in the skin and the adaptation of the skin to its varied functions is needed. Perhaps some enterprising student of comparative anatomy now reading this chapter will undertake such a study.

From a comparative point of view, it should be noted that the ectodermally derived "collagen" of the earthworm cuticle is also arranged in orthogonal plies, and this array is believed to be controlled by the epidermis. The multilayered, lamellate cuticle in insects exhibits a similar array of protein-chitin fibers formed by the single-layered epidermis. The genetic potential to produce such a complex array may be inherent to all ectodermally derived epithelial cells. The fundamental problem remains, however: What are the mechanisms by which chemical changes (patterns in time) are translated into morphological changes (patterns in space)?

**C. Summary of Part 1**

1. The considerations discussed in part 1 above demonstrate the fundamental similarities if not also the underlying homology present in the developmental processes of all vertebrate integuments.

2. The multitude of morphological end-products we see in the mature skins of the various vertebrate classes (fig. 5.3; and part 2) are the results of complex variations on these common developmental themes (epithelial-mesenchymal interactions, delamination, and functional epithelial extinction).

3. These variations are brought about by a finely tuned interplay between the genetic background, cell behavior, and ultimately the functional requirements of the adult form.

4. The primary function of any skeleton is the mechanical and structural support or protection of the functionally related soft tissues of the body.

5. The integumental skeleton includes all hardened and nonhardened components of the vertebrate integument that subserve the primary function of a skeleton.

6. Many derivatives of the integumental skeleton are embryologically composite structures.

# Part 2: Variation in End Products

**D. The Integu-mental-Environ-mental Interface**

1. The Epidermal Surface

The outer surface of the vertebrate skin is the boundary layer between the environment and the organism. In fishes and in unmetamorphosed amphibians, the entire epidermis is alive and metabolically active, even in the surface (apical) layer of cells. These apical cells are usually covered by a *mucous cuticle* (see below). In adult amphibians, reptiles, birds, and mammals, the surface layer is fully or partly keratinized and one or more layers of dead or dying horny cells may be present (see p. 145) as the *stratum corneum.*

The cells that make up the surface layer bear intriguing surface patterns on their outer plasma membranes (fig. 5.1). For example, the surface cells of the lamprey, lungfish, brook trout, California newt, and the human oral mucosa have been found to bear submicroscopic patterns variously described as striations, cilia, microfolds, or microvilli. The surface layer of squamate reptiles bears patterns that, in cross section, are reminiscent of the teeth of a saw (fig. 5.11). These microscopic surface patterns, when first seen in cross section (even by modern researchers) were incorrectly thought to represent microvilli such as are present in intestinal epithelium. However, in the three-dimensional surface view made possible by the scanning electron microscope (SEM), they appear not as microvilli but as *microridges* or folds in the plasma membrane (fig. 5.7). In fishes, at least, these distinct patterns are thought to be species-specific. In many recently published transmission electron micrographs of fish and amphibian skin, the latent microridges can be seen already forming in the upper surface of replacement cells immediately below the superficial cell layer. Such ridges enormously increase the surface area of the living epithelia and therefore may play an important role in the osmoregulatory activities of the lower aquatic vertebrates. It also seems possible that as superficial cells are shed or sloughed the upcoming replacement cells might collectively offer a relatively large cuticle-free surface area to the bathing environment. Whether or not the surface thus exposed can be considered to be a "leak" or temporary "hole" in the osmotic barrier of the epidermis is a matter of speculation. Recent studies of the rat ileum, for example, indicate that the temporary gaps left in desquamated cells in the epithelial mucosa are sufficient to account for the passive permeability of that tissue. None of the many physiological studies of anuran skin have taken into account that projections of the plasma membrane could substantially increase the surface area and potentially affect the osmotic properties. The exact role of the microridges, like that of the mucous cuticle that coats the skin (see below), has yet to be demonstrated.

2. Integumental Glands

In addition to producing the cuticle in fishes and amphibians, the vertebrate epidermis contains a spectrum of *unicellular* and *multicellular* exocrine *glands.* The unicellular glands (fig. 5.8) usually discharge directly onto the epidermal surface through a small pore, as from goblet cells (fig. 5.9). In others—for example, poison cells, alarm substance cells, Leydig cells—the gland cell is not in contact with the epidermal surface

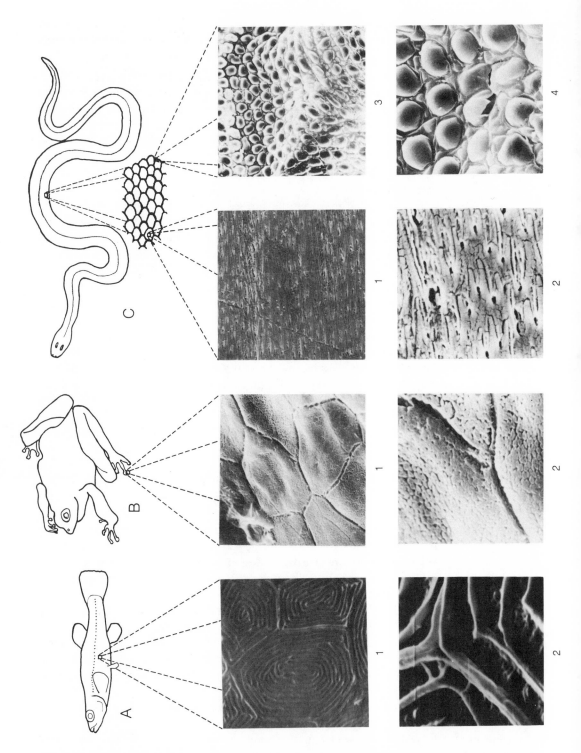

Fig. 5.7.   Patterns in epidermal structure. Scanning electron micrographs of surface features of a fish scale (*A*); frog epidermis (*B*); and snake scales (*C*). Note that in *A*, the surface layer is thrown into a pattern of microridges—such patterns are thought to be species-specific. In *B*, the pattern of the recently sloughed layer is seen still impressed on the new outer cell layer. In *C* the ornamentation on two overlapping snake scales and their hinge region is shown.

in any way and the secretions are released externally only after physical trauma, stress, or pressure, for example, as in biting. Unicellular glands are not found in the integument of reptiles, birds, or mammals.

The *multicellular glands* are formed by ingrowths of the *stratum germinativum* into the dermis. They generally open to the surface by a neck-like tubule of varying length. Multicellular glands are of two general but overlapping types—tubular and alveolar. *Tubular glands* may be simple, coiled, or branched. Spherical dilations of the terminal portions of the tubules, termed *acini* or *alveoli,* distinguish the *alveolar (saccular) glands.* These may be simple, branched, or compound.

## 3. Integumental Glands in Fishes

The greatest variety of skin glands is found in fishes. Most *mucous glands* are unicellular and are classified either by *shape,* for example, goblet cells, club cells; or *content,* for example, mucous cells, granular cells, poison cells, alarm substance cells. Unfortunately, few glands are categorized by *function* for the simple reason that the exact function(s) of many of the varying cell types, or the substances they produce, are yet to be elucidated.

Among the fishes, the agnathan epidermis is characterized by several gland types. In lampreys (fig. 5.8*A*) at least three morphologically and histochemically distinct cell types are present: small mucous cells predominate throughout the epidermis; larger binucleate *club cells* attach to the basal lamina and extend about halfway up through the epidermis; large *granular cells* are found in the upper half of the epidermis. In hagfishes (fig. 5.8*B*) three cell types are also present: small mucous cells predominate in the upper half while large, spherical mucous cells are limited to the basal half of the epidermis; smaller, elongate *thread cells* are found in the basal half also. Hagfishes also possess a multicellular *slime gland* (fig. 5.8*C*) composed of large mucous cells and thread cells.

The glandular cells of sharks and rays have not been well studied. *Goblet* mucous cells are the most predominant cell type found (fig. 5.8*A*). *Pterygopodial glands* are multicellular glands associated with the clasping organs of certain sharks and rays. Their precise function is not well understood. Among the bony fishes, *goblet* cells and *club* cells are the most common cell gland type (fig. 5.8*E*). Fishes with glands bearing known poison substances have been classified as *venomous* if the glandular contents are released through the action of a traumagenic organ, that is, tooth, spine, and so forth, or as *ichthyocrinotoxic* if the glands merely secrete their product into the water without the aid of a mechanical device such as a spine (fig. 5.8*F*).

Perhaps the most thoroughly studied structural gland cell type in fishes is the club cell associated with the production of an "alarm substance" (Schreckstoff) that elicits a fright reaction in fishes belonging to the superorder Ostariophysi. When an ostariophysan fish is experimentally wounded, the injured club cells are reported to spill out an *alarm substance* that elicits a fright reaction in other members of its species and even within and between different genera of the same family. The response, though weaker, is also evident between fishes of related families. Although most families of Ostariophysi examined possess all components of the fright reaction, some few lack both the alarm substance and

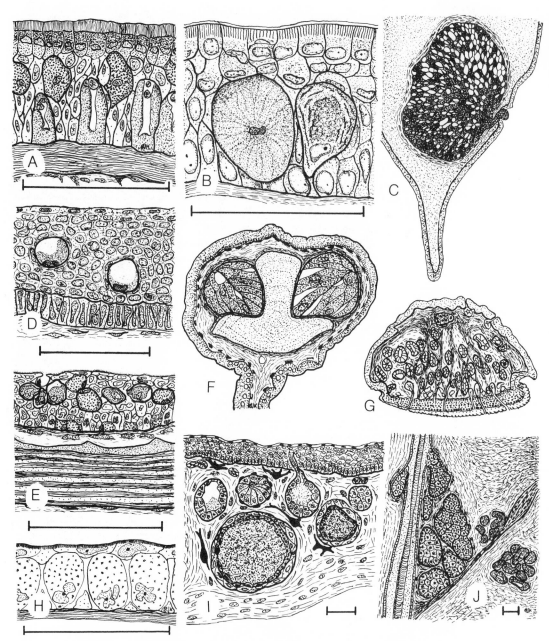

Fig. 5.8. Unicellular and multicellular glands in a variety of vertebrate skins. Note: in panels *A*, *B*, *D*, *E*, *H*, *I*, and *J*, measured bar equals 100 microns; in panels *C*, *F*, and *G*, original magnification is approximately ×20. *A*, granular cells (heavy stipple) and club cells (binucleate) in epidermis of adult lamprey. The "brush border" on surface epithelial cells is probably the microridge pattern. *B*, large mucous cell (left) and thread cell (right) in epidermis of the hagfish, *Myxine glutinosa* (after Schreiner, 1916 [*Arch. Mikr. Anat. Bonn* 89:79]). The slime layer and "striated apical ectoplasm" of original drawing are probably the mucous cuticle and microridge pattern of the surface epithelial cells. *C*, slime gland, in the hagfish ventral fin, is comprised of mucous cells (clear) and thread cells (dark). The entire gland is enclosed in a capsule of connective tissue and circular striated muscle. *D*, simple mucous cells in developing epidermis of embryonic dogfish, *Squalus acanthias*. Notice the columnar shape of the stratum germinativum cells. *E*, mucous cells (fine stipple) and club (alarm substance or "Schreckstoff") cells (heavy stipple) in juvenile gold-

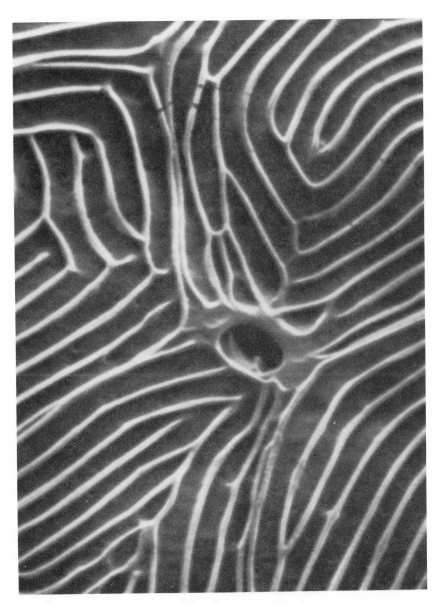

Fig. 5.9. Scanning electron micrograph of goblet mucous cell orifice on skin of the guppy, *Poecilia reticulata*. Notice the arrangement of four adjacent epithelial cells with their microridge patterns. No mucous cuticle is present in this micrograph because of fixation of specimen with glutaraldehyde. Original magnification (×15,000).

fish, *Carassius auratus. F,* frontal section through tenth dorsal spine of the scorpionfish. *Scorpaena guttata,* showing cardiaform multicellular venom gland cluster on either side of anteriomedian ridge of spine (after Halstead and Chitwood, 1970). Small hollow circles in ventral epidermis show location of simple mucous glands. *G,* frontal section through toe of the tree frog, *Hyla regilla,* showing tubular mucous glands separated by connective tissue strands originating on toe cartilage. Note: see fig. 5.13*C* for detail on toe pad epidermis. *H,* Leydig cells in ten-week old larval California newt, *Taricha torosa. I,* mucous glands (smaller) and poison glands (larger) in integument of adult frog, *Rana pipiens. J.* sweat glands (coiled), and sebaceous glands (with follicular opening) in dermis of adult human scalp.

the reaction to it. The ostariophysan electric fishes (order Gymnotiformes) lack the alarm substance and the fright reaction. Some ostariophysans possess the alarm substance but show no reaction to it, for example, the blind Mexican cave fish and certain members of the piranha subfamily.

In the uniquely predatory squawfish, *Ptychocheilus oregonense* (family Cyprinidae), response to the alarm substance is modified by ontogenetic changes in behavior. While the squawfish is part of a juvenile school, it reacts well to the alarm substance. But the maturing squawfish becomes cannibalistic and indifferent to the alarm substance of its own and other cyprinid species. The functionally oriented student should attempt to explain the adaptive significance of this change.

Another phenomenon, possibly related to the fright reaction, is found in members of the freshwater catfish genus *Ictalurus* and the minnow genus *Phoxinus*. Both of these ostariophysans have been shown to recognize individual members of their own species by their smell. In the catfish, the mucus serves as a "social pheromone," that is, as a means of chemical communication between individuals. Trained catfish can discriminate between the mucus secreted by two donor catfish, but after the donors are subjected to experimental stress conditions, their mucus is no longer recognized by the same test fish. Conversely, a test fish whose olfactory epithelium was interfered with no longer recognized the members of its hierarchical community and began to act with increasing aggression toward them. This example especially illustrates the need for interdisciplinary communication when approaching comparative anatomy from a functional viewpoint. Such an unusual function of mucus probably would have gone unsuspected had the behavioral studies not been correlated with the morphological studies. It still remains to be demonstrated, however, whether the "mucus" a catfish smells is that produced by simple goblet cells or by the apical cuticular epithelia.

### a. Comments on the Supposed Function(s) of Mucus in Fishes

The mucous coating of the fish epidermis has traditionally been thought to protect the fish against osmoregulatory stress and against cutaneous abrasion and infection, and to reduce surface friction in swimming. Modern research has introduced not only new functions but also alternative interpretations of traditional ones. Recent transmission electron microscopic studies have demonstrated the existence of an extracellular "cuticle" covering the surface epidermal cells of a variety of fish species and a few larval amphibians. This *mucous cuticle* is of variable thickness but averages about 1 $\mu$m (fig. 5.4) and is composed of mucopolysaccharides. Among fishes, such a noncellular cuticle was previously thought to be present only in the skin of cyclostomes, as in *Lampetra*. A thin film of mucopolysaccharide-staining material has also been reported on the stratum corneum of adult frog skin.

The mucous cuticle stains histochemically not unlike the mucopolysaccharide material found within the goblet mucous cells so universally present in the skin of fishes. Thus when, in the past, an investigator or a fisherman spoke of the "mucous coat" of a live fish (or of the filmy material the student finds on preserved fish specimens in the lab), it was generally accepted that he was referring to the slimy, viscous material supposedly produced by the goblet cells. The mucous cuticle, however, is secreted not from the goblet cells but from or through the plasma membrane of each superficial (apical) epidermal cell. These cells have been conveniently called "cuticularcytes" by some recent authors. If they produce the "mucous coat" of the fish, a problem then arises about the function of the mucous cuticle and about what role simple goblet cells fulfill now that their previously assumed function is in question. It is well known among fisheries biologists and physiologists that even minimum handling of fishes in warm water leads to extensive loss of the "mucous coat" and an imbalance of salts in the fish's bloodstream. Subsequent heavy mortality through infection results unless enough common salt is added to fresh water to slow the outflux of salts from the body. Is the "mucous coat" actually the mucous cuticle, or is it the result of a secretion from the goblet cells or other epidermal glands? The answer is not known at this time.

The goblet cells may begin to function early, at least by the time of hatching. Recent studies have indicated a pituitary involvement in the control of mucous cell production in fish skin, mediated through secretion of prolactin or a prolactin-like hormone. In freshwater fishes, the skin is reported to be more permeable to the outflow of salts in the absence of prolactin. A direct relationship has been assumed between the absence of prolactin, the decrease in the number of goblet cells, and the decrease in mucus synthesis. However, most such experiments have been done without the realization that a mucous cuticle exists in fishes. It is not known whether the mucous cuticle is affected by the absence of prolactin.

A "mucous coat" on the skin of fishes has also been thought to make the skin less permeable to water inflow. Fishes experimentally subjected to high salinities have been found to have fewer mucous cells than their control counterparts in fresh water, yet the former have less influx of water into the body. In hypophysectomized fish, which also have few mucous cells, water influx is also decreased. It is apparent that the relationship(s) between goblet cells, mucus, cuticularcytes, mucous cuticle, and prolactin in the osmoregulatory function of fish skin are yet to be precisely elaborated.

The traditional view that mucus increases locomotory efficiency is based on the assumption that the fish epidermis is perfectly smooth. According to hydrodynamic theory, any departure from a perfectly smooth plane may introduce turbulence into the otherwise laminar flow of the boundary layer between fish and water. The turbulence would create drag and result in decreased swimming efficiency. As stated previously (p. 129), the superficial plasma membrane of the outer epidermal cells of the fish skin is folded into a series of fingerprint-like microridges (fig. 5.7) that are covered by a mucous cuticle. The mucus produced by goblet cells in the skin is supposed to allow the fish to slip more easily through the water. However, any mucus secreted onto the perfectly smooth surface presented by the mucous cuticle might tend to be sloughed off during vigorous swimming movements. This would likely require an almost continuous secretion of mucus in an active fish, a very costly metabolic process. Perhaps the goblet-cell mucus acts only as an emergency lubricant when the fish is being pursued by a predator. It is known, for example, that the application of polyethylene oxide (a long-chained polymer) through vertical slots in the bow of an oceangoing vessel increases the vessel's speed by increasing the flow of liquid over the smooth surface. The structural chemistry of mucin indicates that it is a long-chained glycoprotein molecule, and therefore, its presence could serve a similar hydrodynamic function in fishes.

The cuticle has been found to be considerably thicker on the fin rays of the ventral fins of several bottom-dwelling fishes. This suggests that the cuticle here might serve as a friction pad.

Yet another specialized function of the mucous cuticle might be related to the presence of electroreceptors and the lateralis system in fishes. All of the following receptor organs are formed embryologically as elaborations or invaginations of the epidermis, and all contain a gelatinous mucopolysaccharide component: the cupulae of all neuromasts; the ducts of pit organs and ampullae of Lorenzini in elasmobranchs; and the ampullary electric receptors of the gymnotid, electrophorid, and gymnarchid fishes (see chap. 4). The site of origin of the gelatinous component is unknown. The simplest assumption, as yet untested, is to consider all the gelatinous components as specialized derivatives of the mucopolysaccharide cuticle, secreted by the specialized epidermal cells (otherwise cuticularcytes) that constitute the particular sense organ in question.

The mucous cuticle is probably a normal constituent of the skin of all bony fishes and all unmetamorphosed amphibians. What is perhaps most significant about its discovery in lower aquatic vertebrates is that metabolically active epidermal cells, those that form the interface between organism and environment, are capable of producing and elaborating an external, acellular coat of mucopolysaccharide material. This occurs whether the integument is that of a bony fish, a cyclostome, a tunicate, a mollusk, an annelid, an arthropod, a coelenterate, or, even certain protozoans—as the "pellicle." Indeed, secretion of an extracellular coating appears to be a characteristic property of all epithelia exposed to an aquatic medium (for example, compare it with the "gylcocalyx" of the mucosal epithelia of the gut tube) and it may be a property common to all cells. Perhaps such a secretion served primitively as a method of communication between individual cells or groups of cells?

## 4. Integumental Glands in Amphibians

In amphibians, the types and functions of the integumental glands change at metamorphosis. In larval stages, all the glands are unicellular and merocrine. In mature skin they are multicellular and may be either merocrine or apocrine.

### a. Larval Skin Glands

All larval amphibians possess typical *goblet mucous cells* that are in contact with the epidermal surface. In addition, some urodeles, for example, *Taricha,* possess club-like *Leydig cells* that are interposed between the basal and apical (surface) cells of the epidermis. Some bufonid tadpoles (see below) possess *giant cells* that contain an alarm substance.

All of these unicellular gland cells disappear during metamorphosis.

As in the fishes, the mucous cuticle covering the larval integument appears to be a product of the superficial epidermal cells. Similarly, the function(s) of the simple goblet cells remain unclear. In the larva of the California newt, *Taricha torosa,* for example, simple goblet cells of unknown function are found in the anterior head region. It is possible that, like those in the head region of some anuran tadpoles, these may serve to digest away the egg capsule. The highly mitotic Leydig cells are the major component of the tail fin skin of the mid-larval newt. There is no evidence that these cells ever come in contact with the skin surface nor that they ever release any material or secretion onto the epidermal surface. It has been suggested that, during periods of desiccation, the mucus-like content of these cells is released into extracellular compartments where it serves as an internal fluid reserve beneath the epidermal surface. The surface, meanwhile, is protected by the hardening of the mucous cuticle. As metamorphosis occurs, the epidermis becomes thicker, keratinization occurs in the outermost layer, and the Leydig cells disappear.

Interestingly enough, the schooling larvae of certain bufonid anurans, for example *Bufo bufo* and *B. calamita,* possess giant cells that produce an alarm substance similar in function to that of the ostariophysan fishes (see p. 131). The substance is present in young tadpoles as soon as they become free-swimming, and its release into the water elicits a typical fright reaction. No such cells are present in other anuran species, nor is such a reaction known to occur. The independent occurrence of an alarm substance in such nonrelated vertebrates as bufonid anurans and ostariophysan fishes is a remarkable but not unique evolutionary coincidence. Similar examples of convergent chemical evolution can be found in the production of the highly potent, chemically similar neurotoxins *tarichatoxin* and *tetrodotoxin,* synthesized, respectively, in the embryonic eye vesicle of newts (*Taricha*) and in the ovaries and other tissues of the tropical puffer fishes (family Tetraodontidae). The fright reaction is an important component of social behavior, especially in schooling forms. Some investigators believe its presence in bufonid tadpoles, as well as in ostariophysan fishes, is of selective advantage to the survival of the group rather than of the individual. This group selection phenomenon is a controversial issue in evolutionary studies.

### b. Adult Skin Glands

Most adult amphibians have subepidermal multicellular glands opening to the surface. These are absent in young tadpoles but develop as ingrowths of the basal *stratum germinativum* into the dermis only during or after metamorphosis. Some glands are already mature and functioning by the time metamorphosis is completed. The student should obtain and microscopically examine a typical frog skin slide.

The adult glands generally are of two types and functions. The more numerous *mucous glands* secrete a slimy mucous substance that apparently lubricates the skin in water and keeps it moist on land. A thin coat of smooth muscle tissue, generally surrounding each mucous gland, aids in expelling the contents. Less numerous are the larger *granular* (poison) *glands* that produce a protective apocrine secretion. Excessive excitement or violent stimulation of these glands makes them secrete a noxious fluid that may temporarily inflame the skin, or paralyze and even kill other species. Granular glands have been studied, for example, in *Rana palustris, Xenopus laevis,* and *Hyla vasta.* The true toads, family Bufonidae, possess enlarged "parotid" glands behind the eyes. When frightened or injured, a toad emits a potent poisonous fluid from these glands that irritates the oral epithelium of a would-be predator. When the tiny toad-like Catholic frog, *Notaden bennetti,* of Australia is irritated or roughly handled, a copious milky-yellow, acrid fluid is secreted from the entire dorsal skin.

Among the urodeles, the skin secretions of adult newts (*Taricha*) have been found to contain, among other toxic effects, a potent cardiac poison. California newts have a very low rate of adult mortality, part of which is believed to be accounted for by the protective function of the skin toxins. When roughly handled, the well-named slimy salamander, *Plethodon glutinosus* (family Plethodontidae), secretes a thick, sticky, glue-like substance from its skin, as do many other plethodontid salamanders.

Simple tubular glands appear in the swollen "thumb" pads of male anurans and in the so-called *mental glands* of male plethodontid salamanders. Both are apparently used as aids to amplexus (the mating embrace) during reproductive behavior. Glands in the thickened back skin of the female Surinam toad, *Pipa pipa* (family Pipidae), apparently nourish the developing embryos encysted therein. In hylid (tree) frogs, such as *Hyla regilla,* multicellular, slightly coiled tubular glands resembling mammalian sweat glands are abundant in the toe-pad dermis (fig. 5.8). The mucous secretion of these glands is believed to aid in adhesion during climbing.

## 5. Integumental Glands in Reptiles

Reptiles have few integumental glands, and these are poorly studied. Where glands do occur, they are generally restricted to certain regions of the body. Femoral pores and glands are present in rows or clusters in the posteroventral skin of the thigh regions of lacertid, gekkonid, and iguanid lizards. The glands produce a wax-like secretion not unlike that of the sebaceous glands in mammals. Aside from having some taxonomic value, their function is unknown. The so-called *escutcheon scales* of the gekkonid lizard, *Gonatodes,* have a glandular component of unknown function. A multicellular, gland-like structure is found in the *stratum germinativum* below the horny layer of the blind snake, *Typhlops braminus.* The function of its secretion is unknown. The so-called *nuchodorsal glands* of certain colubrid snakes, for example, *Natrix,* secrete a fluid that irritates oral epithelium. It is thought to have a defensive function.

In crocodilians and certain chelonians, *musk* or *scent glands* are present. These are believed to function during reproductive behavior but have not been well studied. In both sexes of the alligator, a pair of large musk glands open into the cloaca. Another pair, with slit-like openings, emerge on the inner margins of the lower jaw. In the musk or stink-pot turtle, *Sternotherus* (family Chelydridae), a strong-smelling secretion is produced by scent glands in the anterior portion of the hind-limb pockets. Crocodilians also have a row of minute skin glands of unknown function opening between scutes of the dorsum, lateral to the midline.

The *venom glands* of the heloderm lizards and of the various poisonous snake families Colubridae, Elaphidae, Hydrophiidae, and Viperidae are parts of a functional morphological unit that involves modified salivary glands, teeth, and certain bones, muscles, and nerves of the jaw and skull.

## 6. Integumental Glands in Birds

Birds have even fewer glands than reptiles. The only gland of any consequence is the simple branched alveolar oil gland (preen gland, uropygial gland) at the base of the tail. The secretion of this gland, which is squeezed out by the bird's beak, contains fatty acids, fat, wax, and possibly vitamin D. Preening the feathers, which has become a ritualized behavioral pattern in many birds, stimulates the flow of this oily secretion. Preening itself realigns the feather barbules and hooklets (the student should try this with a feather) and thereby redistributes the air spaces uniformly (see fig. 5.10). Applying the oil to the feathers apparently helps to keep them in good condition and to make them impervious to water. Water birds in which the gland has been experimentally removed suffer dramatic deterioration in the quality not only of their feathers, but also of their beaks and scales.

Variation in the external characteristics of the oil gland has some taxonomic value. Oil glands are largest in the water birds. In many birds, the opening of the oil gland may be surrounded by a tuft or circlet of small feathers that gives it a nipple-like appearance. In other birds it may be naked. In the rails, herons, and owls, the external structure of the oil gland orifice varies within species. Oil glands are absent in the ostriches, emus, cassowaries, bustards, frogmouths, mesites, some parrots, some pigeons, and some woodpeckers. Because of the similarity of the oily, holocrine secretions, some authors have compared the preen gland in birds with the sebaceous glands in mammals and the femoral glands in lizards.

No sweat glands are present in birds. This is probably correlated with the fact that an enormous amount of water, and heat, is lost by evaporation from the respiratory passages. In a few gallinaceous birds, for example, the common turkey (*Meleagris gallopavo*), modified oil glands are reputed to be present in the outer ear canal.

## 7. Integumental Glands in Mammals

Compared with the rather sparse occurrence of glands in birds and reptiles, mammals possess a wide variety of integumental glands. There are four main types—*sebaceous, sweat, scent,* and *mammary*. The latter two are thought to be derived from one or both of the former two. *Lacrimal* glands are also present, but they are not confined to mammals.

### a. Sebaceous Glands (*Lat.* sebum = *tallow*)

These glands produce oily or waxy holocrine secretions and are present over most of the skin surface but absent from the palms of hands and soles of feet. They usually occur in association with an opening into hair follicles (fig. 5.8*J*) and function to lubricate and waterproof the hairs much as the secretions of the oil gland in birds maintain the condition of the feathers. There may be more than one gland per follicle. Some sebaceous glands are also present where there are no follicles, as in the angle of the mouth, on the *glans penis,* on the internal surface of the foreskin (preputial glands), on the *labia minora,* and on the mammary papillae. Their function is generally considered to be lubrication. The *tarsal* (Meibomian) *gland,* in the connective tissue plate (tarsus) of the eyelid, secretes an oily film over the surface of the eyeball. This oil also waterproofs the rim of the eyelid so that tears, from the *lacrimal gland,* do not normally overflow onto the face but are guided to the naso-lacrimal duct into the nose. The *wax* (ceruminous) *glands* of the outer ear canal produce an *ear wax* (cerumen) that moistens and lubricates the epithelium of the ear canal and tympanum. Students are encouraged to "clean" their ears gently with a cotton swab and examine the ear wax with a microscope. This wax, along with the hairs therein, also serves to prevent insects or other foreign bodies from wandering into the ear canal where they might injure the sensitive tympanic membrane.

### b. Sweat Glands

These are long, slender tubes of epidermally derived cells. One end of the tube is generally coiled, sometimes branched, and found deep within the dermis (fig. 5.8*J*), while the other end perforates the *stratum corneum* and opens onto the surface or into a hair follicle through a tiny pore.

---

The student should examine a histological section of mammalian skin with the low and medium power lenses of the microscope. Under high power, one might observe that the coiled end of the sweat gland is surrounded by a minute capillary network and also by a thin layer of myoepithelial cells that help expel the serous secretion. Sweat glands occur in most mammals, but they are absent in spiny anteaters, pangolins (scaly anteaters), three-toed sloths, moles, sea cows, whales, and elephants.

---

The outer surface of an animal, whether it is naked or contains feathers or fur, is the interface across which interaction between the animal and its environment occurs. The environment influences the animal through exchanges of energy associated with metabolism, moisture loss or gain, radiation, convection, and conduction. In terrestrial vertebrates considerable amounts of energy are exchanged through evaporation from respiratory or skin surfaces. In mammals, sweating provides a special method of temperature control through evaporative cooling. This function of the sweat glands has been well studied and is the best understood. Other, perhaps more primitive, functions such as skin and hair lubrication or elimination of metabolic wastes have been less well elaborated.

Behavioral traits can often substitute for the sweat glands. Elephants, for example, are nearly hairless and lack both sebaceous glands and sweat glands. They might be expected, because of their relatively low surface-to-volume ratio, to have difficulty in losing heat to the environment by convection. They seem to have survived the lack of sweat gland evaporation by flapping their highly vascularized ears and taking frequent water baths. Pigs also do not always keep cool by evaporative water loss from their sweat glands but induce evaporative cooling by a coat of mud that they spread on themselves by wallowing (pigs experimentally smeared with mud evaporated thirty times as much water per unit surface area as clean pigs). Similarly, the "blood" that the nearly hairless hippopotamus is reported to sweat is in reality a viscous red compound containing watery fluid, mucus, and a red pigment that turns brown when it dries. This sweat compound may serve a cosmetic function by keeping the *stratum corneum* moist and flexible and protecting the skin from sunburn.

Man, another mammal with very few hairs per unit body surface, has nearly the greatest number of sweat glands per unit of skin surface (only the chimpanzee is reported to have more). The sweat glands in man are especially abundant on the palms of the hands and soles of the feet. They do not occur in the lips, eardrums, or nail beds or on the glans penis. In many other mammals, sweat glands are restricted to certain regions. For example, in the duckbill platypus, sweat glands are limited to the snout region; in deer, they are present only at the base of the tail; in mice, rats, and cats, only on the underside of the paws; and in rabbits, around the lips. Various races of man do not differ significantly in number of sweat gland pores per unit skin surface. Recently, it has been found that some pores bear twin glands, and future investigations may reveal that there are many more functional glands than is indicated by simply counting pores.

There are two functional types of sweat glands in mammals—*eccrine* (a type of merocrine gland) and *apocrine* (where the secretion includes parts of the cell itself). The *eccrine glands* are best developed in the catarrhine primates, especially man. They form during fetal life, and no more develop after birth. The infant skin therefore has more sweat glands per unit of body surface area than the adult. Even though sweat gland development is complete two to three months before term, premature babies do not sweat but control their body temperature by increasing heat production in a cool environment and controlling skin blood flow to prevent heat loss. Each sweat gland may increase its activity during childhood.

*Apocrine sweat glands* are found in most mammals, especially in the lemurs and the platyrrhine primates. In man, however, they are confined to the pubic region, the inner surface of the foreskin (prepuce), the labia minora, and the nipple region of the mammary glands. Especially large apocrine glands, about 3–5 mm in diameter, occur in the axillae and the circumanal region. These may well qualify as *scent glands* (see below). The *ciliary glands* (glands of Moll), which open into the base of the eyelashes and onto the margin of the eyelids, are also modified apocrine sweat glands.

Before leaving the topic of sweat glands, each student is encouraged to recall the last time he or she was frightened or under emotional stress. What happened? When a human being is frightened, he often breaks out into a "cold sweat" and the hairs stand on end—he gets "goose bumps." This happens when the myoepithelial cells of the sweat glands and the arector pili muscles of hairs are innervated by the sympathetic portion of the autonomic nervous system. Any mammal in a "flight or fight" behavioral situation secretes adrenalin that stimulates these tiny skin muscles to contract. Thus sweating and goose bumps occur together. This same response, in mammals more hairy than humans, results in a behavioral display that increases the apparent size of the animal (threat display) or directs attention toward a certain part or region of the body.

### c. Scent Glands

Odors or scents are means of communication that keep individuals together or keep them apart. Smell is the most important sense in a majority of mammals and is used in a variety of behaviors from marking territories to laying trails as well as functioning as a means of protection, recognition, or attraction. It is not unexpected then that most mammals have large integumental *scent glands* that produce characteristic scents, pleasant or otherwise. The phylogenetic derivation of scent glands is unknown.

Scent glands may occur in almost every imaginable place on the body: in the eye region as in deer and antelope; in the temporal region, as in elephants (both male and female); on the face, as in bats; on the chest or arms, as in many carnivores; near the opening of urogenital organs, as in rodents; at the base of the tail, as in dogs; on the feet or between the hoofs, as in artiodactyls, for example, sheep, in which they lay a trail that can be followed by other members of the herd; along the abdominal wall, as in musk deer; or along the dorsum, as in kangaroo rats, peccaries, and camels.

Scent glands are most frequently associated with the excretory functions of urination and defecation. However odoriferous the scent of urine or feces itself might be, it is often heightened by the secretion of *preputial* or *anal scent glands. Preputial glands* in many mammals produce a secretion that humans find repugnant, as the urine of rabbits, hares, rats, mice, and boars, or pleasant, as in beavers or musk deer. The latter are prized for their use in perfumes by humans (whose purposes may well rival those of the original producer). *Anal glands* are common and present in both sexes. As with the preputial glands, the odors of some are strongly objectionable whereas others are pleasant to humans and thus are used in perfumery. The large apocrine sweat glands found in the armpit of man may also be classified as scent glands in that, when associated with certain bacteria (in the days before underarm deodorants and antiperspirants), a characteristic odor is produced.

### d. Mammary Glands

These glands are present in both sexes but with few exceptions they are functional only in lactating females. *Milk,* a nutritious fluid containing water, fat, carbohydrate, and protein, is secreted from these glands, whose development and function is intimately controlled by ovarian and pituitary hormones and by stimulation from the newborn suckling animal or infant.

Mammary gland buds form along two ventrolaterally placed ectodermal ridges that in the fetus extend from the axilla to the groin. These are called *mammary ridges or milk lines* and are the developmental analogs to the dental lamina (which occur in the oral epithelia of all vertebrates), the lateral ectodermal ridges of fishes (in which the lateral line neuromasts and first scales occur), the ventral ectodermal tail ridges of mammals (in which occur the buds of tail hairs), and in the feather tracts of birds (wherein feather papillae originate). In a pattern similar to those listed above (see fig. 5.6), the *mammary gland buds* (fig. 5.3*l*) form ducts and nipples along the milk line, and the intervening tissue between the developing buds disappears.

In the juvenile female especially, or in the prelactating adult gland, only tubules or ducts are present. Growth and branching is controlled by estrogen. At puberty, estrogen (and progesterone) similarly control the enlargement of the gland and the formation of secretory ends of the ducts, the *alveoli,* which occur in clusters called *lobules.* Tubules from the alveoli form ducts in the lobules and these in turn unite to form one or more common ducts that lead to the surface. The size of the mature gland is also dependent upon the amount of adipose tissue deposited in the surrounding tissues. At the surface, the common duct(s) may open directly onto the skin surface, usually into a small open depression; into a small depression or *cistern* around which has formed an elevated collar of epidermis, the *teat,* which forms a wide diameter secondary duct through which all milk must flow from the cistern to the surface; or directly to the surface through a raised epidermal papilla called a *nipple.* The nipple is generally surrounded by a halo of highly pigmented tissue, the *areola.* In Prototheria there is no nipple or teat formation and breasts do not form. The milk is exuded from ducts onto the skin where it is lapped up or sucked from small tufts of hair. Both monotreme parents possess functional mammary glands, an exception among mammals.

In Metatheria and Eutheria, nipples and teats are present and the glands enlarge under estrogenic stimulation at puberty, often with fat deposition to form breasts (Lat. = *mammae*). In marsupials, insectivores, and certain rodents, nipples have a single common duct that opens at the surface. In man, and some carnivores, multiple common ducts are present in the nipple. Teats are present in the ungulates. Variation is widespread among mammals.

During pregnancy, milk formation is initiated by prolactin secreted from the anterior pituitary. The suckling of the infant stimulates a neural reflex from nipple to hypothalamus. This causes the release of oxytocin from the posterior pituitary, which in turn causes contraction of the muscles surrounding the alveoli. This contraction causes *milk ejection* ("letdown"). So strong is this particular response in the lactating human female that even the thought of nipple stimulation is enough to initiate release of oxytocin and eventual milk ejection. Students are urged to ask their mothers or acquaintances who have borne babies about this phenomenon.

The number and distribution of mammary glands in a given species varies, generally in proportion to the average number of young born and to the nursing behavior of the parent(s). Students should verify this by examining available laboratory or farm mammals. Make a list of the number and locality of the glands, the number of young born, and the type of nursing behavior exhibited by the mother. Generally, the number of glands may vary from one pair in man to eleven pairs in certain insectivores. The distribution may follow the entire path of the embryonic milk line or only parts of it. Location may be classified as: (1) *ventro-lateral,* as in dogs, cats, pigs, mice, edentates, and other animals that lie on one side when nursing; (2) *thoracic,* as in bats, manatees, dugongs, and primates, especially in tree-dwellers and those in which infants are held in the arms; (3) *axillary* (in the armpits or between the forelegs), as in flying lemurs and elephants; (4) *inguinal* (between the hind legs), as in cetaceans, cattle, and horses; or (5) *combined* (inguinal or thoracic), as in insectivores. Some mammals have their mammary glands in unexpected places, as for example, on the shoulders of male lemurs or on the dorsum in the aquatic nutria, which has four teats on its back so the babies can suckle above water while the mother swims.

It has generally been held that mammary glands are modified sweat glands. Some recent research argues against the traditional view, which is based on the assumption that the muscular tissue of the mammary gland resembles the myoepithelial cells surrounding the sweat glands. The mammary muscle tissue, however, responds to oxytocin and is situated on the dermal side of the basement-membrane complex, whereas the myoepithelial cells of the sweat gland respond to adrenalin and sympathetic innervation and occur on the epidermal side of the basement-membrane complex. This supports the theory that the mammary glands may have been derived from sebaceous glands. Further support is offered by an experiment in which male rabbits, given regular injections of estrogen and progesterone, developed teats. Extra lobes of the mammary gland were formed by proliferation of surrounding sebaceous glands, and hair was lost from the follicles into which these sebaceous glands formerly secreted. The composition of milk is of no use in settling this controversy, since it contains components normally secreted by both sweat glands and sebaceous glands. Furthermore, whether milk is a merocrine or an apocrine secretion is still a matter of debate. Unfortunately, neither sweat glands, sebaceous glands, milk, nor hairs have been preserved in the fossil record. Whatever their precursors in some synapsid (?) reptile, mammary glands, like hairs, are unique characteristics of modern mammals.

## E. The Keratinized Integumental Skeleton

### 1. Keratin and Keratinization: A New Perspective

The term *keratin* has been widely used to denote a spectrum of different cell structures or products ranging from the entire mature *stratum corneum* to all or some of the fibrous and amorphous components of a *keratinocyte*—a keratin-forming cell. Keratinization, however, is a complex process involving self-destruction of the epidermal cell content and concurrent depositing of keratin protein in its place. Depending upon the state of development of the keratinocyte, the protein may be fibrous, amorphous, or a combination of both, but no single keratin protein has been isolated. The plasma membrane of a keratinocyte also imparts much of the integrity of the horny cells or tissues, generally being about twice as thick as the plasma membrane of nonkeratinizing epithelia. The end product "keratin," produced by a maturing keratinocyte, is similar to the end product "enamel" produced by an ameloblast in that both are complexes of fibrous proteins and amorphous matrix that attract and bind calcium salts.

The traditional viewpoint that only in the terrestrial environment did the vertebrate skin begin to express its vast potential of variation, as now seen in a variety of tetrapod derivative structures, is often stated or implied in textbooks and treatises on the integument. Most often, such a view has been incorporated into a discussion of the epidermal "horny layer"—the *stratum corneum*—and its role in waterproofing or otherwise adapting a terrestrial vertebrate to its "dry" environment. This "dry environment theory" holds that keratinization, in a phylogenetic sense, originated in (or is limited to) the terrestrial vertebrates. Those who have held this view were not aware of the great variety of integumental derivatives found in the aquatic lower vertebrates. It should be made emphatically clear that the process of keratinization, though most obvious in the amniotes, is not limited to them, nor is it associated exclusively with dry environments. For example, the vaginal and cervical epithelia of certain mammals undergo cyclical periods of keratinization and mucus production. Furthermore, the fully aquatic oral-pharyngeal cavity of most vertebrates, from fish to mammals, is populated locally by keratinized mucosal tissues, for example, the teeth of lampreys and larval frogs, and the mammalian gingiva and palate.

All metamorphosed terrestrial amphibians have a complete, that is, continuous, stratum corneum of varying thickness, although it is relatively thin compared with that of amniotes. The first (stegocephalian) amphibians had calcified dermal scales, but whether they also possessed a keratinized epidermal layer is not known. Teleosts, though lacking a complete horny layer, have many localized keratinous structures, including tubercles, friction pads, adhesive disks, jaw coverings, and juvenile teeth. Some fishes, such as the algae eaters of home aquariums, have epidermal invaginations that form highly vascularized follicles or pits wherein generations of epidermal cells undergo periodic hypertrophy and keratinization to produce horny epidermal tubercles (fig. 5.15, and p. 159). Such phenomena have heretofore generally been believed to be limited to amniotes—more specifically, to the squamate reptiles—and were associated with the dry environment theory. The rich variety of keratinized epidermal derivatives found in the fish skin and other aquatic epithelia offer a fertile field for future investigations.

A keratinized epidermis has been assumed to have a waterproofing function, but this is not necessarily true. The traditional, insoluble, sulphur-rich "keratin" molecule apparently is hydrophilic, and its reputed role in waterproofing the terrestrial skin is most likely an indirect one provided by the phospholipids bound to its side chains. In mammals, the stratum corneum reduces evaporative water loss from the living cells below. In birds, the evidence is contradictory. Recent experiments comparing pulmocutaneous water loss in a scaleless gopher snake with water loss in a normal individual of comparable age and size, however, suggest that reptilian scales cannot be regarded as adaptations to restrict water loss and that other functions (protection?) should be considered.

From the above introduction, it should be clear that the process of keratinization (cornification) is only one, albeit a well-used, pathway selected and modified for many specific functions from an entire spectrum of potential integumentary sclerifications available to the early, pre-terrestrial vertebrates. Among the alternative pathways, one could include all vertebrate dental and osseous tissues, both cellular and acellular (p. 163), and the vast array of sclerotized and calcified or mineralizable invertebrate tissues.

**2. Shedding, Sloughing, Molting, and Retention**

Within the overall framework provided by the functional epithelial extinction hypothesis (p. 121), the normal fate of a *keratinoblast* maturing into a component of the epidermal horny layer can conveniently, if only approximately, be categorized as follows (examples are summarized in Table 5.1):

Shedding: Continuous loss in small flakes of the most superficial keratinocytes.

Sloughing: Periodic loss in complete sheets, or large fragments, of all keratinocytes in the mature stratum corneum.

Molting: Continuous or periodic loss of specialized keratinous epidermal derivatives other than the stratum corneum itself.

Retention: Continuous or periodic piling up of layer upon layer of firmly bound mature keratinocytes. Loss is limited to mechanical erosion or other environmental degradation.

However distinct these categories appear at first sight, the student should be aware constantly that such definitions are neither all-inclusive nor mutually exclusive. Ambiguities invariably have arisen in the works of different investigators, depending upon how precisely one interprets what is continuous or noncontinuous and what time interval constitutes a period or cycle. Because of such complications and until we better understand the nature of the cycles and periodicities involved, the spectrum of related processes to be studied below might better be considered as parallel long-term solutions to similar functional demands.

**3. Hair and Functional Epithelial Extinction in Mammals**

Hair is commonplace in modern society; we style it, imitate it, wear it, and even write stage plays and songs about it. And though we know much about its growth and function in man, sheep, and laboratory rodents, we know comparatively little about it in other mammals. Hairs, like feathers, are developed in an invaginated epidermal follicle that surrounds a dermal papilla (fig. 5.3C), and are composite structures formed of hundreds of thousands of epidermal cells, most of which are keratinized and dead.

The best way to learn about hairs is to examine them. Pull a hair from your scalp or beard and another from your eyelashes or nostril. Notice that one end often has a slightly swollen collar (part of the *inner sheath*) or tapers to a small *bulb* (part of the follicle surrounding the dermal papilla) or both. Immediately touch this end (before it dries out) and notice that it is soft and flexible, whereas in the collar region above (if the inner sheath has remained), the hair is more hardened and inflexible. This blunt, nonkeratinized end is the growing *root end*, whereas the *tip end* of a new or uncut hair is pointed. If you cannot distinguish which end is which and cannot afford another hair, perform the following experiment. Pull the hair between the forefinger and

TABLE 5.1
The Fate of Mature Keratinocytes in Various Classes of Vertebrates

| Class | Shedding | Sloughing | Molting | Retention |
|---|---|---|---|---|
| Mammals | Most skin including tail scales | In some seals and whales; elephants | Hair; horns | Some horns; nails; claws; armadillo plates |
| Birds | Most body skin; tarsal scales (?) | In some, e.g., pigeons and penguins; others (?); tarsal scales (?) | Feathers; some bills, beaks, and claws; egg caruncle | Most bills, beaks, and claws |
| Reptiles | Skin of "soft" body parts of Chelonia and Crocodilia | Skin of Squamata and Rhynchocephalia | Egg caruncle of Chelonia, Crocodilia, and Rhynchocephalia | "Hard" parts (carapace; beak; plastron) of Chelonia and (dorsal scutes) of Crocodilia; claws |
| Amphibia | (?) | Most; or all (?) | Larval teeth; forelimb nuptial pads | Claws and digital caps |
| Fishes | (?) | (?); May include the mucous cuticle of many Scorpaeniformes, e.g., Scorpaenidae, Congiopodidae, Cottidae and Triglidae; also in Dactylopteriformes and Pegasiformes | Nuptial tubercles; Lamprey teeth; juvenile cyprinid teeth | Nonbreeding tubercles of torrential stream species; horny jaws of herbivorous cyprinids |

thumb, exerting moderate to firm pressure. When passing from the root to the tip, the hair will slide easily and smoothly between the finger and thumb. If pulled from the tip toward the root end, it will offer resistance (often with a squeak). The friction thus produced indicates that the thousands of cells that make up the outer layer of the hair— the *cuticle*—are laid down like overlapping scales with the scale edges pointing toward the tip (fig. 5.10*A*). This directional orientation of the hairs not only prevents tangling of the hair coat, but also allows dead

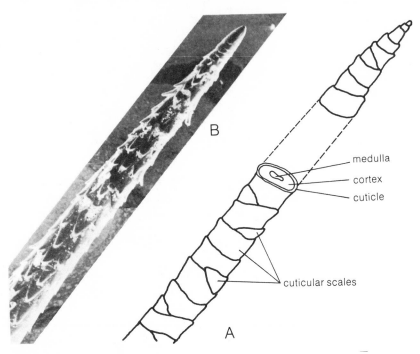

Fig. 5.10.  Hair. *A*, diagram of typical hair tip (uncut) with its cuticular scale edges pointing toward the tip. *B*, scanning electron micrograph of a porcupine quill tip (×500). Note that cuticular scale edges point basally thus forming an effective barbed weapon.

cells (dandruff) and other foreign matter to move away from, rather than toward, the skin. When hair is cut off and the fibers become disoriented, the opposing scale edges catch and cause the hairs to cling together, as in fleece or felt. In a porcupine, however, the scale edges point back toward the root end, an obvious functional adaptation of the barb-like quill (fig. 5.10*B*).

The hair is also characterized by an inner cortex of cells that can be seen with a microscope (also fig. 5.10*A*). When a short piece of cut hair is placed in a whole mount with glycerine, the *cortex* becomes more visible. A central *medulla,* usually occluded by air bubbles, is present in thick hairs but absent in fine hairs, as it is in the root end or cut tips. When a hair is treated with a proteolytic enzyme solution, such as trypsin, the cells separate and can be seen under high-power magnification. The cuticle cells appear flattened and scale-like, the cortical cells elongate and narrow, and the medullary cells as lattice-like meshworks formerly filled with air spaces. The *medulla* is thought to be a structural adaptation that, as in a long bone, allows maintenance of the specific diameter of a cylinder or shaft (cortex) with maximum strength, minimum weight, and at a minimum expenditure of energy. There are many species-specific variations in the dimensions and hence arrangement pattern of cuticular, cortical, and medullary cells.

---

In most mammals there are two kinds of hairs: long, coarse ones, the *guard hairs,* which compose the outer coat of the mammal, and short, finer, softer hairs, the *underhair.* There are many gradations and categories of guard hairs—spines, bristles, awns, and so forth, and of underhair —wool, fur, vellus, and so forth, and even these are further subdivided. The important point, however, is that *guard hairs* grow in *primary follicles,* whereas *underhair* grows in *secondary follicles.* The former develop first and the latter usually form later in specific clusters associated with the primary follicles. Primary follicles are usually accompanied by the development of sebaceous glands, a sweat gland, and an erector (*arector pili*) muscle, whereas the secondary follicles usually have only sebaceous glands associated with them. The number of follicles developed in a given species is apparently genetically determined, but environmental factors such as poor nutrition can also affect the number. This is especially important in mammals grown for wool or fur.

Many mammals have primary hair follicles in a *trio pattern* (central plus two lateral) that become associated with later-forming secondary follicles. But some mammals, such as the cow and horse, have no follicle groupings, the follicles being distributed evenly throughout the skin. In man, there are no follicle groups of the scalp, face, or body, but examine the hairs on your fingers. Are they single or in groups? That such trio grouping occurs in most mammalian follicles has led many researchers to consider it a basic pattern associated with the evolution of mammals from scaled reptiles. That such a trio pattern of hairs is confined to the hinge region between scales of the tails of rodents and marsupials is taken by many as evidence that hairs first appeared between the scales in mammalian precursor reptiles. Without confirmation from the fossil record, and recognizing the variety of keratinized follicular derivatives (figs. 5.3, 5.14, 5.15) present in all vertebrates, I prefer the less tenuous conclusion, expressed earlier, that such derivatives might best be considered long-term solutions, often parallel, to functional demands on the embryonic follicle potential present in all vertebrate integuments.

Hair development occurs in three stages—*anagen, catagen,* and *telogen.* Cells from the *stratum germinativum* adjacent to the dermal papilla begin to proliferate. The forming follicle sinks deeper into the dermis. An *outer root sheath* and *inner root sheath* are formed, the latter as a collar around the base of the forming hair. Germinative cells, induced by the dermal papilla (fig. 5.3C), form the hair itself, which grows out toward the skin surface. It becomes keratinized in a *keratogenous zone* within the inner root sheath. Concurrently, the hair loses water and its diameter is reduced; the final emergent product is a string of keratin-filled cells joined by cell-to-cell junctions. Once a mature hair reaches its maximum length (in noncontinuous growing hairs), the germinative cells of the follicle cease their mitotic activity. This signals the catagen stage, a quickly occurring breakdown process during which parts of the follicle, especially the inner root sheath, are resorbed and remodeled. A swollen "club" forms at the root end of the hair, which is now separated from the dermal papilla. No hair growth occurs in the final, quiescent telogen stage, but a new anagen hair germ forms. The "club" end of the old hair holds it in position until it is molted during the next anagen stage.

The various stages of the hair growth cycle vary in duration among different species, as does the functional epithelial extinction cycle in general. In mammals, any given keratinocyte within the nonspecialized stratum corneum is constantly changing its spatial position with relation to the basement-membrane complex. Except when epidermal growth ceases or is temporarily interrupted, for example, during hibernation, torpor, or telogen (in the case of hairs), old keratinocytes are continuously being shed from the surface as new ones replace them from below. In contrast, a single hair (fig. 5.10) is a complex generation of thousands of keratinocytes that is molted as a single unit either simultaneously or asynchronously with other hairs from adjacent follicles. The growth of a new hair alongside the old is a complex phenomenon involving not only the intrinsic rhythm of growth (genetically determined) and seasonal effects, but the loss of the old hair by molting. Growth cycles, and the accompanying molt, may vary from complete synchrony, wherein the entire coat is molted at one time as in some seals, through bands of synchrony (wave patterns) such as occur in fur-bearing rodents, and mosaic synchrony, wherein small localized areas of synchronized molting occur, to asynchrony where little or no synchrony occurs in the initiation of growth cycles between adjacent follicles. Many factors add to the difficulty of classification. Among these are the proximity of hair follicles to one another, age of the skin, sex of the animal, and time elapsed since previous initiation of hair growth. Wave phenomena, however variable, generally are most pronounced and occur more frequently in the juvenile *pelage* (hair coat) whereas in older animals the wave pattern becomes more diffuse, eventually resulting in small islands of simultaneous hair replacement or even individual asynchronous replacement.

---

Follow the pattern of hair growth in some newborn mice or laboratory rats. If no babies are available, shave a small area of a guinea pig or rabbit back and mark off small areas (square cm or inch) with india ink. The number of hairs, the weight of the fur, and the variation in length and diameter can be determined during a normal school term. All these characteristics except the number could change if environmental factors such as diet were changed. Can you think of an experiment that would allow you to study these effects?

---

Whereas the *maturational molt* of the juvenile pelage may occur continuously or several times during the year, the replacement process in adults is widely variable. In some it occurs *annually,* as in foxes, kangaroo rats, bats, and some rabbits. In others, it is *semiannual,* as in mink, lemming, weasels, moles, pocket gophers, brush mice, and some rabbits. In some it is more *frequent,* as in laboratory rats and small rodents. And in still others, it is *continuous,* as in cats, sheep, and the human scalp. Some studies have demonstrated a correlation between distribution, life history, and type of molt. For example, in the various subgenera of ground squirrels, the more northerly and alpine species hibernate and undergo only one molt a year, whereas the most southerly species, which do not hibernate, undergo two molts each year. The Barrow ground squirrel, however, is active only five months of the year, during which two molts apparently occur. In the pocket gopher, molting is extended over long periods of time and one molt may overlap another. In wood rats the annual molt begins after reproductive activity ends. Thus marked differences in coat color between sexes occur in summer because the reproductive period of the female often extends beyond that of the male and the molt is delayed.

In all considerations of phenomena involving functional epithelial extinction, a distinction should be made between the loss of the stratum corneum itself and other specialized horny derivatives. The cycles are not always in synchrony and may even vary within a given taxon. For example, in most pinniped mammals the hair molt occurs separately from the shedding of the stratum corneum. In the southern elephant seal, the monk seal, and the leopard seal, however, the horny cells of the stratum corneum become fused to the old hair shafts and the two epidermal components are sloughed as a unit. In some whales, the stratum corneum is known to peel off in large sheets twice a year, whereas elephants' habit of "showering" themselves with sand or gravel is thought to help get rid of partially sloughed skin.

4. Feathers and Functional Epithelial Extinction in Birds

Birds are characterized by highly specialized, highly adaptive, multifunctional epidermal derivatives termed *feathers* (fig. 5.11). These form in a follicle (figs. 5.3*B* and 5.15*A*) and, like hair, follow intrinsic growth cycles. However, no catagen stage occurs in feather development. Feather formation, though differing in many minor details (feather germs evaginate and have a dermal core, hair germs invaginate and surround a dermal papilla), follows an inductive sequence similar to hair formation wherein an epithelial-mesenchymal interaction occurs and there is a mitotically active zone of germinal cells adjacent to a dermal papilla. These cells give rise to an outer follicle sheath that surrounds and protects the developing feather, a *shaft* or *rachis,* of which the bottom portion, the *calamus,* surrounds the highly vascularized dermal core. There is also a *keratogenous zone* in the calamus wherein the *shaft* becomes stiffened, as do the off-branching, parallel extensions of the shaft called *barbs* (fig. 5.11*A*). In what remains one of the most fascinating mysteries of precise biological orchestration in the animal kingdom, the barbs interdigitate with their neighbors by means of *barbules* (fig. 5.11*B*), which in turn have interlocking *hooklets* or *barbicels* (fig. 5.11*C, D*). As can be seen only in scanning electron micrographs, the barbicel cells are sequentially arranged much like the cells in the hair cuticle. Observe also (fig. 5.11*C*) that the barbicels and barbules together form small, boxlike cells that entrap air. This mechanism provides not only insulation but also, in water birds, a flotation device. Can you imagine what must happen to those air cells when a water bird is trapped in an oil spill?

Fig. 5.11. The intricate design and external structure of a bird feather. *A*, scanning electron micrograph of feather surface (× 50) showing the relationship of the parallel branching barbs to the rachis (shaft) and to each other. *B*, feather surface showing relationship of barbules between two adjacent barbs (× 250). *C*, box-like air spaces formed between interlocking barbicels (hooklets) of lower barbule and adjacent barbule above (×50) showing terlocking detail of individual barbicel with adjacent barbule (× 3,000). Notice the node-like branching of individual barbicel from the barbule. Compare with cuticular scales on a hair (fig. 5.10). *E*, flight feather. *F*, contour feather (penna). *G*, down feather (plumule). *H*, a hair-like filoplume.

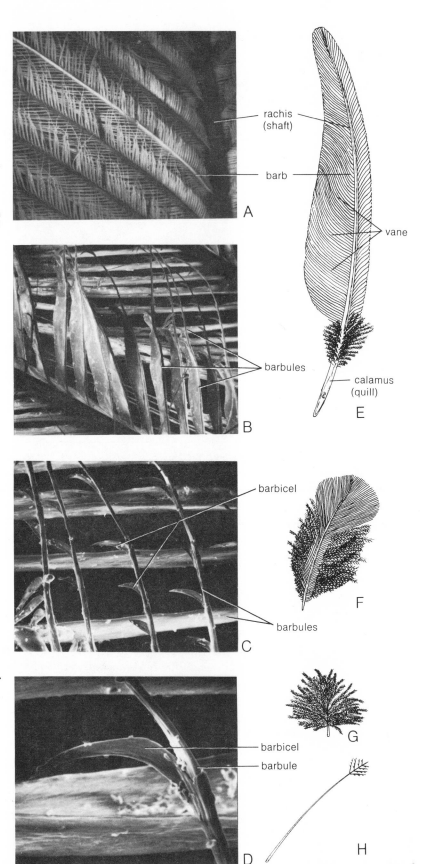

What do you think happens to the barbicels when well-intentioned rescue workers try to remove oil with solvents and detergents? Is it any wonder that survival rates in "rescued" birds are extremely low? Can you think of a way to rescue an oil-soaked bird without adding it to the mortality list?

As feathers wear out, they loosen in the follicle and eventually drop out or are pushed out by the replacement feather below. In many birds, however, the calamus of each old feather is fused to the tip of the *feather sheath* of the replacement feather, and when this sheath ruptures, exposing the new feather, each old feather is individually lost or plucked off by the bird. Penguins offer an interesting contrast to the general rule in that their feather barbules are closely intermeshed and the entire plumage is periodically lost in large pieces, especially from the scaly wing regions.

As in mammals, molting in birds is a variable phenomenon and few precise rules can be laid down. In general, complete renewal of the *plumage* happens once a year, usually after the breeding cycle. In migratory birds, this *postnuptial molt* occurs before the return of migration. In some birds, for example, swallows and the falconiforms, the *annual molt* occurs in the winter. Many birds, however, have a *semiannual molt,* with both a winter plumage and a nuptial plumage. In many of these forms the wing primaries are molted only once a year. Cranes renew their wing primaries every two years.

The duration of the molting process is also highly variable, from a few weeks in the yellow-eyed penguin through two full summers in captive golden eagles. Partial or incomplete molts also occur in some bird species. This results in a composite plumage containing varying proportions of old and new feather generations. The phenomenon is equivalent to the waves of hair growth in mammals, and, similarly, the pattern is most typical of juveniles.

Except for some histochemical descriptions of the epidermis in chicks and adult fowl, pigeons, and penguins, little is known about the loss of the mature stratum corneum. Tarsal scales of the chick are shed about seven to ten days after hatching, concurrently with the first feather molt. In grouse and ptarmigan, the horny covering of the claw is molted in one piece at the end of the breeding season.

5. Epidermal Generations and Functional Epithelial Extinction in Reptiles

Major differences in histochemistry, morphology, and time intervals of keratinocyte loss are found within the reptiles. The shedding of the stratum corneum of the soft body parts, that is, the limbs, neck, and tail of chelonian and crocodilian reptiles, is similar to that of mammals and birds and, similarly, no "fission zone" occurs (see below). As in mammals, epidermal growth apparently ceases during hibernation. The hard body parts—carapace, plastron, dorsal scutes, and so forth, are retained and wear out by erosion or environmental degradation. When compared with that of the Chelonia, Crocodilia, and other amniotes, the stratum corneum of the Squamata and Rhynchocephalia is exceedingly complex, composed of at least five or more distinct strata of cells (fig. 5.12). At regular intervals, the entire germinative layer of the epidermis undergoes synchronous mitotic activity, followed by a maturation process

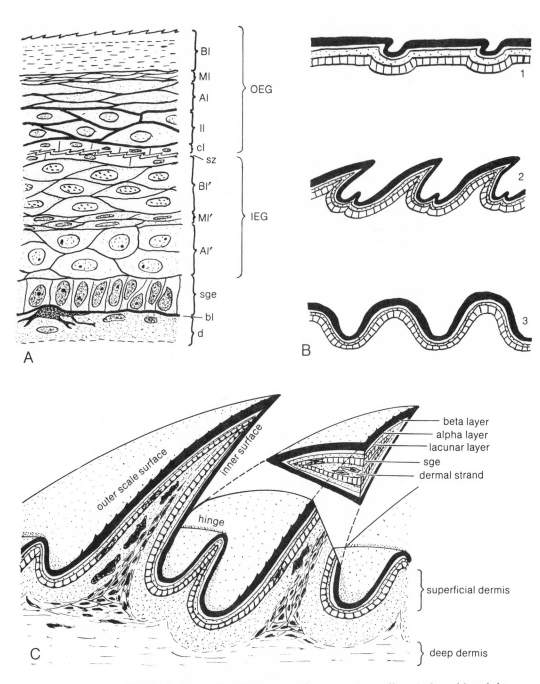

Fig. 5.12. Generalized integument in squamate reptiles. *A*, the epidermis in late, presloughing stage showing the relationship of the multilayered outer epidermal generation (*OEG*), which will be sloughed, and the developing, inner epidermal generation (*IEG*), which will take its place: the "beta" layer (*Bl*) comprised of a superficial "Oberhautchen" layer (note zipper-like profile) and a subjacent layer of keratinized β-cells; the "mesos" layer (*Ml*) of keratinized cells; the keratinized "alpha" layer (*Al*), which overlies the living "lacunar" layer (*ll*); and the "clear" layer (*cl*), which, at sloughing, separates from the underlying inner epidermal generation by a splitting zone (*sz*); the "beta" layer (*Bl'*) of the inner epidermal generation comprised of the "Ober-

during which keratinocytes, at different topographic levels within the stratum corneum, undergo varying degrees of keratinization. Some keratinocytes never reach the surface, as they do in a typical vertebrate stratum corneum. This entire multilayered structure, comprising both the *scale* and the *interscale* (hinge) region, is referred to as an *epidermal generation* (fig. 5.12).

The mature epidermal generation is separated from an underlying, immature epidermal generation by an unkeratinized *clear layer* (= *stratum intermedium*). During the sloughing process, the cells of the clear layer undergo autolysis, a "fission zone" forms, and the mature epidermal generation splits off as an entire unit in large sheets. The subjacent, similarly derived, keratinizing, immature epidermal generation then replaces it as the functional body surface. As its keratinocytes become fully mature, it in turn is eventually sloughed and replaced from below. Fission zones have not been identified in mammals, birds, amphibians, or fishes, and a full understanding of what mechanisms trigger autolysis and how it occurs in the fission zone of reptiles is not yet available. It is known, however, that cell division continues in the interval between the two keratinization cycles and that this produces the unkeratinized, mechanically weak fission zone. This evidence corroborates the earlier contention (p. 121) that the continuance of epidermal proliferation—functional epithelial extinction—is not dependent upon keratinization.

The process of sloughing the outer epidermal generation of the squamate reptiles has been considered quite different from that known to occur in any other vertebrates. Unfortunately, very little is yet known about sloughing in amphibians and fishes. As in other vertebrates, it is influenced by thyroid and pituitary hormones, especially prolactin, but this is a relatively new field of comparative endocrinology in which few generalizations have as yet been proposed, much less documented.

The chelonian shell is an example of *retention* of a specialized horny derivative in which neither shedding nor sloughing occurs. The shell is composed of an inner bony layer and an outer, keratinized layer formed by cells of the underlying epidermis. Young leather-back turtles have small horny laminae in their skin, but these disappear in the adult, which has a smooth leathery skin over a very specialized shell. In two families of turtles (Carettochelydae and Trionychelyidae), the horny laminae are absent.

---

hautchen" layer and keratinizing $\beta$-cells; the keratinizing "mesos" layer ($Ml'$); the presumptive "alpha" layer ($Al'$) of cells, which overlies the stratum germinativum ($sge$), the basal lamina ($bl$), and the upper portion of the dermis ($d$) (after Maderson, 1967 [*Copeia* 1967:743]). *B*, variation in scale types of "resting" (postslough) stage snake and lizard epidermis: flattened head scales and hinge region (1); overlapping body scales and hinge region (2; see *C* for details); and tubercular scales of gekkonid and some lacertid lizards (3) (after Maderson, 1965 [*J. Zool. London* 146:98]). *C*, three-dimensional representation of overlapping body scales of a snake showing the relationship of outer and inner scale surface to the hinge region (redrawn after Maderson, 1964 [*Brit. J. Herpet.* 3:151]). See Fig. 5.7*C* for surface details of scale and hinge region.

6. Sloughing
and Functional
Epithelial
Extinction
in Amphibia

With the exception of a few recent contributions, little is known about keratinization or the sloughing mechanism in amphibians beyond the early contributions and summary of Noble (1931). Amphibian sloughing, like that of other vertebrates, is under endocrine control of thyroid and pituitary hormones, but even less is understood here than in reptiles.

In general, all epidermis except that on unmetamorphosed larvae or persistent larval species has a thin but complete stratum corneum. Periodic sloughing of unkeratinized epidermis in urodeles has been reported but not verified by modern techniques. In most amphibians, the stratum corneum is sloughed periodically or at varying time intervals in large patches or as a single layer. The process may take from a few hours to more than a day. There is no fission zone as in reptiles, and the epidermis is not highly organized except at certain specialized points such as toe pads and warts (fig. 5.13). Once the mature stratum corneum has been sloughed, the underlying layer of transitional keratinocytes requires two or three days to become fully keratinized. During the sloughing process, epidermal glands discharge mucus into horizontal intercellular spaces, and this presumably helps the animal get rid of its old skin. Puffing up the body with air and rubbing the skin against stones are also used to help the sloughing process. With the use of the scanning electron microscope, the imprint of the recently sloughed horny layer can often be seen on the surface of the new replacement layer (fig. 5.7B). The horny body tubercles (warts) found on many amphibians are yet to be analyzed with modern techniques.

7. Sloughing,
Keratinization,
and Functional
Epithelial
Extinction
in Fishes

Among the fishes, keratinized derivatives have been looked upon generally as novelties and until recently little attention has been paid to the processes of epidermal cell loss and renewal. Sloughing of "skin" has been reported or observed in only twenty-four species in fifteen genera of fishes, mostly in the order Scorpaeniformes. But there is no published documentation on whether the sloughed material is actually epidermis or simply a mucous cuticle. I have examined material from several preserved and fresh specimens and found the material to be acellular and composed of mucopolysaccharides, which suggests that the sloughed material is the mucous cuticle. It seems unlikely that, as more fishes are examined or closely observed in aquariums, the scorpaeniform fishes and their close relatives will remain unique in their ability to slough a mucous cuticle or, indeed, any other epidermal derivative. Electron microscopic examination of normal fish skin suggests that the nonkeratinizing epidermal cells are shed individually, in the same manner as the epithelial cells of the oral and alimentary mucosa.

Some juvenile cyprinid fishes (minnows) have horny teeth, and certain herbivorous adult cyprinid fishes have thick horny layers covering their jaws. Horny "adhesive" surfaces occur in certain Asian fishes adapted to torrential stream flow. But the most familiar and well-studied keratinized structures of fishes are the *horny teeth* of cyclostomes and the *nuptial* (breeding) *tubercles* in representatives of five of the eight superorders of teleost fishes currently recognized. The nuptial tubercles are hormonally controlled secondary sexual characteristics that occur primarily, though not exclusively, in male fishes. According to the species,

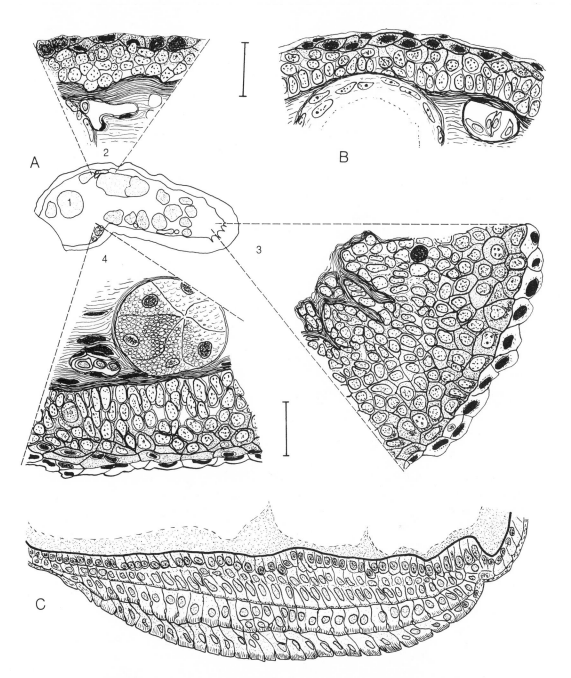

Fig. 5.13. Amphibian skin form and function: *A*, variation in epidermal thickness in different locations of postshedding toe skin of the California newt, *Taricha torosa:* 1. longitudinal section through terminal digit showing precise locations of magnified sections; 2. dorsal skin is thin, heavily pigmented and has no fully keratinized layers. Note: measured bar equals 50 microns in 2, 3, and 4; 3. epidermis from tip of toe pad is thickened but not fully structured (as, for example in *C*), and a single layer of keratinized cells is present; ventral skin has intermediate thickness, the least pigment, a keratinized and a partially keratinized layer of cells. Notice the large mucous gland in the dermis. *B,* tail skin from postshedding *Taricha torosa* (same specimen for comparison). *C,* mature toe pad epidermis of the tree frog, *Hyla regilla* (see fig. 5.8*G* for entire section). Frontal section showing multiple generations of keratinizing epidermis (compare with outer and inner epidermal generations of snake skin, fig. 5.12*A,* and with multiple generations in fish tubercles, figs. 5.14 and 5.15). Note: the bristle-like edges of each of the outer two cell generations is an artifact of light microscopy. Under scanning electron microscopy, the plasma membrane of the outer cell surface is formed into many microscopic cuplike folds.

Fig. 5.14. Horny epidermal tubercles in fishes. *A*, Lateral scale tubercle from the body epidermis in predorsal area of *Clinostomus fundulcides*, an American cyprinid. *B*, a horny opercular tubercle overlying a keratinizing replacement cap in *Hemizyon formosanum*, an Asian homalopterid. *C*, parakeratotic horny layer in the opercular tubercle of *Cheiloglanis brevibarbis*, a stream-dwelling African mochokid catfish. Note: keratinized cap (*kc*); parakeratotic layer (*pkl*); keratinizing layer (*kl*); basement membrane complex (*bm*); stratum germinativum (*sge*); and scale (*sca*). Measured bar equals 100 microns. After Wiley and Collette 1970 (*Bull. Am. Mus. Nat. Hist.* 143:147).

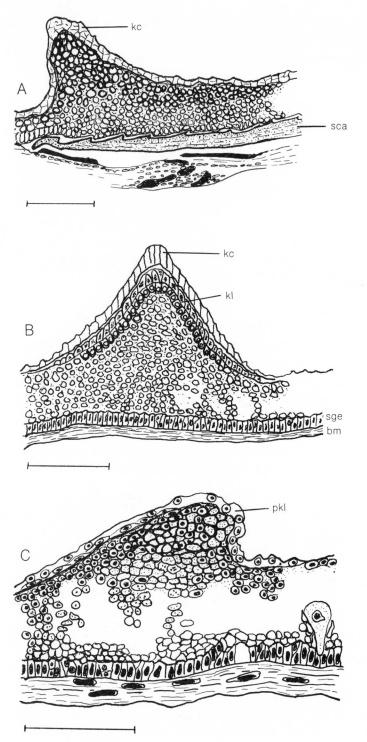

they appear variously on the fins, snout, chin, operculum, and over the entire head or body (fig. 5.14). Detailed behavioral studies have correlated structure with function by determining that in some species the tubercles are used to contact or clasp the female during reproductive behavior.

The morphological variation of tubercles is so great that a concise, descriptive nomenclature is not yet forthcoming. In fact, not all tubercles are "nuptial" in function. For example, in certain catfishes of Africa and the hill-stream fishes of Asia, tuberculation is pronounced all over the body surface in both juveniles and adults. These forms dwell in torrential streams, and it seems likely that natural selection has here utilized keratinization to hydrodynamic advantage in adapting the body to the swift currents.

In those species where tubercle formation is functionally associated with reproductive behavior, little information is available on how tubercles are lost after the breeding season. In at least two genera of loaches, the head and pectoral fin tubercles are so densely distributed that their bases fuse together to form a continuous keratinized sheet. In preserved specimens this sheet can be removed intact, but whether it is sloughed in nature is not known.

In the genus *Gyrinocheilus,* the algae eater of home aquariums, there exists a condition, outside the breeding season, wherein generations of horny tubercles are formed in small "pits" in the skin. These pits are epidermal invaginations that have become surrounded by highly vascularized dermal borders. Each pit is crowned by a mature keratinized tubercle (fig. 5.15). Below this are two presumptive tubercles in differing stages of hypertrophy and keratinization. The surface tubercles, when lost, apparently are replaced by the synchronous distal movement of the underlying tubercles. The convergence of this phenomenon with the formation of an *epidermal generation* in the reptilian skin is striking (p. 153). The evidence in fishes thus far suggests that the phenomenon of epidermal generations' being sloughed as a unit can no longer be upheld as exclusive to squamate reptiles. The ontogeny of these epidermal pits has not yet been investigated; but they have all the general attributes of hair and feather follicles, most likely including epithelial-mesenchymal interaction during early stages of development and, perhaps, during the replacement cycle. Assuming the latter, the exclusive view that "true" follicles are limited to mammals and birds also can no longer be held with firm conviction. Indeed, as early as 1895 F. Leydig remarked on the developmental and morphological similarities between skin tubercles in fishes and mammalian hair. Dare one now speculate that the tubercles contained in these follicles might be the developmental homologs of feathers and hairs?

## 8. Nails, Claws, and Hoofs

*Nails* are broad, slightly curved, keratinized plates on the dorsal surface of mammalian digits as, for example, in primates. *Claws* (talons) are arched, laterally compressed structures that protect the tip as well as the dorsolateral surface of the digit. They are found in many mammals, most birds and reptiles, and some amphibians. *Hoofs* are enlarged, keratinized digital appendages found in the ungulates.

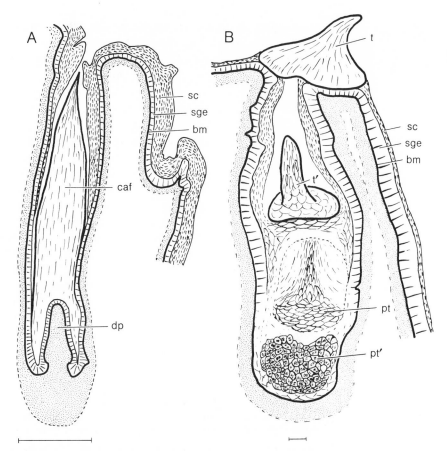

Fig. 5.15. Comparison of follicles in bird and fish. *A*, insunken wing feather follicle in the king penguin, *Aptenodytes patagonica* (after Spearman, 1969 [*Z. Morph. Tiere* 64:361]). *B*, "epidermal pit" in the algae eater, *Gyrinocheilus aymonier* (after Wiley and Collette 1970 [*Bull. Am. Mus. Nat. Hist.* 143:147]). Mature tubercle (*t*) on snout surface, replacement tubercle (*t′*) and two progressively maturing presumptive replacement generations (*pt,* and *pt′*) in the follicle below. Note: dermis (*d*); stratum germinativum (*sge*); basement membrane complex (*bm*); stratum corneum (*sc*); dermal papilla (*dp*); calamus of feather (*caf*). Measured bar equals 100 microns.

The development of amniote nails, claws, and hoofs has been relatively well studied, but not thoroughly understood from a comparative viewpoint involving all vertebrates. More is known about the human fingernail than any other similar structure, and its basic morphology (fig. 5.16 *E,F*) will be used as a reference point against which to compare the horny digital appendages of other vertebrates.

Trim a small piece of one of your fingernails with sharp scissors and then cut it into several squares. Examine it under low and high power of the dissection microscope. With a needle, probe the upper, middle, and lower surfaces and notice that the middle portion is softer. The *nail plate* is composed of three layers. The *dorsal stratum,* owing to its high calcium content, is much harder and tougher than the underlying *intermediate* and *ventral strata.* With a fine forceps, attempt to scrape or tear apart the piece of nail and notice whether it frays or flakes. Repeat the procedures with the claw of a mammal, bird, or reptile. Do the same for an ungulate hoof and a fish scale. Compare. How do they differ? Why? You might wish to test small pieces of each of these for thirty minutes with a solution of proteolytic enzyme such as trypsin and also of weak hydrochloric acid. Compare what happens. Also, you might try boiling small pieces both before and after such treatment. What happens? Why?

---

The trilaminar structure of the human fingernail has been established histologically and histochemically. There are three active growth centers within the continuous underlying germinative epidermis, which has been divided arbitrarily into *root matrix* and *nail bed* portions. The "upper" root matrix gives rise to the dorsal nail plate (stratum); the "lower" root matrix gives rise to the intermediate nail plate; and the nail bed gives rise to the ventral nail plate (fig. 5.16*E*). In dorsal aspect, the boundary between the root matrix and the underlying nail bed is seen as a whitish, crescent-shaped zone, the *lunula* (fig. 5.16*F*).

The stratum corneum forms a *continuous,* albeit flattened, *collar* around the margins of the *nail plate* (fingernail: L = *unguis,* Gr = *onychos*). That portion of the stratum corneum below the distal free edge of the nail is the *hyponychium* (subunguis), and that portion forming the lateral and proximal margins of the exposed nail plate is the *epionychium* (fig. 5.16*F*). This horny fingernail collar is similar to the infolded ridge surrounding the avian or reptilian claw (fig. 5.16*C,D*). Its resemblance to the keratinized *free gingiva* of a normal tooth, and even to the outer edge of a hair follicle, is striking.

In mammals, birds, and reptiles, claws are practically identical in structure, and in each case they partially or completely cover the terminal digital element, the *phalanx.* An early hypothesis described the phylogeny of mammalian claws as they changed from the supposed primitive condition found in the crocodilian reptiles. According to this unchallenged hypothesis, the crocodilian claw is a thimble-like cap that covers the entire tip of each phalanx (fig. 5.16*C*), whereas the mammalian claw has apparently lost the ventral side of the thimble. But this hypothesis neglects the fact that claws and simple thimble-like caps also occur in several amphibian groups (fig. 5.16*A,B*) and therefore such structures are not limited to the amniotes. Their wide appearance in such diverse groups suggests, rather, a phenomenon of evolution wherein one pathway of natural selection probably favored the function of a simple protective keratinized digital cap. On another pathway, functional demands probably favored the gradual modification of the cap into an adaptable hook, a claw-like grasping device, or a hoof-like structure. In birds, the long established diet of a species also has a selective influence on the structure and function of its feeding equipment. The character of the beak and the shape and strength of the feet both determine and are determined by the food a given species may eat. Very often, the type of talon is correlated with the type of beak.

Based on current knowledge of development, I suggest an alternative hypothesis. The *generalized vertebrate claw* may be considered a hori-

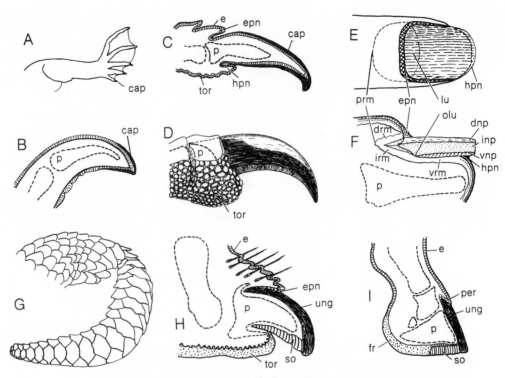

Fig. 5.16. Nails, claws, and hoofs in various vertebrates. *A*, claws on the inner three toes of the right limb of the South African clawed frog, *Xenopus laevis* (after Cochrane, 1961 [*Living amphibians of the world*. Garden City: Doubleday]). Note: similar claws found in the Congo genus *Hymenochirus* but not in *Pipa*. *B*, median section through terminal toe phalanx of *Siren lacertia* (from Maurer, 1913, after Gegenbaur). *C*, median section through terminal toe phalanx of embryonic crocodile (after Maurer, 1913). *D*, claw on first toe (hallux) of the hawk, *Buteo lineatus*. *E*, dorsal aspect of human fingernail. *F*. Sagittal section of human fingernail (after Jarrett and Spearman, 1966 [*Histochemistry of the Skin: Psoriasis*. Princeton: D. Van Nostrand]). *G*, body and tail scales of the giant pangolin, which develop similar to mammalian claws or nails rather than to scales in fishes or reptiles. *H*, claw in the forepaw of the dog. *I*, hoof in the horse. Note: phalanx or digit (*p*); horny cap (*cap*); epionychium (*epn*); hyponychium (*hpn*); dorsal nail plate (*dnp*); intermediate nail plate (*inp*); ventral nail plate (*vnp*); dorsal (upper) root matrix (*drm*); intermediate (lower) root matrix (*irm*); ventral (nail bed) root matrix (*vrm*); proximal limit of nail root matrix (*prm*); lunula (*lu*); outer margin of lunula (*olu*); dermis (*d*); epidermis (*e*); torus or digital pad (*tor*); frog (*fr*); sole (*so*); unguis (*ung*); periople (*per*).

zontally placed cone in which the ventral half has partially collapsed inward, whereas the *generalized nail* would be a completely flattened cone. Rather than postulating the complete loss of the ventral half of a thimble-like cap, one can more easily imagine that the germinative epidermis from the ventral edge of the cone (the presumptive nail bed) migrated dorsally and distally to, or around, the tip of the digit. Here, to varying degrees, it partially (as in a claw) or completely (as in a nail) approximates the lateral and dorsal edges of the cone (the presumptive lower and upper root matrixes).

An analogous reshuffling of the germinative epithelium may also be visualized if one compares the cone-shaped *enamel organ* of normal tooth development (fig. 5.3*D*) with the epidermal portion of the lens- or disk-shaped *scale pocket* of the embryonic fish (fig. 5.3*H*). In all cases, the most active cells are at the edges of the cone, whether it is fully opened (as in an enamel organ) or dorsoventrally collapsed into a disk (as in a scale pocket). The developmental homology with nails and claws is striking, although nothing is yet known about the role of the dermal mesenchyme in influencing nail and claw development.

**F. The Ossified Integumental Skeleton**

**1. Sclerification**

We use the generic term *sclerification* ( = induration) in reference to the processes of hardening that occur in many kinds of biological tissues, plant or animal. In its widest sense, the term includes both those processes that are not mediated by living cells, such as fossilization, and those that *are,* such as keratinization, chondrification, sclerotization, and all types of biological mineralization, especially calcification and ossification. In a narrower sense, sclerification refers to a continuous spectrum of processes that produce skeletal tissues ranging from uncalcified connective tissue fibers to bone.

All biological sclerifications are the result of the vital activity of a generalized cell type, the *scleroblast.* Vertebrate scleroblasts are usually referred to under a variety of locally cytodifferentiated forms—osteoblasts (bone-forming cells), chondroblasts (cartilage-forming cells), odontoblasts (dentin-forming cells), and so forth. Regardless of embryonic origin, the *first phase* of scleroblastic activity results in the formation of an *organic matrix* composed of complexes of fibroprotein and macromolecular amorphous "ground substance." The latter is usually composed of complex protein-polysaccharides or phospholipids.

The proportion of fibrous elements to ground substance within the matrix varies within and among the various phyla of animals. In general, matrix elaboration may occur intracellularly (as in keratinization), or extracellularly (as in ossification, chondrification, and sclerotization). In the latter, scleroblastic activity terminates after the first phase. In other types of sclerification, the process is biphasic, and the second phase is the *mineralization phase.* Where this occurs, inorganic crystals of a variety of mineral salts are formed on or within the mineralizable organic matrix. When the salt is either calcium carbonate or phosphate, the process is called *calcification.*

The process of mineralized tissue (bone) formation in vertebrates is termed *ossification.* The term refers to the formation of a specific calcium phosphate salt, *hydroxyapatite,* on or in an organic matrix containing the fibrous protein collagen. During the first phase of ossification, the scleroblast produces an extracellular organic matrix in which collagen makes up about 90 percent of the organic component, mucopolysaccharides composing the remainder. Early anatomists termed this material "osteoid." By a mechanism as yet little understood, the scleroblast simultaneously produces and arranges the steric molecular configuration of the matrix in a manner that enhances or facilitates localized primary nucleation of the hydroxyapatite crystals. Recent evidence suggests this is an intracellular event.

During the second phase, hydroxyapatite crystals form in this mineralizable matrix at certain specific sites, the *initial calcification loci,* that are visible only ultrastructurally. However, these loci grow radially into distinct spherulitic developmental units, termed *bone nodules,* that can be seen with ordinary light microscopy. The growing bone nodules eventually fuse to form the mature mineralized tissue. Initial calcification loci have been observed in developing membrane bone and fish scales, and also in dentin formation.

**2. Developmental Cell Behavior and Mineralization**

From a functional viewpoint, the various types of scleroblasts are influenced not only by the intracellular and extracellular microenvironments, but also by the function performed by the mature tissue. This is not to deny the role played by the genome in programming a scleroblast to secrete a mineralizable matrix, but rather to emphasize the fact that the behavior of the scleroblasts both during and after matrix formation is especially important. For example, in the formation of primary vascular bone (defined below), a maturing *osteocyte* (bone cell) secretes the organic matrix around itself and becomes trapped within it to form the typical cellular bone pattern (fig. 5.17). Even though surrounded by the matrix, however, a direct contact of the osteocyte with a nutritional source is provided through minute tubules, *canaliculi,* which anastomose with the vascular canal. Within these canaliculi are minute cytoplasmic projections of the osteocyte that are in direct contact with the mineralizing matrix and also directly or indirectly with the vascular canal. Thus, though trapped in their own matrix, mature osteocytes appear to deposit, demineralize, and remodel the bone as required by growth and normal function. It has been estimated that in the fully mineralized skeletal bone of a seventy-kilogram man, the crystal surfaces exposed to extracellular, extravascular fluid in the walls of canaliculi, lacunae, and Haversian canals is between 1,500 and 5,000 square meters. It has not been demonstrated, however, that all of this surface is necessarily available for mineral ion exchange.

During cartilage formation (chondrogenesis), chondrocytes (cartilage cells) become similarly entrapped in the matrix they secrete. Examine a thin section of epiphyseal cartilage and note that, as in bone formation, the chondrocytes are in approximate contact with a rich network of cartilage canals that contain capillary glomeruli. During *endochondral ossification* (defined below) of the epiphysis, the primary centers of ossification occur immediately adjacent to these richly vascularized areas of matrix. Some investigators now believe that the connective tissue of the cartilage canals not only is a source of chondroblasts, but may also be a source of osteoblasts, that is, as osteoprogenitor cells.

Another example of the importance of scleroblast behavior during matrix formation is seen in the development of dentin and enamel. Before tooth formation occurs, part of the oral epithelium differentiates in a band of proliferating cells, the *dental lamina,* which follows the shape of the jaw (both upper and lower). Part of each lamina becomes the inner side of the lip and also the gingiva (gums), whereas the basal portion gives rise to equally spaced epithelial invaginations into the mesenchyme known as *tooth buds* (fig. 5.6*B*). The base of each proliferating tooth bud itself becomes evaginated by a core of mesenchyme tissue that forms the dental papilla (fig. 5.3$D_4$) and the rearranged tooth bud now forms the *enamel organ* (fig. 5.3$D_7$).

During *dentin* formation (odontogenesis), the cells of the dental papilla responsible for producing the dentin are seldom trapped within their own matrix as are the osteocytes of cellular bone, or chondrocytes of cartilage. These *odontoblasts,* having been induced by the overlying enamel organ, "retreat" away from the newly deposited predentinal matrix but remain actively in contact with it by means of long cytoplasmic extensions that fill the characteristic dentinal tubule. Verify the tubular nature of dentin by examining a thin section of a developing tooth. As may be seen in a thin section of tooth from the bowfin (*Amia calva*) if it is available, rarely odontoblasts do not retreat but instead become entrapped in their own matrix. The resulting mature dentin is classified as "cellular dentin."

During *enamel* formation (amelogenesis) the cells of the basal layer of the enamel organ differentiate under the inductive influence of the subjacent dental papilla. These basal cells, now termed *ameloblasts,* differentiate in size, shape, and internal architecture and begin to secrete an organic preenamel matrix. As the ameloblasts retreat, the matrix is left behind, adjacent to the underlying predentinal matrix that has been secreted by the odontoblasts retreating in the opposite direction. In tooth formation, then, the epithelial and mesenchymal  scleroblasts retreat away from their newly deposited matrixes.

In teleost *scale* formation (lepidogenesis) a *dermal scale papilla* forms (fig. 5.3$H$) and the scleroblasts responsible for the mineralization of the *bony ridge layer* are found around the periphery of the dermal portions of the disk-shaped *scale pocket* (fig. 5.3$H_6$). These *lepidoblasts* are derived from and continuous with the scale papilla cells covering both the upper and lower surfaces of the forming scale. The peripherally situated lepidoblasts are thought by some authorities to be under the early inductive influence of certain basal cells in the overlying epidermis. However this may be, as the lepidoblasts lay down the matrix (osteoid) (fig. 5.3$H_5$) for the *bony ridge layer,* they actively retreat centrifugally, away from the *focus* (growth center) of the scale. Recall here the comment made above (p. 164) that whether it is a cone-shaped enamel organ or a disk-shaped scale pocket, the most active cells are found around the inner edges. No cytoplasmic processes are left behind by the lepidoblasts, and therefore the bony ridge layer is acellular. This is true at least in the classical two-layered *cycloid* and *ctenoid scale*s of most teleost fishes. A stratified, collagenous *fibrillary plate* (fig. 5.3$H_7$) forms below the bony ridge layer through the activity of cells split off from the underside of the scale papilla. In more mature scales, these cells are indistinguishable from normal fibroblasts and line the bottom of the scale (fig. 5.3$H_8$). In certain teleost species, the fibrillary plate is known to become secondarily ossified.

**3. Ossified Tissues**

Those connective tissues in which hydroxyapatite nucleates on or in a collagenous matrix are termed *ossified* (bony) *tissues.* In the earliest-known fossil agnathans, ossified tissues were restricted to the dermal plates of the integumental skeleton; that is, the bony carapace of the ostracoderms originated in the skin. The cartilaginous endoskeleton is presumed to have been uncalcified. In all vertebrates, ossified tissues arising in the dermis or in dermal capsules by the process of intramembranous ossification (see below) are termed *dermal bones*. From the earliest records of known vertebrate history, and even in many extant primitive fishes, dermal bones are found fused to and overlain by layers of dentin and enamel. This association has led to many misinterpretations and much speculation over the years. More recently, however, it is considered the strongest evidence that delamination phenomena and epithelial-mesenchymal interactions were operative during the earliest periods of vertebrate history.

During development, bones grow by a continuous, extensive, and complex process of building, remodeling, and rebuilding. In a developing long bone, for example, functional demands placed upon the bone by developing muscle attachments interact with complex growth processes to form a three-dimensional combination of bone deposition and resorption on appropriate outer (periosteal) and inner (endosteal) surfaces. The resulting long bone is a functional compromise between growth in *diameter* (peripheral and circumferential growth), growth in *length* (linear or longitudinal growth), and growth in *space* (positional or translative growth) relative to the soft tissues—muscles it serves and in which it is usually embedded.

Bony tissues form by either *intramembranous* or *endochondral ossifications*. In the former, *membrane bone* (and its variants) forms as a result of the modulation of connective tissue fibroblasts into functional osteoblasts. In the latter, a preformed cartilaginous model is replaced by *endochondral* (replacement) *bone*.

Virtually all bones of the typical vertebrate axial and appendicular skeleton, whether endochondral or dermal in origin, are arranged into an outer *cortical region,* usually possessing compact (dense) bone of various types, and an inner *medullary region* containing varying amounts of cancellous (spongy) bone. The student should recall that hair and feather shafts also have a dense outer cortex and a spongy inner medulla. Can you detect a common functional principle here? Depending upon the particular conditions of growth undergone by a given species, the cortical region of a given bone may be arranged into one or more irregularly stratified zones of variable thickness. Each zone may be composed of either lamellar or nonlamellar bone (defined below).

Histologically, we can define a *basic bone plan* of vertebrates as having three major components: a fibrous matrix; a nutritive supply; and osteocytes. In general, we refer to the tissues that have all three basic components as *cellular bone.* You should begin to examine specimens and prepared thin sections of all bone types.

The most widely encountered type of cellular bone in vertebrates is *primary vascular bone* (fig. 5.17*A*), in which the osteocytes are regularly spaced and arranged in a lamellar pattern (see below). The vascular canals that arise in the developing bone ordinarily are longitudinally oriented and are not surrounded by cylinders of secondarily formed concentric lamellae (osteons).

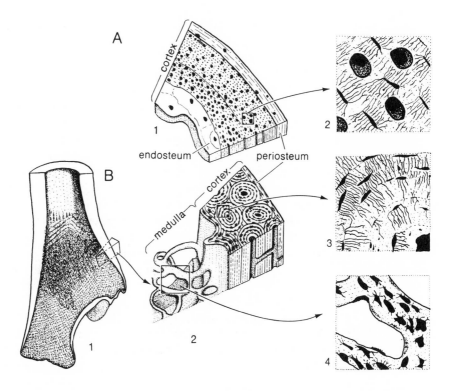

Fig. 5.17. A few examples of the many types of bone. *A*, primary vascular bone: 1. Thin wedge section cut from crocodile femur. Notice that the vascular canals in the cortex are not surrounded by concentric lamellae and that there are relatively few canals in the periosteum or endosteum. The bone in the endosteal portion of the cortex has probably been produced by the conversion of medullary trabeculae into compact bone during earlier growth of this long bone. 2. Magnified thin section of vascular canals and surrounding osteocytes. *B*, Haversian bone (secondary osteonal bone) and coarse cancellous bone: 1. Longitudinal section of the head portion of a mammalian femur. 2. Wedge section from cortical and medullary region of femur showing the relationship of compact to spongy bone. Notice in the cortex that cylinders of concentric lamellae surround the vascular canals to form osteons. 3. Magnified thin sections of portion of an osteon. 4. Magnified thin section of portion of the trabeculae in the coarse cancellous bone.

*Haversian bone* (secondary osteonal bone) is derived from primary vascular bone when, in the process of internal bone reconstruction, the primary canals undergo resorptive enlargement. These eroded resorption canals then become lined secondarily with cylinders of concentric lamellae, termed *osteons* or *Haversian systems* (fig. 5.17*B*). The diameter of a secondary osteonal canal is approximately the same as that of a primary vascular bone canal, but the latter is not lined with concentric lamellae.

There are many variations in the basic bone plan, the most frequent being the partial loss or complete absence of one of the basic components —either of the blood vessels, as in *nonvascular bone,* or of enclosed osteocytes, as in *acellular bone.* Other variations occur in the spatial pattern of the basic bone plan or in its origin rather than in any loss of components. The following bone types are examples, most of which can be seen in a cross section of a limb bone.

*Lamellar bone* involves the most typical pattern of cellular bone, wherein the collagen fibrils of the organic matrix are arranged into a seemingly orthogonal array. This results in the uniform stratification or layering effect of individual osteocytes (fig. 5.3*F,G*). This underlying pattern of stratification is even more visible in decalcified sections. Recall that the basal lamella of developing skin is similarly stratified (figs 5.1 and 5.4), as is also the uncalcified stratum compactum of adult skin.

*Nonlamellar bone* (also variously termed "woven," "fibrous," "membrane," and "coarse-bundled" bone) is a tissue in which the *random orientation* of the collagen fibrils produces a coarse, woven effect in the organic matrix. Although best seen with polarized light optics, this pattern can be distinguished from lamellar bone under ordinary light microscopy by the irregular arrangements and spacing of the osteocytes.

*Cancellous bone,* both coarse and fine, can be included under the category of nonlamellar bone. *Coarse cancellous bone* (fig. 5.17*C*) is usually in the medullary region of a bone. During periods of remodeling, this spongy bone often becomes incorporated into the compact bone of the cortex by the deposition of new bone on the irregular trabecular surfaces. This new endosteal bone has a distinctive, irregular, convoluted structure that differs markedly from the normally stratified appearance of the compact periosteal bone of the cortex.

*Fine cancellous bone* often is deposited in the cortical region, but the spaces in this bone type are larger than primary vascular canals and much smaller than the spaces in coarse cancellous bone of the medulla. Cancellous bone, because of its loose structure and enormous surface-to-volume ratio, not only is readily remodeled but also serves as an immediate source of mineral salts during periods of heavy mineral demand, as during pregnancy or egg-laying.

**4. Functional Bone Biology**

The presence of various bone patterns or types can give the functionally oriented anatomist an important insight into the conditions of growth that prevailed at the time the bone—fossil or recent—was being laid down. For example, lamellar bone typically forms during periods of slow skeletal growth, as during periosteal osteogenesis following either endochondral or intramembranous ossification, or in those parts of a bone that, relative to certain other fast-growing parts, are not deposited rapidly. Nonlamellar bone, however, typically forms in skeletal regions undergoing rapid development or where large amounts of bone are deposited in relatively short periods of time, for example, all membrane bones, developing bones of the mammalian fetus, and the postnatal skeletons of reptiles and mammals. Furthermore, bone that has been laid down in relatively short periods of time tends to contain greater numbers of primary vascular canals than bone that has been laid down slowly. In humans, with the exception of fracture calluses or bone tumors, most nonlamellar bone disappears by four years of age.

Early anatomists did not recognize that the structure of bone is related to the function it is to perform. Haversian bone, for example, was long regarded as the culmination of skeletal tissue evolution in vertebrates. Little or no consideration was given to the fact that it is rarely found where the effects of gravity are not as evident as in the terrestrial environment, as in fishes, unmetamorphosed amphibians, and certain reptiles. Furthermore, avian bones are hollow, and the long bones of cetaceans lack marrow cavities and trabecular bone. Haversian bone is unique only in that it is functionally suited to weight-bearing and the support of terrestrial vertebrates. It is not necessary in fishes. The functional approach makes bone a dynamic rather than a static tissue. Do you recall what happened to the bone of the early astronauts after several days of weightlessness in outer space? What happens to a paralyzed limb if the muscles are not artificially exercised?

Anatomists of the late nineteenth and early twentieth centuries considered that the primary tissues involved in skeletal function—fibrous connective tissue, cartilage, and bone—represented not only an increasingly complex histological and structural order, but also the precise order of their appearance in the phylogeny of bone. Thus cartilaginous fishes were thought to be ancestral to bony fishes; but this leads to difficulties when a modern representative of an early group appears with more cartilage than bone or, worse yet, with no bone. We are forced to postulate that such fishes have lost their bone through some undefined process of evolutionary "reduction." Without fossil evidence to the contrary, it might be simpler and more in keeping with known embryological phenomena to assume that the earlier members of the group had no structural or physiological requirement for bone and therefore did not produce it.

The traditional histological description of bone as "a mineralized fibrous connective tissue arranged mostly in thin plates with enclosed branching osteocytes" is actually limited to a vascularized, lamellar type of bone. More often than not, textbooks have featured human adult Haversian bone as an example of "true bone." Whereas Haversian bone is widespread in the *adult* human skeleton, nonlamellar and non-Haversian bone is present in the *fetal* skeleton. This oversight led many early investigators (and even some modern histologists) to define Haversian bone as a phylogenetic end point. As we have seen above, no such conclusion is justified.

The same early thinking also led to the coining of additional anthropomorphic epithets such as "true enamel" and "true dentin." The former, for example, was believed by some to be formed only by the ectoderm and limited to amniotes. Enamel formation in the "lower" vertebrates supposedly involved only the mesoderm. We know today that all vertebrate enamel is formed by ectodermal ameloblasts that have been induced by underlying mesenchymal odontoblasts. The early interpretations, though perhaps valid on the basis of then-current information, have survived much beyond their usefulness and have led to a restricted concept of mineralized tissue formation and allowed little latitude for broad comparative analyses.

### a. What Then Is Bone?

From the perspective given by our consideration of sclerification, our discussion of cell behavior, bony tissues, and of dental tissues (below), we can look upon the term *bone* in a very wide sense. Variations in bone structure are typical of all vertebrates. In fact, *bone* can be considered to represent *a spectrum of ossified tissue types* found throughout the vertebrates from the earliest fossil agnathans to the modern mammals. Convergence occurs in groups having only distant phylogenetic relationships. For example, cellular bone is found in amphibians, reptiles, birds, mammals, and certain of the earliest fossil fishes—the Osteostraci—whereas the Heterostraci had acellular bone. But there is no overall pattern evident between the classes of vertebrates and there are marked variations within each class. One can generally conclude that the presence or absence of a particular type or pattern of bone is determined for the most part by a dynamic interaction between growth processes and functional demands. Several investigators have independently confirmed the notion that the histology of vertebrate bone is generally more closely related to functional biomechanical demands than to phylogenetic level.

At present, the fishes form an interesting exception. As noted above, the early ostracoderms had either cellular or acellular bone. Among the modern teleosts, some phylogenetically "lower" groups, such as the cypriniform and salmoniform fishes and the clupeomorph fishes, possess cellular bone whereas most of the "higher" groups of fishes have acellular bone. A question arises, therefore, whether skeletal bone histology is in any way related to the evolution of the modern fishes or, for that matter, to the earliest agnathans living in different aquatic environments. It has been found for several fishes that no correlation exists between cellularity or acellularity and the freshwater or marine environment.

Another phenomenon, not yet well studied, occurs when bone that is actually cellular appears to be acellular. For example, in certain localized regions of turtle or crocodile bone, usually isolated or far from a vascular canal, osteocytes undergo a process of necrocytosis (cell death). The resulting bony tissue matrix appears clear and transparent in ground sections and is characterized by the absence of osteocyte nuclei in decalcified sections. Cell death also occurs when vascular canals are occluded by mineral deposits. Such closure results in the degeneration of surrounding osteocytes, and the canaliculi also frequently become filled with the mineral deposit. In extreme cases, the osteocyte lacunae also become filled and, in ground sections, the bone becomes totally transparent, appearing acellular to the noncritical eye.

The examples above indicate how important it is for the student of comparative anatomy not only to understand the physical relationships of existing bony tissues being compared, but also to understand the functional relationships and the developmental phenomena involved during growth and maturation.

**5. Dental Tissue: The Relationship of Teeth and Scales**

Vertebrate *dental tissues,* as here defined, include all nonkeratinous teeth and scales (the horny teeth of the cyclostomes are an exception to what follows). They are *composite tissues* consisting generally of varying combinations of an outer, ectodermally derived, acellular *enamel layer* ("ganoin" of early authors), a mesodermally derived middle layer of *dentin* ("cosmine" of early authors), and a fused inner (basal) layer of *dermal bone* (of one or more types including cementum, "isopedine," etc.). The composite nature of dental tissues strongly reflects the presence and operation of delamination phenomena and of epithelial-mesenchymal interactions during early development (fig. 5.3).

Our current knowledge of dental tissues and the relationship of teeth to scales has been derived from a long and interesting history that is often isolated in the literature from other integumental derivatives but that similarly involves the ectodermal or mesodermal origin of the various end products or layers. Despite a greater knowledge of the fossil record than was available one hundred years ago, many modern textbooks still incorrectly presume that shark teeth are the phylogenetic forerunners of all other vertebrate teeth (and scales). In ignorance of the fact that the oral epithelium has the same developmental potential for epithelial-mesenchymal interactions as the integumental epithelium, the traditional view presumed that shark teeth are simply placoid scales of the skin that somehow turned inward into the oral cavity and assumed another function. Old myths die hard, and it is instructive to examine some of the history of how this one began.

Until the mid-nineteenth century, most anatomists and naturalists held firmly to the Aristotelian belief that fish scales were homologous to feathers and hair in the higher vertebrates. Even the great Louis Agassiz, who classified all fishes into four classes on the basis of scale types—placoids, ganoids, cycloids, and ctenoids—believed that scales were horny or calcareous lamina secreted by the epidermal scale pocket. As we shall come to appreciate, Agassiz's view was at least partly correct. But since mature fish scales did not lie in the epidermis itself, Williamson and other mid-century naturalists scoffed at his idea that fish scales were epidermal products. Indeed, Williamson concluded that the shiny, prismatic layer of nonplacoid scales was not "true enamel," like that of a mammalian tooth, but an enamel-like substance for which he coined the term *ganoin*. Thus in part arose the traditional view that fish scales were entirely mesodermal rather than wholly or even partly ectodermal in origin.

The close similarity of fish scales to vertebrate teeth, however, was recognized very early in the nineteenth century. Then in 1849, Williamson introduced an idea, based on the histology of placoid scales, that all scale types could be phylogenetically derived from the placoid scales of elasmobranch fishes. His theory remained virtually unknown until its rediscovery and promulgation by Goodrich in 1907. Meanwhile, in 1874, Hertwig first demonstrated histologically the presence of what he called a "true enamel organ" (i.e., ectodermally derived) in the development of the placoid scale of a shark (fig. 5.3*E*). On that basis he proclaimed the homology of human teeth with shark denticles. With the imperfect knowledge of the geological record that existed at that time, the view that shark teeth were the phylogenetic forerunners of all other vertebrate teeth soon gained wide acceptance, an uncritical acceptance that has been carried into many modern textbooks.

Only recently has the enamel layer of fish teeth been demonstrated, by modern methods, to be of ectodermal rather than mesodermal origin (although we now know that the underlying mesenchymal papilla induces the overlying ectoderm), and the same is suspected for the surface layer of teleost scales. During the mid-1800s, before Hertwig's discovery, placoid scales were known to be covered with a hard, shiny transparent substance (now known to be enamel) variously described as "modified dentin" or a "combination of dentin and enamel." The underlying scale layer was correctly identified as dentin, but where a similar substance occurred in nonplacoid scales, it was given the name "cosmine." By 1907 Goodrich had recognized that in the cosmine-bearing scales of extinct crossopterygians and dipnoan fishes, the ganoin was thin and nonlamellar whereas in other fossil ganoid fishes it was thick and lamellar. Thereafter a new group, the cosmoid fishes, was recognized as having equal taxonomic status with the other major fish groups. The various scale types of these fishes, as first categorized by Agassiz, were now envisioned as forming a direct phylogenetic sequence—placoid → cosmoid → ganoid → cycloid → ctenoid—the major features of which were differences in growth and progressive loss of certain layers of the scale, especially the dentin and enamel (see fig. 5.3E, F, G, and H). Thus it was not long after that the combined Williamson-Goodrich view, along with that of Hertwig, became entrenched in the literature and has been, with few exceptions, uncritically accepted ever since. The view is partly summarized in table 5.2. It can be seen that, without further elaboration and study, the teleost scales do not "fit" very well into the schema because of the present uncertainty regarding the origin and nature of the various layers of typical cycloid and ctenoid scales (see below also).

*a. Scale Types*

Elasmobranch *placoid scales* (fig. 5.3E), and teeth in most vertebrates (fig. 5.3D, and table 5.2), are characterized by a cap of *dentin* that encloses a pulp cavity (vascularized) from which radiate numerous canaliculi. The outer surface of the dentin is covered by a layer of hard, transparent enamel. Below the dentin is a basal plate that possesses neither dentinal canaliculi nor enclosed osteocytes (see p. 174). Although its nature is still a matter of controversy, it is likely that the basal plate will prove to be simply a plate of acellular dermal bone, formed by intramembranous ossification after delamination from the basement-membrane complex. Fusion with the overlying dental layers likely occurs as development proceeds.

The *cosmoid scale* (fig. 5.3F) is characterized by a thickened dermal bone composed of a lowermost lamellar *bony layer* ("isopedine" of early authors) and a highly canalized *vascular layer* above it; a typical *dentin* ("cosmine") *layer;* and superficially, a thin *enamel* ("ganoin") *layer*. If available, the student should examine the scale from a coelocanth (*Latimeria*). Although not classified as cosmoid scales, the dermal "armor"—the carapace—of the early ostracoderms possesses variations of the same layers as found in the cosmoid scale.

A very much thickened, onion-layered surface layer of enamel ("ganoin") characterizes the *ganoid scale,* and no *dentin layer* is present. Two subtypes are recognized: the more primitive "palaeoniscoid" scale (fig.

**TABLE 5.2**
An Attempt to Synonomize the Various Layers of Vertebrate Teeth and Fish Scales with the Three Main Layers of the Generalized Vertebrate Dental Tissues

| Generalized Dental Tissue | Normal Vertebrate Tooth | Major Fish Scale Type | | | | | |
| --- | --- | --- | --- | --- | --- | --- | --- |
| | | Placoid | Cosmoid | Ganoid: Palaeoniscoid | Ganoid: Lepidosteoid | Cycloid | Ctenoid |
| *Enamel layer* | Enamel | Enamel | "Ganoin" | "Ganoin" | "Ganoin" | Ridges(?)[a] | Cteni(?)[a] |
|   Thick Lamellar | | | | | | | |
|   Thin | | b | b | b | b | b | b |
|   Nonlamellar | b | b | b | b | b | b | b |
| *Dentin layer* | Dentin | Dentin | "Cosmine" | — | — | Bony Ridge Layer(?) "Hyalodentin"(?) | "Hyalodentin"(?) |
|   Tubular | b | b | b | | | | |
|   Atubular | | | b | | | b | b |
| *Dermal bone* | Cementum | Basal plate | "Isopedine" | Bony layer | Bony layer | Fibrillary Plate(?) | Fibrillary Plate(?) |
|   Vascular Cellular | (Pulp cavity) | (Pulp cavity) | b | b | | | |
|   Acellular Lamellar | b | b | b | b | b | b | b |
|   Nonlamellar | b | b | | | | b | b |

[a] Situated in the posterior "exposed" field of the scale, overlain by epidermis.
[b] Noncalcified in most teleosts but bone nodules (Mandl bodies) present in some.

5.3*G*) in which the vascular layer of the dermal bone is present only as a horizontal network of canals; and the later "lepidosteoid" scale, in which both the vascular layer and the dentin are absent.

---

If available, the student should examine a scale of the reedfish (*Polypterus*) as an example of the more complex ganoid scale type, and examples of scales from sturgeon (*Acipenser*) or garfish (*Lepisosteus*) for the less complex ganoid type. If ground sections of this latter scale type are available, note that there are large centrally located canals that pierce the scale vertically through both the bony layer and the enamel layers. These resorption channels are often lined with bony lamellae ("isopedine") similar to that which constitutes the bony layer. Some smaller diameter, vertical and unlined tubules may be seen under high-power magnification. These also penetrate both layers. If your slide section is particularly thin, you might also notice some very minute, transverse fibers (uncalcified collagen) passing through the bony layer. These extrinsic fibers (Sharpey's fibers) serve to anchor the scale in the dermis just as they anchor some vertebrate teeth in the jaw.

---

In the *cycloid* and *ctenoid scales* (figs. 5.3*H,* and 5.18), the major portion of the scale is the lamellar, sometimes calcified, collagenous *fibrillary plate*. In thin section it resembles the lamellar bony layer of the ganoid and cosmoid scales, but it is acellular and generally not calcified. Superficial to the fibrillary plate is a thin, homogeneous surface layer of acellular calcified tissue known as the *bony ridge layer*. Its lack of cells and its transparent appearance in thin sections earned it the name "hyalodentin" from early anatomists who were not aware of the role of the lepidoblasts in the peripheral scale pockets (see p. 165). The bony ridge layer is embossed with concentric grooves and ridges called *circuli* (growth lines). Fishery workers use the arrangement of these growth lines to determine the age of the fish. Under very high magnification, the individual circulus is seen to be a row of fused miniature tooth-like structures called *microcteni*. The *radii* are erosion channels that radiate out from the *focus* (growth center) of the scale across and through the circuli. The radii develop early in the bony ridge layer, that is, after the first few circuli form, through osteoclastic activity. They give flexibility to the otherwise stiff scales during locomotion. The *enamel layer* is generally believed to be absent in cycloid and ctenoid scales.

The main morphological difference between cycloid and ctenoid scales, and perhaps an important clue to the riddle of scale layer homology, resides in the superficial ornamentation of the posterior (exposed) field of the scale, that portion that is covered by a thin layer of skin but not overlapped by other scales. Here the upper surface of the scale is in direct contact with the overlying epidermis that might influence its structural configuration. Perhaps significant in this regard is the fact that the grossly visible, enamel-covered surface sculpturing or ornamentation of all enamel-bearing recent and fossil fish scales (note the surface of the garfish) always occurs in the posterior field of the scale. The posterior field of cycloid scales is covered with low- or high-profile *ridges* (fig.

5.18), whereas that of a ctenoid scale generally possesses slender, tooth-like projections called *cteni*.

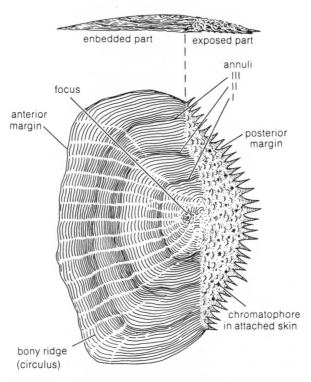

Fig. 5.18. Ctenoid scale of a fish. Semidiagrammatic view of a ctenoid scale of a fish, showing the exposed tooth-like structures (cteni) and the concentric annuli.

---

The student should obtain cycloid scales from a carp or goldfish and compare them with those from a bass or sunfish. There is some evidence that these ridges or cteni actually form a third layer (perhaps produced by the overlying epidermis rather than by the peripherally situated lepidoblasts) that covers the circuli of the posterior field in both cycloid and ctenoid scales. In decalcified scale sections of carp scales, for example, the ridge layer can be peeled off the posterior field and the underside of the peel bears a negative impression of the underlying microcteni on the circuli of the bony ridge layer. Such a layer, if demonstrated to be enamel in composition, would link cycloid and ctenoid scales with ganoid scales and thus favor the view that the bony ridge layer is truly a remnant of the cosmoid or ganoid bony layer. The homology of the layers of cycloid and ctenoid scales with those of earlier scale types is enigmatic.

---

Even though we have made an attempt to categorize scales into the traditional distinctive scale types, the student should be fully aware that there is no such thing as a "typical" scale. Among the teleost fishes, especially, scales come in all sizes and shapes. Compare cycloid and ctenoid scales from locally available species, and from the ichthyology lab or local aquarium shop. If whole specimens are available, also examine the tooth types and body shapes of the fishes above. Can you correlate body shape and scale pattern or tooth type with swimming and feeding habits?

### b. Tooth Types and Patterns

The teeth of vertebrates are an essential part of the feeding apparatus. Their appearance on the jaws of the early vertebrates allowed what was formerly a muscular filter pump and sucking device to also become a mechanism for grasping and manipulating food. The location of a functional grasping device at the oral end of the alimentary canal was accompanied and assisted by the anteriad location of the sense organs that aided in first detecting the presence of food, then exactly locating it. The evolutionary phenomenon of *cephalization* is undoubtedly associated with the appearance of teeth in the vertebrate mouth. As hinted above, the functionally oriented student can find a wide variety of tooth types and patterns in fishes. By examining the body shape and the tooth types and patterns in jaws and oral cavity of a given fish, you can make a reasonable guess about its swimming and feeding habits (see also chaps. 7, 8, 9, and 10). In the tetrapods, where the limbs can aid or completely take over the prey-grasping function, tooth types are not so variable as in fishes, but the pattern and individual shapes still give a good idea of the function. Specific types of teeth, tooth patterns, and tooth replacement patterns are discussed in chapter 8.

## 6. Cartilage

Cartilage is a relatively rigid skeletal tissue that serves a variety of supportive, protective, and locomotory functions. In vertebrates, but not in invertebrates, cartilage often becomes mineralized. Chemically, cartilage is especially characterized by an organic matrix high in collagen, acid polysaccharides, and water. Histologically, cartilage, like bone, occurs in a spectrum of tissue types. As in bone, the scleroblasts or *chondroblasts* that give rise to the organic matrix are subject to intracellular and extracellular influences. The nature of these microenvironmental influences is not fully understood, but, since various cartilage types occur in areas where specific biomechanical forces are present, it is assumed that the predominant extracellular forces affect the behavior of the developing chondrocytes. For example, *hyaline cartilage* is generally found on the articular surfaces of bones, especially where compressive forces predominate, as in the epiphyseal joint and the temporomandibular joint of the infant. *Elastic cartilage* is found in regions where tensile forces predominate, as in the outer ear and larynx. *Fibrocartilage,* sometimes considered an intermediate type of connective tissue because of its high percentage of collagen fibers, is found in areas where both shearing and compressive forces predominate, as in the intervertebral disks, the knee joint, and the juvenile and adult temporomandibular joint.

During the process of *endochondral ossification* (p. 166), cartilage of course forms the model that is eventually replaced by bone. Because it thus generally precedes bone formation during the embryogenesis of vertebrates, in modern times cartilage has been considered an embryonic adaptation of vertebrates that evolved as an ephemeral scaffolding for the construction of the bony skeleton. Furthermore, it has also been considered a tissue unique to vertebrates. The question which is older, bone or cartilage, has been argued from several viewpoints. The presence of bone in the earliest ostracoderm armor weighed heavily for those arguing for the primacy of bone. But the fact that cartilage precedes most endoskeletal bone embryology was used in favor of the primacy of cartilage. Recently the presence of cartilage in many invertebrates has been reconfirmed. The remarkable resemblance of cartilage tissues and plant cell walls has also been pointed out.

**G. Integumental Color**

**1. Introduction**

Color seems to be present everywhere in the vertebrate integument. In addition to being found in the skin, eyes, lining of cavities, and investing membranes of various organs, pigments are found in the scales, fin rays, mouth, meninges, and eggs of fishes; in the mouth, claws, and meninges of amphibians; in the crests, beaks, claws, and dermal scutes of reptiles; in the feathers, beaks, combs, wattles, leg scales, and eggshells of birds; and in the hairs, scales, claws, and nails of mammals.

Color and color changes in all animals are complex phenomena involving physical, chemical, and functional considerations. *Structural (physical) colors* are related to the physical nature, especially the size and shape, of specific surfaces of cells or cellular organelles from which light waves are either reflected or scattered. Light from the sun is a mixture of many colors that we perceive as white or uncolored. The perception of a specific color is produced by subtracting unwanted components from white light. Structural color changes result from differences in the angle from which the object is seen, as with *iridescent colors,* or by the nature of the object from which light waves are scattered, as in *Tyndall effects.* More specifically, structural colors appear because certain components of white light are diverted out of the observer's line of sight.

*Pigmentary (chemical) colors,* on the other hand, are the result of selective light absorption by chemical compounds we call *pigments.* Pigment molecules are so constructed that certain colors are absorbed while others are reflected. They are contained in a variety of cell types generically known as chromatophores (Gr. *chromatos:* color; *-phoreus:* bearer). However, under certain conditions, such as blushing and albinism, the blood pigments play a role in vertebrate coloration.

Pigmentary color changes may be *morphological* (quantitative), that is, those that depend upon a change either in total content of pigment within the chromatophore and associated cells or in the total number of chromatophores within a given unit area; or *physiological* (qualitative), that is, those that depend on rapid changes in the distribution of pigment granules within the chromatophores. Pigmentary colors depend, therefore, on a whole host of directly or indirectly interdependent factors, some of which will be elaborated upon below. Structural color changes, though complex, will be treated in less detail.

**2. Structural Colors and Color Changes**

*Iridescent colors* are the result of a phenomenon known as *interference* that occurs when a light beam passes from a material with one refractive index to another. Some portions of the incident light beam are bent out of the line of sight and others are reflected. The reflected light varies with the angle of incidence, and as the angle of incidence increases from the perpendicular the color shifts from the longer wave lengths (reds) to the shorter wave lengths (greens and blues). The organic materials that possess the different refractive indexes vary, but in general they are found in the shape of submicroscopic intracellular organelles that act as inhomogeneous double films with different refractive properties.

The visualization of iridescent colors depends not only on these cellular organelles, which reflect light waves, but also upon a narrowly defined set of positions involving the object viewed, the observer, and the source of light. The understanding of these interactions is an exercise in three-dimensional geometry. A most lucid explanation of this complex phenomenon is contained in the classic monograph on hummingbirds by Greenewalt (1960).

Among the vertebrates, iridescent colors are most prominent in birds. Specific examples can readily be seen in the feathers of crows, pigeons, starlings, peacocks, hummingbirds, and others. In fishes, amphibia, and reptiles, specialized chromatophores are found in the dermis. These *iridophores* contain ultrastructural reflecting platelets that efficiently reflect incident light and cause interference phenomena. Rarely, as in dove eyes, have these been found in birds. These platelets are crystal-like bodies composed of one or more of the purine pigments—adenine, guanine, hypoxanthine, or even uric acid. Guanine-bearing iridophores are especially prominent in fish larvae and in adults are closely associated with the eyes, scales, and gut membranes. In most cases iridophores act in association with adjacent chromatophores to produce a range of colors (see p. 179 below).

*Scattering colors* are the result of selective deflection or scattering of the shorter (blue) wave lengths of sunlight by submicroscopic particles while the longer wave lengths pass through virtually unchanged. Blue sky is the most common example in nature. Among the vertebrates, blue color produced by scattering can be seen most readily in birds: in the head and neck skin of the turkey and the guinea fowl; from air-filled cavities within the feather barbs of the kingfisher, blue jay, blue tit, bluebird, and indigo bunting. The brilliant blues of some of the tropical marine fishes of the wrasse family (Labridae) are also produced by scattering phenomena, while in other labrids a blue pigment is responsible. A combination of blues produced by scattering and yellows from pigments results in the green coloration of tropical birds such as parrots and parakeets, and also of many reptiles, amphibians, and fishes. A bright green pigment (turacoverdin) found in the feathers of the African turaco (Musophagidae) is reputed to be the only true green pigment known in birds.

3. Pigmentary
Colors and
Color Changes

Pigmentary colors depend upon a whole series of interrelated physiological and morphological factors. On the basis of the pigments they possess, chromatophores generally have been classified as *dark colored,* for example, melanocytes and melanophores, or *bright colored,* for example, xanthophores (= lipophores), which produce yellows; erythrophores, which produce reds; and iridophores (= guanophores), which are light-reflecting. A summary of vertebrate chromatophore types is given in table 5.3.

TABLE 5.3
A Summary of Vertebrate Chromatophores

| Chromatophore Type | Principal Pigments | General Appearance | Principal Organelle |
|---|---|---|---|
| Dark colored | (Occur in all vertebrates) | | |
| Melanophore | Melanin | | Melanosome |
| | Eumelanin, | Black, brown | |
| | Phaeomelanin | Red, orange, yellow | |
| Bright colored | (Occur in all poikilotherms except cyclostomes) | | |
| Iridophore | Purines | Silvery | Reflecting |
| | Adenine, | | platelet |
| | guanine, | | |
| | hypoxanthine | | |
| Xanthophore | Pteridines | Yellow, | Pterinosome; |
| | and | orange, | carotenoid |
| | carotenoids | and red | vesicle |
| Erythrophore | Pteridines | Yellow | Pterinosome; |
| | and | orange, | carotenoid |
| | carotenoids | and red | vesicle |

Xanthophores and erythrophores utilize pteridines and carotenoids as their fundamental pigments. Pteridines are found in cellular organelles, called *pterinosomes,* whereas the carotenoids are found in smaller, ultra-structural vesicles distributed uniformly between the pterinosomes. Less is currently understood about these bright-colored chromatophores than about the dark-colored ones.

The major contributions to our understanding of the dark-colored chromatophores have centered on the process of melanogenesis and the cells it occurs in. There are two principal types of the pigment *melanin:* the eumelanins, which are black or dark brown; and the less well known phaeomelanins, which are red, orange, or yellow. The chemistry of melanogenesis is similar throughout the vertebrate subphylum: the copper-containing enzyme tyrosinase initiates the conversion of the amino acid tyrosine into the polymeric pigment melanin that in turn is associated with a discrete cytoplasmic organelle, the *melanosome.* The process of melanogenesis, briefly, is as follows: in *melanoblasts* (melanocyte precursors derived from the embryonic neural crest), a developing complex lattice of cross-linked proteinaceous fibers, the *premelanosome,* forms a matrix upon which tyrosinase acts on tyrosine to synthesize melanin. Once melanin deposition is complete, the premelanosomal lattice is obscured—becomes electron dense—and tyrosinase activity is absent or minimal. The mature organelle in which melanization is complete is the melanosome. Melanosomes are approximately 0.5 $\mu$ in diameter and they have been found in all the vertebrate melanocytes (see below) thus far examined.

The usage of the terms *melanocyte* and *melanophore* has been associated with a wide degree of confusion depending upon whether the authority happened to have worked on homeotherms, poikilothermic vertebrates, or invertebrates.

The term *melanocyte,* for example, traditionally has been reserved for the spindle-shaped, dendritic, melanin-bearing cells in the epidermis of warm-blooded vertebrates, especially mammals. In addition to their shape, these cells have been characterized by the transport of melanosomes to the associated keratinocytes (Malpighian cells). Conversely, they have not been known to rapidly disperse or aggregate melanosomes intracellularly in response to hormonal stimuli. Recent evidence, however, indicates that some dispersed pigment does occur as a response to hormonal administration in the epidermal melanocytes of man.

In contrast, the term *melanophore* has been traditionally restricted to a melanin-bearing cell found in cold-blooded vertebrates or in invertebrates. A melanophore could previously be defined functionally as a melanocyte that, by intracellular displacement of melanosomes, participated in the phenomenon of rapid color change (with or without other chromatophores). Recently, however, "melanophore" has been designated as the specific term for the *black* pigment-bearing cells in all animals.

Throughout this chapter, I have consistently used the suffix -blast to indicate the undifferentiated embryological precursor to a particular mature cell type identified by the suffix -cyte; for example, keratinoblast—keratinocyte; ameloblast—amelocyte. The terms are embryologically and functionally consistent. Similar logic requires that we should define any cell, vertebrate or invertebrate, that synthesizes melanosomes as melano*cyte*. A melano*phore* is by definition a type of melanocyte. More precisely, if not redundantly, a melanophore is a melanin-bearing cell. To specifically designate the melanophore as the *black* pigment-bearing cell has the advantage of consistently identifying pigment cells by their color. Hence, we have melanophores, xanthophores, and erythrophores. However, it must be recognized that there are exceptions; some epidermal melanophores synthesize red or yellow (phaeomelanin) rather than black (eumelanin) pigments; and some melanophores, such as those that produce the agouti coat color in mouse hair, are capable of synthesizing both kinds of melanin, given the appropriate stimuli.

Keeping these precautions firmly in mind, the present text will follow the terminology just described. The former homeothermic melanocyte, characterized by its epidermal location and its cytocrine activity in hair, feathers, or other epidermal derivatives, is now known to occur in all vertebrates and is now designated as the *epidermal melanophore*. The former poikilothermic melanophore, characterized by its dermal location and by the intracellular migration of its melanosomes, is now designated the *dermal melanophore*. Other melanophores, similarly derived from the neural crest but situated in various organs or on nerves or blood vessels, have not been well studied or specially named. The term *melanocyte,* whenever it appears in this chapter, will broadly refer to any mature cell that synthesizes melanosomes.

### a. Morphological Color Changes

Morphological color changes involve slow changes in the total *quantity* of pigment present or in the number of chromatophores in a given unit area of skin. The basic problem of how the quantity of melanosomes increases is mostly unknown. Understanding hinges on a greater knowledge of the factors involved in the control of tyrosinase activity, which is related to the degree of melanization and the size, age, and physiological condition of the animal. There does not appear to be any relationship between phylogenetic status of a given vertebrate group and its level of tyrosinase activity. The only generalization, of limited validity, is that tyrosinase activity levels are generally higher in the dorsal integument than in the ventral. The anurans, however, form a notable exception in that they show a higher tyrosinase activity in their pale ventral integument than in their darker dorsal integument. The highest levels of enzyme activity have been recorded overall in the amphibians, while the few other vertebrates thus far studied have lower but remarkably similar levels of skin tyrosinase.

Generally, adaptation to background provides the cue to slow color changes: more melanin is developed in animals maintained on dark backgrounds; melanin is lost in animals maintained on light backgrounds. The student should compare the epidermal pigmentation of laboratory frogs that have been exposed to prolonged storage on a dark background with newly received or naturally captured frogs.

In many fishes and amphibians, changes in other chromatophores also affect the overall quantity of pigment. For example, the guanine content in the iridophores of one species of light-adapted fish was four times greater than in dark-adapted controls. Increases in guanine content or iridophores has also been shown to be accompanied by a complementary decrease in melanin content, and vice versa.

In mammals, the intensity of skin pigmentation depends mostly on the color and size of melanosomes and on their numbers and distribution within both the epidermal melanophores and the keratinocytes. Some authors think that these associated epidermal cells acquire melanosomes actively by phagocytosing portions of the melanophore dendritic processes, whereas others believe that the melanosome discharge is the result of cytocrine activity of the melanophore itself. This is yet another example (see p. 112) of the structural and functional integration of cells that transcends that of individual cells. The total amount of pigment borne by any particular keratinocyte depends upon the total time it is in contact with the dendrites of one or more epidermal melanophores. The intensity of pigmentation is dependent not only on the turnover rate of the keratinocytes but also on the rate of melanogenesis within the melanophore. Both rates are hormonally controlled, thus linking morphological color changes to physiological color changes.

### b. Physiological Color Changes

These changes involve relatively rapid movement—seconds to hours —of the pigments within chromatophores, for example, of melanosomes within the melanophores. Pigment mobility is not due to amoeboid activity as was once thought, that is, to expansion or contraction of the chromatophore, but rather to the centrifugal or centripetal movement of cytoplasmic pigment organelles within a relatively stable cell configuration. Pigments widely scattered in the cytoplasm are said to be *dispersed;* those gathered perinuclearly into the central body of the cell are said to be *aggregated.* The precise mechanism(s) that control pigment movement are not fully understood, but the antecedent mechanisms that initiate these movements have been subjected to much investigation. Chromatophores are known to be affected neuronally, by neurohumoral agents, by hormones, and even directly by the environment.

*Neuronal control,* present in many teleosts, is a function of the sympathetic portion of the autonomic nervous system (see chap. 13). The transmitter substance, still unidentified, is adrenergic in function and causes aggregation of melanosomes. Some investigators believe that it may be norepinephrine. The pattern of peripheral innervation of melanophores has not been well studied either, except in the fin rays of fishes, where it has been found that some motor axons, while extending peripherally to control distally located melanophores through normal nerve endings or by neuromelanophoric junctions, also pass over and have presynaptic contact with more proximally located melanophores. Transmitter substances released by these axons appear to cause melanosomal aggregation.

The evidence for the existence of a reciprocal neuronal control of melanophore activity in fishes is controversial. Many investigators believe there is no such system, while others hold that the induction of rapid melanosome dispersal caused by some drugs indicates the presence and active role of parasympathetic involvement.

There is no direct evidence for neuronal control of chromatophores in amphibians. In reptiles, neuronal control of color change has only been inconclusively studied in two or three species of chameleons. *Neurohumoral control* is known in some teleosts and in the horned lizards. In birds and mammals, color control (where it exists) is either structural or hormonal. That only fishes have been shown to have direct nervous regulation of chromatophores may reflect the adaptive value of immediate accommodation to the variety of backgrounds fishes encounter in the aquatic environment.

Most studies on color control in vertebrates have centered on melanophores and on hormonal effects in melanosomal aggregation or dispersal or both. The main hypotheses of hormonal control of color change currently focus on the role of mammalian pituitary MSH (melanophore-stimulating hormone). MSH or MSH-like molecules can cause the dispersal of melanosomes and other pigments in all groups of vertebrates. The structures of the melanin-dispersing hormone in cyclostomes, elasmobranchs, teleosts, and amphibians are similar to that of mammalian MSH. In fishes and amphibians, the hormone is known as *intermedin,* a secretion of the *pars intermedia.*

Although the details of specific steps are not thoroughly understood, the overall phenomenon of vertebrate pigment cell regulation can be summarized basically in the context of a generalized "first messenger–second messenger" model. An environmental cue received by the animal elicits a response of the pars intermedia. MSH (the first messenger), secreted by the pituitary, attaches to a receptor site on the chromatophoral plasma membrane. This receptor is believed to mediate the MSH stimulation and control the intracellular concentration of cyclic AMP (the second messenger), a cyclic nucleotide: adenosine 3,5-monophosphate. Adrenergic receptors, also situated on or in the plasma membrane, respond to other "first messengers" and also control the level of cyclic AMP. MSH stimulates the synthesis of cyclic AMP within chromatophores, as does the so-called *beta* adrenergic receptor. An intracellular increase in cyclic AMP results in a dispersion of melanosomes (within a melanophore) and an aggregation of reflecting platelets (within an iridophore). This complementarity of function between melanophores and iridophores is the basis of the dermal chromatophore unit described below.

The actual mechanism(s) by which pigments or pigmentary organelles move within the cytoplasm are unknown. Recent ultrastructural evidence, however, indicates the presence of microtubules within the melanophores of certain fishes, amphibians, and reptiles and has led to the suggestion of a possible role for microfilaments in melanosome migration.

## 4. Mechanisms of Color Change

Contrary to traditional considerations, it has become increasingly clear that chromatophores do not function independently of other integumental cells. Pigmentary color is the result not only of the physiological and morphological factors described above, but also of integrated activities and interaction of several integumental cell types and of the genetic and environmental factors that influence these component cells.

For example, in amphibians the aggregation of reflecting platelets in iridophores is accompanied by dispersal of melanosomes in melanophores. There appears to be a similar reciprocal relationship between the effects of intermedin on frog iridophores and xanthophores. During intermedin stimulation, it has been demonstrated that the purine content of reflecting platelets is decreased, whereas the pteridine level of xanthophores increases. From such experiments it has been concluded that these components are mediated by the same part of the intermedin molecule; that in each case both chromatophores possess the same receptor site for intermedin.

When a frog is caused to darken by the administration of intermedin, the melanosomes disperse from their perinuclear positions and come to occupy positions in the distally directed dendritic processes of the melanophore. These processes surround the iridophores and interdigitate between the xanthophores. Thus the dispersed melanosomes almost completely obscure the reflecting platelets within the iridophores that have aggregated in response to the intermedin. Although quantitative changes occur in its pteridine content, the role of the xanthophore is essentially passive; that is, before intermedin treatment, the xanthophore acts as a yellow filter for the iridophore, whereas after the melanosomes have obscured the iridophore, the filtering effect of the xanthophore is negated. This association of iridophores and xanthophores with the underlying melanophores has been called the *dermal chromatophore unit*.

Functionally, this combination of a filtering layer, a scattering layer, and an absorbing layer is the primary vehicle for physiological color change among the poikilothermic vertebrates. The following explanation gives a specific example of the functional integration not only of three chromatophore types but also of the role of structural color in determining the final color we visualize. In the anole lizard, short waves of light, reflected off reflecting platelets within iridophores, pass through overlying yellow xanthophores. A green skin color results. When melanosomes of the underlying melanophores become dispersed into the dendritic processes, however, light can no longer be reflected from the reflecting platelets of the now-covered iridophores. A brown color results.

Notwithstanding the interactions demonstrated above, the intensity of the overhead lighting and also the temperature affect the color response directly. For example, high temperatures cause skin darkening in lizards, whereas low temperatures cause skin lightening. Temperature has been demonstrated to affect the ability of melanophores to react to intermedin stimulation, the response being much reduced at low temperatures.

Pigmentation in mammals and in bird feathers (see below) is the result of another integrated interaction—between epidermal melanophores, which synthesize melanosomes, and the epidermal keratinocytes, which acquire their pigment secondarily. An *epidermal melanin unit* has been defined as an epidermal melanophore and the group of keratinocytes with which it maintains functional contact. Originally described in homeotherms, similar units have now been described in reptiles and amphibians and can be considered to be present in all vertebrates. This functional unit may be considered comparable to the series of components that together form the nephron of the kidney. The mechanisms by which genetic and environmental factors influence the component cells of the epidermal melanin unit are little known for most vertebrates.

In the normal melanogenesis of bird skin, melanophores are situated in the epidermal collar of the feather germ. As feather barbs are formed, melanin is transferred from the melanophore either by inoculation into or engulfment by the barb cell. During the course of development of this epidermal melanin unit, feather melanophores, which are highly sensitive to environmental change, leave a permanent record of their responses to such influences in the mature feather pattern. Thus, subtle variations in the structure of feather keratin and in the kind and distribution of pigments result in different feather color patterns. The most obvious color changes in birds occur during the molts, whether from juvenile to adult plumage, seasonal reproductive, or cyclical replacement of old feathers.

Some birds also obtain their color through the incorporation of fat-soluble carotenoids and water-soluble xanthophylls from their diet, as in the flamingos, the roseate spoonbill, and the scarlet ibis. A hen fed a carotenoid-free diet produces nearly colorless yolk in her eggs. Canaries fed a carotenoid-free diet eventually become white-feathered; a reversal to yellow occurs if they are fed on xanthophylls.

Certain fishes, in addition to their normal melanogenic activity, can incorporate dietary carotenoids into their skin coloration. The eel blennies of two Pacific Coast genera (*Anoplarchus* and *Xiphister*), for example, mimic the color of the kelp in which they live by ingesting carotenous invertebrate prey species that live on the kelp.

The seasonal changes in coat color of some mammals occurs in gradual response to changing photoperiods. Initially detected through the eyes, such stimuli elicit specific hormonal responses within the hypophysis, apparently mediated via the hypothalamus. In the varying hare, for example, seasonal molts are correlated with seasonal changes in gonadotropin output, whereas color changes are thought to be related to seasonal changes in corticotropin and MSH. In summer the fur of the entire dorsum is brown (agouti) whereas in winter it appears white. In reality the winter underfur remains brown and only the distal ends of the projecting guard hairs are white. This seems to be the result of the delay in differentiation of guard hair follicular melanophores during the fall hair growth cycle. The lack of melanin synthesis thus results in the apical apigmentation during the winter. MSH has also been implicated in the normal regulation of color in mammalian pelage cycles.

Other hormones have an indirect but distinct effect upon human skin pigmentation and that of other mammals. A marked variation in the sensitivity of melanophores to specific steroid hormones has been demonstrated. For example, in the pregnant human and the pregnant guinea pig, there is evidence for an increase in the number of melanogenic melanophores in response to circulating progesterone and estrogen (and MSH?). This results in an increased pigmentation of: the nipples and areolae; the facial skin (in lesser amounts); the midline of the anterior abdominal wall; and the genitalia. In male mammals, testosterone may also have a specific role in determining certain localized pigmentation. The precise mode of action of specific hormones within the epidermal melanin unit is still obscure. MSH, of course, appears to cause a dispersion of melanosomes and increase in melanogenic activity (morphological color change), and also an increased transport of melanosomes within keratinocytes.

In keratinocytes, melanosomes appear to be localized within lyosome-like vesicles. In human skin, melanosomes within such vesicles are found to occur in groups, as in caucasoids and mongoloids, or singly, as in negroids. Such ultrastructural differences in melanosome distribution may partly account for observed racial differences in skin color.

## 5. Adaptive Value of Color Change

The ability to change color, whether immediately, gradually, or seasonally, is of positive functional and ecological significance. Birds and mammals may change color seasonally, usually in association with reproductive activities. Fishes and reptiles undergo profound, rapid color changes during courtship and reproductive displays. The student should observe the mating displays of a pair of Siamese fighting fish if they are available. Fishes especially, but also some amphibians and reptiles, have the ability to change color almost immediately depending upon the background on which they find themselves. On a pale background, iridophore-reflecting platelets disperse as melanophoral melanosomes aggregate: the animal lightens. On a dark background, iridophore platelets aggregate as melanosomes disperse: the animal darkens. That these reciprocal activities should both occur in response to a single stimulus (intermedin) is a further reflection of the parsimonious manner in which natural selection operates.

The mere possession of chromatophores per se is of considerable adaptive value in preventing potentially damaging effects of solar energy. Sunlight, for example, can kill developing fish embryos and unpigmented cave animals. In many diurnal vertebrates, dark pigments occur around the central nervous system, the gonads, and the internal organs. These are not found in closely related nocturnal forms. Some species of diurnal lizards, such as the desert iguana (*Dipsosaurus dorsalis*), have a heavily pigmented peritoneum that acts as a shield, essential and sufficient to protect organs within the body cavity from solar ultraviolet irradiation. In lieu of black peritoneums, some diurnal lizards have intense fixed concentrations of melanin pigments in their body walls that exclude as much or more incoming light from the body cavity. These examples represent only two alternative evolutionary solutions to the problem of blocking harmful solar radiation.

Reptiles also use color changes for thermoregulation. As a reptile such as the desert iguana warms into the range of temperature necessary for its normal activity, it changes color from the dark phase to the pale. As the body pales, that is, as melanosomes aggregate, light reflectivity from the reflecting platelets of the iridophore is increased and the aggregation of melanosomes in the dermal chromatophore unit allows a greater penetration of light waves that are potentially mutagenic to the central nervous system, the gonads, and other internal organs. Nearly twice as much visible and ultraviolet light penetrates the body wall in the light phase as in the dark phase. As noted above, this is when the black peritoneum becomes a very functional shield.

As in other vertebrates, color in mammals serves a variety of functions. The rump patches (specula) of ruminants, such as the white-tailed deer, alert other members of the species to possible danger. Some colors serve as a warning to other species, as in the skunk. Less conspicuous color marks often subtly express the moods of members of the species to each other, as in the white ear spots of tigers, or the white eyelids of some monkeys. Color patterns or single colors confer protection from predators. Coat colors may be disruptive, as in deer fawns, or concealing, as in young seals. Fawns' irregular spots distract attention from the outline of the animal, often by blending with the shadow pattern of the habitat. Seals born on ice floes generally have white natal fur, whereas those species born on beaches generally have dark brown or black natal fur.

For the most part, mammalian coloration is dull, the colors varying singly or in combinations of orange, brown, gray, black, and white. Few mammals, with the exception of the squirrels and certain primates, such as the mandrills, are brightly colored. The drab coloration of most mammals is correlated with the fact that most species have no or few cones in their retinas and consequently are color-blind or nearly so. Man, on the other hand, has excellent color vision, but his skin coloration is relatively drab.

The pigmented epidermis of man is adapted to latitudinal variation in ultraviolet radiation. This adaptation is correlated with the synthesis of vitamin D (calciferol) in the stratum granulosum. In tropical latitudes, the heavily pigmented upper epidermis of a dark-skinned man screens out more than two-thirds of the incoming ultraviolet radiation and thus prevents overproduction of vitamin D in the underlying layers. In northern latitudes, such as north Germany, England, and Scandinavia, natural selection apparently has favored lightly pigmented skin that maximizes the amount of ultraviolet light penetration, especially in the long winters when the sun angle is low.

Ultraviolet light causes the conversion of 7-dehydrocholesterol into vitamin D, which, along with parathyroid hormone and thyrocalcitonin, is responsible for controlling the level of calcium and phosphorous in the blood. Overproduction of vitamin D may lead to the deposit of calcium salts in the arteries or in the kidneys. Underproduction results in *rickets,* a disease of bone erosion and deformation. Rickets is not necessarily a dietary deficiency disease, although it can be prevented by vitamin supplements, but is caused by a deficiency of sunshine. It was not considered a major disease until the advent of the industrial revolution in England and northern Europe. A geographical relationship was found between the incidence of rickets and northern European cities where industrial smog was heavy. The introduction of smoke pollution to areas where ultraviolet light exposure was already marginal, especially during the winter months, created ultraviolet deficiency and vitamin D thus could not be synthesized by the skin. Any northern European infant born in the fall or winter months was thereby predisposed to rickets because of the lack of sufficient ultraviolet light during early growth. The widespread use of vitamin D supplements has now largely avoided this problem.

The epidermis of northern man, especially, is also adapted to seasonal variations in ultraviolet light. Suntanning, a summertime response of the epidermis to ultraviolet radiation, results in increased pigmentation and keratinization, a shield that prevents the overproduction of vitamin D. Perhaps the seasonal coat color changes of some northern mammals are also correlated with the necessity of increasing exposure to ultraviolet light in the wintertime and decreasing it in summer in order to regulate vitamin D production.

Mammals, birds, reptiles, and amphibians can synthesize vitamin D in certain areas of the body receptive to ultraviolet, such as the skin of man, the ears of rabbits, the feet of birds. Fish, on the other hand, are shielded from ultraviolet radiation by their watery environment, but they can synthesize vitamin D enzymatically and their body oils are rich in it. Arctic mammals, such as seals, polar bears, and Eskimos, living year-round in ultraviolet-deficient latitudes, obtain their vitamin D from a staple diet of fish. Natural selection against rickets in such northern populations appears also to have been toward the bearing of offspring in the springtime. Infants born at this time are exposed to a maximum amount of sunshine during their early growth.

**H. Summary of Part 2**

1. The epidermal surface of the integument of fishes and unmetamorphosed amphibians is metabolically active and covered by a "mucous cuticle." Secretion of an external, acellular mucopolysaccharide coating

appears to be a characterisitc of all metabolically active epithelia exposed to aquatic media. In adult amphibians, reptiles, birds, and mammals, the epidermal surface layers are partly or fully keratinized.

2. The vertebrate integument possesses a spectrum of gland types, both unicellular and multicellular. The greatest variety of skin glands are found in fishes, the least in birds. Among the vertebrates, more attention has been given to gland cell type than to function.

3. The traditional viewpoint, that only with the acquisition of the terrestrial environment did the vertebrate skin begin to express its vast potential of variation (as now seen in a variety of tetrapod structures) is challenged by modern evidence. The process of keratinization, though most obvious in the amniotes, is not limited to them, nor is it associated exclusively with dry environments.

4. The presence of keratinization is of itself no guarantee of the waterproofing function so often attributed to it. Rather, keratinization is but one pathway, selected and modified for many functions, from an entire spectrum of potential integumentary sclerification processes available to the early, preterrestrial vertebrates.

5. The fate of mature keratinocytes, as they undergo the processes of shedding, sloughing, molting, and retention, is reexamined in context of the functional epithelial extinction hypothesis. Rather than being considered all inclusive or mutually exclusive, these processes are better considered parallel, long-term solutions to similar functional demands.

6. In certain fishes the existence of epidermal pits, which possess all the general attributes of hair and feather follicles and which produce generations of horny epidermal tubercles, suggests that "true follicles" are not limited to birds and mammals and that the phenomenon of epidermal generations being sloughed as a unit can no longer be held as exclusive to squamate reptiles.

7. Nails, hoofs, and claws have not been well studied. An early hypothesis suggested that claws originated as thimble-like caps in the crocodilian reptiles and were later modified in the mammals. The presence of both claws and thimble-like caps in amphibians, however, suggests an alternate evolutionary hypothesis wherein one pathway of natural selection favored the function of a simple protective digital cap, whereas on other probable pathways functional demands favored the gradual modification of the cap into an adaptable hook, a claw-like grasping device, a hoof-like structure, or a flattened protective device. Although little is known about the role of dermal mesenchyme in influencing nail and claw development, there is a striking developmental similarity between claws, nails, teeth, and fish scales.

8. Sclerification is a generic term referring to all the processes of hardening that occur in many kinds of biological tissues, plant or animal. Biological sclerifications are the result of the activity of scleroblasts, the various types of which are defined by the type of tissue they produce. Scleroblasts are influenced not only by the intracellular and extracellu-

lar microenvironments, but also by the function to be performed by the mature tissue.

9. In vertebrates, the process of mineralized tissue (bone) formation is termed ossification. It involves a biphasic, scleroblastic production of a collagen-containing organic matrix and the formation of hydroxyapatite salts on or in that organic matrix. The behavior of the scleroblast both during and after matrix formation is especially important.

10. In all vertebrates, membrane bones arising in the dermis or in dermal capsules are termed dermal bones. In the fossil record of fishes, the existence of dermal bones, fused to and overlain by layers of dentin and enamel, is the strongest evidence that delamination phenomena and epithelial-mesenchymal interactions were operative during the earliest periods of vertebrate history.

11. The histologic structure of bone is generally more related to the function it is to perform than to its phylogenetic level. Bone is a dynamic rather than a static tissue. The identification of various bone patterns or types offers an insight into the condition that prevailed during the growth of a given bone. Haversian bone, for example, is not a phylogenetic end point as was previously believed. Rather, it is unique only in that it is functionally suited to weight-bearing and the support of terrestrial vertebrates. Such a bone type is not needed in fishes.

12. Variations in bone structure are typical of all vertebrates. Bone represents a spectrum of ossified tissue types found throughout the vertebrates from the earliest fossil agnathans to the modern mammals. The term "true bone" is as much an anthropomorphic misnomer as the terms "true dentin" and "true enamel."

13. Vertebrate dental tissues are composite tissues of varying combinations of an outer ectodermally derived, acellular enamel layer, a mesodermally derived middle layer of dentin, and a fused inner layer of dermal bone. This composite nature strongly reflects the presence and operation of common developmental themes.

14. Cartilage, long considered a vertebrate embryonic adaptation that evolved as an ephemeral scaffolding for the construction of the bony skeleton (through endochondral ossification), is also found in many invertebrates. It can no longer be considered unique to vertebrates.

15. Color and color changes in animal integuments are complex phenomena involving physical, chemical, and functional considerations. Structural (physical) colors involve the size, shape, and surface of specific cellular organelles from which light rays are either reflected or scattered. Pigmentary (chemical) colors are the result of selective light absorption by pigments that occur in chromatophores.

16. The ability to change color, whether immediately, gradually, or seasonally, is of positive functional and ecological significance. In addition to the adaptive value of various pigment cell strategies, the possession of chromatophores is itself of considerable value in preventing potentially damaging effects of solar radiation.

17. Pigmentary color is the result not only of morphological and physiological factors, but also of the integrated activities and interaction of several integumental cell types and the genetic and environmental factors that influence them. The dermal chromatophore unit, for example, is a functional combination of a filtering layer, a scattering layer, and an absorbing layer of cells. This combination of cell types is the primary vehicle for physiological color change among the poikilothermic vertebrates. The epidermal melanin unit is composed of an epidermal melanophore, which synthesizes melanosomes, and a group of keratinocytes, which acquire the melanosomes secondarily. This functional interaction of two cell types produces the integumental pigmentation pattern in vertebrates.

**References**

American Society of Zoologists. 1972. The vertebrate integument. *Amer. Zool.* 12:13–171. (Twelve articles from a symposium sponsored by the American Society of Zoologists, 27, 28 December 1971, at the A.A.A.S. meeting, Philadelphia, Pa.).

Bagnara, J. T., and Hadley, M. E. 1973. Chromatophores and color change. In *The comparative physiology of animal pigmentation.* Englewood Cliffs, N.J.: Prentice-Hall.

Bagnara, J. T.; Taylor, J. D.; and Hadley, M. E. 1968. The dermal chromatophore unit. *J. Cell Biol.* 38:67–79.

Biedermann, W. 1926. Vergleichende Physiologie des Integuments der Wirbeltiere. I. Die Histophysiologie der typischen Hautgewebe. *Ergebn. Biol.* 1:1–342.

Fitzpatrick, T. B., and Breathnach, A. S. 1963. Das epidermale Melanin-Einheit-System. *Derm. Wochschr.* 147:481–89.

Fleischmajer, R., and Billingham, R. E., eds. 1968. *Epithelial-mesenchymal interactions.* Baltimore: Williams and Wilkins.

Goodrich, E. S. 1904. On the dermal fin-rays of fishes—living and extinct. *Quart. J. Microsc. Sci.* 47:465–522.

———. 1907. On the scales of fish, living and extinct, and their importance in classification. *Proc. Zool. Soc. Lond.* 1907:751–74.

Greenewalt, C. H. 1960. *Hummingbirds.* Garden City, N.Y.: Doubleday (published for American Museum Natural History).

Grinberg, M. M. 1961. On the ontogenesis of the scales of bony fish. (In Russian). *Zool. J.* 40:234–42. (The included literature, especially that of B. S. Matveev, 1932–45, is especially important to the understanding of fish-scale development.)

Hall, Brian K. 1975. Evolutionary consequences of skeletal differentiation. *Amer. Zool.* 15:329–50.

Hertwig, O. 1874. Ueber Bau und Entwickelung der Placoidschuppen und der Zahne der Selachier. *Jena Zeitschr. Naturwiss.* 8:331–404.

Holmgren, N. 1940. Studies on the head in fishes: Embryological, morphological, and phylogenetical researches. Part I. Development of the skull in sharks and rays. *Acta Zool., Stockholm* 21:51–267.

Hyman, L. H. 1942. *Comparative vertebrate anatomy.* Chicago: University of Chicago Press.

Jarvik, E. 1959. Dermal fin-rays and Holmgren's principle of delamination. *Kungl. Sven. Vet. Akad. Handl.,* ser. 4, 6:1–51.

Kollar, E. J. 1972. The development of the integument. Spatial, temporal, and phylogenetic factors. *Amer. Zool.* 12:125–36.

Krejsa, R. J. 1970. Functional epithelial extinction: A synthetic view of epidermal shedding and keratinization in animal integuments. *Amer. Zool.* 10:322 (abstr.).

Ling, J. K. 1970. Pelage and molting in wild mammals with special reference to aquatic forms. *Quart. Rev. Biol.* 45:16–54.

Montagna, W., and Parakkal, P. F. 1974. *The structure and function of the skin.* 3d ed. New York: Academic Press.

Moss, M. L. 1965. Studies on the acellular bone of teleost fish. *Acta Anat.* 60:262–76.

———. 1968. Comparative anatomy of vertebrate dermal bone and teeth. I. The epidermal co-participation hypothesis. *Acta Anat.* 71: 178–208.

Noble, G. K. 1931. *The biology of Amphibia.* Reprinted 1954 by Dover Publications.

Orvig, T. 1967. Phylogeny of tooth tissues: Evolution of some calcified tissues in early vertebrates. In *Structure and chemical organization of teeth,* ed. A. E. W. Miles, 1:45–110. New York: Academic Press.

Person, P. and Philpott, D. E. 1969. The nature and significance of invertebrate cartilages. *Biol. Rev.* 44:1–16.

Sengel, P. 1976. *Morphogenesis of skin.* London: Cambridge University Press. (Covers the important experiments and theories on skin development and cites significant papers through 1973.)

Spearman, R. I. C. 1973. *The integument: A textbook of skin biology.* London: Cambridge University Press.

———. ed. 1977. *Comparative biology of skin.* Symposia of the Zoological Society of London, no. 39. (An interdisciplinary approach to the comparative skin biology of vertebrates and invertebrates.)

Williamson, W. C. 1849. On the microscopic structure of the scales and dermal teeth of some ganoid and placoid fish. *Phil. Trans. Roy. Soc. Lond.* 139:435–76.

Wolpert, L. 1972. Positional information and pattern formation. In *Current topics in developmental biology,* ed. A. A. Moscona and A. Monroy, 6:183–224. New York: Academic Press.

# 6 The Endoskeleton:
## The Comparative Anatomy of the Vertebral Column and Ribs
*David B. Wake*

**A. General Considerations on the Endoskeleton**

**1. The Endoskeleton and Its Parts**

The endoskeleton is the internal supporting system of hardened material characteristic of vertebrates. The endoskeleton of vertebrates consists of cartilage, bone, or mixtures of these two types of tissue, and is mainly mesodermal in origin. The parts of the endoskeleton of vertebrates are: the *skull,* in the head; the *visceral skeleton,* composed of *gill arches* supporting the gills, and their derivatives; the *vertebral column,* occupying the median dorsal region; the *ribs,* projecting from the vertebrae one pair to each vertebra primitively; the *sternum,* occupying the median ventral region of the anterior part of the trunk; the *pectoral girdle,* supporting the anterior paired appendages; the *pelvic girdle,* supporting the posterior paired appendages; and the *skeleton* of the *appendages.* The four parts first named constitute the *axial* skeleton, and other parts constitute the *appendicular* skeleton.

**2. The Notochord**

The *notochord,* or *chorda dorsalis,* is a stiffened rod extending longitudinally in the middorsal region of the chordate body, just beneath the central nervous system. It probably originated as a support for an elongated body of an organism that swam by lateral undulations. In larval vertebrates it is a resilient rod that restores the body to its elongated resting position after flexion. It is formed embryologically from the roof of the archenteron or its equivalent. In tunicates the notochord forms a supporting rod for the "tadpole" larva. In those species that metamorphose, the tail and the notochord are resorbed. This occurs quickly, either through the action of a contractile tail epidermis, in some species, or by the contraction of notochordal cells themselves in other species. In the latter case, the notochord ruptures anteriorly and its contents enter the body cavity of the trunk. In *Branchiostoma* the notochord constitutes the axial support of the body. Recently, early reports that the *Branchiostoma* notochord contains muscle fibers have been confirmed, and it appears that a unique muscular mechanism permits the organ to vary its stiffness. Embryos of all vertebrates have a notochord, and in them it extends anteriorly to the level of the hypophysis. In *Branchiostoma,* however, it extends virtually to the anterior tip of the body. In most vertebrates the vertebral column replaces the notochord as the main axial support during later stages of development. The vertebral column forms around the notochord and gradually squeezes it out of existence or reduces it to remnants that are often restricted to the areas between vertebrae. In some groups the notochord may persist throughout life, however, and may play an important functional role in the adult.

The notochord develops early in vertebrates and at first consists of tightly packed cells that resemble a pile of coins. The cells soon enlarge and become vacuolated. A fully developed notochord consists of enlarged notochordal cells and a relatively complex *notochordal sheath* (fig. 6.1). The superficial cells of the notochord have their nuclei aligned

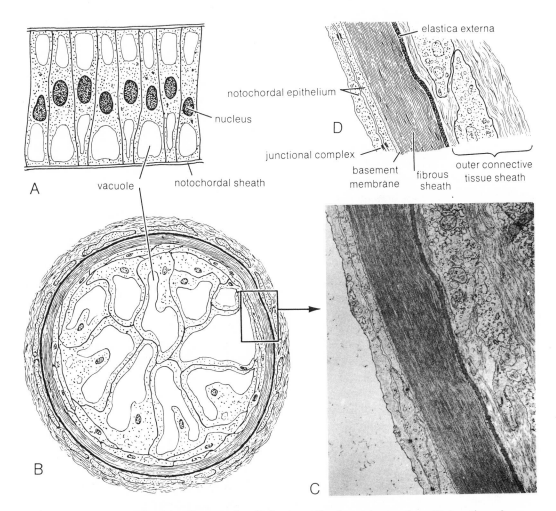

Fig. 6.1. The notochord of a generalized vertebrate. *A*, sagittal section of a developmental stage, at which the cells are layered like a pile of coins. *B*, cross section through the notochord of a tadpole of the African clawed frog, *Xenopus laevis*. *C*, electron micrograph illustrating detail of a portion of the notochordal sheath region, courtesy Richard M. Cloney, University of Washington. *D*, diagrammatic representation of portion of *C*.

near the surface and are often termed the *notochordal epithelium*. All notochordal cells are joined by junctional complexes characteristic of epithelia. The first layer of the sheath is the *basement membrane* of the notochordal epithelium. Surrounding this is a very thick *fibrous sheath* composed of collagen fibers. The next layer is the much thinner *elastica externa*, composed of elastin fibers. Finally, an *outer connective tissue sheath* of alternating strata of flattened fibroblasts and collagen fibers joins the notochord to surrounding tissues. In some accounts the fibrous sheath alone is termed the notochordal sheath, and in early treatments the entire sheath is said to be composed of an *elastica interna* and an elastica externa. These features, based on light microscopy, are not easily reconciled with the above account based on electron microscopy. Recently an elastica interna, lying between the basement membrane and the fibrous sheath, has been found in the distal part of the tail of frog tadpoles.

3. The Skeleto-
genous Regions

The endoskeleton develops from mesenchyme. The mesenchyme for the vertebral column and ribs comes chiefly from the sclerotomes that are formed by the breakdown of the medial sides of the epimeres (somites) into mesenchyme; but contributions from other mesodermal sources also occur. The mesenchyme accumulates around the notochord and neural tube and in certain other *skeletogenous regions*. The arrangement of the latter is largely dependent on the disposition of the myotomes. The myotomes, or muscle segments, which are those portions of the epimeres remaining, grow down between the skin and the digestive tract to form the muscular layer of the body wall. Each myotome is separated from adjacent ones by a transverse partition of mesenchyme, the *myoseptum* or *myocomma*. Each myotome (except in cyclostomes) is further divided into a dorsal (*epaxial*) and a ventral (*hypaxial*) half by a horizontal partition, the *horizontal skeletogenous septum,* which extends from the notochord to the level of the lateral line on the sides of the body. The mesenchyme surrounding the notochord (*perichordal* mesenchyme) and neural tube continues to the median dorsal line as the *dorsal skeletogenous septum,* and similarly from the notochord to the median ventral line (in the tail) as the ventral skeletogenous septum. In the trunk region the *ventral skeletogenous septum* is naturally split into two *ventrolateral* septa which become the outer wall of the coelom (fig. 6.2). The horizontal, dorsal, and ventral septa are continuous longitudinal septa, extending the length of the body. As their name implies, the skeletogenous septa are regions of skeleton formation. *At the inter-*

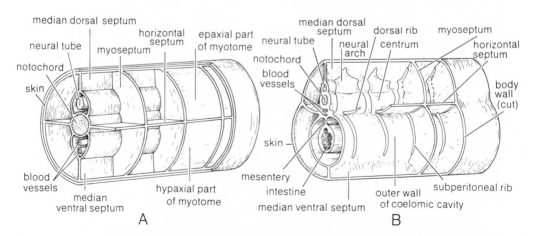

Fig. 6.2. Diagrams to show the skeleton-forming septa in *A*, the tail region, and *B*, the trunk region of a generalized vertebrate.

*section of every myoseptum with the dorsal, ventral, and horizontal septa and with the perichordal mesenchyme, a vertebra arises.* As the myosepta are segmentally repeated, because of the primary segmentation of the myotomes, it follows that the vertebrae are also *segmentally repeated* and that the vertebrae *alternate with the myotomes.*

**B. Anatomy and Embryonic Origin of Vertebrae and Ribs**

**1. Parts of a Typical Vertebra**

The axis of the vertebrate skeleton is the *vertebral column* or backbone, composed of a longitudinal series of bones, the *vertebrae.* A typical vertebra consists of a central cylindrical mass, the *body* or *centrum,* that surrounds and replaces or incorporates the notochord, a dorsal *neural arch* enclosing the spinal cord, and a ventral *hemal arch,* enclosing blood vessels (fig. 6.3A). Neural and hemal arches are commonly prolonged dorsally and ventrally, respectively, into *neural* and *hemal* spines. In the trunk region the hemal arch is typically missing. A vertebra also com-

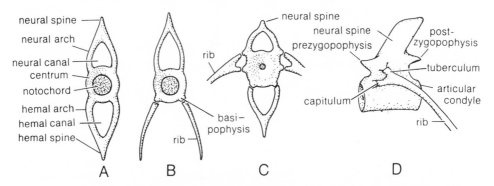

Fig. 6.3. Diagrams of generalized vertebrae. *A,* cross section of tail vertebra of a teleost fish. *B,* cross section of trunk vertebra of the same, showing opening of the hemal arch to form the basapophyses. *C,* cross section of trunk vertebra of a terrestrial vertebrate, showing attachments of ribs. *D,* lateral view of a trunk vertebra of a salamander, showing characteristic processes and attachment of ribs.

monly has a variety of projecting processes, termed *apophyses,* serving for articulation with adjoining vertebrae or with ribs or for muscle attachment. The most common apophyses are:

*a) Zygapophyses,* articulations between successive vertebrae; they are divisible into *prezygapophyses,* anterior projections, and *postzygapophyses,* posterior projections, of the basal region of the neural arch (fig. 6.3).

*b) Basapophyses,* also called hemapophyses or basal stumps, are a pair of ventral projections of the centrum, which in some cases represent the remains of the hemal arch and may serve for rib attachment (fig. 6.3).

*c) Diapophyses,* lateral projections of the vertebra for the attachment of the upper head of two-headed ribs (fig. 6.3).

*d) Parapophyses,* lateral projections of the vertebra for the attachment of the lower head of two-headed ribs (fig. 6.3).

*e) Pleurapophyses,* lateral projections representing the rib attachments of the vertebra plus the fused rib.

*f) Hypapophyses,* midventral projections from the centrum.

The expression *transverse process* for any lateral projections of vertebrae is useful, although it must be understood that it has no exact morphological meaning. Often transverse processes and pleurapophyses are synonymous, but diapophyses alone may also be termed transverse processes. In some groups the diapophysis and parapophysis may fuse to form a *synapophysis;* this composite structure is also often called a transverse process. On the tail vertebrae, transverse processes are formed that cannot be positively identified according to the more specific terminology above.

## 2. Shapes of Centra

The vertebral column is formed and functions as an axial support by the end-to-end placement of the centra of the vertebrae. The shape of the ends of the centra is thus important in the mechanics of the vertebral column (fig. 6.4); generally applied terms include:

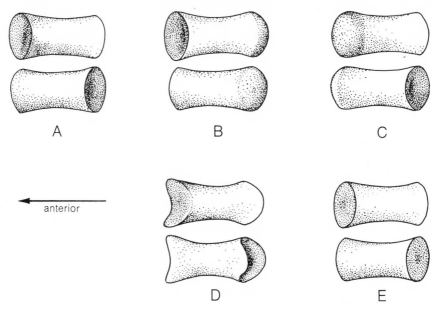

Fig. 6.4. Diagrams of the anterior (*top*) and posterior (*bottom*) ends of the centra of vertebrates, to show the varied articular shapes. *A*, amphicoelous. *B*, procoelous. *C*, opisthocoelous. *D*, heterocoelous. *E*, amphiplatyan.

a) *Amphicoelous*, both ends concave.

b) *Procoelous*, anterior end concave, posterior end convex.

c) *Opisthocoelous*, anterior end convex, posterior end concave.

d) *Heterocoelous*, both ends shaped like the seat of a saddle placed transversely at one end and vertically at the other.

e) *Amphiplatyan*, both ends flat.

These terms are strictly anatomical and relate solely to the bone, without reference to the extremely important fibrous and cartilaginous parts that play important functional roles in joints. Other shapes occur infrequently.

Between the ends of the centra are found the remains of the notochord, intervertebral disks of fibrocartilage or, in some groups, masses of hyaline and fibrous intervertebral cartilage.

3. Development of the Vertebrae

The vertebrae arise from cells that migrate medially, surround the notochord and neural tube, and spread along the skeletogenous septa as mesenchyme. These cells produce a more-or-less continuous *perichordal tube,* whose thickness varies along its length and from group to group. Typically the perichordal tube bears enlargements, *perichordal rings,* that surround the notochord near the midpoint of each segment. Cartilage usually appears in the mesenchyme of the skeletogenous regions. This cartilage may persist, often stiffened by mineralization, or may undergo partial or complete ossification. First appearance of the cartilaginous masses is in intersegmental locations, in the intermyotomal recesses that lie vertically along the notochord and neural tube (fig. 6.5). In

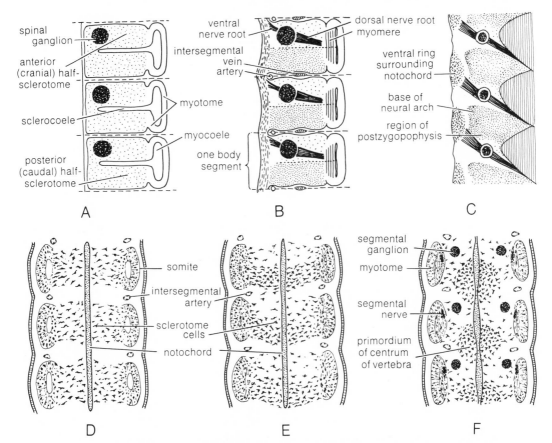

Fig. 6.5. Development of the vertebral column. *A*, highly diagrammatic frontal section through right half of the prevertebral stage, to show the parts of the somites and their relation to the notochord and ganglia. *B*, later stage, to show that the caudal half-sclerotome (sclerotomite) is denser than the cranial half-sclerotome. *C*, start of skeletogenesis, to show that the base of the neural arch forms in close proximity to the skeletogenous ring surrounding the notochord that will become the centrum. *D*, developmental stage similar to *A*, but not so diagrammatic. *E*, developmental stage similar to *B*, to show the general appearance in a less diagrammatic manner. *F*, developmental stage similar to *C*, but not so diagrammatic. *A, B, C*, modified after Remane (1936). *D, E, F*, modified after Patten 1958 (*Foundations of embryology.* New York: McGraw-Hill).

most vertebrates direct ossification of the sclerotomal mesenchyme also occurs, and the vertebra is compounded of parts preformed in cartilage and parts that ossify directly. The vertebral column thus forms around the notochord, which is enclosed inside the centra, where it persists in the more primitive vertebrates. In more advanced vertebrates the notochord is gradually reduced and replaced by the developing centra and persists only between vertebrae as highly modified remnants.

The vertebrae are derived from somites, but in order to function they have an out-of-phase metamerism, relative to the musculature derived from the myotomes. In other words, muscles must run from vertebra to vertebra in order to bend the vertebral column. To accomplish this, development of the major parts of the vertebrae takes place at myotomal borders. This is intersegmental or transsegmental development. Cells from a given sclerotome concentrate at sites in the vicinity of segmental borders. Areas of low cell concentration in the anterior part of a segment (generally termed the *cranial sclerotomite,* fig. 6.5*B,E*) are thought by many workers to contribute to the posterior parts of one vertebra, and the dense column of cells in the posterior part of each segment (termed the *caudal sclerotomite*) contributes in a major way to the anterior and central parts of the next vertebra. Each vertebra therefore seemingly incorporates parts of two adjacent sclerotomes. In amniotes the sclerotome is cell-rich, and distinct sclerotomal units, divided into distinct cranial and caudal sclerotomites, are evident. The caudal sclerotomite is especially evident as a prominent condensation. These two cell masses are usually said to undergo a reorganization to form the vertebrae, but recently this has been questioned, for it seems that very little, if anything, is contributed by the cranial sclerotome (fig. 6.5*C,F*). In fishes and most amphibians the sclerotome is generally scanty and is not clearly divided into parts. Yet the vertebrae form at myotomal borders, and cells from adjacent sclerotomes might contribute to the adult vertebra even though this has not been directly observed. In all vertebrates the original sclerotomal segmentation pattern of the early embryos is lost with the formation of the continuous perichordal tube. The adult vertebrae form from the perichordal tube in transsegmental positions; a resegmentation can be said to occur if one focuses on the fact that segmental muscles run between vertebrae. However, Baur (1969) and Verbout (1974) argued that vertebrae form at the caudal end of each segment and that no reorganization is required. According to this view, vertebrae arise from the beginning at their definitive sites, and there is no shifting of segment parts or regrouping of segment halves. Most illustrations in textbooks are in error in showing blocks of tissue that separate, migrate, and regroup to form definitive vertebrae, but it is still too early to go so far as to deny that vertebrae are formed from components of adjacent segments. This remains an area of controversy.

4. Vertebral
Components

In primitive fishes the vertebral column consists of separate parts that arise independently. These parts are attached to each other in various ways, but individual vertebrae may be difficult to identify. In the past, numerous attempts have been made to explain the development and evolution of the vertebrae of more derived groups in terms of fusions, enlargements, reductions, and elimination of these separate elements. Notable among these attempts are those of Gadow, whose work had great influence for many years. His views have been increasingly criticized, and recently his ideas concerning tetrapods have specifically been challenged. Components are recognizable in all vertebrates, but their homologies are not at all clear, except in a most general way. These basic components are:

### a. Arch Components

Definite paired cartilages (called *arcualia* by Gadow) appear in primitive fishes and give rise to adult vertebral parts. Typically there are four pairs to each vertebral unit (fig. 6.6). The two largest pairs of elements arise from tissue of the caudal part of each sclerotome. The dorsal pair (called *"basidorsals"* by Gadow) produce the neural arch, and in the tail region the ventral pair (*"basiventrals"*) give rise to the hemal arch. Together these arches become the cranial and dominant parts of the adult vertebra. Usually their basal portions rest against the notochordal sheath. Other cartilages may form above the neural arch and below the hemal arch, and these contribute to the formation of neural and hemal spines. Although originally separate entities, the paired cartilaginous elements grow around the neural tube and caudal blood vessels and fuse to form the arches of the adult vertebrae. In addition to these dominant elements, intercalary elements may be present between successive dorsal and ventral arches (fig. 6.6). In some instances both dorsal and ventral intercalary cartilages (termed *"interdorsals"* and *"interventrals"* by Gadow) are present in the posterior parts of the vertebra. These seem to originate from the cranial parts of sclerotomes. In some groups (elas-

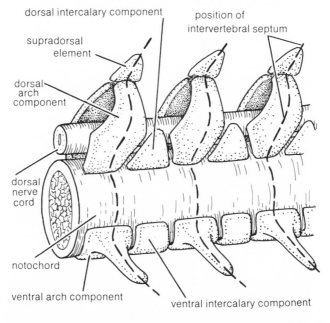

Fig. 6.6. Generalized vertebral column of a primitive fish, viewed laterally, to show major components.

mobranchs) the intercalaries are rather large and may form complete arches. The dorsal intercalary arch may persist as a large, complete arch in the adult, and it often bears the foramen for the dorsal root of the spinal nerves. Ventral intercalaries are less well developed. Dorsal and ventral arches may fuse with the intercalaries in various patterns in different groups of primitive fishes. In teleosts, cartilaginous intercalaries seem to be absent, and neural and hemal arches are the sole arch components. Some attempts have been made to identify embryonic mesenchymatous masses between neural arch rudiments in several teleosts with dorsal intercalaries. These masses give rise to the interarch articulations (zygapophyses), and their homology with dorsal intercalaries is dubious. Only two pairs of arches are ever present in tetrapods, and these are directly comparable with the neural and hemal arches of fishes.

### b. Centrum Components

Three major kinds of developmental "centers" may contribute to the body, or centrum, of the adult fish vertebra (fig. 6.7). All three may be

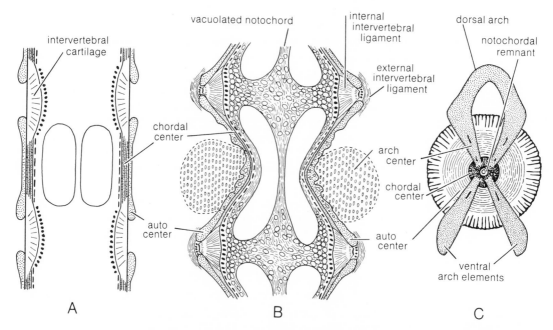

Fig. 6.7. Components of fish centra. *A*, early developmental stage of centrum, showing major components (after Francois 1966), frontal section. *B*, later developmental condition (after Francois 1966). *C*, cross section of later developmental stage, showing interpretation of three developmental centers.

observed in many fish groups. The notochordal sheath itself first undergoes differentiation, with vertebral parts forming *chordal centers* of mineralized, lacunar tissue. Intervertebral parts thicken but remain unmineralized and give rise to parts of the intervertebral ligaments. The *auto center* forms from mesenchyme surrounding the notochord. This tissue produces a series of bony lamellae that are continuous with those covering the cartilaginous arch elements. As a result of this process, the bases of these cartilages are incorporated into the centrum, forming *arch centers*. The three different centers in *Salmo* have three distinct kinds of ossification patterns. The chordal centers are formed of compact acellular bone, or osteoid tissue. The auto centers are the result of perichondrial, membranous ossification. The arch centers, late in development, may undergo endochondral ossification. In other species they may remain cartilaginous.

The most primitive fishes (agnathans, sturgeons) lack distinctive centers and may be termed *acentrous*. Sturgeons have well-developed cartilaginous arches, but dorsal and ventral arches fail to join, except anteriorly, and no perichondrial tissue develops.

Various conditions prevail in chondrichthyans. Many have centra with the same components as in *Salmo*, but some lack auto centers. Some holocephalans also lack arch centers, but these do have chordal centers.

*Polypterus* has vertebrae similar to those of *Salmo* in terms of centrum components, and holosteans have predominant arch centers. Many teleosts are like *Salmo*, in which the auto centers are responsible for most of the centrum, but with large arch centers as well. In various groups of derived teleosts the arch centers may be very reduced or even absent. The chordal centers may be of minor significance and become reduced in size and importance as the large auto center thickens, squeezing the notochord.

Centrum components are essentially absent in lungfishes and *Latimeria*, in which the notochord is essentially unconstricted. What little skeletal tissue is found in the "centrum" of these groups seems to be formed from perichordal mesenchyme, but the cartilaginous arch elements are expanded where they come into contact with the notochord.

All tetrapods have centra that form almost exclusively from perichordal mesenchyme. In most groups the centrum is first laid down in cartilage, which is later replaced by bone, but in some amphibians most of the bony centrum forms directly from mesenchyme. This material, whether preformed in cartilage or not, can be considered an auto center. The bases of the neural and hemal arches may contribute to the adult centrum in amphibians (arch centers).

5. The Ribs

Each vertebra is theoretically provided with a pair of *ribs,* which articulate to various projections of the centrum and extend out into the body wall. Ribs serve to strengthen the body wall and provide muscle attachments. The ribs, like other parts of the axial skeleton, arise in the skeletogenous septa. There are two kinds of ribs, both of which are situated in the myosepta and hence are segmental in arrangement. One type is formed at the intersection of each myoseptum with the horizontal skeletogenous septum. Since the horizontal septum divides the myotomes into dorsal and ventral halves, such ribs lie between the muscles and hence are called *intermuscular* ribs, also *dorsal* or *upper* ribs. The second type of rib arises at the points of intersection of the myosepta with the ventral skeletogenous septa or its derivatives. In the trunk region the ventral septum is split into two lateral septa because of the intervention of the coelom and its contents on the ventral side of the body. The second type of rib appears at the points of intersection of the myosepta with these ventrolateral septa. They typically lie just outside the coelomic lining between the coelomic wall and the muscle layer (fig. 6.8). They are therefore called *subperitoneal* ribs, also *ventral, lower,* or *pleural* ribs. Both kinds of ribs may occur simultaneously on a vertebra, and in fact some fishes may have additional ribs of the category of dorsal ribs. The ventral or pleural ribs are generally thought to be the older phylogenetically.

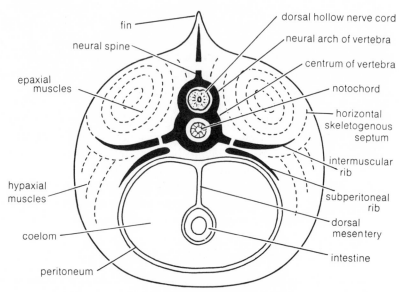

Fig. 6.8. Diagrammatic cross section through the trunk of a vertebrate to show the relation of the ribs to the skeletogenous region, and the positions of the two kinds of ribs.

## C. Vertebral Columns of Some Archaic Fishes

Vertebral columns of the most primitive living vertebrates are rather poorly developed and consist of separate pieces, often cartilaginous in structure. Because of the cartilaginous structure of primitive vertebrae and the poor fossil record for living relicts (e.g., lampreys, primitive actinopterygians), it is difficult to reconstruct the early history of the vertebral column. From all available evidence it appears that elements of the dorsal arch were the first to appear, followed by ventral arches and finally by centra. Thus the early vertebrae, formed of arch parts only, are called *acentrous.*

## 1. Vertebral Column of Cyclostomes

The notochord of lampreys and hagfishes is well developed and has a thick sheath. Vertebral elements of hagfishes are erratically developed and irregular in shape from segment to segment. The tiny dorsal elements lie against the nerve cord, and they are largely restricted in distribution to the tail region. In lampreys, three kinds of dorsal elements are present, and we will call them A, B, and C for purposes of orientation. In the anterior part of the trunk A and B are fused together and are penetrated by a foramen serving the ventral root of the spinal nerve. Between these relatively large cartilages are very tiny cartilaginous pieces, C (fig. 6.9). By the midtrunk region A and B are separated and C is

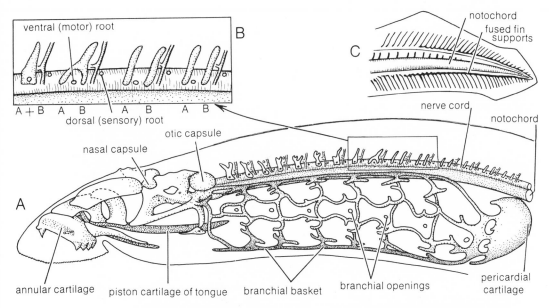

Fig. 6.9. The skeleton of a lamprey. *A*, the anterior end, to show the small, poorly developed vertebral elements lying along the large notochord (after Jollie 1962 [*Chordate morphology.* New York: Reinhold]). *B*, enlargement of the transitional region to show separation of vertebral elements (after Remane 1936). *C*, the tail region to show fusion of bases of fin supports around notochord (after Remane 1936).

absent. Thus in this region there are also two pairs of dorsal cartilages per segment. Posteriorly the development of A and B is erratic, and they may be fused to each other, separated, or one or the other may be absent. Ventral arch elements have been reported in the tails of lampreys, but it is unclear whether these are more than the expanded bases of the cartilaginous fin supports. The fin supports tend to fuse basally in the tail region, and more or less continuous cartilaginous structures extend for some distance in the tail region (fig. 6.9*C*). Homologies of the vertebral elements in cyclostomes with those of fishes are unclear.

## 2. Vertebral Column of the Sturgeon

The vertebrae of sturgeons and related fishes are composed of separate cartilaginous arches that rest on the large notochord (fig. 6.10). There is no centrum. The unconstricted notochord traverses the center of the vertebral column and is covered dorsally and ventrally by cartilage pieces, leaving its lateral exposure uncovered by skeletal elements. Large dorsal arch elements develop in the myosepta and unite above the nerve cord to form a neural arch. The arch is topped by a separate unpaired element, the *neural spine.* Dorsal intercalaries develop in each segment, anterior to the dorsal arch. The intercalaries are variable in number, size, and shape, and there is often more than one pair per segment. Ventral arches and intercalaries have positions corresponding to those of their dorsal counterparts. The ventral arches bear large processes for articulation with ribs, and a small cartilaginous bridge joins them anteriorly. Posteriorly in the trunk and tail the arches join to form the hemal arch.

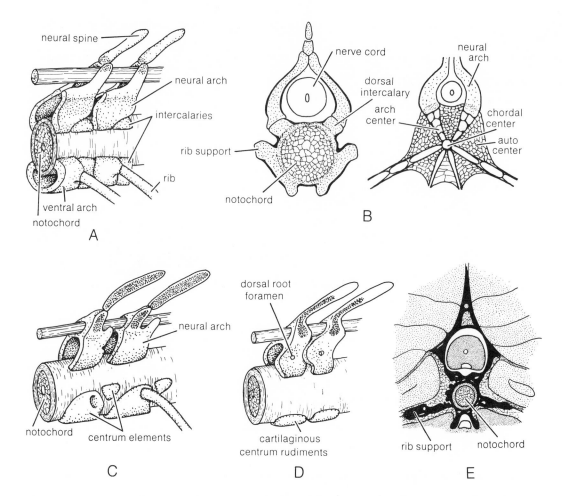

Fig. 6.10. The vertebrae of some archaic fishes. *A*, lateral view of the posterior trunk vertebrae of a sturgeon. *B*, cross sections of developmental (left) and adult (*right*) trunk vertebrae of *Amia* (after Schaeffer 1967). *C*, lateral view of the trunk vertebrae of a lungfish *Neoceratodus*. *D*, lateral view of the trunk vertebrae of *Latimeria*. *E*, cross section of a trunk vertebra of a larval *Protopterus* (after Budgett 1900 [*Trans. Zool. Soc. Lond.*, vol. 16]).

**3. Vertebral Column of *Polypterus***

The fish *Polypterus* has well-developed bony vertebrae formed of dorsal and ventral arches and centra (fig. 6.10). The neural spine, which forms as an independent element, is also joined with these elements to produce the adult vertebra. Ventral arch elements are primarily in the form of ventrolateral processes of the centrum, which serve for rib articulation. Hemal arches are present in the tail region.

**4. Vertebral Column of *Amia***

The centrum of *Amia* is well developed. Arch components play a large role in its formation (Schaeffer 1967). In the abdominal region dorsal intercalaries become incorporated into the adult centrum and form the bases for the neural arch (fig. 6.10). Ventral intercalaries are absent, and the ventral arches themselves are incorporated into the adult centrum, from which they protrude as rib supports (laterally) and small ventral processes. In the caudal region ventral as well as dorsal intercalaries are present in development, and their presence seems associated with the curious arrangement of adult vertebrae in the tail. Centra bearing neural and hemal arches alternate with ones devoid of arches, so that there are two centra to each tail segment. Developmentally the dorsal and ventral intercalaries of a given segment are incorporated into the more anterior centrum (called the *precentrum*), while the bases of the neural and hemal arches are embedded in the more posterior centrum (called the *postcentrum*). This arrangement is probably functionally important in increasing the flexibility of the tail region relative to the rest of the vertebral column.

5. Vertebral Column of Lungfish

Lungfishes have a large, unconstricted notochord with a thick fibrous sheath (fig. 6.10). Ossified neural arches rest on parts of the sheath that may also be ossified. Some lungfishes may have thin, ring-like centra, but they are generally absent. Rib supports also rest on the fibrous sheath, possibly on local areas of ossification. Hemal arches are present in the tail region, and small dorsal intercalaries are present in some species.

6. Vertebral Column of *Latimeria*

The large, unconstricted notochord is surrounded by a thick fibrous sheath and a thinner elastic one (fig. 6.10). No centra are present. The ribs and neural arches rest on the sheath, as do the hemal arches in the tail region. The only vertebral parts that ossify are parts of the neural arches and spines.

**D. Vertebral Column of the Dogfish**

1. Cross Section of the Tail

Obtain a cross section through the tail of the dogfish and study the cut surface, being sure that the section passes through the junction between vertebrae and not through the center of a vertebra. (When the section passes through the center of the vertebra, areas of calcification in the form of rings or rays will be seen.) The center of the section contains the vertebra, composed of clear, relatively soft cartilage. Between the vertebra and the skin is a thick layer of voluntary muscles, composed of a number of leaves, the myotomes or muscle segments, separated from each other by plates of connective tissue, the myosepta. The myotomes appear in whorls because they zigzag in form like those of *Branchiostoma,* and hence a number will be cut across in any cross section. The myotomes are somewhat indistinctly divided into dorsal and ventral portions by a connective tissue partition, the horizontal skeletogenous septum, which extends from the centrum of the vertebra to the skin, where it meets the lateral line. The muscles above the septum are the dorsal or *epaxial* muscles; those below it, the ventral or *hypaxial* muscles.

The vertebra itself consists of a central circular concave portion, the *centrum* or *body;* dorsal to this is the *neural arch,* which encloses a cavity, the *neural canal,* in which the soft, white spinal cord is situated. Ventral to the centrum is the *hemal arch,* which encloses a cavity, the *hemal canal,* containing the *caudal artery* and *vein.* The neural arch terminates in a point, the *neural spine,* and the hemal arch similarly terminates in the *hemal spine.* Observe the connective tissue partitions that extend from the neural spine to the median dorsal line and from the hemal spine to the median ventral line. These are *dorsal* and *ventral skeletogenous septa;* and they, together with the horizontal skeletogenous septum already mentioned, mark the chief sites of skeleton formation.

**2. Sagittal Section of the Tail**

Obtain or make a median sagittal section through the tail of a dogfish. Identify the centra in the section. Each consists of two somewhat triangular pieces, apparently separate, the rounded apexes of the triangles directed toward each other, the whole shaped somewhat like an hourglass. The two ends of the centra are concave, so that diamond-shaped spaces are present between successive centra. These spaces are filled with a soft, gelatinous material, the notochord, which also fills the canal that runs through the center of the centrum. Hence the notochord is constricted by the centra but expands to nearly its embryonic size in the space left between the concave ends of adjacent centra (fig. 6.11). The centra of elasmobranchs are thus biconcave or *amphicoel-*

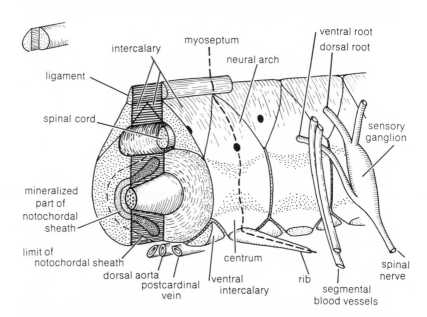

Fig. 6.11. Vertebral column of a shark, partially cut away to show details of the complex notochord-centrum relationship. After Jollie 1962 (*Chordate morphology*. New York: Reinhold).

*ous.* Above each centrum identify the neural arch, arching over the neural canal. Between successive neural arches, and lying therefore dorsal to the diamond-shaped spaces between the centra, observe an extra arch, inverted, however, with apex enclosing the neural canal. This is the *intercalary arch.* Below the centrum is the hemal canal, its sides formed by the hemal arches, rectangular in section.

**3. Cross Section of the Trunk Region**

In a cross section of the anterior part of the trunk region of a dogfish identify the following parts. The muscle segments are arranged as in the tail region, their division into dorsal and ventral masses being well marked by the horizontal skeletogenous septum. The dorsal or *epaxial* muscles above the septum are thick masses, but the ventral or *hypaxial* muscles below the septum form a thin layer enclosing a large cavity, the *body cavity* or *coelom,* lined by a smooth membrane, the *pleuro-peritoneum.* The coelom encloses the viscera, some of which will be observed to be suspended by a delicate membrane, the *dorsal mesentery,* from the median dorsal line of the coelomic wall. The vertebra consists of centrum and neural arch, similar in appearance to those of tail vertebrae; but the hemal arch is absent. It may be represented by a pair of small cartilages at the sides of the ventral part of the centrum. These are the *basapophyses* or *basal stumps,* the apparent homologues of the hemal arch. Examine the horizontal skeletogenous septum carefully and find within it, by picking away the muscles if necessary, a slender cartilage on each side, articulating with the basapophyses. These cartilages are the ribs.

|  |  |
|---|---|
| **4. Preserved Skeleton** | The vertebral column may be conveniently divided into trunk and tail, with all tail vertebrae having hemal arches. The vertebrae are composed of several distinct parts (fig. 6.11). The main part of each vertebra consists of the centrum and the neural arch. The centrum in the trunk region bears a pair of ventrolateral ridges, the basapophyses. The neural arch narrows dorsally to an apex. The spaces between adjacent arches are filled with large intercalary arches which are long dorsally and have a ventral apex on either side of the neural canal. Each neural arch is penetrated by a foramen for the passage of the ventral root of the spinal nerve. The dorsal roots pass through more dorsally located foramina in the intercalaries. Ventrally, small intercalaries lie at the junctions of centra. |

Short ribs articulate with the basapophyses. These extend into the myosepta at the level of the horizontal skeletogenous septum.

|  |  |
|---|---|
| **E. Vertebral Column of Teleosts** **1. The Tail Vertebrae** | Examine a separate, dried tail vertebra of a bony fish. Note that the bony vertebra is very much harder and more opaque than the cartilaginous dogfish vertebrae. Identify the parts already seen in the dogfish vertebrae: the biconcave or amphicoelous centrum, bearing a minute canal in its center for the notochord; the neural arch, terminating in a very long, sharp neural spine; the hemal arch, terminating in a hemal spine (fig. 6.12*D*). The neural canal enclosed within the neural arch is generally smaller than the hemal canal, enclosed by the hemal arch. The spines are directed posteriorly. In some fish there are two neural spines to each vertebra, an anterior and a posterior; the second one probably corresponds to the intercalary arch of the dogfish. |
| **2. The Trunk Vertebrae** | Obtain a separate dried trunk vertebra of a bony fish. Identify, as before, the centrum and the neural arch and neural spine. A small ossification, called a supradorsal, may be present along a ligament that extends through the dorsal part of the neural arch. The hemal arch and spine are missing (fig. 6.12). A pair of projections at the sides of the base of the centrum each bear a long slender rib. |

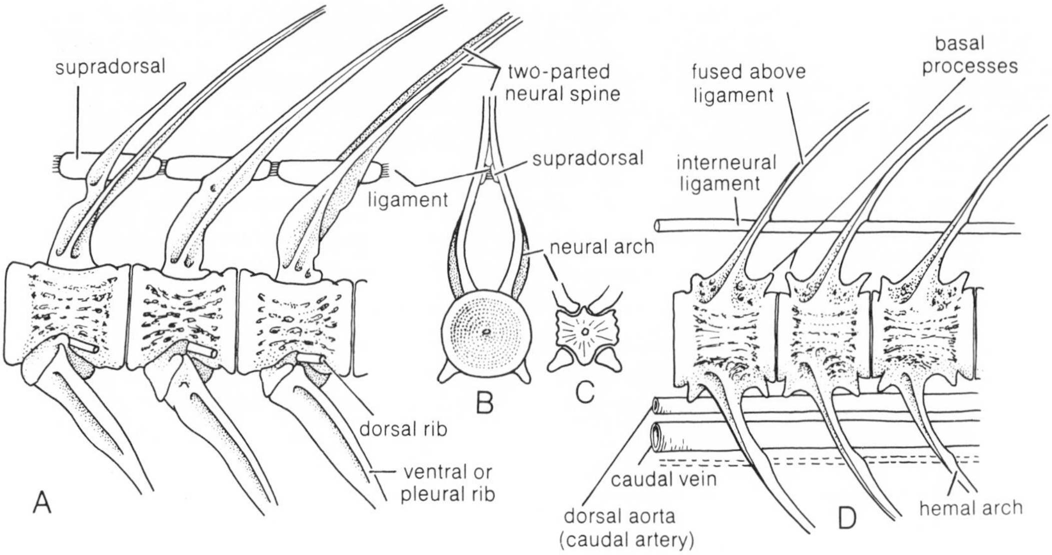

Fig. 6.12. Vertebral structure in a salmon (after Jollie 1962 [*Chordate morphology*. New York: Reinhold]). *A*, lateral view of anterior trunk vertebrae. *B*, anterior view of a trunk vertebra. *C*, cross section through middle of centrum. *D*, lateral view of tail vertebrae.

**3. Section Through the Trunk of a Bony Fish**

Identify the parts already described for a similar cross section of the dogfish. Note the muscle segments, the centrum and neural arch and spine of the vertebra, and the coelom with its lining. Find the ribs, situated just outside the coelomic lining; this location, together with the facts of their development, shows that the ribs of teleosts are pleural (subperitoneal) ribs, like those of elasmobranchs.

**4. Further Study of Ribs**

Some fishes have two (or more) pairs of ribs on each vertebra. Examples are *Polypterus* and many teleosts, including members of the salmon, herring, and pike families. Examine the skeleton of *Polypterus* and note two pairs of ribs attached to each vertebra. The dorsal ribs are articulated to projecting processes of the centrum; the lower or pleural ribs are loosely attached to the ventral surface of the centrum. In the intact fish the dorsal ribs are situated in the horizontal septum, the pleural ribs along the peritoneum. Vertebrae of fishes like the salmon may also be examined, or sections through the trunk of such fishes. Note the pleural ribs; also the dorsal ribs in the horizontal septum. Additional ribs may also be present dorsal to these, articulated with the centrum or neural arch and extending out into the myosepta between myotomes. Teleost ribs approximately equivalent to dorsal ribs may be formed at any level of the myosepta.

**5. The Vertebral Column as a Whole**

Study a mounted skeleton of a bony fish, noting its complete ossification. Observe that the vertebral column is formed by the end-to-end placement of the centra, held together in life by ligaments and muscles. The ends of the centra of fish in general are amphicoelous, and the considerable space left by the concavities of the ends is occupied in life by the remains of the notochord. The vertebral column is divisible into *trunk* and *tail* regions. In the former are the long slender ribs; in the tail region, hemal arches replace the ribs. Observe the transition between trunk and tail regions, noting gradual elongation of the basapophyses and reduction of the ribs toward the posterior end of the trunk. At the beginning of the tail region the reduced ribs finally vanish, and the elongated basapophyses fuse to form the hemal arches. This transition in teleost fishes is considered evidence that the pleural ribs are equivalent to the hemal arches.

Note that the neural arches of successive vertebrae together enclose a continuous neural canal that contains the spinal cord. Similarly, the hemal arches of the tail vertebrae enclose a continuous hemal canal containing blood vessels.

**F. Endoskeletal Fin Supports of Fishes**

The fins of fishes are supported by dermal rays that articulate with endoskeletal supports, the *pterygiophores,* more or less concealed in the animal. Here only the unpaired fins will be considered, since the paired fins are treated in the next chapter.

1. Median Fins

On skeletons of elasmobranch and teleost fish note the endoskeletal supports between the dermal rays of the median fins and the neural and hemal spines of the vertebrae. In elasmobranchs these pterygiophores compose one or more rows of cartilaginous pieces or rods, which are usually larger and fewer in number next to the vertebrae. In teleosts a row of slender bony rods, sometimes flattened, is articulated with the vertebral spines at one end and the dermal fin rays at the other. There may be one or more such pterygiophores to each vertebral spine (fig. 6.13). These pterygiophores seem to be cutoff portions of the vertebral spines.

2. Tail Fin

The dermal rays of the tail fin of fishes are articulated directly to the arches of the vertebrae. Much attention has been paid to the form of this

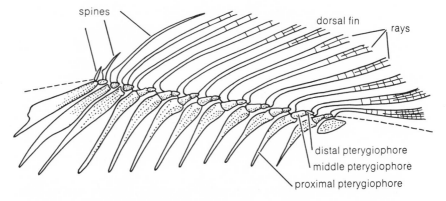

Fig. 6.13. Skeletal supports of dorsal fin of a teleost fish, to show spines, rays, and their supports. Dotted line indicates limit of body of fish. After Jollie 1962 (*Chordate morphology.* New York: Reinhold).

terminal part of the vertebral column, and several types are recognized (fig. 6.14):

### a. Protocercal Tail

Hypothetical primitive type, perfectly bilaterally symmetrical with fin developed equally above and below the straight vertebral (or notochordal) axis; found in cyclostomes and passed through in the development of teleosts.

### b. Heterocercal Tail

Asymmetrical with vertebral axis or notochord bent upward in the tail so that the external fin is larger below than above the axis; characteristic of elasmobranchs and most other lower fishes. Examine the skeleton of an elasmobranch and note asymmetrical tail fin with vertebral column turning upward in it. Observe larger fin expanse below than above the column, and particularly note that the anterior part of the lower fin is expanded into a lobe, which probably represents an originally separate fin now fused to the true tail fin. Note how dermal fin rays articulate to the neural and hemal arches of the vertebrae and that the hemal arches are larger and more expanded than the neural ones. A reversed heterocercal (hypocercal) tail, with vertebral column bending downward and lobe on the upper side, is characteristic of certain ostracoderms.

### c. Diphycercal Tail

Secondarily symmetrical tails, derived by modification from the heterocercal type; seen in present Dipnoi and coelacanth Crossopterygii. They resemble the protocercal type, but intermediate fossil forms show their derivation from the heterocercal condition.

### d. Homocercal Tail

Externally more or less symmetrical but internally like a shortened heterocercal type; found in all higher fishes and derived during embryology from the heterocercal type. Examine the tail of a skeleton of any bony fish. Note upturned end of the vertebral column with last centrum (probably consisting of several fused centra), forming an elongated urostyle that turns sharply upward. The hemal arches accompanying this upturned part of the vertebral column are enlarged and flattened, forming the *hypural* bones, and a few corresponding enlarged neural arches, or *epural* bones, may be present. It is seen that the tail fin is formed largely or wholly of the ventral part of the fin of the heterocercal tail with a great reduction or loss of the dorsal part. This ventral fin has become secondarily bilaterally symmetrical, so that the homocercal tail appears symmetrical externally. Various modifications of the homocercal tail are found, and two extremes, the cod (a simplified tail) and the tuna (highly adapted for speed) are illustrated in figure 6.14.

**G. Vertebral Column of Amphibia**

The transition from an aquatic to a terrestrial existence was accompanied by several modifications of the vertebral column and ribs. Whereas in fishes the major movements of the body were in a horizontal plane accompanying propulsive events, movements in terrestrial vertebrates in-

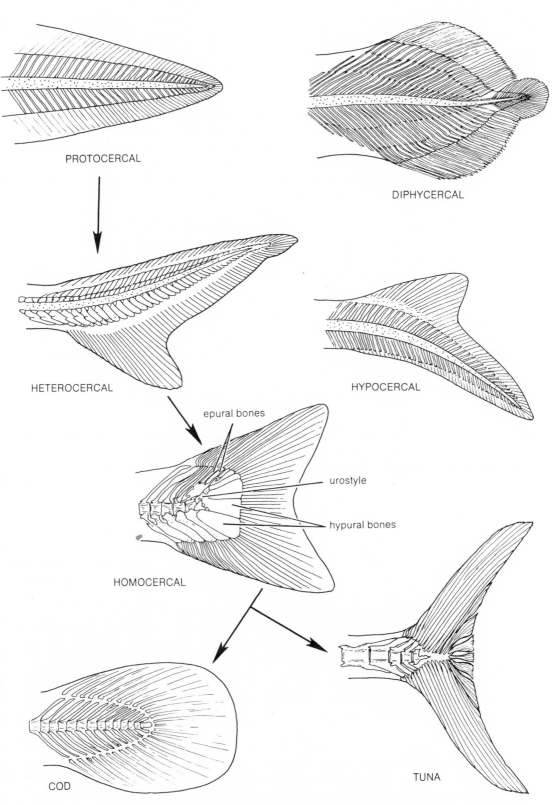

Fig. 6.14. Different kinds of tails found in various groups of fishes. Partly after Jollie 1962 (*Chordate morphology*. New York: Reinhold).

## 1. The Origins of Terrestrial Vertebrates

volve vertical as well as horizontal flexing. Further, the vertebral column becomes a weight-bearing member. As a consequence of these new functions, the notochord is replaced by the sequentially arranged centra as the major axial element, the intervertebral articulations are strengthened, and the ribs are greatly enlarged. The enlargement of the centra is accompanied by a steady decrease in the size and functional significance of the notochord. Zygapophyses, processes that join the neural arches of adjacent vertebrae, are enlarged, and the articulation is increased in extent and complexity. Not only do the zygapophyses increase the total strength of the articulated vertebral column, they enable the trunk to accommodate to the torsions accompanying quadrupedal locomotion. The ribs in the earliest terrestrial vertebrates were massive structures that supported the body during rest on land and prevented collapse of the abdomen while in the air. Probably the earliest terrestrial vertebrates dragged the abdomen and tail along the ground during their forays on land. Suspension of the trunk from the upright limbs and girdles was likely a rather late evolutionary development.

## 2. The Vertebral Columns of Early Amphibians

Almost from their first appearance the amphibians displayed a great diversity in the construction of the vertebral column. In terms of the general structural pattern, with centra, neural arches, ribs, and connections to the girdles, there was uniformity. The greatest diversity occurred in the structure of the centra and their relation to the other vertebral and rib elements. Amphibians are generally considered to have been derived from some group of rhipidistian fishes. Within known members of this latter group a moderate amount of vertebral centrum variation is seen. Usually there is a dominant element in each centrum, but this may be joined by a pair of moderate to small-sized elements that are now usually considered to represent the intercalary arches of archaic fishes. Thus the total centrum may have consisted of a rather large notochord with a single U- or O-shaped bony element surrounding it (fig. 6.15). In front of this element was a pair of intercalary bones having various structural features. In at least some rhipidistians (*Eusthenopteron*) this pair of bones was dorsal and bore traces of grooves that may represent the passage of the dorsal and ventral roots of the spinal nerves. These are considered by some authors to be homologues of the posterior pair of central elements in the early amphibians. Finally, some rhipidistians are known that have only a single central element and no accessory elements for each vertebra. Typically the neural arch is completely separated from the centrum in the fossil remains.

The earliest tetrapods (ichthyostegids) have vertebrae remarkably like those of some rhipidistians, including one dominant centrum element and a pair of accessory elements (fig. 6.15). These are often indicated as posterior centrum elements, but it is likely that they are homologues of the anterior accessory elements of rhipidistians. Many early labyrinthodonts have duplex centra, consisting of a U-shaped, dominant anterior bone and a pair of posterior, dorsally situated elements. The anterior bone is called the *intercentrum,* or the *hypocentrum,* and it usually forms the ventral support for the rib. The posterior pair of bones, the *pleurocentra,* and the neural arch, support the dorsal head of the

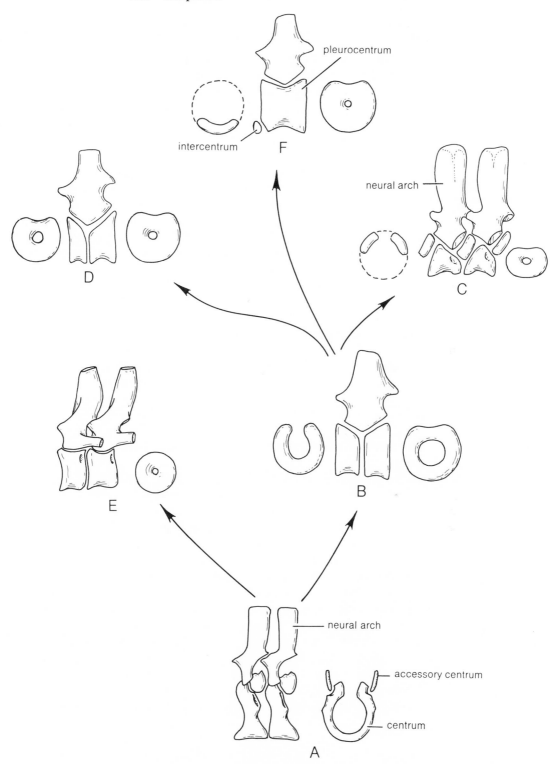

Fig. 6.15. Various tetrapod vertebrae. The vertebrae are shown in lateral view. Centra are also shown in cross section. *A*, a rhipidistian fish, *Eusthenopteron*. *B*, an early diplomerous amphibian. *C*, a rhachitomous amphibian. *D*, an embolomerous amphibian. *E*, a stereospondylous amphibian. *F*, an early reptile.

rib. The neural arch is a separate entity, more closely associated with the pleurocentrum than with the intercentrum.

For many years the classification of early amphibians has been based on the structure of the vertebral centra. The major types of vertebrae recognized are *rhachitomous,* in which a single U- or O-shaped intercentrum and a pair of usually smaller pleurocentra are present (fig. 6.15); *embolomerous,* in which a single U- or O-shaped intercentrum and a single U- or O-shaped pleurocentrum are present (fig. 6.15); *lepospondylous,* in which only a single centrum, not clearly homologous to either the inter- or pleurocentra is present (fig. 6.15); and *stereospondylous,* in which a single cylindrical intercentrum is present (fig. 6.15). The early amphibians with the most generalized structure (temnospondyls) have rhachitomous vertebrae, but in a few specialized groups (Dissorophoidea) the pleurocentra may be the dominant elements. Specialized late lineages of temnospondyls have stereospondylous vertebrae. In early members of the lineage of labyrinthodonts that gave rise to reptiles (anthracosaurs) the circular intercentra were the dominant elements, but in later members of the group the pleurocentra enlarge and fuse so that a duplex centrum is present with an adjacent intercentrum and pleurocentrum for each vertebra. In the ancestors of reptiles the intercentrum is reduced to a small ventral crescent, and the cylindrical pleurocentrum is the dominant element. Recent discoveries suggest that the pattern may not be as simple as presented here, but the major trends seem clear. Certain early amphibians were highly specialized upon their earliest appearance in the fossil record. These aistopods, nectrideans, and some microsaurs have lepospondylous vertebrae similar in general to those of modern salamanders. The centra were thin, hollow cylinders with large anterior and posterior cone-like concavities. Presumably a notochord extended through the centra, and the concavities were filled by intervertebral cartilage that formed the articulations between adjacent centra. In some groups the neural arch was fused to the centrum, but in others (microsaurs) the two units were separate. In some microsaurs a small, crescent-shaped ventral element appears in an anterior position in the tail region, and it has been called an intercentrum. The presence of this element has led some workers to conclude that the dominant centrum in lepospondyls is a pleurocentrum. An extension of this logic has led to the identification of the single centrum of modern amphibians as a pleurocentrum. Supportive evidence is the similar development of the centrum in living amniotes, which clearly have a pleurocentrum based on paleontological grounds, and in the modern amphibians, whose ancestry is unknown. Not all workers accept this reasoning, arguing instead that the questions of ontogeny and phylogeny should be clearly separated until the ancestry of all living groups is clarified. According to this latter view, the centrum of the lepospondyls and modern amphibians, in its unitary state, is the homologue of the entire centrum unit, whether simple or compound, in the rhipidistians, the labyrinthodonts, and the amniotes.

## 3. Vertebral Column of Salamanders

Whole skeletons of salamanders, for example *Necturus, Cryptobranchus* or *Ambystoma,* are needed for this study. The vertebral column consists of five rather poorly differentiated regions. The *cervi-*

*cal* or *neck* region consists of a single vertebra, which lacks ribs and articulates with the skull by means of two large *cotyles* (fig. 6.16). Note that a well-developed process, bearing articular facets, projects into the foramen magnum. This is the *tuberculum interglenoideum,* a feature unique to salamanders. The long *trunk* region consists of vertebrae that are generally similar in construction. All but the last usually bear ribs. From ten to more than sixty trunk vertebrae occur in different species. The *sacral* region consists of one vertebra, the *sacrum,* whose ribs, the *sacral ribs,* attach to and support the pelvic girdle. The *caudo-sacral* region is a transition from sacral to caudal vertebrae. From two to four vertebrae are present. They may bear ribs, but usually do not. The last caudosacral vertebra is recognized as the first vertebra to bear a well-developed hemal arch. This vertebra supports the posterior part of the cloaca and marks the end of the trunk. The *caudal* or *tail* region, composed of vertebrae lacking ribs and having hemal arches, contains from about twenty to more than one hundred vertebrae.

Study individual vertebrae. They are remarkably distinctive, with long, low neural arches fused to rather weak, amphicoelous centra. The tail vertebrae resemble those of fishes, having neural and hemal arches and a centrum that bears projecting lateral processes. On the trunk vertebrae the bicipital or two-headed ribs are attached to two lateral processes. The dorsal process, or *diapophysis,* arises near the midpoint of the neural arch and bears the dorsal rib head, or *tuberculum.* The ventral process, or *paraphophysis,* arises from the dorsal part of the centrum, slightly anterior to the diapophysis, and supports the ventral rib head, or *capitulum.* The ribs appear to be the homologues of the dorsal ribs of fishes. In some species the rib heads are fused together. The tail can be autotomized in many salamander species, and some members of the family Plethodontidae have a marked basal tail constriction. There is no intravertebral autotomy septum and the break occurs between vertebrae.

The vertebrae are articulated to each other by *zygapophyses,* a pair of projections on the neural arch fitting over a similar pair on the anterior end of the succeeding vertebra. Thus each vertebra has a pair of *prezygapophyses* on its anterior end whose articulating surfaces face upward, and a pair of postzygapophyses on its posterior end whose articular surfaces face downward. An *intervertebral cartilage* forms the joint between adjacent centra. This spindle-shaped structure fits into the concavities in adjacent centra. Primitively a large notochord persists in salamanders, and it extends through the center of the centrum and the intervertebral cartilage. In more advanced species the notochord is reduced in size and importance, and the intervertebral cartilage is enlarged. This latter element consists primarily of hyaline cartilage and is firmly anchored into the anterior end of each vertebra. The joint is formed by a disk of fibrocartilage that arches anteriorly from the ends of the centra. Thus a kind of ball-and-socket joint is formed and the vertebrae are functionally *opisthocoelous* (with a convexity anteriorly and a concavity posteriorly in the joint). In newts and some other salamanders the intervertebral cartilages may ossify to produce truly opisthocoelous vertebrae.

**4. Vertebral
Column of
Frogs**

Frogs have a single cervical vertebra that resembles that of salamanders in having two large articular facets and lacking ribs, but it also lacks the tuberculum interglenoideum. The trunk contains from four to eight (usually seven) vertebrae, succeeded by a single sacral vertebra. Behind the sacrum a long *urostyle,* derived from several fused postsacral vertebrae, completes the column; there is no tail (fig. 6.16). No hemal processes are present, and ribs are present in adults only in the most primitive frogs. Typically the vertebrae bear well developed *transverse processes,* and those of the sacrum, termed *sacral diapophyses,* have a variety of shapes. In some species they are rather simple cylindrical structures, while in others they may be extremely large and flattened, with great distal expansion. The sacral diapophyses support the enormously enlarged ilia of the pelvic girdle, which extend posteriorly. Much movement is possible at the sacroiliac joint, which appears to be in the middle of the back (fig. 6.16). Behind this point there is no flexibility and in front of it there is relatively little. What movement does occur is in the vertical plane.

The vertebral column in frogs is much compressed. The neural arches are low and short, with small zygapophyses. The centra have a variety of shapes, but generally they are larger and more solidified than in salamanders. The most primitive species have a persistent notochord and amphicoelous vertebrae, with the intervertebral cartilage forming the joint. A few species have opisthocoelous vertebrae, but the vast majority are *procoelous,* with anterior concavity and posterior convexity (fig. 6.16). There may be variations in the form of the articulation, especially in the immediate presacral region. The centra may develop around the notochord, as in salamanders (*perichordal* development), or they may develop only from tissue that lies atop the notochord (*epichordal*), a unique pattern.

**5. Vertebral
Column of
Caecilians**

Caecilians are limbless and very elongate, with virtually no tail. There is a single cervical vertebra with two large cotyles, which merge ventromedially to form an anterior process quite unlike that in salamanders. There are no cervical ribs. Following this are a series of from about 60 to 285 vertebrae. Most bear bicipital ribs, but these are absent in the cloacal region. There is no sacrum, and the cloacal vertebrae are irregularly shaped lumps of bone. In some features the vertebrae resemble those of salamanders, but they are highly distinctive, with stout centra and high neural arches. There are no hemal arches, save for some anomalous structures in the cloacal region. The pattern of the rib-bearers is especially distinctive. The diapophyses are borne on the anterior part of the neural arch, and the parapophyses are attached to the extreme anterior end of the centrum (fig. 6.16). The centra are amphicoelous, but the joint is formed from a ligament-like structure that joins adjacent vertebrae. This develops from an intervertebral cartilage that is very reduced in adults. The notochord is persistent but is replaced by a mineralized cartilage plug in the midpoint of each vertebra.

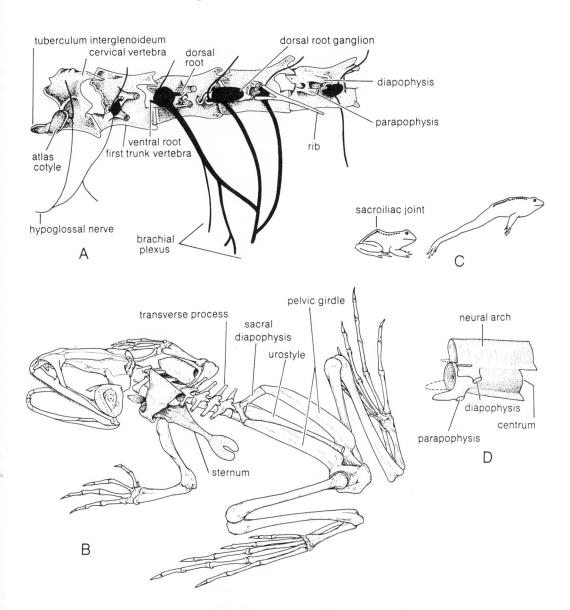

Fig. 6.16. Amphibian vertebral columns. *A*, the anterior part of a vertebral column in a plethodontid salamander, *Eurycea*, to show relation of spinal nerves to vertebrae, and vertebral joints (after Wake and Lawson 1973). *B*, the skeleton of a frog, *Rana*. *C*, a frog at rest (left) and jumping (*right*), to show great movement at the sacroiliac joint. *D*, a trunk vertebra of a caecilian.

6. Development of the Vertebrae in Modern Amphibians

The three orders of living amphibians show few similarities in the mode of vertebral development. Caecilians have the most amniote-like development. Salamanders resemble caecilians in most aspects of development, but they have a very scanty sclerotome. Very little of the centrum is preformed in cartilage in the salamanders and caecilians. By contrast, frogs have much of such preformation. The sclerotome is scanty in frogs, and the mode of vertebral development is unique.

## H. Vertebral Column of Reptiles and Amniotes in General

### 1. Primitive Reptilian and Amniote Vertebrae

The ancestry of reptiles is now reasonably well known, and the stages leading to the development of the typical amniote vertebrae can be documented. Amniotes evolved from a group of anthracosaur labyrinthodonts. This group had a vertebral column in which the pleurocentrum was the dominant centrum element and in which the intercentrum was being reduced in size with time. Thus the proximate amphibian ancestors of reptiles had vertebrae consisting of three distinct parts; the neural arch, a dominant pleurocentrum, and a rather small intercentrum. The pleurocentrum fused with the neural arch, and the intercentrum was reduced to a thin crescent of bone underlying the notochord at the anterior end of each vertebra by the time of the origin of reptiles. Thus the early reptiles, and living amniotes, have essentially monospondylous vertebrae, in which the dominant centrum element is the homologue of the pleurocentrum of ancient amphibians. The intercentrum is usually absent in living amniotes, but it may be present in the tail region, where it contributes to elements of the ventral arch termed *chevron bones.* These always lie at the anterior of the vertebra. Some primitive reptiles have persistent intercentra in the cervical region, but the elements are not typically present in amniotes. Some workers believe that the intervertebral cartilages or intervertebral disks incorporate part of the presumptive intercentrum.

### 2. Evolution of the Atlas and the Axis

The head and trunk of fishes move as a unit, and no specialized joint occurs between the two regions. One of the most striking evolutionary developments of land tetrapods is the craniovertebral joint. Its evolution is doubtless associated with land life, most obviously with sensory perception and feeding, but also with locomotion. Modern amphibians have a well-developed craniovertebral joint with a pair of specialized, convex, occipital condyles on the back of the skull articulating with a pair of enlarged articular facets on the first vertebra. The first vertebra of salamanders has a distinct anteriorly directed process, but the other amphibians lack such a structure. In the ancestors of reptiles the occipital condyle was a single median element, and it retains that morphology in modern reptiles. However, its articulations with the vertebral column are quite different than those of amphibians. The first two vertebrae (termed *atlas* and *axis*) of amniotes are usually modified for the support and movement of the skull (fig. 6.17). The atlas is a ring-shaped element that lacks a typical centrum. It bears one (reptiles and birds) or two (mammals) enlarged, concave articulating facets that receive the knob-like occipital condyle. The axis typically has an anteriorly projecting *odontoid process,* which extends into the base of the foramen magnum and acts as a pivot in the turning of the head. This process thus extends through the relatively short atlas. The most extreme development of this condition is in the mammals, and in many reptiles somewhat intermediate conditions are encountered. There is much controversy concerning exact homologies of the parts of the cervical vertebrae of amniotes. Evidence is good, however, that the definitive centrum of the atlas (the pleurocentrum) is fused to that of the axis to form the odontoid process. The exact composition of the ring-like part of the atlas that surrounds the odontoid is not clear, and it has been suggested that

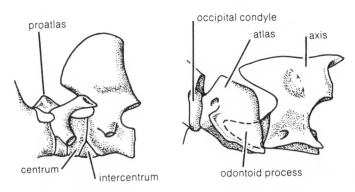

Fig. 6.17. The atlas-axis complex in amniotes (after Romer 1970 [*The Verte-brate body*. Philadelphia: Saunders]). *A, Ophiacodon,* a primitive reptile. *B,* a generalized mammal. The broken line in *B* indicates the odontoid process of the axis, which is thought to be formed from the centrum and intercentrum of the more anterior vertebra of ancestral forms.

either the intercentrum, the ventral arch elements, or both contribute to its formation. Embryological evidence suggests that the odontoid process is an extension of the axis that includes not only the intercentrum of the axis but also the pleurocentrum of the atlas. In some reptiles an additional element, the *proatlas,* may be present between the atlas and the occipital arch. This was long thought to be the remnant of a "lost" vertebra, but it is now thought to be formed by tissue that might have formed the postzygapophyses of the occipital arch of ancestral forms.

### 3. The Tetrapod Rib

Land vertebrates have only a single pair of ribs for each vertebra, and these extend outward from the column, into the lateral trunk muscula-ture, which is increasingly reduced in the higher vertebrates. These are now generally considered to be the equivalent of the dorsal ribs of fishes. Primitively ribs are borne on all vertebrae from the atlas well into the tail. Tetrapod ribs are typically two-headed (bicipital). Primitively the lower head, or *capitulum,* is the largest, and it articulates either directly or with a process (*parapophysis*) of the intercentrum. The upper head, the *tuberculum,* is weaker and articulates with a process (*diapophysis*) attached to the neural arch (fig. 6.18). Accordingly the lower head is ahead of the upper one. The prevailing theory considers the tuberculum and diapophysis to be derived from their more ventral, anterior counter-parts. These structures are considered to have evolved in response to selective pressures associated with supportive requirements for terrestrial locomotion. As these organisms become more and more terrestrial, pres-sure points formed where the ribs were forced upward and toward the midline by the bulk of the abdominal mass. The rib-bearers presumably developed in response to such stimuli.

As the intercentrum was reduced in size, the articulation of the capi-tulum shifted anteriorly, so that it was supported by half-facets on two adjacent centra, or on the intervertebral disk. The typical situation in living tetrapods is the development of a parapophysis on the centrum, al-most directly below the diapophysis, but there are many variations in the mode of rib articulation.

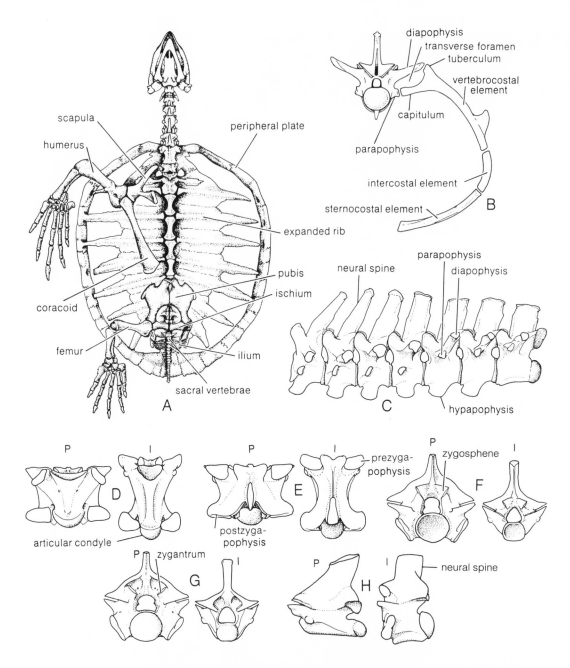

Fig. 6.18. Vertebrae of living reptiles. *A*, the skeleton of a sea turtle, viewed ventrally (after Bellairs 1969 [*The life of reptiles*. London: Weidenfeld and Nicolson]). *B*, the vertebral column of an alligator, in cross section (after Hoffstetter and Gasc 1969). *C*, portions of the neck and trunk of a crocodile, in lateral view (after Hoffstetter and Gasc 1969). *D*, ventral, *E*, dorsal, *F*, anterior, *G*, posterior, and *H*, lateral, views of trunk vertebrae of a snake (*Python*, P) and a lizard (*Iguana*, I) (after Hoffstetter and Gasc 1969).

The early tetrapods had movable ribs along most of the vertebral column, but the more posterior of these fused to their bearers to form pleurapophyses. This pattern has continued into the modern tetrapods, which rarely have postsacral ribs. Usually ribs are small and short in the flexible neck region. Those of the thorax play an important respiratory role in many amniotes, and they are usually imbedded in the body wall musculature. There they curve around to the ventral surface, and articulate with the sternum. They become divided into two or three sections for greater flexibility, usually an upper bony vertebral section and a lower cartilaginous sternal or costal section. Ribs of the lumbar region are incorporated into laterally extending pleurapophyses or, if articulated, greatly reduced in size. The sacral ribs, one or more pairs, attach the pelvic girdle to the vertebral column and are usually short, stout, and nearly immovable.

In modern reptiles two-headed ribs are present only in the crocodilians. The ribs of snakes are very important in locomotion, and they function in conjunction with the elaborately differentiated trunk musculature and enlarged ventral scales. These ribs are long and curved over most of the presacral region. In turtles only ten pairs of ribs are present, and usually the middle eight are distally expanded to contribute importantly to the carapace. The ribs extend from the vertebral column laterally to the dermal skeleton of the shell, forming a flying buttress that strengthens the carapace.

## 4. Abdominal Ribs

*Ventral* or *abdominal* ribs, also called *gastralia,* are imbedded in the ventral wall of the trunk in many fossil and recent amphibians and reptiles. These are often dermal elements that are usually V-shaped, with the apex directed anteriorly. Homologies are not clear, and there have been suggestions that distinction should be made between those that have a typical dermal bone ossification pattern and those that preform in cartilage and even remain so in the adult. These latter elements probably gave rise to the sternum.

## 5. Development of Amniote Vertebrae

Embryos of amniotes are richly supplied with sclerotome, and this purely embryonic tissue plays a dominant role in the formation of the vertebra. The sclerotome is organized in distinct repeating units, corresponding exactly with the myotomal segmentation. These masses of cells should be viewed as dynamic entities, not as blocks of inert material. They fill a large part of the area between the myotomes and the notochord and nerve cord and are separated from each other by areas of markedly lowered cell density rather than by discrete boundaries. The limits of the cell masses are marked by the intersegmental blood vessels, and these may serve as convenient landmarks for study of later stages. An extension of the myocoel—the sclerocoel—has been described in some amniotes, but recent studies have questioned its reality. It is said to lie approximately in the center of each sclerotomal mass, separating the relatively cell-poor cranial half from the cell-rich caudal half.

The first stage in development of the adult vertebra involves a movement of mesenchymal cells around the notochord. These cells multiply to produce condensations near the middle of each segment. The areas between condensations gradually fill with cells and a more or less continuous perichordal tube results that surrounds nearly the full length of the notochord.

After the production of the perichordal tube, a period of rapid differentiation ensues. The areas of condensation in the tube, known as the perichordal rings, are destined to become the intercentrum, or the intervertebral cartilage. The cells that span the old intersegmental fissure will give rise to the adult centrum, the homologue of the pleurocentrum of early tetrapods. Rudiments of the neural arch now appear immediately above the centrum. At first these are represented by basal condensations, but growth is rapid and the condensations prolong around the nerve cord. This tissue typically lies in front of the intersegmental blood vessels and is derived mainly if not exclusively from cells of the caudal sclerotome half. A condensation of cells forms dorsolaterally between adjacent neural arches and unites them. It is in this mass of cells that the zygapophyses form.

Nearly all of the vertebra of amniotes is preformed in cartilage, which is very important as a growth tissue in the rapidly changing element. Differentiation of the intercentra and intervertebral disks proceeds, and the centrum is fully distinguished from them. The centrum chondrifies separately from the neural arch, and a neurocentral suture of rather undifferentiated tissue persists for a long while. The neural arch becomes complete over the nerve cord, and the zygapophyses appear.

Growth of the centrum is much more rapid than growth of the intercentrum and intervertebral cartilage. Incorporation of intervertebral cartilages into the intercentrum is typical of most amniotes. The definitive intervertebral joint forms in the remnant of the intervertebral tissue near the center of each segment.

Ossification of the vertebra is basically a matter of endochondral osteogenesis, but some tissue immediately in front of and behind the cartilaginous core of the neural arch (neural pedicel) ossifies directly. Typically the adult vertebrae are heavily ossified structures containing little cartilage. As in many amphibians, the notochord may persist in some small lizards, and it may contain plugs of cartilage at the midvertebral level. In addition, in *Sphenodon* and in some lizards the intervertebral cartilage remains in a cartilaginous state. When the cartilage does not mineralize, the vertebrae are said to be amphicoelous, even though the presence of the cartilage makes most vertebrae functionally procoelous. Most reptiles are procoelous, though there are many modifications. Birds and mammals have special patterns of intervertebral articulation.

**6. Vertebral Column of Crocodilians**

Regional differentiation is rather well marked in crocodilians. There are nine cervical vertebrae, distinguished by ribs that fail to reach the sternum. The first two cervical vertebrae, the atlas and the axis, consist of separate components. The atlas is a ring composed of four bones, the paired neural arches and dorsal (proatlas) and ventral (intercentrum) connecting pieces. The proatlas is probably the remnant of an ancient zygapophyseal joint between the occipital and atlantal arches. The intercentrum bears a pair of long movable ribs with single heads. Although the centrum of the atlas is apparently absent, embryological evidence indicates that it has fused with that of the axis, forming the odontoid process. The craniovertebral joint is formed by the unpaired occipital condyle and the hollowed parts of the intercentrum and neural pedicels of the atlas. The axis has no free intercentrum. Strong neural arches and elongated neural spines characterize the axis and succeeding cervical vertebrae, and the ribs are rather short and double-headed. From the short ribs of the axis to the long ones of the last cervical vertebra, rib length steadily increases. Distinctive, short midventral projections (*hypapophyses*) are present on the centrum of the cervical and anterior trunk vertebrae (fig. 6.18*C*). Ligaments extending between these suspend the neck and are important in support of the massive head.

The next fifteen vertebrae form a distinct functional unit, the trunk. The first eight or nine of these bear trisegmented ribs that reach the ventral surface of the body and form the thoracic basket. Ribs of the first two arise from bearers that are well separated, as on the cervicals, and connect directly with the sternal plate. The bearers and rib heads enclose a large opening, the *transverse foramen* (fig. 6.18*B*). The successive openings form the *vertebrarterial canal,* in which blood vessels to the head are situated. Ribs of the next three vertebrae consist of one or two segments only. The trunk vertebrae that bear ribs are often called thoracic vertebrae, and the last three or four, which have long transverse processes but no free ribs, are termed lumbar vertebrae.

Crocodilians have two sacral vertebrae, each of which bears stout ribs that articulate with the ilia. The first postsacral vertebra differs from all other vertebrae in having a biconvex (most are procoelous) centrum. It also lacks a hemal arch and is thus a transitional or caudosacral vertebra. The first eighteen caudal vertebrae have intercentral hemal arches, formed by the chevron bones, and transverse processes. The last few caudal vertebrae are strongly compressed and lack transverse processes. The shape of the caudal vertebrae reflects the use of the tail in locomotion, which is accomplished by strong lateral movements. Differences in the orientation of the zygapophyses can also be noted, and those of the trunk tend to restrict movement and increase rigidity, while those of the tail favor lateral bending.

## 7. Vertebral Column of Turtles

All turtles have eighteen presacral vertebrae. The cervical vertebrae, always eight in number, are movable and usually have only vestigial ribs or no ribs at all. Most turtles that fold their necks in a vertical plane (Cryptodira) have a very generalized atlas-axis complex. The atlas typically consists of three separate parts, the intercentrum, the centrum, and the neural arch. In the side-necked turtles (Pleurodira), which fold their necks in a horizontal plane, and in some others, the separate elements tend to fuse, and a distinctive, large atlas may result. The cervical vertebrae behind the axis are very flexibly articulated, with well-developed ball-and-socket joints that are differently organized in different families. These differences reflect differences in the mode of neck movement. The trunk, or dorsal, vertebrae are intimately associated with the carapace, each corresponding to a medial dermal shield (fig. 6.18*A*).

Neural arches tend to be displaced forward to an intercentral position, to agree in conformation with the original segmentation pattern. Ribs of the middle ten trunk vertebrae join the carapace in all but one family (Dermochelydae). The ribs are usually intercentral in position. They expand and fuse to the inner surface of the costal plates of the carapace. Ordinarily there are no special articulations between the trunk vertebrae. The nineteenth and twentieth vertebrae are the sacrals in most turtles, but the region is not well defined in pleurodires. The sacral ribs, and in some groups the last one or two ribs of the trunk vertebrae, meet the ilia. The tail vertebrae articulate freely and are usually procoelous. They are small and few in number. A few anterior vertebrae in the tail may bear ribs. In some species the neural arch is separated from the centrum by a narrow cartilage band. Small hemal arches and in some instances chevron bones are present.

8. The Vertebral Column of *Sphenodon*

This genus has the most primitive reptilian vertebral column among living forms. The notochord is persistent and the vertebrae are amphicoelous. A small intercentrum is present in all but the cervical vertebrae. There are eight cervical, seventeen trunk, two sacral, and twenty-nine to thirty-six caudal vertebrae. The elements of the atlas remain separate, and there is a small proatlas. The centrum of the atlas is fused with that of the axis, which in turn fuses or articulates with intercentra of the first three vertebrae. The last five cervical vertebrae bear ribs of increasing length. The trunk vertebrae bear well-developed ribs that are modified so as to be not quite two-headed. The first three trunk ribs reach the sternum, and some of the more posterior ribs attach to the gastralia by connective tissue. Ribs of the last three trunk vertebrae are short. The result of this pattern is the formation of a thoracic cage anteriorly and a slight tendency toward a waist posteriorly. The sacral vertebrae have strong ribs. Anterior caudal vertebrae are well developed but bear no ribs. The first three postsacral vertebrae are transitional in morphology and have normal intercentra. In more posterior vertebrae the intercentra are modified to form chevron bones, which are part of the hemal arch.

The tail can be autotomized. This is accomplished by a break in specialized tail vertebrae. The region of the break corresponds to the embryonic borders of the sclerotomes. The border region remains unossified, forming a relatively weak transverse autotomic septum. This first appears on the eighth caudal vertebra.

9. The Vertebral Column of Lizards

Lizards are a diverse group with great variation in the numbers of vertebrae. There is a general structural similarity, however, and lizard vertebrae are rather distinct from those of other reptilian groups. Many geckos have amphicoelous centra, but the great majority of lizards are procoelous. Intercentra are typically present only in the cervical region. The number of presacral vertebrae seems to be about twenty-four in most lizards, but dwarf chameleons have as few as sixteen, and some limbless lizards may have as many as 116. Many lizards have eight cervical vertebrae, the last five of which bear increasingly elongate ribs. The first trunk vertebra is the one whose ribs contact the sternum. The atlas is a ring formed of the neural arch articulated to a hypapophysis (ventral process of a centrum) or to an intercentrum. The axis includes the centra of both the atlas and axis. Typically intercentra are present in the cervical region but not in the trunk region. Two intercentra may be fused to the axis, or they may be free. Ribs of about the first five trunk vertebrae are joined to the sternum or xiphisternum. Most trunk vertebrae bear a pair of single-headed ribs, but a few immediately in front of the sacrum may lack them. Some families of lizards have secondary articulation surfaces on the dorsal part of the neural arch, between the zygapophyses. These are termed *zygosphenes* (on anterior end of vertebra) and *zygantra* (facets on the posterior end of the vertebra, which receive the zygosphenes) (fig. 6.18). There are two sacral vertebrae, usually fused. The enlarged sacral transverse processes articulate directly with the ilia. They incorporate embryonic rib rudiments. Usually one to four *pygal* or caudosacral vertebrae lie between the sacral and caudal regions. The first caudal vertebra bears a hemal arch in the form of a chevron bone, which incorporates the intercentrum. There are no free postsacral ribs. Most lizard families have autotomic septa in several to many caudal vertebrae and are capable of tail autotomy. The break occurs midvertebrally as in *Sphenodon*, but the details differ in a taxonomically significant way from family to family (fig. 6.19).

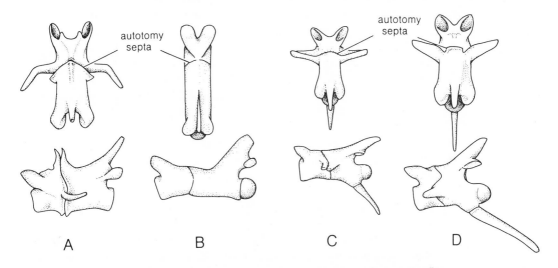

Fig. 6.19. Caudal vertebrae of lizards, to show autotomy septa. Dorsal (*top*) and lateral (*bottom*) views (after Hoffstetter and Gasc 1969 and Etheridge 1967). *A,* a gecko. *B,* an iguanid. *C,* a scincid lizard, *Scincus. D,* a scincid lizard, *Lygosoma.*

## 10. The Vertebral Column of Snakes

The vertebral column of snakes is rather similar in structure to that of *Sphenodon* and lizards, but in its lack of precise regional differentiation and its complex articulations it reflects its role as the major locomotor organ. Musculature of the snake body is extraordinarily complex, with precisely organized tendinous attachments to the vertebrae that vary greatly from one taxonomic group to the next. It is the combination of multiple joints (many vertebrae and long, double-headed ribs) and complex musculature that enables snakes to move in such varied and intricate ways. The atlas-axis complex is much like that of lizards. Other cervical vertebrae cannot be distinguished from trunk vertebrae. From 160 to more than 400 vertebrae are present, and the great majority are precloacal. The individual vertebrae typically are stout and well developed (fig. 6.18). The centrum ranges from very short to very long, relative to lizards. Centra are strongly procoelous, with stout, protrusive posterior condyles. Zygosphenes and zygantra are well developed. Frequently, well-developed hypapophyses are present on the ventral surface of the centrum. Ribs are typically long, robust, and well ossified. There is no sternum, and all ribs have free ends, with rather stumpy tips. Ribs in the cloacal region are forked. There are no postcloacal ribs. Hemapophyses appear in the cloacal region and are distinctive in that they are separated distally. Thus the hemal arches are incomplete. Only a very few species are capable of tail autotomy, and they have autotomic septa, similar to those in lizards, in a few caudal vertebrae.

Compare these features among reptilian skeletons available for study.

## I. Vertebral Column of Birds

Birds display great similarity in the structure of their vertebral columns, which are highly specialized as a result of adaptations for flight. Reduction of vertebral flexibility, except in the neck, is a striking feature of the bird vertebral column. This reduces intervertebral movement during flight, hence reducing frictional loss of energy. The most striking feature is the fusion of numerous vertebrae with the pelvic girdle to form a weight-bearing synsacrum in the posterior part of the body. The column is rather short, with a total of nearly forty to more than sixty vertebrae (fig. 6.20). No free intercentra occur in adults, but embryonic rudiments contribute to the atlas, axis, and pygostyle. The atlas is very similar to that of mammals but seems to have been independently derived. It is a ring-shaped element that lacks a centrum. Embryological evidence indicates that the centrum of the atlas fuses with that of the axis to form the odontoid process. The atlas and the axis articulate with the unpaired occipital condyle. In hornbills the atlas and axis are fused, but this is an exceptional situation. The cervical vertebrae are those anterior members of the column that lack ribs, or whose ribs do not articulate with the sternum. The number is variable (from eleven to twenty-five, but usually thirteen to sixteen). Generally the longer-necked birds have high numbers of cervical vertebrae. The thoracic vertebrae bear ribs that reach the sternum, and they usually number four to five (from three to ten). Varying degrees of fusion of the thoracic vertebrae occur, and frequently the middle three form a single unit. In some species one or two cervical vertebrae may be fused to this mass. The synsacral mass incorporates a number of vertebrae (ten to twenty-two), and several subregions, based on vertebral components, can be recognized. The thoracic synsacrum fuses to the ilia by means of dorsal and ventral rib-bearers, and the first of the series usually bears a small rib. Elements of the lumbar synsacrum are fused to the ilia by the ventral rib-bearers only, the dorsal bearers having been lost. The primary synsacrum is composed of from zero to three vertebrae that are fused to the ilia by dorsal and larger ventral rib-bearers. The ventral bearer incorporates a

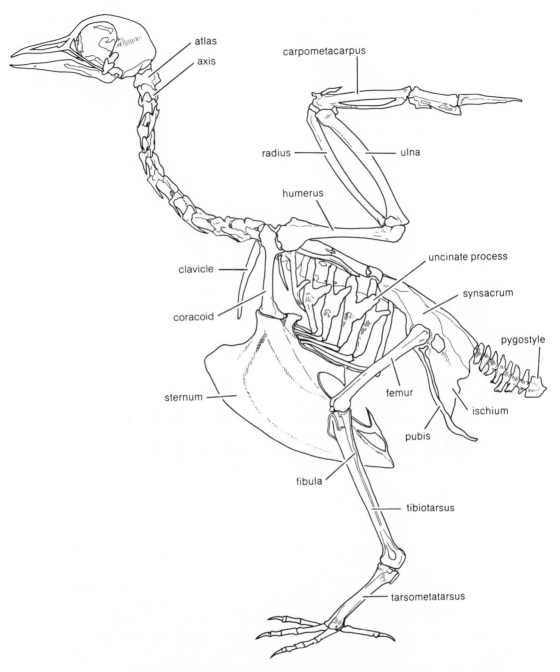

Fig. 6.20. Skeleton of a pigeon. After Pettingill 1956 (*Ornithology*. Minneapolis: Burgess).

sacral rib. The caudal synsacrum includes two to ten vertebrae, united to the ilia by often rather stout transverse processes. The caudal vertebrae usually number six (three to twelve) that are free, and from four to seven additional vertebrae that are fused to form the pygostyle. The pygostyle serves as the base of attachment of the long tail feathers, or retrices, that are of great locomotory and behavioral importance in birds. In ancient birds (*Archaeopteryx*) a long tail, composed of regular caudal vertebrae, was present, but in modern birds this is reduced to a small, but very important, stump. Hemal arches are absent on all caudal vertebrae.

Although most of the bird vertebral column is characterized by rigidity, that portion of it that is vertical, the neck, has great flexibility. Most birds can turn their heads nearly 180° in either direction. This flexibility results from the saddle-shaped, or heterocoelous, centrum ends (fig. 6.4), which should be examined on isolated neck vertebrae. These articulations are unique among vertebrates in the extent of movement permitted.

The ribs of birds are divisible into the usual vertebral and sternal sections. The vertebral portions bear posteriorly directed uncinate processes that are a characteristic avian feature. These processes form bridges between adjacent ribs and strengthen the rib cage.

---

## J. Vertebral Column of Mammals

The vertebral column of mammals is differentiated into five regions, and the typical vertebrae of each region are sufficiently distinct to be readily identifiable when isolated. The following description applies primarily to articulated skeletons and isolated vertebrae of cats and rabbits.

### 1. The Cervical Vertebrae

Typically mammals have seven cervical vertebrae, the first two of which are differentiated to form the atlas and axis. Sloths have six to nine and sirenians have six cervicals, but even such long-necked forms as giraffes retain the generalized number of seven. In aquatic mammals with short necks, such as whales and dolphins, the vertebral centra are very short and often more or less fused. Fusions also occur in armadillos and in such ricochetal rodents as kangaroo rats and jerboas (fig. 6.21). Some believe that fusions in these rapidly moving rodents reduce head-bobbing during locomotion. The atlas is a ring-shaped bone with wide, wing-like lateral projections. These projections represent ribs, and they are perforated by the vertebrarterial canal. In monotremes the sutures between the cervical ribs and the processes of the centrum are evident in young animals, and this is also true of some young dogs. Typically, however, the sutures disappear. The low, flat neural arch of the atlas is perforated for the passage of the first spinal nerve, a situation also encountered in lower vertebrates. The anterior surface of the atlas has a pair of large, curved concavities that articulate with the paired occipital condyles of the skull. As in reptiles, the atlas consists mainly of the bases of the neural arch, but the median ventral region is produced by the intercentrum rudiment. A separate intercentrum occurs in the atlas of some marsupials, but in most it is replaced by a ligament.

The axis has a large and elongated neural arch with a forward-projecting neural spine. Zygapophyses first appear on its posterior border. The odontoid process extends from the anterior end of the centrum of the axis into the ring of the atlas and is important in rotation of the head on the neck. The odontoid develops from the centrum of the atlas embryologically, as in reptiles. The axis and succeeding cervical vertebrae have transverse processes that are formed in part of ribs fused to their rudimentary bearers. All but the last are pierced by the

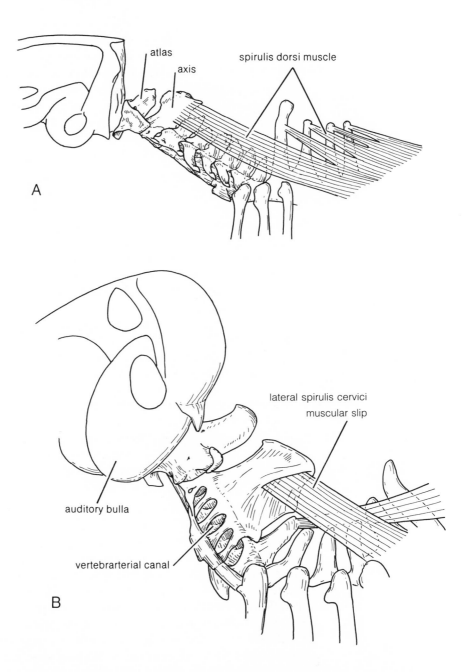

Fig. 6.21. Neck region of mammals (after Hatt 1932 [*Bull. Am. Mus. Nat. Hist.* 63:599–738]). *A, Rattus,* to show the general mammalian condition. *B, Jaculus,* a Jerboa, to show the fused cervical vertebrae found in some richochetal rodents.

vertebrarterial canal, a feature that distinguishes cervical from thoracic vertebrae.

The remaining cervical vertebrae are similar in general form, with well-developed neural arches and spines, pre- and postzygapophyses and transverse processes. The prezygapophyses are distinguished from the postzygapophyses by dorsal, as opposed to ventral, orientation of the articular surfaces.

## 2. The Thoracic Vertebrae

These rib-bearing vertebrae are recognized by the presence of costal facets for rib articulation. Cats have thirteen, and rabbits usually twelve, thoracic vertebrae; the number varies from group to group. The commonest numbers are twelve to fifteen (twelve in man). Cetaceans have the fewest (nine) and sloths the most (up to twenty-five). Often the anterior thoracic vertebrae have very tall neural spines that are directed caudad, short centra, small pre- and postzygapophyses, and short, stout transverse processes. The ribs articulate with the transverse processes, or diapophyses, by their upper heads. The lower heads of the ribs articulate between the centra, where the intracentra of ancestral forms occurred, but partial facets for articulation are found on the anterior end of the centra and the epiphyses. The epiphyses of two articulating vertebrae, plus the *meniscus,* or articular pad, between them, seem to be homologues of the intervertebral disk of lower vertebrates, and possibly also of the intercentra.

The last thoracic vertebrae differ somewhat from the others in having reduced neural spines and transverse processes, more prominent zygapophyses, a single rib facet, and an extra process from the prezygapophysis, termed the *metapophysis* or *mammillary process.*

## 3. The Ribs

The ribs of mammals typically consist of a bony vertebral rib and a cartilaginous sternal rib or costal cartilage. The tuberculum diminishes in size posteriorly, and the last ribs (three in the cat) have only capitular heads. The narrowest part of the rib between the two heads is termed the *neck;* the remainder of the rib, the *shaft;* and the point of greatest curvature of the shaft, the *angle.* Ribs that have an independent attachment to the sternum are *true ribs;* those that join the preceding ribs or are unattached ventrally are called *false ribs;* the unattached false ribs are known as *floating ribs.* Cats have nine true and four false ribs, of which the last is floating; man and rabbits have seven true and five false ribs, three floating in rabbits and two in man. In sirenians and whalebone whales all but the first one to three ribs are floating. Only monotremes and some sloths have three-sectioned ribs, and it is not clear that these correspond to the divisions of reptilian ribs. Uncinate processes are lacking in mammals but are present in many of their therapsid ancestors. Anteaters, some armadillos and a few slow-moving arboreal primates (lorisids) have expanded, overlapping ribs. These increase the stability of the thorax, and hence of the vertebral column. This is important in trunk stabilization during underground activity, or in methodical arboreal locomotion. Some arboreal anteaters anchor themselves with their hind limbs and tail and extend the trunk to reach an adjacent branch with outstretched forelimbs; this requires much trunk stability.

**4. The Lumbar Vertebrae**

From four to seven (the latter typical of cats and rabbits) lumbar vertebrae are usually present in mammals, but monotremes and some edentates have from two to four, and as many as twenty-one occur in cetaceans. The lumbar vertebrae are large and stout, with prominent neural spines and long, anteriorly directed transverse processes. These processes typically include rib rudiments and are hence called *pleurapophyses*. A prominent metapophysis projects above the prezygapophyses, and a spine-like *anapophysis* (*accessory process*) is seen below the postzygapophyses.

**5. The Sacrum**

The sacrum is composed of a variable number (three in cats, usually four in rabbits, five in man) of fused vertebrae that articulate with the ilia. From six to eight sacral vertebrae occur in some perissodactyls, and up to thirteen are found among edentates. Typically the first assumes the greatest part of pelvic girdle support, and it has large lateral expansions. These incorporate the sacral ribs, which are indistinguishably fused to the vertebra.

**6. The Caudal Vertebrae**

A highly variable number of caudal vertebrae occur in mammals, usually corresponding with tail length (up to fifty in long-tailed forms, but usually only three to five, fused to form the coccyx, in man). Neural arches, transverse processes, and zygapophyses diminish caudally, and the last vertebrae consist only of centra. Chevron bones are commonly present, and very small ones, often lost in skeletal preparation, occur in cats. Some mammals are capable of tail autotomy to a slight degree, but regeneration is limited to completion of the broken vertebra and wound healing, in contrast to the situation in reptiles and amphibians.

---

**7. Functional Consideration**

The work of Slijper has provided new perspectives for studies of the functional morphology of the vertebral column in mammals, and the two paragraphs that follow, quoted from his work, admirably summarize his outlook (see fig. 6.22).

"The body-axis (vertebral column and spinal musculature) of mammals may neither be compared to an arched roof, nor to a bridge. In the first place it forms part of the construction of the whole trunk-skeleton. This construction may be considered as an elastic bow (pelvis and body-axis of the trunk) bent in the dorsal direction (ventral concave) by a string (sternum, abdominal muscles, linea alba, extrinsic muscles of the legs). The head and neck may be compared to a loaded beam supported at one end only. On the other hand, the whole body-axis may be compared to such a beam if the animal stands or sits on its hind quarters only, a posture that is attained by every mammal now and then. Thus the principal static function of the body-axis is to resist bending in the dorsal direction. The elastic resistance is caused by the strength of the intervertebral discs and ligaments, but chiefly by the tonus of the epaxial musculature.

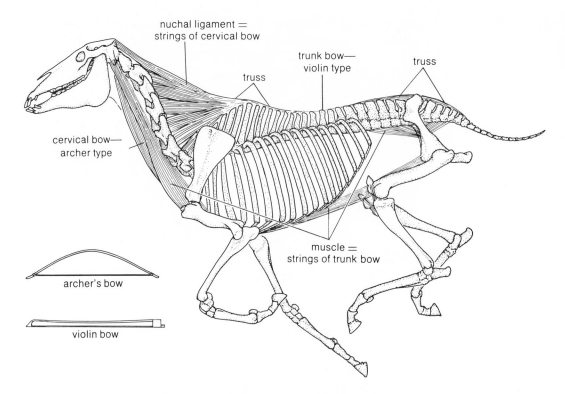

Fig. 6.22. A running horse, to show features of the functional dynamics of the spine in relation to the head and limbs.

"In the second place the body-axis is an organ of locomotion. It has to transmit the locomotive power from the hind quarters to the forehand (this means chiefly to resist bending in the dorsal direction) and it has to bend and extend the back in the sagittal plane, especially when the animal moves in a leaping-gallop."

Mammals differ from fishes, amphibians and reptiles in the reduction of lateral flexion as an important component of the locomotory pattern. The vertebral column of a generalized mammal, such as the cat, is far more flexible in the sagittal plane than in the frontal plane. Further, secondarily aquatic mammals, unlike secondarily aquatic reptiles, never use side-to-side swimming motions of the body. Rather, they engage in a dorsoventral movement of the body axis.

Among cursorial terrestrial mammals two extremes of adaptation in the vertebral column are observed. Compare with the adaptations mentioned in chapter 7. These correspond to the two extremes of running patterns. The highly flexible body axis of such forms as the cheetah is characteristic of sprinters. The lumbar region is long and the neural spines of more anterior vertebrae are relatively short. Much dorsoventral movement is possible. In contrast, the vertebral column of cursorial ungulates, which have considerable staying power while running, has a relatively short lumbar region, tall anterior neural spines, and almost no flexibility.

There are functional reasons for these differences. A sprinting mammal expends much energy to attain high speed quickly. The flexible back of the cheetah is in effect an extra set of muscles and an extra joint segment to the hind legs, which in the cheetah are the primary propellers. By flexing and extending its trunk in addition to accelerating its legs, the cheetah moves most of its mass through a great distance with each stride. It pays a high price in terms of energy expenditure.

In contrast, running ungulates require endurance and have evolved adaptations that permit high speed with relatively low energy expenditure. The rigidity of the back precludes moving the body mass great distances in the sagittal plane, and this saves energy. The forelimbs are principally involved in steering, but propulsive power, though generated mainly by the hind limbs, comes from them as well.

Richochetal (leaping) mammals have vertebral columns similar to those of sprinters, in part because the main propulsive force is from the hindquarters. However, the anterior parts of the vertebral column tend to be much reduced and the cervical vertebrae may fuse, in contrast to the situation seen in such sprinters as cats.

Burrowing mammals tend to have highly flexible vertebral columns, for they must be able to turn around in their burrows. Arboreal mammals that are very active (monkeys) have a flexible column, but those that are slow and deliberate (sloths, lorises) have very rigid backs.

## K. Summary

1. The vertebral column develops around the notochord, a stiffened rod that extends from the region of the braincase to near the end of the tail. The complex notochordal sheath includes several layers of fibers surrounding the vacuolated notochordal cells. The notochord has enough resilience to serve as the main axial support and locomotor organ in *Branchiostoma,* in some primitive fishes, and in the early larval stages of modern amphibians. In most vertebrates the notochord is either replaced by or incorporated into the developing vertebrae.

2. Vertebrae form at the intersections of myotomes, and they have a pattern of repetition that is about one-half cycle out of phase from the segmentation of the body. This pattern is imposed by the functional necessity to bend the body axis, thus creating waves for locomotion. It is not necessary to postulate any highly mechanistic explanation for the formation of vertebrae in particular positions. The neural arch develops somewhat independently from the centrum, especially in lower vertebrates. In fishes the centrum forms from a series of centers, some originally serving as bases for the neural arch, others forming from notochordal materials, and others forming from mesenchyme in the area. These centers are interrelated in different ways in different vertebrates, with some having all centers and others only one. Most higher vertebrates show only so-called auto centers, which form in mesenchyme surrounding the notochord.

3. Ribs are formed along the margins of most vertebrae in fishes and along the trunk vertebrae of tetrapods. Both intermuscular ribs, which lie at the intersection of each myoseptum with the horizontal skeletogenous septum, and subperitoneal ribs, which lie against the coelomic lining, are present in many fishes. In other fishes and in tetrapods, only the first group is present.

4. The vertebral columns of cyclostomes and some archaic fishes are very poorly developed. In some instances it is evident that the skeleton is less well developed than in ancestral forms, and there seems to be a trend in the direction of skeletal reduction. Such is the case in lungfishes and *Latimeria,* for example.

5. Cartilaginous fishes have well-developed vertebral columns, composed of complexes of skeletal plates that form a continuous skeletal tube.

6. Modern bony fishes have vertebrae that differ strikingly from those of archaic and cartilaginous fishes. The vertebrae are typically very well ossified and have a variety of specialized processes. Each group has vertebrae that have evolved specialized structure in relation to mechanical demands resulting from locomotor adaptations.

7. Related to the vertebral column are the endoskeletal fin supports of the median and tail fins in fishes. These form two extremes of structure, rigid spines and flexible rays. The rays of the tail fin are arranged in a variety of patterns, reflecting phylogenetic history and function.

8. Early amphibians initially had vertebrae not much different from those of lobe-finned fishes, but as terrestrial adaptation proceeded the vertebral centra diversified greatly. Zygapophyses evolved between the neural arches, and a variety of articulations between adjacent centra appeared. The centra consisted of two major parts, often called intercentrum and pleurocentrum, in different proportions among different groups. In some groups, notably the lepospondyls, there is only a single element in the centrum.

9. The modern amphibians are similar in vertebral structure in that a single unitary vertebra forms, with no separation of the neural arch and the single centrum. There are many differences among the three major groups. Salamanders have the most generalized vertebrae, with a distinct tail and paired ribs along the trunk vertebrae. Caecilians have only a vestigial tail and the ribs have a curious anterior placement on each of the very numerous trunk vertebrae. Frogs have extremely shortened vertebral columns, with rudimentary ribs present in only a few primitive species; a specialized urostyle forms during development from vertebral rudiments. The development of the vertebrae is distinctive for each of the three groups.

10. Amniotes all share a similar pattern of vertebral development, but it is a matter of controversy whether a distinct resegmentation occurs. Rudiments of a second centrum are found in the chevron bones of the tails of some living reptiles, but the main centrum is thought to be homologous with the pleurocentrum of ancient amphibians and reptiles.

11. A distinct neck region is present in all amniotes, and a specialization of the first two vertebrae to form the atlas and axis has occurred independently in several different lines. The odontoid process of the atlas forms in a very different manner than does the similarly shaped tuberculum interglenoideum of modern amphibians.

12. Tetrapod ribs are thought to be the equivalent of the intermuscular or dorsal ribs of fishes. Characteristically each rib is two-headed. Some modern and many fossil reptiles and amphibians also have abdominal ribs, termed gastralia. These are dermal elements embedded in the ventral body wall that develop independently of the vertebral column.

13. Each of the living reptilian groups has distinctive vertebral columns. This is especially true of the turtles, in which the trunk vertebrae are incorporated into the carapace of the shell. In many lizards and in snakes there are accessory articulations between the adjacent vertebrae associated with specialized locomotory modes. Specialized autotomy planes are present within the tail vertebrae of many different lizards and amphisbaenians, and some snakes.

14. Birds' vertebral columns feature variable numbers of neck vertebrae and fusions of other vertebrae into complexes of bone associated with attachments of enlarged muscles related to locomotion.

15. The mammalian vertebral column is well differentiated into five regions, and vertebrae from each are easily identified by their shapes as to position and function. Mammals with specialized modes of locomotion may show modifications of parts of the vertebral column. For example, the cervical vertebrae of different jumping rodents are fused. Ribs of some climbing, slow-moving species overlap, thus providing stability for the thoracic region. Ribs are borne on thoracic vertebrae, and these are followed by a series of ribless lumbar vertebrae. Different numbers of vertebrae are incorporated into the relatively complex sacral region, and tails vary greatly in length.

16. Vertebral columns play important functional roles in static support as well as dynamic support during movement. In many groups the vertebral column is the organ of locomotion. Thus in lower vertebrates waves are propagated at the anterior end of the vertebral column and these travel down the body axis, applying force to the medium or substrate and propelling the animal forward. Usually this wave is generated simply by differential contraction of the segmental musculature. However, in lizards and especially in snakes there are many additional, highly specialized muscles that are involved in bending the vertebral column and transmitting force to the external scalation. In addition to traveling waves, terrestrial vertebrates are capable of propagating waves that "stand" in one position on the column, causing one part of the body to oscillate. These waves are found at the junctures of the pectoral and pelvic girdles to the vertebral column where they cause the girdles, and their associated limb elements, to move in anterior-posterior planes. In this manner waves propagated along the vertebral column contribute directly to increased stride length. In many groups of vertebrates (birds, frogs, many mammals) wave propagation is insignificant or does not occur. However, movements in other planes may be significant. Dorsal-ventral movements of the column contribute to the jumping mechanisms of frogs and to the increased stride length associated with relatively great speed of such mammals as the cheetah. Finally, there are groups in which the vertebral column becomes a rigid supporting member during

locomotion in a way that suggests that the rigidity itself is an adaptation. This seems to be the case with large mammals, where lateral shifts of the massive bodies would require great energy expenditures.

**References**

Alexander, R. McN. 1968. *Animal mechanics.* London: Sidgwick and Jackson.

Baur, R. 1969. Zum Problem der Neugliederung der Wirbelsäule. *Acta Anat.* 72:321–56.

Bruns, R. R., and Gross, J. 1970. Studies on the tadpole tail. I. Structure and organization of the notochord and its covering layers in *Rana catesbeiana. Am. J. Anat.* 128:193–224.

Carroll, R. L. 1969. Problems of the origin of reptiles. *Biol. Rev.* 44: 393–432.

DeVillers, C. 1954. Structure et évolution de la colonne vertébrale: Les côtes. In *Traité de zoologie, ed.* P. P. Grassé, 12:605–97. Paris: Masson et Cie.

Etheridge, R. 1967. Lizard caudal vertebrae. *Copeia* 1967:699–721.

Francois, Y. 1966. Structure et développement de la vertèbre de *Salmo* et de téléostéens. *Arch. Zool. Exp. Gen.* 107:287–328.

Gadow, H. F. 1933. *The evolution of the vertebral column.* Cambridge: At the University Press.

Gray, James. 1968. *Animal locomotion.* London: Weidenfeld and Nicolson.

Hildebrand, M. 1959. Motions of the running cheetah and horse. *J. Mammal.* 40:481–95.

Hoffstetter, R., and Gasc, J-P. 1969. Vertebrae and ribs of modern reptiles. In *Biology of the Reptilia,* ed. C. Gans, A. d'A. Bellairs, and T. S. Parsons, 1:201–310.

Jenkins, F. A., Jr. 1970. Anatomy of expanded ribs in certain edentates and primates. *J. Mammal.* 51:288–301.

Kemp, T. S. 1969. The atlas-axis complex of the mammal-like reptiles. *J. Zool. Lond.* 159:223–48.

Laerm, J. 1976. The development, function, and design of amphicoelous vertebrae in teleost fishes. *Zool. J. Linnean Soc.* 58:237–54.

Moffatt, L. A. 1974. The development and adult structure of the vertebral column in *Leiopelma* (Amphibia: Anura). *Proc. Linnean Soc. New South Wales* 98(3): 142–74.

Mookerjee, H. K. 1930. On the development of the vertebral column in Urodela. *Phil. Trans. Roy. Soc. Lond.,* ser. B, 218:415–46.

———. 1931. On the development of the vertebral column of Anura. *Phil. Trans. Roy. Soc. Lond.,* ser. B., 219:165–96.

Panchen, A. L. 1977. The origin and early evolution of tetrapod vertebrae. *Linnean Soc. Symp. Ser.* 4:289–318.

Parrington, F. R. 1967. The vertebrae of early tetrapods. *Coll. Inter. CNRS, Paris* 163:269–79.

Remane, A. 1936. Wirbelsäule und ihre Abkömmlinge. In *Handbuch de vergleichenden Anatomie der Wirbeltiere,* eds. L. Bolk, E. Goppert, E. Kallius, and W. Lubosch, 4:1–206. Berlin and Vienna: Urban und Schwarzenberg.

Romer, A. S. 1966. *Vertebrate paleontology*. 3d ed. Chicago: University of Chicago Press.

Schaeffer, B. 1967. Osteichthyan vertebrae. *J. Linn. Soc. Lond., Zool.* 47:185–95.

Sensenig, E. C. 1949. The early development of the human vertebral column. *Contrib. Embryol. Carn. Inst.* 214:23–41.

Slijper, E. J. 1946. Comparative biologic-anatomical investigations on the vertebral column and spinal musculature of mammals. *Verh. Kon. Nederl. Akad. Wet.* 42:1–128.

Smit, A. L. 1953. The ontogenesis of the vertebral column of *Xenopus laevis* (Daudin) with special reference to the segmentation of the metotic region of the skull. *Ann. Univ. Stellenbosch.* 29:79–136.

Thomson, K. S., and Vaughan, P. P. 1968. Vertebral structure in Rhipidistia (Osteichthyes, Crossopterygii) with description of a new Permian genus. *Postilla* 127:1–19.

Verbout, A. J. 1974. The early embryonic development of the vertebral column of sheep, with a critical evaluation of the "Neugliederung" theory. Thesis, University of Leiden.

———. 1976. A critical review of the "Neugliederung" concept in relation to the development of the vertebral column. *Acta Biotheoretica* 25:219–58.

Wake, D. B. 1970. Aspects of vertebral evolution in the modern Amphibia. *Forma et Functio* 3:33–60.

Wake, D. B., and Lawson, R. 1973. Developmental and adult morphology of the vertebral column in the plethodontid salamander *Eurycea bislineata,* with comments on vertebral evolution in the Amphibia. *J. Morph.* 139:251–300.

Werner, Y. L. 1971. The ontogenetic development of the vertebrae in some gekkonid lizards. *J. Morph.* 133:41–92.

Williams, E. E. 1950. Variation and selection in the cervical central articulations of living turtles. *Bull. Amer. Mus. Nat. Hist.* 94(9):505–62.

———. 1959. Gadow's arcualia and the development of tetrapod vertebrae. *Quart. Rev. Biol.* 34:1–32.

# 7

# The Endoskeleton:
# The Comparative Anatomy of
# the Girdles, the Sternum,
# and the Paired Appendages
*George Zug*

**A. General
Considerations**

**1. Definitions**

The girdles are crescent-shaped portions of the endoskeleton lying within the trunk. Their main mass occupies a ventral position, and their ends extend dorsally. The girdles bear articular surfaces for the endoskeleton of the paired appendages and provide the major sites of appendicular muscle attachments. In the lower fishes the girdles are cartilaginous, but in bony fishes and land vertebrates they are more or less ossified. The *pectoral* girdle supports the anterior appendages; the *pelvic* girdle supports the posterior ones.

The *sternum* or *breastbone* is an elongated structure lying midventrally in the anterior part of the trunk and articulating with the pectoral girdle. It is lacking in fish and first appears among amphibians. Primitively, it is a simple cartilaginous plate; in later forms it usually consists of a chain of cartilages or bones or both. In amniotes the sternum articulates with the pectoral girdle and ribs. The ribs fail to reach the sternum in all present amphibians, although they may have done so in extinct forms. The combination of pectoral girdle, ribs, and sternum forms the anterior trunk into a noncollapsible chamber that protects the anterior viscera and aids air-breathing by lungs.

The paired appendages, fins in fishes and limbs in land vertebrates, are supported by an endoskeleton. The anterior appendages, *pectoral fins* or *forelimbs,* articulate with the pectoral girdle through the glenoid cavities; and the pectoral girdle transmits their support to the body by muscular attachment, since the girdle is rarely directly articulated to the vertebral column. The posterior appendages, *pelvic fins* and *hind limbs,* articulate to the pelvic girdle through the *acetabula* and transmit their support of the body through the girdle.

**2. Locomotion
and the Origin
of Girdles and
Paired
Appendages**

The origin of chordates and particularly vertebrates is inextricably linked with the evolution of motile, somatic animals from sessile, visceral animals. Improvement and diversification of locomotor patterns has been at a high selective premium in all vertebrate lineages, as can be seen in the diversity of the locomotor structures—vertebral column, girdles, and appendages. The shape and arrangement of the endoskeleton components of these structures reflect their role in locomotion; so structure cannot be divorced from function.

The protovertebrates were aquatic, and thus axial locomotion—propulsion by undulatory movement passing caudad through the body—appears to have been the original mode of locomotion. Efficiency in aquatic locomotion is reduced by the motion deviations of yawing, pitching, and rolling—side-to-side seesawing, up-and-down seesawing, and rotation around the longitudinal body axis (fig. 7.1). Both paired and

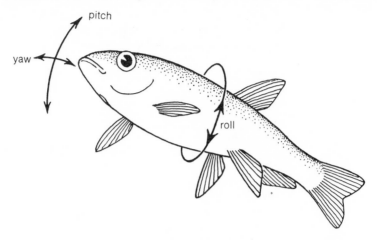

Fig. 7.1. The three motion deviations encountered by swimming animals. From Alexander 1967 (*Functional design in fish*. London: Hutchinson).

unpaired lateral extensions of the body probably evolved to reduce these motion deviations. The paired lateral extensions were the precursors of the paired fins (*ichthyopterygium;* pl., *ichthyopterygia*), which were in turn the precursors of limbs (*cheiropterygium*). Girdles likely appeared after, not simultaneously with, the paired fins as internal struts for the bilateral support of the fins.

Two main theories have been proposed to provide the lateral body extension precursors for the paired fins. According to the *fin fold theory*, the ancestral vertebrates possessed a pair of continuous, ventrolateral fin folds, which fused behind the anus to a single median fin extending around the tail and along the dorsal midline. This hypothetical condition resembles *Branchiostoma* (amphioxus), with its paired metapleural folds and its continuous caudal and middorsal fins. The paired fins and single median fins of present fishes are supposed to have arisen through the persistence of certain regions of the fin folds. In contrast, the *body-spine theory* postulates that the early fishes had two or more pairs of ventrolateral spines. Fins develop from the spines first as flaps of skin extending backward from the spine to the body wall and then later by elaboration of the spine into an endoskeleton framework. There is no conclusive set of evidence to support one theory over the other. See figure 7.2.

No matter which theory is accepted, the evolution of paired and unpaired fins was largely influenced by their role in locomotion. In the fin fold theory, the ventrolateral folds would have acted as stabilizers from the beginning, whereas body spines were probably initially protective or bottom support devices. Increased mobility of the early fish would favor the development of bilateral hydroplanes, or paired fins, to reduce pitch and roll.

The origin of girdles was secondary to the origin of paired fins. As the locomotor function of the fins increased, there would have been a need

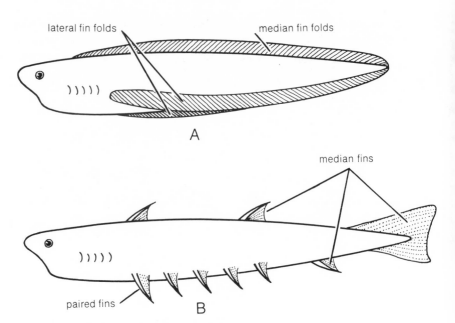

Fig. 7.2. Diagrams to illustrate the origin of paired appendages. *A*, hypothetical finfold condition of an early fish. *B*, hypothetical ventrolateral series of spines on an early fish.

for mechanical strengthening of their internal bracing. Although the actual origin of this bracing, the development of girdles, is unknown, there are clues that it arose from two sources anteriorly and a single one posteriorly. The pectoral appendages may have initially gained internal bracing from the dermal thoracic armor. As the fish became more mobile, there was a tendency to reduce the dermal armor, but portions of it remain associated with the fins. As the dermal armor was reduced, a basal pterygiophore of each side enlarged and extended medially to come in contact on the ventral midline with its opposite mate, thereby creating a transverse ventral base. This hypothesis explains the presence of both dermal and endochondral elements in the pectoral girdle. The medial extension of basals from the pelvic fins explains the existence of only endochondral elements in the pelvic girdle (see fig. 7.3).

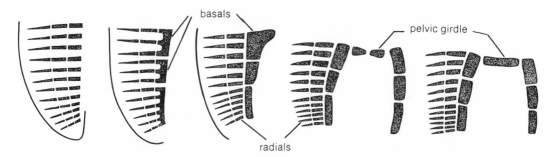

Fig. 7.3. Morphological sequence showing the formation of the endochondral part of the limb girdles from basal pterygiophores.

**3. Parts of the Primitive Limbs**

The tetrapod limb is divisible into five fundamental segments: *propodium, epipodium, mesopodium, metapodium,* and *phalanges* (fig. 7.4). The propodium contains either the *humerus* in the upper arm or the *femur* in the thigh. The epipodium contains a pair of bones in both the

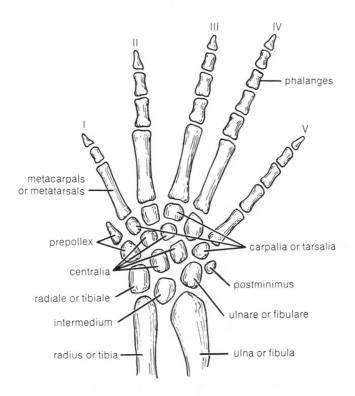

Fig. 7.4. Endoskeletal elements of the primitive autopodium.

fore- and hind limb, the *radius* and *ulna* of the forearm and the *tibia* and *fibula* of the shank. The radius and tibia are on the preaxial, primitively anterior, border of the limbs, the ulna and fibula on the postaxial border. The hand or foot possesses the mesopodium, *carpus* (wrist) or *tarsus* (ankle), the metapodium, *metacarpus* (palm) or *metatarsus* (sole), and the *phalanges* (fingers or toes). The mesopodium contains a series of small bones, which originally seem to have been arranged as a proximal row of three bones—*radiale* or *tibiale, intermedium, ulnare* or *fibulare;* a central row of four bones—the *centralia;* and a distal row of five *carpalia* or *tarsalia.* Each phalanx or digit consists of linear series of bones also termed phalanges. Primitively the hand and foot were pentadactyl, that is, with five digits, and possessed the *phalangeal formula* 2, 3, 4, 5, 4 for the number of phalanges from first to last digit. Extra wrist or ankle bones occur in some tetrapods, a *prepollex* or *prehallux* on the preaxial side and a *postminimum* on the postaxial side. Although these elements are occasionally interpreted as vestigial phalanges, implying that primitively the hand and foot had seven digits, they are more likely to be mesopodial elements. Similarly, the first metacarpal appears to be derived from a carpalium and the fifth metacarpal from a phalanx. This indicates a marked asymmetry of the primitive hand, corresponding to the asymmetry of the crossopterygian fin, with extension along the pre-axial side and crowding together along the postaxial side (fig. 7.5). Specializations of the hand and foot tend to be loss of phalanges and loss or fusion of mesopodial elements.

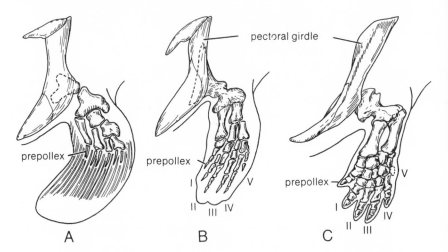

Fig. 7.5. Skeletal changes from fin to limb. Anterolateral view of left pectoral girdle and appendage of *A, Eusthenopteron. B,* hypothetical intermediate. *C, Eryops.* After Gregory and Raven 1941.

## 4. Evolution of the Fish Fin into the Tetrapod Limb

The ancestral tetrapod could have been derived only from the fish groups Dipnoi and Crossopterygii, since they alone have internal nares, functional lungs (swim bladders) connected by a duct to the ventral pharyngeal wall, and lobed fins containing a series of endoskeleton elements transmutable into limb bones. Holmgren and a few associates have argued for a diphylectic origin of the tetrapods, salamanders from the dipnoans and anurans and the reptilian lineages from rhipidistian crossopterygians. If this were true the tetrapod limb would have evolved twice, once from a monoaxial fin of dipnoans and once from the biserial fin of crossopterygians. But most of the facts indicate a close relationship between salamanders and anurans and their derivation from crossopterygians.

The sequential development of limbs and terrestrial locomotion may have been much like the following hypothesis. Initially the tetrapod ancestors used their paired fins as braces to hold their bodies upright when they came out of the water and rested on the bank. Movement on land was restricted to undulatory movements of the body, the fins simply preventing the animal from falling on its side. At the next stage, the appendages became accessory anchor points. Although the ventral body surface provides the main frictional surface for undulatory locomotion, the lateral projecting appendages would have acted as pivotal points for forward progression. The appendicular skeletons of *Eusthenopteron* and *Sauripterus* illustrate this stage (fig. 7.5). In their appendages, the basal elements form single stout proximal bones supporting the articulation of two enlarged radials. The linear arrangement of the endoskeleton elements, while providing a straight axis for the transmission of support from the tip of the fin to the body, also developed a transverse hinge (elbow or knee) between the propodial and epipodial elements. The "foot" and "ankle" were not differentiated, simply the curved distal tip of the fin.

Quadrupedal walking was the final stage in development of the tetrapod limb. The sequence of limb movement in quadrupedal walking—right forelimb, left hind limb, left forelimb, right hind limb, and so on—is related to the undulatory body movement. As an undulatory wave passes caudad, the body alternately takes on an S and a reversed S shape. The limbs on the convex surface are moved anteriorly and those on the concave surface posteriorly, thus generating the walking gait. The foot—that is, the mesopodium, metapodium, and phalanges—differentiated from the radials concurrently with the development of the walking gait. The propodial segment of the limb projected perpendicularly from the body, the epipodial extended almost perpendicularly downward, and the foot flexed through the mesopodials and metapodials so that the phalanges formed the main contact surface. The highly mobile epimesopodial joints did not appear until later in the reptilian lineages.

Structurally, the development of a limb endoskeleton from a fin skeleton is more complex than the above suggestion of elaboration or growth of radials and basals and the appearance of joints between the segments. Both the pectoral and the pelvic fins of *Eusthenopteron* were monobasic, for the anterior two basals, propterygium and mesopterygium, had been lost along with most of their radials. Assuming the ancestors of the tetrapod had monobasic fins, the metapterygium forms the propodium in each limb. The first two radials become the epipodial pair, and the succeeding rows of radials become the mesopodial, metapodial, and phalangeal elements. These bony elements shift from a closely interlocking framework providing a rigid foundation for the dermal rays to an articulated lever arm capable of wider range of movements (fig. 7.5).

## 5. General Function of Paired Appendages

The pectoral and pelvic appendages are neither homologous nor strictly analogous. Both serve locomotor roles and possess similar form and segmentation owing to similarities in selective pressures and morphogenetic fields. Yet their specific locomotor roles differ, and so do their detailed morphologies.

Primitively, in fishes the fins are broad-based, as in sharks, and largely immobile. They project perpendicularly from the body and act as hydroplanes to stabilize the body by reducing the tendency of the fusiform piscine body to roll (rotate around its anteroposterior body axis) and to pitch (seesaw up and down with the fulcrum at the body's center of gravity). At this functional stage, the pectoral appendages tend to be larger than the pelvic ones, since they are anterior to the center of gravity and can be effective stabilizers to counter movement deviation generated by the caudal fin. In the higher fishes the fins retain their role as stabilizers but take on the more active role of steering and occasionally of propulsion. The base of each fin becomes narrower, thereby greatly increasing its mobility. The fins become more flexible and are able to open and close like fans. The pectoral fins remain as the major stabilizing and steering appendages. The importance of fins anterior to the body's center of gravity is emphasized by the tendency of the teleost fishes to have shifted their pelvic fins anteriorly so that pectoral and pelvic fins lie in nearly the same transverse plane.

The forelimbs retain their steering role in tetrapods. The pectoral girdle loses its attachment with the cranial skeleton and gains increased mobility. The hind limbs become the propulsive mechanism. In order to transmit the force of forward locomotion from the hind limbs to the body with a minimum loss of energy, the pelvic girdle develops a strong articulation to the vertebral column by way of the sacral ribs. As you already learned in chapter 6, the region of the vertebral column that bears the sacral ribs and hence supports the pelvic girdle is called the sacrum. With the shift from an aquatic to a terrestrial existence, the appendages and girdles take on the new function of support. In water, gravity is counterbalanced by buoyancy, and the need for support is negligible; however, on land gravity must be counterbalanced by structural means, hence the more massive development of the bony endoskeleton of the girdles and limbs.

**B. The Pelvic Girdle and the Posterior Appendages**

**1. The Pelvic Girdle and Pelvic Fins of Fishes**

The three groups of extant fishes—cyclostomes, Chondrichthyes, and Osteichthyes—are less closely related to one another than are the tetrapod groups and illustrate distinct levels of girdle and appendicular development. The cyclostomes have neither girdles nor paired fins. They are lethargic animals and move either by swimming close to the bottom or by hitching rides on more active fishes. However, their more active ancestors provided the finfolds or spines that became the paired fins of the Chondrichthyes and Osteichthyes.

---

The pelvic girdle of most Osteichthyes shows a pair of ventral pelvic cartilages or bones that suggest the medial extension of basals. Note the pelvic girdle in *Acipenser;* it looks like basal cartilages with a row of radials attached to each. The pelvic girdle and fins are primitively situated immediately anterior to the anus. In osteichthyian fishes, there has been an evolutionary trend to shift the pelvic girdle and fins to a more anterior position so that in some fishes it lies anteroventral of the pectoral fin. This forward shift is associated with a functional shift from a stabilizer to a rudder. In the extreme anterior position, the pelvic fins cooperate with the pectoral fins to increase the fish's maneuverability, since both fins lie anterior to the body's center of gravity. The fins are extremely flexible and can be folded like fans or adpressed tightly to the body. This flexibility obtains from a size reduction of the basals and radials and their restriction to the proximal bases of the fins. Compare fin bases in *Acipenser* or *Amia* with that of *Perca*. The fins' internal supports are dermal fin rays.

In elasmobranchs, the pair of pelvic plates has fused at the midline to form a cartilaginous pelvic girdle of one continuous piece. The fins have broad bases and little flexibility; they act as stabilizers. Study preserved skeletons of elasmobranchs. The pelvic girdle is the bar of cartilage, the *puboischiac bar,* across the ventral side at the end of the trunk. Laterally the puboischiac bar has small dorsally projecting ends, the *iliac processes,* and at the base of each process is an articular surface bearing the fins with an endoskeleton of pterygiophores or cartilaginous fin rays. The pterygiophores are arranged in a proximal series of one or two (in some forms three or five) large cartilages, the *basals,* and a distal series of one or more rows of small, rod-like cartilages, the *radials.* The dermal fin rays extend from the radials to the edge of the fin. There is usually a single basal, the *metapterygium,* a long, curved cartilage extending along the whole medial border of the fin; a small *propterygium* may be present anterior to the metapterygium. The basals have probably arisen by the fusion of a number of pieces similar to the radials. In males large radials support the claspers, the male copulatory organs.

## 2. The Tetrapod Pelvic Girdle

The shift from an aquatic to a terrestrial life necessitated a functional shift of the girdle from the minor role of supporting the pelvic fins to supporting the body. The puboischiac plate must have been strengthened and enlarged rapidly to transmit the support of the walking legs to the body and to provide larger surfaces for muscle attachment. Typically, two pairs of ossification centers appear in the puboischiac plate, anteriorly the *pubes* and posteriorly the *ischia,* and an ossification center in each iliac process, the *ilium.* All three pairs of pelvic bones are endochondral bones. A *pelvic symphysis* is formed medially where the left and right sides of the girdle are united through a cartilaginous union. A cup-shaped depression, the *acetabulum,* receives the head of the femur; all three pelvic bones meet at the acetabulum and contribute to the depression. The ilium is attached along its upper edge to the sacral ribs, so that the sacrum and pelvic girdle form a skeletal circle through which the urogenital and digestive systems must exit to the exterior.

Primitively in tetrapods, the pelvic girdle is an ossified ventral plate of the pubes and ischia with a pair of broad ossified ilia extending dorsally, such as in *Cacops* (fig. 7.6). Primitively an obturator nerve passes through the pubic foramen in each pubis. The next opening, the *puboischiac* foramen, appears between the pubis and ischium and frequently enlarges and extends to the symphysis. The pubic and puboischiac foramen are separate in primitive reptiles, such as *Sphenodon* (fig. 7.6). In more advanced reptiles and in mammals they join to form a single obturator foramen.

Minor elements of uncertain homology occur among various tetrapods. A single or paired anteromedial projection from the pubis also found in turtles is the *epipubis;* it may be cartilaginous or osseous. A single or paired posteromedial projection from the ischium, *hypoischium,* occurs in some reptiles (fig. 7.6). A separate acetabular bone lies in the acetabulum of many mammals.

## 3. Pelvic Girdle and Hind Limb of Urodeles

The largely unossified urodele girdle, although primitive in appearance, represents a derived condition because the primitive amphibian girdle consisted of large plates of bone. Study preserved skeletons of

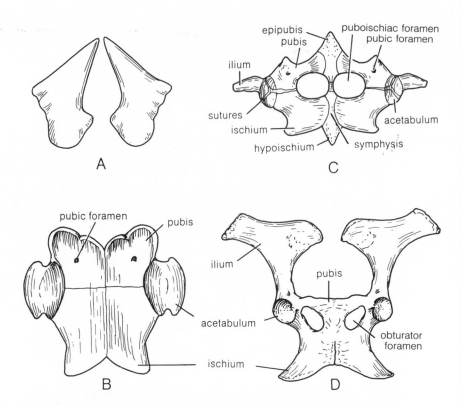

Fig. 7.6. Primitive to advanced pelvic girdles. Ventral views. *A, Eusthenopteron*, a single osseous plate on each side. *B*, extinct embolomerous amphibian *Cacops*, showing three primary divisions of osseous plate with only small pubic foramina (after Williston 1925 [*The osteology of the reptiles*. Cambridge: Harvard University Press]). *C, Sphenodon*, fenestration by puboischiac foramina. *D, Equus*, fusion of bones and foramina.

*Necturus* or *Cryptobranchus*. Both of these salamanders are aquatic, thus their body support role is largely negated by the buoyancy of the water; the feeble appearance of the girdle and appendicular skeleton reflects this reduced function. The girdle consists of a ventral puboischiac plate; the anterior part is the pubic region, which remains wholly cartilaginous. A pair of rounded ossification centers in the posterior part of the plate represent the ischia. An ilium extends dorsally from each side of the plate and articulates with the end of the sacral rib. The acetabulum occurs at the junction of the pubic cartilage, ischium, and ilium; the femur inserts there by a ball-and-socket joint.

The posture of the urodele's hind limb remains similar to the primitive tetrapod posture (see p. 243). The limb is divided into a proximal segment, the thigh, with a single bone, the femur; a middle segment, the shank or crus, with two parallel bones, a preaxial tibia and a postaxial fibula; and a distal segment of the ankle and foot. The ankle or tarsus consists of several small bones, usually impossible to differentiate; it is followed by four elongate bones, the metatarsals, forming the sole; and these bear the toes, each made of a row of phalangeal bones. What is the phalangeal formula of the hind foot? *Necturus* and a few other urodeles have four toes, although most amphibians have five; it is probably the fifth that is missing.

---

**4. Pelvic Girdle and Hind Limb of Reptiles**

The reptilian girdle contains little cartilage. In most reptiles, extinct and extant, the pelvic plate is relatively flat, with a large foramen between the pubis and ischium and a large dorsally projecting ilium. Striking changes in form and size of the girdle occur only in those reptilian lineages that have shifted from quadrupedal locomotion to axial locomotion, such as the loss or great reduction of pelvis in snakes and ichthyosaurs.

---

The turtle pelvic girdle and limbs are representative of the quadrupedal condition. The girdle consists of three pairs of stout bones—anteroventrally the pubes, posteroventrally the ischia, and dorsolaterally the ilia. Anteriorly the pubes and posteriorly the ischia meet their mates to form pubic and ischiac symphyses. Between the ischium and pubis of each side is the large obturator foramen, separated in life from its fellow by cartilage continuous with that of the symphyses. An epipubic cartilage extends anterior from the pubic symphysis, and each pubis has an anterolateral projection, the pectineal process. Note that all three pelvic bones contribute to the acetabulum.

In spite of the shell, the turtle limb posture retains the primitive reptilian stance. Identify the femur (note the proximal *head* that fits into the acetabulum), tibia, and fibula; the fibula is the smaller of the two shank bones. The ankle bones are reduced in number from the primitive plan through extensive fusion of elements. Next to the shank is a large, transversely elongated bone composed of two or more fused pieces (tibiale, intermedium, fibulare, and one or more centralia). In some turtles the fibulare, at the distal end of the fibula, is separate. Distal to this compound bone is a row of four tarsalia (the fourth is a fusion of the fourth and fifth tarsalia and possibly one or more centralia). Distal to the tarsalia are the five metatarsals, followed by the phalanges, of which the terminal ones are provided with horny claws. The general phalangeal formula 2,3,3,3,3 of turtles is also considered typical for later tetrapods.

In extant reptiles, the functional ankle joint is a *mesotarsal* or *intratarsal* joint; that is, the functional foot is composed of the distal tarsalia, metatarsals, and phalanges, while the proximal tarsal bones act as a functional extension of the epipodial segment. As seen in the turtle, the proximal tarsal bones have tended to fuse; the *astragalus* is probably composed of the intermedium, tibiale, and centrale. The *calcaneum* or fibulare usually remains separate although it is sometimes fused with the astragalus. Primitively the ankle joint is a tarsometatarsal joint with the functional foot composed of metatarsals and phalanges. In mammals, the crurotarsal joint develops between the epipodial (tibia and fibula) and the mesopodial (tarsus) segment. However, as evolution from plantigrade (flat-footed) to unguligrade (on the toes, primarily) locomotion occurs, there is a distal shift of the functional ankle joint from crurotarsal to intraphalangeal.

**5. Pelvic Girdle and Hind Limb of Birds**

Modifications of the pelvic girdle and limbs were a major adaptive fore-runner of flight in birds, for without the structural shift permitting bipedal locomotion, the forelimbs would have remained legs instead of becoming wings. Birds still possess the anteroposterior elongation of the girdle, elongation of hind legs, particularly the more distal segments, and broadening of foot contact area, which are typically associated with bipedalism, although these structural modifications are overlaid by adaptations associated with flight, landing, and perching, such as pneu-matization of bones and fusion of girdle to sacrum (fig. 7.7).

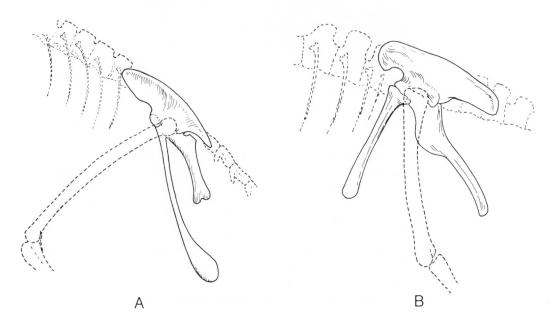

A                                        B

Fig. 7.7. Comparison of pelvic girdle of primitive bird to a bipedal dinosaur. *A, Archaeopteryx. B, Coelurus.* After Romer 1966 (*Vertebrate paleontology.* Chicago: University of Chicago Press).

Examine isolated backbones with girdle attached or study whole mounted specimens. The pelvic girdle consists of three pairs of bones, as in reptiles—*ilium, ischium,* and *pubis.* All three are fused on each side to form a continuous broad bone, the *innominate bone.* The ilium is the largest and most dorsal part of the innominate bone, forming a thin elongated plate—concave in front, convex behind—extending from the last thoracic vertebra to the tail region, fused along its entire length with the synsacrum, from which it is bounded by a suture. (In the bird embryo the ilium is articulated to only two vertebrae, the true sacral vertebrae.) The ischium is the part of the innominate bone below the rear half of the ilium, sepa-rated from the latter by the large oval ilioischiac foramen. The pubis is the long, slender bone along the ventral border of the ischium, from which it is separated by a more or less distinct suture and by a slit-like obturator foramen, which is sometimes subdivided. Ilium, ischium, and pubis meet in the acetabulum, which is perforated by an acetabular foramen, as in crocodilians. Embryologically the ilium, ischium, and pubis originate as separate bones. The pubis begins anterior of the acetabulum but later growth is posteriad. Note the absence of pubic and ischiac symphyses. The two innominate bones are widely separated ventrally, probably an adaptation for the laying of large eggs.

Note the posture of the hind limbs. They extend directly ventral from the girdle; the primitive sprawled posture is replaced by a vertical stance. The femur has a large head fitting into the acetabulum and a prominent projection lateral to the head, the great *trochanter.* The distal end of the femur is shaped like a pulley, consisting of a central depression with curved ridges—the *condyles*—on either side. Over the joint between the thigh and shank is an extra small bone, the *patella* or *kneecap,* not found in lower vertebrates. The patella is a sesamoid bone; a bone developed in a tendon at a zone of stress. Sesamoid bones are common in the limbs of higher vertebrates. The shank is composed of two bones, a medial large one (tibiotarsus) and a lateral short rudimentary one (fibula). The tibiotarsus consists of the tibia fused at its distal end with the proximal tarsal bones. Proximally, the tibiotarsus has two *condyles* for articulation with those of the femur and bears in front two diverging elevations or crests for muscle attachments. The distal portion of the fibula is atrophied. Distally the tibiotarsus bears a pulley-like articular surface, the *malleolus,* for articulation with the next limb bone, the *tarsometatarsus.* The tarsometatarsus arises from the fusion of the distal tarsal bones and three metatarsals, which are visible as three ridges distally. The three metatarsals fused in the tarsometatarsus are the second, third, and fourth; a remnant of the first articulates with the distomedial border of the tarsometatarsus. Each metatarsal articulates with its respective clawed digit; fifth metatarsal and digit are absent. Birds are digitigrade; thus the functional ankle joint lies between the metatarsals and phalanges.

---

**6. Pelvic Girdle and Hind Limb of Mammals**

The mammalian pelvic girdle consists of the usual three bones indistinguishably fused into an *innominate* or *hip* bone on each side. The ilium, the most dorsal and largest of the three, articulates with the sacrum and terminates anteriorly and dorsally in a curved border, the *crest* of the ilium. The ischium is that part of the dorsal region of the girdle posterior to the acetabulum. The posterior ends of the two ischia form prominent projections in the rabbit or a rough curved surface in the cat, called the *ischial tuberosities,* and extend toward the midventral line as the rami of the ischia, which meet to form the ischial symphysis. The anteroventral part of the innominate is the pubes, and the two pubes also have *rami* meeting to form the pubic symphysis. Both ischial and pubic symphyses are fibrocartilage. Between the pubic-ischial rami is the large obturator foramen. The ilium, ischium, and pubis meet in the acetabulum and are part of it.

On a demonstration specimen of a kitten or any other young mammal note that the ilium, ischium, and pubis are separate. Note further in the acetabulum a small acetabular bone that forms a part of the acetabular fossa which would otherwise be occupied by the pubis; its phylogenetic significance is uncertain.

The femur has a head, a greater trochanter lateral to the head (continuing in the rabbit to a small posterior projection, the third trochanter), and a lesser trochanter below the head. These trochanters serve for muscle attachments. The large articulating surfaces at the distal end of the femur are condyles (medial and lateral), and they bear additional elevations or roughened areas, the epicondyles. At the knee joint a patella is present. The shank is composed of a stout tibia and a slender fibula; in the rabbit the latter is fused with the tibia for the greater part of its length. The anterior face of the tibia presents a crest; its proximal articulating surfaces are known as condyles; its distal ones as malleoli. The bones of the ankle are identical with those of the human ankle and are designated by the same names, which unfortunately are somewhat fanciful and not based upon comparative anatomy. The name derived from comparative anatomy is given in parentheses after the name derived from human anatomy. The ankle consists of seven bones (cat) or six (rabbit). The largest and most conspicuous of these, which projects backward as the heel, is the *calcaneus* (*fibulare*). Articulating with the malleoli of the tibia and fibula is the *astragalus* or *talus* (believed to represent the *tibiale,* the *intermedium* and several *centralia*). Directly in front of the astragalus is found the *navicular* or *scaphoid* (representing one or two *centralia*), a curved bone reaching to the medial side of the foot. Directly in front of the calcaneus is the *cuboid* (*fourth* and *fifth tarsalia* fused), articulating with the fourth and fifth metatarsals. Medial to the cuboid is the *third* or *lateral cuneiform* (*third tarsale*), articulating with the third metatarsal. Medial to this is the second or *intermediate cuneiform* (*second tarsale*), articulating with the second metatarsal. In the cat there is a first or *medial cuneiform* (*first tarsale*) along the medial border of the anterior part of the ankle in front of the navicular. It articulates with the small rudimentary first metatarsal, which lies directly in front of it. In the rabbit the first cuneiform is fused to the proximal end of the second metatarsal. The homology of these ankle bones with those given in the primitive vertebrate plan is evident. The sole consists of four long *metatarsals* and one rudimentary one (the first) on the medial or ventral side of the proximal end of the second metatarsal. The terminal phalanges of the digits are curiously beak-shaped for the support of the horny claws, and in the cat they have sheaths at their bases into which the claws fit.

The ankle joint of mammals is a crurotarsal joint; that is, it lies between malleoli of tibia and fibula and the proximal tarsal bones. The gait of the cat and rabbit is chiefly digitigrade, although the hind legs assume the plantigrade posture when the animals sit down. Review page 243 on the torsion of the mammalian hind limbs.

**C. The Pectoral Girdle, the Sternum, and the Anterior Paired Appendages**

**1. Phylogeny of the Sternum**

The sternum occurs only in tetrapods. In amphibians and reptiles, the cartilaginous sternum seldom ossifies and is rarely fossilized. Thus, we do not know when it appeared, its initial morphology, or its initial function. There are three main hypotheses, deriving the sternum from the ribs, the pectoral girdle, and the true abdominal ribs. The rib hypothesis implies that the functional development of the sternum resulted from the additional support and protection given to the anterior viscera; the girdle hypothesis implies a midventral strengthening of the girdle or, perhaps, increased pectoral girdle mobility. The rib hypothesis, based on embryology, is that the ventral ends of the ribs fuse to form a rod-shaped cartilage on each side and then these rods unite to become the sternum. This explanation conflicts with the absence of sternum-to-rib contact in amphibians. To avoid this objection, the amniote sternum has been suggested as a rib derivative (neosternum), whereas the amphibian sternum is derived from the pectoral girdle (archisternum). The pectoral girdle hypothesis postulates that the sternum is the separate median region of the primitive pectoral girdle. This hypothesis relates the sternum to the condition in fishes but has little supporting evidence. The third hypothesis derives the sternum from cartilages forming in the ventral myosepta. Such cartilages are the same as the true abdominal ribs (*parasternals*) previously mentioned. Fusion of the most anterior parasternals would form a sternum, initially independent of ribs or pectoral girdle. At present this hypothesis is regarded as the most probable. It is further supported by the presence of a series of parasternals directly continuous with the sternum (fig. 7.8) in several lizard forms.

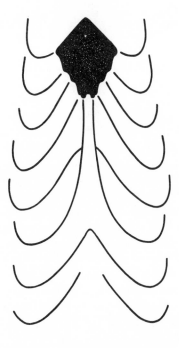

Fig. 7.8. Sternal elements of a lizard. Note midventral fusion of some parasternals. After Etheridge 1965.

2. Evolution of
the Pectoral
Girdle

The cartilaginous or endochondral part of the pectoral girdle originated, like the pelvic girdle, in the enlargement and midventral fusion of a pair of basal pterygiophores (fig. 7.3). The dermal part of the pectoral girdle was probably derived from the dermal thoracic armor. These dermal bones strengthened the girdle, and a number of them are still present in fishes, such as *Polypterus* (fig. 7.9); the principal ones from the ventral

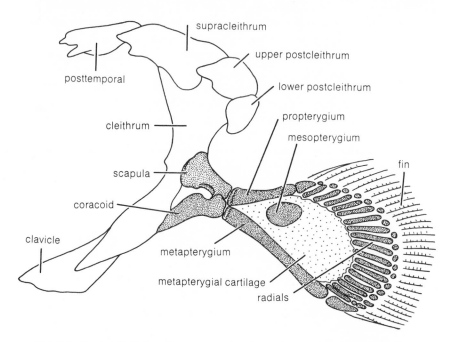

Fig. 7.9. Pectoral girdle of *Polypterus*, to show large number of dermal bones in girdle of an osteichthyian fish. Viewed from the inside (dorsal view). Dermal bone, blank; endochondral bone, close stippling; cartilage, open stippling. After Goodrich (in part 9 of E. R. Lankester, *Treatise on zoology.* Courtesy of the Collier-Macmillan Co., London: A. and C. Black).

side moving dorsally are on each side: *clavicle, cleithrum, supracleithrum,* and *posttemporal;* the posttemporal in nearly all Osteichthyes attaches the pectoral girdle to the skull. This dermal component of the pectoral girdle persists in all tetrapods, and its history is easily followed. In the earliest embolomerous amphibians, as in *Eogyrinus* (fig. 7.10*A*), the dermal pectoral girdle is practically identical with that of fishes, including even a probable skull attachment by way of the posttemporal; but from the start in tetrapods there was an added element, an unpaired midventral *interclavicle.* In the further evolution of tetrapods there has been a gradual *reduction and loss of the dermal elements of the pectoral girdle* (fig. 7.10). Interclavicle and paired clavicles and cleithra are seen among many extinct amphibians and reptiles, but the cleithrum is soon lost and does not occur in any living amniote. Interclavicle and clavicles persist among present reptiles and birds; but the interclavicle is lost in mammals, being present only in monotremes (fig. 7.10*G*). Clavicles are also frequently absent in mammals.

The history of the endochondral part of the pectoral girdle is less clear,

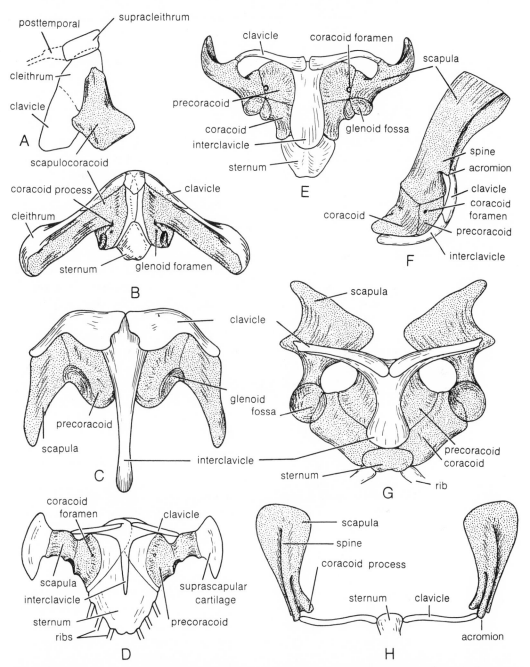

Fig. 7.10. Series of tetrapod pectoral girdles. *A*, half of the pectoral girdle of the embolomerous amphibian *Eogyrinus*, seen from the side, showing fish-like series of dermal bones (after Watson 1926 [*Phil. Trans. Roy Soc. Lond.* 214:189]). *B*, pectoral girdle of the rhachitomous amphibian *Eryops*, cleithrum still present, ossified scapulocoracoid in one piece (after Miner 1925 [*Bull. Amer. Mus. Nat. Hist.* 51:145]). *C*, pectoral girdle of *Seymouria*, cleithrum lost, suture separating scapula and precoracoid (after Williston 1925 [*The osteology of the reptiles.* Cambridge: Harvard University Press]). *D*, pectoral girdle of *Sphenodon*, retains same elements as *Seymouria*, clavicles reduced. *E*, pectoral girdle of extinct mammal-like reptile, showing two bones in the coracoid region (after Broom 1932 [*Mammal-like reptiles of South Africa.* London: H. F. and G. Witherby]); note that coracoid foramen is through the anterior bone, indicating that this is the single coracoid element of present reptiles. *F*, same type of girdle of the monotreme *Echidna* (after Broom 1932 [*Mammal-like reptiles of South Africa.* London: H. F. and G. Witherby]); note identity with reptilian girdle of parts *E* and *F*, except that precoracoid is completely excluded from the glenoid fossa. *H*, pectoral girdle of a placental mammal (muskrat); precoracoid completely lost, coracoid reduced to process.

and homologies are still questionable. The original single pectoral bar or plate became paired, forming on each side a *scapulocoracoid* cartilage, having near its middle a depression, the *glenoid fossa,* for articulation with the humerus. The part above the glenoid fossa is the *scapular* region; that below, the *coracoprecoracoid* region. Among early fossil amphibians the endoskeletal girdle ossifies on each side into a single scapulocoracoid bone; presumably considerable areas of unossified cartilage remained, but these are not preserved as fossils. In early reptiles, also, the endoskeletal girdle consists of a scapulocoracoid on each side; but this becomes divided by a suture into two bones—a *scapula* or shoulder blade above the glenoid fossa and a *precoracoid* below it, as in *Seymouria* (fig. 7.10*C*); both bones contribute to the glenoid fossa. The endochondral girdle remains in the *Seymouria* condition, consisting of a scapula and a precoracoid (usually called coracoid) on each side in present reptiles and in birds, and the precoracoid appears to be homologous throughout these forms. But in some primitive reptiles, and especially in the extinct reptilian groups ancestral to mammals, there occurred two ossifications in the coracoprecoracoid cartilage instead of one, resulting in two bones below, an anterior precoracoid and a posterior *coracoid* (also called *metacoracoid*); the scapula, of course, was also present. Among the mammal-like reptiles the anterior bone, the precoracoid, gradually became reduced in importance and excluded from the glenoid fossa, whereas the posterior bone or true coracoid came to surpass it in size (fig. 7.10*E*). A continuance of this process has resulted in the loss of the precoracoid in all mammals except monotremes (where it is usually called *epicoracoid*). The coracoid is fully developed in monotremes (fig. 7.10*G*), but in other mammals it is reduced to a projection of the scapula, the *coracoid process* (fig. 7.10*H*).

According to the above account (which follows the views of Broom, Watson, and Romer), then, there was in early tetrapods one ossification in the coracoprecoracoid cartilage, and this formed the precoracoid bone, which is the single coracoid element of present reptiles and of birds; but in the extinct reptiles ancestral to mammals two ossifications developed in the cartilage, resulting in an anterior precoracoid and a posterior coracoid. Both bones are seen in the reptiles in question and in monotremes, but in other mammals only the posterior one survives. Thus the so-called coracoid of reptiles and birds is not homologous to the bone called coracoid in monotremes; the former is really the precoracoid and hence will be so termed in the following discussion. This view is not accepted by all vertebrate anatomists. The alternative view is that the original single coracoid ossification is equivalent to the two later ones; that the reptilian and avian coracoid elements represent both precoracoid and coracoid. This conclusion contradicts the rule that separate centers of ossification represent originally distinct bones.

It is thus seen that throughout the tetrapods the pectoral girdle is compounded of dermal and endochondral bones. It was previously mentioned that dermal bones, really part of the exoskeleton, may participate in the endoskeleton; and we here meet them for the first time, associated with the pectoral girdle. The dermal bones never take part in the articulation with the forelimb, whereas the cartilage bones (scapula, precoracoid, and coracoid) regularly do so. The coracoid elements of the girdle frequently articulate with the sternum, whereas the scapula rarely has any connection with the vertebral column (exceptions: skates, pterosaurs). In this respect the pectoral girdle contrasts with the pelvic girdle, which, as already seen, is always in land vertebrates firmly articulated with the sacrum.

Some of the original cartilage of part of the pectoral girdle is likely to persist unossified. Various names have been given to such cartilages of the girdle, but no particular importance can at present be attached to such names.

| | |
|---|---|
| 3. Pectoral Girdle and Pectoral Fin of Fishes | As with the pelvic girdles and fins of fishes, the pectoral girdles and fins show different levels of specialization. The cyclostomes lack the pectoral girdle and fins as well as the pelvic ones. In chondrichthyian and osteichthyian fishes, the pectoral fins play a more active role in locomotion, that is, steering. Unlike the passive stabilizing role of the pelvic fins, the pectoral fins are used to turn the body to the left and right or direct the body upward or downward in addition to stabilizing it in forward locomotion. |

In cartilaginous elasmobranchs the pectoral girdle is a curved cartilaginous bar nearly encircling the anterior part of the trunk. The median portion between the bases of the two fins is the *coracoid bar;* the long processes extending dorsally above the articulation of the fins are the *scapular processes;* the ends of the scapular processes may consist of separate pieces, the *suprascapular cartilages.* The *pectoral* fin is supported at its base by a series of cartilaginous pterygiophores and distally by dermal fin rays. The pterygiophores consist of a proximal row of enlarged *basals* and several distal rows of smaller *radials.* There are generally three basals: an inner or posterior one, the longest, the *metapterygium;* a middle one, the *mesopterygium;* and an outer or anterior one, the *propterygium.*

In bony fishes, the pectoral girdle contains many components. On a skeleton of a teleost note a series of bones extending from the ventral to the dorsal side in front of the pectoral fins. The uppermost bone of the series, the *posttemporal,* usually a forked bone, is attached by one fork to the rear of the skull. Extending from the posttemporal to the ventral midline are the *supracleithrum, cleithrum,* and *clavicle.* The endochondral bones lie more or less concealed by these bones and consist on each side of a scapulocoracoid, usually divisible into a lower coracoid and upper scapular region (fig. 7.9). These may be cartilaginous or in many bony fishes are partially ossified into two bones, coracoid and scapula. The homology of these with the bones of the same names in tetrapods is, however, uncertain.

| | |
|---|---|
| 4. Pectoral Girdle and Forelimb of Amphibians | The importance of the extinct embolomerous and rhachitomous amphibians in the story of the evolution of the pectoral girdle was noted above (p. 252). They originally had a series of bones similar to that of bony fishes plus a midventral interclavicle (lacking in all fishes), and the post- |

temporal was attached to the skull in at least the earliest forms. Some of the bones were soon lost, so that the dermal girdle of primitive tetrapods is composed of an interclavicle, clavicles, and cleithra (fig. 7.10). The endochondral girdle of these early amphibians consists of a single scapulocoracoid on each side (fig. 7.10*B*).

In urodele amphibians, as an adaptation to aquatic life, where less support is needed, the dermal girdle has been wholly lost and the endochondral remains in a largely cartilaginous condition, probably through the evolutionary process of neoteny. Study *Necturus* and *Cryptobranchus*. The girdle consists of two halves, the scapulocoracoids. The ventral coracoid region forms a flat plate of cartilage, often overlapping its fellow. Dorsally above the glenoid fossa is seen the scapula, the only ossified part of the girdle; its dorsal border bears the *suprascapular cartilage*. In *Cryptobranchus* and a number of other urodeles, the sternum is present as a cartilaginous piece between the rear parts of the coracoid cartilages. In *Necturus* the sternum is represented by two or three transverse cartilages in the ventral myosepta.

Note, again, the primitive posture of the limbs (p. 242). Identify the bones of the forelimb; the small wrist elements are impractical to distinguish (there are six or seven cartilages in three rows).

In association with their terrestrial locomotion and jumping ability, the girdle of anuran amphibians is better ossified, and a more definite sternum is often present. In primitive frogs the cartilaginous portions of the coracoid region overlap and the sternum is limited to a piece between the rear parts of these. In common frogs (*Rana*) note the firm construction and the union of girdle and sternum. The girdle consists ventrally on each side of an anterior clavicle and a posterior coracoid; the clavicle covers the precoracoid, which remains cartilaginous. The sternum consists of two parts: a *prezonal* part in front of the girdle, termed the *omosternum* and composed of ossified and cartilaginous sections; and a *postzonal* part or *xiphisternum* behind the girdle, also partly ossified and terminating in a rounded cartilage. The dorsal part of the girdle is composed of the scapula, participating in the glenoid fossa, and the suprascapular cartilage above the scapula; this cartilage is partially covered by a thin, flat bone, now believed to represent the cleithrum.

Note the lack of attachment of the pectoral girdle of amphibians to the vertebral column. There are always four fingers in present amphibians; it is now accepted that the fifth is lacking. A prepollex is present in Anura.

**5. Pectoral Girdle, Sternum, and Forelimb of Reptiles**

Primitively, the girdle resembles that of amphibians, having a dermal part, composed of an interclavicle, clavicles, and often traces of cleithra, and an endoskeletal part, consisting of scapula and precoracoid in the lines of reptiles that led to present reptiles and birds, and of scapula, coracoid, and precoracoid in the lines that led to mammals (fig. 7.10*C*–

*E*). Among present reptiles *Sphenodon* has a representative girdle (fig. 7.10*D*). The dermal part is composed of the T-shaped interclavicle and slender clavicles; the endoskeletal girdle is largely cartilaginous; and the sternum between the coracoprecoracoid cartilages is wholly so. In the coracoprecoracoid cartilages there is a single bone on each side, the precoracoid, having a coracoid foramen; the scapula is also ossified and bears on its upper end a large suprascapular cartilage. A similar pectoral girdle is found among lizards, except that the ossified portions are scalloped to form openings.

---

The pectoral girdle of the alligator consists of the midventral dagger-shaped interclavicle, a stout ventral bone on each side, the precoracoid, and a stout dorsal bone, the scapula. Clavicles are lacking. The sternum is a plate of cartilage between the ventral ends of the coracoids, underlain by the interclavicle. The sternum is drawn out posteriorly into long, curved cartilages, the *xiphisternal* horns, behind which is the series of gastralia, or ventral ribs. Note attachment of the sternal ribs to the sternum and xiphisternal horns.

The turtle lacks a sternum—no doubt because it has the plastron; and its pectoral girdle is peculiar in that it has moved to the inside of the ribs. The dermal part of the turtle's girdle has been incorporated into the plastron. The entoplastron and epiplastra represent the interclavicle and clavicles, respectively, and their fusion with the dermal osteoderms; the other elements of the plastron may include the gastralia. The endoskeletal part of the girdle is a tripartite bony structure. There is a ventral coracoid, an elongated dorsal scapula, reaching the carapace, and an anterior projection from the scapula, the *acromion process,* believed to be homologous to the acromion process of mammals. Identify the bones of the forelimb of the turtle. The carpus of turtles is in a fairly primitive condition and should be studied if good specimens are available. At the distal end of the ulna are two bones, an outer ulnare and an inner intermedium. The center of the carpus is occupied by a long bone, composed of fused radiale and centrale and probably other centralia, fused with the ulnare and intermedium. A row of five carpalia articulates with the metacarpals.

---

**6. Pectoral Girdle, Sternum, and Forelimb of Birds**

Requirements for flight have resulted in substantial remodeling of the girdle, sternum, and forelimb of birds. The bones have been lightened through a general thinning and pneumatization. The pectoral girdle acts as a triradiate strut to anchor and support the wing against the body. A long sword-like scapula lies above the ribs; a stout precoracoid extends ventrally to the sternum; and the clavicles fuse anteroventrally with the interclavicle to form the *wishbone* or *furcula*. Precoracoid and scapula participate in the glenoid fossa. The sternum is an enlarged bone bearing a strong ventral projection, the *keel* or *carina,* which serves as the attachment of the powerful flight muscles. As in reptiles, the ribs join the sternum through their costal cartilages. The anterior end of the sternum has short *costal* processes, posteriorly each side has two long xiphisternal processes.

On a demonstration specimen of an ostrich or other flightless
bird note the smooth convex sternum lacking a carina. Such birds
are called *ratite* birds, whereas flying birds that have the keel are
called *carinate* birds.

The forelimb of birds is adapted for flight by the concentration of
its distal elements. The stout humerus fits into the glenoid by a convex head, to either side of which are prominent projections, the
greater and lesser tuberosities, for muscle attachment. The lesser
tuberosity continues distally as a sharp ridge, the *deltoid ridge,*
which marks the preaxial or radial side of the limb and so indicates
the dorsal rotation the humerus has undergone. The greater tuberosity is postaxial and bears on its under surface a large hole, the
*pneumatic foramen,* the entrance into the air space in the humerus.
The radius is more slender than the ulna, which has at its proximal
end a projection, the *olecranon process* or *elbow,* here met for the
first time. The wrist is greatly altered, consisting of only two separate
carpals, the homologies of which are unclear. The remaining wrist
bones are fused to the metacarpals to form the carpometacarpus,
which includes three metacarpals—two long ones and a short hump
on the radial side of these. These three metacarpals are now generally considered to be the first, second, and third; they bear the corresponding digits. The second digit normally has two phalanges. A
claw is frequently present on the first finger. *Archaeopteryx,* the
fossil "bird" that retained a number of reptilian characteristics, had
separate metacarpals and claws on all three fingers.

## 7. Pectoral Girdle, Sternum, and Forelimb of Mammals

As already explained, the reptilian ancestors of mammals had two bones
in the coracoid region, precoracoid and coracoid. This reptilian condition of the pectoral girdle is retained in monotremes (fig. 7.10*G*), which
have a dermal girdle of interclavicle and clavicles and an endoskeletal
girdle of precoracoid (called epicoracoid in older accounts), coracoid,
and scapula on each side. In all placental mammals the interclavicle and
the precoracoid have vanished and the coracoid is reduced to a projection on the scapula. Clavicles also are frequently reduced or wanting, so
that the scapula becomes the most important part of the mammalian
pectoral girdle.

Study the pectoral girdle of the rabbit or cat. It consists, on each
side, of a clavicle and a scapula. The clavicles are small, slender
bones imbedded in muscle; and, since they are not articulated to
any other part of the skeleton, they generally fall off in prepared
skeletons. They will be seen in place later, when the muscles are
dissected. The scapulae or shoulder blades are the large, flat, triangular bones above the anterior ribs. The mammalian scapula has
certain characteristics that differentiate it readily from the scapulae
of other tetrapods. It is triangular in form, with the apex articulating
with the humerus. On its outer surface there is a prominent ridge,
the *spine* of the scapula, whose ventral end terminates in a pointed
projection, the *acromion* process, usually serving to receive one end

of the clavicle. Above the acromion is a backward projection, the *metacromion,* very long in the rabbit. The apex of the scapula is concavely curved, forming the *glenoid fossa.* From the anterior side of the glenoid rim there projects medially a small beaklike *coracoid process,* the vestige of the coracoid bone, which embryologically forms by a separate center of ossification. In order to facilitate the description of muscle attachments, the various surfaces and borders of the scapula are named as follows: the part of the external surface anterior to the spine is the *supraspinous fossa;* the part posterior to the spine, the *infraspinous fossa;* the whole of the internal surface is the *subscapular fossa;* the dorsal border is the *vertebral border;* the anterior margin, the *anterior border;* the posterior margin, the *axillary border.*

The sternum consists of a longitudinal series of pieces, the *sternebrae*—eight in the cat, six in the rabbit. The first sternebra is called the *manubrium* and articulates with the first thoracic rib at its center. The next six (cat) or four (rabbit) sternebrae constitute the *body* of the sternum. The last piece is called the *xiphisternum* and terminates in a *xiphoid* or *ensiform* cartilage. Note points of articulation of the sternebrae with the ribs.

The forelimb is fairly typical. The *humerus* has a large, rounded *head* fitting into the glenoid fossa and *greater* and *lesser tuberosities* at the sides of the head. The anterior surface of the humerus below the tuberosities is slightly elevated into *ridges* or *crests* (two in the cat, one in the rabbit), which serve as points of muscle attachment. The lower end of the humerus is rounded for articulation with the bones of the forearm and is divided into two portions—an outer mass, the *capitulum,* and a medial mass, the *trochlea.* Above the capitulum is a projecting ridge, the *lateral epicondyle;* and a similar *medial epicondyle* is situated above the trochlea. Near the medial epicondyle the bone is pierced by an opening, the *supracondyloid foramen,* absent in the rabbit.

The forearm consists of the smaller *radius* and larger *ulna;* the proximal end of the ulna forms a prominent projection, the *olecranon* or *elbow.* Distal to this is a deep semicircular concavity, the *semilunar notch,* which articulates with the trochlea of the humerus. The distal border of the notch forms another projection, the coronoid process. Observe that the radius crosses obliquely in front of the ulna, and review pages 241–44 regarding the cause of this condition.

The wrist is composed of a number of small bones, arranged in two transverse rows. The proximal row consists of four pieces in the rabbit, three in the cat; the distal row, of five in the rabbit, four in the cat. Articulating with the distal end of the *radius* is the large *scapholunar* bone in the cat, separated in the rabbit into a medial *navicular* (*radiale*) and a lateral *lunate bone* (*intermedium*). Lateral to the lunate portion or bone and articulating with the ulna is the *triquetral bone* (*ulnare*). The *pisiform* is the element projecting prominently lateral to the triquetral bone in the cat or behind it in the rabbit; the pisiform is a sesamoid bone, formed in a tendon. The distal row of pieces beginning at the medial side and proceeding laterally are:

the *greater multangular* (*first carpale*), the *lesser multangular* (*second carpale*), the *central* (*centrale,* missing in the cat), the *capitate* (*third carpale*), and the *hamate* (*fourth* and *fifth carpalia* fused). These carpales are situated at or near the proximal ends of their respective metacarpals. These are five *metacarpals,* of which the first is very much reduced, and five digits whose terminal phalanges support the horny claws.

**D. Some Variation of the Appendicular Skeleton in Tetrapods**

1. Limb Proportions

Gravity imposes limitations on the size and form of terrestrial tetrapods. These limitations are readily observed in the effect of increase in body weight upon the stoutness of limb bones. Large animals tend to have proportionately shorter and stouter limbs than those of smaller relatives, and this is directly attributable to surface-volume relationships. The strength of a limb is proportionate to the area of its cross-section. If a terrestrial tetrapod doubles its linear dimension, its volume or weight is the cube of the linear dimensions and will thus be eight times its original weight, whereas the cross-sectional area of the limb increases as the square of the linear dimension and thus will be only four times its original area. Thus, the stoutness of the limb obtained by a simple proportionate increase is insufficient for support of the new weight. The limb must and does exhibit a disproportionate increase to maintain enough strength to support the increase in weight (fig. 7.11).

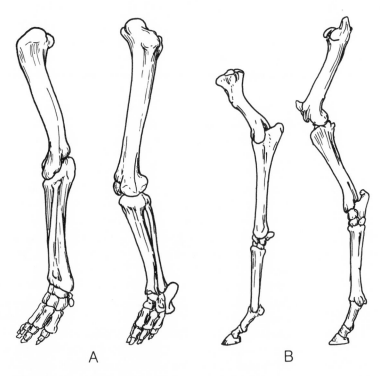

A                                B

Fig. 7.11. Comparison of stoutness and segment length in different-sized mammals. *A,* a graviportal elephant. *B,* an unguligrade horse (limbs drawn to same length).

Relative limb length reflects different locomotor patterns. In general, the hind limbs are slightly longer than the forelimbs, since they are the major propulsive agents and bear the larger muscle masses. In bipedal saltators, such as kangaroos, the hind limb may be nearly twice as long as the forelimb. The converse situation occurs in brachiators, such as gibbons.

## 2. Fossorial Tetrapods

Tetrapods have adapted to a subterranean existence in two strikingly different ways. Caecilians and most fossorial reptiles have lost or greatly reduced their limbs and, thus, the supporting appendicular skeleton, whereas anurans and mammals have enlarged one pair of limbs as digging organs. In reptiles, the shift toward limblessness is associated with an increase in number of trunk vertebrae and initially only a size reduction of the appendicular skeleton. Later stages include a loss of digits and mesopodial elements, loss of interclavicle and sternum, and reduction of pubis and ischium. The ilium and scapula tend to be the most persistent girdle elements during limb reduction. These limbless reptiles burrow by alternately shoving the head into the soil and then pulling the body up, continually repeating this sequence, or by swimming through friable soil with serpentine movements of the body. In mammals the animal actively digs a burrow with its forelimbs. The humerus becomes a stout but short bone with large crests for muscle attachment. The epipodial and other distal bones become enlarged and flattened to form a shovel-like foot. The phalanges are large and flat, and the ultimate phalanx of each digit bears a long, sharp claw.

## 3. Aquatic Tetrapods

Of the many tetrapods living in water, only a few have been able to completely free themselves from periodic returns to the land. It is in these completely aquatic tetrapods, such as ichthyosaurs and whales, that the appendicular skeleton is highly modified, although the trends are discernible in the less aquatic species. The most apparent modifications are reduction in overall length of the limb; relative increase in size, both width and length of the foot, by a broadening and elongation of the phalanges, increase in number of phalanges per digit (hyperphalangy), increase in number of digits (hyperdactyly), thereby producing a flipper or fin-like structure; loss of movable articulation in joints; and a tendency toward the loss of hind limbs.

## 4. Cursorial Tetrapods

Reptiles and birds adapted for running are bipedal, at least while running, whereas the cursorial mammals are quadrupedal. Although the locomotor modes of these two cursorial groups are different, the modifications of the limb skeleton show similar trends: a reduction in relative length of propodial skeleton and a concentration of the appendicular muscle mass in this region; an increase in relative length of the epipodial, mesopodial, and metapodial bones and a reduction in their weight; and the development of a vertical limb posture with a concomitant reduction in lateral swing of the limb. Cursorial mammals also show the loss or great reduction of three or four digits so that in the final unguligrade condition the animal is supported on two toes (artiodactyls) or on one toe (perissodactyls).

**Summary**

1. Paired appendages may have been derived from continuous ventro-lateral finfolds or from a series of ventrolateral spines. Their initial appearance probably reduced the motion deviations of pitching and rolling during swimming.

2. Girdles arose secondarily to the origin of fins and probably from the fusion in the midline of enlarged basal pterygiophores.

3. The pelvic girdle was derived entirely from the basal and is primitively composed of bars or plates of cartilage in which ossification subsequently occurred with the formation of endochondral bone.

4. The pelvic girdle of tetrapods contains three pairs of bones: pubis, ischium, and ilium. At first these are closely attached, but later a large opening arises between pubis and ischium. In birds and mammals the three bones become fused on each side.

5. The pelvic girdle is not attached to the vertebral column in fishes. In tetrapods the pelvic girdle articulates with the vertebral column through the sacral ribs.

6. The pectoral girdle contains cartilage or endochondral components derived from the pterygiophores and dermal components derived from the bony thoracic armor.

7. The endochondral part of the girdle, consisting in early tetrapods of a scapulocoracoid cartilage or bone on each side, may give rise to either two or three endochondral bones, a dorsal scapula and one or two ventral bones, an anterior precoracoid and a posterior coracoid. Only one ventral bone, believed to be the precoracoid, ossifies in early tetrapods and in extant reptiles and birds. Two bones, precoracoid and coracoid, are present in extinct mammal-like reptiles and monotremes. In placental and marsupial mammals the precoracoid is lost and the coracoid reduces to a projection on the scapula.

8. The dermal part of the pectoral girdle overlies the endochondral part in fishes and also attaches to the skull. In tetrapods, the attachment to the skull is lost and the number of dermal bones reduced. The typical dermal girdle of primitive tetrapods consists of a median unpaired interclavicle and paired clavicles and cleithra. The cleithra are soon lost; the interclavicle persists in present reptiles, birds, and monotremes but is lost in marsupial and placental mammals. In tetrapods, the pectoral girdle never articulates with the vertebral column.

9. The mammalian scapula is distinguished by the presence of a spine and of the coracoid process mentioned in paragraph 7.

10. The sternum is limited to tetrapods and originally consists of a cartilaginous plate; later this has one to several ossifications in a linear arrangement. Ribs articulate with the sternum in most amniotes.

11. The phylogenetic origin of the sternum is disputed; the most acceptable hypothesis derives it from the fusion of the anterior members of a series of true abdominal ribs (parasternals).

12. The bones of the limbs are probably derived from the cartilaginous fin supports (pterygiophores).

13. Each limb is divided into five segments: propodium, containing one bone, femur or humerus; epipodium, containing two parallel bones, radius and ulna or tibia and fibula; mesopodium, ankle or wrist; metapodium, sole or palm; and phalanges or digits. Ankle or wrist consists primitively of twelve small bones; a proximal row of three bones: tibiale or radiale, intermedium, and fibulare or ulnare; a middle group of four centralia; and a distal row of the five tarsalia or carpalia. Sole or palm consists of five metatarsals or metacarpals. There are typically five digits. Each digit has a skeletal axis of a row of phalanges; the original number of phalanges for the five digits from the preaxial to the postaxial side was 2, 3, 4, 5, 4.

14. Loss or fusion of tarsals or carpals, metatarsals or metacarpals, and digits and phalanges is common. These losses begin from both preaxial and postaxial sides and proceed toward the middle. Increase in the number of phalanges (hyperphalangy) occurs in aquatic (extinct) reptiles and aquatic mammals.

**References**

Cracraft, J. 1971. The functional morphology of the hind limb of the domestic pigeon, *Columba livia. Bull. Amer. Mus. Nat. Hist.* 144(3): 173–267.

Charig, A. J. 1972. The evolution of the archosaur pelvis and hindlimb: an explanation in functional terms. In *Studies in vertebrate evolution,* eds. M. S. Joysey and T. S. Kemp. Edinburgh: Oliver and Boyd.

Eaton, T. H. 1951. Origin of tetrapod limbs. *Amer. Midl. Nat.* 46:245–51.

Etheridge, R. 1965. The abdominal skeleton of lizards in the family Iguanidae. *Herpetologica* 21:161–68.

Evans, F. G. 1957. *Stress and strain in bones.* Springfield, Ill.: Charles C. Thomas.

Gambaryan, P. P. 1974. *How mammals run.* New York: Wiley.

George, J. C., and Berger, A. J. 1966. The avian skeleton. In *Avian myology,* eds. J. C. George and A. J. Berger. New York: Academic Press.

Gray, J. 1968. *Animal locomotion.* London: Weidenfeld and Nicolson.

Gregory, W. K., and Raven, H. R. 1941. Studies on the origin and early evolution of paired fins and limbs. *Ann. N.Y. Acad. Sci.* 42:273.

Holmgren, N. 1949. On the tetrapod limb problem—again. *Acta Zool.* 30:485–508.

Jenkins, F. A. 1971. Limb posture and locomotion in the Virginia opposum (*Didelphis marsupialis*) and in other non-cursorial mammals. *J. Zool.* 165:303–15.

Nursall, J. R. 1962. Swimming and the origin of paired appendages. *Amer. Zool.* 2:127–41.

Osburn, R. C. 1906. Adaptative modifications of the limb skeleton in aquatic reptiles and mammals. *Ann. N.Y. Acad. Sci.* 16:447–482.

Preuschoft, H. 1970. Functional anatomy of the lower extremity. *Chimpanzee* 3:221–94.

Romer, A. S. 1956. *The osteology of the reptiles.* Chicago: University of Chicago Press.

Schaeffer, B. 1941. The morphological and functional evolution of the tarsus in amphibians and reptiles. *Bull. Amer. Mus. Nat. Hist.* 78: 395–472.

Stokely, P. S. 1947. Limblessness and correlated changes in the girdles of a comparative morphological series of lizards. *Amer. Midl. Nat.* 38:725–54.

Sukhanov, V. B. 1974. *General system of symmetrical locomotion of terrestrial vertebrates and some features of movement of lower tetrapods.* New Delhi: Amerind.

Sy, M. 1936. Funktionell-anatomische Untersuchungen am Vogelflugel. *Ornith.* 84:199–296.

# 8

# The Endoskeleton:
# The Comparative Anatomy
# of the Skull and the Visceral Skeleton
## *Herbert R. Barghusen and*
## *James A. Hopson*

**A. Introduction: Concept of the Skull**

The *skull* is that part of the axial skeleton that encloses the brain, houses organs of special sense, and forms structures intimately involved in feeding. The skull is the most complex of all parts of the skeleton because it is derived from several different sources and is involved in many different functions. It consists of three embryonic components: the *chondrocranium,* part of the endoskeleton that forms around the brain and organs of special sense; the *splanchnocranium,* originally forming in the pharyngeal region as endoskeletal gill supports; and the *dermatocranium,* consisting of membrane bones that, in the majority of vertebrates, superficially invest the endoskeletal portions of the skull. Each of these three components has had a long, complex evolutionary history, and the relative contribution of each to the skull has varied greatly in different groups.

In the material that follows, we will attempt to understand the skull by studying its three components separately. Be sure to review the sections in chapter 4 on developmental considerations and chapter 5 on the ossified integumental skeleton. That information will improve your understanding of the evolutionary and ontogenetic development of the skull and visceral skeleton as they are considered in this chapter.

**B. The Chondrocranium**

1. Development

The best way to understand the complex structure of the chondrocranium is to examine how it is gradually built up from various components during embryonic development. The formation of the chondrocranium has been studied in many vertebrates and found to differ considerably in detail among different groups; here only a generalized account can be given.

The early developmental events in the normal formation of the chondrocranium take place in relation to, and are perhaps dependent upon, the embryonic brain and sense organs and the anterior part of the notochord (fig. 8.1*A*). In the head mesenchyme below the brain there appear two pairs of elongate cartilages: the *parachordals,* on either side of the notochord; and the *trabeculae,* anterior to these (fig. 8.1*A, B*). In a number of vertebrates a pair of *polar* cartilages is present between parachordals and trabeculae; in others they seem to be represented by the rear ends of the trabeculae. Cartilaginous capsules also develop from mesenchyme around the chief sense organs of the head: *nasal capsules* around the olfactory sacs, *optic capsules* around the eyes, and *otic* or *auditory capsules* around the internal ears. The optic capsules do not participate in the formation of the chondrocranium because the eyes

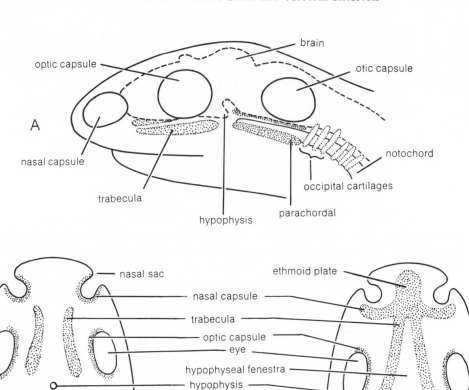

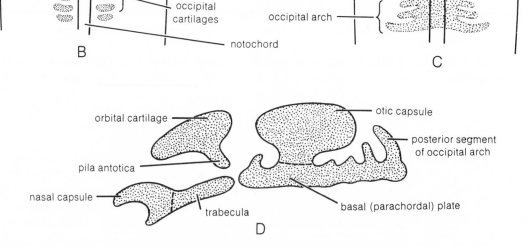

Fig. 8.1. Diagrams of the development of the chondrocranium. *A*, relationship of the main elements of the early developing chondrocranium to structures in the head. *B*, early stage, and *C*, later stage of the developing chondrocranium. *D*, shark chondrocranium before complete fusion of the main embryonic components. Modified from Goodrich 1930.

must remain freely movable, but the other two sense capsules become immovably incorporated into it. Behind the parachordals appear a small but variable number of *occipital* cartilages; these are derived from the sclerotome of the first few postotic somites and are homologous with the rudiments of vertebrae forming farther back behind the head. Together they form the *occipital arch* of the definitive skull.

The parachordals fuse medially to form the *basal plate,* which encloses the notochord (fig. 8.1*C*). The basal plate unites with the otic capsules on each side and with the occipital cartilages behind. The anterior ends of the parachordals usually become connected by a transverse bar, the *acrochordal* cartilage; and in this process an opening, the *basicranial fenestra,* is usually left behind the acrochordal bar (fig. 8.1*C*). Polar cartilages (when present) and trabeculae unite with each other and with the anterior ends of the parachordals. The anterior ends of the trabeculae fuse to form the *intertrabecular* or *ethmoid plate.* The ethmoid plate fuses with the nasal capsules or, more commonly, the trabeculae themselves form much of the nasal capsules, though the capsules are usually completed by independently arising cartilages. The ethmoid plate also forms a vertical *internasal septum* between the nasal capsules. The posterior ends of the trabeculae remain separate, leaving a large space, the *hypophyseal fenestra,* between them and the acrochordal bar behind; this opening transmits the hypophysis and the internal carotid arteries.

The fusions described above result in the formation of an incomplete cartilaginous trough beneath the brain. In the region between the eyes, side walls of the chondrocranium develop from a pair of *orbital* cartilages (fig. 8.1*D,* not to be confused with the optic capsules), which connect with the trough by three main outgrowths. From back to front these are the *antotic* pillars, connecting with the acrochordal region just in front of the otic capsule (fig. 8.1*D,* 8.2); the *metoptic* or *postoptic pillars,* fusing with the trabeculae behind the optic nerve (fig. 8.2); and the *preoptic pillars,* fusing with the trabeculae in front of the optic nerve. Of these, the first is the most important and constant. Outgrowths from the upper part of the orbital cartilages extend forward (the *anterior marginal taenia* or *sphenethmoid commissure*) to join the nasal capsules and backward (the *posterior marginal taenia*) to join the otic capsules (fig. 8.2). The former may fuse to form a flat median plate below the anterior part of the brain, the *planum supraseptale.* In the space below the planum a vertical midline septum, the *interorbital septum,* may form between the orbits from the fusion of the preoptic pillars with upgrowths of the trabeculae; it is continuous anteriorly with the internasal septum.

In most vertebrates, there is no roof to the chondrocranium except in the occipital region, where one or more vertebra-like elements of the occipital arch are completed above by a roof (*tectum*). The roofing arch may form a connection between the otic capsules (thus forming a *synotic tectum*) or it may lie behind this (*posterior tectum*), or both may be present. The large opening surrounded by the occipital arch is the *foramen magnum,* through which the brain is continuous with the spinal cord.

A cartilaginous chondrocranium resulting from the fusion of originally separate cartilages, as described above, occurs in the embryo of all verte-

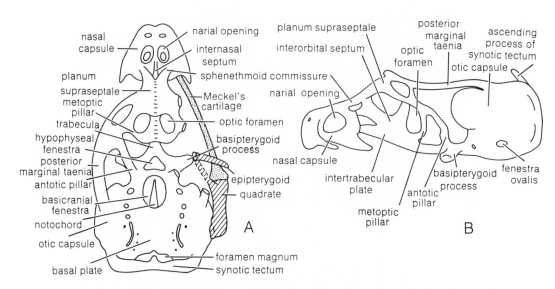

Fig. 8.2. Diagrams of the fully developed chondrocranium of the lizard *Lacerta* (from Goodrich 1930, and deBeer 1937). *A*, dorsal view with portions of the orbital cartilage removed on the right side to show the relations of the mandibular arch elements to the chondrocranium in amniotes. Palatoquadrate elements indicated by diagonal lines, Meckel's cartilage by stipple. *B, l*ateral view.

brates. It typically has no roof except at the rear end. In the adult, with the exception of agnathans and chondrichthians, it usually becomes partly or almost completely ossified by the formation of centers of ossification; further, it becomes roofed and floored by dermal bones. In sharks, where an unossified chondrocranium forms the adult skull, upgrowths of cartilage form a nearly complete roof.

The chondrocranium of the crossopterygian fishes ancestral to tetrapods possessed a movable transverse joint that divided it into anterior and posterior divisions (fig. 8.6*A*); the anterior half formed in the trabecular region, the posterior half in the parachordal region. This intracranial joint was an important part of the mechanism allowing movements within the skull that were significant in feeding. In the earliest tetrapods, this joint was already lost, though its former position is marked by a suture (fig. 8.6*C*).

## 2. Chondrocranium of the Dogfish

Study specimens preserved in fluid. The chondrocranium is a cartilaginous mass without divisions, broad and flat above, narrower and more irregular below. Its anterior trough-like region is termed the *rostrum.* On each side at the base of the rostrum is a *nasal capsule;* in the middle of the chondrocranium is a large lateral depression, the *orbit,* which in life holds the eye; behind the orbits the region on each side is the *otic capsule.*

### a. Dorsal Surface

The rostrum contains a large cavity, the *precerebral cavity,* filled in life with a gelatinous material; this is continuous posteriorly with the *precerebral fenestra,* which opens into the *cranial cavity* (interior of the chondrocranium occupied by the brain), though in life it is closed by a membrane. The nasal capsules are rounded, thin-walled structures whose posterior wall is continuous by way of a ridge, the *antorbital process,* with a thick projecting shelf, the *supraorbital crest,* that forms the dorsal wall of the orbit. The posterior end of the crest continues into a lateral projection, the *postorbital process,* triangular as seen from above. Along the base of the supraorbital crest runs a row of openings, the *superficial ophthalmic foramina,* for the passage of superficial ophthalmic branches of the trigeminal and facial nerves; additional foramina for other branches are seen in the roof of the nasal capsules. In the median line just behind the rostrum is an opening, the *epiphyseal foramen,* through which a brain projection, the epiphysis or pineal body, extends. In the median line behind the level of the postorbital processes is a rounded depression, the *endolymphatic fossa,* in which are situated two pairs of openings, foramina for the passage of the *endolymphatic* (smaller, more anterior holes) and *perilymphatic ducts* of the internal ear. These ducts connect the fluid-filled chambers of the ear with the skull surface. The massive region to each side of the endolymphatic fossa is the *auditory* or *otic capsule* incorporated into the chondrocranium. The terminal opening shortly behind the endolymphatic fossa is the *foramen magnum,* through which the brain continues into the spinal cord.

### b.  Ventral and Lateral Surfaces

The rostrum bears a midventral keel, the *rostral carina;* to each side of the rear end of this keel is a *rostral fenestra,* through which one may look into the cranial cavity. The nasal capsules are situated lateral to these fenestrae; because of their thin walls they are liable to breakage, but when complete they are nearly spherical, with a ventrally directed opening (*external naris*) divided in two by a cartilaginous bar. Behind each nasal capsule the antorbital process already mentioned forms the anterior wall of the orbit. The walls of the orbit are pierced by nerve foramina, of which the largest, ventrally located, is the *optic foramen* for the passage of the optic nerve. Projecting into the orbit behind the optic foramen is a mushroom-shaped structure, the *optic pedicel,* which in life supports the eyeball. The narrow ventral surface of the chondrocranium between the two orbits bears a pair of rounded lateral projections, the *basitrabecular* or *basipterygoid processes,* which furnish articulation for the orbital processes of the upper jaw. Behind the basitrabecular processes, the ventral surface expands to a broad, flat region, the *basal plate,* continuous on each side with the otic capsules. The median streak in the basal plate is the notochord; its anterior end turns dorsally into the cartilage at about the level of a small median foramen through which the united internal carotid arteries pass. The posterior end of the basal plate projects slightly on either side of the notochord as an *occipital condyle,* which articulates with the first vertebra.

**C. The Splanchnocranium**

The *splanchnocranium* or *visceral* skeleton is that part of the endoskeleton that in the most primitive vertebrates serves to support the gills and furnish attachment for the respiratory muscles. It is called the visceral skeleton because the gills, as we will see later, are part of the wall of the digestive tract. The visceral skeleton forms from mesenchyme derived not from mesoderm, but from neural crest ectoderm that migrates downward from the forming neural tube. In fishes the visceral skeleton consists of a longitudinal series of crescentic elements situated in the pharyngeal wall between the gill slits. Each such element is termed a gill *arch* or skeletal visceral arch; it is usually divided into a number of separate pieces in fishes, though not in agnathans, where it is unjoined. Although the gill arches arise in complete independence of the chondrocranium in all vertebrates, they early become associated with it, and in the course of evolution the most anterior ones become incorporated into the skull. There are typically seven visceral arches in fishes (fig. 8.3*A*), although some sharks have as many as nine and the primitive agnathan number was probably greater.

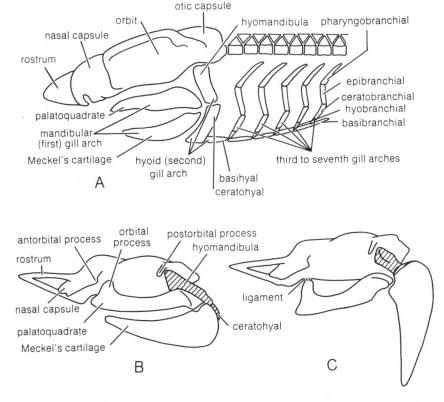

Fig. 8.3. *A,* diagram of the chondrocranium, vertebral column, and gill arches of an elasmobranch to show the parts and relations of the seven gill arches. *B, C,* skull of an advanced hyostylic shark (*Carcharhinus*) to show the freely movable palatoquadrate (after Moss 1972). *B,* with the jaws closed, and *C,* with the jaws maximally opened and palatoquadrate fully protracted and depressed. Further movement of the palatoquadrate limited by ligament binding orbital process to chondrocranium.

| | |
|---|---|
| **1. The Visceral Skeleton of the Dogfish** | Obtain a specimen in which the visceral skeleton has been left attached to the chondrocranium, and study it carefully. The seven gill arches form a series of curved cartilages ventral to the posterior part of the chondrocranium and extending posteriorly to the pectoral girdle. The first gill arch, the *mandibular arch,* is the largest and most modified of the series. When viewed from below, it is seen to consist of dorsal and ventral halves. Each side of the dorsal half is called the *palatoquadrate* or *pterygoquadrate* cartilage; in profile it will be seen that this cartilage is closely applied to the ventral surface of the chondrocranium, to which in life it is attached by a ligament. It sends up a well-developed *orbital* process into the orbit. The palatoquadrate cartilages bear teeth and *constitute the upper jaw of the animal.* The ventral half of the mandibular arch consists of two halves, each of which is known as *Meckel's cartilage.* These bear teeth and together *constitute the lower jaw of the dogfish.* The wide gap between the two jaws is the mouth opening. At their posterior ends the palatoquadrate and Meckel's cartilages join at an acute angle, forming a hinge joint that permits opening and closing of the mouth. |

The second or *hyoid* arch is more slender than the mandibular arch, to the posterior face of which it is closely applied. It consists of a ventral median piece, the *basihyal;* a slender bar, the *ceratohyal,* on each side of the basihyal; and a stout piece, the *hyomandibular,* dorsal to each ceratohyal. The hyomandibular articulates with the otic region of the skull and is bound by ligaments to the articular region of the mandibular arch; it thus acts as a *suspensor* of the jaws. The hyomandibular bears on its posterior margin some slender projections, the *gill rays,* which in life support the gills.

The remaining arches, known simply as *gill* or *branchial arches,* are similar to each other. Each typically consists of five pieces, named from the dorsal side ventrally: *pharyngobranchial,* the most dorsal piece, elongated and directed posteriorly; *epibranchial,* the succeeding, much shorter piece; *ceratobranchial,* another elongated piece; *hypobranchial,* curved ventral pieces, of which there are but three pairs to the five branchial arches; and the two *basibranchials*—an anterior small one situated between the medial ends of the first and second pairs of hypobranchials and a large posterior piece between the bases of the fifth ceratobranchials and terminating in a caudally directed point. Epi- and ceratobranchials bear gill rays. Note that the gill arches are not attached to the vertebral column.

| | |
|---|---|
| **2. The Origin of Jaws and Jaw Suspension** | A great deal of interest, and controversy, has arisen over which of the gill arches of the agnathan ancestors of gnathostomes gave rise to the jaws or mandibular arch. It has been variously argued that one or even two premandibular gill arches and slits were lost, so that the mandibular arch is actually homologous with the second or third arch of the pregnathostome ancestor. Recent work has shown that this theory is probably not true. The most anterior gill slit in fossil agnathans and larval |

lampreys is that of the hyoid arch; therefore the anteriormost gill arch was probably always the mandibular arch.

With the enlargement of the first gill arch to form biting jaws, the need of supporting or buttressing the jaws against the chondrocranium became imperative. As the mandibular arch became elongated backward, to allow an increased gape, it impinged upon the arch behind; the second, or hyoid, arch was then "recruited" to serve as an accessory buttress of the jaw mechanism, supplementing the direct contacts of the palatoquadrate with the chondrocranium. In most fishes, direct fusion of the upper jaws to the braincase would be disadvantageous because they must be free to rotate outward to aid in respiratory movements (and, in some fishes, feeding movements as well). Thus the palatoquadrate of fishes typically has only a few points of contact with the chondrocranium, usually no more than three; from back to front these are (fig. 8.4A):

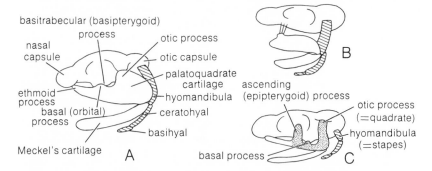

Fig. 8.4. Diagram of types of jaw suspension. A, amphistylic type of jaw suspension showing primitive articulating processes of the palatoquadrate. B, hyostylic type. C, autostylic type of tetrapods with the ascending process, a process found only in tetrapods and sarcopterygian fishes. Amniote condition of palatoquadrate indicated by stipple. Hyoid arch elements indicated by diagonal lines.

### a. Otic Process

Upward projection of the rear end of the palatoquadrate behind the orbit that contacts the otic region; this becomes the quadrate bone of tetrapods.

### b. Basal Process

Usually a low projection articulating with a projection of the chondrocranial base termed the basitrabecular or basipterygoid process. The basal process is probably identical with the orbital process of sharks and is also found in primitive bony fishes and most tetrapods. It is the most persistent of the palatoquadrate attachments.

### c. Ethmoid Process

Forward projection of the anterior end of the palatoquadrate making contact or fusing with the ethmoid region of the skull; found in bony fishes and some amphibians. It is never present in amniotes, where the anterior portion of the palatoquadrate does not develop (fig. 8.4C).

### d. Ascending Process

In sarcopterygian fishes and tetrapods there is an additional connection in front of the otic process and lateral to the basal process; it becomes the epipterygoid bone (fig. 8.4C).

3. Jaw
Suspension
Types

Three broad types of jaw suspension, that is, palatoquadrate support, are recognized, though numerous subtypes can be distinguished.

### a. Amphistylic

The palatoquadrate is attached by ligaments to the chondrocranium directly at the basal and otic processes (and sometimes ethmoid process as well) and indirectly by means of the hyomandibular which contacts the otic capsule and is ligamentously bound to the region of the jaw articulation; found in extinct acanthodians and crossopterygians and in primitive fossil and living sharks (fig. 8.4A).

### b. Hyostylic

Otic contact of the palatoquadrate is absent, and so jaws are suspended primarily by way of ligamentous attachments to the hyomandibular, which is attached to the otic region; basal or ethmoid contact may be retained anteriorly. Hyostyly is characteristic of advanced elasmobranchs and all actinopterygian fishes and is important in the protrusible jaw mechanisms evolved independently in these groups (figs. 8.3B, C, and 8.4B).

### c. Autostylic

Processes of the palatoquadrate articulated or fused with the chondrocranium; hyoid arch does not participate in jaw suspension. Found in lungfishes and tetrapods (fig. 8.4C), where it is clearly secondarily derived from an amphistylic ancestor. A distinct type of autostyle is found in Holocephali; termed "holostyly" by some authors, it is characterized by indistinguishable fusion of the palatoquadrate with the chondrocranium. Another peculiar form of autostyly occurred in fossil placoderms in which the palatoquadrate was fused to the dermal armor covering the cheek.

It is generally believed on theoretical grounds that the primitive type of jaw suspension was autostylic and that the hyoid arch became involved in jaw support somewhat later. Until recently, the fossil record has not provided unequivocal evidence to support this theory. The oldest gnathostomes, the acanthodians, are now known to have been amphistylic, not autostylic as was formerly thought. The autostyly of holocephalans is most likely primitive, but this has been disputed. The problem has now been settled by the recent discovery of primitive fossil elasmobranchs, in which the hyoid arch resembles a normal branchial arch and lacks all contact with the mandibular arch. This undoubtedly represents the unmodified primitive autostylic condition and, further, provides corroboration for the primitive nature of the autostylic condition seen in holocephalans.

The hyomandibular of tetrapods, once freed from support of the mandibular arch, became the *columella* or *stapes,* a sound-conducting element in the middle ear.

**D. The
Dermatocranium**

**1. Origin of the
Dermal Bones
of the Skull**

In addition to the chondrocranium and visceral arches, still another component participates in the definitive skull. This consists of the dermal or membrane bones, collectively termed the *dermatocranium*. As noted above, the chondrocranium in most vertebrates lacks a roof and has imperfect sides. Roof and sides are completed by these dermal bones, which thereupon become the *superficial bones of the skull*. Similarly, bones encase the palatoquadrate and Meckelian cartilages, which are the primitive upper and lower jaws, and became the superficial bones of the jaws. Finally, superficial bones form in the dermis of the roof of the mouth and cover the undersurface of the chondrocranium. In this fashion the chondrocranium and jaws become sheathed in dermal bones, originating in the dermis of the skin and hence really belonging to the exoskeleton. These encasing bones are known as *dermal, membrane,* or *investing* bones; they develop directly in the mesenchyme. Early stages of the dermal skull are best seen in primitive bony fishes, such as extinct crossopterygians, but a good idea of them may be gained from examining a living holostean actinopterygian, the bowfin (*Amia*).

**2. Dermal Bones
of a Primitive
Bony Fish**

Study the head bones of the bowfin, *Amia,* and compare with figure 8.5*B*. For comparison, an early fossil actinopterygian skull pattern is shown in figure 8.5*A;* note that certain bones have been lost in *Amia*. Note the general similarity in the arrangement of skull bones in these actinopterygians with that of a crossopterygian near the ancestry of tetrapods (fig. 8.5*C*). The larger head bones are probably really homologous between the major groups of bony fishes and early land vertebrates (see fig. 8.7), but the smaller bones vary so greatly in pattern, even between members

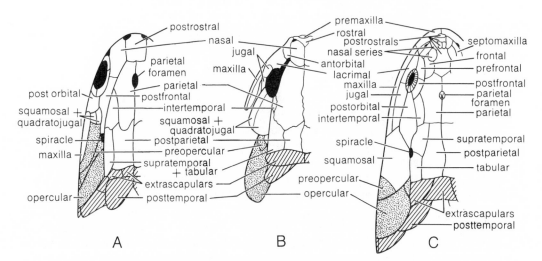

Fig. 8.5. Dermal bones of the skull of three primitive fishes. *A,* an extinct actinopterygian (*Moythomasia*). *B,* the bowfin (*Amia*). *C,* an extinct crossopterygian (*Eusthenopteron*). The opercular series (stippled) and bones of the neck region (diagonal lines) are lost in tetrapods. *A* and *C* after Moy-Thomas and Miles 1971.

of the same species, that their homologies are very uncertain. The names currently applied to the head bones of crossopterygians are based on their reasonably certain homology with the head bones of tetrapods; except in the snout, where the pattern of bones is extremely variable, there is a nearly one-to-one correspondence between the skull bones in the two groups. Of course, in the transition to land life certain of the fish bones have been lost, especially those of the operculum and the posterior part of the skull roof (see fig. 8.7). The terminology based on the tetrapod pattern of skull bones is applied here to the actinopterygians, though with less assurance of strict homology than exists for crossopterygians.

Identify on the skull of *Amia* anterior and posterior nostrils and the orbits. Between the anterior nostrils projects the small triangular *rostral* bone; in front of it are the tooth-bearing *premaxillae*. In the space between the four nasal openings are the paired *nasals*. Posterior to them, in the medial line, are the large *parietals* (usually called "frontals" in actinopterygians), and behind them the *postparietals* (usually called "parietals"). The posteriormost dorsal elements, the *extrascapulars* and *posttemporals,* are actually posterior to the skull. Lateral to the nasals are the *antorbitals* and behind them the large *lacrimals* bordering on the front of the orbit. Below each orbit are two small bones which appear to correspond to the *jugal* of crossopterygians. Behind the orbit are two large bones corresponding to the *squamosal* and *quadratojugal*. A small bone forming the posteromedial border of the orbit is probably the *intertemporal* and the larger element directly behind it is the *supratemporal* plus *tabular* of crossopterygians. The sides of the upper jaws are formed by the tooth-bearing *maxillae*. The operculum is supported by several opercular bones.

Note in figure 8.5 that a number of the skull bones found in crossopterygians and tetrapods are absent in *Amia* but are still present in the more primitive actinopterygian. The tendency toward loss of bones in the skull is a common one in vertebrates. Note also that although *Amia* lacks a parietal foramen in the large bones between the orbits, the homology of these elements with the parietals of crossopterygians and tetrapods is established by the presence of the parietal foramen in the topographically identical bones in the fossil actinopterygian (see fig. 8.5*A*).

3. Demonstration of the Origin of Chondrocranium and Dermal Bones

Examine the demonstration specimen of the head of *Amia* in which the encasing dermal bones have been loosened from the underlying chondrocranium. Remove the sheath of dermal bones, noting that they are situated in the dermis of the skin, and note the chondrocranium, similar to that of the dogfish, lying within the sheath. Such a specimen shows in a particularly graphic fashion the origin of the skull from these two sources (the membrane bones and the endoskeletal chondrocranium) because the braincase of *Amia,* like that of other primitive living bony fishes, has lost much of the ossification found in earlier fossil actinopterygians. Although the absence of bone in the skeleton of chondrichthians may be primitive, the absence or reduction of ossification in the endoskeleton of bony fishes as *Amia,* sturgeons, and lungfishes is definitely secondary.

**E. The Compo-**
**sition of the**
**Complete Skull:**
**Characterization**
**of the Primitive**
**Tetrapod Skull**

The skulls of tetrapods are built up from the same components—chondrocranium, splanchnocranium, and dermatocranium—as those of fishes already studied, although the parts are usually welded together to form a more solid unit than exists in fishes. Nevertheless, the contributions of the three components can be distinguished and the composite nature of the tetrapod skull can be demonstrated. It is instructive to compare the components of the skull of a primitive labyrinthodont amphibian (figs. 8.6*B, D,* and 8.7), as representative of a generalized tetrapod, with the skull components of a crossopterygian fish (figs. 8.5*C* and 8.6*A, C*) and note both the fundamental similarities between them and the important differences. In both, the skull consists of cartilage bones ossified in the chondrocranium and mandibular and hyoid arches with a sheath of dermal bones covering the cartilage bones everywhere except on the

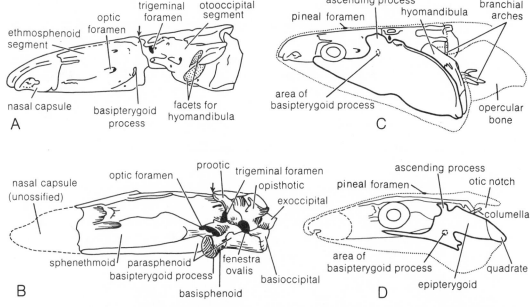

Fig. 8.6. Chondrocranium and splanchnocranium of a crossopterygian and a labyrinthodont amphibian showing differences in proportions and modifications of the palatoquadrate and hyomandibular in the shift from fish to tetrapod. *A,* chondrocranium of the crossopterygian *Eusthenopteron,* with position of intracranial joint marked by arrow. *B,* chondrocranium of the labyrinthodont *Palaeoherpeton,* with position of former intracranial joint marked by arrow. *C,* chondrocranium of *Eusthenopteron* with splanchnocranial components of the skull added. *D,* chondrocranium of *Palaeoherpeton* with splanchnocranial components of the skull added. In *C* and *D,* the positions of the eyes, parietal foramen, and basipterygoid joint are indicated; the outline of the dermatocranium is indicated by dotted lines. (*A, C* modified from Moy-Thomas and Miles 1971; *B,* after Panchen 1964; *D* modified from Panchen 1964, and Thomson 1967).

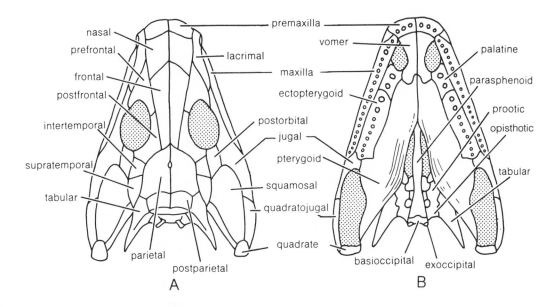

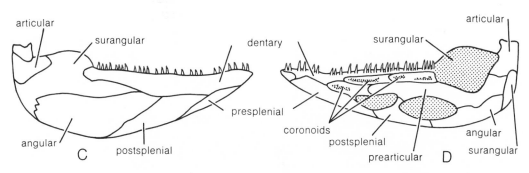

Fig. 8.7. The skull in *A*, dorsal and *B*, ventral view and the lower jaw in *C*, lateral and *D*, medial views of the extinct labyrinthodont amphibian *Palaeoherpeton*. Modified from Panchen 1970.

posteroventral and the posterior surfaces of the skull. The primitive tetrapod skull differs from that of the ancestral crossopterygian most prominently in the absence of a fully developed series of branchial arches and the opercular series of dermal bones, changes associated with the shift from gill-breathing to lung-breathing. Also lost are the large dermal bones of the rear of the fish skull, associated with the development of a flexible neck in tetrapods. There have been important changes in the feeding apparatus that are reflected in the very different proportions of the two skulls (see fig. 8.6) and the great reduction of intracranial mobility within the skull of the early tetrapod. The joint within the braincase, characteristic of crossopterygians, is lost in amphibians (although its former location is sometimes indicated by a suture; see fig. 8.6*B*). The palatoquadrate is reduced in size in tetrapods and becomes more solidly enclosed within the dermal skull so that it lacks the extreme mobility present in fishes. Jaw suspension in tetrapods is autostylic; the hyomandibula no longer participates in supporting the mandibular arch and has been freed to serve as a sound-conducting element in the middle ear, the columella or stapes (fig. 8.6*D*). Finally, the number of bones in the dermal skull, especially in the snout region, has been reduced in the transition from fish to tetrapod, and the pattern of bones has become more standardized. As a result, it is possible to homologize the individual bones of the skull throughout the tetrapods with a great degree of confidence, though in some cases uncertainties still remain.

The skull of an extinct labyrinthodont amphibian, *Palaeoherpeton,* is described below in order to illustrate the morphological pattern from which the skulls of all tetrapods were derived. As will be seen, the adaptive radiation of land-living vertebrates in many cases involved profound modifications of this primitive structural pattern.

## 1. Bones of the Chondrocranium

The chondrocranium of primitive amphibians is a single structure, in contrast to that of crossopterygians, which is subdivided into two halves by a transverse intracranial joint (fig. 8.6*A, B*). It is made up of a series of discrete ossifications that in the adult skull form distinct bones separated from one another by lines of apposition or *sutures;* in some parts of the braincase, however, several such bones may fuse to form a single mass of bone in which individual ossifications cannot be recognized, except in the immature, incompletely ossified skull.

An occipital group of bones is typically ossified around the foramen magnum in the parachordals and tecta at the rear end of the chondrocranium (fig. 8.6*B,* 8.7*A, B*). They are the *supraoccipital* above; the two *exoccipitals,* one on each side; and the *basioccipital* below. Modern amphibians lack supra- and basioccipitals, but these bones were present in labyrinthodonts. Primitively, the basioccipital forms most of the *occipital condyle*—a single projecton for articulation with the vertebral column. The contribution of the exoccipitals is minor in fishes and primitive tetrapods; however, in modern amphibians and in mammals, reduction of the basioccipital and enlargement of exoccipital projections result in the formation of two occipital condyles by the exoccipitals.

The floor of the braincase immediately anterior to the basioccipital is formed by the *basisphenoid,* ossified in the trabeculae and acrochordal cartilage; conspicuous features of the basisphenoid are the *basitrabecular* or *basipterygoid* processes (figs. 8.6*B,* 8.7*B*), which extend laterally to form a movable joint with the epipterygoids (described below) and in some cases the pterygoids as well. In front of the basisphenoid the floor of the braincase is formed by a single extensive ossification, the *sphenethmoid,* which ossifies in the anterior part of the trabeculae. The side walls of the braincase in the region between the otic and nasal capsules are formed by paired wings of the sphenethmoid, ossified in the orbital cartilages and extending upward to enclose the anterior half of the brain. In living reptiles and mammals, the interorbital region of the chondrocranium is usually not well ossified, but several bones may form here, presumably remnants of the single sphenethmoid ossification of primitive tetrapods. In reptiles there may form a *laterosphenoid* in the antotic pillars and an *orbitosphenoid* in the metoptic pillars. In mammals, a median ventral *presphenoid,* ossified in the trabeculae anterior to the basisphenoid, and paired lateral orbitosphenoids, ossified in the orbital cartilages, seem to correspond to the primitive sphenethmoid. In addition, some mammals possess an unpaired *mesethmoid* ossified in the internasal septum.

The otic capsules of the chondrocranium are well ossified and intimately associated with the occipital group of bones. There are typically two main ossifications in the otic capsule of tetrapods, an anterior *prootic* and a posterior *opisthotic,* though an additional element, the *epiotic,* is sometimes formed. When the otic bones are indistinguishably fused, as they are in mammals, the single ossification of the otic capsule is designated as the *periotic* or *petromastoid* bone.

As previously explained, the optic capsules never fuse with the skull because of the necessity for retaining free movement of the eyes. The optic capsule becomes the sclerotic cartilage in most vertebrates (though not in mammals, with the exception of monotremes); in some vertebrates these cartilages ossify to become a ring of sclerotic bones around the pupil. Sclerotic bones occurred in labyrinthodont amphibians (inherited from crossopterygian ancestors) but are not found in living amphibians. They are common, however, in both extinct and living reptiles and in birds.

The nasal capsules of tetrapods typically fail to ossify during ontogenetic development. In mammals, however, the *turbinals* or *conchae,* scrolled bones in the side walls of the nasal capsule, are ossified; similar outgrowths of the nasal capsule occur in other tetrapods, though they are not so well developed as in mammals and do not ossify.

## 2. Bones of the Splanchnocranium

### a. In the Palatoquadrate Cartilages

These cartilages are, as we have seen, the dorsal halves of the first or mandibular pair of gill arches that form the primitive upper jaws. Whereas in primitive fishes, such as extinct crossopterygians, most or all of the palatoquadrate ossifies (fig. 8.6C), in tetrapods much of the palatoquadrate remains unossified or may degenerate. The ossified parts consist of two bones on either side: the *epipterygoid,* which ossifies in the ascending process and the *quadrate,* which ossifies in the otic process (see fig. 8.4C). The epipterygoid maintains the primitive basal articulation with the braincase via the basipterygoid process of the basisphenoid. In the primitive amphibian, the epipterygoid is a large element that forms much of the medial wall of the temporal fossa (see below) and extends back to make contact with the quadrate. In modern amphibians the epipterygoid does not ossify. It is reduced to a slender rod in lizards and is rudimentary in turtles; it is absent in snakes, crocodilians, and birds. In mammals, the epipterygoid becomes sutured into the braincase and is called the *alisphenoid.*

The quadrate contacts the otic capsule dorsally and, except in mammals, bears the articulating surface for the lower jaw ventrally; jaw suspension is thus of the autostylic type. It is sutured to the dermal skull in labyrinthodonts so that it is immovably fixed, or *monimostylic;* a fixed quadrate occurs in living amphibians, turtles, crocodilians, and *Sphenodon.* In some vertebrates, notably lizards, snakes, and birds, the quadrate is freed so that the bone is movable, a condition termed *streptostylic.* The quadrate of extinct mammal-like reptiles was also streptostylic, a condition that persists in highly modified form in mammals (see p. 313).

### b. In Meckel's Cartilages

These are the lower halves of the mandibular arches and they function as the lower jaws in elasmobranchs. In bony fishes and tetrapods they become sheathed in dermal bones that bear the teeth and serve as the functional lower jaws; Meckel's cartilage either remains as a cartilaginous core inside the dermal sheath or it atrophies during development, leaving only its rearmost portion to serve as the articulation with the palatoquadrate. This part that forms the lower half of the jaw joint is the *articular* bone. This quadrate-articular type of jaw articulation is characteristic of all tetrapods except mammals; in the latter, both bones have become greatly reduced and have become bones of the middle ear.

### c. In the Hyoid and Other Gill Arches

The dorsal part of the hyoid arch is the hyomandibular cartilage of elasmobranchs and ossifies into a bone, the *hyomandibula,* in bony fishes. This element serves as a suspensor of the mandibular arch (hyostylic type) in most fishes. In crossopterygians, in which the palatoquadrate was freely movable on the braincase, it also served to suspend the jaw; however, in the transition to tetrapods, in which the upper jaw lost a great deal of its former mobility by becoming more firmly buttressed against the braincase and more tightly encased by the outer sheath of dermal bones, the hyomandibula was no longer required to serve as a jaw-supporting structure. Because of its proximity in crossopterygians (fig. 8.6A, C) both to the inner ear (via its contact with the otic capsule) and to the outer surface of the head (via its contact with the opercular bone; see below), the hyomandibula was in the optimal location to conduct sound vibrations from the surface of the head to the inner ear in the earliest land-living tetrapods. Thus it became the *columella* or *stapes* of the tetrapod middle ear (see below). In labyrinthodont amphibians (fig. 8.6B, D), this bone has a dorsolateral orientation, extending from the fenestra ovalis in the otic capsule to the eardrum, which was held in an emargination in the posterior border of the cheek (otic notch).

The rest of the hyoid arch in tetrapods associates with the remaining gill arches, much reduced and simplified, to form the *hyobranchial skeleton.* This consists of the *hyoid apparatus* and the cartilages of the larynx. The hyoid apparatus is a plate or rod of cartilage or bone, situated in the throat and having projecting processes extending to the otic region; a connection to the columella (upper element of the hyoid arch) is present early in development but in the adult the attachment usually shifts directly to the otic capsule. It acts as a support for the tongue and larynx, serves for muscle attachment, and in amphibians plays an important role in buccal respiration. The larynx or voice box is a chamber at the top of the windpipe whose walls are supported by cartilages derived from the gill arches (see chap. 10). The exact composition of the hyoid apparatus and the laryngeal cartilages varies in different vertebrates.

## 3. The Dermal Bones of the Skull and Lower Jaws

### a. The Skull Roof

Since the chondrocranium is typically roofless in tetrapods, there are never any cartilage bones on the dorsal surface of the skull; the skull roof is formed by dermal bones. The general arrangement of these roofing bones was already seen in the study of the skull bones of *Amia* (fig. 8.5*B*), but the exact correspondence of these with the dermal bones of the tetrapod skull remains uncertain despite the identity of the names applied to them. The correspondence between the dermal bones of extinct crossopterygians and primitive tetrapods is much closer and homologies are made with far greater confidence (compare figs. 8.5*C* and 8.7*A*). The dermal skull of a labyrinthodont amphibian (fig. 8.7) is described here because it serves as the pattern from which the skulls of all tetrapods were derived and, furthermore, provides a good connecting link to the fish skull as seen in early crossopterygians.

The dermal skull roof may be somewhat arbitrarily divided into a number of topographic regions to aid in description; these are: the *snout,* situated anterior to the orbits and through which the external nares open; the *margins of the orbits;* the *skull table,* which forms the covering for the braincase and roofs the temporal fossa (which houses jaw-closing muscles) on each side; the *cheeks,* which form the lateral walls of the temporal fossae; and the marginal *tooth-bearing bones* below and anterior to the orbits. A conspicuous emargination bordered by the posterodorsal edge of the cheek and the posterolateral margin of the skull table is the *otic notch*—a hallmark of the labyrinthodont amphibian skull which in life held the eardrum.

Bones of the dorsal surface of the primitive tetrapod snout include: the *nasals,* on either side of the midline; the *lacrimals,* lateral to the nasals and bordering on the external nares; the medially situated *frontals,* which extend back between the orbits to meet the skull table; and the *prefrontals,* lateral to the frontals and contributing to the margins of the orbits. A small element, the *septomaxilla,* lies on the margin of the external narial opening in primitive tetrapods.

Around each orbit is a series of bones: the already mentioned prefrontal; the large ventrolateral *jugal,* which also contributes substantially to the cheek; the *postfrontal,* forming the medial rim of the orbit; and the *postorbital,* forming its posterior rim.

The medial portion of the skull table behind the level of the orbits is formed by paired *parietals,* which enclose a median *parietal foramen;* behind the parietals are paired *postparietals.* The lateral margin of the skull table is formed on either side by *intertemporal, supratemporal,* and *tabular* bones; the last is greatly elongated posteriorly and forms much of the dorsal margin of the otic notch.

The bones forming the cheek are the jugal, mentioned above, and behind it the *quadratojugal* (below) and the *squamosal* (above). The squamosal forms the inferior margin of the otic notch.

The upper jaw of tetrapods is formed by marginal bones of the dermal skull that sheath the remnants of the palatoquadrate cartilages. These marginal bones are termed the *maxillary arch* and include on each side, from front to back: the tooth-bearing *premaxilla* and *maxilla,* plus the jugal and quadratojugal, already mentioned, which form the lower border of the cheek.

Typical of the primitive tetrapod skull (and lower jaw) is the presence of well-defined dermal sculpturing covering the outer surfaces of the superficial bones (not indicated in the figures). This sculpturing indicates the close adherence of skin to bone, for ridges are produced in the bone as a stress response to the connective tissue attachment points for the skin (see alligator discussion, p. 295). Impressions of lateral line canals are also commonly visible on the labyrinthodont skull, providing evidence that many of these early amphibians were aquatic.

### b. The Palate

The ventral surface of the skull also becomes sheathed in dermal bones that conceal the cartilage bones of the true floor of the cranial cavity. These dermal bones form the roof of the oral cavity and hence constitute the *palate.* In crossopterygians they are closely applied to the palatoquadrate, forming a sheath below the primitive upper jaw cartilage. With 'he reduction in importance of the palatoquadrate in tetrapods, the palatal bones become more firmly tied to the dermal roofing bones and the marginal tooth-bearing elements. The ventral surface of the braincase is also sheathed by a dermal bone, the unpaired median *parasphenoid* (figs. 8.6*B,* 8.7*B*). Two pairs of conspicuous openings are seen in the palate; anteriorly, the internal openings of the nostrils, *internal nares,* open into the oral cavity; posteriorly, the large *subtemporal fossae* form openings through which adductor (jaw-elevating) musculature passed from its origin in the temporal fossa to its insertion on the mandible. The premaxillae and maxillae (bones of the dermal roof that constitute the functional upper jaws) form the anterior and lateral margins of the palate and bear the marginal dentition. Internal to these elements are a series of three bones: anterior *vomers,* lying on the midline between the internal nares; *palatines;* and *ectopterygoids* (=*transpalatines*). Large tusk-like palatal teeth are characteristically found on the palatines and ectopterygoids (and sometimes the vomers) in labyrinthodont amphibians. Behind the vomers are a pair of very large *pterygoids* that form the greater part of the palate and extend posterolaterally to meet the quadrates and form the medial margins of the subtemporal fossae.

### c. The Lower Jaw

As many as nine dermal bones sheath each lower jaw in primitive amphibians. On the outer surface (fig. 8.7*C*) these bones include: the *dentary,* bearing the marginal dentition; the *splenial;* the *postsplenial;* the *angular;* and the *surangular.* Forming the medial surface (fig. 8.7*D*) are: the splenial and postsplenial; the dentary, forming the anterior union or *symphysis* of the two lower jaws; three *coronoid* bones, bearing denticles; the *prearticular;* and the angular and surangular. The articulation with the upper jaw is formed, as already noted, by the articular, a cartilage bone ossified in Meckel's cartilage.

As seen in medial view, a large *adductor fossa* is found at the posterior end of the jaw in front of the articular. It is bounded laterally by the surangular and medially by the prearticular. In life, the majority of the jaw-closing musculature passed from the temporal fossa through the subtemporal fossa to insert on the margins and walls of the adductor fossa. Two large *Meckelian fenestrae* are also seen anterior to the adductor fossa; these fenestrae open into a channel running the length of the jaw within the dermal sheath. Meckel's cartilage probably occupied the inferior part of this channel and an intramandibularis muscle may have occupied the superior part.

In the evolution of tetrapods, a number of dermal elements are lost from the lower jaw. The postsplenial and two of the coronoids are lost very early in reptiles, but the remaining six dermal bones are typically retained. Birds lose the coronoid but retain the remaining reptilian bones, though these are typically fused to form a single unit. In mammals the lower jaw is formed entirely by the dentary, the reptilian elements having either disappeared or shifted into the middle ear (see below). The articulation of the lower jaw with the skull in mammals is between the dentary and the squamosal.

### d. The Hyoid and Other Gill Arches

No dermal bones occur in connection with the hyoid or branchial arches.

**F. The Teeth**

Although teeth are functionally part of the digestive system, they are phylogenetically part of the skeleton because, as already learned, they form as a result of delamination and epimesenchymal interactions (see chap. 5). They are analogous to scales in the sense that scale formation is the result of similar developmental processes occurring in the skin tissues covering the body. Teeth or denticles are formed in the skin of the oral cavity; with the origin of jaws the denticles on the margins of the mouth became enlarged to serve as grasping structures to aid in the capture of prey.

1. Teeth of
Elasmobranchs

Examine a variety of available specimens. In many elasmobranchs the biting surface of the jaws resembles the external skin, having a mosaic of similar small or minute, rounded, diamond-shaped, or hexagonal teeth. A great enlargement of the teeth is seen in those species that eat hard food, such as mollusks or crustaceans; examples are the Port Jackson shark (*Heterodontus*), which has small pointed teeth in front and very large flattened ones farther back, and certain skates, in which the middle teeth are elongated, the lateral teeth small polygons. Among the larger predaceous sharks the teeth are enlarged formidable cutting weapons with one or more sharp points and often serrated edges. They are seen to be arranged in rows, of which only one or a few rows are in use, standing upright on the jaw edges, whereas the others lie flat along the inner surface of the jaws. Replacement occurs by the teeth in the back rows moving up, new teeth developing at the inner-most ends of these rows.

The teeth of elasmobranchs are not fastened directly to the jaw cartilages but are held in place only by strong connective tissue bands from the dermis.

2. Teeth of Bony Fishes

When the jaw cartilages and palate become sheathed in dermal bones, the teeth are borne upon the dermal bones as a consequence of their mode of development. Consequently, in bony fishes teeth may occur on any of the palate or jaw bones. Furthermore, in many teleosts teeth may also occur on the gill arches. The teeth of bony fishes vary greatly in shape and arrangement though a simple, pointed form predomi-nates; available materials may be examined. In primitive bony fishes, such as *Amia,* teeth occur on both the maxilla and premaxilla of the upper jaw, but in teleosts the maxilla is usually toothless and in some advanced teleosts the entire marginal dentition is lost. The teeth of bony fishes are usually ankylosed, that is, immovably fixed to the un-derlying bones by way of intervening bone of attachment. In some teleosts the teeth may be attached by fibrous bands and are some-times movably hinged to the bones.

A type of tooth structure termed *labyrinthine* occurred among extinct crossopterygians; the enamel is infolded along longitudinal grooves, often making a complicated pattern in the interior. Similar teeth are in-herited from crossopterygian ancestors by early amphibians, whence the name Labyrinthodontia applied to the latter.

3. Development, Structure, and Arrangement of Tetrapod Teeth

As already indicated in chapter 5, the development of the teeth of bony fishes and of tetrapods is similar to that of a placoid scale. The stratum germinativum of the lining of the mouth cavity differentiates to form the *enamel organ,* whose epithelium, one cell thick, takes on a cup shape (fig. 8.8*A*). The interior of the cup is filled with mesenchyme derived from neural crest, forming the *dental papilla.* The epithelial cells of the enamel organ, termed *ameloblasts,* secrete enamel on their under sur-face; the layer of superficial cells of the dermal papilla, termed *odonto-blasts,* secretes dentin on its outer surface. The combined secretions be-come a tooth, and the remains of the dental papilla persist as the pulp which nourishes the tooth.

Obtain a longitudinal section of a simple tetrapod tooth and identify the following parts (see fig. 8.8*B*) : the *crown,* or shiny upper part of the tooth, which in life projects above the gum or gingiva; the *root,* or dull lower part, which sets into the jaw; the *pulp cavity,* the interior space, filled in life with the *pulp,* consisting of connective tissue, nerves, and blood vessels; the neck, or junction of crown and root; the *dentin,* the bone-like material of which most of the tooth is composed; the *enamel,* the thin, shiny layer covering the dentin of the crown; the *cementum,* a thin layer of bone that coats the dentin of the root.

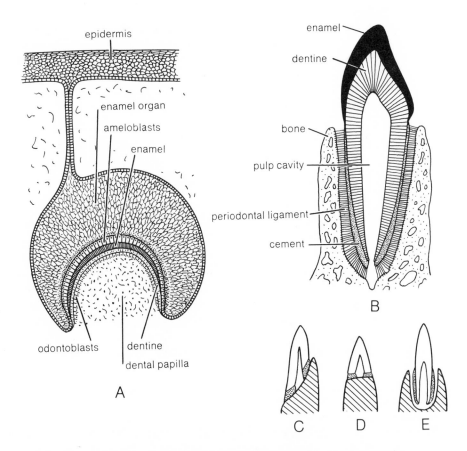

Fig. 8.8. *A*, diagram of the development of a tetrapod tooth, showing the enamel organ. *B*, diagram of a mammalian tooth, showing anchorage of the root in the socket by means of the periodontal ligament. *C*, pleurodont, *D*, acrodont, and *E*, thecodont attachment of the tooth to the jaw bone; stipple indicates cement.

The dentition of labyrinthodont amphibians consisted of numerous small, simple, conical teeth along the jaw margins and a very few large fang-like teeth on the bones of the palate. In modern amphibians the teeth are usually very small and numerous and may occur both on the palate and along the margins of the jaws. They are confined to the upper jaws in most frogs and are entirely absent in some toads.

Simple conical teeth predominate in modern reptiles, but various fossil groups show a wide variety of tooth shapes. Palatal teeth occur in modern lizards, snakes, and *Sphenodon* but are absent in crocodilians. Among the therapsid reptiles ancestral to mammals, palatal teeth are lost and the marginal teeth become progressively differentiated into several functional types, as in mammals (see later). Teeth are lost in turtles and in birds, being functionally replaced in both groups by horny beaks developed from the epidermis.

**4. Terminology of Teeth**

When all the teeth of an animal are alike in form, the dentition is said to be *homodont;* when they vary in form in different parts of the mouth, the term *heterodont* is applied. Heterodonty is characteristic of mammals but, as noted above, is also seen in many fishes; it is also seen to varying degrees in some living snakes and lizards and in some extinct reptiles such as therapsids and certain dinosaurs. Animals with a continuous succession of teeth throughout life, such as fishes, amphibians, and reptiles, are termed *polyphyodont;* those in which a first evanescent set of teeth is replaced by a second permanent set, as in most mammals, are *diphyodont;* and those with a single set of teeth that cannot be replaced, such as toothed whales and a few other specialized mammals, are *monophyodont.* The teeth of tetrapods are immovably fixed to the bones of the skull and jaws in three principal ways (fig. 8.8*C–E*): when the teeth are ankylosed to the inner side of the supporting bones as in most lizards, the term *pleurodont* is applied; when ankylosed to the summit of the bones, as in *Sphenodon,* the teeth are *acrodont;* and when set in sockets, as in crocodilians and mammals, they are *thecodont.* In the first two cases attachment to the bone is direct, by means of intervening *cementum,* and the teeth cannot fall out of the dried skull. In thecodont teeth, the root of the tooth is held in place within the socket by means of collaginous fibers, termed the *periodontal ligaments,* that are anchored in the bone of the jaw at one end and in the cement covering the root at the other end; in the dried skull, the teeth are loosely held in their sockets and thus may sometimes fall out.

**G. The Modern Amphibian Skull Exemplified by Necturus**

The skull of living amphibians is greatly modified from that of the early labyrinthodont amphibian illustrated in figures 8.6 and 8.7; in fact, the modern amphibian skull is among the most specialized of all tetrapod skulls. It is usually moderately to greatly flattened and the braincase has lost much ossification; furthermore, many dermal bones are lost, notably in the temporal roof. Such trends are especially notable in those frogs and salamanders that are paedomorphic (that retain larval features in the adult); these forms are usually permanently aquatic. Although the modern groups of Amphibia differ in many ways, they nevertheless share certain specialized features that suggest that they are closely related and may be classified together as the subclass Lissamphibia. The Anura (frogs and toads), Urodela (salamanders), and Apoda or Gymnophiona (caecilians) have teeth that are bicuspid and pedicellate (possessing an uncalcified zone of weakness between the crown and the base), and a middle ear apparatus of a unique type (described below).

---

*Necturus* is a permanently aquatic and paedomorphic salamander; its skull is poorly ossified. The skull of *Necturus* is profoundly specialized away from the primitive condition as exemplified by *Palaeoherpeton* (figs. 8.6, 8.7). Ossification of the chondrocranium and palatoquadrate cartilage is very incomplete and the number of dermal bones is considerably less than was possessed by labyrinthodonts. The temporal roof (lateral part of the skull table) and cheek bones that primitively enclose the temporal fossae are absent in *Necturus;* moreover, major landmarks of the primitive tetrapod skull are missing—one cannot discern the external and internal nares or the orbits because there are no bony margins delimiting these openings. Nevertheless, a study of *Necturus* is valuable in that it permits the student to evaluate the degree of modification possible in vertebrate skulls while being able to easily observe the basic components of the tetrapod skull.

For the study of *Necturus*, complete skulls, preferably preserved in fluid, and chondrocrania from which the dermal bones have been removed should be at hand.

**1. General Regions of the Skull**

The skull is partly bony, partly cartilaginous; in fact, there is just enough ossification present to encase the brain and to support the feeding apparatus. The bony part exists in the form of distinct bones, separated by jagged lines, the *sutures*. The dermal bones are somewhat distinguishable from the cartilage bones by their more superficial positions. The skull is divisible into a median portion, encasing the brain and inner ears, and marginal regions formed of the upper jaws and palatal bones and bearing teeth. Identify the *foramen magnum* and, below this on each side, an *occipital condyle*, a smooth projection for articulation with the first vertebra. The expanded region on each side of the posterior part of the skull is the *otic capsule*, in front of which a lateral projection, formed in part by the *quadrate* bone, bears an articular facet for the lower jaw. Above and anterior to the otic capsule is the *temporal fossa*, completely open above and to the side and filled in life by adductor jaw musculature (the entire posterior two-thirds of the dorsal surface of the skull is covered by jaw musculature). Anteriorly, a laterally projecting cartilage (generally missing in dried skulls), the *antorbital process*, marks the position of the eyeball that lies just above it. In front of and medial to the antorbital process, an oval cavity marks the location of the nasal capsule.

**2. Bones of the Skull**

*a. Dermal Bones of the Dorsal Side*

From front to rear are found the V-shaped *premaxillae*, the elongated *frontals* covered anteriorly by one limb of the V, and the *parietals* partly covered by the frontals. Note the absence of nasals, maxillae, and bones around the orbit. In life the external nares enter the soft tissue of the head anterolateral to the premaxillae; the location of the nasal passage is marked by a groove cut into the frontal bone immediately in front of the oval opening in which the nasal capsule lay. On each side of the parietal is a slender *squamosal* bone, extending from the quadrate to the lateral margin of the otic capsule.

*b. Dermal Bones of the Ventral Side*

The chief bone here is the large median *parasphenoid,* extending from the occipital condyles anteriorly between a pair of tooth-bearing bones, the *vomers,* which lie behind the tooth-bearing lateral wings of the premaxillae. Between the vomers, at the anterior end of the parasphenoid, there is exposed a bit of the cartilage of the chondrocranium, the *ethmoid plate.* At the side behind the vomers are the *pterygoids,* bearing teeth on their anterior ends and articulating posteriorly with the quadrates. The internal nares enter the oral cavity at the level of the vomer-pterygoid suture just lateral to the tooth row.

### c. The Teeth

The teeth are arranged in two rows—a short outer row on the premaxillae and a longer inner row on the vomers and pterygoids. The teeth of the lower jaw fit into the space between these two rows when the jaws close. The teeth are homodont and pleurodont.

### d. Cartilage Bones Ossified in the Chondrocranium

Only the rear end of the chondrocranium is ossified. The *exoccipital* bones bear the occipital condyles and constitute the lateral walls of the foramen magnum. The chondrocranium is completed above by a strip of cartilage, the *synotic tectum.* (Basioccipital and supraoccipital are absent in living amphibians.) The otic capsules are partially ossified, each containing two cartilage bones, the *opisthotic* and the *prootic.* The opisthotic, dorsolateral to the exoccipital, is a cone-shaped bone forming the posteriorly projecting angle of the skull on each side of the occipital region. The opisthotic meets the parietal and squamosal above and the parasphenoid below. In front of the opisthotic on the dorsal side is an area of cartilage, and in front of this is the prootic bone, wedged between the squamosal and the parietal. Examine the otic capsule from the side and note a small rounded bone, the *columella,* in its wall between the opisthotic and the prootic. A projection of the columella, termed the *stylus,* meets a columellar projection extending from the squamosal. The round base of the columella, called the footplate, fits into an opening, the *oval window* or *fenestra ovalis,* in the cartilage between the prootic and the opisthotic. The window leads to the internal ear, situated within the prootic bone. The columella is a sound-transmitting structure and part of the middle ear. The morphology of the middle ear of *Necturus* represents the continuation into the adult of a morphology characteristic of larval amphibians in general (see fig. 8.9). This probably represents the persistence into the paedomorphic adult of the larval system that is adapted to hearing in an aquatic environment.

### e. Middle Ear of Primitive Tetrapods

The middle ear originated in the earliest tetrapods and is retained in its complete form among living amphibians only in some frogs, being reduced in urodeles and caecilians. It is an air-filled chamber interpolated between the internal ear (lodged in the otic capsule) and the skin and is derived from the first or spiracular gill pouch of fishes (see chap. 10). It is closed externally by the *eardrum* or *tympanic membrane,* a double-layered membrane of which the inner layer consists of the endodermal lining of the spiracular pouch and the outer layer is the skin. Primitively the eardrum is flush with the outer surface of the head, and current evidence suggests that it corresponds in position with the opercular bone of crossopterygians. Across the *tympanic cavity* (cavity of the middle ear) there stretches a rod-like element, the *columella auris,* which is a persisting remnant of the upper part of the hyoid arch, that is, the hyomandibular bone of fishes. The columella consists of a proximal bony piece, usually termed the *stapes,* and a distal cartilaginous piece, the *extra-*

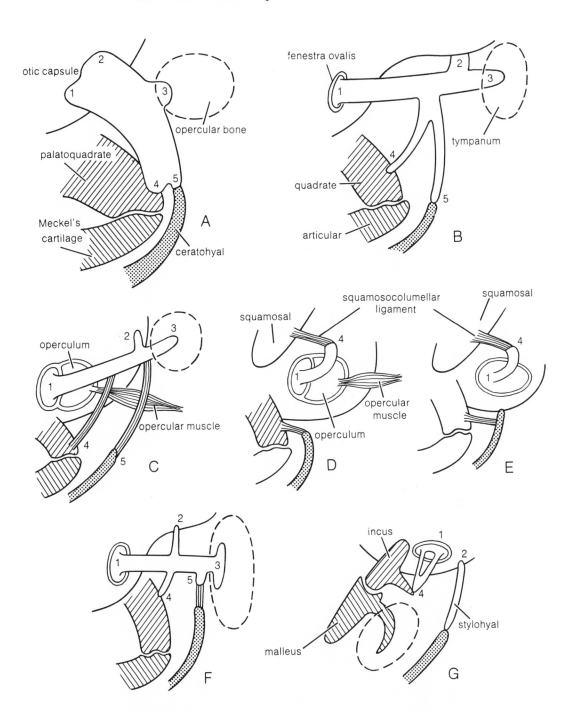

Fig. 8.9. Diagrams of the hyomandibular bone and its homologue, the columella, in a series of vertebrates, with the homologous parts of the bone indicated by numbers: *1*. ventral process-footplate; *2*. dorsal process; *3*. opercular process-tympanic process; *4*. internal or quadrate process; *5*. hyoid process. *A*, crossopterygian; *B*, labyrinthodont amphibian; *C*, frog; *D*, terrestrial salamander (*Ambystoma*); *E*, *Necturus*; *F*, lizard; *G*, mammal.

*columella* or *extrastapes*. The distal end of the columella is attached to the eardrum; its proximal end is expanded to form the footplate that fits into the fenestra ovalis.

The middle-ear mechanism amplifies the pressure of sound waves in air (which are some one-sixty-fourth the strength of sound waves in water) so that they will displace the fluid in the inner ear (where hearing takes place). Thus, weak airborne sound pressures cause the large eardrum and the attached columella to vibrate, and these displacements are transmitted from the vibrating columellar footplate to the inner-ear fluid. The whole system acts as a hydraulic lever, transforming a small pressure acting on a large surface (the eardrum) to a large pressure concentrated on a small surface (the footplate).

The columella becomes greatly modified in various tetrapod groups, but the nature of the changes can be best understood in terms of modifications away from the primitive morphology as represented by the hyomandibula of crossopterygians (fig. 8.9A). The hyomandibulae possess five processes, each with a characteristic attachment point to an adjacent structure. These are: (1) the *ventral process* = columellar footplate, meeting the otic capsule lateral to the inner ear in crossopterygians and fitting into the fenestra ovalis in tetrapods (see fig. 8.6A, B); (2) *dorsal process,* meeting the otic capsule dorsolateral to the inner ear; (3) the *opercular process = tympanic process,* meeting the opercular bone in crossopterygians and the eardrum in tetrapods; (4) the *internal* or *quadrate process,* meeting the quadrate bone; and (5) the *hyoid process,* meeting the ceratohyal or ventral element of the hyoid arch. In the evolution of tetrapods, some of these contacts are lost or are reduced to ligamentous connections.

The typical middle ear of primitive tetrapods as described above occurred in the extinct labyrinthodonts (fig. 8.6D, 8.9B) and persists in many frogs and toads (fig. 8.9C), though certain processes of the columella are reduced to ligaments. In urodeles and caecilians, however, the eardrum and the entire middle-ear cavity are absent, though a columella is usually present. The middle ear of modern amphibians is peculiar in that there are usually two elements fitting into the fenestra ovalis; an anterior columella (in frogs sometimes called the "plectrum") and a posterior *operculum* (no relation to the fish operculum implied), which is a new element unique to living amphibians (fig. 8.9C, D). The latter develops as an independent rudiment in the membrane covering the fenestra ovalis; it is thought to be derived from the wall of the otic capsule and hence is not a hyoid-arch derivative. A muscle extends from the operculum to the shoulder girdle. The operculum appears at metamorphosis, hence it is absent in larvae and in paedomorphic forms such as *Necturus* (fig. 8.9E). In urodeles and caecilians, the columella typically has a rod-like process, the *stylus,* that makes its primary contact with the squamosal though it may secondarily extend to the quadrate. This process may best be homologized with the dorsal process of primitive amphibians (see fig. 8.9B).

The function of the columella-operculum apparatus of amphibians is not well understood. Recent experimental work on frogs suggests that contraction of the "opercular" muscle binds, or couples, the operculum and columellar footplate so that the system is optimally adapted for transmission of low-frequency sounds; relaxation of the muscle uncouples the two bony elements and the system then transmits high-frequency sounds best. In urodeles and caecilians, which lack a tympanic membrane, sound vibrations are probably picked up by the surface of the head and transmitted via the squamosal or quadrate attachment of the columella to the inner ear.

---

*f. Cartilage Bones Ossified in the Palatoquadrate Cartilages*

The only bone so formed is the quadrate, already noted as occupying the projection between squamosal and pterygoid and bearing an articulating surface for the lower jaw. Medial to the quadrate a considerable area of cartilage, extending up to the parietal bone, represents unossified portions of the palatoquadrate cartilage. In well-prepared wet skulls the *ascending process* of the palatoquadrate cartilage can be seen extending dorsally to contact a lateral projection of the parietal. The ascending process contributes to the medial wall of the temporal fossa and branches of the trigeminal (fifth cranial) nerve emerge from the braincase just behind it. It is ossified in labyrinthodonts and reptiles as the epipterygoid bone.

**3. Chondro-cranium**

When the skull of *Necturus* is soaked in warm soap solution, the dermal bones can easily be lifted off, revealing the chondrocranium beneath. This consists largely of cartilage, with a few cartilage bones already noted. The otic capsules are inseparably incorporated into the chondrocranium, but the nasal capsules have very thin walls and are not firmly attached to the chondrocranium, so that they are usually lost in preparation.

Compare the chondrocranium with figure 8.10. Its posterior part consists chiefly of the rounded otic capsules connected dorsally by the narrow *synotic tectum* and ventrally by a still narrower bridge between the exoccipitals. This ventral bridge apparently represents all there is of the basal plate. Thus there is very little fusion of the parachordals in *Necturus*. From the anterior end of each otic capsule a slender, curved *trabecular* cartilage extends forward and fuses with its fellow to form the ethmoid plate already noted. The trabeculae enclose a very large fenestra formed by the fusion of the basicranial and hypophyseal fenestrae. (In the complete skull the fenestra is covered by the frontals and parietals and below by the parasphenoid). From the ethmoid plate a slender process continues forward on each side, the *trabecular horn* or *cornua trabeculae*. (In front of the otic capsule is the quadrate bone and accompanying cartilage; the latter sends processes to the otic capsule and trabecula. These are not, of course, parts of the chondrocranium proper.) The laterally projecting *antorbital* cartilage is loosely attached to the chondrocranium and may be missing. The boundaries of prootic, opisthotic, and columella can be seen better than in the complete skull.

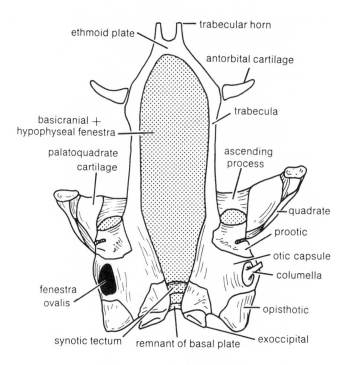

Fig. 8.10. Chondrocranium and palatoquadrate of *Necturus* with associated cartilage bones.

## 4. Lower Jaw

The lower jaw consists of a pair of Meckel's cartilages, united in front and sheathed for the greater part of their course in dermal bones. As in the skull, many dermal bones contributing to the formation of the primitive tetrapod lower jaw are missing in *Necturus,* which retains only three dermal bones for each half. The outer surface of each half of the lower jaw consists of the *dentary,* a dermal bone bearing teeth. The inner surface is formed by two dermal bones probably representing the *coronoid* and *prearticular* bones of other vertebrates. The coronoid is a small bone situated at about the middle of the inner surface, bearing the last group of teeth. The prearticular covers the remainder of the inner surface, except anterodorsally, and passes onto the outer surface at the extreme posterior end of the jaw. The articulating surface of the lower jaw is composed of cartilage representing the posterior end of Meckel's cartilage. (Note that the posterior end of Meckel's cartilage is not ossified to form the articular bone commonly found in tetrapods.) Meckel's cartilage forms the core of the lower jaw running almost its entire length, concealed except posterodorsally beneath the dermal bones. It can be revealed by removing these bones.

## 5. The Remaining Gill Arches

The hyoid and three succeeding gill arches are present in *Necturus* and are almost completely cartilaginous. Preserved material is necessary for their study. The hyoid arch is a broad, somewhat V-shaped cartilage situated just posterior to the lower jaw in the floor of the mouth cavity. On each side it is divisible into two cartilages—a small anterior *hypohyal* and a much larger posterior *ceratohyal*. The third gill arch (first true gill-bearing arch) is more elongated and slender than the hyoid arch and is likewise divisible into two pieces—an anterior *ceratobranchial* and a posterior, slightly longer, *epibranchial*. Between the median ventral portions of the hyoid and third arch is a triangular *copula* representing a *basibranchial* piece. The fourth and fifth gill arches are short, curved rods of cartilage on each side, composed of *epibranchials*. At the anterior end of the epibranchial of the fourth arch is a small *ceratobranchial*. In the median ventral line is a slender bone, the second *basibranchial*. It will be seen that the gill arches of *Necturus* are reduced both in number and in the pieces of which they are composed, compared with the gill arches of the elasmobranchs. For the typical condition of the gill arches in elasmobranchs, see figure 8.3 and compare with the condition in *Necturus*.

**H. The Reptil-ian Skull Exem-plified by the Alligator or Caiman**

**1. Distinguishing Features of the Reptilian Skull and Lower Jaw**

Reptiles retain many basic characteristics of the primitive tetrapod skull and lower jaw but show marked tendencies to superimpose their own specializations on this primitive organizational plan. Characteristic reptilian features include: single occipital condyle, elimination of the amphibian otic notch and secondary development of new otic notch on the quadrate in crocodilians, turtles, and lizards; tendencies for reduction in size and number of bones that in primitive amphibians form the lateral margin of the skull table (temporal roof); tendencies for development of a partial secondary or "false" palate in some turtles and a few lizards, and the formation of a complete secondary palate in crocodilians; elimination of palatal tusks and greater emphasis on the marginal dentition in all reptiles (except turtles, which completely eliminate teeth); development of massive pterygoid flanges (secondarily reduced in turtles) related to the differentiation of posterior pterygoid adductor jaw musculature; development of a streptostylic (movable) quadrate in lizards and snakes with important implications for their feeding behavior; and development of a coronoid process of the lower jaw for attachment of adductor jaw musculature in most groups (crocodilians are an exception).

One obvious distinguishing feature of the reptilian skull is the tendency to modify the roof and side wall of the temporal fossa (fig. 8.11). Primitively, the reptilian temporal fossa is completely enclosed, as in fossil amphibians, by the temporal roof above and the cheek at the side. This is the *anapsid* condition. In every major line of reptilian descent, however, strong tendencies develop for reduction of the dorsal or lateral bony walls of the temporal fossa. This is possibly related to changes in size, differentiation, and orientation of jaw-closing musculature housed in the temporal fossa.

Some living turtles (e.g., sea turtles) retain the anapsid condition almost intact; but many turtles eliminate much or all of the cheek and temporal roof by *emargination* (elimination of bone proceeding dorsally from the ventral margin of the cheek or anteriorly from the posterior margin of the temporal roof, or both). In other reptilian groups, openings or *temporal fenestrae* develop in the cheek or temporal roof or both. The *euryapsid* condition, characteristic of extinct plesiosaurs and ich-

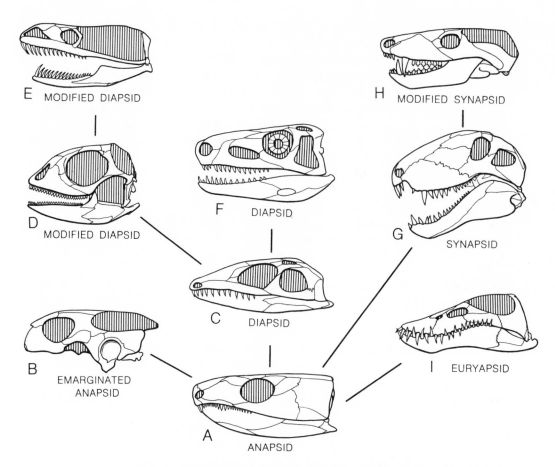

Fig. 8.11. Modification of the temporal region in reptiles from the primitive anapsid condition. *A*, the cotylosaur, *Captorhinus; B*, the turtle, *Emys; C*, the eosuchian, *Youngina; D*, the lizard, *Iguana; E*, the snake, *Python; F*, the primitive archosaur, *Euparkeria; G*, the pelycosaur, *Dimetrodon; H*, the advanced therapsid, *Thrinaxodon; I*, the plesiosaur, *Styxosaurus. F*, after Romer 1966 (*Vertebrate paleontology*. Chicago: University of Chicago Press); all others after Romer 1956.

thyosaurs, consists of a dorsally situated *superior temporal fenestra* on either side of the midline. The *synapsid* condition, found in fossil mammal-like reptiles, consists of a laterally situated *inferior temporal fenestra* on each side. When both superior and inferior temporal fenestrae are present, as in *Sphenodon* and crocodilians, the condition is *diapsid*. Lizards and snakes possess a highly modified diapsid temporal region. In lizards, the inferior temporal arch, forming the lower margin of the inferior fenestra, is eliminated. In snakes, both inferior and superior temporal arches are missing; thus all vestiges of the temporal roof and cheek are eliminated.

The skull of the alligator or caiman is almost completely ossified and because of its large size is convenient for introducing the student to the names, locations, and relations of the skull and lower jaw bones of living tetrapods. However, because of its specialization for an aquatic life, the crocodilian skull is greatly modified in the nasal, palatal, and otic regions from the generalized reptilian condition.

**2. General Topography of the Skull**

Obtain a skull and study its structure. It is composed of a number of separate bones united along jagged lines, the *sutures*. The various components of the vertebrate skull are here closely knit into a single structure. The bones forming the dorsal and lateral surfaces of the skull are rough and pitted, reflecting the close adherence of skin. The parietal foramen, normally found in reptiles, is absent in the Crocodilia. At the anterior end of the dorsal surface are the two nasal openings or *external nares*. The posterior half of the skull roof is dominated by the orbits, and posterior to them is the temporal region, with two pairs of openings, termed *temporal fenestrae*. As noted above (p. 293), the temporal region of the skull of early tetrapods was completely roofed; however, in reptiles this condition is greatly modified by emargination or fenestration. The alligator or caiman illustrates the diapsid condition of the temporal region, in which there are two temporal fenestrae on each side. The upper or *superior* temporal fenestrae are the openings in the roof on each side of the midline of the posterior part of the skull. The lower or *inferior* (lateral) temporal fenestrae are situated immediately behind the orbits, from which they are separated by a bony rod, the *postorbital bar*. Both the superior and the inferior temporal fenestrae open into a large cavity, the *temporal fossa,* situated posterior to the orbit; in life the temporal fossa houses adductor (jaw-closing) jaw musculature. The projecting ledge lateral to the superior temporal fenestra is called the *superior temporal arch,* and the bony bar below the inferior temporal fenestra is called the *inferior temporal arch*. Beneath the superior temporal arch is an elongated space through which the middle ear or tympanic cavity can be seen. In life the lateral part of this space is occupied by valve-like skin flaps that seal off the *external auditory meatus* from the exterior. These flaps lie just lateral to the eardrum or *tympanic membrane*. The various openings seen on looking deep into this space are found within the tympanic cavity. Dorsally in the tympanic cavity you will see a canal leading to the tympanic cavity of the other side. Immediately ventral to this canal is another canal through which the *stapedial apparatus* passes as it extends from the tympanic membrane to the *fenestra ovalis*. The stapedial apparatus is generally missing in prepared skulls. Turn the skull so as to view its posterior end. Note the two slitlike *posttemporal fossae* just beneath the rear edge of the skull roof; the *foramen magnum;* the single *occipital condyle* characteristic of reptiles below the foramen magnum; the quadrate bones forming the side projections and each bearing an articular surface for the lower jar; and the large *subtemporal fossae* below and anterior to the quadrates, continuous anteriorly with the orbits and above with the temporal fossae. On looking into the foramen magnum you will see a pair of bulges; these are the *otic capsules* containing the internal ears. The brain occupies the relatively small cavity seen around and in front of the otic capsules.

Study the ventral surface and identify: a small opening near the tip, the *premaxillary fenestra;* a pair of larger openings below the orbits, the *palatal* or *suborbital vacuities;* the *internal nares,* a pair of openings in the medial region near the posterior end of the skull; and the subtemporal fossae through which the jaw-closing muscles descend to the lower jaw. Note that the internal nares are very far posterior compared with their position in amphibians and most reptiles; this is a specialized condition caused by the development of a secondary palate separating the nasal passageways from the oral or mouth cavity.

Most of the small openings in the rear part of the skull are nerve foramina; some serve for the passage of blood vessels. The small openings on the midline just behind the internal nares are for the eustachian tubes, which connect the tympanic cavities with the throat region.

Examination of the general features of the crocodilian skull so far should reveal that the head is dorsoventrally flattened, with the external nares, orbits, and ear openings located high on the skull. Thus crocodilians are able to lie inconspicuously in water with only their main sense organs and air intake openings exposed above the surface. A secondary palate, separating nasal and oral cavities, enables the animal to breathe in this position for extended periods of time even though the oral cavity is submerged and, presumably, not completely sealed off from the water.

**3. Dermal Bones of the Dorsal Side**

These are the *premaxillae,* in front of and at the sides of the external nares; the *nasals,* the pair of long median bones behind the external nares; the *maxillae,* the large bones to the sides of the nasals; the *prefrontals,* to the sides of the rear ends of the nasals and contributing to the anterior wall of the orbits; the *lacrimals,* lateral to the prefrontals, also bounding the orbit anteriorly; the *frontal* bone, the median bone between the orbits, single in the adult but paired in the embryo; the *parietal* bone, behind the frontal and between the superior temporal fenestrae, also single in the adult but paired in the embryo; the *postorbital,* forming the anterior part of the superior temporal arch and sending down a process that constitutes the upper half of the postorbital bar; the *squamosal,* directly behind the postorbital and forming the posterior part of the superior temporal arch; the *jugal (malar),* the elongated bone below the orbit, bearing a projection that forms the lower half of the postorbital bar; and the *quadratojugal,* a slender, obliquely placed bone forming the posterior margin of the inferior temporal fenestra. The *quadrate,* a cartilage bone, is wedged between the quadratojugal, which it parallels, and the squamosal; its lower end has a saddle-shaped surface for articulation with the lower jaw, and its upper end supports the tympanic membrane. The ear region is closed behind by a ventral process of the squamosal that descends to meet the quadrate. A small bone, the *palpebral* (eyelid), often lost in prepared skulls, occurs in the eyelid.

The premaxillae, maxillae, jugal, quadratojugal, and quadrate bones form the maxillary arch and are really the upper jaw inseparably incorporated into the skull. All are dermal bones except the quadrate, which is ossified in the otic process of the palatoquadrate cartilage.

Compare the bones of the alligator or caiman skull with those of a primitive tetrapod (fig. 8.7). Which are missing?

The closure of the otic notch by the squamosal is a specialized feature of the Crocodilia that was not yet complete in the earliest fossil crocodilians; in these forms the otic notch was open behind and bore a superficial resemblance to that of labyrinthodont amphibians.

4. Palate and Ventral View of the Maxillary Arch

Study the ventral surface and identify premaxillae and maxillae, both bearing teeth. Behind the maxillae are the palatines, forming the inner boundaries of the palatal vacuities. In life, a major mass of adductor jaw musculature, the anterior pterygoideus muscle, lies in the suborbital region just dorsal to the palatal vacuity. Behind the palatines are the broad *pterygoids;* these enclose the internal nares and laterally form prominent ventrally directed pterygoid flanges that serve as attachment areas of powerful jaw-closing muscles, the posterior pterygoideus musculature. The posterior edges of the pterygoid flanges form the anterior margins of the subtemporal fossae. The *ectopterygoids* extend from the pterygoids laterally to meet the maxillae and jugals and form the posterior and lateral boundaries of the palatal vacuities. Identify also quadrates forming the medial and posterior margins of the subtemporal fossae and quadratojugals.

In the Crocodilia, the maxillae, palatines, and pterygoids have put out horizontal shelves that, growing medially, have met in the midventral line and completely hidden the primary palate of primitive tetrapods (see fig. 8.7). This secondary palate has caused the internal nares to lie far back on the skull. The vomers, which are exposed ventrally in the typical reptilian palate, now are hidden in the interior of the nasal passage. A parasphenoid bone, also a typical palate bone of amphibians and reptiles, is lacking in crocodilians although it is represented by a vestige in the embryo.

5. Occipital Region

This region forms the posterior end of the skull and consists of four cartilage bones ossified in the basal plate and synotic tectum. The foramen magnum is bounded by three bones, the median *basioccipital* below and, on each side, an *exoccipital* (fused with the otic capsule so its exact limits are not determinable). The basioccipital forms most of the large rounded occipital condyle, but processes from the exoccipitals participate. Above and between the exoccipitals is the triangular *supraoccipital,* which articulates above with the parietal; between the supraoccipital and the squamosals are the posttemporal fossae, already mentioned. Between the exoccipital and the quadrate on each side is a canal through which the seventh nerve and several blood vessels leave the middle ear region (see below); the inner end of this passage can be seen close to the opening into the tympanic cavity as viewed from the side. In the exoccipitals, on the side of the region taking part in the condyle, are two small foramina for the twelfth cranial nerve and a large foramen for the ninth, tenth, and eleventh cranial nerves; directly below the latter is a passage for the internal carotid artery into the tympanic cavity.

In more typical reptiles there is an elongate space bounded by the braincase medially and the quadrates and pterygoids laterally, through which pass the seventh nerve and several blood vessels. In crocodilians, the quadrates and pterygoids closely parallel the side of the braincase, and this elongate space is reduced to the circular canal described above.

6. Posterior Sphenoid Region

This region is seen in front of the occipital region when the skull is oriented ventral side up with the posterior end toward you; it forms the floor of the cranial cavity. The *basisphenoid* appears as a vertically oriented bone in the median region between the basioccipital and the pterygoids. A single median eustachian canal is found at the basioccipital-basisphenoid suture. Most of the basisphenoid is concealed from view by the pterygoids and quadrates, but its anterior end projects as a thin median plate, the *rostrum,* into the space between the orbits. In life this space is closed by the interorbital septum, which remains cartilaginous except for the rostrum of the basisphenoid. The rostrum is best seen by looking from the side into the subtemporal fossa. Posterior and superior to the rostrum on each side is a *laterosphenoid* bone, ossified in the antotic pillar, which ascends to the skull roof, where it articulates with the frontal, postorbital, and parietal. Between the base of the laterosphenoid and the quadrate, a conspicuous oval opening, the *trigeminal foramen,* provides passage for the maxillary and mandibular divisions of the fifth cranial nerve. The two laterosphenoids fail to meet in the median line, leaving a cleft through which pass the optic nerves and the olfactory tracts.

Like turtles, crocodilians possess a totally akinetic skull. The quadrate is firmly sutured to the rest of the skull, whereas in lizards and snakes it is mobile. Notable is the loss, in crocodilians, of the basipterygoid joint that in primitive fossil reptiles and amphibians exists as a movable articulation between the pterygoid and epipterygoid on the one hand and the basisphenoid on the other. *Sphenodon* and lizards retain a movable basipterygoid joint. Crocodilians lack an epipterygoid, this bone being replaced by a downgrowth of the laterosphenoid that is suturally joined to the pterygoid immediately in front of the trigeminal foramen. This downgrowth is separated by a narrow vertical slit from the main part of the laterosphenoid. The ophthalmic branch of the fifth cranial nerve passes forward through this slit.

7. Orbitosphe-nethmoid Region

There are no ossifications in the chondrocranial floor anterior to the basisphenoid; the cartilaginous interorbital septum is continuous anteriorly with the cartilaginous internasal septum.

8. Otic Capsule
and Median
Sagittal Section
of the Skull

The otic capsules, as already noted, can be seen by looking into the
foramen magnum but can be studied only in sagittal sections of the
skull. In such a section, locate the capsule as the rounded swelling in
the side wall close in front of the foramen magnum. Study of their
embryology shows that three bones ossify in the otic capsule: *epiotic,
prootic,* and *opisthotic.* A vertical suture in the capsule marks the
boundary between opisthotic and prootic. The opisthotic is behind the
suture, fused to the exoccipital; the prootic, in front of the suture, remains as an independent bone. Through its lower part passes the trigeminal foramen, noted above; and the prootic can be seen in the external view of the skull as the posterior boundary of this foramen. The
dorsal part of the otic capsule above the suture represents the epiotic
fused to the supraoccipital. The row of foramina below the capsule
beginning behind the trigeminal foramen serves for the exit of the
seventh to the twelfth nerves; the slit-like foramen below the capsule
gives passage to the ninth to eleventh; and those behind this are for the
twelfth. Below the trigeminal foramen is the foramen for the sixth
nerve.

In the parietal bone note the large canal cannecting the tympanic
cavities of the two sides. Identify laterosphenoid and basisphenoid in
the section. The rounded depression in the basisphenoid between laterosphenoid and rostrum is the *sella turcica,* which lodges the pituitary.
The brain occupies the cavity from the foramen magnum to the anterior
limits of the laterosphenoids. The upright bar in front of the interorbital space is formed chiefly of the prefrontal. Observe formation of
the long nasal passage by the pterygoid, palatine, and maxilla. In front
of the long partition belonging to the pterygoid is seen the thin *vomer.*

---

9. The Middle
Ear of Reptiles

In reptiles, as in labyrinthodont amphibians, the columella auris stretches
from the eardrum to the fenestra ovalis and is derived wholly from the
hyoid arch (fig. 8.9*F*). It typically consists of an ossified stapedial part,
having a broad footplate, and a cartilaginous extrastapedial part, having
an expanded tympanic process attaching to the eardrum and bearing an
*internal process* (passing to the quadrate) and a *dorsal process* (passing
to the opisthotic); in addition, the lower part of the hyoid arch may retain its primitive attachment to the columella ( = the hyomandibula), as
in *Sphenodon,* though normally it becomes detached. An eardrum is
present in turtles, crocodilians, and most lizards, being held in a deep otic
notch in the posterior surface of the quadrate. Because the early anapsid
ancestors of modern reptiles lacked an otic notch, it is believed that the
otic notch of the living reptiles is not homologous with that of labyrinthodont amphibians. Recent studies indicate that the tympanum (and the
tympanic process of the columella) of reptiles (see fig. 8.9*F*) may also
have arisen independently of these features in labyrinthodonts. An eardrum is absent in some lizards, especially burrowing forms, *Sphenodon,*
and all snakes. In these reptiles, sound vibrations may be picked up
through the surface of the cheek and transmitted to the inner ear via the
internal (quadrate) process of the columella. The eardrum of turtles is
always flush with the side of the head; that of lizards may also be super-

ficial or may be set in from the surface to a slight extent (although much less so than in mammals). In crocodilians, the eardrum is set well in from the surface and is covered by a movable flap of skin which closes the outer ear opening when the animal submerges.

10. The Lower Jaw

The lower jaw or mandible is composed of two halves or *rami,* united in front by a *symphysis.* Each ramus consists of six separate bones. Near the posterior end of the lateral surface of the jaw is a large vacuity, the *external mandibular fenestra,* and in front of this on the inner surface a smaller *Meckelian fenestra.* The posterior third of the jaw is dominated by a large fossa, the *adductor fossa,* leading to the interior of the jaw and situated immediately below and lateral to the subtemporal fossa of the skull. In life adductor jaw musculature originating in the temporal fossa of the skull descends and enters the adductor fossa to insert on its margins and internal surfaces. The bones of each ramus are: the *dentary,* bearing teeth and forming the outer surface of the anterior two-thirds of the ramus; the *splenial,* of about the same shape and size as the dentary and in the same position on the inner surface; the *angular,* below the external mandibular fenestra and in part separated from the splenial by the Meckelian fenestra; the *surangular,* forming the lateral wall of the adductor fossa above the external mandibular fenestra; the *coronoid,* a small bone on the inner surface, between the anterior ends of the angular and surangular; and the *articular,* above the posterior end of the angular and bearing a concavity for articulation with the quadrate. The posterior portions of the angular and articular form a large process, the *retroarticular process,* posterior to the jaw articulation. The dorsal surface of this process serves, in part, for the insertion of depressor mandibulae (jaw-opening) musculature and the medial and ventrolateral surface for the insertion of posterior pterygoideus musculature.

A cavity exists in the interior of the ramus extending forward between the dentary and the splenial. This cavity is occupied in life by Meckel's cartilage (and by an intramandibularis muscle in crocodilians). The posterior end of this cartilage has ossified into the articular, which is the only cartilage bone of the lower jaw. The articulation between upper and lower jaw is by way of the articular and the quadrate, a condition found in the majority of vertebrates with the notable exception of mammals.

11. The Hyoid Apparatus

As has already been explained, the hyoid apparatus is derived from the hyoid arch and remaining gill arches. It is generally missing on dried skeletons, and for its study preserved specimens are necessary. It consists of a broad cartilaginous plate, the body of the hyoid, and a pair of processes or horns (cornua) extending posteriorly and dorsally from the body, one on each side. The cornua are partially ossified. The body of the hyoid apparatus is derived from the bases of the second and other arches, and the horns are remnants of the third arch. The second (hyoid) arch persists as a tendon in the quadrate and articular bones. In lieu of alligator material the hyoid apparatus may also be studied on turtle skeletons. In turtles the hyoid apparatus consists of a median ventral plate, the body of the hyoid situated in the floor of the mouth cavity, and two pairs of posteriorly extending processes, the anterior and posterior horns of the hyoid; they are portions of the third and fourth gill arches.

**12. The Teeth**

The teeth of crocodilians occur in a single row on the margins of the premaxillae, maxillae, and dentary bones; in other reptiles teeth may also be borne on the palatine and pterygoid bones. Note that the teeth are of simple form and are all alike (homodont) except that certain ones, the grasping teeth, are considerably larger than the others. The teeth are thecodont, set into sockets in the jaw and held by the periodontal ligament; in other modern reptiles tooth attachment is either pleurodont or acrodont. On an isolated tooth note the large pulp cavity widely open at the base; this condition is related to the ability of reptiles to replace the teeth an indefinite number of times. Numerous developing teeth lie within the sockets below the open roots of the old teeth.

---

The teeth of reptiles are usually similar in different parts of the jaw, but in many reptiles a degree of heterodonty is present, although not as much as occurs in mammals. In *Sphenodon* and some lizards the anterior teeth are enlarged and incisor-like while the posterior teeth may possess cutting edges and resemble, to a slight degree, the molars of mammals. The dentitions of extinct dinosaurs and mammal-like reptiles exhibited a degree of heterodonty very comparable to that of mammals. In poisonous snakes there are special teeth termed poison fangs, which are greatly enlarged and provided with a groove or a canal (formed by the closing of the groove) for conducting the venom. The venom comes from glands of the nature of salivary glands. The vipers and rattlesnakes have the canal type of poison fang. There is a large poison fang ankylosed to each maxilla and folded back against the roof of the mouth when not in use. By means of a movable chain of bones in the palate, the maxilla can be rotated so that the fangs are "erected," that is, brought to a vertical position for striking into the prey. In another group of poisonous snakes, the elapid snakes, which includes the cobras, the poison fangs are immovable and permanently erect.

**I. The Mammalian Skull Exemplified by the Opossum and the Cat**

**1. Distinguishing Features of the Mammalian Skull and Lower Jaw**

The mammalian skull is completely ossified except for a small part of the ethmoid region. The number of bones in the mammalian skull is greatly reduced, as compared with the number in primitive tetrapods (fig. 8.7), partly through the loss of dermal bones and partly through extensive fusions, especially among the cartilage bones. Other characteristic features include the completely enclosed and greatly expanded cranial region containing the brain; paired occipital condyles; the formation of a secondary palate, causing the internal nares to open behind the mouth cavity, and a diphyodont and markedly heterodont dentition, related to mastication of food; the greatly enlarged temporal fenestra formed by the reduction of the temporal roof and loss of the postorbital bar as the temporalis muscle was differentiated; the formation of the zygomatic arch (inferior temporal arch of reptiles) bowed outward from the lower jaw, related to the differentiation of the masseter muscle; the great reduction in size of the pterygoid bones; the great differentiation of the internal ear; the reduction of the quadrate and articular to middle ear bones; the great enlargement of the dentary, with the formation of conspicuous coronoid and angular processes related, respectively, to the differentiation of the temporalis and masseter muscles; the elimination of all bones except the dentary from the lower jaw, and the formation of a new jaw articulation between the squamosal and the dentary.

The phylogenetic development of most of the features typical of the mammalian skull can be observed in the extinct mammal-like reptiles of the order Therapsida (see figs. 8.13, 8.14). These reptiles were of the synapsid type; the large temporal opening of mammals is the result of the expansion of the reptilian inferior temporal fenestra.

2. General
Features and
Regions
the Skull

The opossum and the cat are presented as examples of the mammalian skull. The opossum is a member of the order Marsupialia, and its skull is an excellent example of a primitive, or generalized, mammalian skull; as such, it can be used as a standard of comparison for studying more specialized types. Moreover, there is less fusion of individual bones in the opossum than in most mammals, which facilitates study of the basic components of the skull. If an opossum skull is not available, a cat skull may be used. The cat is a highly specialized member of the order Carnivora, and its skull is greatly modified from the primitive mammalian condition; furthermore, extensive fusions between individual bones has taken place in this species.

The adult skull is a hard, bony case composed of separate bones, immovably jointed in dovetail fashion along the sutures. The *facial* region supporting the nose and eyes is distinguishable from the posterior *cranial* region with its expanded braincase and greatly enlarged temporal fossae. At the anterior end of the facial region are seen the two external nares, separated in life by a cartilaginous partition, which constitutes part of the *septum of the nose.* At the side of the posterior facial region, the circular orbit is confluent with the temporal fossa behind it. The temporal fossa is filled in life with the temporalis muscle. The ventrolateral boundary of the orbit and temporal fossa is formed by a flaring arch, the *zygomatic arch,* massive in the opossum, more slender in the cat. The boundary between the orbit and the temporal fossa is marked by two small *postorbital* processes, one projecting dorsally from the zygomatic arch and the other projecting laterally from the roof of the skull; the postorbital processes are small in the opossum, prominent in the cat.

The cranial portion of the skull that encompasses the brain presents the following features. At the posterior end is the large *foramen magnum;* on each side of this is a projection, the *occipital condyle,* which articulates with the atlas. On the side of each condyle in the opossum is a large ventrally projecting process, the *paroccipital (exoccipital)* process, serving for the attachment of digastric (jaw-opening) and neck muscles; an abbreviated *mastoid* process is located just lateral to the paroccipital process. In the cat, the paroccipital process is short and broad and is pressed against the posterior surface of a hollow bulbous expansion, the tympanic bulla (described below); the mastoid process lies against the lateral surface of the bulla. As viewed from above, the anteromedial boundary of the temporal fossa is marked by a *temporal crest* indicating the anterior boundary of the attachment of the temporalis muscle. The temporal crests of both sides pass medially from the postorbital process to meet in the midline to form a *sagittal crest,* very conspicuous in the opossum, less so in the cat, serving for the attachment of the temporalis muscle. Posteriorly the sagittal crest connects with the *lambdoidal* (*nuchal*) *crest* of either side. The lambdoidal crest separates the temporal fossa from the occipital surface and thus marks the posterior boundary of the attachment of the temporalis muscle and the anterior edge of neck muscle attachment.

On the ventral surface, the anterior part of the skull is occupied by the *hard* (*secondary*) *palate.* Dorsal to this are the nasal passages, which open at the posterior end of the hard palate by the *internal nares* or *choanae.* Anteriorly, the hard palate contains a pair of *incisive foramina;* posteriorly, in the opossum, are two pairs of *palatine fenestrae* (characteristic of marsupials). The flattened ventral border of the zygomatic arch marks the area of attachment of much of the masseter muscle. At the posterior end of the zygomatic arch is a depression, the *glenoid fossa,* for the articulation with the lower jaw. Immediately behind the fossa is the *postglenoid process,* which acts as a backstop for the lower jaw. The concavity between the postglenoid and mastoid processes is the *external auditory meatus.* An expanded *bulla* is present in both species: that of the opossum lies directly medial to the postglenoid process and only partially encloses the *tympanic cavity* (cavity of the middle ear); the bulla of the cat is greatly expanded and completely encloses the tympanic cavity.

3. Dermal Bones of the Roof of the Skull

Beginning just behind the external nares are the paired *nasals* roofing the nasal cavity; next posterior, the paired *frontals;* then the paired *parietals;* and last, the small unpaired *postparietal* (also called the *interparietal* or *dermosupraoccipital*). In the opossum, the frontals, parietals, and postparietal contribute to the formation of the sagittal crest; in the cat, only the parietals and postparietal contribute to the sagittal crest. In both species the postparietal forms the medial half of the lambdoidal crest. The frontals form the upper postorbital processes. In the anterior wall of the orbit is the *lacrimal* bone, which is pierced by the *lacrimal canal.* The *nasolacrimal duct,* which in life drains tears into the nasal cavity, passes through the lacrimal canal.

4. Bones of the Upper Jaw and Palate

The maxillary arch consists, on each side, of the following elements: the *premaxilla,* ventral to the external nares and bearing teeth; the *maxilla,* forming the sides of the facial region and also bearing teeth; the *jugal* (or *malar*), forming the anterior root and almost the entire lower border of the zygomatic arch; and the *squamosal,* which completes the zygomatic arch, forms the glenoid fossa and postglenoid process, and forms the lateral half of the lambdoidal crest. The premaxillae form the anteriormost part of the hard palate by means of their *palatine processes,* which meet in the median ventral line and include the incisive foramina. The maxilla is the main bone of the facial region. It extends to the frontal bone above by its *frontal process;* its *palatine process* meets its fellow in the median ventral line and contributes to the hard palate; its *alveolar process* bears teeth; and its *zygomatic process* unites with the jugal to form the anterior root of the zygomatic arch.

The rearmost part of the hard palate is formed by the *palatine bones,* which therefore form the ventral border of the internal nares. A *lateral wing* of the palatine contributes to the medial wall of the orbit. The unpaired *vomer,* which forms part of the primary palate in primitive tetrapods, lies dorsal to the maxillae and palatines and forms the base of the nasal septum. In the opossum it is completely hidden from view by the secondary palate, but its extreme posterior tip is visible in the cat.

Observe that the quadrate is absent from the upper jaw. Because of the absence of the quadrate, all the bones of the mammalian upper jaw are of dermal origin.

## 5. The Occipital Region

The *occiput* or posterior surface of the skull extends from the lambdoidal crest above to the base of the foramen magnum below. Surrounding the foramen magnum is a single *occipital* bone. In the opossum it is composed of paired *exoccipitals* that lie on either side of the foramen magnum and bear the paired occipital condyles and the ventral median *basioccipital;* all three bones are fused in the adult but are separate in embryonic and young animals. The exoccipitals meet above the foramen magnum in the opossum; in the cat and most other mammals the dorsal border of the foramen magnum is formed by a median *supraoccipital* that also fuses into the occipital bone. In the opossum the supraoccipital lies above the exoccipitals and is fused with the dermal postparietal. The basioccipital forms the ventral margin of the foramen magnum between the occipital condyles and extends forward on the undersurface of the skull to a point between the bullae. The occipital condyles are derived from the single one of reptiles by the retrogression of the basioccipital contribution and dorsolateral expansion of the exoccipital contribution. Lateral to the occipital condyles, the exoccipitals bear prominent paroccipital processes, elongate in the opossum, short in the cat. Lateral to each exoccipital is exposed the mastoid portion of the *petromastoid* bone, which forms the small mastoid process on each side; in the opossum, the anterior face of the mastoid process is formed by the squamosal, which also forms the lateral border of the occiput.

## 6. The Otic Capsules and Middle Ear

The bones of the otic capsule of mammals are fused to form a single *periotic* or *petromastoid* bone; the separate prootic and opisthotic bones can be seen only in early embryonic stages. The mastoid portion of this bone is visible on the occipital face of the skull where it contributes to the formation of the mastoid process. The petrous portion of the petromastoid bone encloses the inner ear. It lies within the tympanic bulla of the cat and is therefore not easily seen in external view. In the opossum it is visible as a pear-shaped bone lying just in front of the paroccipital process and medial to the external auditory meatus. In some opossum specimens, the *tympanic* bone is preserved as a delicate horseshoe-shaped structure lying posterolateral to the cavity in the bulla. In life, the tympanic ring supports the tympanic membrane (eardrum) which separates the tympanic cavity from the external auditory meatus. In the cat the tympanic bone is fused into the lateral part of the bulla where it forms an expanded and somewhat irregular surface, contrasting markedly with the smooth surface of the medial part of the bulla which is formed by an *entotympanic* bone. The tympanic cavity of mammals contains the middle ear and its three little bones, the *auditory ossicles;* of the chain of three ossicles which traverse the tympanic cavity, the *malleus, incus,* and *stapes,* the malleus may be preserved attached to the tympanic ring in the opossum or lying just inside the bulla in the cat. If specimens are available, identify them as follows: the malleus, having an enlarged *head* for articulation with the incus and a slender handle or *manubrium* that abuts against the eardrum; the incus, having an expanded *body* with a saddle-shaped surface for articulation with the malleus and two pointed processes, one of which meets the stapes, a tiny stirrup-shaped bone with a *head* for articulation with the incus, a large central *stapedial foramen,* and an expanded *footplate (base)* that fits into the fenestra ovalis of the otic capsule.

---

The middle-ear cavity in mammals is enclosed to varying degrees by a bulla formed by contributions from various surrounding bones. In marsupials, the tympanic bone is usually unexpanded and a partial (as in the opossum) or complete bulla is formed by the alisphenoid. In placental mammals, the bulla may contain contributions from surrounding skull bones (especially the alisphenoid and basisphenoid), but is usually formed by the *tympanic* (often called *ectotympanic*), the *entotympanic,* or both. The tympanic is a dermal bone that primitively forms a horseshoe-shaped ring supporting the eardrum; it may expand to enclose the middle ear cavity and may also contribute to the bony external auditory meatus (as in man). It is homologous with the angular bone of the lower jaw of reptiles. In therapsids the angular possesses a hook-like *reflected lamina,* which in cynodont therapsids bears a striking resemblance to the tympanic bone of primitive mammals (fig. 8.12C, D). The entotympanic is a cartilage bone that appears to have no homologue in lower vertebrates. In some placentals the various components of the bulla remain distinct; generally, as in man, they fuse with one another and with the adjacent squamosal bone to form a compound temporal bone.

The homologies of the auditory ossicles were worked out through embryological studies and have since been verified by fossil discoveries. The malleus, which ossifies from the posterior end of Meckel's cartilage, represents the reduced articular bone; the manubrium is homologous with the downturned retroarticular process of therapsids, and a small dermal component of the malleus (the *goniale*) represents the reptilian pre-

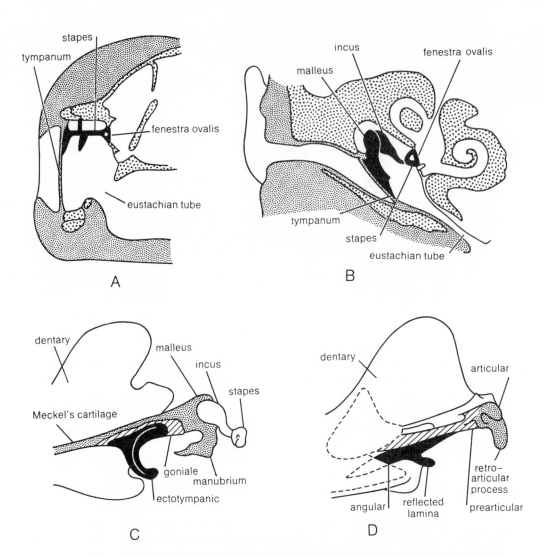

Fig. 8.12. *A, B,* semidiagrammatic representation of the ear region of *A,* a reptile (lizard) and *B,* a mammal (man) in transverse section showing main features of the middle ear (from Hopson 1966, after Goodrich 1930). *C, D,* medial view of posterior end of lower jaw and auditory ossicles of *C,* pouch young of the opossum, *Didelphis,* compared with *D,* posterior end of the lower jaw of the advanced cynodont therapsid *Diademodon* (from Hopson 1966). *C* modified from Goodrich 1930; *D* modified from Crompton 1963.

articular (fig. 8.12*C, D*). The incus is homologous with the reduced quadrate, and the stapes is the remnant of the stapedial apparatus of reptiles and hence is derived from the hyoid arch. Stages in the reduction of the quadrate and lower jaw bones to form the mammalian auditory ossicles and tympanic ring can be followed in a series of mammal-like reptiles and early fossil mammals (fig. 8.13). The joint between the head of the malleus and the incus represents the articulation of the lower jaw with the quadrate seen in nonmammalian vertebrates. The complete shift of the bones forming the reptilian jaw articulation into the middle ear became possible only after a new joint between what remains of the

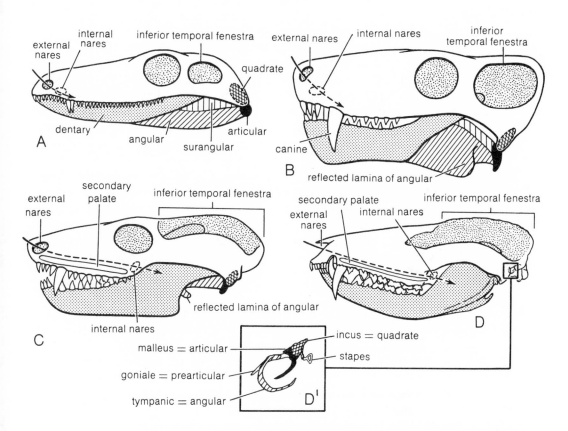

Fig. 8.13. Evolution of the skull from primitive mammal-like reptile to mammal: *A*, a primitive pelycosaur mammal-like reptile; *B*, a primitive therapsid reptile; *C*, an advanced (cynodont) therapsid; and *D*, a mammal. Illustrated is (1) the progressive enlargement of the inferior temporal fenestra and temporal fossa; (2) the development of a secondary palate in advanced therapsids separating the nasal passages from the oral cavity; (3) the development of a heterodont dentition beginning with greatly enlarged canines as seen in *B*; (4) the progressive enlargement of the dentary (*stipple*) to form the entire mammalian lower jaw; (5) the reduction of the postdentary bones, and transformation of the reptilian angular (*diagonal lines*), articular (*black*) and quadrate (*cross-hatched*) bones into the mammalian tympanic ring, malleus, and incus, respectively. Also illustrated is a dermal bone, the goniale, ossified in association with the malleus and believed to represent the reptilian prearticular.

lower jaw of mammals (namely, the dentary) was established with the posterior end of the dermal upper jaw (namely, the squamosal). This dentary-squamosal articulation is, as already noted, a distinguishing feature of mammals. In some advanced cynodonts and in the earliest mammals of the Mesozoic, the articular-quadrate joint, though greatly reduced, functioned in the jaw articulation beside the newly established joint between the dentary and squamosal.

---

**7. The Posterior Sphenoid Region**

On the ventral surface of the skull in front of the basioccipital is the *basisphenoid*. On either side of and partially (opossum) or completely (cat) fused to the basisphenoid are the *alisphenoids*. In lateral view, each alisphenoid meets the squamosal behind (in the opossum) or laterally (in the cat), the parietal and frontal above and the orbitosphenoid and palatine (by means of a process extending forward beneath the orbitosphenoid) in front. The ventral edge of that part of the palatine that forms the internal nares meets the thin, ventrally directed *pterygoid,* often missing in prepared skulls. The pterygoids are drawn out posteroventrally into processes, the *hamuli,* that serve for attachment of part of the internal pterygoid jaw musculature.

The mammalian alisphenoid is homologous to the epipterygoid bone of reptiles, absent in crocodilians but present in lizards and *Sphenodon.* This was already stated to be a derivative of the palatoquadrate cartilage and not a true bone of the cranial wall (p. 291). It became incorporated into the braincase in advanced therapsids (fig. 8.14C). The parasphenoid is lacking in mammals, though a vestige of it has been identified in the embryos of several species.

| | |
|---|---|
| 8. The Anterior Sphenoid Region | In the median region in front of the basisphenoid is the slender *presphenoid.* The presphenoid is fused on each side with the *orbitosphenoids,* best seen in lateral view. Each orbitosphenoid is contacted by the alisphenoid behind, the frontal in front, and the palatine below. In the cat it contains the large optic foramen; in the opossum it forms the anterior border of a large opening, the confluent optic foramen and orbital fissure. |

Many of the individual bones described in the opossum skull are commonly fused with one another in the adults of other mammals, though their separate identities can be observed in the embryonic or young individual. The supraoccipital, exoccipitals, and basioccipital fuse to form a single occipital bone. In the human skull the postparietal, fused to the supraoccipital, is also considered part of the occipital bone. The squamosal, tympanic, and periotic bones fuse to form the squamous, tympanic, and petrous parts of the compound temporal bone characteristic of many mammals. In man the basisphenoid, alisphenoids, presphenoid, orbitosphenoids, and pterygoids are fused into one bone, termed the sphenoid.

| | |
|---|---|
| 9. The Ethmoid Region and the Sagittal Section of the Skull | The ethmoid region can be studied only in sagittal section. There should be available sections cut to one side of the median line, so that one half retains the septum of the nose and the other half exposes the turbinals. The section shows that the interior of the cranial portion of the skull is occupied by a cranial cavity, divisible into three regions of unequal size. The most posterior cavity, enclosed by the occipital, otic, and postparietal bones, is the *posterior* or *cerebellar fossa* of the skull. Its anterior boundary is marked in the opossum by a slight ridge of bone passing up and back across the lateral wall of the cranial cavity. In the cat and many other mammals, this boundary is marked by a prominent ridge or shelf of bone, the *tentorium,* which in life is completed by a membrane. In the lateroventral wall |

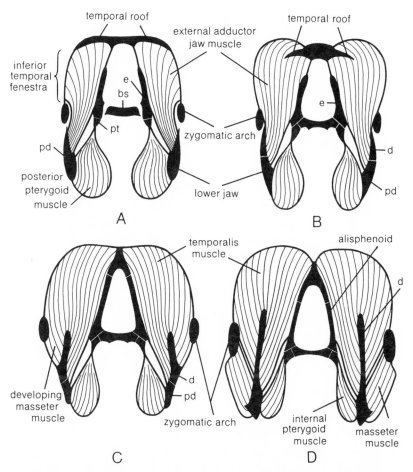

Fig. 8.14. Progressive stages in the evolution of the mammalian skull, lower jaw, and associated adductor jaw musculature illustrated by diagrammatic cross sections through the temporal region of *A*, a primitive pelycosaur mammal-like reptile; *B*, a primitive therapsid mammal-like reptile; *C*, an advanced (cynodont) therapsid; and *D*, a primitive mammal. In *A*, the temporal fossa and enclosed adductor jaw musculature are covered above by the temporal roof; the inferior temporal fenestra isolates a zygomatic arch at the side; the post-dentary bones (*pd*), forming the posterior part of the lower jaw, serve for insertion of the adductor jaw musculature; the epipterygoid (*e*) fails to meet the dermal skull roof; and the basipterygoid joint between the basisphenoid (*bs*) and epipterygoid is open. In *B*, the temporal roof is partially reduced owing to the invasion of musculature onto its dorsal surface; the dentary (*d*) serves for the insertion of part of the adductor jaw musculature through the development of an abbreviated coronoid process; and the basipterygoid joint is closed by a sutural contact. In *C*, the temporal roof is completely eliminated, with most of that part of the musculature occupying the temporal fossa now equivalent to the mammalian temporalis muscle; the zygomatic arch is bowed laterally away from the lower jaw and the space created is occupied by the developing masseter muscle; the epipterygoid (= mammalian alisphenoid) meets a ventral extension of the dermal skull roof to begin complete closure of the braincase. In *D*, the dentary forms the entire lower jaw; the masseter muscle is fully differentiated (as it also is in the most advanced therapsids); and, part of the reptilian posterior pterygoid muscle forms the mammalian internal pterygoid muscle.

of the cerebellar fossa is a rounded area of very hard, compact bone
bearing a small oval opening below and, in the opossum, a larger
circular depression above; this is the *petrous* part of the petromastoid
bone and encloses the internal ear. The greater part of the cranial
cavity comprises the *middle* or *cerebral fossa,* lying anterior to the
cerebellar fossa. Its roof and walls are formed by the parietals, frontals, alisphenoids, and orbitosphenoids; its floor by the basisphenoid.
In the floor of the cerebral fossa is a marked saddle-shaped depression, the *sella turcica,* which in life lodges the pituitary body. A foramen opening into the sella turcica of the opossum is for the internal
carotid artery. In the anterior end of the cranial cavity is the small
*anterior* or *olfactory fossa* situated between the more anterior parts
of the frontal bones. The olfactory fossa is separated from the nasal
cavities, which lie below and in front of it, by an oblique plate of bone
perforated by numerous holes, the *cribriform plate* of the ethmoid.
This plate is best seen in the intact skull by looking through the foramen magnum. The plate, pierced by holes like a sieve, is then seen
closing the anterior end of the cranial cavity. Our study of the sagittal section shows that the floor of the skull is composed of a chain
of cartilage bones—basioccipital, basisphenoid, presphenoid, and
ethmoid—on which the brain rests. These bones, as has already
been explained, are derived from the chondrocranium.

The nasal cavities or nasal fossae are enclosed partly by cartilage
bones and partly by dermal bones. The roof of the cavities consists
of the nasal bones and a small part of the frontals. Behind the nasal
cavities of the cat (but not of the opossum) are extensions of these
spaces into the frontal bones above the cranial cavity and the sphenoid bone below. These *sinuses* occupy "dead" space within the
skull between the walls of the cranial cavity and the external surface
of the skull that otherwise might be occupied by bone; their presence
decreases the weight of the skull. The two nasal cavities are separated
by a median, vertical, bony partition, the *perpendicular plate* of the
ethmoid; in the living state this is continued to the anterior nares by a
cartilaginous plate. The two together constitute the *septum* of the
nose. Dorsally the septum meets the nasal and frontal bones; ventrally
it meets the *vomer,* an elongated bone dorsal to the maxillae and
palatines. The posterior end of the septum meets the cribriform
plate.

On the half of the skull where the septum is missing, the *turbinal
bones* or *conchae* may be studied. They are peculiar, delicate,
grooved, and folded bones that occupy the lateral walls of the nasal
cavities and fill most of the interior. The most posterior of these
bones is the *ethmoturbinal,* situated below and just in front of the
cribriform plate and composed of folds that enclose spaces known as
the ethmoid cells. In front of the ethmoturbinal is the smaller *maxilloturbinal,* borne on the inner surface of the maxilla. A single elongated
bony ledge, the *nasoturbinal,* occurs on the uppermost scrolls of the
ethmoturbinals. The function of the turbinals is to increase the respiratory and olfactory surface of the nose. The ethmoturbinals are covered by olfactory epithelium, whereas the maxillo- and nasoturbinals

serve to moisten and strain the inspired air. Turbinals occur in reptiles and birds and present much variation among mammals; homologizing them in different groups has so far proved impractical.

---

10. The Foramina of the Skull

The skull is pierced by numerous openings for the passage of nerves and blood vessels and sometimes other structures. These are listed below for convenient reference.

a. Incisive Foramina: Anterior end of secondary palate between maxillae and premaxillae; connect roof of mouth with nasal cavities.

b. Infraorbital Foramen: Large opening in the maxilla anterior to the zygomatic arch; for the passage of certain branches of the fifth cranial nerve and blood vessels.

c. Lacrimal Canal: Lacrimal bone in anterior border of the orbit and passing through the maxilla into the nasal cavity; contains the nasolacrimal duct for the draining of the tears.

d. Sphenopalatine Foramen: In that part of the palatine bone that extends into the orbit, posterior to the lacrimal; for the passage of branches of the fifth nerve into the nasal cavity.

e. Palatine Canal: The posterior end of this canal or foramen is posteroventral (opossum) or ventral (cat) to the sphenopalatine foramen; its anterior end is on the posterolateral part of the secondary palate, on the contact between the maxilla and palatine; for the passage of a branch of the fifth nerve to the palate.

f. Optic Foramen and Orbital Fissure: A single large foramen in the opossum on the contact between presphenoid and orbitosphenoid, divided into two foramina in the cat, and most other mammals: an optic foramen for the passage of the optic nerve and a large orbital fissure behind it for the third, fourth, and sixth cranial nerves to the muscles of the eyeball, and the first branch of the fifth nerve.

g. Foramen Rotundum: In the alisphenoid immediately behind the preceding foramen; transmits the second branch of the fifth nerve.

h. Transverse Foramen: Foramen in the opossum (absent in the cat) on ventral surface of skull passing anteromedially into the basisphenoid; for the passage of a vein.

i. Carotid Canal: Foramen in the opossum directly behind transverse foramen, enters the sella turcica; transmits internal carotid artery.

j. Foramen Ovale: Oval foramen in the alisphenoid of the cat and in anterior face of the alisphenoid bulla of the opossum, enters the cranial cavity immediately in front of the petromastoid bone; transmits the third branch of the fifth nerve.

k. Canal for the Eustachian Tube: In the medial wall of the bulla of the opossum immediately posteromedial to the foramen ovale and separated from it by a short ridge or spine of the alisphenoid; for the passage of the eustachian tube from the pharynx into the tympanic (middle ear) cavity.

l. Jugular Foramen: Foramen on posteromedial side of petromastoid bone medial to the paroccipital process (opossum)
or on medial side of the posterior end of the bulla (cat); for
the passage of the ninth, tenth, and eleventh cranial nerves.

m. Hypoglossal Foramina: One (cat) or two (opossum) small
foramina on the ventral surface of the occipital bone anterior
to the occipital condyle (opossum) or in the medial side of
the jugular foramen (cat); for the passage of the twelfth
cranial nerve.

n. Postglenoid Canal: Canal in the opossum with one or two
external openings in the squamosal behind the postglenoid
process, opens into cranial cavity above petromastoid bone;
transmits a vein.

o. Primitive Facial Foramen: Anteriormost of a row of three
very small foramina in the petromastoid bone within the tympanic cavity of the opossum (hidden by the bulla in the cat);
transmits the seventh nerve from the cranial cavity; the nerve
passes back across the tympanic cavity to exit via a small
notch or foramen, the *stylomastoid* foramen in the depression
between the mastoid and paraoccipital processes. In the cat,
the stylomastoid foramen lies at the ventral tip of the mastoid
process.

p. Fenestra Ovalis: Second foramen of the row of three in the
petromastoid of the opossum; an oval foramen, the "oval
window," opening into the inner ear; the footplate of the
stapes sits in this opening.

q. Fenestra Rotunda: Third foramen of the row in the opossum;
a circular foramen, the "round window," covered in life by a
membrane; serves to relieve pressures on fluid of inner ear.

r. Internal Auditory Meatus: In the lower half of the petromastoid bone of the opossum, seen in sagittal section, below a
larger depression; in the cat opening in the center of the
petromastoid bone, for the passage of the auditory nerve from
the internal ear to the brain; the seventh nerve also begins its
exit in common with this foramen and appears within the
tympanic cavity through the primitive facial foramen.

## 11. The Lower Jaw

The lower jaw or *mandible* consists of a single pair of bones, the
*dentaries,* connected in front by a *symphysis* that in life allows some
independent movement of the dentaries relative to each other. All
other bones seen in the lower jaw of reptiles have vanished (except
the articular and angular, which have become the malleus and tympanic ring respectively). The anterior horizontal part of the mandible bearing the teeth is named the *body.* The posterior vertical part,
for the attachment of adductor jaw musculature, the *ramus.* (In
lower vertebrates, the term ramus refers to each half of the lower
jaw.) The ramus extends dorsally to form a strong *coronoid* process
that projects into the temporal fossa of the skull. Both the lateral
and the medial faces of the coronoid process serve for insertion of
the temporalis muscle. The articulating surface of the mandible is

borne on the *condylar* process. The depressed area in the lower part of the lateral face of the ramus is the *masseteric* fossa, which marks the attachment of the deep components of the masseter muscle. The ventral surface of the ramus bears the *angle* or *angular process*. In the opossum, as in most marsupials, the angular process projects medially to form an *inflected angle*. In the cat, as in all placental mammals, the angular process projects ventrally rather than medially. The ventral (external) surface of the angle serves for insertion of the superficial component of the masseter muscle, the dorsal (internal) surface, for insertion of the internal pterygoid muscle. Near the anterior tip of the mandible on the outer surface is the *mental* foramen, through which the nerve (a branch of the mandibular division of the fifth cranial nerve) of the lower jaw exits. A second smaller mental foramen lies behind the first. Near the posterior end of the inner surface is the large *mandibular* foramen, through which the nerve enters and pursues a course in the interior of the mandible forward to the mental foramen.

In the absence of the quadrate and of all of the bones of the lower jaw except the dentary, the articulation of the lower jaw to the skull is *between the dentary and the squamosal*. This feature distinguishes mammals from all other jawed vertebrates, for in the latter the articulation is between the articular and the quadrate. The condition found in mammals is, however, closely approached by those therapsid reptiles directly ancestral to mammals.

## 12. The Teeth

The teeth of mammals are thecodont, heterodont, and diphyodont; some of them are of complicated form, having several cusps or crests (ridges) and more than one root. Because of their complex morphology, which is closely correlated with feeding habits, the teeth (particularly the molars) are important in working out phylogenetic relationships. Teeth are the most commonly fossilized parts of extinct mammals because enamel is the hardest and most durable tissue in the body; as a result, much of our knowledge of the relationships and probable habits of fossil mammals is derived from study of their teeth. The opossum provides an example of the teeth of a relatively unspecialized mammal with a mixed diet of plant and animal food. The cat, on the other hand, has the teeth of a highly specialized carnivorous mammal.

*Opossum:* The teeth of the opossum possess a very primitive mammalian pattern very much like that from which the teeth of all marsupial and placental mammals have been derived. At the tip of the jaws is a series of small, simple teeth—the *incisors*. In the upper jaw there are ten incisors, five on each side, and in the lower jaw, eight incisors, four on each side. Behind the incisors on either side is a *canine,* a long, pointed but simple tooth. Between the upper incisors and the upper canine is a gap or *diastema* into which the lower canine fits when the jaws are closed. Behind the canine on each side are seven teeth in both upper and lower jaws. These teeth are more complicated than the preceding in having more than one *cusp* or pointed projection and more than one root. They are known as *premolars* and *molars*. The first three teeth are premolars; they are much simpler than the molars, each having a single large main cusp and only a small posterior cusp or heel lying close to the level of the alveolus. The first premolar is very small and may be missing in some skulls. There are four molars; the upper ones are triangular in crown view and have three main cusps on the inner part of the tooth and a row of smaller cusps along the outer edge. The lower molars are elongate, with a high triangle of cusps on the front half of each crown and a low heel behind with three low cusps surrounding a broad basin. In old individuals the cusps may be obliterated through wear. Note that the upper incisors are borne on the premaxillae; the other teeth of the upper jaw on the maxillae.

*Cat:* At the tip of the jaws are six small, simple teeth, called *incisors.* On either side of the incisors is a *canine,* a long, sharp but simple tooth. Between the upper incisors and canine is a gap or *diastema* into which the lower canine fits when the jaws close. Behind the canine on each side are four teeth in the upper jaw and three in the lower. These teeth are more complicated than the preceding, having more than one *cusp* or pointed projection and more than one root. In the upper jaw the first three are *premolars,* in the lower jaw the first two. The last tooth in each jaw is a *molar.* The last upper premolar and the lower molar are enlarged to form cutting blades called *carnassials,* a characteristic feature of most carnivorous mammals. Note that the upper incisors are borne on the premaxillae, the other upper teeth on the maxillae.

---

From the foregoing it is seen that the teeth of mammals are differentiated into four kinds: incisors, canines, premolars, and molars. The incisors are usually simple teeth with rounded or cutting edges. In the opossum they are very small and apparently used for little more than holding. In man and many other mammals the incisors are used for cutting. The canines are usually enlarged, sharply pointed teeth adapted for piercing; they are the main instruments for capturing and killing prey as well as the principal weapons of defense. The premolars and molars are often spoken of together as *cheek teeth.* They are distinguished on the basis of replacement; premolars have milk predecessors, molars do not. They are not always distinguishable on the basis of morphology, although in general premolars are simpler than molars and serve more for puncturing than for cutting and grinding. The molars usually have curved crests (ridges) joining the cusps that serve as shearing (cutting) surfaces. It is evident that the increase in complexity of teeth from front to back is correlated with the nature of food processing in the mouth, which increases progressively toward the rear.

Tooth replacement in mammals is greatly reduced over the polyphyodont condition seen in reptiles. Primitively, mammals have two sets of teeth: the juvenile or milk dentition, which is shed early in life, and the permanent dentition that replaces the milk teeth. The molars, which are not replaced, erupt after the milk dentition is complete and may be thought of as a continuation of this first set of teeth that is not replaced by the second set. In many mammals some or all of the milk dentition fails to develop and only the permanent teeth erupt. In marsupials the last premolar is the only tooth to have a milk predecessor. On the other hand, in some mammals parts of the permanent dentition fail to erupt and the milk teeth are retained; the first premolar of placental mammals is thought to be a retained milk tooth.

Mammalian heterodonty is often described by a dental formula, which expresses the number of each different kind of tooth in each half-jaw from the anterior median line to the posterior end of the jaw. The upper teeth are placed in the numerator of the formula; the lower teeth in the denominator. The complete dentition of the ancestral therian mammal is not retained in any living mammal, but it is inferred to have been 5/4 incisors, 1/1 canine, 4/4 premolars, and 4/4 molars in each half jaw for a total of 54. It is written thus: $\frac{5 \cdot 1 \cdot 4 \cdot 4}{4 \cdot 1 \cdot 4 \cdot 4}$. In the marsupial line that split off from this ancestral therian, the dentition was reduced by the loss of one premolar in each jaw to give a formula of $\frac{5 \cdot 1 \cdot 3 \cdot 4}{4 \cdot 1 \cdot 3 \cdot 4}$, which is still retained in the opossum. In the line leading to placental mammals, the dentition was reduced from the ancestral therian formula by the loss of several incisors and one molar to give a formula of $\frac{3 \cdot 1 \cdot 4 \cdot 3}{3 \cdot 1 \cdot 4 \cdot 3}$. Further reductions have characterized later evolution in mammals; for example, the formula in the cat is $\frac{3 \cdot 1 \cdot 3 \cdot 1}{3 \cdot 1 \cdot 2 \cdot 1}$; in the sheep, $\frac{0 \cdot 0 \cdot 3 \cdot 3}{3 \cdot 1 \cdot 3 \cdot 3}$; and in man, $\frac{2 \cdot 1 \cdot 2 \cdot 3}{2 \cdot 1 \cdot 2 \cdot 3}$.

The fossil record indicates how the complex, multicusped molar teeth of mammals evolved from the simple conical teeth of reptiles (fig. 8.15). Additional cusps first appeared as small upgrowths close to the base of the crown anterior and posterior to the main cusp (similar teeth are found in some lizards). These cusps, and the low crests connecting them, increased in prominence to become important parts of the crown. Additional cusps appeared and formed a *cingulum,* which is a ridge around the base of the crown. Initially, the main cusp and two accessory cusps and their connecting crests lay in an anteroposterior row on the crown, but in the earliest therian mammals there was a shifting of the accessory cusps relative to the main cusp so that the three cusps formed a triangle. The teeth are so placed that when the jaws are closed the lower triangles fit between the upper triangles and the crests shear past one another. In more advanced therians, the upper molars developed an additional internal cusp from the cingulum which fits into a basined heel formed behind the main triangle of the lower molars (fig. 8.15G, H). This introduced a crushing function to the formerly exclusively shearing molars. The fossil record demonstrates that the triangles of the upper teeth of modern therians are not the same as the original triangles. In the modern therian, or *tribosphenic,* molar, the main cusps in the upper triangle, or *trigon,* are called *protocone, paracone,* and *metacone;* in the lower triangle, or *trigonid,* the cusps are *protoconid, paraconid,* and *metaconid* (fig. 8.16A, B). The paracone and protoconid represent the original reptilian cusps. The *heel* or *talonid* of the lower molar bears additional cusps; the *hypoconid, hypoconulid,* and *entoconid* (fig. 8.16B). Among

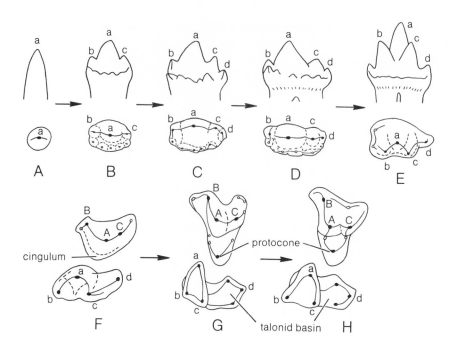

Fig. 8.15. Origin of the complex molar teeth of therian mammals from the simple conical tooth of primitive therapsid reptiles. *A–E*, right lower tooth in internal (*above*) and crown (*below*) views: *A*, simple cone (*a*) of primitive therapsid; *B*, addition of small accessory cusps (*b,c*) in early cynodont therapsid (*Procynosuchus*); *C*, elongation of crown and addition of small heel (*d*) in advanced cynodont (*Thrinaxodon*); *D*, further elongation of the crown and subdivision of root in early nontherian mammal (*Morganucodon*); *E*, shifting of accessory cusps (*b, c*) to form triangle with main cusp (*a*) in early therian mammal (*Kuehneotherium*). *F–H*, crown views of left upper (*above*) and right lower (*below*) molars to show origin of the protocone and talonid basin: *F*, primitive therian (*Amphitherium*) with internal cingulum on upper and enlarged heel (*d*) on lower from which protocone and talonid basin develop; *G*, more advanced therian (*Pappotherium*) with protocone and basined talonid moderately well developed; *H*, primitive placental mammal (*Didelphodus*) with protocone and basined talonid well developed. Homologous cusps denoted by same letter; capitals in upper teeth, lower case in lowers. Modified from Crompton 1971; Hopson and Crompton 1969; and original.

omnivorous or herbivorous mammals, such as man, the molars develop square, four-cusped crowns by addition of a *hypocone* in the uppers and reduction or loss of the paraconid and hypoconulid in the lowers (fig. 8.16*C*). Further adaptive modifications of the mammalian dentition are discussed below.

---

**13. The Hyoid Apparatus**

As already explained, this is the remnant of the hyoid and other gill arches. It is generally absent on prepared skeletons, and isolated specimens should be provided for study.

*Opossum:* The hyoid apparatus of the opossum is greatly reduced from that of more primitive tetrapods. It consists of a bony plate lying in the root of the tongue just in front of the larynx; this plate is

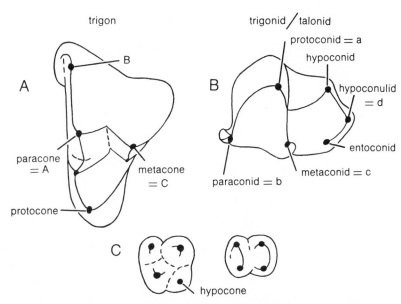

Fig. 8.16. *A*, upper and *B*, lower molar of extinct therian mammal (*Didelphodus*) to show the parts of the primitive tribosphenic molar. *C*, development of square, four-cusped molars in herbivorous mammals by addition of hypocone in the upper (*left*) and loss of paraconid and hypoconulid in the lower (*right*). *A* and *B* after Crompton 1971.

called the *body* of the hyoid, or *basihyal*. From it extend two pairs of processes or horns. The short anterior horns consist of *hypohyals;* they are attached to the base of the skull by ligaments. The posterior horns, called *thyrohyals,* are actually derived from the ceratobranchials of the third gill arch. The remaining gill arches are represented in the cartilages of the larynx, studied in a later chapter.

*Cat:* The hyoid of the cat consists of a bony bar placed in the root of the tongue just in front of the larynx; this bar is called the *body* of the hyoid, or *basihyal*. From it extend two pairs of processes or horns. The anterior horns consist of a chain of four pieces, called *hypohyal, ceratohyal, stylohyal,* and *tympanohyal,* of which the last is attached to the tympanic bulla just ventral to the stylomastoid foramen. The groove it occupies can generally be seen on the side of the bulla. The posterior horn consists of a single piece (*thyrohyal*) attached to the larynx. The body and anterior horns belong to the hyoid arch, of which an additional piece, as already learned, forms the stapes in the middle ear; the posterior horns are remnants of the third gill arch. The remaining gill arches are represented in the cartilage of the larynx, studied in a later chapter.

**14. Comparison of the Skull and Lower Jaw of a Carnivore (Cat) with That of an Herbivore (Sheep)**

To carry out this study, cat and sheep skulls and lower jaws are needed.

The skulls of advanced placental mammals, as represented by the cat and the sheep, show certain differences from the skulls of the more primitive marsupials, such as the opossum. Note: the greater expansion of the braincase, the greater fusion of individual bones and the greater expansion of the bulla in the placentals. On the other hand, cursory examination shows that there are also substantial differences between the cranial skeletons of the cat and the sheep. Many of these relate to differences in the type of food eaten and in the way it is masticated. Others relate to differing ways the head is used in intra- and interspecific competitive and defensive behavior.

Make the following comparisons between the skulls of the cat and the sheep, noting the enlarged pointed canines in the cat as opposed to the absence of canine-like teeth and upper incisors in the sheep; well-developed postglenoid process and the smaller, but still prominent, ventrally directed process bordering the glenoid cavity anteriorly in the cat as opposed to the presence of a poorly developed postglenoid process and absence of an anterior process in the sheep; the relatively large temporal fossa with a midsagittal crest developed posteriorly in the cat as opposed to a relatively small temporal fossa failing to reach the middorsal line in the sheep; the location of the articular condyle of the lower jaw at the level of the lower tooth row and a coronoid process extending relatively high above the jaw joint in the cat as opposed to the location of the articular condyle well above the level of the tooth row and a coronoid process extending only a short distance above the jaw joint in the sheep; a relatively small angle on the lower jaw of the cat as opposed to a greatly expanded angle in the sheep. In what follows, these and other differences in skull morphology between cat and sheep are related to differences in how the heads of these animals are used in life-sustaining activities.

### a. Feeding Behavior—Food-gathering

Most mammals (humans are a notable exception) use their heads as organs of prehension. However, the physical demands of prehension vary greatly according to the type of food eaten.

In those skeletal parts seen so far, the cat illustrates features typical of mammalian carnivores. The pointed canines obviously serve to grab, hold, and kill prey. While loads imparted to the skull and lower jaw of a domestic cat during prehension may be slight because of the small size of its usual prey, in other carnivores, such as the large felids (lion, leopard, etc.), which prey on large animals, these loads are substantial. The skull and lower jaw must be able to withstand impact that may tend to force the open lower jaw back against the skull; they must also be able to withstand the effects of the weight and struggles of the prey that may pull the lower jaw forward and rotate it downward. Note that the jaw joint of a cat is much more secure than that of a sheep. Forces that drive the lower jaw back against the skull are resisted by the well-developed postglenoid process. Forces that pull the lower jaw forward are resisted to some extent by the anterior process of the glenoid cavity; moreover, the line of action (direction of pull) of the relatively large temporalis muscle (as reflected in the relatively large temporal fossa) is appropriately directed to resist forward and downward displacement of the lower jaw. This may help explain the presence of a relatively large temporalis muscle acting with a relatively long lever arm (roughly equivalent to the height of the coronoid process above the jaw joint) in mammalian carnivores. Also note that the postglenoid process will resist tendencies for posterior displacement of the lower jaw created by the direction of pull of the powerful temporalis muscle.

In contrast, the sheep is a grazing animal that pulls up grasses. The absence of enlarged canines in sheep reflects total lack of predatory behavior; it also reflects the development of types of competitive and defensive behavior that do not employ canines either for display or for actual biting. Furthermore, the loss of large canine teeth is also a consequence of the development of considerable transverse movement of the lower jaw when chewing. It should be apparent that the physical demands of food procurement are much less on the skull of a grazing or browsing animal than on a carnivore; insofar as food procurement is concerned, a large temporalis muscle and a relatively secure jaw joint are thus not needed for procuring food.

### b. Feeding Behavior—Chewing

Unlike the great majority of other vertebrates, most mammals chew their food. The effect of chewing is to reduce food items to a size suitable for swallowing, as in carnivores, or to reduce great amounts of tough plant material of low nutritive value to small bits in order to break down nondigestible or poorly digestible cell walls and to increase the surface area available for digestive juices to work on.

In carnivores, flesh is cut into pieces mainly by the shearing action of the carnassial teeth (last premolar above, first molar below). Demonstrate on a cat skull with articulated lower jaw that this shearing action mostly involves upward rotation of the lower jaw. Slight transverse movement of the lower jaw to the side on which food is being cut (as well as slight rotation around the long axis of that half of the mandible on the chewing side) probably also occurs in order to press the cutting surfaces of the carnassials tightly together and ensure effective shearing. Note that the jaw joint of the cat will allow upward rotation of the lower jaw and slight sliding from side to side. It will not readily allow forward and backward sliding of the lower jaw, for reasons already considered in relation to prehension.

Chewing in the sheep, on the other hand, employs a highly mobile jaw joint. The ridges (lophs) on the lower molars and premolars move sideways across the ridges on the upper teeth. The intervening food is ground and cut to pieces in the process. To accomplish this, the lower jaw swings (rotates) and slides from side to side; actions which involve forward, backward, and transverse movements of the articular condyles. As is typical of mammalian herbivores, the relatively large size of the angular region of the lower jaw reflects the fact that the masseter and internal pterygoid muscles are the dominant jaw adductors. The temporalis muscle is much smaller, as indicated by the small size of the temporal fossa. The masseter and internal pterygoid muscles also have a substantially larger lever arm than the temporalis because the articular condyle is raised far above the level of tooth row. (Note that in carnivores the temporalis muscle is the largest and also has the greatest leverage.) The former muscles are those mainly responsible for the chewing movements described.

Numerous suggestions have been made as to why many mammalian herbivores have increased the relative size and leverage of the masseter and internal pterygoid muscles at the expense of the size and leverage of the temporalis muscle. The primitive mammalian condition, retained by the opossum and living carnivores, is one in which the temporalis is the largest muscle with the greatest leverage. One suggestion has been that once the demands of predatory behavior are eliminated, the development of a condition like that in the sheep results in a substantial reduction of force exerted at the jaw joint over that which exists when the temporalis muscle dominates. It is possible, therefore, that selection has operated in herbivores to reduce joint wear and tear in animals that chew for extensive periods of time.

### c. Inter- and Intraspecific Aggressive and Defensive Behavior

In addition to procuring and processing food, the head is important in defensive activities against enemies and in aggressive activities toward members of the same species for space (territory), social position, or mates.

Carnivores typically display their well-developed canines, exhibiting them as potential weapons of aggression or defense. When actual fighting occurs, either within or between species, biting is a common occurrence. This type of fighting, involving the teeth and jaws, utilizes the same adaptations that carnivores possess for bringing down prey; qualitatively, at least, the physical demands on the skull and lower jaw are similar. It also appears that those herbivores that retain conspicuously developed canines (e.g., gorillas, pigs) do so for their value as threat or weapons in combat.

Sheep represent herbivores that have eliminated canines (in this case, by modifying the lower canines to function as incisors) and that utilize head-to-head butting as the major method of intraspecific combat, especially among males competing for mates. In the domestic sheep, some osteological modifications for this type of behavior can be seen. Note the great breadth of the frontal bone in the interorbital region. In the adult male (especially in some wild species) this region is massively developed and expanded through development of sinuses within the frontal bone itself. This region forms the structural base for support of the horns, which may be subjected to great impact forces. If the specimen you are examining is a juvenile, note the developing horn cores on the posterior slope of the frontal bone. Also, note the downwardly rather than backwardly facing foramen magnum, which is so oriented that when the head is lowered and the horns and frontal region are facing forward, the forces created by the impact of butting are directed into the neck, which therefore forms a buttress against these forces.

---

**15. Some Variants of the Mammalian Skull and Teeth**

### a. Primitive Therians—the Opossum

The primitive therian mammal skull is much like that of the opossum; both the snout and temporal region are long, the braincase is rather narrow, and the zygomatic arch is complete and well-developed. The lower jaw possesses a large coronoid process, slender body, and an articulation slightly above the level of the teeth. Canines are large, incisors small and unspecialized, premolars simple, and tritubercular molars suited both for shearing and grinding; the entire dentition is adapted for a generalized carnivorous-omnivorous diet. From this type of skull, that of all later therian mammals is probably derived.

### b. Insectivores

Primitive insectivores, like hedgehogs and tree shrews, are similar to the opossum in general skull morphology; most groups show an enlargement of certain incisors and reduction of the canines. The zygomatic arch is lost in tenrecs and shrews.

### c. Carnivores

The long-snouted skulls of generalized fissiped carnivores, such as civets and dogs, differ little in appearance from that of the opossum. In the cats and weasels, the facial reigon is extremely shortened. The jaw articulation of most carnivores lies close to the level of the lower tooth row. All fissipeds are characterized by large canines and by the specialization of the last upper premolar and the first lower molar as a set of enlarged flesh-shearing blades called *carnassials*. In extremely carnivorous forms, such as cats, the number of molars is greatly reduced; in more omnivorous forms, like the bears and raccoons, the molars are expanded for grinding and the carnassials are reduced.

In the fish-eating pinniped carnivores (seals, sea lions), the skull is usually similar to that of fissipeds, but the cheek teeth are simpler and all alike and there are no carnassials. The skull of the mollusk-eating walrus is extremely massive; the cheek teeth are flat-crowned crushing surfaces, and the upper canines are greatly elongated to form the "clam-rake" tusks.

### d. Edentates, Pangolins, and the Aardvark

The edentates (armadillos, sloths, and anteaters) are a diversely adapted group characterized by the degeneration of the teeth, which are simple pegs lacking enamel in armadillos and sloths and are entirely absent in anteaters. Incisors are lacking and enlarged anterior teeth in sloths are not certainly homologous with canines. In some armadillos the number of teeth is greatly increased over the usual placental number. Except in sloths, the skull is long and tubular, notably so in the giant anteater. The distantly related pangolins and aardvark are convergent upon true anteaters because of their similar ant- and termite-eating habits. Pangolins are toothless, but the aardvark (which may be derived from primitive ungulates) possesses peculiar enamelless cheek teeth in which the dentin forms numerous columnar prisms, each with a tubular pulp cavity.

### e. Rodents and Lagomorphs

These two groups are characterized by the absence of canines and the possession of an enlarged pair of rootless, ever-growing upper and lower incisors separated from the cheek teeth by a diastema; otherwise they are quite different and are not at all closely related. Lagomorphs (rabbits, hares, and pikas) are easily distinguished by the possession of a second pair of small upper incisors behind the enlarged pair and by numerous perforations (fenestrations) of the snout. Mastication is by transverse movements of the jaws, and the skull and mandible therefore resemble those of ungulates in proportions (see below). The skull of rodents tends to be low and little arched, unlike that of rabbits, and the lower jaw has a small coronoid process and a well-developed angular process. Anterior premolars are lost and those present are usually molariform; the cheek teeth have a complicated morphology with well-developed cross ridges. Chewing is by fore-aft movements of the mandible. The masseter muscle invades the face: (1) by forming a trough in front of the zygomatic arch (squirrel or sciuromorph condition); (2) by passing forward through a greatly enlarged infraorbital foramen (porcupine or hystricomorph condition); or (3) by utilizing both of these ways (rat or myomorph condition).

### f. Cetaceans

Perhaps the most remarkable variants of the mammalian skull are seen among the whales and dolphins. The cranial region is short and broad, and the face has a beak or rostrum formed by elongated maxillae and premaxillae. The external nares have moved far back and open upward by way of the "blowhole"; the very reduced nasals lie behind them. The bones surrounding the blowhole of toothed whales are asymmetrical in size and shape; this is most pronounced in the sperm whale. The greatly enlarged supraoccipital spreads forward and almost meets the nasals, so that the parietals and frontals have little or no dorsal midline exposure. Toothed whales possess homodont, conical teeth that may greatly exceed the normal placental number and that do not undergo replacement. In the sperm whale they are confined to the lower jaw. The whalebone whales are devoid of teeth, although tooth germs occur in the young; instead, there is along each side of the upper jaw a series of frayed horny plates (the "whalebone" or baleen), which hang vertically and are used to sieve plankton from the water.

### g. Ungulates

These are the herbivorous hoofed mammals of the orders Perissodactyla (horses, tapirs, and rhinos) and Artiodactyla (pigs, hippos, camels, deer, cattle, and a great variety of other "cloven-hoofed" types). In all, the cheek teeth tend to become square and the cusps tend to become complicated. In the perissodactyls, the premolars become molariform and the cheek teeth develop ridges connecting the main cusps; this is called a *lophodont* pattern. The most complicated pattern of lophs is seen in the horse. Primitive artiodactyls, like the pigs and hippos, have molars with separate bulbous cusps—a *bunodont* pattern. Advanced artiodactyls, the ruminants, possess molars in which the cusps are crescentic—a *selenodont* pattern. In artiodactyls, the premolars tend not to be fully molariform.

The skulls of most ungulates share a number of common features that are related to the way plant food is procured and processed. With few exceptions, the face is elongated and very deep and the cranial region is short; the orbit is closed behind by a postorbital bar; canines are reduced and incisors are specialized for cropping (in advanced ruminants, the lower canine has the morphology and function of an incisor); there is a diastema in front of the cheek teeth; the lower jaw has a deep angle, a small coronoid process, and an articulation elevated well above the tooth row; weapons are often developed on the skull, solid keratinous horns in rhinos, bony antlers in deer, true horns (keratinous sheath covering a bony core) in cattle, antelopes, and their allies. Only a few ungulates retain enlarged canines that serve as weapons: the curved, ever-growing tusks of pigs and hippos and the saber-like upper canines of tragulids and musk deer.

### h. Primates

The higher primates—monkey, apes, and man—are characterized by skulls with a large, rounded cranium, a relatively short face, the complete separation of the forwardly facing orbits from the temporal fossae by a bony wall, and the shift of the occiput and foramen magnum to a more ventral position below the braincase. The molar teeth of apes and man are bunodont, suitable for an herbivorous or omnivorous diet. The teeth of Old World monkeys are somewhat lophodont. There are only two, instead of the usual three, incisors. The canines are enlarged in male apes and monkeys, usually small in females; in man they are small in both sexes. The apes and man are characterized by extensive fusions of the skull bones.

## J. Summary

1. The vertebrate skull is formed from the fusion of three originally separate components: the cartilaginous cranium or chondrocranium, the splanchnocranium, and the dermatocranium.

2. The chondrocranium, which forms around the brain and organs of special sense, arises through the fusion of originally separate cartilages—the parachordals alongside the notochord and the trabeculae anterior to these—with the nasal and otic capsules and with lateral cartilages, the orbital cartilages. Its roof is typically open and its walls imperfect, but one or more vertebra-like arches roof the rear end.

3. The ossification of the chondrocranium produces most of the cartilage bones of the skull, which are limited to its ventral and ventrolateral surfaces and rear end. The principal cartilage bones are the supraoccipital, exoccipital, basioccipital, basisphenoid, laterosphenoid, orbitosphenoid (or sphenethmoid), the otic bones, and some bones of the nasal capsules. Two bones, opisthotic and prootic, usually ossify in the otic capsules of tetrapods, though a third bone, the epiotic, is often present; in mammals

they are fused to form a single periotic or petromastoid bone. One or two additional bones may occur in the basicranial axis of mammals, the presphenoid and mesethmoid.

4. The number of cartilage bones tends to be reduced in higher tetrapods, especially in mammals, through the fusion of originally separate bones.

5. The splanchnocranium consists of the visceral (gill) arches, typically seven in number. Of these, the first or mandibular arch forms the upper and lower jaws. The upper jaw is braced against the chondrocranium, while the lower jaw forms a movable joint with the upper. The second or hyoid arch contributes to the bracing of the mandibular arch in fishes; the remaining five or branchial arches support the gills. In land vertebrates the hyoid arch and remaining five arches contribute to the hyoid apparatus and the laryngeal cartilages.

6. The dermatocranium consists of dermal bones that encase the chondrocranium and jaws and become part of the definitive skull. In early (extinct) tetrapods they form a complete roof for the skull, but in the evolution of higher tetrapods they are greatly reduced in number, chiefly through loss. Vacuities also tend to be formed in the posterior part of the roof, and these temporal fenestrae are important in the evolution of the various reptilian groups and of birds and mammals.

7. The upper and lower jaw cartilages ossify in fishes and also become ensheathed in dermal bones. The upper jaw (palatoquadrate) and its dermal sheath become incorporated into the skull as the bones of the palate and the maxillary arch; only two cartilage bones formed in the palatoquadrate cartilage persist through the vertebrates: the quadrate at its posterior end and the epipterygoid further forward. The epipterygoid in mammals is called the alisphenoid, a cartilage bone fused into the side wall of the braincase. The lower jaw cartilage (Meckel's cartilage) gives rise to one main cartilage bone, the articular, at the rear end of the lower jaw. The upper end of the hyoid arch forms the hyomandibula, typical of fishes.

8. In all tetrapods except mammals the lower jaw articulates with the skull by way of the articular and quadrate bones. In mammals, these two bones are greatly reduced and a new articulation has formed between two dermal bones, the dentary and the squamosal.

9. In primitive tetrapods there are a large number of dermal bones in the lower jaw; these become reduced in number in many higher tetrapod groups. In mammals all bones but the dentary are lost from the lower jaw, which thus consists of a single pair of bones, the dentaries.

10. The middle ear contains a sound-transmitting apparatus that in all tetrapods except mammals consists of one element or a chain of elements, termed the columella auris, which is a remnant of the upper element of the hyoid arch, the hyomandibula of fishes. This persists in mammals as the stapes. In addition, the mammalian middle ear contains two more small bones, the malleus and the incus, representing the greatly reduced articular and quadrate, respectively.

11. The teeth of vertebrates develop in a manner similar to that of the dermal denticles of the scales of primitive fishes, similar to the individual placoid scales of elasmobranchs. Usually of simple form and all alike among lower tetrapods, they become heterodont, that is, differentiated into several kinds, each with its distinct function, in mammals; in most lower tetrapods the teeth are ankylosed to the bone of the jaw by a bone-like cement, but in mammals and some reptiles (such as crocodilians) they are set in sockets in the jaw bones, termed the thecodont condition. Borne in lower vertebrates on various jaw and palatal bones, they become limited in higher tetrapods to the jaw margins. The complicated form of mammalian molar teeth is best explained as the result of elaboration on a primitive three-cusped tooth derived from mammal-like reptile ancestors.

**References**

Allin, E. F. 1975. Evolution of the mammalian ear. *J. Morph.* 147:403–38.

Barghusen, H. R. 1972. The origin of the mammalian jaw apparatus. In *Morphology of the maxillo-mandibular apparatus,* ed. G. H. Schumacher, pp. 26–32. Leipzig: VEB Georg Thieme.

Beer, G. R. de. 1937. *The development of the vertebrate skull.* London and New York: Oxford University Press.

Carroll, R. L. 1969. Problems of the origin of reptiles. *Biol. Rev.* 44: 393–432.

Crompton, A. W. 1963. On the lower jaw of *Diarthrognathus* and the origin of the mammalian lower jaw. *Proc. Zool. Soc. Lond.* 140: 697–753.

————. 1971. The origin of the tribosphenic molar. In *Early mammals,* ed. D. M. and K. A. Kermack, supplement no. 1, *Zool. J. Linnean Soc.* 50:65–87.

Crompton, A. W., and Hiiemae, K. 1969. How mammalian molar teeth work. *Discovery* (Magazine of the Peabody Mus. Nat. Hist., Yale Univ.) 5:23–34.

Edmund, A. G. 1969. Dentition. In *Biology of the Reptilia morphology A,* vol. 1, eds. C. Gans, A. d'A. Bellairs, and T. S. Parsons, pp. 117–200. London and New York: Academic Press.

Goodrich, E. S. 1930. *Studies on the structure and development of vertebrates.* London: Macmillan.

Gregory, W. K. 1933. Fish skulls: A study of the evolution of natural mechanisms. *Trans. Amer. Philos. Soc.* 23:75–481.

————. 1934. A half century of trituberculy: The Cope-Osborn theory of dental evolution, with a revised summary of molar evolution from fish to man. *Proc. Am. Philos. Soc.* 73:169–317.

Hopson, J. A. 1966. The origin of the mammalian middle ear. *Amer. Zool.* 6:437–50.

Hopson, J. A., and Crompton, A. W. 1969. Origin of mammals. In *Evolutionary biology,* eds. Th. Dobzhansky, M. K. Hecht, and W. C. Steere, 3:15–72. New York: Appleton-Century-Crofts.

Jarvik, E. 1954. On the visceral skeleton in *Eusthenopteron,* with a discussion of the parasphenoid and palatoquadrate in fishes. *K. Svenska Vetensk Acad. Handl.* (4) 5:1–104.

Jollie, M. 1962. *Chordate morphology*. New York: Reinhold.

Kingsbury, B. F., and Reed, H. D. 1909. The columella auris in Amphibia. *J. Morph.* 20:549–627.

Lombard, R. E., and Straughan, I. R. 1974. Functional aspects of anuran middle ear structures. *J. Exp. Biol.* 61:71–93.

Maynard Smith, J., and Savage, R. J. G. 1959. The mechanics of mammalian jaws. *School Science Rev.* 141:289–301.

Miles, R. S. 1971. The Holonematidae (placoderm fishes), a review based on new specimens of *Holonema* from the Upper Devonian of Western Australia. *Phil. Trans. Roy. Soc. Lond.* ser. B, 263:101–234.

Moss, M. A. 1972. The feeding mechanism of sharks of the family Carcharhinidae. *J. Zool.* 167:423–36.

Moy-Thomas, J. A., and Miles, R. S. 1971. *Palaeozoic fishes*. 2d ed. Philadelphia: W. B. Saunders.

Panchen, A. L. 1964. The cranial anatomy of two Coal Measure anthracosaurs. *Phil. Trans. Roy. Soc. Lond.,* ser. B, 247:593–637.

―――. 1970. *Anthracosauria*. In *Handbuch der Paläoherpetologie,* ed. O. Kuhn, Part 5A. Stuttgart: Gustav Fischer Verlag.

Parrington, F. R. 1967. The identification of the dermal bones of the head. *J. Linn. Soc. (Zool.)* 47:231–39.

Parrington, F. R., and Westoll, T. S. 1940. On the evolution of the mammalian palate. *Phil. Trans. Roy. Soc. Lond.,* ser. B, 230:305–55.

Parsons, T. S., and Williams, E. E. 1963. The relationships of the modern Amphibia: A re-examination. *Quart. Rev. Biol.* 38:26–53.

Peyer, B. 1968. *Comparative odontology*. Chicago: University of Chicago Press.

Romer, A. S. 1956. *Osteology of the reptiles*. Chicago: University of Chicago Press.

Scapino, R. P. 1972. Adaptive radiation of mammalian jaws. In *Morphology of the maxillo-mandibular apparatus,* ed. G. H. Schumacher, pp. 33–39. Leipzig: VEB George Thieme.

Schaeffer, B., and Rosen, D. E. 1961. Major adaptive levels in the evolution of the actinopterygian feeding mechanism. *Amer. Zool.* 1:187–204.

Thomson, K. S. 1966. The evolution of the tetrapod middle ear in the rhipidistian-amphibian transition. *Amer. Zool.* 6:379–97.

―――. 1967. Mechanisms of intracranial kinetics in fossil rhipidistian fishes (Crossopterygii) and their relatives. *J. Linn Soc. (Zool.)* 46:223–53.

Westoll, T. S. 1943. The hyomandibular of *Eusthenopteron* and the tetrapod middle ear. *Proc. Roy Soc. Lond.,* ser. B, 131:393–414.

Wilder, H. H. 1903. The skeletal system of *Necturus maculatus*. Rafinesque. *Mem. Boston Soc. Nat. Hist.* 5:357–439.

Zangerl, R., and Williams, M. E. 1975. New evidence on the nature of the jaw suspension in Palaeozoic anacanthous sharks. *Palaeontology* 18:333–41.

Ziegler, A. C. 1971. A theory of the evolution of therian dental formulas and replacement patterns. *Quart. Rev. Biol.* 46:226–49.

# 9 The Comparative Anatomy of the Muscular System
## Leonard Radinsky

**A. Introduction**

**1. Kinds and Origin of Muscle**

The muscles of the vertebrate body may be divided into two general classes, the *involuntary* and the *voluntary*. The involuntary or *smooth* muscles occur in the walls of the digestive tract and other viscera and in the skin and certain of its derivatives. They originate through the transformation of mesenchyme cells, which may be of various origins, but typically they come from the splanchnic mesoderm of the hypomere, since in development the hypomere closes around the archenteron and its derivatives (see chap. 4). Cardiac muscle is a special type of smooth muscle in which the cells have fused as a *syncytium*. Adult heart musculature is a continuous network of dividing and recombining strands with prominent bands, termed *intercalated disks,* subdividing the network into short units. The voluntary or *striated* muscles, on the other hand, arise from the myotomes (also called muscle plates), with certain exceptions noted below. The myotomes, you will recall, are those portions of the epimeres remaining after differentiation of the sclerotomes and dermatomes. From their original dorsal position the myotomes grow down between the hypomere and the skin, and left and right sides meet in the midventral line. Review pages 102–3. In this way there is produced a complete coat of voluntary muscles over the body, lying beneath the skin. This muscle coat is divided into dorsal and ventral parts by the horizontal skeletogenous partition (also called lateral septum), which extends from the vertebral column to the lateral line. The muscles dorsal to the septum are called *epaxial* and those below the septum, *hypaxial*. It should be noted, however, that the horizontal septum is absent in *Branchiostoma* (amphioxus) and cyclostomes and hence apparently is not a primitive chordate feature.

The muscles originating from the myotomes are called *parietal* or *somatic* muscles, but not all the voluntary muscles are of myotomic origin. In the gill region of vertebrates a system of voluntary muscles is developed for moving the gill arches. The gills are of endodermal origin, and the muscles in the walls of the gill region come from the mesoderm of the hypomeres. For this reason the gill arch musculature is frequently called the *visceral* musculature, although, unlike the musculature of the viscera, it is voluntary and striated. A better term, and one that avoids confusion, is *branchial* or *branchiomeric* musculature; this will be used here. There are consequently voluntary muscles of two kinds of embryonic origin: the parietal or somatic muscles, derived from the myotomes of the epimere, covering most of the body and innervated by spinal nerves; and the branchiomeric muscles, derived from the hypomere, limited to the gill region and innervated by cranial nerves.

The presence of the gill apparatus blocks the downward growth of muscles from the head and neck myotomes. Midventral musculature in that region comes from myotomes behind the gills. These myotomes grow downward, turn forward, and extend anteriorly to the jaw. The muscles so originating are termed the *hypobranchial* musculature and are innervated by spinal nerves, like the rest of the myotomic musculature. There is also an *epibranchial* musculature above the gill region, but this is of slight importance, being limited to elasmobranchs.

The terms muscle and musculature used in this chapter refer only to the voluntary muscles, since only these are studied in vertebrate dissection; the study of the involuntary muscles belongs properly to histology. It is assumed that the student understands the histological difference between smooth and striated muscle.

**2. Terminology of Muscles**

The major function of most striated muscles is to move the skeleton. (Exceptions are certain subcutaneous muscles that move or wrinkle the skin or close orifices, such as the mouth or anus, and the muscles that move the eyes.) Muscles are attached to bone by connective tissue. When the connective tissue attachments form flat sheets, these are termed *fasciae* or *aponeuroses;* when they are concentrated into a band, this is termed a *tendon.* A muscle is attached at its two ends and is more or less free in the middle. The place of attachment that in any particular movement remains fixed when the muscle contracts is called the *origin;* that which is caused to move is the *insertion;* and the in-between free part of the muscle is the belly. When a muscle has more than one origin, these are called *heads.* Several points of attachment segmentally arranged are termed *slips.* Persistent myosepta in the course of a muscle that extends over more than one body segment are called *inscriptions.* The specific movement produced by a muscle is called its *function (action)* and is brought about by the shortening of the muscle.

One method of naming muscles is to combine the origin and the insertion into a compound word in which the origin precedes, as *pubo-femoralis,* which would mean a muscle originating on the pubis, inserting on the femur and causing the femur to move toward the pubis on contraction. However, many muscle names are based on shape, function, or fanciful resemblances. In many muscles origin and insertion are interchangeable; that is, either end may be held fixed and the other end moved.

A single muscle rarely, if ever, acts alone to produce a specific movement, and a given muscle may play a role in several different functions. One or more muscles that are the main movers of a given joint are called the *prime movers* (or *agonists*) of the action or function. The action of the agonist(s) is usually counteracted by another set of prime movers called the *antagonists* (which pull in the opposite direction). Contraction of the prime mover can produce the required effect only if the bone from which it acts is relatively fixed. Muscles that anchor, steady or support a bone or joint in order that a prime mover may have a firm base from which to pull are called *fixation muscles.* The term *synergist* has been used in so many different ways that its meaning has become ambiguous. Generally synergists act along with prime movers to help them produce a specific movement.

The chief actions of muscles are: flexion, extension, adduction, abduction, rotation, elevation, depression, and constriction. Flexion is the bending of a joint, and the antagonistic action of extension is the straightening out of a joint; these terms are applied chiefly to limbs, and muscles with such action are named flexors and extensors. Adduction means drawing a limb toward the midline; abduction, away from the midline; and the corresponding muscles are called adductors and abductors. Muscles that rotate a limb segment about its long axis are termed rotators, pronators, and supinators. *Protractors* pull a part forward, and *retractors* pull a part backward. Elevation and depression are illustrated by the closing and opening of the lower jaw, and the muscles concerned are called *elevators* and *depressors*. Constrictor muscles act by compressing a part; when encircling an aperture such as the mouth or anus, they are termed *sphincters*.

**3. Muscle Topography and Its Functional Significance**

To understand the functional significance of the muscular system, it must be studied in conjunction with the skeletal and articular systems. The bones individually or collectively serve as rigid bars or *levers* even though some may have peculiar shapes. A joint (or articulation) is a *fulcrum* around which movements occur. The *power* or force is exerted by muscles. The result is the overcoming of a *resistance,* weight, or load.

The pull of muscles often causes bones to rotate about their joints. A measure of the effectiveness of a muscle in rotating a bone is its *moment* or torque. The moment of a force about any point is equal to the magnitude of the force multiplied by the perpendicular distance from the line of action of the force to that point. Thus the moment of a muscle is equal to the pull exerted by the muscle multiplied by the perpendicular distance from the line of action of the muscle (its direction of pull) to the fulcrum. The moment of the resistance may be similarly calculated (see fig. 9.1). The perpendicular distance from line of action of force or resistance to the fulcrum is known as the *moment arm*.

The moment of a muscle, which is a measure of how effective it is in producing a given movement, may be increased by increasing its force (magnitude of pull) or by increasing its moment arm (perpendicular distance from the joint, or fulcrum). An increase in the moment arm can greatly increase the moment, even if the force is not enlarged. Animals that have specialized for powerful movements often have the relevant muscles situated far from the joint, for a muscle with a relatively large moment arm requires less force to overcome a resistance than does a muscle of the same physical properties situated closer to the joint. It is commonly stated that muscles with small moment arms have the advantage of increased speed and range of movement compared to similar muscles with larger moment arms. However, in reality the situation is complicated by such factors as the angle of attachment, which changes as the bones move, and the length of the muscle at any given time compared with its optimal length (see below). Thus it is an oversimplification to talk of muscles adapted for speed and range at the expense of

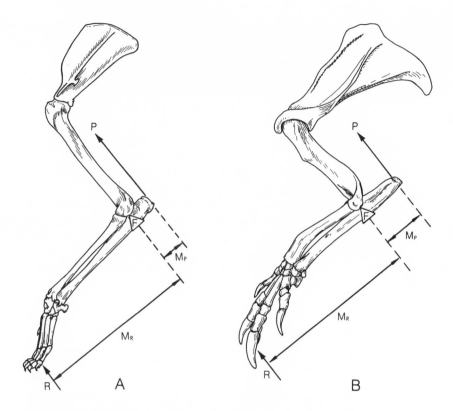

Fig. 9.1. Partial analysis of action of triceps muscle. *A,* walking cat. *B,* digging armadillo. *F,* fulcrum; *P,* line of action of triceps muscle; *R,* resistance; *Mp,* moment arm of triceps muscle; *Mr,* moment arm of resistance.

power, or vice versa, solely from considerations of moment arms. The angle of pull of a muscle affects its power for only at an angle of 90 degrees is the entire force of the muscle acting to rotate the bony lever around its fulcrum. At angles greater or less than 90 degrees, some of the muscle energy is dissipated.

The biological meaning of many features of the skeletomuscular system can be understood by considering the mechanical advantages with which the muscles work. One can often deduce from biomechanical analysis whether natural selection has favored great power or, alternatively, speed and a greater range of movement (such as has been shown in running animals). Of course, in many instances a compromise is the optimal solution, and it occurs frequently in nature.

**4. Muscle Architecture**

The architectural arrangement of muscle fibers within a muscle is an important anatomical parameter when comparing muscles within or between individuals of one or more taxa. A muscle is composed of fibers arranged in a three-dimensional pattern between two simple or complex surfaces of connective tissue. Each muscle fiber is surrounded by a delicate connective tissue sheath, the *endomysium*. Muscle fibers are grouped into *fasciculi,* which are enveloped by *perimysium*. A muscle as a whole consists of many fasciculi and is enclosed by *epimysium,* which is closely associated with an aponeurosis and is sometimes fused with it.

In *parallel-fibered muscles* the fibers may run parallel from surface to surface so that the length of a single fiber gives an estimate of the length of the contracting part of the whole muscle (fig. 9.2*A*). Arrangement of fibers in parallel provides the greatest possible excursion of the insertion

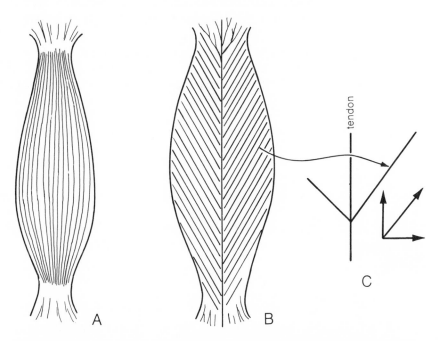

Fig. 9.2. Two types of muscle architecture. *A,* parallel-fibered muscle; *B,* pinnate-fibered muscle; *C,* vector analysis of force exerted by a pinnate muscle fiber. The effective force is represented by the arrow parallel to the central tendon.

with minimum nonphysiological loss of power during contraction. Because the distance of contraction represents a smaller fraction of the fiber length in this parallel arrangement than in any other, the fibers will also remain in a more *isometric* condition (when a muscle develops tension without changing its length) so that the force per fiber will be maximum. Parallel forces can be summed by simple algebraic addition of their magnitudes.

In *pinnate-fibered muscles* the fiber position lies at an angle to the direction of exerted force or induced motion (fig. 9.2*B*). The fibers can be arranged either in a parallel pinnate pattern, remaining parallel but lying at an angle to the direction of tendon motion, or in a radially pinnate pattern, whereby the fibers converge on a tendon from a fan-shaped (or cone-shaped) organization. An important parameter of a pinnate muscle is the angle of pinnation between muscle fiber and direction of tendon motion. Because of the angle of pinnation, the contractile force of a fiber must be resolved into an effective component parallel to the direction of the tendon axis and a component that does not contribute to the useful work of the muscle, since it is at right angles to the tendon axis (fig. 9.2C). This loss of force is the fundamental disadvantage of fiber angling. The advantage of the pinnate arrangement is that more muscle fibers, and thus more power, may be packed into a given volume.

It is well known that there are "red" and "white" muscles among vertebrates. The chicken pectoralis muscle is often referred to as white meat (muscle) and the chicken thigh muscles are referred to as dark (red) meat (muscle). Most vertebrate muscles are composed of different proportions of at least three functionally and morphologically distinct fiber types. The three types may be designated as R (type 1), W (type 2), and I (intermediate between R and W). The R fibers are red and have a small diameter, dense vascularization, high myoglobin and fat content, and numerous mitochondria and are capable of sustained and slow contraction. The W fibers, on the other hand, are white and possess a large diameter, sparse vascularization, low myoglobin and fat content, and a small number of mitochondria and are capable of short and fast contraction. Some muscles are composed of a mixture of various proportions of W, I, and R fibers, others contain a mixture of R and W fibers, and certain muscles in certain taxa are made up of either I fibers or R fibers. No vertebrate muscle is known to have only W fibers.

5. Kinds of Muscle Contraction

Contraction refers to the development of tension within a muscle. It does not necessarily mean that any shortening of the muscle takes place. When a muscle develops tension without changing its length, it is said to be in *isometric contraction. Isotonic contraction* takes place when a muscle develops tension and actually shortens. The force produced by a single muscle fiber reaches a maximum when the fiber is at an optimum length (or "resting" length) during isometric contraction. The force drops off slowly when the fiber shortens to not more than 80% of its optimum length and declines very sharply thereafter. Although there is considerable intermuscular and interspecific variation, in the vast majority of vertebrates whole muscles cannot shorten by more than 30% ( = to 70% of resting length). The force produced by a muscle also decreases when the fiber is stretched beyond its optimum length. There are definite limits on the extent to which a whole muscle can be stretched. Although the figures vary both intermuscularly and interspecifically, we may regard a stretch of 130% of the optimum length of a muscle as a maximum value.

6. Determination of Muscle Function

Several approaches are available to determine the function of a muscle. Most of the functions ordinarily attributed to muscles have been determined by the anatomical method, which deduces function from the origin and insertion of a muscle. The derivation of function by an origin-insertion approach, using preserved material, is an oversimplified and sometimes erroneous procedure, because fiber direction, muscle topography, shape, and size may change after death and fixation. The anatomical method can be made more accurate by including the nature of the attachment and the natural position of muscle, the lever system, the permissible excursion, the normal loading, the physiological cross section, muscle volume or dry weight of the muscle, and types of muscle fibers (R, W, or I). However, these parameters tell us what a muscle can do, not necessarily what it actually does. The traditional textbook approach of assigning one specific function to a given muscle is often erroneous, being based on the assumption that the whole animal is no more than a simple sum of its isolated parts. A muscle often has multiple functions when considered in the framework of a major biological function in which it participates, such as locomotion, feeding, or respiration. Some functions will not be detected if the muscle is studied as an isolated part.

Another approach used in determining muscle function is *electrical stimulation*. Electrical stimulation of a muscle causes contraction. This approach has some of the same disadvantages as the anatomical approach. Furthermore, the results of electrical stimulation may be misleading. For example, in man, when the deltoid muscle, which is attached to the scapula and the humerus, is stimulated, both bones move toward each other and the scapula is depressed. When the deltoid acts naturally in coordination with other muscles of the scapula, the latter is fixed and stabilized so that the humerus can be abducted, which is one of the true functions of the deltoid. A muscle function determined as an isolated entity may not be the correct or natural one.

Finally, muscle function can be determined by the electromyograph. The mechanical twitch of a muscle fiber is preceded by a conducted electrical impulse. The electrical activity of the fibers of a muscle can be detected by electrodes within the muscle. The recorded response constitutes an electromyogram (EMG). Records can be made from several muscles simultaneously and under natural conditions. This makes electromyography the most accurate method of determining muscle function, especially when it is combined with synchronous cinematographic recording of the motions produced by the muscles under study.

Analysis of muscle function is usually done to study the mechanics of a particular activity, such as locomotion or feeding. The results of such analyses are presented here in summaries of the mechanics of locomotion and feeding in fish and in mammals. The shark and the cat are emphasized, for you will likely be dissecting them. You should compare and contrast these models and keep them in mind as you do your dissections.

### a. Locomotion in Primitive Vertebrates—Swimming

Fishes swim by producing lateral flexures of the trunk and tail, and the fish moves forward by transverse motion of the tail and body through the water. As the fish moves through the water, it pushes water away from its surface in a backward and sideways direction relative to the axis of the body. However, the inertia of the water resists this movement and consequently the fish's body is acted upon by a force equal, but opposite, to the force that the body exerts on the water (fig. 9.3). This force, opposite to the force that the tail applies to the water, can be resolved into a lateral force and an anteriorly directed force that drives the fish forward. The lateral force is neutralized by the water pressure against the mass of the anterior end of the fish.

The side-to-side movements of the trunk and tail of a fish are waves caused by serially contracting and relaxing parietal muscles on the right

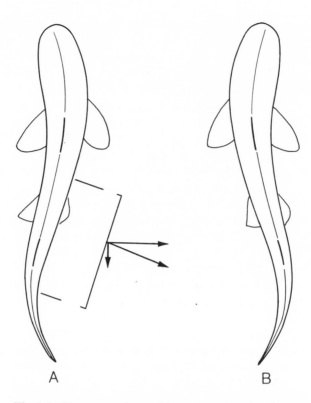

Fig. 9.3. Two successive positions of a swimming shark. As the wave of contracting myomeres passes from front to back, the segment of trunk indicated is moved from left to right against the surrounding water. Because it is obliquely oriented, there is a backward component of thrust (see inset showing resolution of forces), against which the inertia of water exerts an equal but opposite force, which propels the fish forward. The lateral component of thrust is neutralized by water pressure at the head end of the shark.

and left sides of the vertebral column. The waves are initiated anteriorly and pass caudally with increasing amplitudes.

The muscle fibers of the parietal muscles are oriented anteroposteriorly and arranged in a complicated zigzag pattern of blocks (myomeres), associated with cone-shaped myosepta (see your dissection). All jawed fishes possess myomeres that are W-shaped externally, with the upper edge of the W facing anteriorly (fig. 9.4). During unilateral contraction the longitudinal muscles cause a bending of the body, because the vertebral column is longitudinally incompressible but laterally flexible and acts as a fulcrum. Because the front end of the fish is so much more massive than the tail, the side-to-side movements of the tail have minimal influence on the head end, which travels forward in a straight line while the tail end is sweeping back and forth.

The fish turns by generating a unilateral wave of contraction of larger amplitude, so that the head is turned in a new direction, the area just behind the head serving as a fulcrum. At about the halfway point along the body the relationship switches, so that the posterior body swings into a position on the new course, using the anterior body as the lever against which to turn.

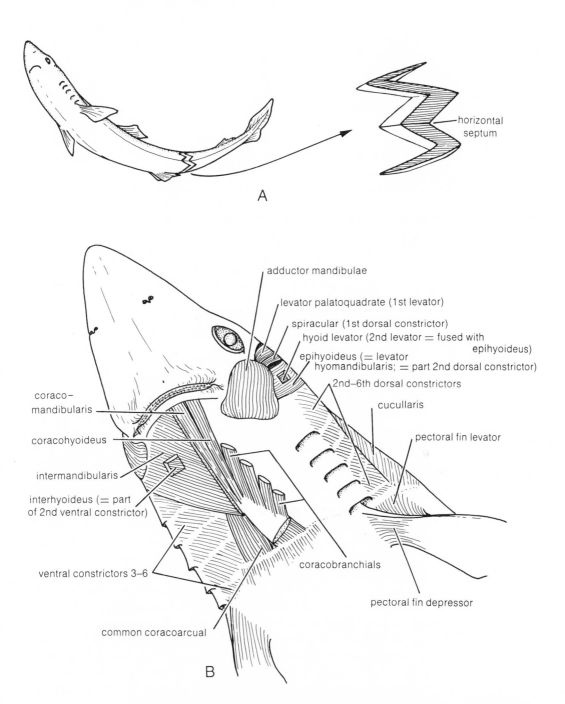

Fig. 9.4. Dogfish shark musculature. *A*, an individual trunk myomere, to show three-dimensional folding. *B*, branchial and hypobranchial musculature. Ventral constrictors removed on left side to show deeper muscles.

The complex and folded arrangement of the myomeres serves several functions. The obliquity of the myosepta extends the pull over several vertebral joints, because a given myomere extends anteriorly and posteriorly over several body segments and the multitude of short, parallel muscle fibers can exert their force over a relatively great distance. Any vertebral joint is acted upon by several myomeres at once, because any one myomere extends over several segments and overlaps several neighboring myomeres. Such overlapping in structure, and therefore in phases of myomere contraction, reinforces and smoothes the movements. (For more detailed analysis of fish locomotion, see this chapter's bibliography.)

The greatest proportion of the body musculature of fishes is made up by white fibers (W-type), while muscles with a high proportion of red fibers (R-type) are found along the horizontal septum. Red-fibered muscles occur much more abundantly in fishes with great staying power for continuous swimming (e.g., salmon).

The median and paired fins serve to stabilize the body against pitch, yaw, and roll (see also chap. 7). Pitch, or rotation about a transverse axis, is counteracted by the paired (pectoral and pelvic) fins. Yaw, or rotation around a vertical axis, is counteracted by the median fins and the caudal fin. Roll, or rotation around the long axis of the body, is countered by both paired and median fins.

Many problems related to the complex of anatomical features involved in the different types of successful fish locomotion remain to be studied. For example, the essential functional anatomical differences between fast and slow swimmers, between hunters and herbivores, and between fishes able to sustain high speeds continuously and high speeds for only a short burst, are either not known or insufficiently known and explained.

### b. Locomotion in Mammals

There have been few in-depth studies of locomotion in mammals, and several of those done have been on primates. Most functional studies of locomotion in other mammals have emphasized the skeleton, with little discussion of muscles and little data from electromyography. With a few exceptions, such as Davis's outstanding monograph on the anatomy of the giant panda, most studies involving mammalian locomotor musculature have been largely descriptive and at best provide data on muscle weights.

There have been some recent studies on locomotion in the cat. Manter (1938) analyzed the mechanics of movements in a walking cat but offered little specific mention of muscle action. Hildebrand (1961) analyzed movies of a running cheetah, a close relative of the cat, and Jenkins (1971b) provided data on limb posture in a walking cat, but neither discussed the locomotor musculature. Goslow and his associates have begun an extensive study of locomotion in the cat. They are particularly interested in muscle performance and nervous system control during movement. Several techniques are used in these analyses. Movements are filmed, and angles and joints and lengths of bones are measured from the films so that differences in walking, trotting, galloping, jumping, and landing can be analyzed. Activity of specific muscles is analyzed electromyographically. In this way gaits can be analyzed and compared, and a comparison with other animals can be made. For example, comparison of Goslow's data on the cat with Hildebrand's for the cheetah, the fastest-running cat, can give considerable insight into what bone-muscle relationships provide for increased speed of running.

Other methods are used to determine the means of control of locomotor activities. Goslow's group has done extensive dissections to determine muscle-fiber orientations, origins and insertions, and relationship to tendons and aponeuroses. Experimental procedures include sectioning nerves to various muscles in order to assess the activity of muscles that retained innervation and to test the force produced by muscles at various joints. In addition, the properties of motor units were measured by analyzing contraction time and tension for motor units in various muscles. Among the conclusions reached by Goslow are the following: the cat is adapted primarily for running, and its foot is specialized. Digit number and length are reduced, metatarsals are elongated, plantar musculature is reduced, and digitigrade posture is assumed (see chap. 7). This has resulted in a "division of labor" among flexor and extensor muscles.

The four phases of the cat step cycle have been carefully analyzed to see how joint angles and muscle activity change as speed of locomotion increases and gaits change. A number of highly specific changes take place in muscle contractile length and other properties and these changes differ from muscle to muscle, depending on their morphologies, functions, and neural controls. These studies continue and are providing much information about the mechanics and neural control of locomotion. Keep this discussion in mind when you dissect the muscles of your specimen, and pay attention to muscle morphology, including fiber direction and attachment and the relationship of muscles to bones and joints. Move the limbs of your specimen to simulate gaits in order to get an idea of possible muscle activity. Compare, if you can, this passive movement with the activity of a living, freely moving cat.

### c. Feeding and Respiration in Primitive Vertebrates

The major functions in which the cephalic musculature of the shark is important are the production of a respiratory water current to ventilate the gills, the initial capture of food, and swallowing (deglutition) of food.

The respiratory current is produced by changes in volume of the orobranchial and parabranchial cavities (see pp. 401–2). Water enters the respiratory system through both the mouth and the spiracle during expansion of the orobranchial cavity, and after passing across the gills it enters the parabranchial cavities before being expelled to the outside through the five pairs of external gill slits.

The principal muscles and skeleton of the dogfish head are depicted diagrammatically in figure 9.5. The direction in which movement occurs when a given muscle contracts is shown by an arrow. Electromyographic analysis has revealed that most muscles (all parts of all constrictors synchronously, the levator palatoquadrate, levator hyomandibulae, interarcualis dorsalis, adductor branchialis, and adductor mandibulae) are active during expiration, which is brought about by compression—that is,

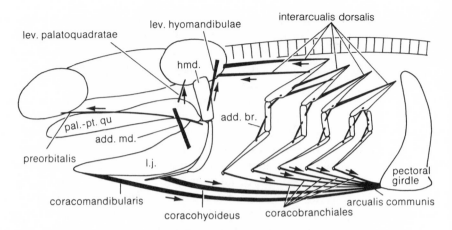

Fig. 9.5. Diagram of the skeleton and main muscles of the dogfish head. Superficial constrictors are not shown (except for the levator hyomandibulae = epihyoideus). The direction in which movement occurs when a given muscle contracts is shown by the arrow; *add br.*, adductor branchialis: *add. md.*, adductor mandibulae; *pal.-pt. qu.*, palatopterygoquadrate. After Hughes and Ballintijn 1965.

decrease of volume—of the orobranchial and parabranchial cavities. Expiration, therefore, takes place actively. Inspiration takes place by the expansion of the orobranchial and parabranchial cavities. During the expansion of the orobranchial and parabranchial cavities only one muscle is in tonic contraction: the adductor mandibulae muscle. Inspiration, therefore, takes place almost passively by relaxing all head muscles except the adductor mandibulae. The expansion phase is caused by elastic recoil of the cartilaginous visceral skeleton of the dogfish's head. The branchial region is compressed by contraction of most of the head muscles and on their relaxation the elastic properties of the skeleton and ligaments are sufficient to account for the expansion of the orobranchial and parabranchial cavities. Because of the ligamentous connections between jaws, the hyoid and other visceral arches, the elastic forces responsible for the expansion phase also exert a traction on the jaw. If the jaws are partially closed by tonic contraction of the adductor mandibulae muscles, they serve as an anchor against which the branchial arches can rotate and move to cause the expansion phase. When the shark is swimming, variations in the degree of contraction of the adductor mandibulae muscles control the volume of water entering the mouth. It should be emphasized that the hypobranchial muscles (coracomandibularis, coracohyoideus, coracobranchials; see fig. 9.4) are not active during all phases of the normal respiratory cycle of sharks. An unexpected aspect is that the levator palatoquadratae and levator hyomandibulae muscles function in reducing the volume of the orobranchial cavity, although the levator series is generally concerned with expansion movements.

An active expiration and a passive inspiration seem to represent a phylogenetically primitive state. In lampreys, electromyographic studies have shown a passive inspiration and an active expiration. But in the holostean *Amia* and many teleosts the hypobranchial muscles become involved in active inspiration. In more advanced fishes both inspiration and expiration take place actively by muscular actions.

The shark captures prey or food by a rapid and often extensive lowering of the mandible. As in all fishes, depression (lowering) of the mandible has to take place against the resistance of the water. Unlike the pattern during respiratory movements, feeding in sharks involves active participation of all hypobranchial muscles. Rapid and extensive lowering of the mandible is accomplished by strong contractions of the coracomandibularis and the coracohyoideus muscles (fig. 9.5). Closure (adduction) of the jaws is accomplished by contraction of the preorbitalis and adductor mandibulae muscles, immediately followed by contractions of the levator palatoquadratae and levator hyomandibulae muscles that constrict the oral cavity to force the food posteriorly toward the pharynx. Once the food has been transported to the pharynx, the pharyngeal phase of swallowing (deglutition) is activated by a peristaltic wave traveling posteriorly toward the esophagus. The pharynx is widened by contraction of the coracobranchialis muscles, which enlarge the angles between the epi- and ceratobranchials. Generally there is a peristaltic relationship in which the anterior parts of the coracobranchiales contract before those more posteriorly placed. Active expansion of the pharynx is followed by a peristaltic wave of compression involving the interhyoideus, the interarcualis dorsalis, cucullaris, and adductor branchialis muscles (fig. 9.5). The interarcualis dorsalis muscles reduce the distance between the dorsal elements of adjacent arches, constricting the branchial cavity. The levator branchiales and adductor branchiales muscles reduce the angle between epi- and ceratobranchial, reducing the volume of the branchial cavity.

Numerous adaptations have evolved in the jaw mechanisms of fishes. Much comparative anatomical research must yet be done to acquire a clear understanding of the biology and evolution of feeding mechanisms of chondrichthyes and bony fishes. In most teleosts two to four different mechanisms to open the jaws against the resistance of water may be present in the same individual, involving not only the hypobranchial muscles but also the muscles associated with the opercular and palatopterygoid bones.

## 7. Analysis of Jaw Movement

There is a considerable body of literature that attempts to analyze jaw movement in mammals, but most of those studies focus on the hard parts—teeth and jaws—rather than on muscles (see discussion, p. 319). There also exists a large literature on mammalian jaw musculature, most of which is descriptive and at best gives rough average fiber direction and muscle weights. Turnbull (1970) has provided detailed descriptions of the jaw muscles of several mammals, plus a summary of published jaw muscle weights of a wide variety of mammals, plus an excellent bibliography of previous literature. Hiiemae and Houston (1971) produced an exemplary study of rat jaw muscles that involved histological, cineradiographic, and detailed fiber direction analyses in addition to the more usual approaches.

In the jaw musculature of the cat, as in most specialized carnivores and in what appears to have been the primitive condition, the temporalis comprises the main part of the jaw musculature. It is larger than the masseter and internal pterygoid combined. Several recent studies have analyzed neural control of biting, tooth shearing patterns, and the pattern of function of individual muscles. In ungulates (e.g., horses, cows, sheep) and many other specialized herbivores, the masseter and internal pterygoid have enlarged and make up more than half of the total jaw musculature. The large temporalis of carnivores can counter large stresses on the incisor and canine teeth, such as those involved in subduing prey and ripping off pieces of meat, since it pulls the jaw back as well as up, while the masseter and internal pterygoid pull forward and up. For many specialized herbivores, transverse movement of the jaws is important for shredding grass between the complicated cheek teeth, and the masseter and internal pterygoid muscles are better positioned for lateral jaw movement than is the temporalis, since they are on the same side of the fulcrum (the mandibular condyle) as the resistance (the food). For further discussion of jaw movement, see p. 318.

8. Relation of
Muscle and Nerve
and the Question
of Muscle
Homology

A muscle is the end organ of a motor nerve and is caused to contract by a nervous impulse originating in the central nervous system and transmitted to the muscle along its nerve. If the nerve is cut across, the animal is no longer able to make the muscle contract, although contraction can be elicited by artificially stimulating the nerve stump attached to the muscle. A denervated muscle eventually atrophies.

Primitively, the chordate muscular system consists of an axial series of myotomes from anterior to posterior end. Each myotome is supplied by a corresponding nerve issuing from between two adjacent vertebrae, for, as already learned, vertebrae and myotomes alternate. In general the evidence indicates that this primitive relation of a nerve of a given body segment to the myotome of that segment is relatively constant and that the same nerve continues to supply the muscles derived from that myotome through all their evolutionary changes. Similarity in nerve supply reflects similarity in pattern of development. Hence, nerve supply is a valuable criterion for determining muscle homology throughout the vertebrate classes. But this principle, which seems simple enough in itself, is not so easy to apply in practice because of the branching and anastomoses of nerves and the difficulty of tracing fine branches. Other criteria of muscle homology are the origins and insertions and the topographical relations to other muscles. The former are subject to considerable shift, especially when the bones to which they were originally attached are lost; for example, the disappearance of the cleithrum forced muscles that were fastened to it to find new attachments.

Ontogenetic origin is an important indicator of muscle homology. Groups of individual muscles in the adult can be traced back to larger masses of muscular or premuscular tissue in the embryo. The mode by which these larger masses subdivide gives valuable evidence of homologies.

Nevertheless, the task of homologizing muscles among the vertebrates is extremely difficult. Muscles are very plastic structures and change their shape and attachments with adaptive structural modifications of the animal. Connecting groups are extinct and are known chiefly by their skeletons, but it is possible to partially reconstruct probable muscle arrangements by means of the marks left on bones by muscle attachments. Recent workers have made much progress in such reconstructions of limb and cranial muscles of extinct amphibians and reptiles. But the whole subject of muscle homology is too difficult for adequate presentation here. To the natural difficulties of the subject has been added an immense confusion of terminology; the same muscle in the same animal has been called different names by different investigators and in different animals; nor have modern investigators reached complete uniformity as they attempt to produce a terminology that will express homologies among vertebrate groups.

The following account does not try to give all the details of muscle origins and insertions or to describe every muscle of the animals dissected; nor should the student memorize every muscle of the animals dissected. The intention is to give an idea of the major aspects of structural evolution of vertebrate musculature (excluding that of teleost fishes and birds) and to follow some of the accompanying major functional changes.

**B. Generalized Vertebrate Musculature as Exemplified by the Dogfish**

The musculature of the dogfish serves as a convenient starting point for the study of vertebrate musculature, although, of course, it has undergone some modification from the primitive chordate plan. The latter is presumably exemplified by *Branchiostoma* (amphioxus), where there is a complete set of myotomes from anterior to posterior end. These myotomes in *Branchiostoma* nearly cover the gill region, which has an intrinsic musculature of hypomeric origin. A horizontal septum is absent in primitive chordates.

It is wise to review skeletal morphology *before* beginning to dissect the muscles of any specimen.

The musculature of the dogfish will be considered under the headings of somatic, fin, branchial, and hypobranchial musculature.

**1. The Parietal or Somatic Muscles**

Strip the skin from one side of the dogfish tail, starting just behind the pelvic fins and extending back about 3 inches; make a cut through the skin, and with your fingers carefully separate the cut edge of the skin from the underlying muscles and pull back the skin without the further use of the knife. (Trying to cut the skin from the body usually results in cutting into the muscles.) Note that the body is completely sheathed in a coat of muscles, the parietal muscles, which are of direct myotomic origin. They consist of a series of zigzag *myomeres* separated from one another by white connective tissue partitions, the *myosepta* or myocommata. The myosepta reach inward to attach to the vertebral column. In the middle of the side of the body there is a white longitudinal line, the outer edge of the horizontal skeletogenous septum, more briefly termed *lateral septum*. This divides myomeres into dorsal or epaxial portions and ventral or hypaxial portions. The epaxial muscles form the dorsal longitudinal bundles, usually two in number, as can be seen in cross sections (make a transverse cut completely through the tail in the region you have stripped to expose a cross-section view). The hypaxial muscles are also divided into two longitudinal bundles; a *lateral* one, just below the lateral septum and easily distinguishable by its darker color; and a *ventral* longitudinal bundle, which in cross sections can be seen to be subdivided into two bundles. Thus in most elasmobranchs and most teleost fishes the fibers of the myomeres are arranged into five longitudinal bundles on each side. The muscle fibers are oriented in a basically anteroposterior position in each myomere.

In the midventral line note a white partition, the *linea alba,* which separates the myomeres of the two sides of the body. Muscles do not cross the middorsal or midventral lines and hence are always paired, even when they look single.

2. Fin Musculature

Strip off the skin from both surfaces of a pelvic fin, preferably in a female, since the muscles of the clasper introduce complications in the males. On the dorsal side of the fin you will see a mass of muscles extending into the fin; it originates on the fascia (connective tissue) of the myomeres and on the iliac process of the pelvic girdle and may be regarded as an abductor, a levator, or an extensor of the fin. When you cut through this mass, the myomeres will be revealed beneath it; this suggests that the fin muscles are outgrowths of the myotomes, as in fact they are. The cut also reveals a deeper part of the muscle that originates on the metapterygial cartilage and radiates into the fin. Both parts of the muscle insert by way of connective tissue on the pterygiophores and ceratotrichia. On the ventral side of the pelvic fin, the muscle mass, which acts as a depressor, an adductor, or a flexor, is divisible into two parts, one extending from the linea alba and puboischiac bar to the metapterygium, and the other from the latter into the fin, inserting like the above.

Similarly skin the base of the pectoral fin. On the dorsal side of the fin you will see the fanlike extensor or levator muscle, originating on the scapular process of the pectoral girdle and adjacent fascia. A similar ventral flexor or depressor mass originates on the coracoid bar of the pectoral girdle. Both muscles radiate into the fin in bundles and insert on the pterygiophores.

Thus the musculature of the paired fins consists essentially of a dorsal extensor mass and a ventral flexor mass. Both masses have an obviously segmental appearance, and embryology shows that the fin musculature originates by budding from the myotomes. Typically there are two buds from each myotome (see fig. 4.11C); each of these again subdivides, and the upper two become dorsal fin musculature, the lower two ventral fin musculature.

3. The Branchial Musculature (figs. 9.4, 9.5)

Make a midventral incision in the skin of the head from the pectoral girdle forward and strip off the skin on one side in an upward direction over the gill slits up to the middorsal line. Note that the epaxial part (dorsal longitudinal bundles) of the myomeres continues unaltered above the gill region and attaches to the rear part of the skull. The lateral longitudinal bundle is seen to terminate at the scapular process. Lateroventrally the longitudinal bundles continue unaltered to the pectoral girdle, to which they are attached. The region in front of the pectoral fin is seen to be occupied by a set of muscles separate from the myomeres. These are the branchial muscles, which operate the gill arches and jaws (modified gill arches). In the evolution of vertebrates, with the loss of gill respiration the branchial muscles undergo striking changes but continue to serve the jaws and remnants of the gill arches, as well as spreading into new territory, and continue to be supplied by the same motor nerves (cranial), that serve them in the dogfish.

### a. Constrictor Series

The musculature covering the gill region from the pectoral fin to the eye and mouth is termed the *superficial constrictor* musculature and consists of six *dorsal* constrictors above and six *ventral* constrictors below the external gill slits. Only the most posterior part of each of the last four constrictors is exposed on the surface, and these exposed parts are separated from one another by vertical lines of connective tissue (*raphes*). Most of each constrictor is concealed under the preceding constrictor, as can easily be seen by slitting along a raphe upward and downward from a gill slit. The innermost part of the constrictor supports the wall bearing the gill plates.

The second dorsal constrictor, belonging to the hyoid arch, is very large and extends from the raphe above the second gill slit forward to the muscle mass at the jaw angle. It continues above this mass as the *epihyoidean* muscle ( =levator hyomandibulae), which originates on fascia and the otic capsule and inserts on the hyomandibula. (From behind the spiracle there is seen emerging the hyomandibular trunk of the seventh cranial nerve (facial), which slants downward across the lower part of the epihyoidean muscle.) Immediately in front of the hyoidean constrictor is the small *spiracular* (*first dorsal constrictor* or *craniomaxillaris*) muscle, which passes in front of the spiracle, originating on the otic capsule and inserting on the palatoquadrate cartilage. (The muscle immediately in front of the spiracular is the levator palatoquadratae, and belongs to the levator series [see below]). The large muscle mass at the jaw angle is termed the adductor mandibulae; it originates on the rear part of the palatoquadrate cartilage and inserts on the mandible ( =Meckel's cartilage). The adductor mandibulae plus the dorsal constrictor are part of the first constrictor, which belongs ontogenetically to the mandibular arch. The *preorbitalis muscle* (also called *suborbitalis* and *levator labialis superioris*) will be found by dissecting deeply between the upper jaw and eye. This will uncover the slender *labial cartilage,* which bends around the mouth angle; there are also *extrabranchial* cartilages under the skin over the gill pouches; the significance of these cartilages external to the regular gill arches is unknown. The preorbitalis is a cylindrical muscle that originates on the midventral surface of the chondrocranium, extends along each side of the upper jaw, and narrows posteriorly to a tendon that joins the adductor mandibulae muscle.

In the ventral view, the ventral constrictors of the last four gill pouches are seen to be similar to the dorsal ones. The first ventral constrictor is the very broad sheet originating on a midventral raphe and slanting forward on both sides to insert on the mandible as the *intermandibularis muscle*. The most posterior fibers attach to the fascia of the *adductor mandibulae* muscle. By carefully cutting through the middle of the intermandibularis of one side, you will find another thin sheet immediately above it; this is the rest of the second ventral constrictor, belonging to the hyoid arch and properly termed *interhyoideus* (*constrictor hyoideus*).

### b. Levator Series

On turning again to the dorsal side of the animal, directly in front of the spiracular (first dorsal constrictor) muscle, but a little deeper, you will find the first levator or *levator palatoquadratae,* which originates on the otic capsule and inserts on the palatoquadrate cartilage next to the adductor mandibulae. It acts to raise the upper jaw. The second or *hyoid levator* is under the epihyoideus, to which it is fused. The remaining levators (3–8) are represented by the *cucullaris* (often called the *trapezius,* although its homology with the tetrapod trapezius is very doubtful), an elongated muscle mass lying between the dorsal longitudinal bundles and the upper ends of the last gill pouches. It looks like a continuation of the lateral longitudinal bundle in front of the scapular process. The cucullaris originates on the fascia of the dorsal longitudinal bundle and inserts on the epibranchial cartilage of the last gill arch and also on the scapular process of the pectoral girdle, hence acting as a levator and protractor of the scapula. The anterior end of the cucullaris thins out to a point. The whole levator series, termed *levatores arcuum,* acts in general to raise the gill arches.

### c. Interarcual Series

The upper ends of the gill pouches can easily be separated from the dorsal longitudinal bundles and the cucullaris (the space above the gill pouches so exposed is the anterior cardinal sinus (see p. 504). By spreading these structures apart as much as possible, you will see the cartilaginous gill arches; running between their upper ends and slanting forward are the short *interarcual* muscles. They extend chiefly between the pharyngobranchial cartilages of the gill arches and can draw the arches together and so expand the pharynx.

**4. The Hypobranchial Musculature**

This group of muscles occupies the region between the coracoid bar and the mandible, passing dorsal to the mandibularis and intermandibularis muscles. They strengthen and elevate the floor of the mouth cavity and strengthen the walls of the pericardial cavity. Immediately in front of the coracoid bar in the midventral region is a triangular area containing a pair of diverging muscles, the *common coracoarcuals,* which originate on the coracoid bar and insert on fascia just above the rear edge of the mandibularis. Slit up the midventral raphe forward to the middle of the mandible and note just above the constrictor layer a median (really paired) muscle, the *coracomandibular* (also called the *geniocoracoid* or *geniohyoid*), extending forward to the mandible. Dorsal to this is a pair of large muscles, the *coracohyoids* ( = *rectus cervicis* of tetrapods), running obliquely forward to insert on the basihyal. Just behind the center of the lower jaw, between the anterior parts of the coracomandibular and coracohyoids, is a flat dark-colored mass, the *thyroid gland.* On cutting through the middle of the coracohyoids and deflecting them, you will see dorsal to them the *coracobranchials,* a muscle sheet that extends obliquely laterally, with slips of insertion on the ceratohyal cartilage and on various parts of the last five gill arches. (Throughout this book typical gill arches will always be referred to as the third to seventh arches; many books call them the first to fifth; see also chap. 10.) To avoid cutting important blood vessels, do not trace the details of these muscles at present.

The hypobranchial musculature arises by forward growth of the ventral ends of the myotomes behind the branchial region and may be regarded as a forward continuation of the ventral longitudinal bundles. The hypobranchial musculature, like the body myotomes, is innervated by spinal nerves, whereas the branchial musculature is innervated by cranial nerves. The coracobranchial series in elasmobranchs is of hypobranchial origin.

**C. Early Tetrapod Musculature, Exemplified by Necturus**

When vertebrates evolved from an aquatic to a terrestrial mode of life, the musculature underwent many alterations. With loss of gill respiration, parts of the branchial apparatus and its musculature were lost, but some of the gill arches and muscles persist with altered functions. More complex movements were required of the jaws, with greater specialization of mandibular arch muscles. Limbs for support and walking require a heavier musculature with broader attachments than do fins, and their greater mobility and varieties of movement necessitated the splitting of the simple muscle masses of fish fins into a number of specific muscles. The trunk musculature underwent less change but became more covered by the expanded musculature of the limb girdles

The muscles of salamanders serve to illustrate the transition from aquatic to terrestrial vertebrates, and *Necturus* is the most convenient salamander to dissect, because of its ready availability and relatively large size. However, it should be noted that *Necturus* is secondarily specialized for an adult aquatic life and therefore differs from early tetrapods in having relatively smaller and weaker limbs and in retaining functional gills throughout life.

Make a middorsal incision extending the length of head and trunk. Loosen the cut edges of skin with the fingers, noting in the cut surfaces the flask-shaped cutaneous glands, which secrete slime. Then with the fingers separate the skin from the muscles in a ventral direction until you have removed the skin from head, trunk, and appendages, leaving the gills in place. The white fibrous material between the skin and muscles is termed the subcutaneous connective tissue or superficial fascia. Confine the dissection of the muscles to one side of the animal.

### 1. The Muscles of the Trunk and Tail

These muscles preserve the generalized arrangement typical of primitive and embryonic vertebrates. They consist, as in the dogfish, of a series of myomeres, separated by myosepta. The myomeres are long, nearly rectangular blocks extending from the middorsal to the midventral line. Their narrowed dorsal ends slant forward. Note their division into *epaxial* and *hypaxial* portions by the horizontal septum. Although the trunk muscles appear to be unmodified, they are in reality already separating into layers. On cutting into the epaxial muscles, you will see that they constitute a mass whose fibers are all directed forward; this is the dorsal muscle mass, most of which forms the *dorsalis trunci,* whose fibers are attached to the myosepta and, deeper down, to the transverse processes of the vertebrae. A deeper portion of the dorsal muscle mass runs between the vertebrae (*interspinalis*). That portion of the dorsalis trunci that passes between the transverse processes of the vertebrae may be distinguished by the name *intertransversarius*. The general function of the dorsal muscle mass is lateral flexion of the body.

The hypaxial musculature is divisible into the *rectus* and the *oblique* groups. The oblique muscles form the sides of the body wall and, although seemingly one mass below the horizontal septum, are really split into three layers. The outer layer, or *external oblique* muscle, has fibers directed obliquely caudad and ventrad. On cutting through this rather thick layer, you will find internal to it the thin *internal oblique* muscle, with fibers directed obliquely craniad and ventrad. When this is cut through, the thin *transverse* muscle is revealed. These three layers, typical of tetrapods in general, serve to compress the viscera and have other subsidiary actions according to their attachments. There is a *rectus lateralis* of slight girth running longitudinally just ventral to the horizontal septum.

On either side of the linea alba the hypaxial musculature is, as in tetrapods in general, differentiated into the *rectus abdominis* muscle, a narrow band of longitudinal fibers interrupted by regular segmental *inscriptiones tendineae* ( = myosepta). The rectus extends from the anterior edge of the pubis to the sternum and hyobranchial apparatus and acts to compress the viscera, curve the body ventrally, and retract the hyobranchial apparatus.

The epaxial and hypaxial muscle masses continue into the tail and act to bend the tail from side to side. On skinning the tail, note the numerous mucus-secreting cutaneous glands in the tail fin.

### 2. General Features of the Limb Musculature

The urodeles exhibit a simple type of limb musculature, but there is good evidence from the fossil record that this reflects secondary reduction rather than retention of the primitive condition. The shoulder girdle has lost its primitive dermal components, and there are associated specializations of the shoulder muscles.

Note, particularly for the forelimb, the fanlike muscle sheets that originate on trunk and girdle and converge toward the limb, on both dorsal and ventral sides; these sheets illustrate the spread of limb musculature over the myomeres. A similar arrangement is seen on the ventral side of the hind limb but here is less evident dorsally. Infer actions of the dorsal and ventral muscle sheets on the limb as a whole. Within the limb itself the general arrangement comprises a set of muscles from each part of the limb to the next more distal part, for example, from upper to forearm, forearm to wrist and hand, and so forth. The muscles of the original upper surface of the limb sections have an extensor action; those of the lower surface, a flexor action. Note the great similarity of the muscles of the distal parts of both limbs and less similarity of the proximal parts.

**3. The Muscles of the Pelvic Girdle and Hind Limb**

Here, as in the dogfish, the continuity of the muscles of the pelvic girdle and hind limbs with the body myotomes is more or less evident. The limb musculature obviously consists of dorsal abductor and ventral adductor that have become split up into a number of separate muscles.

The ventral surface of the pelvic girdle is covered by a flat, fan-shaped sheet of muscle divisible into two parts along a transverse line of connective tissue about the middle of the mass (fig. 9.6). The most anterior muscle is the *puboischiofemoralis externus,* originating

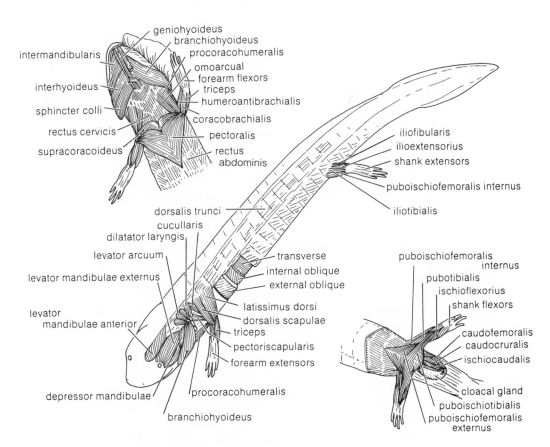

Fig. 9.6. Muscles of *Necturus.*

on the median line of the girdle and inserting on the middle of the ventral surface of the femur. It covers the rear end of the rectus abdominis, and by lifting its anterior edge you can trace the latter muscle to its attachment on the anterior end of the pubic cartilage. The posterior part of the puboischiofemoralis externus is covered by the second of the two muscles mentioned above, the *puboischiotibialis,* which originates on the median region of the pelvic girdle and passes to the tibia.

Around the cloaca is a massive glandular mass, the *cloacal gland,* and on scraping this away, you will see a set of muscles running between the tail and the pelvic girdle. The median one is the *ischiocaudalis,* from the posterior border of the ischium into the tail; it flexes the tail laterally; lateral to this is the *caudocruralis* (or *caudopuboischiotibialis*), originating on the fascia of the puboischiotibialis; it can retract the thigh. On turning the animal sideways and separating the hypaxial mass from the caudocruralis, you will see between them a slightly deeper muscle, the *caudofemoralis;* originating on the femur, it can retract the thigh.

The dorsal girdle muscles are less extensive than the ventral ones. Dorsally the preaxial border of the thigh is formed of the *puboischiofemoralis internus,* which originates on the dorsal surface of the puboischial plate and inserts along a considerable length of the femur. Behind this is the *iliotibialis* and immediately posterior to this the *ilioextensorius;* both originate on the base of the ilium (noticeable as a curved bone between the myomeres) and are inserted on the tibia; the second is continuous with the aponeurosis over the knee and along the shank. Immediately behind the ilioextensorius is the slender *iliofibularis,* from the base of the ilium to the proximal end of the fibula. These muscles abduct the thigh and extend the shank.

On the ventral side of the thigh four muscles are visible. The most anterior is the *puboischiofemoralis internus* seen from above. Next to this is the *pubotibialis,* from the puboischiac plate to the proximal part of the tibia. Next comes the distal part of the puboischiotibialis already noted (this part is sometimes called *gracilis*), and the most posterior muscle is the slender *ischioflexorius.* This blends with the puboischiotibialis at its origin but is distinct distally, where it forms the posterior edge of the thigh and inserts on the fascia of the shank. The foregoing muscles serve chiefly to flex the shank; they also adduct the thigh.

The shank musculature comprises a dorsal extensor mass and a ventral flexor mass; both operate the feet. Twist the shank so that it is in line with the thigh. The extensor mass, on the dorsal side, is divisible into three muscles; a preaxial *extensor tibialis;* a median and largest mass, the *extensor digitorum communis,* which send three conspicuous slips to the digits; and the postaxial *extensor fibularis.* Note the short extensors (*extensores breves*) on the proximal phalanges. On the ventral or sole surface of the foot the main muscle mass is the *flexor primordialis communis,* which continues distally into an aponeurosis, the *plantar fascia,* having tendinous slips to the digits.

**4. The Muscles of the Pectoral Girdle and Forelimb**

On the ventral surface, the broad oblique sheet from the linea alba to the humerus is the *pectoralis;* note the continuity of its fibers with the rectus abdominis and external oblique muscles. The anterior part of the pectoralis, where the fibers run transversely to the upper arm, is somewhat separable as the *supracoracoideus;* it overlies the rounded coracoid part of the pectoral girdle. The pectoralis and supracoracoideus adduct the humerus and thus are important in keeping the anterior part of the body off the ground. In front of the supracoracoideus and extending anteriorly on each side of the neck is the elongated procoracoid cartilage; it is covered by the *procoracohumeralis* muscle, which proceeds from the surface of the cartilage to the humerus, lateral to the insertion of the supracoracoideus. It serves to protract the limb. The procoracohumerales enclose between them a large area of the neck, covered by the *sternohyoid* muscle. This is segmented by inscriptions, is obviously a forward continuation (above the pectoralis) of the rectus abdominis, and hence is more properly termed *rectus cervicis*.

Turn the animal sidewise, push the gills forward, and note the series of shoulder muscles from the fascia, suprascapula, and scapula to the humerous (fig. 9.6). The most posterior one is the *latissimus dorsi* (or *dorsohumeralis*), originating by separate slips on the fascia of the epaxial muscle mass and inserting on the posterior border of the humerus. It is an important retractor of the limb. In front of this is the *dorsalis scapulae* (representing the later *deltoid*), extending from the suprascapular cartilage to the anterior border of the humerus, and serving to protract the humerus. Anterior to this is the *cucullaris,* originating on fascia and inserted on the girdle at the shoulder joint; it draws the scapula anterodorsally. The origin of the cucullaris is partly covered by the most posterior of a series of muscles above the gills (see below). Anterior to the cucullaris is the *pectoriscapularis* (*coracoarcualis, omohyoid*). This muscle slants obliquely backward to its origin on the scapula next to the insertion of the cucullaris and serves to draw the scapula forward. Between the pectoriscapularis and the procoracohumerales, the lateral part of the rectus cervicis is exposed on the side of the neck. Some separate strands of the rectus cervicis attached to the dorsal surface of the procoracoid cartilage are sometimes regarded as a distinct muscle (*omoarcual*).

In ventral view, the upper arm (brachium) has a medium slender muscle, the *humeroantibrachialis,* which extends from the humerus to the radius. Behind this on the inner surface of the upper arm is the larger *coracobrachialis*. These muscles are flexors of the arm. The outer or back side of the arm is covered by a large extensor mass, the *triceps,* divisible into four parts; it originates on shoulder muscle and humerus and inserts on the fascia over the elbow.

The musculature of the forearm (antebrachium) closely parallels that of the shank. On the dorsal or upper surface is an extensor group consisting of a median *extensor digitorum communis,* flanked on the preaxial side by the *extensor antibrachii et carpi radialis,* and on the postaxial side by the *extensor antibrachii et carpi ulnaris*. On the ventral surface of the forearm is the flexor group, a median *flexor primordialis communis,* a preaxial *flexor antibrachii et carpi radialis,* and a postaxial *flexor carpi ulnaris,* of which the antibrachial part is deeper down.

**5. Analysis of Locomotion**

The definitive studies of primitive tetrapod locomotion, particularly based on cineradiographic and electromyographic techniques, remain to be done, although several studies (see references) have provided important insights. The relative simplicity of the limb musculature of *Necturus* makes it a convenient subject to study for general understanding of how tetrapod limbs function in walking. With references to the locomotor sequence shown in figure 9.7, list the muscles (ap-

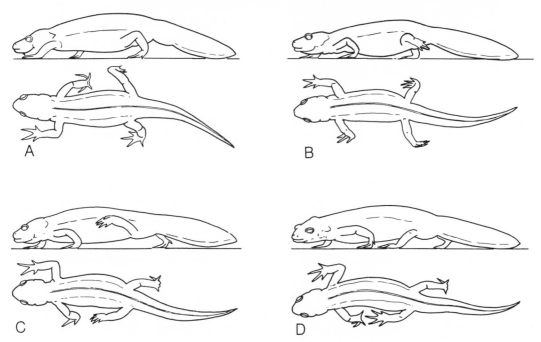

Fig. 9.7. Limb movements during locomotion of a salamander, *Triturus pyrogaster,* traced from a motion picture. Left front limb beginning propulsive phase in *A,* near end of propulsive phase in *C,* and beginning recovery phase in *D.* Left hind limb at end of propulsive phase in *A,* beginning recovery phase in *B,* and beginning propulsive phase in *D.* After Schaeffer 1941.

pendicular and axial) that are active in the propulsive phase and the recovery phase for a forelimb and a hind limb. Note the importance of body undulations in moving limbs forward in the salamander, and the diagonal pattern of limb movement. It is worthwhile to take time at this point to understand the operation of limbs in *Necturus,* for it will greatly facilitate understanding the function of the much more complicated limb musculature of the cat (to follow).

**6. Branchial and Hypobranchial Musculature**

As already learned, these are the muscles of the head, jaws, gill arches, floor of the mouth cavity, and throat. Since *Necturus* retains three pairs of gills and the corresponding arches, the arrangement of the branchial musculature bears considerable resemblance to that of the dogfish.

Above the gills is seen a series of thin, flat muscles, originating on the skin and fascia and passing to the three ceratobranchial cartilages of the gill arches; these are the *levatores arcuum* or *branchiarum,* the gill levators. The thin, narrow muscle behind them and in series with them, covering the origin of the cucullaris, is the dilatator laryngis, which acts to dilate the larynx and open the glottis.

On the dorsal surface of the head the dorsalis trunci is seen to terminate by attaching to the skull. The large muscle mass that covers the roof of the skull is the *levator* (or *adductor*) *mandibulae,* functioning to close the lower jaw. It is divisible into two main portions: the elongated mass to either side of the middorsal line, termed the *levator mandibulae anterior,* and the large mass lateral to this, lying behind the eye, called the *levator mandibulae externus.* The large mass between the latter and the gills is divisible into two muscles, of which the anterior is the *depressor mandibulae,* or opener of the jaw, and the posterior one, next the gills, the *branchiohyoideus.* The latter is traced further below. The levator mandibulae inserts at about the middle of the mandible, below the eye, and the depressor mandibulae inserts on the rear end of the mandible at its articulation with the quadrate region of the skull.

The ventral side of the head and the throat are covered with a sheet of transverse fibers, as in the dogfish. The anterior part of this sheet is the *intermandibularis,* extending from the median raphe to the mandible on each side. Behind this is the *interhyoideus,* which also inserts on the median raphe and originates chiefly on the fascia of the depressor mandibulae and branchiohyoideus. The interhyoideus is divisible into anterior and posterior parts; the latter is also called *sphincter colli* and *gularis.* Some of its fibers attach to the skin of the gular fold.

Cut through the intermandibularis and interhyoideus slightly to one side of the median raphe and deflect them. The branchiohyoideus can now be followed forward as it curves along the cartilage of the hyoid arch on which it inserts. To either side of the median line is seen the narrow *geniohyoid* muscle, which extends from the mandible to the second basibranchial cartilage ( = urobranchial). The rectus cervicis also inserts on this cartilage; and, in fact, the geniohyoids may be regarded as a forward continuation of the median part of the rectus cervicis. At the sides of the urobranchial cartilage and somewhat under the branchiohyoid muscle is a small glandular mass, the thyroid gland. When you cut through and deflect the geniohyoids and press the branchiohyoid to the side, the gill arches are exposed. The *subarcual* muscles are now visible—short strap-like muscles running longitudinally from one gill to the next; the first one from the hyoid to the third arch and the second from this to the fourth arch are easily seen. The *depressores arcuum* or gill depressors are muscles that run along the gill cartilages into the gills and can be seen lateral to the subarcuals. The attachment of the rectus cervicis to the hyobranchial skeleton is now exposed. Deep down between the rectus cervicis and the last gill cartilage you will see the transverse fibers of the *transversi ventrales* series.

The levator mandibulae and the intermandibularis are mandibular arch muscles, homologous to the first constrictor set of elasmo-branchs, and are innervated by the fifth cranial nerve (trigeminus). The depressor mandibulae and the interhyoideus are hyoid arch muscles, homologous to the hyoid levator and ventral hyoid constrictor, respectively, of elasmobranchs and are innervated by the seventh cranial nerve (facial). The levators and depressors of the gills, the branchiohyoid, the dilatator laryngis, the subarcuals, the transversi ventrales, and the cucullaris belong to the remaining gill arches and are innervated by the ninth (glossopharyngeal) and tenth (vagus) cranial nerves. The geniohyoids, rectus cervicis, and pectori-scapularis are hypobranchial muscles with a spinal innervation.

**D. Reptilian Musculature**

Early reptiles, like early amphibians, had relatively heavy, clumsy bodies, with short stumpy limbs held out in a sprawling posture. It appears that a relatively massive musculature was required at the pectoral and pelvic girdles just to hold the body up off the ground. During the Mesozoic era, great evolutionary radiations produced flying reptiles (pterosaurs), swimming reptiles (e.g., ichthyosaurs, plesiosaurs), and a variety of both quadrupedal and bipedal running reptiles (various dinosaurs); and from the fossil skeletons, great changes in locomotor musculature may be inferred. In addition, many reptiles became specialized herbivores, and in some the jaw muscula-ture was modified to meet the requirements of chewing tough plant material (see references). Unfortunately, only a few groups of rep-tiles survive today, and except for snakes they present no great modi-fications in feeding or locomotor muscle systems over what is seen, in relatively simple pattern, in *Necturus*.

**E. Avian Musculature**

In many ways the musculature of birds is similar to that of reptiles. Some attention should be paid to the modifications of muscles to per-mit flight. Peel the feathers and skin from the left side of the body, but not from the wing. Place the specimen on its back in a dissecting tray. The *pectoralis major* covers the entire breast area, originating on the furcula and sternum and inserting on the ventral surface of the humerus. Note the extent of the sternal origin and its relation to the large sternal keel. The pectoralis major produces the downward part of the wing stroke during flight. Cut through the left pectoralis major parallel to the sternal keel. Try to cut only that muscle, not those just below it. Reflect the pectoralis major to expose the *axillary air sac* and the *pectoralis minor* muscle. Note that the air sac connects to the pneumatic foramen of the humerus. The pectoralis minor originates on the sternum and keel, and its tendon inserts on the upper side of the humerus. This muscle raises the wing. The coracobrachialis is posterior to the upper part of the pectoralis minor. It originates on the anterior ribs and inserts on the humerus, and it retracts the wing. Ex-amine the skin, now still attached to the left wing. Within the fold of skin on the anterior edge of the bend of the wing, notice a short muscle at right angles to the humerus, the *tensor patagii brevis*.

Dissect the skin from the ventral surface of the wing. The *biceps* is anterior and median to the humerus, and the *triceps* is posterior to the humerus. Their origins and insertions are similar to those of other tetrapods, and they are used in flexion and extension of the wing, especially at the beginning and end of flight. The muscles of the forearm are also similar to those of other tetrapods, many inserting via tendons on the carpi or phalanges. Compare these muscles in the pigeon with those of the cat or rabbit and consider their morphology relative to their function.

**F. Mammalian Musculature**

Generalized mammalian musculature differs from that of primitive tetrapods in several major respects, all of which correlate with the development of homeothermy in mammals. First, for more efficient digestion, necessary to maintain a relatively high and steady metabolic rate, mammals chew their food more than do other vertebrates. This required elaboration of tongue musculature and development of lips and muscular cheeks to control food position in the mouth, and remodeling of the jaw musculature to provide for powerful chewing. Second, lateral undulation of the trunk is not important in mammalian locomotion, and the axial musculature is reduced in relative bulk and its primitive segmental arrangement further obscured. Third, with the change in limb posture for more efficient locomotion—compared with primitive tetrapods, mammal limbs are held more under the body, with elbows rotated back and knees forward—the massive shoulder and hip girdle musculature needed to keep the primitive sprawling tetrapod body off the ground is reduced, and shoulder and hip muscles are reorganized in mammals to provide fore and aft rotation of the limbs (i.e., a shift in emphasis from adduction to protraction and retraction). An example of this reorganization is the migration of the supracoracoideus, a chest muscle that adducts the arm in primitive tetrapods, up onto the lateral face of the scapula to form the mammalian supraspinatus and infraspinatus muscle, which help rotate the arm. Of the mammals generally available for class dissection, the cat is selected for detailed treatment because its musculature is relatively unspecialized and it is large enough to allow relatively easy muscle dissection.

**1. The Dermal or Integumental Muscles**

These are muscle sheets that have acquired insertion on the skin and that act to move the skin or skin structures such as hairs, scales, and bristles. Although present in all amniotes, they are particularly developed in mammals, where they accomplish such movements as shivering of the skin, erection of the fur, movements of vibrissae, and changes of facial expression. They are extensions of trunk musculature, especially of the pectoral muscle, or of branchial muscles, such as the trapezius and hyoid constrictor. The last named is the source of the *platysma,* the dermal muscle of the head, with numerous subdivisions acting to move the lips, cheeks, eyelids, ears, and so forth.

Skin the animal and, as you do so, note the structures named below. In skinning, make a middorsal incision from the base of the tail to the back of the head, being sure to cut only through the skin. Make incisions through the skin around the throat, ankles, and wrists, and along the outer surface of the limbs. Connect these with the middorsal incision. Loosen the skin along the incisions and gradually work the skin loose from the muscles, using your fingers and the back of the scalpel. Avoid, as far as possible, cutting the skin from the body, since one is liable to cut into the muscles by that procedure. Work from the dorsal side toward the ventral side. Leave the skin on the head and the perineal region for the present.

The following points should be noted during the skinning. The skin is connected with the underlying muscles by a loose weblike material, the subcutaneous connective tissue or superficial fascia, often impregnated with fat. Below this is the much firmer and tougher connective tissue on the surface of the muscles, forming the deep fascia. Passing from among the muscles into the skin at regular intervals, which represent the segments of the body, you will see a slender cord, composed of an artery, a vein, and a sensory nerve. These may be severed. Other blood vessels, not segmentally arranged, will also be seen passing onto the under surface of the skin, from anterior and posterior regions toward the middle. The arteries are colored by an injection mass and are readily recognized. The veins are usually of a very dark reddish brown color. All vessels to the skin should be severed.

When the skin has been loosened from the sides of the animal, you will note a thin layer of muscle fibers on its undersurface, looking like a fine striping. Toward the chest and shoulder region this assumes the form of a thin sheet. This muscle is a dermal or skin muscle, the *panniculus carnosus* or *cutaneous maximus*. It covers the entire lateral surface of the thorax and abdomen, being more prominent anteriorly. As you continue to skin forward and ventrally, you will find that the muscle takes its origin from the outer surface of muscle (latissimus dorsi) situated posterior to the shoulder, from the axilla, and from the linea alba and various points on the ventral side of the thorax. These points of origin should be cut through and the cutaneous maximus removed with the skin to which it generally adheres. The muscle is inserted on the skin and serves to shake the skin; it is degenerate in man. The cutaneous maximus is chiefly an outgrowth of the pectoral muscles and has various parts in different mammals.

The other chief dermal muscle, the *platysma,* will be found on the undersurface of the skin of the neck and head and consists of many different parts, bearing separate names. Some of these will be seen later, but studying the parts of the platysma is almost impossible in any but freshly killed specimens. The platysma muscle is inserted on the skin of the ears, eyelids, and so forth, and moves them; in man it constitutes the muscles of facial expression. The platysma is a branchial muscle, derived from the muscles of the hyoid arch by extension.

On the undersurface of the skin of the ventral side in females, you will note *mammary* or *milk glands,* spread out as a thin, irregular layer. They may be removed to expose the underlying muscles.

When the skin has been removed, clean away fascia and fat from the surface of the muscles. There is generally a large mass of fat at the base of the hind legs. You will now see that the exposed surface consists in part of muscles, pinkish masses composed of parallel fibers, and in part of the deep fascia, which forms very strong white sheets. The posterior half of the back is covered by such a sheet, known as the *lumbodorsal fascia*. In the median ventral line is the linea alba. The angle between the base of the thigh and the abdominal wall is known as the *inguinal* region. The angle between the upper arm and chest is called the *axilla,* or *axillary fossa.*

In studying the muscles it is necessary to separate each muscle from its neighbors by searching carefully for the white lines of connective tissue that mark the boundaries of muscles and slitting along these lines with the point of the scalpel. Observing the direction in which the fibers run will also aid in separating muscles, since the fibers in one muscle generally run in the same direction, which may be different from that of the neighboring muscles. After freeing the margins of a muscle, work your fingers under the muscle until it is separated from its fellows and can be traced from origin to insertion. Since each muscle is enclosed in a connective tissue sheath, it will separate smoothly from its neighbors; rough edges indicate that the muscle itself has been cut. Avoid using sharp instruments to free the muscles. In case it is necessary to cut through a muscle in order to reveal another muscle beneath it, always cut through the center of the muscle so that original relationships may be reconstructed later.

In the following dissection of the muscles the dissection is to be confined strictly to the left side, leaving the right side intact for the dissection of other systems.

**2. Muscles of the Pectoral Girdle and Forelimb (figs. 9.8, 9.9)**

*a. Superficial Chest Muscles*

These are differentiations of the *pectoralis* mass. They originate on the sternum and insert on the humerus. Their main action is adduction (pulling the arm toward the chest) and, for the most posterior fibers, retraction (pulling the arm back).

1) *Pectoantibrachialis.* Anterior and most superficial of the chest muscles. Origin, manubrium; insertion, by a flat tendon on the fascia of the forearm.

2) *Pectoralis major.* Next posterior to the preceding and extending anteriorly dorsal to the preceding, which should be cut across. Origin, sternum and median ventral raphe; insertion, pectoral ridge on ventral side of humerus.

3) *Pectoralis minor.* Next posterior to the preceding and covered in large part by the pectoralis major. The latter should be cut through and the extent of the pectoralis minor noted. In the cat p. minor is larger than p. major; the names come from human anatomy. Origin, sternum; insertion, ventral side of humerus, proximal to preceding.

4) *Xiphihumeralis.* A thin, flat, long muscle, passing from the xiphoid process of the sternum, its anterior part passing dorsal to

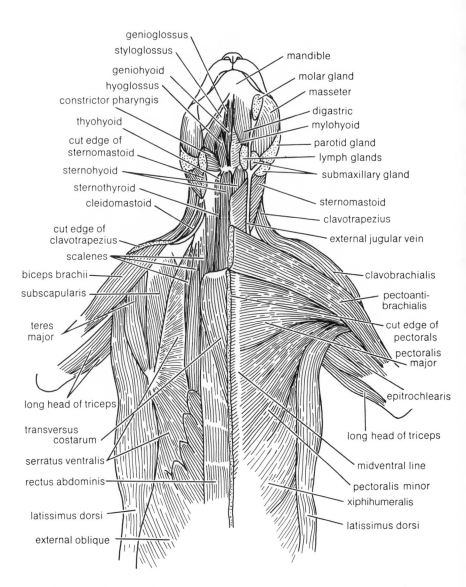

Fig. 9.8. Ventral view of anterior part of cat to show the muscles. All dermal muscles have been removed. Superficial muscles removed from right side to show deeper muscles.

the posterior part of the pectoralis minor, and inserted on the ventral side of the humerus with the pectoralis minor.

### b. Shoulder Muscles

1) *Latissimus dorsi.* A large sheet, partly covered at its cranial end by the spinotrapezius. Origin, from the neural spines of the last thoracic and most of the lumbar vertebrae and from the lumbodorsal fascia; insertion, by a tendon on the medial surface of the humerus; action, retraction of the humerus.

2) *Trapezius muscles.* There are three trapezius muscles in the cat; they are the thin, flat muscles covering the back and neck anterior to the preceding. The posterior trapezius or spinotrapezius originates

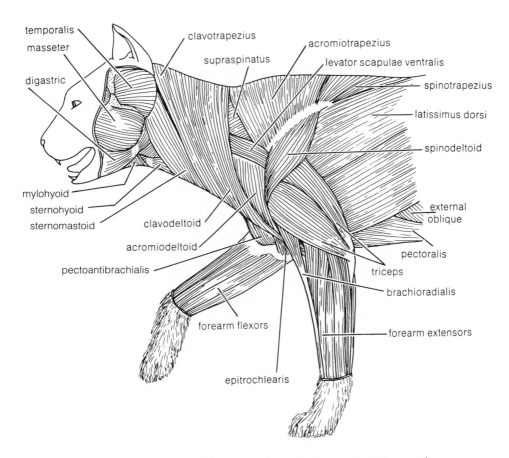

temporalis
masseter
clavotrapezius
acromiotrapezius
supraspinatus
levator scapulae ventralis
digastric
spinotrapezius
latissimus dorsi
spinodeltoid
mylohyoid
sternohyoid
sternomastoid
clavodeltoid
external
oblique
acromiodeltoid
pectoralis
pectoantibrachialis
triceps
brachioradialis
forearm flexors
forearm extensors
epitrochlearis

Fig. 9.9.  Lateral view of anterior part of cat to show the muscles.

from the spines of the thoracic vertebrae and passes obliquely forward, covering part of the latissimus, to be inserted on the fascia of the scapula; action, draws the scapula dorsad and caudad. In front of this is the middle trapezius or acromiotrapezius. Origin, neural spines of cervical and first thoracic vertebrae; insertion, metacromion process and spine of the scapula and fascia of the preceding muscle; action, draws scapula dorsad and holds the two scapulae together. The anterior trapezius or clavotrapezius originates on the superior nuchal line (lambdoid crest) and median dorsal line of neck; passes obliquely ventrally to be inserted on the clavicle, which is imbedded on its surface; it is continuous with the clavobrachial muscle. Action, draws the clavicle dorsad and craniad.

3) *Levator scapulae ventralis.* Along the ventral border of the acromiotrapezius and apparently continuous with it is seen a flat, band-like muscle that passes anteriorly, diverging from the acromiotrapezius and passing internal to the clavotrapezius, which should be cut across. Origin, transverse process of the atlas and occipital bone; insertion, metacromion process and neighboring fascia; action, draws the scapula craniad.

4) *Rhomboideus.* Cut across the middle of the bellies of the spino- and acromiotrapezius muscles. A thick muscle will be seen beneath

them extending from the vertebral border of the scapula to the mid-dorsal line; this is the rhomboideus. Origin, neural spines of the vertebrae and adjacent ligaments; insertion, vertebral border of the scapula; action, draws scapula dorsad and medial. The most ventral portion of this muscle is a practically separate muscle, the rhomboideus capitis, which extends as a slender band forward to originate from the superior nuchal line; insertion, scapula; action, draws the scapula craniad or raises head.

5) *Deltoids.* There are three deltoids in the cat. The *clavodeltoid* (also known as the *clavobrachialis*) originates on a raphe at the level of the clavicle with the clavotrapezius, of which it appears as a continuation. The two muscles together are also known as the *cephalohumeralis.* Insertion, ulna; action, protracts arm or flexes forearm. The *acromiodeltoid* originates on the acromion process and inserts on the spinodeltoid and brachialis muscles and on the deltoid ridge of the humerus. Action, retracts and abducts humerus. The spinodeltoid is a thick muscle that crosses the levator scapulae ventralis at right angles, originating on the spine of the scapula and inserting with the acromiodeltoid on the deltoid ridge of the humerus. Action, with the acromiodeltoid, retracts and abducts the humerus.

6) *Supraspinatus.* On turning back the ventral half of the cut acromiotrapezius, a stout muscle is seen occupying the supraspinous fossa of the scapula. Origin, whole surface of the supraspinous fossa; insertion, greater tuberosity of the humerus, next to the insertion of pectoralis minor; action, extends (protracts) the humerus.

7) *Infraspinatus.* Cut across the belly of the latissimus dorsi. Bring the anterior parts of the latissimus and the spinotrapezius forward so as to expose the posterior part of the scapula. Here you will see two large muscles. The anterior one fills the infraspinous fossa of the scapula, from whose surface it takes its origin, and is inserted on the greater tuberosity of the humerus, the insertion being concealed by the deltoids, which may be cut across to see it. This muscle is the infraspinatus. Action, rotates the humerus outward (abduction). The embryological studies of Cheng (1955) have showed the supraspinatus and infraspinatus muscles to be derived from the supracoracoideus muscle of reptiles.

8) *Teres major.* The stout muscle immediately behind the preceding, its fibers running in the same direction. Origin, glenoid border of the scapula and fascia of neighboring muscles; insertion, in common with latissimus dorsi on the medial surface of the humerus; action, rotates the humerus and retracts it.

9) *Teres minor.* Carefully separate the infraspinatus from teres major and separate the former from the deltoids and the muscles of the upper arm. On the posterior border of the infraspinatus and somewhat covered by it is the small teres minor. Origin, glenoid border of the scapula; insertion, greater tuberosity of humerus; action, assists the infraspinatus.

10) *Subscapularis.* Place a finger under the anterior border of the scapula and clear away connective tissue from the undersurface of the scapula. The subscapular fossa is seen to be occupied by a

muscle, the subscapularis, that covers the inner or medial surface of the scapula. Origin, subscapular fossa; insertion, lesser tuberosity of the humerus; action, pulls the humerus medially. Posterior to the subscapularis will be found part of teres major, which extends onto the medial surface of the scapula.

11) *Serratus ventralis.* Cut through the rhomboideus and loosen the scapula. The serratus ventralis is a large fan-shaped muscle that originates by slips from the first nine or ten ribs and the anterior part (sometimes called levator scapulae) from the transverse processes of the last five cervical vertebrae and inserts on the scapula near the vertebral border; action, anterior part draws the scapula craniad; main part draws scapula ventrad, that is, supports the trunk.

### c. Upper Arm Muscles

1) *Triceps brachii.* The great extensor mass on the back of the upper arm, called anconeus in lower tetrapods, is named triceps in mammals, because it usually has three heads. These are practically distinct muscles in the cat. Separate the heads. The *long* head is the large mass on the back of the upper arm; origin, scapula, from the glenoid border. The *lateral* head is on the lateral surface of the upper arm, ventral to the preceding; origin, greater tuberosity and deltoid ridge of the humerus. The *medial* head is in contact with the humerus. To see it, spread the other two heads apart and look deep between them or cut through the middle of the belly of the lateral head; origin, along the dorsal surface of the humerus. All three heads insert on the olecranon process of the ulna and are the great extensors of the forearm.

There is also in the cat a small fourth part of the triceps, a small triangular muscle at the elbow joint, covered by the distal end of the lateral head, which should be deflected. It is sometimes called anconeus in cat anatomy texts. Origin, distal end of humerus; insertion, lateral surface of the ulna; action, strengthens elbow joint, which it covers.

2) *Epitrochlearis or extensor antibrachii.* On the medial side of the long head, a thin sheet, taking origin from the latissimus dorsi and inserted on the olecranon; action, in common with the triceps, tending also to rotate the ulna.

3) *Biceps brachii.* Beneath the pectoralis, which must be cut near its insertion and reflected. The biceps originates on the supraglenoid tubercle of the scapula, extends down the anteromedial surface of the humerus, and inserts on the bicipital tuberosity of the radius. (The name is derived from human anatomy, where the muscle has two heads.) Action, flexes forearm and assists in supination of forearm (i.e., rotates hand upward).

4) *Brachialis.* Lateral to the biceps, in contact with the lateral head of the triceps. Origin, lateral surface of the humerus; insertion, ulna; action, with the biceps.

5) *Coracobrachialis.* A very small muscle on the medial surface of the shoulder joint. Origin, coracoid process of scapula; insertion, proximal end of humerus; action, adducts humerus. In reptiles and

amphibians with sprawling posture the coracobrachialis is a much larger muscle and is important in raising the body off the ground.

### d. Forearm Muscles

These consist of long, strap-like extensors on the outer or lateral surface of the arm and similar flexors on the inner or medial surface. Remove the tough fascia from the surface of the muscles as you proceed. At the wrist the tendons to the metacarpals and fingers pass under tough ligaments of the wrist, which must be removed if the tendons are to be traced.

1) *Extensor carpi ulnaris.* Begin on the outer surface of the forearm next to the ulna. The olecranon process of the ulna is exposed at and below the elbow, and the first muscle preaxial to the ulna is the extensor carpi ulnaris. Origin, lateral epicondyle of the humerus, and semilunar notch of the ulna; insertion, proximal end of fifth metacarpal; action, extends fifth digit and ulnar side of wrist.

2) *Extensor digitorum lateralis.* This is next to the preceding, going toward the preaxial side. Origin, lateral surface of humerus above the lateral epicondyle; insertion, tendon passes internal to wrist ligaments and then splits into three or four tendons somewhat underlying the tendons of the next muscle; these tendons go to three or four digits, which they extend.

3) *Extensor digitorum communis.* Next to the preceding muscle. Origin, just above the preceding; tendon and action like the preceding.

4) *Brachioradialis or supinator longus.* Next to the preceding on the preaxial border of the forearm, but loose and standing away from the underlying extensor (5). Origin, about middle of humerus; insertion, lower end of radius and adjacent ligaments; action, rotates hand to supine position.

5) *Extensor carpi radialis.* This underlies the brachioradialis and extends onto the inner or medial surface of the arm. The forearm should now be turned so that the medial side faces the student. The extensor carpi radialis is divisible into a long part (*longus*) and a shorter part (*brevis*); the former is the part seen on the medial surface. Origin, humerus, near other extensors; insertion, second and third metacarpals; action, extends hand.

6) *Pronator teres.* Proceeding on the medial surface of the forearm toward the ulnar side, the pronator teres is next to the extensor carpi radialis longus and somewhat under it. Its broader upper end slants proximally toward the radius, to which the edge of its narrower lower part is fastened. This muscle is continuous with a strong fascia overlying the tendon of the next muscle. Origin, medial epicondyle of the humerus; insertion, radius; action, rotates radius to prone position (palm down).

7) *Flexor carpi radialis.* Lies next to, and mostly under, the preceding. Origin, medial epicondyle of humerus; muscle narrows to a tendon that is bound to adjacent ligaments and finally passes through a deep groove between the capitate (third carpale) and the first met-

acarpal, inserting on second and third metacarpals; action, flexes hand.

8) *Palmaris longus.* Flat muscle forming outer surface of forearm next to preceding. Origin, medial epicondyle of humerus; insertion, flat tendon passes through wrist ligaments and divides into four or five tendons, which pass to pads and phalanges; action, flexes digits.

9) *Flexor digitorum profundus.* This complex muscle lies under the preceding and projects to the radial side of it. It has five parts, originating on the ulna and humerus and converging to a broad flat tendon under the tendon of palmaris longus. This tendon divides into five, inserted on the basal phalanges. Action, general flexing of the fingers.

10) *Flexor carpi ulnaris.* This consists of two nearly separate muscles which form the ulnar border of the forearm from the medial view. One head, somewhat under the palmaris longus, originates on the medial epicondyle of the humerus in common with one of the heads of the flexor digitorum profundus. The other head lies alongside the exposed olecranon process of the ulna, from which it originates. Near the wrist both heads join to a tendon that inserts on the pisiform bone of the wrist. Action, flexes ulnar side of wrist.

The muscle along the ulna between the flexor carpi ulnaris and the extensor carpi ulnaris is part of the flexor digitorum profundus.

## 3. Trunk Muscles

### a. Hypaxial Muscles

In the trunk region the hypaxial musculature of mammals is similar to that of *Necturus* and reptiles. The abdominal wall is supported by three layers of muscles plus paired longitudinal bands on each side of the midventral line. All function to compress or constrict the abdomen, and the last also retracts the sternum and ribs.

1) *External oblique.* Covers the whole abdomen from the base of the hind leg, as well as part of the thorax. Fibers run mostly caudoventrally. Origin, lumbodorsal fascia and posterior ribs; insertion, linea alba, from sternum to pubis.

2) *Internal oblique.* Very carefully cut through the middle of the belly of the external oblique, in a longitudinal direction, and separate it from the underlying muscle, which is the *internal oblique*. This separation is often difficult. Fibers run mostly cranioventrally. Origin, lumbodorsal fascia and border of pelvic girdle; insertion, linea alba.

3) *Transversus abdominis.* On cutting through the preceding and separating edges, a third muscle layer, very thin, will be found. This is the *transversus abdominis*. Its fibers are directed ventrally and slightly posteriorly. Origin and insertion are similar to the preceding.

4) *Rectus abdominis.* A pair of long, slender muscles on either side of the linea alba, found inside and between the aponeuroses of the preceding muscles. Its fibers run longitudinally and are crossed at regular intervals by white lines, the myosepta or inscriptions. Origin, pubis; insertion, sternum and costal cartilages. Internal to the trans-

verse and rectus muscles is the peritoneal membrane or lining of the coelom.

There are two derivatives of the hypaxial musculature that lie beneath the vertebral column and will not be exposed at this time. The *psoas minor* originates on the centra of the posterior thoracic and anterior lumbar vertebrae, inserts on the pelvic girdle, and acts to flex the back. The *quadratus lumborum* originates on the centra of the posterior thoracic and lumbar vertebrae and the transverse processes of the lumbar vertebrae, inserts in the ilium, and acts to bend the body laterally.

In the thoracic region the hypaxial musculature is modified to move the ribs.

5) *Serratus dorsalis.* The anterior part of this muscle arises by a number of fleshy slips from the ribs near their angles. The short slips soon pass into a thin aponeurosis that overlies the epaxial muscles of the thorax and is fastened to the median dorsal line. The posterior part of this muscle consists of a few slips lying under the latissimus dorsi and appearing like a forward continuation of the internal oblique. These slips insert on the last ribs and originate by means of an aponeurosis from the median dorsal line. Action, draws the ribs forward.

6) *Scalenes.* Raise up the pectoralis muscles from the chest wall by passing the fingers under them. If necessary, their posterior parts may be cut into. Several long muscles will be seen in the chest wall ventral to the origin of the serratus ventralis and in front of the anterior boundary of the external oblique. These muscles are scalenes; they originate on the ribs and pass forward in a nearly straight course to be inserted on the transverse processes of the cervical vertebrae, uniting anteriorly into one band, which you will readily see by looking immediately ventral to the origin of the anterior part of the serratus ventralis. Action, draw the ribs forward and bend the neck.

7) *Intercostals.* The intercostals are a set of muscles extending from one rib to the next. The external layer is called the *external intercostals;* you will see them in part by looking at the chest wall between the origins of the serratus ventralis and dorsalis. Their fibers run obliquely backward and downward. On cutting through some of them, another layer, the *internal intercostals,* will be seen inside of the external layer. The fibers of the internal intercostals run obliquely forward and downward. Near the median ventral line the external intercostals are lacking, so that the internal ones are exposed by cutting through the scalenes. Action, external intercostals bring the ribs forward, internal intercostals draw them back again. The intercostals are the chief respiratory muscles of the thoracic wall.

### b. Epaxial Muscles

These are similar to those of reptiles, although somewhat more differentiated, and considerably more complex than in *Necturus*. Remove the lumbodorsal fascia over the extreme posterior part of the back, finding beneath it a great thick mass of muscle enclosed in a tough, shining fascia; this is the epaxial muscle mass, sharply separated from the hypaxial muscles in the abdominal region by a furrow corresponding to the position of the horizontal skeletogenous septum of lower vertebrates. The epaxial mass in the lumbar region is divisible into a slender narrow median portion, the *multifidus spinae,* and a very thick lateral portion, the *sacrospinalis.*

In the thoracic region the multifidus terminates. Cut through the aponeurosis of the serratus dorsalis to expose the sacrospinalis, which here is divided into three parts of about equal width: a dorsal part, next to the median dorsal line, the *semispinalis dorsi* (also known as the *spinalis dorsi*); a median part, the *longissimus dorsi;* and a lateral part, the *iliocostalis.* The first two dorsiflex the back, and the last, which consists of separate bundles extending across several ribs, draws the ribs together.

In the neck region the largest epaxial muscle is the *splenius,* which lies anterior to the rhomboideus and internal to the clavotrapezius and rhomboideus capitis. Origin, from the middorsal line and fascia; insertion, superior nuchal line (lambdoidal crest) of occiput; action, raises or turns the head. Ventrolateral to the splenius, and possibly partly fused with it is the *longissimus capitis.* Origin, last four cervical vertebrae; insertion, mastoid process; action, turns the head. The splenius and longissimus capitis appear to be continuations of the longissimus dorsi. Also in the neck region, the semispinalis dorsi becomes the *semispinalis cervicis et capitis,* and is divisible into a medial part, the *biventer cervicis,* and a lateral part, the *complexus.*

| | |
|---|---|
| 4. Head and Neck Muscles (figs. 9.8 and 9.9) | Slit the skin up the center of the throat to the tip of the lower jaw and loosen it so as to fully expose the lower jaw. In doing so, note parts of the platysma muscle on the under side of the skin. It sweeps from the median dorsal line of the neck around the sides of the head to face and ears, and portions of it generally are attached near the anterior end of the sternum. In dissecting the throat muscles, work on one side only, leaving the other side intact for the study of other parts. Avoid cutting any blood vessels. A large vein, the *external jugular vein,* runs in the superficial muscles of the throat. At the angle of the jaw is a rounded pinkish body, the *submaxillary salivary gland,* crossed by the posterior facial vein. Other small bodies are *lymph glands.* |

### a. Neck Muscles

1) *Sternomastoid.* This is the superficial muscle of the ventral side of the neck. The external jugular vein crosses its surface at an angle to the direction of its fibers. Origin by two parts, from the median raphe and the manubrium of the sternum, the first-named origin lying ventral to the second, so that the muscle appears divisible into two muscles. From its origins the muscle passes obliquely away from the median ventral line around the sides of the neck and is inserted on the skull from the lambdoidal ridge onto the mastoid process. The muscle passes internal to the submaxillary gland and the *parotid* gland; the latter is a mass at the base of the ear. The insertion on the

mastoid process is by means of a thick tendon. Action, singly turn the head, together depress head on neck.

2) *Sternohyoid.* The anterior ends of these muscles are visible between the two sternomastoids as the latter diverge from the median raphe. Slit the raphe of the sternomastoids to the manubrium of the sternum, thus exposing the full length of the sternohyoids. They extend in the median ventral line from the first costal cartilage to the body of the hyoid bone, the two being closely united in the median line. Action, draw the hyoid posteriorly.

3) *Cleidomastoid.* Lateral to the sternomastoid is a long muscle passing from the head to the upper arm. Loosen this up and find, internal to it, a narrow, flat muscle, the cleidomastoid. Extends from the clavicle to the mastoid process, dorsal to the insertion of the sternomastoid. Action, pulls clavicle craniad or turns head acting singly, or lowers head on neck. Origin and insertion are thus interchangeable.

4) *Masseter.* The great, thick muscle covering the angle of the jaws, situated in front of the submaxillary and parotid gland. It is covered by a very tough shining fascia. Origin, zygomatic arch; insertion, posterior half of the lateral surface of the mandible ( = angular process of mandible); action, elevates the lower jaw.

5) *Temporalis.* Remove the skin from the side of the head up to the median dorsal line. You will see a great mass of muscle covered by a strong shining fascia occupying the temporal fossa of the skull, dorsal to the ear. Origin, from the side of the skull from the superior nuchal line to the postorbital process of the frontal bone, and from part of the zygomatic arch; insertion, coronoid process of the mandible; action, elevates the jaw, in common with the masseter. Though both masseter and temporalis act as adductors or elevators of the lower jaw, it should be noted that the former pulls up and forward whereas the latter pulls up and back. Those differences have been significant in the specialization of jaw musculature in various groups of mammals.

6) *Digastric.* The muscle lying along the medial surface of each half of the mandible. It extends posteriorly internal to the submaxillary gland. Origin, jugular and mastoid processes of the skull; insertion, mandible; action, depresses the lower jaw.

7) *Mylohyoid.* The thin transverse sheet passing across between the two digastrics from one half of the mandible to the other. Origin, mandible, the origin concealed by the digastrics; insertion, median raphe; action, raises floor of the mouth and brings hyoid forward.

8) *Geniohyoid.* Cut through the median raphe of the mylohoid. This exposes a pair of long, slender muscles, the geniohyoids, lying in the median line. Origin, mandible near the symphysis; insertion, body of the hyoid; action, draws the hyoid forward.

9) *Sternothyroid.* Separate the two sternohyoids in the median line. This exposes the *trachea* or *windpipe,* a tube stiffened by rings of cartilage. At the top of the trachea is a chamber with cartilaginous walls, the *larynx.* The chief cartilage of the larynx is the large, shield-shaped *thyroid* cartilage, forming the ventral walls of the larynx.

Just in front of the thyroid cartilage the body of the hyoid is felt as a bony bar. The sternothyroid muscles are located, one on each side of the trachea, dorsal to the sternohyoids. Origin, sternum in common with the sternohyoid; insertion, thyroid cartilage of the larynx; action, pulls the larynx posteriorly.

10) *Thyrohyoid.* Short, narrow muscle on each side of the thyroid cartilage, from which it takes its origin; insertion, posterior horn of the hyoid; action, raises the larynx.

11) *Internal (or medial) pterygoid.* A large muscle on the medial surface of the caudal end of the mandible, with fibers that roughly parallel those of the masseter on the other side of the mandible. To expose it, muscles of the throat must be cut and deflected. Origin, pterygoid process and adjacent region of skull; insertion, medial surface of the angular process of the mandible; action, elevates the jaw, in common with the masseter.

A small muscle that need not be exposed in this dissection is the *external (or lateral) pterygoid,* which extends from the pterygoid region of the skull to the medial side of the condyle of the mandible and assists in jaw movement.

The foregoing muscles of the neck and throat belong chiefly to the branchial and hypobranchial groups of muscles. The temporalis and the masseter are considered to be derived from the adductor mandibulae externus of nonmammalian tetrapods. There are also two *pterygoideus* muscles found along the inner surface of the mandible that belong to the general adductor mandibulae mass. The mylohyoid is homologous with the intermandibularis. All these muscles are branchial muscles derived from the original musculature of the first or mandibular arch and innervated by the fifth cranial nerve. The digastric muscle is so called because in most mammals it is a compound muscle with a tendon dividing the belly into two parts; the anterior part is innervated by the fifth nerve and hence is a mandibular arch muscle; the posterior belly is innervated by the seventh nerve and hence is a derivative of the musculature of the hyoid arch. The posterior belly is not identical with the depressor mandibulae of reptiles and amphibians but comes from the same embryonic muscle mass as that muscle. Other mammalian muscles derived from the mandibular arch musculature are the tensor palati and tensor tympani (not seen in the dissection).

The posterior belly of the digastric and the various parts of the platysma muscle of mammals are derivatives of the hyoid arch musculature. An interhyoideus muscle is present in monotremes.

The cucullaris muscle of lower tetrapods is subdivided in mammals into one or more trapezius muscles and the sternomastoid, sternocleidoid, and so forth. These are also branchial muscles. A ceratohyoid muscle, homologous to the first subarcual muscle of *Necturus,* is found in many mammals, including the cat; it runs between the two horns of the hyoid. Monotremes have, in addition, an interthyroid muscle homologous to the third subarcual. The striated pharyngeal and laryngeal muscles are also remnants of branchial arch musculature.

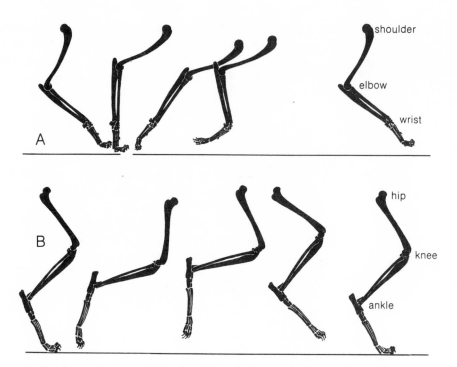

Fig. 9.10. Limb movements of a running cheetah, traced from a motion picture. Right front limb going through propulsive phase in first three frames, beginning recovery phase in fourth frame. Right hind limb passes through propulsive phase in first two frames, recovery phase in last three frames. After Hildebrand 1960.

The hypobranchial musculature, formed from growths of somatic myotomes into the region below the branchial musculature and innervated by the eleventh cranial nerve, includes the geniohyoid and various other muscles of the tongue and hyoid apparatus. The rectus cervicis of lower tetrapods is split up in mammals into the sternohyoids, sternothyroids, and thyrohyoids, with the addition of omohyoids in some mammals.

**5. Muscles of the Pelvic Girdle and Hind Limb (fig. 9.11)**

*a. Muscles of the Hip and Thigh*

1) *Tensor fasciae latae*. Examine the lateral (outer) surface of the thigh. The anterior part of this is covered by a tough fascia, the fascia lata. In the dorsal part of this is a muscle, the *tensor fasciae latae,* a thick triangular muscle. Origin, ilium and neighboring fascia; insertion, fascia lata; action, tightens the fascia lata.

2) *Biceps femoris*. This is the large muscle on the lateral surface of the thigh posterior to the fascia lata and covering more than half the surface of the thigh. It has only one head in the cat. Origin, tuberosity of the ischium; insertion, patella and tibia by a tendon, and the fascia of the shank; action, abducts the thigh and flexes the shank.

3) *Caudofemoralis*. Clean away the fascia from the back in front of the base of the tail, as far forward as the anterior end of the pelvic girdle. Muscles will be found between the median dorsal line and

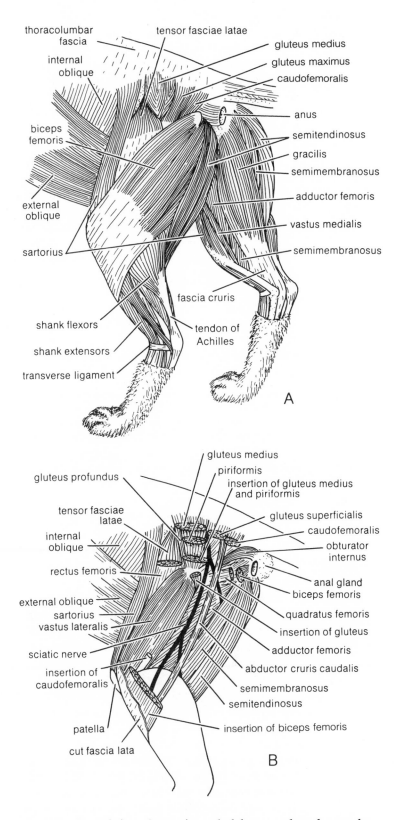

Fig. 9.11.　Lateral view of posterior end of the cat to show the muscles.

the thigh. The most posterior of these is the narrow, flat *caudofem-oralis,* passing from the side of the root of the tail toward the dorsal end of the biceps femoris. Origin, transverse processes of the second and third caudal vertebrae; insertion, the muscle passes ventrally, concealed by the anterior margin of the biceps femoris. This should be lifted up and the caudofemoralis followed to its tendon; the latter is very long and passes to the patella, on which it is inserted. Action, abducts the thigh, extends the shank.

4) *Gluteus maximus.* A rather thin, flat muscle immediately anterior to the preceding. It is imbedded in the fascia and is continuous with the tensor fasciae latae anteriorly. Origin, from the fascia and from the transverse processes of the last sacral and first caudal vertebrae; insertion, fascia lata and to a slight extent on the greater trochanter of the femur; action, in common with the gluteus medius.

5) *Gluteus medius.* The very large, triangular muscle immediately in front of the preceding and partly covered by it. The gluteus maximus should be cut across to see it. Origin, adjacent fascia, crest of the ilium and lateral surface of the ilium, and transverse processes of the last sacral and first caudal vertebrae; insertion, by a strong tendon on the great trochanter of the femur; action, abductor of the thigh. Along the anterior border of this muscle the origin of the tensor fasciae latae passes internally toward the ilium.

6) *Sartorius.* This muscle forms the anterior margin from the lateral view of the thigh. It is folded over the margin and, on following it to the medial or inner surface of the thigh, you will find it covers the anterior half of the medial surface. Origin, crest and ventral border of the ilium; insertion, proximal end of the tibia and the patella and the fascia and ligaments between; action, adducts and rotates the thigh and extends the shank.

7) *Vastus lateralis.* Cut through the fascia lata by a longitudinal slit extending to the patella. Separate well the sartorius from underlying parts. The tensor fasciae latae is now well exposed. The large, stout muscle that was covered by the fascia lata is the *vastus lateralis.* Origin, greater trochanter and surface of the femur; insertion, see below.

8) *Rectus femoris.* At its anterior margin the vastus lateralis will be found partly separable from a stout muscle lying on its medial side and covered externally by a sartorius. The sartorius may be cut across the middle. The muscle in question is the *rectus femoris.* Origin, ilium near the acetabulum.

9) *Vastus medialis.* This is on the medial side of the thigh posterior to the rectus femoris, which its anterior margin partly covers. It also is covered externally by the sartorius. Origin, femur.

10) *Vastus intermedius.* On widely separating the rectus femoris from the vastus lateralis, this muscle will be seen, deep down, next to the shaft of the femur. Origin, surface of the femur.

The rectus femoris and the three vastus muscles are more or less united to each other and constitute the great *quadriceps femoris* muscle. Its origins have been described; all its parts are inserted on the patella and adjacent ligaments; action, extends the shank.

11) *Gracilis*. This is the large, flat muscle forming the posterior half of the medial surface of the thigh. Origin, ischial and pubic symphyses; insertion by an aponeurosis that passes to the tibia; action, adducts the leg.

12) *Adductor longus and adductor femoris*. Cut through the middle of the gracilis and separate each half from the underlying muscles. The latter consist of three muscles passing from the median ventral line to the femur. The most anterior of the three is very small; this is the *adductor longus*. Origin, pubis; insertion, femur; action, with the next. The middle muscular mass is the large *adductor femoris* (corresponding to adductor magnus and brevis of other mammals). Origin, pubis; insertion, femur; action, adducts the thigh.

13) *Pectineus*. Anterior to the adductor longus is the *pectineus*. Origin, pubis; insertion, femur; action, adducts the thigh.

14) *Semimembranosus*. The large posterior part of the mass that was covered by the gracilis. Origin, ischium; insertion, medial epicondyle of the femur and proximal end of the tibia; action, extends the thigh. The muscle is more or less divisible into two parts.

15) *Semitendinosus*. The most posterior muscle of the thigh, posterior to the preceding. Origin, ischial tuberosity; insertion, tibia; action, flexes the shank.

16) *Tenuissimus*. Turn to the lateral surface of the thigh. Cut through the middle of the biceps femoris. Beneath it you will note a very narrow, long muscle, the *tenuissimus*. Origin, transverse process of the second caudal vertebra, in common with the caudofemoralis; insertion, on the same fascia as the insertion of the biceps.

Separating the biceps from the underlying muscles reveals them as extensions of muscles already identified on the medial surface. The adductor femoris is seen in contact with the femur posterior to the vastus lateralis; the semimembranosus comes next, and the semitendinosus is again the most caudal of the thigh muscles.

There is a series of deep pelvic muscles that is difficult to dissect and may be considered an optional task.

17) *Piriformis*. Cut and reflect the gluteus maximus, gluteus medius and caudofemoralis to expose a thin triangular muscle, the *piriformis*. Origin, sacral and caudal vertebrae; insertion, greater trochanter of femur; action, abducts thigh.

18) *Gluteus minimus (gluteus profundus) and gemellus superior*. These muscles are fused and difficult to separate. They lie beneath the piriformis, which must be cut and reflected to expose them. The former originates on the ventral half of the ilium and the latter on the ilium and ischium. Both insert on the greater trochanter of the femur and act to rotate and abduct the thigh.

19) *Obturator internus*. Lift the caudal edge of the gemellus superior to expose the *obturator internus*. Origin, ischial ramus insertion, trochanteris fossa of femur; action, abducts thigh.

20) *Quadratus femoris*. Ventral to the obturator internus and beneath the origin of the biceps femoris is the quadratus femoris. Origin, ischial tuberosity; insertion, lesser trochanter; action, retracts and rotates thigh.

21) *Gemellus inferior*. Mostly beneath the obturator internus lies the gemellus inferior. Origin, lateral surface of ischium; insertion, tendon of obturator internus; action, abducts thigh.

22) *Obturator externus*. Lies deep between the adductor femoris and the quadratus femoris. Origin, pubis and ischium; insertion, trochanteric fossa of femur; action, rotates and protracts thigh.

### b. Muscles of the Shank

An exposed edge of the tibia separates an extensor group, on the anterolateral side, from a flexor group, on the posteromedial side.

1) *Tibialis anterior*. Clean away the tough fascia of the shank and also the insertions of the biceps and the gracilis. Examine the lateral (outer) surface of the shank. The most ventral muscle, whose ventral border is in contact with the tibia, is the *tibialis anterior*. Origin, proximal parts of tibia and fibula; insertion by a strong tendon, which should be traced into the foot, where it will be found to pass obliquely to the medial side of the foot to be inserted on the first metatarsal; action, extends (or dorsiflexes) the foot.

2) *Extensor digitorum longus*. This is the muscle next dorsal to the preceding on the lateral surface of the shank. It is placed so close to the preceding that it appears part of it, but the line of separation will be found by a little searching. Origin, lateral epicondyle of the femur; insertion, by a stout tendon which if followed into the foot is found to diverge into four tendons, one of which is inserted on each digit; action, extends the digits and dorsiflexes the foot. A set of intrinsic foot muscles, the extensor digitorum brevis, is associated with the tendons of insertion of the extensor digitorum longus.

3) *Peroneus muscles*. These are next dorsal to the preceding, originating on the fibula. There are three of them, more or less fused to each other. Each ends in a tendon; the three tendons pass over the lateral surface of the lateral malleolus of the fibula and over the calcaneus and are inserted on the metatarsals and digits. Action, extensors of the foot.

The following muscles make up the flexor group, which is considerably larger than the extensor group because the flexors are involved in the propulsive stroke (backward thrust) in locomotion, whereas the extensors are involved only in the recovery phase (bringing the foot forward).

4) *Gastrocnemius*. This is the large muscle forming the posterior or caudal surface of the shank. It is divisible into two large portions, one on the medial surface, the other on the lateral surface of the shank. The lateral head is subdivisible into four heads. Origins, from the surface fascia, the femur, and the tendon and fascia of the plantaris muscle (see below); insertion, by a strong tendon (Achilles tendon), which passes to the heel bone (calcaneus), on which it is inserted. Action, ventroflexes the foot (technically, extension).

5) *Soleus*. On carefully separating the lateral head of the gastrocnemius, a muscle, the soleus, will be found internal to it. It is a flat muscle in contact with the peroneus muscles ventrally; it tapers abruptly to a tendon that joins the tendon of the gastrocnemius. Ori-

gin, fibula; insertion, calcaneus; action, with the gastrocnemius, of which it is sometimes considered a part.

6) *Plantaris.* On carefully separating the medial head of the gastrocnemius a large muscle will be found internal to it, lying between the two heads of the gastrocnemius, which practically enclose it. It is fused to a considerable extent to the lateral head but is quite separable from the medial head, being covered on the medial side by a shining aponeurosis. Origin, patella and femur; insertion, by a thick tendon that passes in the middle of a sort of tube formed by the tendon of the gastrocnemius and soleus onto the ventral surface of the calcaneus. Here it broadens and finally divides into four slips, each attached to a digit. Action, flexes the digits; extends the ankle.

7) *Popliteus.* A triangular-shaped muscle, originating on the lateral epicondyle of the femur, passes behind the knee joint and inserts in the proximal one-third of the tibia. Action, flexes and rotates the shank.

8) *Flexor digitorum longus.* Immediately beneath the insertion of the popliteus and medial to the exposed edge of the tibia is the *flexor digitorum longus,* which consists of two parts, somewhat separated. The other part is more lateral and in contact with the peroneus muscles. Separate the part of the flexor that appears on the medial surface from the tibia by a cut, and lift it up. Internal to it you will see a long tendon, and on the other side of this tendon is the other part of the flexor, this part corresponding to the *flexor hallucis longus* of man. Both parts of the flexor terminate in slender tendons that unite distally into a broad tendon, which eventually divides into four tendons inserted on the digits. Origin, tibia, fibula, and adjacent fascia; action, flexes the digits. A *flexor digitorum brevis* group is an intrinsic foot muscle associated with the distal tendon of the plantaris.

9) *Tibialis posterior.* The long tendon between the two parts of the flexor digitorum longus was noted above. It is the tendon of the *tibialis posterior,* and on following this tendon proximally you will locate the belly of the muscle. Origin, fibula, tibia, and fascia; insertion, scaphoid and medial cuneiform of the ankle; action, ventroflexes the foot.

From consideration of your dissection of the cat, and with the information provided in fig. 9.10 or from your own observations of a walking cat, list the main muscles involved in propulsive and recovery phases of locomotion of the forelimb. After you have completed the dissection, do the same for the hind limb.

---

**F. Summary**

1. The voluntary muscles are of two kinds: the somatic or parietal muscles, derived from the epimeres, which constitute most of the musculature; and the branchial or visceral muscles, derived from the hypomere, which are limited to the gill region. Primitively, the branchial muscles operate the gill arches, but in tetrapods they may spread over a considerable area. They come from the entire hypomere, since in the gill region the hypomere does not divide into somatic and splanchnic layers.

2. In primitive vertebrates the somatic muscles exist in the form of an axial succession of muscle segments or myotomes, separated by connective tissue partitions, the myosepta. Each myotome extends from middorsal to midventral line.

3. Beginning with fishes, the myotomes are divided into dorsal or epaxial halves and ventral or hypaxial halves by the horizontal skeletogenous septum. Although the division of the musculature remains, the septum becomes indistinct in amniotes.

4. The girdles and paired appendages interrupt the series of myotomes. The musculature of the girdles and appendages is derived in lower vertebrates from muscle buds sent out by the adjacent myotomes. Although

TABLE 9.1
Muscle Derivatives

| Salamander | Mammal |
| --- | --- |
| *Forelimb Derivatives of Primitive Dorsal Muscle Mass* | |
| Latissimus dorsi | Latissimus dorsi, teres major |
| Subcoracoscapularis | Subscapularis |
| Dorsalis scapulae | Spinodeltoid, acromiodeltoid |
| Procoracohumeralis longus | Clavodeltoid |
| Procoracohumeralis brevis | Teres minor |
| Triceps | Triceps |
| Forearm extensors and supinators | Forearm extensors and supinators |
| *Forelimb Derivatives of Primitive Ventral Muscle Mass* | |
| Pectoralis | Pectoralis |
| Supracoracoideus | Supraspinatus, infraspinatus |
| Humeroantibrachialis | ——— |
| ——— | Biceps |
| Coracobrachialis | Coracobrachialis |
| Forearm flexors and pronators | Forearm flexors and pronators |
| *Hind Limb Derivatives of Primitive Dorsal Muscle Mass* | |
| Ilioextensorius | Rectus femoris |
| ? Iliotibialis | Sartorius |
| ——— | Vastus group |
| Puboischiofemoralis internus | Iliacus, psoas, pectineus |
| Iliofemoralis, ? Iliofibularis | Gluteus group, tensor fascia lata |
| Peroneus group | Peroneus group |
| Other extensors | Other extensors |
| *Hind Limb Derivatives of Primitive Ventral Muscle Mass* | |
| Puboischiofemoralis externus | Obturator externus, quadratus femoris |
| Ischiofemoralis | Obturator internus, gemelli |
| Adductors | Adductors |
| Puboischiotibialis | Gracilis |
| Ischioflexorius | Semimembranosus, semitendinosus |
| ——— | Biceps femoris |
| Caudofemoralis | Pyriformis, crurococcygeus |
| Other flexors | Other flexors |

SOURCE: After Romer 1970.

this mode of origin can no longer be seen in the embryos of higher vertebrates, their girdle and limb musculature must be regarded as of myotomic origin phylogenetically.

5. In the evolution of tetrapods the muscles of the girdles and limbs increase in size and importance and spread over the segmented musculature. The latter more or less loses its segmental arrangement.

6. The epaxial musculature tends throughout vertebrates to split up into longitudinal systems—some long, some short and segmental. In tetrapods there are typically three longitudinal systems: the transversospinalis (divisible into the spinalis and semispinalis systems), the longissimus, and the iliocostalis, from the middorsal line downward. These extend in general from the sacrum to the back of the skull and also have continuations into the tail; they have attachments on vertebrae and ribs. The short segmental muscles include several systems between various parts of the vertebrae and ribs, and each member extends over only one body segment.

7. The hypaxial trunk musculature in tetrapods is divisible into the oblique system on the body sides and the rectus system near the midventral line. The former splits into three layers; the latter forms the rectus abdominis in the trunk, rectus cervicis in the neck, and geniohyoid in front of the hyoid apparatus.

8. The appendicular muscles consist in fishes of a dorsal or abductor mass and a ventral or adductor mass. In tetrapods these differentiate into a number of muscles from girdle to limb and between sections of the limb. The accompanying table shows salamander and mammalian derivatives of the dorsal and ventral masses.

9. The branchial musculature, derived from the hypomere and innervated by cranial nerves, originally functions to move the gill arches for respiratory purposes. Very early in vertebrate history, however, the mandibular arch became the upper and lower jaws, and its musculature took on masticatory functions. The musculature of the remaining gill arches, with the loss of gill respiration in tetrapods, became reduced and altered but continues to move those structures derived from gill arches, namely the hyoid apparatus and the laryngeal cartilages. The following attempts to give the main points in the history of the branchial musculature, but some things are still controversial.

*a. General*

Primitively, probably each gill arch had a levator and a constrictor muscle. In living vertebrates there are usually present in addition to the levator and constrictor series, the following ventral series: coracobranchials, subarcuals, and ventral transversals (transversi ventrales). The last two series are lacking in elasmobranchs, and their coracobranchials are usually considered to be hypobranchial muscles, not homologous with the coracobranchials of other fishes. Their levator series is also imperfect. The ventral branchial series does not occur in connection with the mandibular and hyoid arches.

### b. Mandibular Arch Muscles

The dorsal part of the mandibular arch constrictor is generally represented in vertebrates by the dorsal constrictor (craniomaxillaris) and the adductor mandibulae. The former persists in reptiles but not in amphibians or mammals; the latter is the chief masticatory muscle of vertebrates, acting to close the lower jaw. The adductor mandibulae in mammals differentiates into two groups of masticatory muscles, one including the temporalis, masseter and external pterygoid, and the other including the internal pterygoid, tensor tympani, and tensor palati. The tensor tympani and tensor palati are small muscles associated with the middle ear, a result of the reduction of the quadrate and articular to middle ear ossicles.

The ventral part of the mandibular constrictor is the intermandibularis, termed mylohyoid in mammals; other derivatives are the mandibularis, when present, and the anterior belly of the mammalian digastric muscle.

The mandibular arch muscles through all their evolutionary changes are innervated by the fifth cranial nerve (trigeminus).

### c. Hyoid Arch Muscles

The hyoid constrictor sheet forms in tetrapods chiefly the interhyoideus and the constrictor colli. The interhyoideus becomes in mammals the posterior belly of the digastric. The digastric is the opener or depressor of the jaw in most mammals. The constrictor colli is highly developed in reptiles and birds and is usually considered to develop into the platysma and other muscles of facial expression of mammals. The second or hyoid levator typically becomes the depressor mandibulae, but in some tetrapods the depressor may come from the constrictor sheet. In most mammals the hyoid levator becomes the stapedius, a small muscle of the middle ear.

The hyoid arch muscles are innervated throughout all their changes by the seventh (facial) cranial nerve.

### d. Other Branchial Arch Muscles

The levators persist in gill-bearing urodeles like *Necturus* and also occur in larval stages of other amphibians. The cucullaris belongs to the levator series but has acquired attachments to the pectoral girdle. In higher tetrapods it subdivides into the trapezius and sternocleidomastoid or some variant thereof. The branchial constrictors appear to have no representatives in tetrapods. The coracobranchials are limited to fishes. The subarcual series occurs in urodeles and anuran larvae; and the first subarcual is found in reptiles, birds, and mammals as the branchiohyoid or branchiomandibularis muscle.

The constrictor esophagi, the laryngeal muscles, and the striated pharyngeal muscles of mammals are probably derivatives of some of the ventral branchial series.

10. The hypobranchial musculature is the ventral musculature below the gill region and is formed by forward growth from the ventral ends

of the myotomes behind the branchial muscles. It may be regarded as in series with the midventral systems of the hypaxial trunk musculature. It includes the rectus cervicis (sternohyoid, sternothyroid, thyrohyoid, and omohyoid) in the neck, the geniohyoids in the throat, the muscles from the skull, mandible and hyoid apparatus into the tongue (genioglossus, styloglossus, hyoglossus), the intrinsic tongue musculature, and muscles of the epiglottis.

**References**

American Zoologist Symposium. 1961. Evolution and dynamics of vertebrate feeding mechanisms. *Amer. Zool.* 1:177–234. (six articles.)

Bakker, R. T. 1971. Dinosaur physiology and the origin of mammals. *Evolution* 25:636–58.

Ballintijn, C., and Hughes, G. M. 1965. The muscular basis of the respiratory pumps in the trout. *J. Exp. Biol.* 43:349–62.

Barghusen, H. 1968. The lower jaw of cynodonts (Reptilia, Therapsida) and the evolutionary origin of mammal-like adductor jaw musculature. *Yale Peabody Mus. Postilla* 116:1–49.

Basmajian, J. 1972. Electromyography comes of age. *Science* 176:603–9.

Cheng, C. 1955. The development of the shoulder region of the opossum, *Didelphys virginiana*, with special reference to the musculature. *J. Morph.* 97:415–71.

Crompton, A. W. 1963. On the lower jaw of *Diarthrognathus broomi* and the origin of the mammalian jaw. *Proc. Zool. Soc. Lond.* 140:687–750.

Crompton, A. W., and Hotton, N. 1967. Functional morphology of the masticatory apparatus of two dicynodonts (Reptilia, Therapsida). *Yale Peabody Mus. Postilla* 109:1–51.

Davis, D. D. 1964. The giant panda: A morphological study of evolutionary mechanisms. *Fieldiana: Zool. Mem.* 3:1–339.

Galton, P. M. 1969. The pelvic musculature of the dinosaur *Hypsilophodon* (Reptilia: Ornithischia). *Yale Peabody Mus. Postilla* 131:1–64.

Gans, C. 1974. *Biomechanics: An approach to vertebrate biology.* Philadelphia: J. B. Lippincott.

Gans, C., and Bock, W. 1965. The functional significance of muscle architecture—a theoretical analysis. *Rev. Anat. Embryol. Cell Biol.* 38:115–42.

Goslow, G. E.; Stauffer, E. K.; Nemeth, W. C.; and Stuart, D. G. 1972. Digit flexor muscles in the cat: Their action and motor units. *J. Morph.* 137(2):335–52.

Goslow, G. E.; Reinking, R. M.; and Stuart, D. G. 1973. The cat step cycle: Hind limb joint angles and muscle lengths during unrestrained locomotion. *J. Morph.* 141(1):1–42.

Grand, T. E. 1968. The functional anatomy of the lower limb of the howler monkey (*Alouatta caraya*). *Amer. J. Phys. Anthropol.* 28:163–82.

Hiiemae, K., and Houston, W. J. B. 1971. The structure and function of the jaw muscles in the rat (*Rattus norvegicus* L.). (3 pts.). *Zool. J. Linnean Soc.* 50:75–99; 101–9; 111–32.

Hildebrand, M. 1960. How animals run. *Sci. Amer.* 202(5):148–57.

———. 1961. Further studies on locomotion of the cheetah. *J. Mammal.* 42:84–91.

Hughes, G. M., and Ballintijn, C. M. 1965. The muscular basis of the respiratory pumps in the dogfish (*Scyliorhinus canicula*). *J. Exp. Biol.* 43:363–83.

Jenkins, F. A., Jr. 1970. Limb movements in a monotreme (*Tachyglossus aculeatus*): A cineradiographic analysis. *Science* 168:1473–75.

———. 1971*a*. Limb posture and locomotion in the Virginia opossum (*Didelphis marsupialis*) and in other non-cursorial mammals. *J. Zool.* 165:303–15.

———. 1971*b*. The postcranial skeleton of African cynodonts. *Yale Peabody Mus. Bull.* 36:1–216.

Jouffroy, F. K., and Lessertisseur, J. 1967. Correlations musculo-squelettiques de la ceinture scapulaire chez les reptiles et les mammifères: Remarques sur un probleme de paleomyologie. In Problemes actuels de paleontologie. *Colloq. Int. Centre Nat. Rech. Sci.* 163:453–73.

Liem, K. F. 1973. Evolutionary strategies and morphological innovations: Cichlid pharyngeal jaws. *Syst. Zool.* 22:425–41.

Liem, K. F., and Osse, J. W. M. 1975. Biological versatility, evolution, and food resource exploitation in African cichlid fishes. *Amer. Zool.* 15:427–54.

Manter, J. T. 1938. The dynamics of quadrupedal walking. *J. Exp. Biol.* 15:522–40.

Miner, R. W. 1925. The pectoral limb of *Eryops* and other primitive terapods. *Bull. Amer. Mus. Nat. Hist.* 51:145–312.

Nursall, J. R. 1962. Swimming and the origin of paired appendages. *Amer. Zool.* 2:127–41.

Osse, J. W. M. 1969. Functional morphology of the head of the perch (*Perca fluviatilis* L.): An electromyographic study. *Neth. J. Zool.* 19:289–392.

Ostrom, J. H. 1966. Functional morphology and evolution of the ceratopsian dinosaurs. *Evolution* 20:290–308.

Oxnard, C. E. 1963. Locomotor adaptations in the primate forelimb. *Zool. Soc. Lond. Symp.*, No. 10, pp. 165–82.

Romer, A. S. 1922. The locomotor apparatus of certain primitive and mammal-like reptiles. *Bull. Amer. Mus. Nat. Hist.* 46:517–606.

———. 1927. The pelvic musculature of ornithischian dinosaurs. *Acta Zool. Stockholm* 43:225–75.

———. 1970. *The vertebrate body*. Philadelphia: W. B. Saunders.

Roos, P. R. 1964. Lateral bending in newt locomotion. *Verh Kon. Nederl. Acad. Wetensch., Amsterdam,* 67c:223–32.

Schaeffer, B. 1941. The morphological and functional evolution of the tarsus in amphibians and reptiles. *Bull. Amer. Mus. Nat. Hist.* 78(6): 395–472.

Schaeffer, B., and Rosen, D. E. 1961. Major adaptive levels in the evolution of the actinopterygian feeding mechanism. *Amer. Zool.* 1:187–204.

Slijper, E. J. 1946. Comparative biologic-anatomical investigations on the vertebral column and spinal musculature of mammals. *Verh. Kon. Nederl. Akad. Wetensch., Amsterdam,* 42(5):1–128.

Smith, J. M., and Savage, R. J. G. 1956. Some locomotory adaptations in mammals. *J. Linnean Soc. Zool.* 42:603–22.

————. 1959. The mechanics of mammalian jaws. *School Sci. Rev.* 141:289–301.

Snyder, R. C. 1954. The anatomy and function of the pelvic girdle and hindlimb in lizard locomotion. *Amer. J. Anat.* 95:1–46.

————. 1962. Adaptations for bipedal locomotion of lizards. *Amer. Zool.* 2:191–203.

Turnbull, W. D. 1970. Mammalian masticatory apparatus. *Fieldiana: Geol.* 18:149–356.

Tuttle, R. H. 1969. Quantitative and functional studies on the hands of the Anthropoidea. I. The Hominoidea. *J. Morph.* 128:309–64.

The above references are, with a few exceptions, recent works and provide an introduction to the older literature. An excellent source of information on musculature of living vertebrates is P. P. Grasse, ed., *Traité de zoologie,* vols. 1–17 (Paris: Masson). See relevant volumes for each vertebrate class.

# 10 The Comparative Anatomy of the Coelom and of the Digestive and Respiratory Systems
## Ronald Lawson

**A. Introduction**

The coelom is a cavity associated with both the digestive and the respiratory systems, and a special portion of the coelom (the pericardial cavity) surrounds the heart. Its functional importance is that it allows the organs of the systems it encloses to move relatively freely within it and thus facilitates their action. This chapter includes information on the embryological origin of the coelom and its mesenteries, integrated with similar data for the digestive and respiratory systems. Details of the development and functional morphology of the oral cavity and the pharynx are given, and special attention is devoted to the special visceral muscle of the pharyngeal wall. The structure and mode of action of the gills is described and linked with the evolution of the lungs and swim bladders. The alimentary canal and the liver and pancreas are described in the final section on the digestive system (fig. 10.1). Directions for dissection follow these discussions.

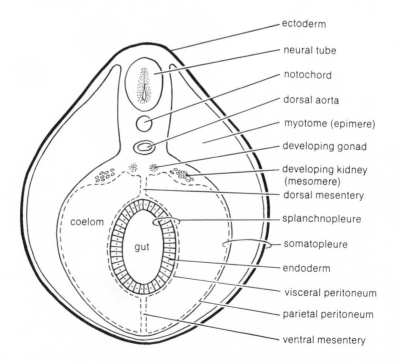

ectoderm

neural tube

notochord

dorsal aorta

myotome (epimere)

developing gonad

developing kidney (mesomere)

dorsal mesentery

splanchnopleure

coelom

gut

somatopleure

endoderm

visceral peritoneum

parietal peritoneum

ventral mesentery

Fig. 10.1. Diagrammatic transverse section of vertebrate to show the relations of the mesoderm. After Arey, in Romer 1949.

**B. The Coelom and Mesenteries**

**1. Embryological Origin**

The *coelom* or *body cavity* of vertebrates is the cavity of the hypomere (see chapter 4) and is never segmented, as it may be in invertebrates. The *somatic* or *parietal peritoneum,* the epithelial sheet lining the coelomic cavity, is derived from the outer wall of the hypomere, and this layer, together with the tissues of the body wall and the overlying ectoderm, forms the *somatopleure.* The inner wall of the hypomere on each side produces the *visceral peritoneum* or *serosa,* which, together with the gut wall and the endoderm, forms the *splanchnopleure.* It is important to remember that cells derived from the walls of the hypomere not only form the peritoneal membranes but also contribute extensively to the smooth muscle of the gut wall and the striated muscle of the body wall and limbs. Above and below the gut, the walls of the hypomere form a double-layered membrane, the *primary mesentery.* The portion of the mesentery between the dorsal body wall and the gut is the *dorsal mesentery* and that below the gut is the *ventral mesentery.* In the early stages of development the coelom is divided into right and left halves by the primary mesentery, but later the coelomic compartments may become confluent by the breakdown of the mesentery (fig. 10.2). In most cases it

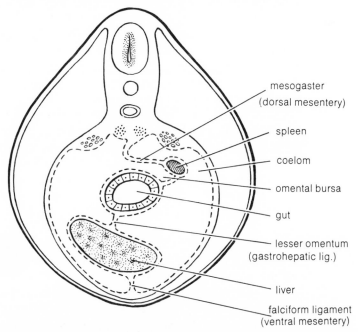

Fig. 10.2. Diagrammatic transverse section through a mammal in the region of the stomach, liver, and spleen. After Langman 1963.

is the ventral portion of the primary mesentery that is eliminated, while the dorsal portion usually remains more or less intact along its entire length. The dorsal mesentery is often strengthened by fine bundles of collagen fibers and carries the arteries, lymphatics, and nerves of the gut. The portion of the dorsal mesentery attached to the stomach is the *mesogaster* or *greater omentum,* that associated with the duodenum is the *mesoduodenum.* The *mesointestine* supports the small intestine, the *mesocolon* relates to the colon, and the *mesorectum* is connected to the rectum. In mammals the dorsal mesentery is often a site for fat deposition. During development in most jawed vertebrates, because of differential growth of parts of its wall and changes in position of surrounding organs such as the liver, the stomach moves toward the left side of the body. As the stomach moves it pulls the mesogaster with it; hence the right part of the coelom tends to encroach on the left side of the body. This is exaggerated by the fact that the greater omentum does not descend directly from the dorsal body wall but is folded to the left so that the cavity of the right coelom in this area is essentially a pouch, the *omental bursa*. As indicated above, the ventral mesentery is greatly reduced in size and persists only as a few specialized ligaments. The *falciform ligament* attaches the anterior end of the liver to the ventral body wall; the *gastroduodeno-hepatic ligament* (*lesser omentum*) connects the stomach to the liver and the duodenum; the *median ligament* lies between the urinary bladder and the ventral body wall. (The mesenteries and peritoneal membranes serve to support and cover organs and hence exclude such structures from the coelom.) Clearly none of the visceral organs actually lie *within* the coelom; they are *retroperitoneal* (external to the coelomic lining) or lie between the two layers of the mesentery. Two notable exceptions are the *anterior funnels of the oviduct* and the *anterior kidney tubules* in some vertebrates (e.g., the larval lamprey). In some cases developing organs project toward the cavity of the coelom and push the peritoneum ahead of them. The peritoneum in turn converges behind them to form a lateral mesentery, such as the *mesovarium* and the *mesorchium,* which support the ovary and the testis, respectively.

In *Branchiostoma* the mesoderm arises from the archenteron as a series of pouches that later combine to form the definitive coelom (i.e., enterocoelic). Although the coelom in modern vertebrates does not arise in this way, the enterocoelous manner of origin of the mesoderm and the coelom is generally considered ancestral for the vertebrates. If this is the case it suggests that the vertebrates are most closely linked to those invertebrates that are enterocoelic, such as the echinoderms.

**2. Divisions of the Coelom**

In its early stages of development the coelom is a continuous cavity along the entire length of the trunk, divided into right and left halves by the dorsal and ventral mesenteries. In the adult of all vertebrates the coelom is divided into at least two more cavities by the development of a *transverse septum*. This septum is formed as the common cardinal veins from each side of the body push across to enter the sinus venosus and carry the folds of the coelomic wall with them. The transverse septum divides the coelom into a small anterior *pericardial cavity,* which contains only the heart, and an extensive *pleuroperitoneal cavity,* which houses all the other viscera. In most vertebrates the transverse septum is complete, but in hagfishes, elasmobranchs, and some ray-finned fishes there is a small opening, the *pericardioperitoneal canal.* The production of this septum separating the pericardial and peritoneal cavities is associated with the liver, which during development fuses widely with it. In many lower vertebrates the main attachment of the liver is to the transverse septum, but in most higher forms the liver tends to draw away from the septum and remains attached to it by a narrow *coronary ligament.* The pericardial cavity in fishes and urodeles lies directly anterior to the pleuroperitoneal cavity and the transverse septum lies across the body. In other forms the pericardial cavity lies ventral to the anterior end of the pleuroperitoneal cavity and the transverse septum therefore assumes a sloping position. In anurans and some reptiles this is accomplished by the forward growth of the anterodorsal portion of the pleuroperitoneal cavity, which forms the *pleural pouches* around the lungs (fig. 10.3).

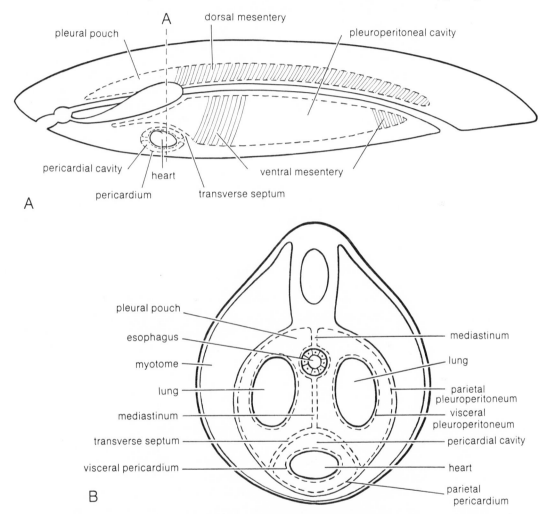

Fig. 10.3. Representation of the coelom in a primitive reptile: *A*, lateral view; and *B*, transverse section through A. After Smith 1960.

The area between these pouches is the *mediastinum,* in which the esophagus is embedded. In other reptiles (including crocodiles, snakes, and some lizards) and in birds the lungs are enclosed in chambers that are separated from the remainder of the body cavity. This is accomplished by the formation of a new septum that isolates the former pleural pouches (as the *pleural cavities*) from the remainder of the body cavity, now termed the *peritoneal* or *abdominal cavity.* The septum that isolates the pleural cavities from the peritoneal cavity is the *oblique septum,* derived partly from the transverse septum and partly from a double-walled membrane (*pleuroperitoneal membrane*) infolded from the dorsal and lateral parietal pleuroperitoneum. In birds the presence of a complex respiratory system of lungs and air sacs has resulted in a complicated subdivision of the body cavity. The oblique septum shuts off the lungs into two individual pleural cavities. A second septum forms a pair of lateral cavities that house the air sacs, and a third horizontal septum runs on either side of the liver to form separate coelomic cavities above and below this organ. The development of the oblique septum in reptiles and birds is paralleled by the presence of a *diaphragm* in mammals (fig. 10.4). In this case the origin of the septum is complex, for as well as involving the transverse septum and the pleuroperitoneal membranes, it includes muscular elements. The origin of the muscle in the diaphragm is from cells derived from the third, fourth, and fifth cervical segments and also from some thoracic somites. The cervical origin of some of the muscle of the diaphragm is demonstrated by the fact that the diaphragm is largely innervated by the phrenic nerve derived from the third, fourth and fifth cervical segments. The muscle of thoracic origin is supplied by *intercostal nerves*. The diaphragm in mammals is an important respiratory structure. The respiratory system in mammals is further elaborated by the enlargement of the pleural cavities which invade the area between the heart and the ventral thoracic wall (fig. 10.5).

**C. The Digestive Tract and Associated Structures**

The primitive intestine or *archenteron,* produced in the embryo by a process of *gastrulation,* has the form of a simple tube with a single opening to the exterior, the *blastopore* (see chapter 4). The endodermal tube persists as the lining or *mucosa* of the adult digestive tract. The connective tissue of the submucosa and the muscle layers of the gut wall are added from the inner wall of the hypomere (see above). At the extreme anterior and posterior ends of the digestive tube the endodermal lining gives way to an ectodermal one produced during development by invaginations of the ectoderm. Initially these take the form of shallow indentations that become deeper as the flexure of the head and tail increases, to become conspicuous pockets. The anterior one is the *stomodeum,* which eventually forms the *oral cavity,* and the posterior pocket forms the *proctodeum,* or *anal cavity.* A plate of tissue (the *pharyngeal membrane* or *stomodeal plate*) is formed by the apposition of the stomodeal ectoderm and the foregut endoderm. A similar membrane (the *cloacal membrane* or *proctodeal plate*) is formed posteriorly. In due course both of these plates rupture, and the primitive gut acquires communication with the exterior (fig. 10.6).

**1. Embryological Origin**

**2. Oral or Buccal Cavity**

The oral or buccal cavity is bounded by the jaws in front and at the sides, the palate above and the tongue below. Lips and cheeks develop in higher forms, and the space between the lips and the teeth is usually termed the *vestibule.* In *Branchiostoma* the oral cavity is enclosed by the oral hood and is demarcated from the pharynx by the *velum.* The position of the velum probably corresponds to the boundary between the oral cavity and the pharynx in gnathostomes. In adult gnathostomes, however, there is no such landmark and it is difficult to decide where the buccal cavity

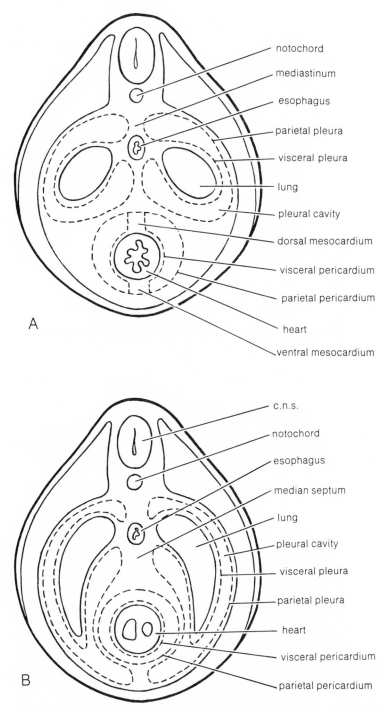

Fig. 10.4*A–B*. Diagrammatic transverse section of the thorax of a mammal showing the development of the lungs and pleural cavity. After Torrey 1963.

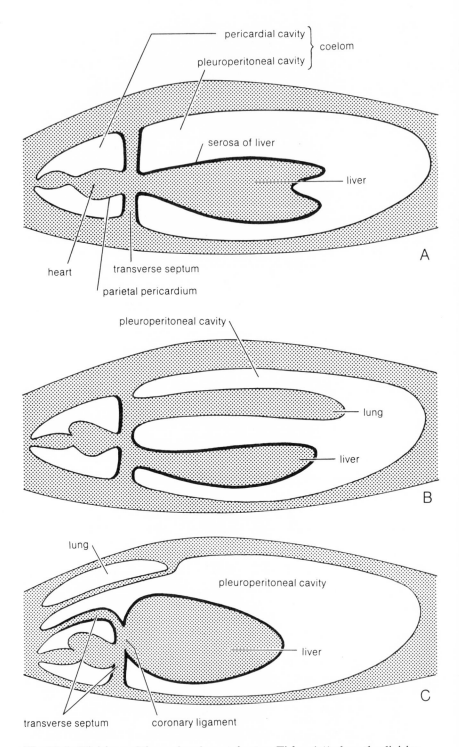

Fig. 10.5. Divisions of the coelom in vertebrates: Fishes (*A*) show the division of the coelom into pericardial and peritoneal cavities (which may be connected by a pericardioperitoneal canal in some forms) separated by a transverse septum. Urodeles (*B*) are similar except for the addition of the lungs that project into the pleuroperitoneal cavity. In the turtle (*C*) the pericardial cavity has descended posteriorly until it lies ventral to the anterior portion of the pleuro-

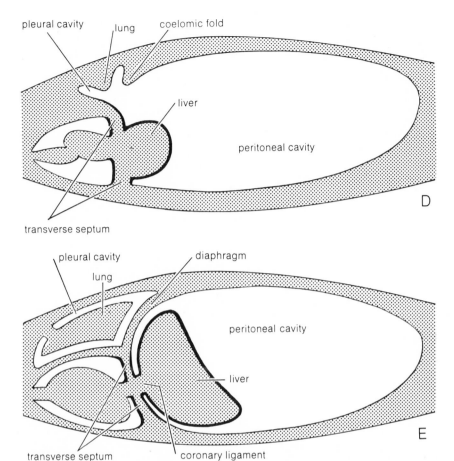

peritoneal cavity. The anterior face of the transverse septum becomes part of
the peritoneal sac; the lungs are retroperitoneal. In the early stages of development the mammal (*D*) shows the beginning of the coelomic fold descending
from the dorsal body wall. Later (*E*) the coelomic fold and the transverse septum unite to form the diaphragm which separates the peritoneal and pleural
cavities.

The relations of the liver to the transverse septum are complex. The liver is a
diverticulum from the small intestine that grows out ventrally into the transverse septum. It lies first within the mesenchyme of the septum between the
two walls of the septum. Since the great veins enter the heart by way of the
transverse septum, the liver also acquires important relationships to these
veins. The liver increases rapidly in size so that it can no longer be contained
within the limits of the septum. It consequently bulges posteriorly, carrying the
posterior wall of the septum, which thus becomes the peritoneal covering, or
serosa of the liver. Later the region where the liver bulges from the septum
narrows on the dorsal and lateral sides, leaving anterior and ventral connections
between the liver and the septum. The anterior connection is the coronary
ligament and is, as its name implies, a circular ligament by which the liver in all
vertebrates is suspended from the transverse septum or its derivative. The
ventral partition formed by the constriction of the liver from the septum is
the falciform ligament of the liver which is part of the ventral mesentery.
Thus the transverse septum has the following parts: the posterior wall of the
pericardial cavity, the anterior wall of the pleuroperitoneal cavity, the falciform
and coronary ligaments, the serosa of the liver, and the mesoderm of the liver.

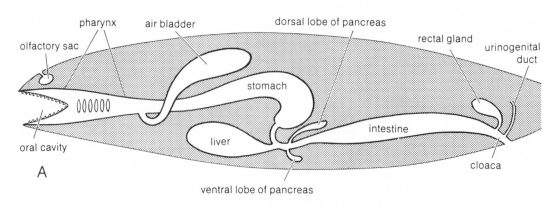

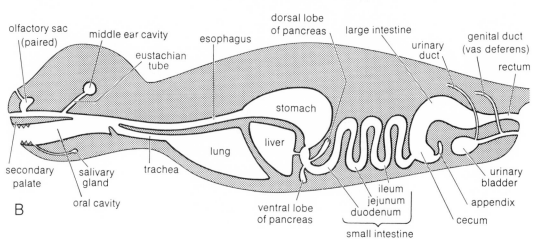

Fig. 10.6*A–B*. Diagram of alimentary systems of fishes and mammals. After Smith 1960.

ends and the pharynx starts. The epithelium of the oral cavity is derived mainly from the stomodeal invagination, but most investigators believe that the lining of the oral cavity has an endodermal component. In the Dipnoi (lungfishes), most crossopterygians, and all tetrapods the nasal cavities open into the roof of the buccal cavity. It is convenient to consider the nasal apparatus at this point. The nasal organs are initially blind sacs produced by the invagination of the paired epidermal *olfactory placodes* (single in cyclostomes; fig. 10.7). Each sac is lined by an epithelium that eventually contains *olfactory cells,* nonsensory *supporting cells,* and *gland cells* (e.g., *Bowman's glands* in tetrapods). Impulses are relayed to the brain by extensions of the *bipolar olfactory cells.* The *nasal cavities* of the chondrichthyes and most actinopterygians lie outside the limit of the oral cavity, the walls of the *nasal sacs* are sensory, and the sacs end blindly. The external opening of the nasal sac, the *external naris,* is usually subdivided into *incurrent* and *excurrent apertures* facilitating the flow of water over the nasal epithelium (fig. 10.8). In the lungfishes the excurrent opening is within the oral cavity, and, in view of the affinities of the lungfishes with the rhipidistians (an-

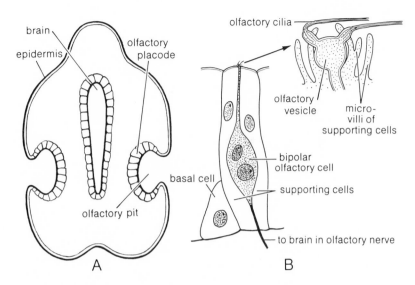

Fig. 10.7. Development of the olfactory apparatus: *A*, development of olfactory placode; *B*, portion of fully differentiated olfactory epithelium.

cestors of the tetrapods) and the tetrapods, many authors have regarded this opening as the *internal nostril* or *choana.* However, there is no evidence that the nasal apparatus in lungfishes ever transmits air into the buccal cavity; and seems to be used like that of other fishes. The opening of the excurrent aperture of the naris into the buccal cavity is also found in certain actinopterygian fishes. There is now general agreement that the internal opening of the nostril in lungfishes does *not* correspond to the choana of tetrapods, which is probably a neomorphic structure—that is, a structure newly acquired and peculiar to the tetrapods. Embryological evidence suggests that the choana in tetrapods is usually developed from an internal process of the nasal sac, together with a small diverticulum from the gut.

### a. Nasal Apparatus

The nasal apparatus in tetrapods typically consists of three parts. The major cavity is the central *cavum nasi proprium,* the *primary* or principal nasal cavity, which is usually an enlarged chamber lined in part with sensory olfactory epithelium. Anterior to this is a tubular *vestibulum* that leads to the external naris, and posteriorly a *nasopharyngeal duct* leads to the *choana.* The term nasopharyngeal duct is used here in a broad sense, since many authors restrict the term to describe the channel above a well-developed *secondary palate* (see below). Projections from the lateral wall into the cavum nasi proprium are termed *conchae.* In many tetrapods an accessory nasal organ, the *vomeronasal organ,* or *Jacobson's organ,* is present, and this may be either a ventrolateral outpocketing of the primary nasal cavity or more simply a special area of nasal epithelium, *vomeronasal epithelium.*

There is, however, considerable variation in the basic structure of the nasal apparatus. In urodeles it is usually relatively simple—the cavum nasi proprium is a large ovoid chamber and there is no well-developed

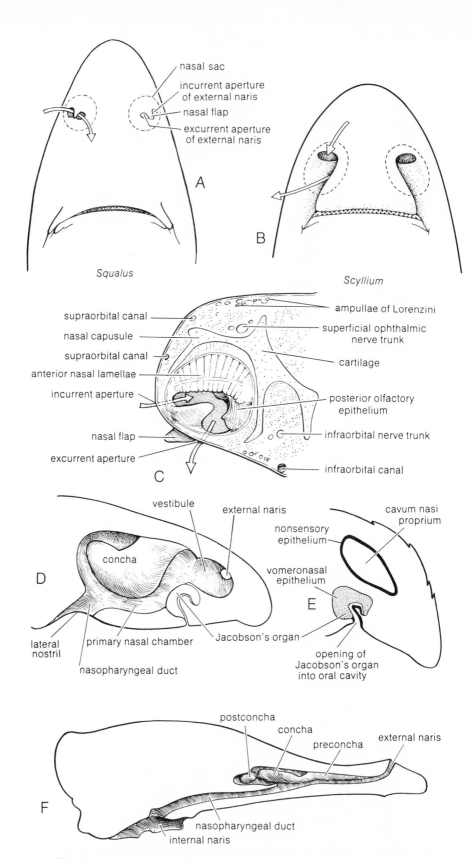

Fig. 10.8. *A, B,* external view, external naris of *Squalus* and *Scyllium; C,* cross section of nasal region of *Squalus* (after Walker 1965); *D,* medial view of lateral wall of nasal cavity of *Lacerta; E,* transverse section through Jacobson's organ of *Anguis; F,* medial view of lateral wall of the nasal cavity of *Alligator* (after Parsons 1959).

vestibulum or nasopharyngeal duct. A ventrolateral groove-like extension, the *lateral nasal sinus,* posteriorly bears the vomeronasal epithelium. In gymnophionans, anurans, and most reptiles the nasal apparatus is more complex than in urodeles, and a distinct Jacobson's organ is usually present. In all cases the choanae open into the anterior portion of the oral cavity, which then forms a common passageway for air and food, as in the urodeles. In amphibians the roof of the mouth is relatively flat, whereas in reptiles and birds the roof is vaulted and has a pair of longitudinal *palatal folds.* These folds help form a passage from the choanae above the tongue to the pharynx to aid the passage of air. In crocodiles the nasal cavities are complex, with a large number of accessory sinuses and recesses opening into the main chamber. From approximately the middle of each cavum nasi proprium arises a long nasopharyngeal duct that opens into the posterior part of the oral cavity via the choana. Crocodiles have thus formed a *secondary palate* so that choanae open into the posterior end of the oral cavity and hence largely separate the respiratory and food passages. Jacobson's organ is absent (fig. 10.9). The olfactory apparatus in mammals is variable, and some groups such as whales and bats have reduced nasal organs. In most cases the nasal cavities are well developed, and within the principal nasal cavity are several conchae usually termed the turbinals (*maxilloturbinals, nasoturbinals,* and *ethmoturbinals*). Also, several accessory sinuses are present within the bones of the nasal region. A well-developed secondary palate extends posteriorly as a thick membrane, the *soft palate.* Vomeronasal organs are usually present as small tubular structures along the nasal septum and may communicate with the nasal cavity, oral cavity, or both (fig. 10.10).

The function of the nasal apparatus in aquatic forms is primarily sensory as it samples the water passing through the nasal sac. In terrestrial animals the nasal apparatus has developed a connection with the oral cavity and in most cases serves a respiratory as well as a sensory function. In many cases the sensory function of the nasal apparatus is enhanced by the development of the vomeronasal organ.

The nerve fibers that carry impulses from the sensory olfactory epithelium run in the olfactory nerve to the main olfactory bulb of the brain. Those from the vomeronasal organ enter a ventrolateral accessory lobe, absent from those forms lacking such an organ.

### b. Mouth

A variety of hypotheses have been advanced concerning the homologies of the vertebrate mouth, and the immediate ancestors of the chordates have been sought in almost every invertebrate group, especially the annelids and the arthropods. It is now generally believed that the chordates are more closely related to the echinoderms than to any other invertebrate group and that the chordates and the echinoderms evolved from a common ancestor. Therefore the mouth in chordates is homologous with that of their immediate ancestors, not a newly evolved structure as it would have to be if the chordates had evolved from an annelid or arthropod stock. The mouth in *Branchiostoma* originates on the left side in a series of gill apertures and only later moves to a median

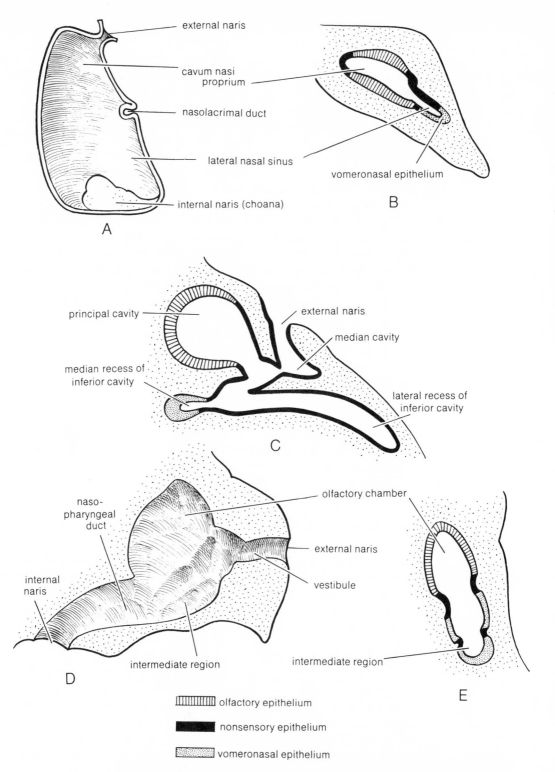

Fig. 10.9. Tetrapod nasal cavities: *A*, dorsal view of ventral half of nasal cavity of *Triturus; B*, transverse section of nasal cavity of *Triturus; C*, transverse section of nasal cavity *Alytes*. The cavum nasi proprium is complex and divided into principal, median, and inferior cavities. *D*, medial view of lateral wall of the nasal cavity of *Chelonia; E*, transverse section of nasal cavity of *Chelonia*. Cavum nasi proprium divided into dorsal olfactory region and ventral intermediate region.

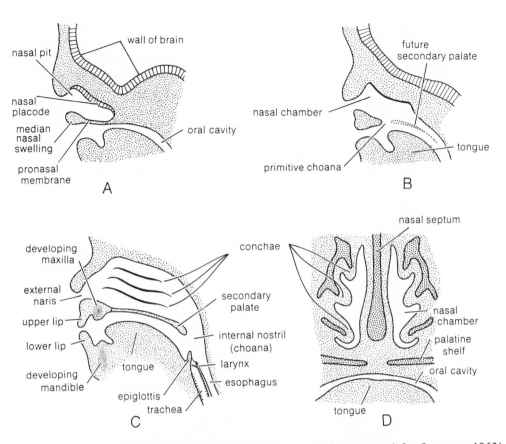

Fig. 10.10. Development of human nasal apparatus (after Langman 1963):
*A–C*, sagittal sections showing three stages in development of the nasal region
in the human embryo. *D*, frontal section of the nasal region of human embryo.

position. The mode of development of the mouth in *Branchiostoma*
lends some support to the idea that the mouth in chordates arose by the
fusion of gill slits. The asymmetrical development seen in *Branchiostoma* is a peculiarity of the protochordates and has no phylogenetic
significance.

### c. Tongue

Fishes have a small *primary tongue* that is simply the portion of the
floor of the buccal cavity covering the anterior end of the hyobranchial
apparatus. It is demarcated anteriorly by a fold. In cyclostomes the
name "tongue" is also given to a protrusible oral structure covered with
horny teeth. This organ has been evolved in parallel and is not homologous with the tongue of other groups. In amphibians a crescentic gland
field usually develops in front of the primary tongue or *copula* and fuses
with it to form the *definitive tongue*. In amniotes the gland field is termed
the *tuberculum impar* and fuses with the pair of lateral swellings (*lingual
swellings*) as well as with the primary tongue to form a complex definitive tongue. The contributions of these various swellings to the tongue
are not the same in all groups. In fact, the swellings form only the
epithelial and mucosal portions of the tongue; the muscular elements are

derived from the *hypobranchial musculature*. As indicated elsewhere, this muscle is derived from the segments immediately posterior to the skull (*occipitocervical somites*) and is innervated by nerves (from these segments) that grow ventrally with the muscle. In fishes and amphibians the nerve involved is a spinal nerve, usually termed the *hypoglossal* or *first spinal nerve,* whereas in the amniotes the nerve issues from the cranium as the *twelfth cranial nerve* or *hypoglossal nerve*. In addition to its motor innervation from the hypoglossal nerve, the tongue is also richly supplied with sensory fibers, mainly in the fifth and seventh cranial nerves. The primary function of the vertebrate tongue is handling food, but this organ has many modifications in different groups. In frogs, toads, and urodeles the tongue may be used in the capture of food, and in the chameleons the tongue may be rapidly everted and withdrawn for the same purpose. Several mammals such as the anteaters also have long, protrusible tongues. The tongue in many lizards and snakes is used to pick up airborne particles that are then placed against the opening of the chemosensory Jacobson's organ.

### d. Oral Glands

The inner lining of the oral cavity in vertebrates may contain numerous glands. These glands are integumental, and it is often impossible to homologize them within the various groups. Broadly speaking, there are two types of glands, *mucous* and *serous*. The former secrete mucus, used primarily in lubrication, and the latter produce enzymes and in some cases poisonous substances. Because of the difficulties encountered in homologizing the oral glands, they are often named according to their position, such as *lingual, sublingual, palatine, dental,* and *labial glands.* Enlarged labial glands are present in many groups and are often referred to as *salivary glands.* The *poison* or *venom* glands found in snakes are examples of salivary glands and have their opening near the base of a poison fang. The venom passes into the wounded prey via a duct or groove in the tooth. The presence of large salivary glands is particularly pronounced in the mammals, where it is thought that they are related to the chewing habits of many members of this group. The products of the salivary glands (*parotids, submaxillary, sublingual,* and *infraorbital*) are primarily concerned with lubrication for swallowing; only a few mammals have saliva that contains starch-splitting enzymes.

### e. Rathke's Pouch

During the early stages of development the roof of the oral cavity produces an epithelial evagination in the form of a small pouch known as *Rathke's pouch*. This grows dorsally toward a ventral extension of the diencephalon, the *infundibulum,* eventually losing contact with the oral cavity and fusing with the infundibulum to form the *hypophysis* or *pituitary body*. During the development of the hypophysis the cells in the anterior wall of Rathke's pouch actively proliferate and form the anterior lobe of the hypophysis, and *adenohypophysis*. Later a small extension of this lobe grows along the infundibulum and eventually surrounds it as the *pars tuberalis*. The posterior wall of the pouch forms the *pars intermedia*. The infundibulum becomes differentiated into a stalk, surrounded by the *pars tuberalis,* and a *pars nervosa* or *posterior* lobe, and these together constitute the *neurohypophysis*. The neurohypophysis is composed of cells that differentiate into *pituicytes* and also contains numerous nerve fibers that link it to the hypothalamic area of the diencephalon.

The hypophysis or pituitary forms an important part of the endocrine system; the secretions of the pituitary often regulate the activity of other ductless glands via complex biological feedback mechanisms.

3. Pharynx

The oral cavity in chordates leads into the pharynx, which is characterized by *pharyngeal slits* or *gill slits*. Such perforations of the pharyngeal wall may often be lacking in adult gnathostomes, but they make at least a transient appearance during development. In *Branchiostoma* and the ammocoete larva of cyclostomes the anterior limit of the pharynx is clearly marked by the velum, but in most gnathostomes the anterior boundary is difficult to fix. It is usually taken as the region of the first gill slit or its derivative. The gill slits in chordates develop as a series of *visceral* or *pharyngeal pouches* in the inner wall of the pharynx. As each pouch approaches the surface a corresponding invagination occurs, which forms a *pharyngeal* or *visceral groove* (furrow). Experiments show that the visceral furrows will not form without the visceral pouches. The absence of the visceral pouches also appears to prevent the formation of certain parts of the skeleton of the gill region. It is concluded that the visceral pouches have an inductive function that affects the visceral furrow areas and also the area between the successive pouches (fig. 10.11).

It is impossible to say how many pharyngeal pouches were present in the ancestors of the modern vertebrates. It is likely that it was a fairly large number, since some modern myxinoids possess fourteen such pouches and the paleozoic ostracoderms frequently had as many as ten. The commonest number of gill pouches in modern gnathostomes is six, and in many cases the first of this series, which is anterior to the hyoid arch, forms the *spiracle*. The original function of the pharynx was food collection, and it acted as a large elaborate filtering device. In modern protochordates the pharynx is extremely large, and the feeding currents are mainly produced by the action of cilia. The food is screened off and usually trapped in a thread of mucus. With the evolution of jaws the role of the pharynx changed, since the animals no longer depended on filtering small organisms from the water, and in these circumstances the pharynx became involved in respiration. The walls of the gill pouches became highly vascularized gills and formed areas for gaseous exchange.

*a. Pharyngeal Pouches*

The tissue that forms the dorsal and ventral areas of the pharyngeal pouches invariably proliferates to form a variety of structures that are mainly glandular and are often associated with the lymphatic system. The number and disposition of these structures is variable, and in some cases the function of the pharyngeal pouch derivatives is uncertain. A summary of the pharyngeal pouch derivatives is shown as Table 10.1. The anterior pharyngeal cleft in amphibians and amniotes is associated

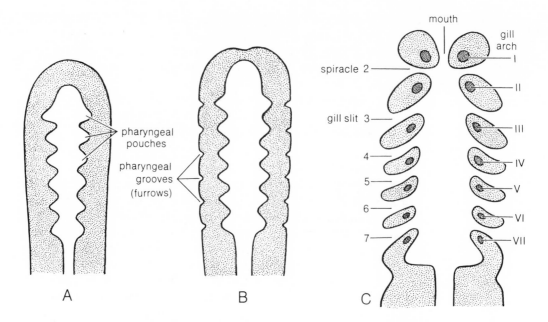

Fig. 10.11. Diagrams to illustrate the formation of pharyngeal pouches (*A–B*) and gill slits (*C*). The gill arches are numbered I–VII, and the gill slit *anterior* to each arch carries the same number as that arch; e.g., gill slit 7 lies anterior to gill arch VII. In some cases, other authors do not number the spiracular slit, and hence the next posterior slit becomes number 1, in which case the slit anterior to arch VII is number 5.

with the otic region when the dorsal part forms a *tubotympanic recess*. This grows out laterally to form a saclike *tympanic* or *middle ear cavity,* which remains connected to the pharynx by a narrow tube, the *pharyngotympanic* or *eustachian tube*. In some reptiles and in birds and mammals a *pharyngeal furrow* corresponding to the tubotympanic recess meets the wall of the *tympanic membrane*. The passage formed by the furrow is the external *auditory meatus,* which hence marks the position of the spiracular gill cleft (fig. 10.12). The *thymus gland* is an important pharyngeal structure that primitively derives from a number of pouches but becomes progressively more compact until in the mammals it is derived from a single pouch. The thymus gland is a constant feature of vertebrates and is primarily concerned with immunity responses. It is particularly important in young animals, and in many mammals just before birth the thymus gland makes up about 1% of the total body weight. The size of the gland usually decreases with age. Removing the thymus from newborn mammals causes death from infections within weeks, owing to low production of antibodies. It is believed that the thymus is responsible for the production of lymphoid cells and possibly certain substances that are transported by the blood to the other lymphoid tissues of the body such as the lymph nodes, which actually produce the antibodies. The *tonsils* and the *parathyroid* glands are also produced from pharyngeal pouches; the former are of lymphoid function and the latter are concerned with the metabolism of calcium and phosphorus. The *carotid bodies,* present in the carotid arteries of all tetrapods, are

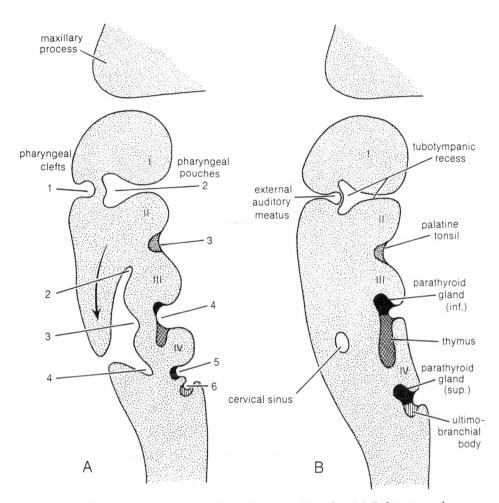

Fig. 10.12. Development of the pharyngeal pouches (*A, B* show stages in gland establishment) in human embryo. After Langman 1963.

largely concerned with detecting the oxygen tension in the blood. The *aortic body* has a similar function. The *postbranchial* or *ultimobranchial* bodies arise from the posterior gill slit; their function is not well understood. In fishes they are probably homologous to the C cells in the thyroid gland of other vertebrates and are a source of the hormone calcitonin. However, the precise role of calcitonin in the regulation of body calcium levels in fish remains problematic. In the cyclostomes the ventral portion of each pouch produces an epithelial corpuscle of unknown function.

### b. Thyroid Gland

As indicated above, the majority of the glandular structures found in the pharynx are proliferated from the pharyngeal pouches. The thyroid gland is an exception, since it develops as an epithelial outgrowth from the *floor* of the pharynx at the level of the hyoid arch. The *thyroid primordium* moves ventrally and posteriorly and forms a bilobed structure. During its migration the developing thyroid remains connected to the pharynx by

TABLE 10.1    Pharyngeal Pouch Derivatives in Vertebrates

| Pharyngeal Pouch | Lamprey | Elasmobranch | Urodele | Anuran | Reptile | Bird | Mammal | Position |
|---|---|---|---|---|---|---|---|---|
| Spiracular | Thymus | Spiracle | Tubotympanic recess | Tubotympanic recess | Tubotympanic recess | Tubotympanic recess | Tubotympanic recess | Dorsal |
| Pouch of hyoid arch Arch II | Epithelial | Spiracle | — | Obliterated by development of tongue | | | | Ventral |
| First branchial or Arch III | Thymus | Thymus | Tonsil | Thymus | Thymus | Tonsil | Tonsil | Dorsal |
|  | Epithelial corpuscle | — | Carotid body | Carotid body | Carotid body | Carotid body | Carotid body | Ventral |
| Second branchial or Arch IV | Thymus | Thymus | Thymus | Tonsil | Thymus | Thymus | Parathyroid | Dorsal |
|  | Epithelial corpuscle | — | Parathyroid | Parathyroid | Parathyroid | Parathyroid | Thymus | Ventral |
| Third branchial or Arch V | Thymus | Thymus | Thymus | Tonsil | Tonsil | Tonsil | Parathyroid | Dorsal |
|  | Epithelial corpuscle | — | Parathyroid | Parathyroid | — | Parathyroid | — | Ventral |
| Fourth branchial or Arch VI | Thymus | Thymus | Thymus | Tonsil | Tonsil | Tonsil | Tonsil | Dorsal |
|  | Epithelial corpuscle | — | Aortic body | Aortic body | Aortic body | Aortic body | Aortic body | Ventral |
| Fifth branchial or Arch VII | Thymus | Thymus | Tonsil | Tonsil | Tonsil | Tonsil | Tonsil | Dorsal |
|  | Epithelial corpuscle | Ultimo-branchial body | Ultimo-branchial body | Ultimo-branchial body | Ultimo-branchial body | Ultimo-branchial body | Ultimo-branchial body | Ventral |

a narrow canal, the *thyroglossal duct* (fig. 10.13). The thyroid gland is an important endocrine gland necessary for the control of metabolic activities, normal growth, metamorphosis, molting, and sexual development.

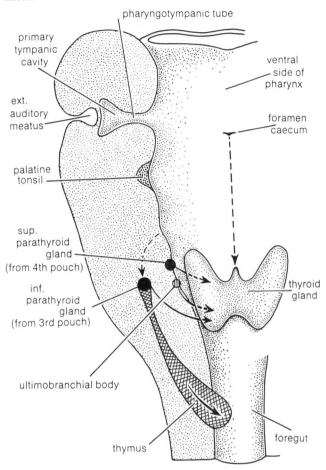

Fig. 10.13. Migration of pharyngeal pouch derivatives and the thyroid. After Langman 1963.

### c. Pharyngeal Muscle

In vertebrates the successive gill slits are separated by pharyngeal or *branchial* arches, each of which contains a skeletal support, the *gill arch*.[1]

---

1. The term has been used in the past with some ambiguity and may refer in various cases to: (*a*) The *gill arch* or *gill skeleton* (also termed a *visceral arch, pharyngeal arch,* or *branchial arch*), which supports the gills or branchiae. The jaws are believed to have evolved from an anterior gill arch and the more posterior arches evolve, in terrestrial vertebrates, into a variety of pharyngeal structures as the larynx. (*b*) The *total* tissue that separates the successive gill pouches, hence including the gill skeleton and the connective tissue, blood vessels, and the *intrabranchial septum.* (*c*) An arterial blood vessel that leaves the heart or the ventral aorta, such as the *carotid arch* or *pulmonary arch.* In this volume the term *gill arch* refers specifically to the major portion of the gill skeleton, typically composed of pharyngo-, epi-, cerato-, and hypobranchial elements.

Extending outward from each gill arch there is usually a vertical sheet of connective tissue that forms the *gill septum.* This is particularly well developed in elasmobranchs but is much reduced in teleosts. In the embryo the pharyngeal arches contain a portion of the coelom. This cavity is lost in the adult, but the mesodermal cells derived from the coelomic wall contribute extensively to the formation of the *pharyngeal muscle.* Thus the pharyngeal muscle separating the gill slits develops from the splanchnic mesoderm, not from the somites as does the comparable musculature of the postpharyngeal region. A distinction may therefore be made between the pharyngeal muscle, whose origin is similar to that of the gut (see p. 382) and may be considered *visceral,* and the more posterior muscle of the body wall, which is derived from the somites and is *somatic.* It should be noted however, that despite its origin from the hypomere the pharyngeal muscle is striated (not unstriated like the rest of the visceral muscle) and is regarded as a special area of *striated visceral muscle.* The pharyngeal muscle is split into a series of columns by the development of the gill slits and therefore assumes a segmental arrangement. There is considerable disagreement on whether the segmentation of the visceral pharyngeal muscle corresponds to that of the somatic muscle forming the myotomes. There is certainly no a priori reason why it should. Many authors believe that the segmentation of the two muscle groups does not correspond and use the term *branchiomeric* to describe the pharyngeal muscle and *metameric* for the somatic segmentation. A recent report suggests that the anterior branchiomeric segments do correspond to metameric segments, except more posteriorly in the region behind the ear (*postotic*). In the latter case the correspondence between the segmental relationships is assumed to break down because the postotic pharyngeal muscle develops as a single mass. This mass later proliferates, then becomes divided in such a way that the segments produced have no relation to the metameric segments.

**4. Gill Arches and Their Derivatives**

In fishes the *gills* are supported by a series of skeletal elements, the gill arches,[2] which typically consist of two major elements: a dorsal *epibranchial* and a ventral *ceratobranchial.* The epibranchial supports a *pharyngobranchial* element and the ceratobranchial supports a *hypobranchial* element. In addition, a median ventral *basibranchial* or *copula* is often present. A number of anterior gill arches and their pharyngeal muscle have been concerned with the evolution of the jaw, and in those land vertebrates where the gills have lost their respiratory function the more posterior gill arches have evolved into a series of elements that support the tongue (*glossal skeleton*), the larynx, and the trachea. The evo-

---

2. It has been the practice of many authors to assign numbers to the gill arches and gill slits. Unfortunately not all authors have used the same numbers. In most cases the mandibular arch, forming the jaws, is arch I, and the hyoid arch, posterior to the hyoid or spiracular slit, is arch II. The next arch, the first branchial arch, is arch III, and so on. I adopt this practice here and consider the mandibular arch as number 1 (since it is common to all gnathostomes). Similarly, the gill pouch preceding each gill arch is assigned the number of that gill arch. Hence the spiracle becomes pouch 2, that in front of the first branchial arch (arch III) becomes pouch 3, and so on (see fig. 10.12).

lution of the jaws, the tongue skeleton, and the larynx is discussed in chapter 8.

**D. Respiratory Structures**

1. Gills

The gills that form on the pharyngeal pouches are *internal gills* and are common in aquatic gnathostomes. In some cases *external* gills may be formed, and these arise as filamentous outgrowths of the superficial epithelium of the gill arches. They are found in the larvae of lungfishes and some actinopterygians. External gills are also common in amphibian larvae; in some cases they are retained in the adult and as such are regarded as paedomorphic characters—larval characters that persist in an otherwise adult animal. The size of the external gills in a particular individual may vary with changes in the nature of the water. When the water is well oxygenated the gills are small and retracted, whereas in water with little oxygen the gills may be large and extended.

The internal gills or *branchiae* typically consist of a double row of gill filaments. The spaces between successive filaments are not simple, since the upper and lower surfaces of the filaments are thrown into a series of folds or *secondary lamellae*. The secondary lamellae have thin epithelial walls separated by specialized *pillar* or *pilaster cells*. In life the filaments are splayed out in the form of a V by the elasticity of the gill rays, and the tips of the filaments of adjacent rays are in close contact so that the path of water across the gills is mainly between the secondary lamellae. The presence of filaments and lamellae greatly increases the gill surface available for gaseous exchange. Such surface areas are correlated with the habits of the fish. For example, slow swimmers have gill lamellae that measure about 151 mm$^2$ per gram of body weight, whereas highly active swimmers may have a value as high as 1,240 mm$^2$ per gram of body weight.

A gill arch with lamellae on both anterior and posterior faces is a *holobranch;* that with lamellae on one face only is a *hemibranch* (fig. 10.14). In sharks, which usually have six gill slits, the last gill is a hemibranch since it lacks lamellae on its posterior surface. The hyoid gill, which is invariably reduced in size to form a spiracle, is also a hemibranch. In addition the spiracle receives oxygenated blood from a vessel that has already passed through a gill; this is therefore not usually considered a "true gill" and is termed a *pseudobranch*. In many teleosts the pseudobranch is totally without lamellae and becomes a glandular structure. This gland is concerned with the production of the enzyme *carbonic anhydrase,* which is important in the release of gases into the swim bladder. The pseudobranch may also be involved in salt excretion and the production or activation of a hormone, since experimental removal of the gland may cause the skin to darken. In some of the fast-swimming elasmobranchs the spiracle is much reduced in size, but in the bottom-living skates and rays it is relatively enormous. In this case it functions as an entrance for water passing into the pharynx, since the gill slits and the mouth are situated ventrally. In the close relatives of the elasmobranchs, the chimaeras, the spiracle and the last gill pouch are absent. In the teleosts the gills are covered externally by a bony *operculum,* and the gill septa and flaps found in the elasmobranchs are absent. The development of the operculum in the bony fishes is paral-

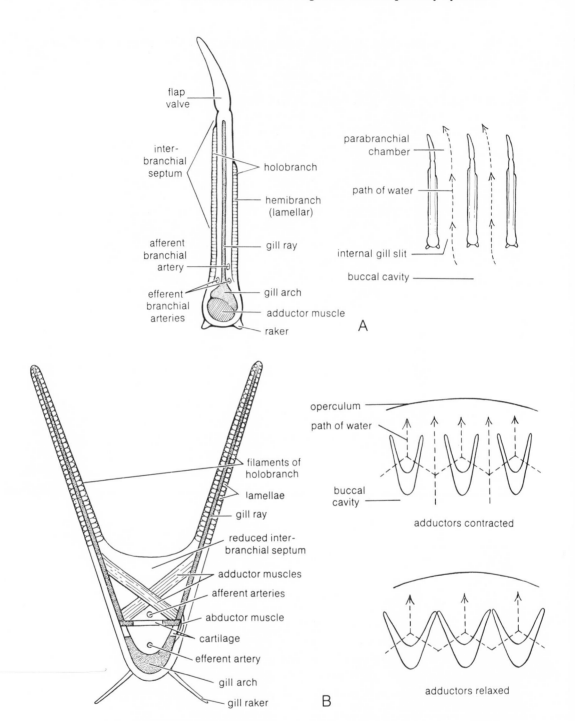

Fig. 10.14. Diagrammatic gill structure sections through gill of an elasmobranch (*A*) and a teleost (*B*).

leled in the chimeras, which also have a covering protecting the opening of the gills. In the lungfishes the gills are reduced in size and in some cases the blood vessels pass through the gill without breaking up into capillaries. The gills in many fishes are used as an area for the removal of salts and ammonia.

It has long been thought that the ventilation of the gills is accomplished by a *single pump mechanism* that forces water across the resistance produced by the gills. The whole of the buccal cavity and the opercular cavity expand, drawing water through the mouth. The water is then forced across the gills and out of the operculum by the almost simultaneous contraction of the buccal and opercular chambers. However, recent investigations have shown that the mechanism of gill ventilation in the teleosts is best considered as a *double pump mechanism* that ensures that the flow of water across the gills is almost continuous despite the intermittent entry and exit of water. Pressure measurements show that a gradient from the buccal cavity to the opercular cavity is maintained throughout the respiratory cycle, except for a very short period when some reversal of flow may take place. The *buccal pressure pump* forces water across the gills, whereas the opercular chamber acts as a *suction pump,* and together they maintain an almost continuous flow. The pump mechanisms work in conjunction with valves that guard the mouth and the opercular opening. These valves are passive and move relative to the pressure gradient against them. When the mouth opens, thin flaps of skin projecting from behind the upper and lower jaw are displaced inward. As the buccal cavity begins to decrease in volume, the increase in pressure causes the valve to close and prevents water from leaving the mouth. The outer rim of the operculum has a thin sheet of tissue that comes into contact with the posterior border of the opercular cavity as it expands and thus prevents the entry of water. Refer to figure 10.15 and compare with figure 10.14*B.*

In sharks the gill slits are covered externally by projections of the gill septa that form a series of flap valves, and a series of *parabranchial cavities* are thus formed external to the gills. Water enters the mouth

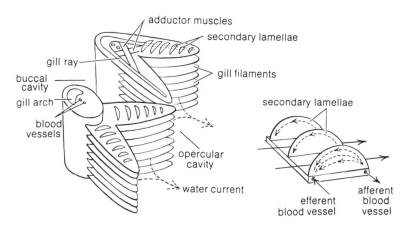

Fig. 10.15. Gill filaments in teleosts (after Hughes 1965): gill filaments and associated structures on two gill arches (*left*) and countercurrent flow of blood and water (*right*).

when the volume of the orobranchial cavity is increased by lateral expansion and the lowering of the floor. These movements are accomplished largely by the elastic recoil of the head skeleton, and the hypobranchial muscles are apparently used when hyperventilation is required and when the branchial region expands before biting. The valves on the gill septa are closed and water enters the mouth and spiracles. It has been demonstrated that the water entering the mouth leaves largely by the more posterior gill slits and that entering the spiracle leaves by the more anterior slits (fig. 10.16). When the volume of the buccal cavity is de-

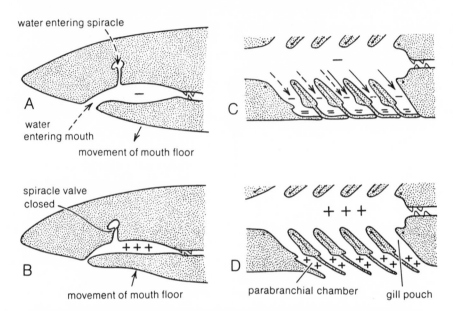

Fig. 10.16. Gill ventilation in the dogfish. Lateral views *A* and *B;* frontal sections *C* and *D.* Relative pressures indicated by + and −; solid arrows indicate flow of water entering the mouth and broken arrows mark the wakes entering spiracle. After Hughes, in Walker 1965.

creased by the action of the constrictor muscles, water is forced across the gill filaments, effecting the *pressure pump.* The *suction pump* is due to changes in the volume of the separate *parabranchial cavities,* and the fall in pressure within them is greater than that within the buccal cavity as it expands. Therefore, as in the teleosts, a differential pressure gradient across the gills is present throughout the whole of the respiratory cycle and results in the almost continuous passage of water across the gills.

Some fishes do not make active respiratory movements but rely on a current of water entering the mouth as a result of forward movement. This mechanism is found in some sharks and in teleosts such as the mackerel (*Scomber*). The secondary lamellae of the gills are so arranged that the direction of blood flow is the *reverse* of that of water. This arrangement, known as a *countercurrent,* facilitates the exchange of gases between the blood and the water. If the blood and water flow in the same direction (*concurrent*), then the oxygen in the blood can reach only tensions approximating those of the expired water. In a coun-

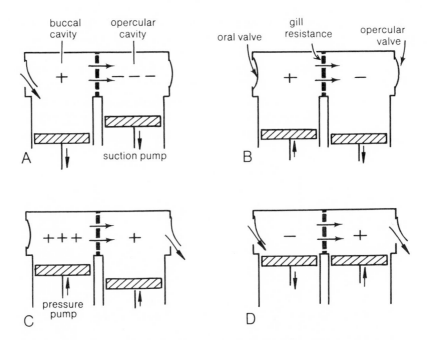

Fig. 10.17.  Gill ventilation in fishes—double-pump mechanism. Water passes almost continuously across the gills (*A, B,* and *C*) in phase *D* (which is very short). Some reversal of flow may occur. Hughes 1965.

tercurrent system much higher blood oxygen tensions can be achieved since the blood can come to equilibrium with the high oxygen tension of the inspired water.

**2. Swim Bladders and Lungs**

The lungs in gnathostomes arise as a central invagination from the floor of the pharynx behind the gill region. The sac formed soon divides into two halves that grow posteriorly while still retaining their contact with the pharynx. It is probable that the lungs have evolved from a pair of posterior gill pouches. A number of fishes such as *Polypterus* have ventral lung-like structures, while in other forms such as the lungfishes the similar organs are placed dorsally. Many fishes have a dorsally placed swim bladder, which may retain a connection with the pharynx via a *pneumatic duct*. This condition is termed *physostomatous (physostomous)*. In the *physocleistous* state the pneumatic duct is lost and hence the swim bladder is a separate structure having no connection with the pharynx.

The evolutionary relationship between the lungs and swim bladder has been a matter of some controversy, and it has often been supposed that since the lungs are essentially a tetrapod structure they must have evolved from the swim bladder. This seems unlikely, since lungs are very ancient structures and were probably present in some placoderms (Devonian fishes) as ventral outgrowths of the pharynx. Paleontological evidence suggests that lungs evolved in vertebrate groups inhabiting environments that were prone to periods of drought, and indeed modern lungfishes survive in areas where there is considerable likelihood of drought and water stagnation.

The development of lungs—however they have arisen in the course of evolution—presents physiological problems related to the concentration of carbon dioxide in the surroundings. The blood of modern fishes that live in fast-moving streams, and also that of marine fish, is extremely sensitive to changes in the concentration of carbon dioxide. So great is the effect of high concentrations of carbon dioxide that the blood will not reach an effective level of saturation with oxygen even in water that is fully oxygenated (fig. 10.18). It therefore follows that invasion of,

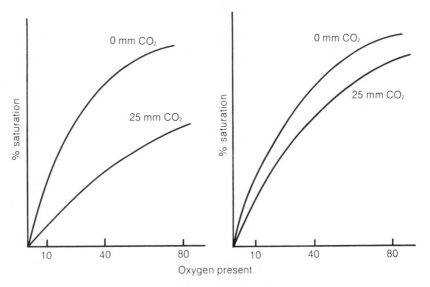

Fig. 10.18. Dissociation curves for river and swamp fish in the presence and absence of carbon dioxide (river fish, *Myleus setiger* [Paku]; swamp fish, *Electrophorus electricus*). After Willmer and Carter, in Lawson 1964.

or survival in, a habitat with a high level of carbon dioxide might be associated with the evolution of blood that became less sensitive to carbon dioxide. It has frequently been demonstrated that in lungs a complete exchange of gases rarely, if ever, takes place, and as much as 7% of carbon dioxide is found in human lungs. The value for fish lungs is about 3%. It is also known that fish that inhabit areas of high carbon dioxide concentration have developed blood that is relatively insensitive to carbon dioxide. In view of these facts it is possible that this kind of physiological change also accompanied the evolution of lungs in the Devonian. Thus, on the basis of paleontological evidence and evidence from living forms, it seems that both lungs and swim bladders have evolved from a pair of ventral pharyngeal diverticulae that arose early in the history of the vertebrates. There are, however, certain anatomical objections to this hypothesis, the chief one being that the swim bladder is a dorsal structure, frequently having an attachment to the pharynx, whereas the lungs are ventral and connected to the ventral portion of the pharynx. In *Erythrinus* (Characoidea) the swim bladder is connected to the pharynx laterally and is supplied with some of its blood by a branch of the coeliac artery that arises *below* it. In addition, the lungs in the lungfishes, although they are situated dorsally, are supplied with blood by vessels that arise from the last afferent branchial artery, pass below the gut, and then run dorsally to the lung. Thus it is apparent that lungs evolved as a ventral structure that in some forms migrated dorsally and either retained its respiratory function (as in the lungfish) (fig. 10.19) or became a swim bladder.

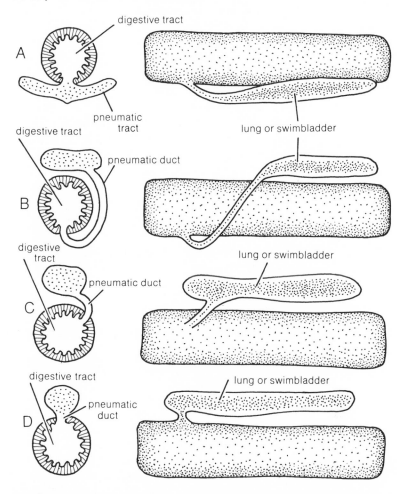

Fig. 10.19. Diagrammatic representation of lungs and swim bladders in various fish: *A, Polypterus; B, Ceratodus; C, Erythrinus; D,* sturgeons and most teleosts. After Goodrich, in Lawson 1964.

In those fishes where the dorsally placed diverticulum does not function as a lung, it assumes the function of a hydrostatic organ that alters the overall specific gravity of the animal in response to changes in the density of the environment. Thus the fish may attain *neutral buoyancy* and remain motionless in midwater without sinking or rising. Changes in the overall density of the fish are produced by varying the amounts of gases in the swim bladder. The gases involved are mainly oxygen, carbon dioxide, and nitrogen. The secretion of gases into the bladder usually involves a special area known as the *gas gland,* which is highly vascular owing to a mass of blood vessels, the *rete mirabilia.* During secretion there is considerable vasodilation in the rete, and the first gas entering the bladder is usually carbon dioxide produced from the breakdown of glycogen. The release of carbon dioxide from the blood is facilitated by the enzyme *carbonic anhydrase,* found in high concentration in the gas gland. Some of the carbonic anhydrase is produced by the pseudobranch (modified spiracular pouch) and carried by the blood to the gas gland. The general increase in level of carbon dioxide in the secretory area of the swim bladder causes a sharp increase in dissociation of oxygen from the blood into the swim bladder. The reabsorption of gases from the swim bladder is a passive process, since the concentration of gases is higher in the bladder than in the blood. In some cases a reabsorption chamber or *oval* is separated from the main chamber of the bladder, from which it may be closed off by a sphincter muscle. The swim bladder is best developed in active midwater fishes and may be lost in bottom-dwelling forms. In deepwater fishes the bladder is often converted into a fat store. In some fishes the swim bladder may perform functions that are not related to the regulation of density. In the group of fishes known as the Ostariophysi the swim bladder is linked to the ear region by a series of bones termed the *Weberian ossicles.* The bladder responds to vibrations that are then transmitted to the ear by the ossicles. In the family Balistidae the bladder acts as a resonating chamber for sounds produced by the stridulation of clavicles and the postcleithrum and the grating of the pharyngeal teeth. In other cases the wall of the swim bladder may be vibrated by special muscles to produce sounds, while some forms produce noises by releasing bubbles of gas down the pneumatic duct.

Primitively the lungs are ventral sac-like structures connected to the pharynx by a short tube or *trachea,* which opens into the *glottis.* In *Polypterus* the inner surface of the lung is relatively smooth except for a few small furrows. In *Lepisosteus* and *Amia* the surface is more complex and thrown into a series of folds. This is also true of the lungfishes. The lungs in amphibians may have smooth walls or in some forms may have a number of ridges forming shallow pockets. In reptiles the inner surface of the lung is more complex, and the lung reaches its maximum complexity in the mammals and birds. In these groups the trachea is very long and leads into a pair of *primary bronchi.* These lead to the lungs, where they subdivide into secondary bronchi, then further subdivide to form the *bronchial tree.* The fine "twigs" of the tree are the respiratory bronchioles, which lead into the *alveoli* in mammals. In birds, the alveoli are represented by tubes, the *air capillaries,* that arise from tubular *parabronchi.* Each primary bronchus is continued through the lung as a single *mesobronchus,* each of which gives rise to a cluster of up to sixteen *posterior secondary bronchi,* then terminates in an abdominal air sac. Each posterior secondary bronchus communicates with a large number of *tertiary bronchi* or *parabronchi.* In fact, the bulk of each lung is made up of a relatively solid mass of parabronchi that branch and form a system of through-tubes. The close packing of the parabronchi is facilitated by their hexagonal shape, and each tube may measure 1 mm across, with a bore of 0.5 mm. Opening into the bore are a large number of finely branching, blind tubes to *air capillaries.* These capillaries are profusely supplied with blood, and here respiratory exchange takes place as it does in the alveoli of the mammalian lung. The intrapulmonary circuit is completed by the eventual fusion of the parabronchi into a small number of *anterior secondary bronchi,* which in turn, along with the anterior air sacs, open into the anterior position of each *mesobronchus.* In birds the lungs thus consist of a complex system of through-tubes. The lungs adhere to the ribs and are relatively inelastic compared with those of mammals. Associated with the lungs of birds are a number of air sacs (fig. 10.20), which are often connected to cavities in adjacent bones such as the *humerus, sternum, coracoid, furcula, scapula, ribs,* and *pelvis,* which are hence known as *pneumatic bones.*

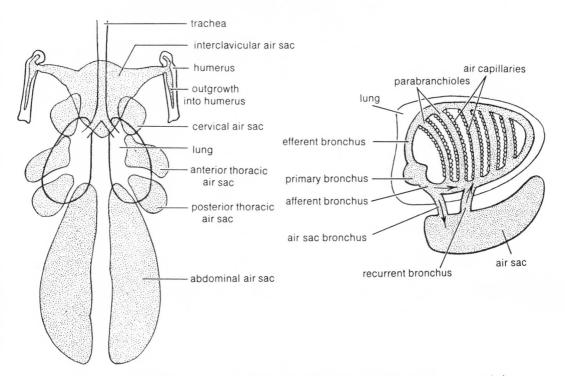

Fig. 10.20. Avian air sacs (after Goodrich, in Smith 1960): *A,* ventral view of air sacs and lungs; *B,* cross section of air sac on one side of the body.

Ventilation of the lungs in amphibians such as the frog largely involves movements of the floor of the buccal cavity. Air is habitually moved into the buccal cavity, with the glottis closed, by relatively small movements of the buccal floor. These movements of air are now believed to be primarily associated with sampling the air and hence to be "smelling" movements. Less frequent ventilatory movements occur when the glottis opens after the floor of the buccal cavity has been depressed. Recent work has shown that the stream of air into the buccal cavity from the lungs does not mix appreciably with that already drawn in via the nostrils. The nostrils are then closed, the subbuccal muscles contract, and the air is forced into the lungs. The glottis closes near the peak of the buccal pressure, leaving the air in the lungs substantially above atmospheric pressure. This type of ventilation mechanism is usually termed the "force pump" method. The evolution of the force pump method of respiration in the Amphibia and its evolutionary relationship with the irrigational mechanism present in fish has aroused considerable specu-lation. It has often been suggested that the force pump method is a secondary feature of the Amphibia and is associated with the loss or re-duction of ribs that has accompanied their evolution. However, recent experiments on ventilation in the lungfish, *Protopterus,* suggests that there are definite similarities between its mode of respiration and that of mod-ern amphibians. In particular they both possess the unusual feature of inhalation into the buccal cavity *before* expiration from the lung (see above), except that in *Protopterus* the air is taken directly into the mouth rather than entering by the nostrils. The experiments on *Protop-terus* also suggest that the amphibian ventilatory mechanism could have evolved, with a few minor changes, from gill irrigation cycles basically similar to those of elasmobranchs and teleosts. The importance of these observations is that they suggest that in the change from aquatic to aerial respiration the machinery was already available and necessitated only small changes in the buccal-opercular pumping mechanism.

In reptiles the force pump method of respiration is replaced by *costal respiration,* which involves the use of the ribs. This method is also used to some extent in birds, where expansion of the thoracic cage is produced by the action of the *intercostal muscles,* which move the ribs forward and outward, while the sternum is slightly depressed. With the conse-quent reduction of the pressure in the body cavity and the pleural cavity, air is drawn into the system. Expiration results mainly from the relaxation of the intercostal muscles but there is evidence that it is aided by the active contraction of the *internal intercostals* and the *costopulmonary muscles.* The latter run from the ribs to a membrane below the lungs which is derived from peritoneum and is the *pulmonary aponeurosis.* During flight the rhythmic raising and lowering of the sternum by the flight muscles ventilates the lungs. The lining of the air sacs in birds is generally smooth, and the sacs are of little value as respiratory surfaces. But they are important in the circulation of air through the lungs, since they frequently exhibit a bellows-like action (produced by the con-traction of surrounding muscles) that forces the air through the lungs. The important feature of the lungs in birds is that they have a straight-through circulation of air and are not composed of dead-end alveoli as they are in other vertebrates.

It is perhaps surprising to learn that there is no general agreement on the precise nature of the air pathways within the lungs of birds, since it is not yet clear whether the flow of air through the parabronchioles is uni-directional or whether the direction is reversed during expiration and inspiration. However, it is clear that the combined ventilation of the air sacs and lungs in birds enables them to achieve high values for oxygen uptake. The exceptional nature of these values is further emphasized when it is remembered that the lungs in birds are relatively small. For example, the ratio of the lung area (cm$^2$) to body weight (g) in the pigeon is approximately 0.5, whereas in the frog the figure approaches 8.5 and in man it is about 11.0.

In mammals the *ribs, intercostal muscles, abdominal muscles,* and *diaphragm* all aid in respiration. Contraction of the diaphragm and the *external intercostal  muscles* enlarges the *pleural cavity,* and air enters the lungs. Relaxation of the diaphragm and the rib muscles causes expiration, which is therefore passive. A forced expiration may be produced by the contraction of the *abdominal muscles* and the internal intercostals.

The extreme complexity of the lung in mammals and the resulting high ratio of lung volume to body weight (see above) is also characterized by extensive air passages that are not concerned with gaseous exchange. Such spaces are referred to as the *dead space*. In man the dead space accounts for an appreciable proportion (about 28%) of the air brought into the lungs at a single inspiration when the subject is at rest. This volume of air (*resting tidal volume*) is approximately 500 ml, and hence the volume of the dead space is approximately 140 ml. However, increased demands for oxygen may invoke respiratory movements that produce inspiratory volumes of up to 3.6 liters, and hence the value for the dead space becomes insignificant.

The development of compartmentalized lungs, especially the alveolar lungs of mammals, produces physical problems related to the surface tension at the gas/fluid interface within the lungs. Such tensions tend to make the lung more difficult to inflate and more likely to collapse when deflated. It has been shown that these problems are largely overcome by the presence of *pulmonary surfactants* that aid inflation of the lung and prevent its collapse when deflated. These substances, which are proteins combined with lipid (lipoprotein), occur widely and have been isolated from the lungs of mammals, birds, amphibians, reptiles, and lungfish.

In a number of vertebrates the lungs may be of unequal size or one may be absent. In *Polypterus* the right lung is smaller than the left, and in snakes the right lung is very large whereas the left lung is rudimentary. This condition is also found in most caecilians and is undoubtedly related to the elongate nature of the body in this group, which parallels that of the snakes. Many vertebrates make use of accessory respiratory areas in addition to the gills or the lungs. The catfish *Clarias* uses the epithelium of the branchial cavity, the climbing perch *Anabas* a special subopercular cavity, and the teleost, *Plectostomus,* the stomach. The amphibians use the skin as a respiratory area and the plethodontids (lungless salamanders) rely entirely on cutaneous respiration. The uptake of oxygen in the buccal cavity of amphibians (*buccal respiration*) has always been considered of considerable importance. Recent evidence suggests that this is not so and that the movement of air into the buccal cavity is largely a sampling process usually involving a very small gaseous exchange. In the plethodontids, however, the buccal respiration may be relatively important. Bucco-pharyngeal respiration is also important in some aquatic turtles, and in others irrigation of the cloaca accounts for considerable gaseous exchange.

**E. The Alimentary Canal**

**1. Esophagus**

The pharynx is connected to the stomach by a narrow muscular tube, the *esophagus,* which is lined with *stratified epithelium* similar to that of the buccal cavity. In a number of elasmobranchs, amphibians, and reptiles, part of the esophagus may be ciliated, but in most vertebrates food transport is accomplished by waves of muscular contraction that pass along the esophagus. This process is termed *peristalsis,* and the esophagus contains both striated and unstriated muscle. In some cases the esophagus has accessory structures such as the *crop* or *ingluvies,* a distensible sac for the temporary storage of food. The crop may secrete a milky material that is regurgitated to feed the young. In some mammals the lower portion of the esophagus is modified as part of an elaborate *gastric apparatus* (see below).

2. Stomach

In most vertebrates the stomach is a muscular pouch-like expansion of the foregut lying in the anterior part of the abdominal cavity. Internally the stomach may be divided into a number of regions. The *esophageal region*—the proximal end of the stomach—is the continuation of the esophagus and is lined with nonglandular stratified epithelium similar to that of the esophagus. This is followed by the *cardiac region,* lined with columnar epithelium and nondigestive glands. The *fundus* or *body* of the stomach also has columnar epithelium and carries the digestive glands, while the *pylorus,* which is again nondigestive, is the distal portion of the stomach leading into the *small intestine.* The regions of the stomach may be distinguished on histological grounds, but there is a good deal of variation and one or more regions may be absent. There is no close correlation between the external form of the stomach and the internal structure. Further, there is considerable variation in the gross form of the stomach, and it is evident that the stomach evolved primarily for food storage and assumed its digestive function secondarily. In some forms such as *Branchiostoma* and the cyclostomes there is no distinct stomach, whereas in birds the stomach is divided into a glandular *proventriculus* and a muscular portion, the *gizzard.* The gizzard has a hard, cornified lining, and in the grain-eating birds it often houses small stones and is used as a grinding mill. The most pronounced modifications of the stomach are seen in the ungulates, especially the ruminants, or cud-chewing forms. In the pigs the base of the esophagus is swollen to form the *"forestomach"*, a small saccular outgrowth. In horses and tapirs the forestomach is relatively larger and in the hippopotamus it is clearly separated from the rest of the stomach and has a pair of large diverticula. In camels the forestomach is divided into a large *rumen* or *paunch* and a smaller *reticulum* or *honeycomb.* In these chambers part of the wall is partitioned into a number of water-storing cavities, the *water cells.* In typical ruminants such as sheep and cattle the stomach is elaborate and consists of four chambers (fig. 10.21). The *rumen* is the largest compartment and may hold up to 300 liters. The *reticulum* is a smaller chamber whose walls have a series of hexagonal folds (*reticular folds*) and is followed by the *omasum,* which is composed of a large number of parallel internal ridges. The final chamber is the *abomasum,* which leads into the small intestine. It is evident that the abomasum is the *true stomach,* since it carries the characteristic gastric epithelia and produces the digestive enzymes, whereas the other chambers are essentially elaborations of the base of the esophagus. In ruminants after a minimum of chewing the food is passed into the rumen, along with copious amounts of saliva. Within the chamber the food is mixed with the saliva and *microfauna* that ferment the food. The fermentation process produces gases, especially carbon dioxide and methane, which are drawn back up the esophagus

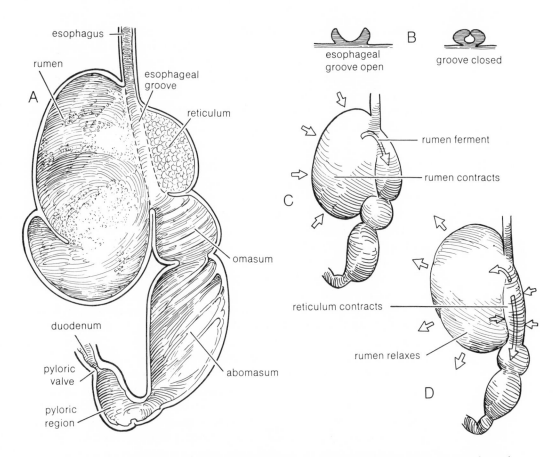

Fig. 10.21. Stomach of sheep: *A*, general view of chambers; *B*, esophageal groove; *C*, *D*, emptying of rumen.

together with a quantity of food, which is then chewed and swallowed again, along with more saliva. This process of rumination and chewing the cud may take place even when the animal is asleep. The rumen periodically contracts and some of its contents are passed into the reticulum, which in turn contracts and forces some of the food back into the rumen and some into the omasum. By this process the rumen is gradually emptied. In the omasum a large portion of fluid is removed from the partially digested food, which then passes on to the abomasum to encounter the enzymes produced by this chamber. The rumen produces no digestive enzymes, and the digestion that occurs within this chamber is accomplished by the microfauna, a complex mixture of bacteria and protozoa able to break down the plant material that forms the ruminant's food. The cellulose and the plant starch are converted to sugars and eventually to fatty acids, which are absorbed by the wall of the rumen. The plant protein is largely converted to ammonia, most of which is used by the microfauna to produce microbial protein. This protein is released as the microbial cells are themselves digested in the abomasum and intestine. The rumen never empties completely but always retains a small reservoir of microbial cells that multiply rapidly in an almost ideal environment, when the rumen next fills with food. In many ruminants an *esophageal groove* is present that remains open during grazing but may be closed when drinking to allow water to pass directly to the omasum.

The presence of an elaborate stomach in the ruminants is related to the inability of mammals to digest cellulose, which is broken down by the microfauna that inhabit the rumen. In other herbivorous mammals such as the horse and rabbit this role is largely taken over by the *caecum* (see next paragraph).

3. Intestine

In mammals the gut posterior to the stomach may be divided into a number of areas that are relatively distinct. The stomach, whose exit is guarded by a *pyloric sphincter muscle,* leads into the *small intestine,* which in some cases may be composed of three portions—the *duodenum, jejunum,* and *ileum.* In many cases, however, the distinctions between them are not well marked and the terms may be applied in a very arbitrary manner. The small intestine in mammals is coiled and its length usually averages seven to eight times the length of the animal, though it is often much greater than this in herbivorous forms. In most mammals the small intestine leads, by a valve, into the *large intestine* or *colon,* which is a tube of considerable diameter, usually with a series of outpocketings and pronounced longitudinal muscle bands. The proximal portion of the large intestine is the *ascending colon,* which runs anteriorly along the right side of the body. After a flexure the *descending colon* runs posteriorly along the left side of the body. In primates a *transverse colon* connects the other two colonic limbs, and in most herbivores the colon may be coiled. At the junction of the small and large intestines there is a diverticulum, the *caecum,* that terminates in a *vermiform appendix.* The caecum varies considerably in size and complexity and in some forms may be vestigial, while in others, such as horses and rodents, it is very large and is important in digestion (see above). The colon leads into the *rectum,* which terminates at the *anus.*

The intestine in other groups of vertebrates is less well differentiated and of much simpler structure. In elasmobranchs the intestine is not clearly divisible into segments but consists of a "*spiral intestine*" that runs along the whole length of the body cavity to terminate in a short *rectal* portion that is supplied by a *rectal gland.* This gland secretes mucus and is also thought to be concerned with salt secretion. The spiral intestine is so named because it contains a *spiral fold (valve),* a fold of epithelium that twists in spiral fashion as it runs along the length of the intestine. The spiral fold is especially well developed in some sharks and it is undoubtedly a primitive feature, since it is also present in lungfishes and primitive actinopterygians. Paleontological evidence suggests that some placoderms had a similar fold. The spiral fold greatly increases the internal surface area of the intestine. In teleosts the intestine is without major internal folds but is usually much longer and coiled. Additional area for digestion and absorption is usually supplied by a number of *pyloric caeca*—diverticula from the proximal portion of the intestine. In amphibians and reptiles there is no trace of a spiral valve, and the gut is of a relatively simple nature. However it is often possible to distinguish a *small intestine* which is separated from the large intestine (colon) by a valvular constriction. The small intestine is usually coiled, and the proximal portion of the large intestine bears a small outpocketing that resembles the caecum of mammals.

4. Cloaca

Primitively the alimentary canal and urogenital ducts open into a common chamber, the cloaca, whose exterior opening is the anus. The cloaca is frequently divided by a ridge into a dorsal chamber, the *coprodeum,* which receives material from the rectum, and a ventral chamber, the *urodeum,* which receives the urogenital ducts. In many fishes and in mammals, with the exception of the monotremes, the gut and the urogenital ducts open separately and there is no cloaca (see chap. 12).

5. Liver

The liver develops as a sacculation from the floor of the embryonic gut immediately posterior to the stomach (fig. 10.22) and remains connected

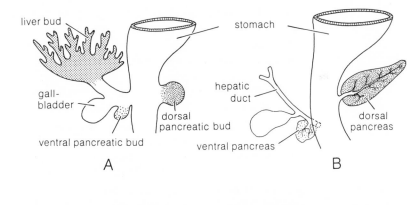

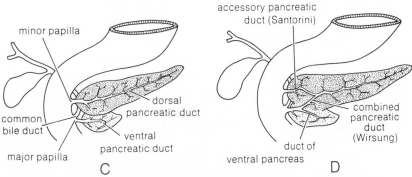

Fig. 10.22. Development of the human pancreas and liver. After Langman 1963.

to the gut by a narrow band of tissue that eventually forms the *bile duct.* The liver develops in the ventral mesentery and remains in contact with the stomach by the *lesser omentum.* Below the liver a persistent portion of the ventral mesentery (the *falciform ligament*) ties it to the body wall. The distal end of the liver diverticulum proliferates into the mass of liver tissue, and the hollow proximal portion forms the common bile duct that receives numerous *hepatic ducts* from the substance of the liver. The *gallbladder* arises as a secondary outgrowth of the liver diverticulum and empties into the bile duct through a *cystic duct*. In some mammals and in birds the gallbladder may be absent, and others may have several bile ducts.

The liver functions in the production of *bile,* which collects in minute *canaliculi* between the liver cells (*hepatic cells*) and is then transferred to larger *bile tubules,* then to the *hepatic duct,* and finally into the *common bile duct.* The liver functions as a storage organ and is the center for a large number of both anabolic and catabolic activities. The liver has a direct blood supply from the arterial system in the *hepatic artery,* but it also receives blood directly from the gut in the *hepatic portal vein* and hence has first call on the absorbed products from the gut.

6. Pancreas

The pancreas is common to all vertebrates and arises as a *dorsal pancreas* from the dorsal side of the gut and a *ventral pancreas* from the primordium of the common bile duct (fig. 10.22). The dorsal and ventral pancreas rudiments may remain separate, as in many fishes, but in the amphibians and the amniotes the two units fuse. The dorsal portion is usually responsible for the greater portion of the adult gland. In man the duct from the ventral pancreas taps the duct of the dorsal pancreas, which then atrophies, and the ventral duct opens into the common bile duct, hence becoming the definitive pancreatic duct. In pigs and cows the dorsal duct rather than the ventral duct remains, and in the horse and dog there are two ducts. The pancreas is a *mixed gland* and has an *exocrine function* involving the production of digestive secretions, or *pancreatic juice,* which is poured into the duodenum. The pancreas is also an endocrine gland and secretes hormones such as *insulin* and *glucagon* into the blood. The hormones are produced by special tissue termed *islet* tissue.

---

**F. The Coelom and the Digestive and Respiratory Systems of Elasmobranchs**

**1. The Body Wall and the Pleuroperitoneal Cavity**

The elasmobranchs illustrate a generalized condition of the coelom and the respiratory system. The heart is far forward beneath the posterior limit of the gill region. The *transverse septum* separates the pericardial and *pleuroperitoneal cavities*.

Directions apply to the spiny dogfish and the skate.

Make an incision from the left side of the cloaca forward through the pelvic girdle slightly to the left of the midventral line up to the pectoral girdle. The incision should cut through the skin first and then extend through the muscle layer. Make a transverse incision on each side to extend from approximately the middle of the longitudinal cut to the lateral line. The four flaps of body wall may then be turned back to reveal the body cavity. In the skate, cut along the left side of the cloaca and then along both lateral borders of the body cavity, but not anteriorly, leaving the flap of body wall adhering to the pectoral girdle. Note the layers of body wall through which you have cut. The outermost is the *skin,* followed by an inconspicuous layer of *connective tissue,* the *hypaxial musculature,* and finally the smooth, shiny lining of the *pleuroperitoneal cavity,* the *parietal peritoneum*.

## 2. The Viscera of the Pleuroperitoneal Cavity

At the anterior end of the cavity is the large, brownish or grayish *liver*. In the spiny dogfish this consists of long *left* and *right lobes* and a small *median lobe,* in which is the long, greenish *gallbladder.* In the skate the liver is composed of *right, median,* and *left lobes* of equal length and size, and the *gallbladder* is situated in the angle between the right and median lobes. Dorsal to the liver on the left side is the large J-shaped *esophagus-stomach,* often distended with food. (In some specimens this is everted into the oral cavity and should be pulled back into the pleuroperitoneal cavity by exerting a gentle traction upon it.) There is no external demarcation between esophagus and stomach, but the anterior part of the organ is esophagus. The stomach continues straight backward from the esophagus to a point somewhat posterior to the caudal ends of the liver lobes; it then makes a sharp bend, decreasing considerably in diameter, and extends anteriorly, terminating in a constriction, the pylorus. Along the posterior margin of the bend of the stomach (or in the skate on the dorsal side of the bend) is a dark-colored, triangular organ, the *spleen.* From the pylorus the short *intestine* extends to the anus. The first part of the intestine beyond the pylorus is called the *duodenum.* It extends for a short distance to the right and then curves posteriorly. The long, stout *bile duct* is easily seen descending from the gallbladder to enter the duodenum shortly caudad of the bend. The bile duct, accompanied by some blood vessels, runs in a strip of mesentery. It passes to the dorsal side of the duodenal wall and runs for a short distance caudad, embedded in the wall, before it penetrates the cavity of the duodenum. In the curve of the duodenum lies the ventral lobe of a white gland, the *pancreas.* The *dorsal* lobe of the pancreas is long and slender in the spiny dogfish, reaching to the spleen, and should be located by raising the stomach and duodenum. The *duct of the pancreas* is somewhat difficult to find in the dogfish, less difficult in the skate; it lies embedded in the tissue of the pancreas near the posterior margin of the ventral lobe and may be exposed by picking away the pancreas tissue in this region. Beyond the duodenum the intestine widens considerably, and its surface is marked by parallel rings. These rings are the lines of attachment of a spiral fold, the *spiral valve,* which occupies the interior of the intestine. (A portion of the intestine often protrudes through the anus and should be drawn back into the coelom by grasping the portion in the cavity and exerting a gentle pull.) The part of the small intestine occupied by the spiral valve is called the *valvular intestine* or *spiral intestine.* Caudal to this, the narrower large intestine or *colon* proceeds to the *anus.* Attached to the colon by a duct is a small cylindrical body, the *rectal* or *digitiform gland,* which secretes mucus and salt. The terminal chamber of the colon is the *cloaca,* opening by the anus.

Cut open the esophagus-stomach and wash out its contents. Partly disintegrated squid and fish are commonly found in the organ. The anterior part, having projecting *papillae,* is the esophagus; the remainder, with lengthwise folds or *rugae,* is the stomach. Cut open the small intestine along one side midway between the large blood vessels that run lengthwise along its wall and observe the spiral valve in the interior. It consists of a fold of the intestinal wall spirally coiled so as to make a series of overlapping cones; it serves to increase the digestive and absorptive surface of the intestine.

The reproductive organs and their ducts may also be identified in part at this time. By raising and moving apart the lobes of the liver it is possible to see the *gonads*—soft bodies lying adjacent to the dorsal body wall. In mature females the oviducts are prominent structures on each side against the dorsal coelomic wall.

**3. The Mesenteries**

The viscera are held in place by delicate membranes, the *mesenteries,* whose mode of origin was explained in the introduction to this section. In studying them, lift and spread each organ as it is mentioned. The *dorsal mesentery* extends from the median dorsal line of the coelom to the digestive tract but is incomplete because there is a gap in the region of the small intestine. The portion of the dorsal mesentery supporting the stomach is called the *mesogaster;* in the skate it is limited to the anterior part of the stomach. The mesogaster encloses the spleen between its two walls, and that portion of the mesogaster from the spleen to the stomach is the *gastrosplenic ligament.* The area of the dorsal mesentery that supports the intestine is referred to simply as the *mesentery.* This is absent in the skate. In the spiny dogfish there is a fusion between the mesentery and the mesogaster, so that a pocket is formed dorsal to the bend of the stomach. In the dorsal wall of this pocket is situated the greater part of the pancreas. The dorsal mesentery begins again in the region of the rectal gland, this portion of the mesentery being named the *mesorectum.*

The ventral mesentery is represented in these animals, as in all vertebrates, by remnants only. Such a remnant is the *gastrohepato-duodenal ligament* extending from the right side of the stomach to the liver and duodenum, also called the *lesser omentum.* It may be roughly divided into two portions: the *hepatoduodenal ligament,* from the liver to the duodenum and containing the bile duct and blood vessels; and the *gastrohepatic ligament,* extending from the stomach to the liver and duodenum and in dogfishes occupying also the angle formed by the bend of the stomach. Another remnant of the ventral mesentery is the *suspensory* or *falciform ligament* of the liver, found at the anterior end of the liver, extending from the ventral surface of the liver to the midventral line of the body wall. In mature females the *mouth of the oviduct* will be noticed in the falciform ligament as a funnel-shaped aperture. After seeing the falciform ligament, you may cut off the flap of body wall left in the skate.

The anterior end of the pleuroperitoneal cavity is closed by a partition, the *transverse septum,* whose posterior face is clothed by the *parietal peritoneum.* The liver is attached to the septum by the strong *coronary ligament,* which is in fact a portion of the septum. In its early development the liver is enclosed in the transverse septum and subsequently, because of increased size, projects posteriorly from the septum, which then narrows around the anterior end of the liver and forms the coronary ligament.

The pleuroperitoneal cavity communicates with the exterior by means of the *abdominal pores*. These will be found one on each side of the anal opening (in the skate, posterior to the anus), somewhat concealed by a fold of skin. Probe into them and note that they lead into the pleuroperitoneal cavity. Their purpose is obscure.

**4. The Peri-cardial Cavity**

With forceps gently strip off the layers of hypobranchial muscle on the ventral side of the throat in front of the pectoral girdle until you have exposed a membrane, the *parietal pericardium*. Slit it open and see that it encloses a cavity, the *pericardial cavity,* in which the heart is situated. To reveal this cavity more fully, cut laterally along the anterior face of the girdle, keeping your instrument in contact with the girdle. Portions of the girdle may be sliced away, but the heart must not be injured. The pericardial cavity is thus revealed as a conical cavity lined by the parietal pericardium and containing the heart. By gently lifting up the heart note that it is attached at its anterior and posterior ends. At these places the pericardial lining is deflected from the walls of the pericardial cavity and passes over the surface of the heart as a covering layer, the *visceral pericardium,* which is indistinguishably fused with the heart wall. With the heart lifted, note that the posterior end of the heart is a fan-shaped chamber, the *sinus venosus,* whose walls are continuous with the partition that forms the posterior wall of the pericardial cavity. This partition is the *transverse septum,* whose posterior face you have already seen. The septum is thus seen to be a partition whose anterior wall is composed of parietal pericardium and the posterior wall, of parietal peritoneum.

**5. The Oral Cavity, Pharynx, and Respiratory System**

Insert one blade of a scissors into the left corner of the mouth and make a cut through the angle of the jaws across the ventral parts of the gill slits through the pectoral girdle, so that you emerge to the left side of the stomach. A flap is thus formed, which should be turned over to the right. A large cavity is revealed that at its posterior end converges into the esophagus, which may be slightly slit to aid in opening the flap.

The anterior part of the cavity enclosed by the jaws and gill arches is the *buccal* or *oral cavity*. On the floor of the mouth lies a *primary tongue,* a flat, practically immovable projection. It is surrounded by the *second* or *hyoid gill arch,* which should be felt within it and identified on the cut surfaces. The skate lacks a tongue because of the reduction of the hyoid arch.

The posterior and greater part of the cavity is the *pharynx,* whose wall is perforated by six *internal gill slits*. The most anterior is the spiracle, a rounded opening in the roof of the pharynx immediately posterior to the jaws. The remaining five slits are elongate, and their openings into the pharynx are protected by a series of papillae-like projections, the *rakers,* that prevent particles from entering the gills.

The internal gill slits open into large cavities, the *gill* or *branchial pouches,* which in turn open onto the surface by a series of *external gill slits.* The tissue between successive gill pouches is the *intra-branchial septum,* at the base of which cross sections of the main gill skeleton the *gill arch* are visible. The *gill rays* extend from the gill arch into the *intrabranchial septum,* which is extended laterally to form a flap-like structure that forms a *valve* covering the opening of the gills. The portions of the gill apparatus between the extensions of the intrabranchial septa are the *parabranchial chambers,* which are important in maintaining the respiratory currents during respiration. The gill pouches are covered with a number of plate-like *primary lamellae.* Examine these with a hand lens and note that they have a series of closely packed *secondary lamellae* extending perpendicu-larly from the surface. The lamellae on one face of an intrabranchial septum constitute a *hemibranch,* and those on both sides of the same septum form a *holobranch.* Much of each intrabranchial septum is composed of the *branchial muscles.* A large *adductor muscle* lies medial to each arch, and an *intrabranchial muscle* and *superficial constrictor muscle* extend into the septum. Close examination will also reveal several blood vessels. An *afferent branchial artery* lies near the middle of the septum just lateral to the gill arch, and an *efferent branchial* lies at the base of the primary lamellae on each side. Note the fine branches of these vessels in the lamellae.

Examine the external opening of the spiracle. Small parallel ridges that represent a reduced gill or *pseudobranch* will be seen on a fold of tissue that is separated from the anterior wall of the spiracular passage by a deep recess. The fold of tissue is a *spiracular valve* that may be closed to prevent water from leaving the spiracle.

**G. The Respira-tory and Digestive Systems of a Bony Fish**

The bony fishes show interesting adaptive modifications of piscine structure, especially in the gill apparatus.

**1.  Contents of the Pleuroperi-toneal Cavity**

The coelom of bony fishes, like that of elasmobranchs, is divided into an anterior pericardial cavity and a posterior pleuroperitoneal cavity, and the two are separated by a transverse septum. Open the pleuroperitoneal cavity by a lengthwise slit from anus to pectoral girdle and enlarge, if necessary, by lateral cuts. Identify liver, esophagus, stomach, and pylorus. Esophagus and stomach are not externally delimited. The stomach varies in form in different teleosts and may simply be U-shaped or may have a sac-like posterior pouch. Note finger-like projections, the *pyloric caeca,* encircling the duodenum. There is no definite division of the intestine into small and large intestine, but the greater part of it, more or less looped, undoubtedly represents the small intestine. Cut open the intestine: Is there a spiral valve? Observe the apparent absence of a pancreas; it is present as diffuse tissue recognizable only by microscope inspection.

The entire dorsal wall of the pleuroperitoneal cavity is occupied by a large *swim bladder,* readily detected by pressing on the region indicated. Note that it is outside the parietal peritoneum. Slit open the swim bladder; the interior is divided by partitions in some fishes.

The swim bladder is connected by a duct to the dorsal wall of the pharynx or esophagus in some teleosts, but lacks such a duct in others.

**2. Oral Cavity, Pharynx, and Respiratory System**

As noted in the study of the external anatomy, the bony fishes lack external gill slits; instead the gill apparatus is covered by an operculum, supported by thin bones. Below the operculum is the *branchiostegal membrane,* supported by bony rays. Cut away one operculum. Note that each gill arch is bony and that the intrabranchial septum, which in elasmobranchs extends to the surface to bound the external gill slits, has been lost. Note on each arch the *gill rakers,* supported by bone, projecting to the interior, and the gills borne on the outer surface of the arch. Notice the absence of a spiracle and the reduced number of gill arches.

Open the mouth and cut through the angle of the jaws on one side and across the middle of the gill arches, thus opening up the oral cavity and the pharynx. In the roof of the oral cavity study the distribution of teeth, usually borne on the premaxilla, maxilla, vomer, and dentary. Teeth may also be present on the gill arches. At the tip of both jaws behind the teeth there is a thin transverse membrane, the *oral valve,* which prevents water from flowing out of the mouth during respiration. Note the inflexible primary tongue.

**H. The Coelom and the Digestive and Respiratory Systems of Necturus**

These systems in *Necturus* are similar to those of fishes except for the addition of lungs. The pericardial cavity is anteriorly placed, and the transverse septum has a transverse orientation.

Place the specimen in a wax-bottomed dissecting pan, fastening it ventral side up by pins through the legs.

**1. Pleuroperitoneal Cavity**

Make a longitudinal incision through the body wall a little to the left of the midventral line, from the left side of the anus through the pelvic girdle to the pectoral girdle. Spread the incision apart. The large cavity is, as in fishes, the pleuroperitoneal cavity, lined by the parietal peritoneum. The body wall consists of skin, muscle, and peritoneum, as can be seen on the cut surface. In the midventral line on the inner surface of the abdominal wall runs a large vein, the *ventral abdominal vein.*

Examine the viscera. The *liver* is the large greenish or brownish organ occupying the anterior half of the pleuroperitoneal cavity. Its margins are divided into several scallop-like *lobes* by shallow indentations. It is united to the median ventral line by a mesentery, which should not be disturbed at this stage. On raising the left side of the liver, you will see the elongated *stomach* dorsal to the liver. Along the left side of the stomach is the dark colored *spleen.* The long, slender *left lung* lies dorsal to the spleen and terminates some distance posterior to the liver. The *right lung* occupies a similar position on the other side of the body. Follow the stomach posteriorly. It is a straight tube, somewhat shorter than the liver, terminating at a constriction, the *pylorus.* The *small intestine* begins at the pylorus and makes an anterior loop, the duodenum, before entering a number of convolutions. A white gland, the *pancreas,* rests in the anterior loop. Part of the gland is in contact with the intestine, part with the liver and one part, the *tail,* passes dorsal to the stomach and almost reaches the posterior tip of the spleen. In the case of females the intestine is coiled on the ventral surface of the large *ovaries,* on the surface of which the eggs will be noted. (The size of the ovaries varies with the sexual state of the animal.) To each side and dorsal to the ovaries is a large, white, much-coiled tube, the *oviduct.* Trace the intestine, posteriorly pressing the ovaries away from the median line. The small intestine widens near the *anus* into the short *large intestine.* In female specimens it lies between the posterior termination of the two oviducts. At the posterior end of the pleuroperitoneal cavity ventral to the large intestine is a sac, generally collapsed and shriveled—the *urinary bladder*—that is attached to the ventral side of the intestine. That part of the intestine to which the urinary bladder is attached (and into which the ducts of the kidneys and gonads also open) is the *cloaca,* which terminates at the anus. In the male a pair of large elongate *testes* lie in a dorsal position in the anterior intestinal region, and the kidneys lie dorsal and lateral to the testes. Each kidney has a conspicuous duct along its lateral border.

2. The Mesenteries

The digestive tract is attached for most of its length to the mid-dorsal line of the coelom by the *dorsal mesentery.* This should be noted by pressing other organs away from the median line. It is missing in the pyloric region of the stomach. The portion of the dorsal mesentery supporting the stomach is the *mesogaster.* The spleen is enclosed in the mesogaster; that portion of the mesogaster that extends from the spleen to the stomach is the *gastrosplenic ligament.* A considerable portion of the left lung is connected to the mesogaster by a short *pulmonary ligament.* A similar structure supports the right lung, which also has an accessory mesentery, the *hepatocavopulmonary ligament,* passing to the dorsal surface of the liver. The large postcaval vein approaches the liver through the posterior margin of this ligament. That part of the dorsal mesentery supporting the small intestine is the *mesentery,* in the limited sense; that part supporting the large intestine is the *mesocolon.* The ventral mesentery is present only in the region of the liver and urinary bladder. It forms the long mesentery extending between the midventral line of the body wall and the midline of the ventral face of the liver. This *falciform ligament* of the liver contains a number of blood vessels that pass from the ventral body wall into the substance of the liver (where they join the hepatic portal vein). In the free posterior margin of the falciform ligament the ventral abdominal vein crosses from the body wall to the liver. On raising the liver you will see the *gastrohepatic ligament* extending from the anterior part of the stomach to the dorsal face of the liver. In the region of the pancreas the *hepatoduodenal ligament* joins the duodenum and liver and encloses the greater part of the pancreas. The tails of the pancreas, however, are situated in the *mesentery* of the small intestine. The posterior part of the ventral mesentery extends from the *urinary bladder* to the midventral line of the body wall; this is the *median ligament* of the bladder.

The falciform ligament should now be severed, without cutting through the ventral abdominal vein. On raising the right side of the liver you may identify the right lung dorsal to it. The small gallbladder will be seen on the dorsal surface of the right side of the liver. Its duct, surrounded by pancreas tissue, may be readily traced to the duodenum. The pancreas opens into the duodenum by a number of fine ducts, which are usually too small to be seen. The anterior end of the pleuroperitoneal cavity is closed by the *transverse septum.* The liver is attached to this by the *coronary ligament,* which is continuous posteriorly with the falciform ligament.

**3. The Pericardial Cavity**

In *Necturus* the pericardial cavity is farther forward than in most amphibians and occupies an essentially piscine position in front of the pectoral girdle and below the posterior hypobranchial muscles. Carefully remove the muscle from this region to expose a membrane, the *parietal pericardium.* Cut through this membrane, exposing the *pericardial cavity,* in which the heart is situated. Widen the opening into the cavity by cutting laterally along the anterior margin of the pectoral girdle. The muscles between the pericardial cavity and the forelimbs may also be split. On gently raising the heart the posterior wall of the pericardial cavity is seen to be formed by the transverse septum. The transverse septum is pierced by two *hepatic veins,* which extend forward and enter the sinus venosus, the most dorsal chamber of the heart.

**4. The Oral Cavity and the Pharynx**

Open the mouth and cut through the angle of the jaws on each side so that the jaws can be spread widely, carrying your cuts back to the gill arches. The cavity thus exposed consists of an anterior *oral cavity* and a posterior *pharynx.*

The oral cavity is bounded externally by the well-developed lips, within which are the small conical teeth. There are two rows of teeth on the roof of the mouth. External to the last teeth of the posterior row on each side is a slit—an *internal naris,* or internal opening of the nasal passage. Probe one of the *external nares* and note emergence of the probe through the *internal naris.* The floor of the oral cavity bears a single row of teeth; on closing the mouth you will find they fit between the two rows on the roof. Posterior to the teeth is the *tongue,* supported by the strongly developed hyoid arch palpable inside it. The tongue of *Necturus* is similar to that of fishes.

The walls of the pharynx are pierced by two pairs of gill slits; probe through them and note that the probe emerges between the external gills. Note the cartilaginous gill arches supporting the gill slits. The gill pouches, unlike the condition in elasmobranchs, are devoid of gill lamellae, since *Necturus* has only external gills. The pharyngeal cavity narrows posteriorly into the *esophagus,* and on passing a probe into the latter you will find it dorsal to the pericardial cavity and to continue into the *stomach.*

5. The Larynx
and Lungs

Although *Necturus* resembles fishes in retaining the gill apparatus, it also has in its typical primitive state the air-breathing apparatus characteristic of land vertebrates. In the floor of the pharynx, midway between the second gill slits, is a short slit, the *glottis*. Its walls, as you should determine by feeling them with a fine forceps, are stiffened by a pair of delicate cartilages, the *lateral cartilages,* representing reduced gill arches. These are the first of the laryngeal cartilages to appear in the phylogenetic series. The small cavity enclosed between the two lateral cartilages, into which the glottis leads, is the larynx. Cut across the gill slits to the left side, so that the pharyngeal cavity can be opened more widely, and slit the glottis posteriorly. The larynx is thus seen to lead into a narrow, flattened passage, the *trachea*. The posterior end of this is widened and receives two openings; probe each and note emergence of the probe into a lung. Slit open one of the lungs and note the smooth interior, not subdivided into air spaces.

**I. The Coelom, and the Digestive and Respiratory Systems of the Turtle**

No trace remains in the adult of gills and gill slits, although certain derivatives of the gill arches persist. The heart and other viscera have descended posteriorly, so that the esophagus and trachea are elongated, the transverse septum assumes an oblique position, and a considerable part of the pleuroperitoneal cavity is situated dorsal to the heart. This cavity, however, remains undivided, as in fishes and amphibians. The large lungs have become spongy through the partitioning of their walls into air spaces. The respiratory mechanisms are much modified in relation to the presence of the shell.

Remove the plastron by sawing through the bridges on each side, lifting up the plastron and separating it with a scalpel from the surrounding skin and underlying membrane.

1. The Divisions
and Relations of
the Coelom

Removing the plastron exposes a membrane, the *parietal peritoneum,* that covers and conceals the *viscera.* Note that the muscle layer that is normally present between the skin and the peritoneum is completely lacking in the ventral body wall of the turtle, because of the presence of the plastron. The ventral body wall in turtles therefore consists of only two layers; the skin with its contained exoskeleton, and the peritoneum. Because of this circumstance the parietal peritoneum can be easily separated from the inside of the body wall, a procedure that is difficult or impossible in other vertebrates. Note, however, the usual muscles in connection with the girdles and limbs.

In the median line in the anterior part of the parietal peritoneum shortly posterior to the pectoral girdle is situated a triangular membranous sac, the *pericardial sac,* which encloses the heart. It will be noticed that the heart is much more posterior in position than in the fishes and *Necturus*. The membranous sac covering the heart is, as in the dogfish, the *parietal pericardium*. In the turtle it takes the form of an isolated pericardial sac, whereas in fishes and *Necturus* it formed the lining of a chamber surrounded by the body wall. The space between the pericardial sac and the heart is the *pericardial cavity,* a portion of the coelom. The ventral face of the pericardial sac rests, in the natural position, against the internal surface of the plastron, and its dorsal face is fused to the parietal peritoneum. Cut into the ventral wall of the pericardial sac, thus exposing the pericardial cavity and the contained *heart*.

Two conspicuous veins, the *ventral abdominal* veins, run longitudinally in the parietal peritoneum between the pericardial sac and the pelvic girdle. Cut through the peritoneum halfway between the heart and pelvic girdle by a transverse cut that severs both of the abdominal veins. The large cavity thus exposed is the *pleuroperitoneal cavity,* whose walls are lined by the *parietal peritoneum.*

The coelom of the turtle, like that of the fishes and *Necturus,* consists of two parts, a small *pericardial cavity* and a much larger *pleuroperitoneal cavity*. Note that whereas in the lower forms the pericardial cavity is anterior to the pleuroperitoneal cavity and separated from it by the transverse septum, in the turtle the pericardial cavity is ventral to the pleuroperitoneal cavity, and the transverse septum seems to have disappeared. We may explain this change as follows. In its posterior descent the heart must necessarily carry with it the transverse septum and the parietal pericardium. The latter, to move posteriorly, must separate from the body wall. It does this and so becomes an independent sac, the *pericardial sac*. This splitting of the pericardial sac from the body wall is aided by the forward invasion of the pleuroperitoneal cavity. The pericardial sac may be thought of as sliding posteriorly below the ventral wall of the pleuroperitoneal cavity. The pericardial sac thus comes to lie ventral to the anterior part of the pleuroperitoneal cavity. The posterior wall of the pericardial sac is still the anterior face of the transverse septum, the posterior face of the latter, as in lower forms, being placed between the pericardial sac and the liver. Thus the transverse septum in the turtle forms part of the partition between the pericardial and the pleuroperitoneal cavities, and the remainder of the partition is composed of the rest of the parietal pericardium, which is now the pericardial sac.

## 2. The Viscera and Their Mesenteries

With bone scissors cut away the margins of the carapace on each side between fore- and hind limbs so as to gain easy access to the pleuroperitoneal cavity. Masses of greenish yellow fat will be found in various places and may be removed. Lift up the edges of the cut already made in the peritoneum, widening this if necessary, and look inside. Identify in the anterior part of the pleuroperitoneal cavity the large brown *liver* lying on each side of the heart. Posterior to the liver are the coils of the *intestine*. In female specimens the *ovaries*, containing eggs of various sizes, are conspicuous in the lateral and posterior part of the pleuroperitoneal cavity. Running alongside each ovary is the coiled oviduct. Just in front of the pelvic girdle is the large, bilobed *urinary bladder*.

The liver consists of right and left lobes, whose lateral margins curve dorsally to fit the curves of the carapace. The pericardial sac rests in a depression between the two lobes that are united by a narrow bridge passing dorsal to the heart. Posterior to the heart the liver is united to the parietal peritoneum by very short mesenteries, corresponding to the falciform ligaments of other vertebrates. In these mesenteries the ventral abdominal veins leave the peritoneum and pass into the liver. Trace the parietal peritoneum anteriorly from this region. It passes along the dorsal face of the pericardial sac, to which it is inseparably fused. This compound membrane between the heart and liver is the transverse septum, which has assumed an oblique position owing to the descent of the heart. The liver is, as usual, attached to the transverse septum by the *coronary ligament*. Continue to trace the parietal peritoneum to the anterior end of the pleuroperitoneal cavity. On the posterior face of the pectoral girdle it turns dorsally to form the inner lining of the carapace. Similarly trace the parietal peritoneum posteriorly by lifting the posterior cut edge of the membrane. It curves dorsally, following the anterior surface of the pelvic girdle, and passes to the inner surface of the carapace.

Press both lobes of the liver forward against the pectoral girdle and look on the dorsal surface of the liver. The elongated *stomach* will be found curving above the lateral border of the left liver lobe. On following the stomach anteriorly you will find the narrow *esophagus* entering the stomach. The stomach passes along the dorsal surface of the left lobe of the liver, and is attached to the middle of it along its entire length by the short *gastrohepatic ligament*. Opposite the bridge connecting the two lobes of the liver the stomach passes into the *small intestine,* the first part of which is the *duodenum,* united to the middle of the dorsal surface of the right lobe of the liver by the *hepatoduodenal ligament*. In this ligament is a long white gland, the *pancreas*. About one-quarter of an inch behind the right end of the pancreas a *pancreatic duct* passes from the pancreas into the duodenum and may be revealed by picking away the substance of the pancreas at this point. On the dorsal surface of the right lobe of the liver, near its lateral border, is the large *gallbladder,* connected to the duodenum by a short, stout *bile duct*. Beyond the entrance of the bile duct the small intestine turns sharply posteriorly and is then thrown into a number of coils. In female specimens you will generally need to remove one of the large egg-bearing ovaries before you can conveniently trace the intestine. Lifting the coils of the small intestine, note the *dorsal mesentery* that attaches it to the median dorsal line of the coelom; this part of the dorsal mesentery is the *mesentery* proper. Follow the dorsal mesentery forward and note the portions of it that support the duodenum and the stomach—the *mesoduodenum* and *mesogaster,* respectively. The mesoduodenum is fused to the hepatoduodenal ligament so that the two look like one, but the megogaster is distinct from the gastrohepatic ligament. Trace the small intestine posteriorly, noting the coiling of the mesentery corresponding to the coils of the intestine. Find on the right side the entrance of the small intestine into the *large intestine* or *colon*. At the junction of the small and large intestine is as light projection, the *colic caecum*. The colon generally crosses the pleuroperitoneal cavity transversely and then turns posteriorly and runs straight to the cloaca. The colon is supported by the *mesocolon,* which in the transverse part of the colon is fused to the mesogaster. In the mesocolon on the dorsal side of the colon, shortly beyond the caecum, is a rounded red body, the *spleen.* Trace the colon to the point where it disappears above the pelvic girdle. At this point, ventral to the colon, will be found the large thin-walled, bilobed *urinary bladder.* It is generally greatly distended with urine, but in some specimens it may be contracted to a small mass. The bladder has no ligaments, for the peritoneum leaves the body wall around the stalk of the bladder and passes over its surface to form its visceral investment.

Cut away the pelvic girdle by making a cut through each side with the bone scissors and removing a median piece. The large intestine can then be traced into a tube, the *cloaca,* which proceeds dorsal to the girdle to the *anus.* At the point where the large intestine enters the cloaca you will find the urinary bladder, attached to the ventral surface of the cloaca by a stalk. In females you will also see a large white *oviduct* entering the cloaca on each side of the stalk of the bladder. Attached to each side of the cloaca in both males and females are *accessory urinary bladders.*

## 3. The Respiratory System

Pry open the jaws of the turtle and cut through the angles of the jaws, nearer the lower than the upper jaw, revealing the *oral cavity* and the *pharynx.* The oral cavity is bounded by the jaws, which have no teeth but are clothed with horny beaks of epidermal origin. These beaks extend as plates into the mouth cavity. In the roof of the mouth, posterior to each plate, is an elongated opening, the *posterior naris.* Probe them and determine that they connect with the anterior nares by passages that run through the nasal cavities. The floor of the mouth cavity is occupied by the fleshy, pointed tongue, which is of the definitive type.

In the pharynx note that neither gills nor gill slits are present, although, as we shall see shortly, the gill arches are represented. Behind the base of the tongue is an elevation, the *laryngeal prominence,* and in the center of it is an elongated slit, the *glottis.* Feel the pair of small *arytenoid cartilages,* one on each side of the glottis; they are derived from the sixth gill arch. On each side of the roof of the pharynx, posterior to the muscles that connect the skull and lower jaw, is the opening of the *auditory* or *eustachian tube,* a canal leading from the pharynx to the cavity of the middle ear. (The opening may have been destroyed when you cut the jaws apart.) The auditory tube and also the cavity of the middle ear are outgrowths from the first gill pouch. Posteriorly the pharynx narrows into the esophagus.

Cut through the skin in the midventral line of the neck and peel the skin from neck to throat. Separate the muscles in the midline of the neck and find a tube stiffened by rings of cartilage, the trachea or windpipe. Trace it forward until it disappears into the pharynx. In front of this note the hard body of the *hyoid,* and by cleaning away muscles find also two pairs of horns of the hyoid extending posteriorly. The hyoid and its horns are derivatives of the second, third, and fourth gill arches. Open the mouth and make a cut around the laryngeal prominence, freeing it from its position on the dorsal surface of the body of the hyoid. The structure thus freed is the larynx, an expanded chamber at the anterior end of the trachea. Find in the lateral walls of the larynx the arytenoid cartilages, small cartilages supporting the two triangular flaps that bound the glottis. Posterior to the glottis is a ring-shaped cartilage, the *cricoid,* much wider ventrally than dorsally.

Now trace the trachea posteriorly. Note the esophagus, a soft tube, lying dorsal to or to one side of the trachea. Just anterior to the heart the trachea bifurcates into the two *bronchi,* which proceed to the lungs. Raise the right and left lobes of the liver and the stomach and find dorsal to them, against the carapace, a large spongy *lung* on each side. Each bronchus is accompanied by a pulmonary artery and a pulmonary vein that can be traced into a lung. Study the relation of the lung to the pleuroperitoneal cavity. Note that the lung is in contact with the inner surface of the carapace and that the parietal peritoneum passes over the ventral surface of the lung, leaving the lung outside the membrane. Such a relation to the peritoneum is spoken of as *retroperitoneal.* The posterior end of the lung, however, projects into the pleuroperitoneal cavity and is clothed with the peritoneum. Cut open the lung and observe its extremely spongy texture; cords of connective tissue divide the interior into air spaces or *alveoli.*

**J. The Coelom and the Digestive and Respiratory Systems of the Pigeon**

A major difference between birds and reptiles is that in birds the pleuroperitoneal cavity is divided into a pair of anterior pleural cavities, each containing a lung, and a posterior peritoneal cavity. The respiratory system of birds has a number of peculiarities, but these are not of phylogenetic significance. Place the specimen in a dissecting plan and remove the feathers. In plucking the feathers from the neck region, take care to avoid tearing the skin.

**1. The Oral Cavity and the Pharynx**

Open the mouth wide by cutting through the angles of the jaws. An anterior *oral cavity* and a posterior *pharynx* are thus revealed.

*a. Oral Cavity*

The roof and floor of the oral cavity are bounded laterally by horny beaks, of epidermal origin, which encase the jaws. Teeth are absent, as in all living birds. The roof of the mouth cavity bears a pair of elongated *palatal folds* with free fimbriated margins. These palatal folds probably correspond to the secondary or hard palate of mammals but differ in that they do not meet in the median line, leaving a deep *palatal fissure.* In the roof of the mouth cavity, above the palatal folds and concealed by them, are the *posterior nares,* found by bending aside or cutting away the palatal folds. Probe to verify the connection of anterior and posterior nares. The floor of the oral cavity is occupied by the pointed *tongue,* which has a free fimbriated posterior border. The tongue is of the compound definitive type but is not very muscular. The numerous glands opening into the oral cavity are too small to study in gross dissection.

### b. Pharynx

Note that, as in all adult vertebrates above urodeles, gill slits are absent from the lateral walls of the pharynx. In the roof of the pharyngeal cavity just posterior to the caudal ends of the palatal folds is a median aperture, the opening of the paired *auditory tubes*. Each auditory tube extends from this opening to the cavity of the middle ear. In birds, unlike other vertebrates, the two auditory tubes unite to one at the point of communication with the pharynx. Posterior to this opening, the roof of the pharynx bears a pair of folds with fimbriated borders, which hang down like a curtain into the pharyngeal cavity. These folds constitute the *soft palate*. In the floor of the pharynx, immediately posterior to the caudal end of the tongue, is a hardened elevation, the *laryngeal prominence,* bearing in its center an elongated opening, the *glottis*. The margins of the glottis are also fimbriated, and immediately posterior to the glottis on each side is a fringed fold. In the walls of the glottis the supporting laryngeal cartilages are readily felt.

**2. The Hyoid Apparatus, the Larynx, the Trachea, and the Esophagus**

Make a median ventral longitudinal incision in the skin of the neck from the throat to the anterior end of the sternum. Deflect the skin on each side of the incision. The *trachea* or *windpipe,* a tube with walls stiffened by rings of cartilage, is immediately exposed. Dorsal to it or to one side of it is the soft *esophagus*.

Trace the trachea forward to the glottis and clean away the muscles that cover its anterior end. At the same time cut to the sides of the tongue, so that it may be pulled down ventrally from the mouth cavity. You can now study the *hyoid apparatus,* which consists of remnants of the hyoid (second) and third gill arches. It is composed of three median elements, arranged in a longitudinal series, and two pairs of *horns* or *cornua*. The most anterior of the three median pieces is the *entoglossal cartilage*. It is situated inside the tongue and may be revealed by dissecting off the covering membrane of the tongue. It represents the two fused *ceratohyals*. From its posterior end there projects posteriorly on each side a small cartilage that occupies the caudal point of the tongue, already noted. These two cartilages constitute the *anterior horns* of the hyoid and consist of the free ends of the two ceratohyals, whose anterior portions fused to form the entoglossal cartilage. Posterior to the entoglossal cartilage is a median bony piece, the *basihyal*. Posterior to this is the *basibranchial* of the third gill arch. From the point of junction of basihyal and basibranchial projects, on each side, the long *posterior horn* of the hyoid, consisting of portions of the third gill arch. On following the posterior horns you will find that they tend toward the ears and are divided into a proximal longer portion, the *ceratobranchial,* and a distal shorter rod, the *epibranchial*.

Cut around the laryngeal prominence, freeing it so that it can be drawn ventrally. Also free the hyoid apparatus from the ventral surface of the larynx. The larynx is the expanded chamber thus revealed at the top of the trachea and opening into the pharyngeal cavity by way of the glottis. By dissecting in the margins of the glottis on each side, expose a slender, curved partially ossified *arytenoid cartilage.* On the ventral side of the larynx note the enlarged triangular *cricoid cartilage.* Follow this around to the dorsal side, where it terminates by much-narrowed ends. Between the two dorsal ends of the cricoid cartilage is a median cartilage, the *procricoid,* which is in contact with the posterior ends of the arytenoids and is simply a separated piece of the cricoid.

Examine the "cartilages" of the trachea. They are broad, hard, and bony ventrally, but narrower, softer, and cartilaginous dorsally. There is consequently a somewhat soft strip along the dorsal side of the trachea, which lies against the cervical vertebrae.

Trace the esophagus posteriorly. Shortly in front of the sternum it widens into an enormous bilobed sac, the *crop,* in which the food, swallowed whole, is detained for a time and may be subjected to muscular and enzymatic action. It is passed on to the stomach in small quantities. In some birds, such as the pigeon, it is regurgitated from the crop and fed to the young. The crop is best developed in grain-eating birds and is small or absent in many birds. The crop should be carefully loosened on all sides.

3. The Anterior Air Sacs and the Pectoral Muscles

*It is advisable to have a prepared skeleton available during this portion of the dissection.* The respiratory system consists not only of the lungs but also of a number of *air sacs* among the viscera and *air spaces* in the bones. The air sacs and air spaces communicate with the lungs by branches of the *bronchi.* The system not only aids in decreasing the specific gravity of the bird but also insures a more complete exposure of the lung tissue to the air; for the residual air is retained in the air sacs, not in the lungs as in other vertebrates, and the air in the lungs is consequently completely renewed at each inspiration. Because of the delicacy of the air sacs you may not be able to locate all of those mentioned below, but you will see some of them. They are best studied in freshly killed specimens in which they have been inflated through the trachea. Free a small portion of the trachea, in the middle of the neck, separate it from the surrounding tissue, and place a ligature around it ready for tying. Just anterior to the ligature make a small slit in the trachea and insert a blowpipe. Blow down the pipe and observe that the abdomen becomes distended and the sternum raised. Draw the ligature tight around the blowpipe and with the abdomen distended quickly remove the blowpipe. Tighten the ligature as the pipe is removed, taking care not to pull it too tight so the trachea will not be severed.

Separate the crop from the pectoral muscles by carefully pulling it away with the fingers. It is often very delicate. Dorsal to the crop, in the angle formed by the two halves of the *furcula* or *wishbone,* is the *interclavicular air sac.* Its delicate ventral wall is in contact with the dorsal wall of the crop. It consists of two lobes, one on each side of the median line. Puncture the interclavicular air sac and find, dorsal to it on each side, another sac, the *cervical air sac.* The *keel* of the sternum can be seen in the midventral line projecting between the muscles.

Extend the median ventral incision in the skin as far as the anus.
Separate the skin from the underlying muscle on each side of the
chest and abdomen. The great *pectoral muscles* are revealed immediately below the skin and occupy the angle between the *keel* and the
*sternum*. With a sharp scalpel cut the pectoral muscles away from
the keel by a vertical cut as close as possible to the bone on one side.
Separate the muscles without cutting into the substance of the muscle.
Between the pectoralis major and the pectoralis minor lies the *axillary air sac*. Immediately behind this air sac are the large *pectoral
arteries* and *veins*. Take care to avoid cutting these vessels.

## 4. The Divisions of the Coelom and the Posterior Air Sacs

Cut through the posterior ventral abdominal wall to the right of the
median line. Beneath the skin are thin layers of abdominal muscles
corresponding to those of mammals, and internal to this is the
*parietal peritoneum,* generally impregnated with streaks of fat. Cut
through this and extend the incision anteriorly, cutting through the
sternum. Laterally the cut must be made through the attachment of
the ribs to the sternum and through the ventral part of the coracoid
passing ventral to the pectoral blood vessels. Keep the scissors in
contact with the bone to avoid injury to the internal parts. Anteriorly
the sternum can readily be separated from the fused ends of the
clavicles and then be lifted off.

The small cavity posterior to the sternum is the *peritoneal cavity*.
Note in it the *liver* dorsal to the posterior end of the sternum, the
closely coiled *intestine,* and to the left the large firm *gizzard*. From
the gizzard a mesentery extends to the ventral body wall to the left
of the midline. This may be designated the *ventral* ligament of the
gizzard. It is continuous with the *falciform ligament* of the liver,
which extends from the median ventral region of the liver to the
midventral line of the body wall and inner surface of the sternum.
The falciform ligament and ventral ligament of the gizzard together
constitute a partition that divides the peritoneal cavity into a large
right portion and smaller left portion. This division is not found in
other vertebrates. In the partition courses a small vein, extending
from these mesenteries to the liver.

Immediately dorsal to the sternum is the delicate *pericardial sac*
containing the *heart*. The ventral wall of the pericardial sac will
probably have been opened in cutting through the sternum. The
heart, as in the turtle, has descended posteriorly, and a pericardial
sac has been formed from the anterior face of the transverse septum
and the parietal pericardium, as described in connection with the
turtle. The space between the pericardial sac and the heart is, as
before, the *pericardial cavity,* a portion of the coelom. The pericardial sac is in contact on its ventral surface with the inner surface
of the sternum, and anteriorly and laterally is also in contact with the
inner surface of the body wall. Hence, only the posterior part of the
pericardial sac is freed from the body wall. From the points where
the pericardial sac meets the lateral body wall, a membranous partition extends obliquely posteriorly on each side. This partition is
called the *oblique septum*. It contains a large air sac and stretches

across from the lateral body wall to that part of the pericardial sac that is derived from the transverse septum, thus dividing the pleuroperitoneal cavity into anterior and posterior portions. The part of the original pleuroperitoneal acvity that remains anterior to the oblique septum consists of the two *pleural cavities,* one on each side of the pericardial cavity. The part of the pleuroperitoneal cavity posterior to the oblique septum is the *peritoneal cavity,* already mentioned.

Inside the oblique septum, enclosed between its anterior and posterior walls, is a large air sac, the *posterior intermediate air sac.* Immediately anterior to this, lying to each side of the heart, is the small *anterior intermediate air sac.*

In the peritoneal cavity, cut through the falciform ligament and ligament of the gizzard at their line of attachment to the ventral body wall. On each side of the viscera and slightly dorsal to them find the large *abdominal air sac.*

## 5. The Peritoneal Cavity and Its Contents

As in other vertebrates, the peritoneal cavity is lined by the parietal peritoneum, which is deflected at certain points to form mesenteries and continues over the surface of the viscera as the visceral peritoneum.

The viscera of the peritoneal cavity may not be studied in more detail. At the anterior end is the large *liver,* consisting of right and left lobes; the former is the larger. The pericardial sac rests between the two lobes of the liver. The liver is attached to the pericardial sac (that portion of it derived from the transverse septum) by the *coronary ligament.* The falciform ligament of the liver has already been noted and severed. To the left and slightly covered by the left lobe of the liver is the *gizzard.* On raising the left lobe of the liver you will see the *gastrohepatic ligament* passing between the gizzard and the liver. The *mesogaster* connects the gizzard with the dorsal body wall. The ventral ligament of the gizzard has already been noted and cut. On breaking through the gastrohepatic ligament you will find the soft proventriculus extending anteriorly from the gizzard, dorsal to the liver. From the stomach, at the junction of *proventriculus* and gizzard, the *small intestine* arises. The duodenum makes a long U-shaped loop posteriorly, and its beginning is attached to the right lobe of the liver by the *hepatoduodenal ligament.* Between the two sides of the duodenal loop stretches the *mesoduodenum,* a portion of the mesentery of the intestine in which is the *pancreas,* lying between the two limbs of the loop. From a deep depression in the dorsal surface of the right lobe of the liver, the two *bile ducts* emerge and pass into the duodenum. Note the absence of a gallbladder. The *left bile duct* is the shorter and stouter of the two and enters the lift limb of the duodenum about half an inch beyond the gizzard. The more slender *right bile duct* joins the right limb of the duodenal loop. There are three *pancreatic ducts,* which pass from the right side of the pancreas into the right limb of the duodenal loop. One of these arises from the anterior part of the pancreas and passes obliquely forward, entering the duodenum near the anterior termination of the right limb of the loop. The other two ducts emerge from the middle of the pancreas and pass across to the right limb of the duodenum. The ducts are easily seen by spreading out the mesentery.

Trace the small intestine posteriorly from the duodenum. It is much coiled and is supported by the *mesentery,* which, owing to the small space into which the intestine is packed, is fused in many places. Near its termination the small intestine turns toward the midline, widens slightly, then runs posteriorly. At about the middle of the peritoneal cavity it passed without enlargement into the *large intestine*. The point of junction of large and small intestine is marked by a pair of small lateral diverticula, the *colic caeca*. The large intestine is short and leads into the *cloaca*. Because of the absence of public and ischial symphyses in birds, the cloaca does not pass through the ring of the pelvic girdle but may be traced directly to the *anus*. A *urinary bladder* is absent. In females the *single left oviduct* may be seen entering the left side of the cloaca. The single left ovary is situated in the anterior part of the peritoneal cavity, dorsal to the gizzard.

The gizzard and proventriculus may now be freed from the adjacent air sacs and the mesenteries. On turning the gizzard far forward you will find, between the proventriculus and the anterior end of the right limb of the duodenal loop, a rounded red body, the *spleen*. The gizzard may now be cut open along its posterior margin. The interior contains small stones and probably partially digested food. Note the thick muscular walls and hard horny lining of the gizzard. Cut from the gizzard into the proventriculus and note the soft glandular walls of the latter. The gizzard compensates for the absence of teeth in birds and grinds up the food into small pieces; hence it is best developed in birds that eat grain and other hard foods.

**6. The Pleural Cavities and Their Contents**

The posterior intermediate air sac, situated in the oblique septum, may now be punctured if you have not already done so. The two walls of the septum are now more evident. The anterior intermediate air sac may also be punctured. Against the dorsal wall of the pleural cavity of each side will be found a reddish, spongy, flattened *lung*. The openings of some of the air sacs into the lungs may be visible on some specimens. On cutting into the lung you will find that the organ is solid, not hollow as in the preceding animals.

The cavity in which each lung is contained is, as already explained, a *pleural cavity* and is lined by a coelomic membrane, the *pleura*. As the lungs are flattened against the dorsal wall of the pleural cavity, the pleura passes over their ventral faces, leaving them outside, so to speak. The pleura, furthermore, passes over the surface of the pericardial sac and lines the inner surface of the body wall.

**7. The Syrinx**

Examine the posterior part of the *trachea*. Two slender muscles, the *sternotracheal muscles,* diverge from their insertion on the ventral surface of the trachea to their origin on the sternum. These muscles should be severed. The trachea disappears dorsal to the heart and to the great blood vessels that enter and leave the heart. These blood vessels must not be injured. Loosen the trachea and pull it forward so you can see the bifurcation of the trachea into the two *bronchi* dorsal to the heart. Cut across the bronchi with a fine scissors and draw the trachea forward. At the point where the trachea forks into the two bronchi, an expanded chamber, the *syrinx,* is present. The voice of birds issues from the syrinx, not from the larynx. Along each side of the trachea extending from the point of insertion of the sternotracheal muscles to the lateral walls of the syrinx is a muscle, the *intrinsic syringeal muscle.* The walls of the syrinx are supported by the last tracheal rings and the first bronchial half-rings. The last two tracheal rings are widely separated from each other but are connected in the midventral line by a narrow bridge of bone; the thin tracheal wall to either side of this bridge forms the *external typaniform membranes.* On the inner side of the bronchi at their junction with the trachea are similar thin internal typaniform membranes. Make a slit in the ventral wall of the syrinx and spread the cut edges apart. In the dorsal wall of the interior is a slight vertical fold, the *semilunar membrane,* supported by a bony ridge, the *pessulus.* The voice is produced by the vibrations of the pessulus and the tympaniform membranes. Large thickenings in the lateral walls of the syrinx also play a role, and the syringeal and sternotracheal muscles aid by changing the shape of the syrinx.

## K. The Coelom and the Digestive and Respiratory Systems of Mammals

In mammals the pleural cavities are separated from the peritoneal cavity by a muscular partition, the *diaphragm,* which consists partly of the transverse septum and plays an important role in respiratory movements. Larynx and lungs also reach a high degree of differentiation, and the digestive tract shows many variations in correlation with various modes of life.

The directions apply to the cat and rabbit, and any differences between the two animals will be specifically mentioned.

## 1. Oral Cavity and Pharynx

### a. The Salivary Glands

The salivary glands are large masses of glandular tissue that are outgrowths of the oral epithelium; the stalk of each outgrowth remains as a salivary duct. The glands are situated among the muscles of the head and throat. They should be located and their ducts followed as far as practicable. The dissection should be carried out on the same side of the head as that on which the muscles were dissected.

The *parotid gland* is below and slightly in front of the base of the *pinna* of the ear, just under the skin. Remove the skin from this region and find the pinkish gland. Its duct passes across the external surface of the *masseter muscle* and penetrates the mucous membrane of the upper lip opposite the last premolar (cat) or the first molar (rabbit). Accessory glandular areas are frequently found along the length of the duct. The *submaxillary gland* has already been noted as a roundish mass at the angle of the jaw near the posterior margin of the masseter. Do not confuse it with the *lymph nodes* in this area, which are smaller and have a smoother texture. Loosen it and find the duct from its internal surface. In the cat the beginning of this duct is surrounded by the elongated *sublingual gland.* Trace the *submaxillary duct* forward (in the cat it is accompanied by the sublingual duct). The duct passes internal to the *digastric muscle.* This muscle should be severed. The duct (or two ducts in the cat) will then be seen to pass internal to the *mylohyoid muscle.* This in turn should be cut and the duct traced forward. In the rabbit the small flattened *sublingual gland* lies in the path of the submaxillary duct, which is situated just external to the lining of the mouth cavity, runs forward nearly to the symphysis of the mandible, and then penetrates the lining. In the rabbit the sublingual gland opens into the mouth cavity by several short ducts that are difficult to find. The parotid, submaxillary, and sublingual glands are the most common salivary glands of mammals, but others are present in some species. The *molar gland,* present in the cat only, is situated between the skin and the external surface of the mandible, just in front of the masseter muscle. It will be found by deflecting the skin in this area. It opens onto the inside of the cheek by several small ducts, impractical to locate. The *infraorbital gland* in both cat and rabbit lies in the floor of the *orbit* and will be seen later when the eye is dissected. The rabbit also has elongated *buccal glands* beneath the skin of the upper and lower lips.

### b.  The Oral Cavity

Cut through the skin at the corners of the mouth and see that the skin is well cleared away over the angles of the jaws. Cut through the masseter and other muscles attached to the lower jaw at the angle of the jaws. It should then be possible to pull the lower jaw down. Pry open the mouth, grasp the lower jaw, and exert a strong traction. The jaw will generally yield; but if it does not, cut through the ramus of the mandible with the bone scissors. The anterior part of the cavity thus releaved is the *oral cavity,* bounded by the *lips* and *cheeks.* That part of the oral cavity lying between the teeth and lips is called the *vestibule* of the mouth. The teeth were described in connection with the skull.

The anterior portion of the roof of the oral cavity is occupied by the *hard palate,* the posterior part of the *soft palate.* Feel the difference between the hard and the soft palate. The hard palate is supported by the *premaxillary, maxillary,* and *palatine bones;* the soft palate lacks bony support. The mucous membrane of the hard palate is thrown into a number of roughened transverse ridges. At the anterior end of the hard palate, just behind the incisor teeth, are a pair of openings of the *nasopalatine ducts,* which connect the mouth and nasal cavities by way of the *incisive foramina* of the maxillary bones. Look for the opening of the duct of the parotid gland on the inside of the cheek opposite the second upper premolar in the rabbit, opposite the last cusp of the third upper premolar of the cat, in which animal it is situated on a slight ridge: the openings are difficult to identify with certainty.

The floor of the oral cavity is occupied by the *tongue,* a fleshy muscular organ, more mobile in mammals than in most other verte-

brates. The anterior attachment of the tongue to the floor of the oral
cavity has the form of a vertical fold, the *lingual frenulum.* Halfway
between the lower incisors and the frenulum in the rabbit you will
find the two small, slitlike openings of the ducts of the submaxillary glands, the two being about an eighth of an inch apart. In the
cat a fold runs forward from the frenulum on each side just within
the teeth and terminates anteriorly in a well-marked, flattened papilla
that bears the openings of the ducts of the submaxillary and sublingual glands.

Cut through the floor of the mouth on each side; keeping the
scalpel next to the mandible, cut through the junction, the *symphysis,*
of the lower jaws. The tongue can now be pulled down and out between the two halves of the lower jaws. The cuts may be continued
on each side at the base of the tongue back to the level of the submaxillary glands so that the tongue can be pulled well down. The
surface of the tongue may now be examined in detail. In the rabbit
the tongue is divisible into two portions: an anterior softer portion,
covered with minute pointed elevations, the *fungiform papillae;* and
a posterior elevated, smoother, and harder portion. At the posterior
end of the latter on each side is situated a *vallate papilla,* consisting
of a round elevation set into a pit. In front of each vallate papilla on
the side of the tongue is an oval area of considerable size marked
by numerous fine parallel ridges, the *foliate papilla.* In the cat the
anterior part of the tongue is covered with the *filiform papillae,*
many of which are hard and spine-like, pointed posteriorly; the
remainder of the tongue is provided with *fungiform* papillae; among
the fungiform papillae are four to six vallate papillae arranged in a
V-shaped row. At the sides of the vallate papillae are some very large
fungiform papillae. The papillae are provided with microscopic
taste buds.

### c. The Pharynx

The *pharynx* is that portion of the cavity lying posterior and dorsal to
the soft palate. Pull the tongue well forward and examine the soft
palate, which descends like a curtain across the posterior end of the
oral cavity. Find its free posterior margin, arching above the base of
the tongue. (The margin may be concealed by a leaf-shaped structure, the *epiglottis,* which projects from the base of the tongue. If so,
the epiglottis should be pressed out of the way.) The opening formed
by the free border of the palate is known as the *isthmus of the fauces.*
This opening leads into the cavity of the pharynx. Shortly anterior to
the free border of the soft palate on each side is a pit, the *tonsillar
fossa,* which contains a small mass of lymphoid tissue, the *palatine
tonsil.* The tonsillar fossa is bounded in front and behind by low
folds, an anterior *glossopalatine arch* and a posterior *pharyngopala-
tine arch.* Now slit the soft palate forward along its median line. A
cavity, the *nasopharynx,* is revealed dorsal to the soft palate. At the
anterior end of the nasopharynx are the two *posterior nares* or
*choanae,* the internal ends of the nasal passages. Posterior to them on
the lateral wall of the nasopharynx are a pair of oblique slits, the

openings of the *auditory* or *eustachian tubes,* which connect the pharynx with the cavity of the middle ear.

The pharynx narrows posteriorly into the *esophagus*. The entrance of the respiratory tract lies anterior to that of esophagus. The opening of respiratory tract is guarded by an *epiglottis,* which if not already identified will be seen on pulling the tongue well forward. In the pharynx the paths for food and air are crossed.

2. Hyoid Apparatus, Larynx, Trachea, and Esophagus

Press the tongue dorsally against the lower jaw and find on its external surface at its base a bone, the body of the *hyoid*. This is a stout bone in the rabbit, a narrow bar in the cat. Expose the hyoid by clearing away the muscles from its surface. Note the *horns* and *cornua* that extend from its sides. In the rabbit the horns are short processes that are connected by slender, tendinous muscles with the *jugular process* of the *occipital bone*. In the cat the anterior horn is long and slender and consists of a chain of four bony pieces, the last of which articulates with the *tympanic bulla;* the posterior horn is short and is united to the *larynx*. The hyoid supports the base of the tongue and serves for the origin and insertion of muscles.

In the median ventral line posterior to the body of the hyoid is a chamber with cartilaginous walls, the *larynx,* which constitutes the projection in the throat popularly known as the Adam's apple. By making a cut through the base of the tongue and gently severing the muscle attachments the larynx may be freed and lifted forward. At the top of the larynx is a large opening, the *glottis,* from whose ventral margin the *epiglottis* projects. Dorsal to the glottis and bound with it by muscles is another opening, generally collapsed and concealed from view by portions of the larynx. This opening should be located by probing; the probe will enter a soft tube that proceeds posteriorly dorsal to the larynx. This tube is the esophagus.

The structure of the larynx should now be examined in detail. The ventral wall of the larynx is supported by a large shield-shaped cartilage, the *thyroid cartilage*. A short distance posterior to this is the *cricoid cartilage,* which forms a ring around the larynx. The dorsal rim of the glottis is supported by a pair of projecting triangular shaped cartilages, the *arytenoids*. On looking into the glottis you will see a pair of folds, the *vocal cords,* extending from the arytenoid cartilages to the thyroid cartilage. They almost occlude the opening. In the cat, in addition to these true vocal cords, there is a pair of false vocal cords, situated lateral to the former and extending from the tips of the arytenoid cartilages to the base of the epiglottis. The vocal cords are not really cords but folds of the lateral wall of the larynx. Dissecting away the esophagus from the dorsal side of the larynx exposes the dorsal side of the cricoid cartilage. It is much broader than the ventral side. By cleaning away the mucous membrane covering it, the two arytenoid cartilages which rest on the anterior extremity of the dorsal part of the cricoid will be exposed.

From the larynx the *trachea* proceeds posteriorly. Its walls are stiffened by cartilaginous tracheal rings, which are incomplete dorsally, leaving a soft strip into which the esophagus fits. On each side of the trachea, lying against the trachea and internal to the muscles, is a flattened elongated body, one of the lobes of the *thyroid gland*. The anterior end of each lobe is at a level with the cricoid cartilage. The caudal ends of the two lobes are connected by a median portion, the *isthmus,* which crosses the ventral side of the trachea. In the cat the isthmus is usually very narrow and in some cases may be absent, whereas in the rabbit it is prominent. The trachea is not to be traced farther posteriorly at this time.

## 3. The Pleural and Pericardial Cavities

The trunk of mammals is divided into an anterior *thoracic region* and a posterior *abdominal region*. Each of these regions contains cavities that are portions of the coelom. The thoracic region has three coelomic cavities: the two *pleural cavities,* laterally situated, and the median *pericardial cavity,* situated between the two pleural cavities.

Cut through the ribs one-half inch to the left of the sternum, extending the cut the length of the sternum. At each end of this, cut laterally and dorsally between two adjacent ribs at right angles to the first cut, forming a flap in the chest wall. Open the flap and bend it dorsally so that you can look within. The cavity thus revealed is the left *pleural cavity* or *pleural sac;* a similar sac exists on the right side. Each pleural sac contains a *lung*. In the median region under the sternum lies the large *heart*. Note the delicate partition that stretches from the heart to the ventral midline. This partition, called the *mediastinal septum,* consists of the two medial walls of the right and left pleural sacs in contact with each other. At the level of the heart the two walls separate so that the heart and its pericardial sac are enclosed between them. This space between the two walls of the mediastinal septum is called the *mediastinum*. The posterior walls of the pleural sacs are formed by a muscular dome-shaped partition, the *diaphragm*. Each pleural sac is lined by a smooth moist membrane, the *pleura,* divided into *parietal* and *visceral* parts. The *parietal pleura* lines the inside of the pleural cavity, covers the anterior face of the diaphragm, and together with the medial wall of the other pleural sac forms the mediastinal septum. The visceral pleura is that part of the pleura that passes over the surface of the lung, to which it is indistinguishably fused. Examine the left lung. It is a soft spongy organ divided into three lobes: a smaller *anterior* lobe, and larger *middle* and *posterior* lobes. The anterior lobe is very small in the rabbit. The large posterior lobe fits neatly on the convex surface of the diaphragm. Cut into the lung; it appears solid but is really composed of innumerable minute "air cells" (*alveoli*).

Now carefully cut through the mediastinal septum ventral to the heart and look into the right pleural cavity. The diaphragm may be slit along its left side so it is easier to spread apart the thoracic walls. The right pleural cavity is similar to the left cavity. The right lung is somewhat larger than the left and is also divided into anterior, middle, and posterior lobes. The large posterior lobe is subdivided into *medial* and *lateral* lobules. The medial lobule projects into a pocket formed by a dorsally directed fold of the mediastinal septum. This *caval fold* supports a large vein, the *postcaval vein,* which ascends from the liver to the heart and will be found enclosed in the free dorsal margin of the caval fold.

Examine the heart and the *pericardial sac*. The pericardial sac or parietal pericardium is a sac of thin tissue enclosing the heart but not attached to it except at the anterior end, where the great vessels enter and leave the heart. The heart is freely movable inside the pericardial sac. The narrow space between the pericardial sac and the heart is the *pericardial cavity,* a portion of the coelom. Cut through the pericardial sac to expose the heart. The surface of the heart is invested by a thin membrane, the *visceral pericardium,* inseparably adherent to the heart wall. The visceral pericardium is continuous with the pericardial sac at the anterior end, where the blood vessels enter and leave the heart. Since the heart with its pericardial sac is situated in the mediastinum, it is evident that there are three coelomic layers surrounding the heart: the visceral pericardium, closely adherent to the heart wall, the parietal pericardium or pericardial sac, separated from the heart by the pericardial cavity, and the parietal pleura of the mediastinal septum, closely fused to the pericardial sac.

In the mediastinum, in the median line ventral to the anterior part of the heart and extending forward, will be found a mass of glandular tissue, the *thymus*. It is larger in younger specimens. In searching for it do not injure the large blood vessels.

Now press the heart and the left lung over to the right. The lung is attached by a narrow region, the *radix* or *root,* which encloses an artery, a vein, and a *bronchus*. In the cat, the lung is also attached along most of its length to the dorsal thoracic wall by the *pulmonary ligament,* a fold of the pleura. Note that dorsal to the root of the lung the pleura continues onto the dorsal and lateral surfaces of the pleural cavity and that certain structures can be seen lying between the dorsal portions of the two walls of the mediastinal septum and consequently in the mediastinum. The most conspicuous of these is the *dorsal aorta,* a large vessel that arches away from the heart to the left and descends toward the *diaphragm*. About one-half inch ventral to the aorta is the *esophagus,* also lying in the mediastinum: trace it posteriorly to the place where it penetrates the diaphragm. The *phrenic nerves* will also be seen in the central portion of the mediastinum as a white strand on each side of the pericardial cavity. These nerves lie ventral to the roots of the lungs and run to the diaphragm. The right phrenic nerve closely follows the postcaval vein, and the left passes through the posterior part of the mediastinal septum. These nerves arise from the ventral rami of the fourth (rabbit) and the fifth and sixth (cat) cervical nerves. The anterior origin of the phrenic nerves is indicative of the cervical derivation of some of the muscles of the diaphragm.

The diaphragm is a curved sheet forming the posterior wall of the thoracic cavity and completely separating it from the abdominal cavity. The center of the diaphragm consists of connective tissue forming a circular *central tendon*. The rest of the diaphragm is muscular. The diaphragm is pierced at several points to allow important structures such as the esophagus to pass through. The diaphragm is a structure peculiar to mammals. It is formed in part of the transverse septum and in part of other coelomic membranes; it then becomes invaded by muscle buds from the adjacent cervical myotomes and intercostal muscles.

**4. The Peritoneal Cavity and Its Contents**

Make a longitudinal slit through the abdominal wall, a little to the left of the midventral line from the inguinal region up to the diaphragm. Widen the opening by a transverse slit in the middle of the left abdominal wall. A large cavity, the *abdominal* or *peritoneal cavity,* is exposed. Its anterior wall is formed by the concavely arched diaphragm, which completely separates the peritoneal cavity from the pleural cavity. Posterior to the diaphragm and shaped so as to fit the concave surface of the diaphragm is the large, lobed *liver,* generally grayish brown in preserved specimens. Posterior to the liver the peritoneal cavity is filled by the coils of the intestine. In the cat the intestine is covered ventrally by a thin membrane impregnated with streaks of fat, the *greater omentum.* This membrane is present also in the rabbit but is much smaller and less conspicuous. At the posterior end of the peritoneal cavity, note the pear-shaped *bladder,* generally distended with fluid. On raising the liver and looking dorsally and to the left of it you will find the stomach, with the *spleen* attached to its left border. On the dorsal wall of the peritoneal cavity at about the level of the posterior ends of the liver lobes are the *kidneys;* to see them, gently lift the coils of the intestine. In female specimens, especially pregnant ones, you will note the horns of the *uterus* as a tube on each side in the posterior part of the peritoneal cavity.

The peritoneal cavity is lined by a membrane, the *peritoneum.* As in all coelomate animals, that portion of the membrane on the inside of the body wall is the *parietal peritoneum.* In both dorsal and ventral regions the peritoneum is deflected from the body wall and passes over the surface of the viscera, forming a covering layer, the *visceral peritoneum* or *serosa,* for all the viscera. In passing to and from the body wall to the viscera the peritoneum forms double-walled membranes, the *mesenteries* or *ligaments.* The *dorsal mesentery* is present intact in mammals, and the *ventral mesentery* persists in the region of the liver and bladder as in other vertebrates.

Examine the *stomach* by raising the liver and pressing it craniad. Slitting the diaphragm on the left side helps expose the stomach. The stomach is large and rounded in the rabbit, smaller and more elongated in the cat. Find where the esophagus emerges from the diaphragm and enters the anterior surface of the stomach. The area of junction of the stomach and esophagus is called the *cardia,* and the region of the stomach adjacent to the junction is the cardiac end of the stomach. The shorter, slightly concave, anterior surface of the stomach from the cardia to the pylorus is the *lesser curvature;* the larger, convex, posterior surface, the *greater curvature.* The sac-like bulge of the stomach to the left of the cardia is known as the *fundus;* the remainder of the stomach is the *body* or *corpus.* At the right, the stomach passes into the *small intestine;* the point of junction, the *pylorus,* is marked by a constriction beyond which the small intestine makes an abrupt bend. Along the left side of the stomach lies the *spleen,* a rather large organ in the cat but smaller in the rabbit.

The relations of the stomach to the peritoneum are somewhat complicated. Raise the fundus and note the *mesogaster* extending from the dorsal wall to the stomach. Only a small portion of the mesogaster passes directly to the stomach; the greater part of it first descends posteriorly as the *greater omentum.* This is a very large and extensive sheet in the cat, covering the intestine ventrally and often containing a good deal of fat. In the rabbit it is much smaller. The greater omentum is to be thought of as formed in the following way. Suppose one should grasp the mesogaster and pull it posteriorly, drawing it into a sac. Such a sac would have two walls, each double —that is, composed of the two layers of the mesogaster; the sac would contain a cavity known as the lesser *peritoneal sac* or *omental bursa* and would open anteriorly (this opening will be seen later). By manipulating the omentum, determine that it consists of two separate walls. Having formed the greater omentum, the mesogaster returns to the stomach wall and passes onto the stomach along the greater curvature. The spleen is enclosed in the ventral wall of the greater omentum just before the latter passes to the stomach. The portion of the greater omentum between the spleen and the stomach is called the *gastrosplenic* (or *gastrolienal*) ligament. Posterior to the spleen, near the left kidney, a secondary fusion, the *gastrocolic ligament,* has formed between the mesogaster and the mesentery of the intestine.

The liver has a convex anterior surface, fitting against the posterior surface of the diaphragm, and a concave posterior surface, fitting over the stomach and first part of the small intestine. The liver is divided into *right* and *left lobes,* each of which is subdivided into two lobes, a *median* and a *lateral.* The left lateral and right median lobes are larger than the others. In the cat the right lateral lobe is deeply cleft into two *lobules.* The large elongated *gallbladder* is embedded in the dorsal surface of the right median lobe in the rabbit, in a cleft in this lobe in the cat. On raising the liver and looking between it and the stomach you will see another small lobe, the *caudate lobe,* situated between the two layers of the lesser omentum. The *lesser omentum,* or *gastrohepatoduodenal ligament,* is the ligament passing from the lesser curvature of the stomach to the posterior surface of the liver. It is divisible into two portions: the *gastrohepatic ligament* from the lesser curvature to the liver, and the *hepatoduodenal ligament* from the liver to the first part of the small intestine. The portion of the gastrohepatic ligament that contains the caudate lobe of the liver forms a sac that extends anteriorly the cavity of the greater omentum. In the hepatoduodenal ligament runs the bile duct, which should be traced from the gallbladder by carefully dissecting the ligament. Note the *cystic duct* from the gallbladder and the *hepatic ducts* from the lobes of the liver. The hepatic duct from the right lateral lobe is especially large. The cystic and hepatic ducts unite to form the *common bile duct,* which passes to the intestine in the hepatoduodenal ligament. It should be traced to the duodenum by cleaning away the connective tissue around it. To the right and dorsal to the bile duct, also in the hepatoduodenal ligament, lies the large hepatic portal vein. This must not be injured. Immediately dorsal to this vein, posterior to its branch into the right lateral lobe of the liver, the hepatoduodenal ligament has a free border that forms the ventral rim of an opening or slit of some size, the *foramen epiploicum* or entrance into the cavity of the omentum. It can be identified with certainty by making a slit into the cavity of the omentum and probing through the slit toward the right, in the direction of the spot just described.

The lesser omentum extends to the middle of the posterior face of the liver, where it becomes the *serosa* of the liver; here its two walls part and, enclosing the liver between them, pass to the anterior face of the liver, where they again unite to form ligaments. The *falciform ligament* extends from between the two median lobes of the liver to the median ventral line as a thin sheet with a concave posterior border. Anteriorly and dorsally it is continuous with the *coronary ligament,* a stout ligament that attaches the liver to the central tendon of the diaphragm. The coronary ligament is circular in form, and its ring of attachment to the liver bounds a small space on the anterior face of the liver that is free from serosa.

Now trace the intestine from the pylorus. Its first portion, the *duodenum,* is bound to the liver by the hepatoduodenal ligament. The duodenum curves abruptly caudad. In the rabbit it is very long and forms a loop. The part of this loop that descends posteriorly is named the *descending limb;* the short turn at the most posterior part of the loop is the *transverse limb;* and the part that ascends anteriorly toward the stomach is called the *ascending limb*. The duodenum of the cat descends caudad for about two inches, then turns to the left. The duodenum is supported by a part of the dorsal mesentery, the *mesoduodenum*. It is also attached to the right kidney by a mesenterial fold, the *duodenorenal ligament*. Situated in the mesoduodenum is the *pancreas,* which can be seen by spreading the mesentery. In the rabbit the pancreas consists of streaks of glandular tissue scattered in the mesentery, chiefly along the courses of blood vessels. In the cat the pancreas is a definite, compact, pinkish gland that extends to the left into the dorsal wall of the greater omentum, dorsal to the greater curvature of the stomach. In the rabbit the *pancreatic duct* enters the duodenum about an inch or an inch and a half anterior to the beginning of the ascending limb of the duodenum. This location of the pancreatic duct is unusual in mammals. In the cat there are two pancreatic ducts. The principal one joins the common bile duct at the point where the latter enters the duodenum. By picking away the substance of the pancreas at this point you can readily find the duct. The swollen chamber where bile and pancreatic ducts unite is known as the *ampulla of Vater*. The second or accessory pancreatic duct in the cat enters the duodenum about three-quarters of an inch caudad of the principal duct but is not easy to find.

From the duodenum trace the coils of the remainder of the small intestine. It is supported by a part of the dorsal mesentery, the *mesentery* proper. The first portion of the small intestine beyond the duodenum is called the *jejunum,* and the remainder the *ileum,* but there is no definite boundary between the two. Note the coils of the mesentery accompanying the intestine and the frequent fusions that occur (especially in the rabbit) between these coils. In the cat you will have to withdraw the greater omentum from the coils of the intestine; the omentum may then be cut across near the spleen and discarded. Follow the small intestine to its enlargement into the *large intestine*. In doing this it may be necessary to tear the fusions of the mesentery slightly, but the structures should be kept as intact as practicable.

*Rabbit:* At the point of juncture of the large and small intestine there is an enlargement, the *sacculus rotundus,* a few inches from which extends an enormous blind sac. The first portion of this is very large and is known as the *caecum;* the last few inches are reduced in diameter and constitute the *vermiform appendix.* Both caecum and appendix are very much longer in the rabbit than in most mammals. The wall of the caecum is marked by a spiral line that denotes the position of an internal spiral fold of the lining membrane of the caecum. The large intestine beyond the caecum is named the *colon.* The colon is supported by a part of the dorsal mesentery called the *mesocolon.* The first part of the colon, the *ascending colon,* is rather long and pursues a winding course. At first the wall bears three longitudinal muscular bands, the *bands* or *taeniae* of the colon. Between the taeniae the wall of the colon is greatly puckered, forming little sacculations, the *haustra.* Beyond the ascending colon, the colon runs for a short distance transversely across the peritoneal cavity from right to left and is then the *transverse colon.* At the left it turns abruptly posteriorly as the *descending colon.* At this turn the mesocolon is fused to the mesogaster. The descending colon passes straight posteriorly and disappears dorsal to the urogenital organs.

*Cat:* The posterior portion of the gut in the cat is similar to that of the rabbit except that the caecum is much smaller and the vermiform appendix is almost absent.

The *urinary bladder* (see chap. 12) is a sac occupying the posterior end of the peritoneal cavity, immediately internal to the body wall and ventral to the large intestine. From the ventral surface of the bladder a mesentery, the *median ligament* of the bladder, extends to the midventral line and here continues forward for some distance. Near the exit of the bladder from the peritoneal cavity there is on each side a slightly developed ligament, the *lateral ligament* of the bladder.

The terminal portion of the descending colon is the *rectum.* Both the rectum and the duct of the bladder pass to the exterior through the ring formed by the pelvic girdle and vertebral column. They will be followed later in connection with the urogenital system. For the present it may be stated that the rectum and urogenital ducts are completely separated, so there is no cloaca.

The small bodies that may have been noted in the mesentery, usually buried in fat, are *lymph glands.* Small portions of lymphatic tissue called *lymph nodules* are also abundantly present in the wall of the intestine. Aggregations of lymph nodules known as *Peyer's patches* can often be seen as thickened oval spots on the surface of the intestine, slightly different in color from the rest of the intestinal wall. In the rabbit they occur along the entire small intestine on the side opposite that attached to the mesentery; there is a larger patch at the junction of sacculus rotundus and caecum. The walls of the sacculus rotundus, caecum, and vermiform appendix are composed almost entirely of lymph nodules. In the cat, Peyer's patches occur as oval light-colored spots along the colon but are best seen from the inside.

Slit open various parts of the digestive tract along the side opposite the attachment of the mesentery. Wash out the interior. In the cat note the marked ridges or *rugae* in the wall of the stomach; these are very slight in the rabbit. Cut through the pylorus and note the thickened ridge or *pyloric sphincter* at this place. In the wall of the small intestine observe the velvety appearance due to the *villi* (finger-like projections of the mucous membrane). Find also the depressions marking the positions of the lymph nodules and Peyer's patches. In the rabbit slit open the sacculus rotundus, caecum, and appendix and note the spotted appearance of the interior, owing to the lymph nodules composing the walls. In the interior of the caecum is a spiral ridge. Cut through the junction of large and small intestine and note in both animals an elevation, the *ileocolic valve,* projecting into the ileum.

L. Summary

1. The coelom develops as a cavity in the nonsegmented lateral plate mesoderm (hypomere).

2. Cells from the inner (visceral) layer of the hypomere produce the smooth muscle of the gut wall as well as the striated pharyngeal muscle.

3. During development the gut is supported by dorsal and ventral mesenteries derived from the visceral wall of the hypomere. The dorsal mesentery is extensive and suspends the gut as well as carrying major blood vessels, nerves, and lymphatics. The ventral mesentery largely disappears but persists in adult vertebrates as specialized ligaments associated with the liver and the bladder.

4. The visceral organs do not lie within the cavity of the coelom but are covered by its membranes and are thus retroperitoneal.

5. The coelom arises as a continuous cavity, but a transverse septum develops that subdivides it into an anterior, paricardial cavity, and a posterior pleuroperitoneal cavity. In reptiles, birds, and mammals the pleuroperitoneal cavity becomes further subdivided to form a cavity surrounding the lungs (pleural cavity) and a body cavity (abdominal or peritoneal cavity) around the digestive system.

6. The coelom greatly facilitates the functioning of the organs it encloses by allowing limited movements and changes in relative position.

7. The nonmuscular portion of the digestive system is differentiated into a large number of glands concerned with the digestion and absorption of food. The gut can often be divided into a number of regions on the basis of function and histological appearance. Such divisions are not necessarily reflected in the gross external appearance of the gut.

8. In *Branchiostoma* the oral cavity is separated from the pharynx by the velum, and it is assumed that this point also corresponds to the boundary between these two areas in gnathostomes.

9. Embryologically, the nasal organs are derived from epidermal sacs, which develop from the nasal placodes. The nasal sacs become lined with sensory cells that are concerned with olfaction and nonsensory cells that are often glandular.

10. The inner portion of the nasal apparatus in lungfishes does not transmit air into the oral (buccal) cavity and does not correspond to the internal nostril (choana) of tetrapods.

11. Many tetrapods have an accessory sensory structure, the vomeronasal organ.

12. The mouth of chordates is homologous with that of their ancestors. The chordates and the echinoderms share a common ancestor.

13. The tongue in fishes is a portion of the floor of the buccal cavity.

14. In other groups the tongue may be more elaborate, more mobile, and used in the capture of food.

15. In fishes and amphibians the motor innervation of the tongue is by the hypoglossal nerve, which is a spinal nerve. In amniotes the hypoglossal is a cranial nerve.

16. The pharynx is characterized by a number of gill slits that develop from a series of pharyngeal pouches. In amniotes the gill slits appear in the embryo but close over during development.

17. The tympanic cavity arises as an outgrowth of the first pharyngeal pouch.

18. The endodermal tissue derived from the pharyngeal pouches proliferates to form a variety of glandular structures such as the thymus, parathyroid, aortic body, carotid body, tonsil, and ultimobranchial bodies.

19. The pharyngeal muscle is regarded as a "special" visceral muscle that is striated. The segmentation of the pharyngeal muscle is imposed by the development of the gill slits and is branchiomeric. This segmentation may or may not correspond to that of the muscle blocks (somites) of the rest of the body, which is metameric.

20. The jaws have evolved from a modified gill arch, the mandibular arch. There is no general agreement about the number of arches involved with the mandibular arch in the evolution of the jaws and their associated structures. The derivation of the jaws, trabeculae cranii, and gill arches from the neural crest material suggests that the trabeculae are also essentially modified gill arches representing an arch anterior to the mandibular arch. Such an arch may be termed the premandibular.

21. The presence of a nervus terminalis in many vertebrate groups suggests that an arch anterior to the premandibular may have existed in the ancestors of the gnathostomes. Such an arch (if it existed) was lost in the dawn of vertebrate history.

22. The arch posterior to the mandibular, the hyoid arch, is closely associated with jaw suspension, and its gill slit is reduced to a spiracle.

23. In tetrapods the tongue skeleton has evolved from the more posterior gill arches. Evolution has tended to produce simplification of the tongue apparatus, and in amniotes such material forms the laryngeal and tracheal skeletons.

24. Gills are concerned with gaseous exchange, and their surface area is enhanced by the presence of gill filaments and lamellae.

25. Blood flow through the gill lamellae is in the opposite direction from that of the water passing over the gills. This countercurrent relationship facilitates a high level of gaseous interchange between the water and the blood.

26. Gill ventilation in fishes involves a double-pump mechanism that ensures that a stream of water is passing over the gills for the major portion of each respiratory cycle.

27. Lungs and swim bladders are thought to have evolved from ventral outgrowths of the pharynx in the early jawed vertebrates. Such areas acted as accessory areas of gaseous exchange in environments where there were periodic reductions in the available oxygen in the water. The pharyngeal extensions evolved into the lungs in tetrapods and Dipnoi, whereas in many fishes they migrated dorsally and formed the swim bladder.

28. Swim bladders control the overall specific gravity with reference to the surrounding medium and can produce neutral buoyancy. Swim bladders are most well developed in active midwater fishes and are reduced, or act as a fat store, in deepwater forms.

29. In some fishes the swim bladder is sensitive to vibrational changes relayed to the ear region by a series of bones, the Weberian ossicles. In other groups the swim bladder may act as a resonating chamber or be vibrated to produce sounds.

30. Lung ventilation in amphibians is accomplished by a "force-pump" method that involves movements of the floor of the buccal cavity, whereas reptiles use the ribs in costal respiration.

31. In amphibians the inner surface of the lung is relatively smooth or thrown into a series of shallow pockets, but in reptiles the inner surface is more elaborate; the lung reaches its maximum complexity in the birds and mammals.

32. In birds and mammals the trachea is long and divides into a pair of primary bronchi that lead into the lungs. The respiratory surface is vastly increased by the presence of air capillaries in birds and alveoli in mammals. Associated with the lungs in birds are a number of air sacs that are often connected to cavities in the pneumatic bones.

33. Lungs in birds have a "straight-through" circulation, but there is no agreement on whether the flow of air is unidirectional or is reversed at inspiration and expiration.

34. Ventilation of the lung in birds is associated with the intercostal muscles and the movements of the sternum during flight. The air sacs aid the flow of air and are not important areas of gaseous exchange.

35. The extreme complexity of lungs in mammals produces a very high ratio of lung area to body weight when compared with other vertebrates.

36. Mammalian lungs are also characterized by an extensive series of air passages that constitute a respiratory dead space. This space accounts for about a quarter of the volume of inspired air during resting respiration but falls to a very small percentage during large respiratory movements.

37. The ribs, intercostal muscles, and diaphragm are used to produce respiratory movements in mammals.

38. Many vertebrates have accessory areas of respiration such as the skin, buccal cavity, and cloaca.

39. The digestive tract is essentially a muscular tube with a variety of glands embedded in its wall. In addition to these glands, the gut is associated with the liver and the pancreas.

40. The esophagus is lined with stratified epithelium similar to that found in the buccal cavity, and its walls contain both smooth and striated muscle. The esophagus may have accessory structures such as the crop.

41. The stomach is a muscular pouch evolved primarily as a storage area and secondarily as a digestive chamber.

42. Major modifications occur in the stomach region, and these are most pronounced in the ruminants, where the stomach is divided into rumen, reticulum, omasum, and abomasum. It is generally accepted that the rumen, reticulum, and omasum are essentially elaborations of the base of the esophagus, whereas the abomasum, which contains the digestive glands, is the true stomach.

43. The morphology of the alimentary canal, especially the caecum and the appendix, varies greatly in relation to the diet and mode of life.

**References**

Alexander, R. McN. 1966. Physical aspects of swim bladder function. *Biol. Rev.* 41:141–76.
————. 1973. *The Chordates*. Cambridge: Cambridge University Press.
Bertmar, G. 1969. The vertebrate nose: Remarks on its structure, functional adaptation and evolution. *Evolution* 23:131–52.
Blaxter, K. L. 1966. *The energy metabolism of ruminants*. London: Hutchinson.
Briggs, M. H. 1964. Digestion in the ruminants. *Science Progress* 52:219–31.
De Beer, G. R. 1937. *The development of the vertebrate skull*. London: Oxford University Press.
DeJongh, H. J., and Gans, C. 1969. On the mechanism of respiration in the bullfrog *Rana catesbiana:* A reassessment. *J. Morph.* 127:259–90.
Field, H. E., and Taylor, M. E. 1969. *An atlas of cat anatomy*. 2d ed. Chicago: University of Chicago Press.

Gans, C. 1970. Strategy and sequence in the evolution of the external gas exchangers in ectothermal vertebrates. *Forma et Functio* 3:61–104.

Gans, C.; DeJongh, H. J.; and Farber, J. 1969. Bullfrog (*Rana catesbiana*) ventilation: How does the frog breathe? *Science* 163:1223–25.

Gaunt, A. S., and Gans, C. 1969. Mechanics of respiration in the snapping turtle *Chelydra serpenta. J. Morph.* 128:195–228.

Goodrich, E. S. 1935. *Studies on the structure and development of the vertebrates.* New York: Dover Publications.

Hildebrand, M. 1974. *Analysis of vertebrate structure.* New York: Wiley.

Hughes, G. M. 1965. *Comparative physiology of vertebrate respiration.* London: Heinemann.

Hungate, R. E. 1966. *The rumen and its microbes.* New York: Academic Press.

Jollie, M. 1968. Some implications of the acceptance of the delamination principle. In *Current problems in lower vertebrate phylogeny: Nobel Symposium,* ed. Tor Orvig, 4:89–107. Stockholm: Almquist and Wiskel.

————. 1971. A theory concerning the evolution of the visceral arches. *Acta Zoologica* 52:85–96.

Jones, J. D. 1972. *Comparative physiology of respiration.* London: Arnold.

Kluge, A. G., et al. 1977. *Chordate structure and function.* 2d ed. New York: Macmillan.

Langman, J. 1963. *Medical embryology.* Baltimore: Williams and Wilkins.

Lawson, R. 1964. Lungs and swim bladders. *School Science Review* 45:386–90.

McMahon, B. R. 1969. A functional analysis of the aquatic and aerial respiratory movements of the African lungfish, *Protopterus aethiopicus,* with reference to the evolution of lung ventilation mechanisms in vertebrates. *J. Exp. Biol.* 51:407–30.

Miles, R. S. 1964. A reinterpretation of the visceral skeleton of *Acanthodes. Nature, Lond.* 204:457–59.

————. 1965. Some features in the cranial morphology of acanthodians and the relationships of the Acanthodii. *Acta Zoologica* 46:235–55.

————. 1968. Jaw articulation and suspension in *Acanthodes* and their significance. In *Current problems in lower vertebrate phylogeny: Nobel symposium,* ed. Tor Orvig, 4:109–27. Stockholm: Almquist and Wiskel.

Moore, J. A., ed. 1964. *Physiology of the Amphibia.* Vol. 1. New York: Academic Press.

Morton, J. 1967. *Guts.* Studies in Biology, no. 7. London: Arnold.

Nickel, R., et al. 1973. *The viscera of domestic animals.* New York: Springer-Verlag.

Panchen, A. L. 1967. The nostrils of choanate fishes and early tetrapods. *Biol. Rev.* 42:374–420.

Parsons, T. S. 1959. Nasal anatomy and phylogeny of the reptiles. *Evolution* 13: 175–87.

————. 1967. Evolution of the nasal structure in lower tetrapods. *Amer. Zoologist* 7:397–413.

Piiper, J., and Schumann, D. 1967. Efficiency of $O_2$ exchange in the gills of the dogfish, *Scyliorhinus stellaris*. *Resp. Physiol.* 2:135–48.

Romer, A. S. 1949. *The vertebrate body*. 1st ed. Philadelphia: W. B. Saunders.

———. 1960. *Vertebrate paleontology*. 3d. ed. Chicago: University of Chicago Press.

Romer, A. S., and Parsons, T. S. 1977. *The vertebrate body*. 5th ed. Philadelphia: W. B. Saunders.

Schmidt-Neilsen, K. 1971. How birds breathe. *Sci. Amer.* 225:72–79.

Smith, H. M. 1960. *Evolution of chordate structure: An introduction to comparative anatomy*. New York: Holt, Rinehart and Winston.

Thomson, K. S. 1964. The comparative anatomy of the snout in rhipidistian fishes. *Bull. Mus. Comp. Zool.* 131:313–57.

———. 1965. The nasal apparatus in Dipnoi with special reference to *Protopterus*. *Proc. Zool. Soc. Lond.* 145:207–38.

———. 1969. Gill and lung function in the evolution of the lungfishes (Dipnoi): An hypothesis. *Forma et Functio* 1:250–62.

———. 1971. The adaptation and evolution of early fishes. *Quart. Rev. Biol.* 46:139–66.

Torrey, T. W. 1963. *Morphogenesis of the vertebrates*. New York: Wiley.

Walker, W. F. 1965. *Vertebrate dissection*. 3d ed. Philadelphia: W. B. Saunders.

Weston, J. A. 1970. The migration and differentiation of neural crest cells. In *Advances in morphogenesis,* ed. M. Abercrombie et al., 8:41–114. New York: Academic Press.

Young, J. Z. 1950. *The life of the vertebrates*. London: Oxford University Press.

# 11 The Comparative Anatomy of the Circulatory System
*Ronald Lawson*

**A. Introduction**

The vertebrate vascular system is essentially a system of closed tubes which permeate the tissues of the body. The blood-vascular system is augmented by an elaborate series of blind-ended vessels forming the lymphatic system. The lymphatic system collects fluid which has seeped out of the capillaries into the intercellular spaces and returns this to the general circulation. The efficient functioning of the blood-vascular and lymphatic systems ensures that the tissues are well supplied with oxygen and the necessary nutrients and that carbon dioxide and other waste products are removed. In addition these systems are intimately concerned with removing cellular debris and combating invading organisms, such as bacteria.

Details of the anatomy and embryology of the vascular system are correlated here with their functional morphology and with their relevance to establishing the evolutionary relationships between the various groups of vertebrates. Special emphasis is placed on the heart and its bearing on the evolution of a double circulation. In addition to a general description of the vascular system a section on specialized adaptations is included.

**B. Parts of the Circulatory System**

The circulatory system of vertebrates is composed of two networks of branching tubes which enclose the circulating fluids, the *blood-vascular* (or *cardiovascular*) system and the *lymphatic system*. The blood-vascular system is the more conspicuous and is the one referred to as the "circulatory" or "vascular" system when the term is used without qualification. The blood-vascular system is a *closed* system which consists of a series of branching *vessels* forming a continuous circuit in which the enclosed fluid travels. It consists of the *heart, arteries, arterioles, capillaries, venules* and *veins*. The heart is a contractile organ composed of a special type of striated muscle, *cardiac muscle,* lying in the mid-ventral region of the anterior part of the body. The *blood* consists of a colorless fluid, *plasma,* containing the microscopic *blood cells* or *corpuscles*. The blood cells are of two main types: *red cells* or *erythrocytes* and *white cells* or *leucocytes*. The red blood cells carry the respiratory pigment, hemoglobin, which is responsible for the transport of oxygen (and, to some extent, carbon dioxide) and they may or may not possess a nucleus. Although there are a number of different types of white blood cell, all have a nucleus. The body is constantly producing blood cells of both types to replace those that become exhausted, and in normal conditions the rate of production balances the rate of breakdown. The blood carries a large number of red cells and in the human the recognized normal value is approximately 4.6 million per cubic millimeter. The value for the white cells is around 5,000 to 9,000 per cubic millimeter. The life of the blood cells is quite short, and in man the red cells survive for some 120 days, the white cells between 8 and 15 days.

Blood vessels cannot be defined on the basis of the type of blood they carry (i.e., *oxygenated* or *deoxygenated*) since, for example, the pulmonary artery carries deoxygenated blood and the pulmonary vein oxygenated blood. They must therefore be classified in terms of whether they carry blood *toward* or *away from* the heart. On this basis arteries carry blood from the heart and veins carry blood toward the heart. The blood in vertebrates leaves the heart under pressure that is quite considerable, and the major arteries therefore have thick walls with considerable *elastic tissue* and *smooth muscle*. The walls of the veins are much thinner than those of the corresponding arteries, although the internal diameter is usually much greater. Venous elastic tissue is usually inconspicuous and some veins completely lack muscle cells.

A characteristic feature of blood-vascular systems is the existence of a *pressure gradient* in which the pressure is high in the major arteries and decreases in the smaller vessels. The flow of blood through the vascular system is retarded by *resistance* resulting from the viscosity of the blood itself, friction, and changes in the diameter of the blood vessels. The total resistance to flow is the *peripheral resistance* and the net blood pressure must exceed this resistance so that the blood may circulate. Arteries divide into smaller arterioles, which in turn become capillaries which join the venules and then the veins. This results in the formation of an extensive series of vessels which permeate the tissues as the *microcirculation*. The total cross-sectional area of the microcirculation is enormous since, when any vessel in the circulation divides, the combined cross-sectional area of the two branches is always greater than that of the original vessel. Taken together the capillaries form the largest "organ" and in man, for example, their total bulk is more than twice that of the liver. If all the capillaries were open at the same time, they would contain the whole of the blood supply of the animal. Clearly this does not happen, and hence there must be some method for regulating blood flow through the capillary network. As the arteries branch into the tissues their muscular sheaths become thinner and in the arterioles they may be only one cell thick. The *thoroughfare* channels that arise from the arterioles are sparsely supplied with isolated muscle cells. The major portion of the capillary bed arises as side branches from the thoroughfare channels and at the point where each capillary leaves its thoroughfare channel a prominent ring of muscle cells forms a *precapillary* sphincter. Such sphincters regulate the flow of blood into a capillary, and by opening and closing periodically they ensure irrigation of first one part of the capillary network and then the other (fig. 11.1).

The dimension of the larger vessels in the vascular system is largely controlled through *vasomotor centers* in the medulla of the brain. The influence of these centers is mediated via the autonomic nervous system and, in general, stimulation of the sympathetic system produces narrowing of the blood vessels (*vasoconstriction*) whereas the parasympathetic

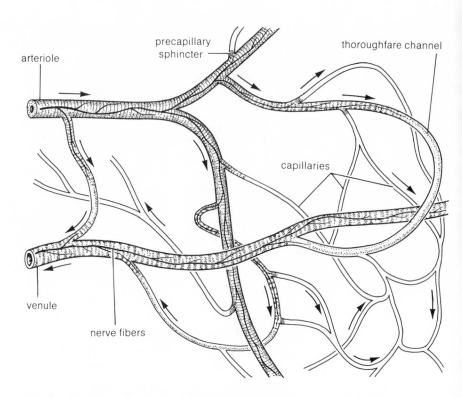

Fig. 11.1. Microcirculation of blood. Portion of capillary network. The blood
enters the capillary bed through the arteriole and leaves via the venule. From
the arteriole blood passes into the thoroughfare channels and then into the capillaries, which then return it to the channels. The blood flow into the thoroughfare channels is regulated by rings of muscle, the precapillary sphincters. After
Zweifach, 1959 (*Sci. Amer.* 200:54–60).

provokes the opposite effect (*vasodilation*). Many of the muscle cells in
the capillary network have no nerve connections and capillary circulation is largely controlled by chemical means. The precise nature of vaso-
control at the microcirculation level is not well understood, but hormones
from the adrenal cortex (corticosteroids) as well as epinephrine and norepinephrine from the adrenal medulla are involved. Such substances affect the capillary muscles and hence the diameter of the vessels. However, it has been shown recently that a number of other chemical compounds, especially compounds containing sulphur (sulphydryls), as well
as substances carried by special cells known as *mast cells,* are important.
Such compounds probably do not act directly on the muscles but modify
the way in which the muscles respond to epinephrine and other
hormones.

The blood-vascular system is essentially closed, only in the liver and
spleen do gaps occur in the *endothelial walls* of capillaries that allow the
blood to establish direct contact with the cells. In other areas the cells
are bathed in a fluid, the *tissue fluid* or *extracellular fluid,* which acts as
an intermediary between the tissue cells and the blood. Tissue fluid is
formed by the movement of fluid from capillaries (fig. 11.2) and has a
composition similar to that of plasma except that it contains only approxi-

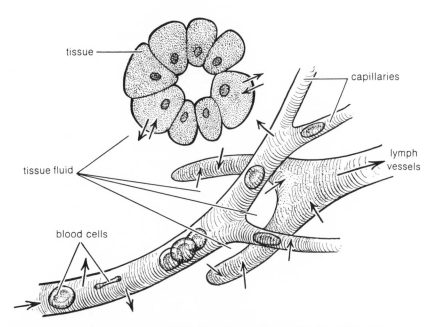

Fig. 11.2. Formation of tissue fluid. Fluid leaves the capillaries in the tissues and is collected by the capillaries or the lymph vessels.

mately 0.1% protein whereas plasma usually has around 7%. Much of the fluid that bathes the tissues is returned to the blood-vascular circulation via the *lymph vessels,* or *lymphatics.* The lymphatic system is, however, an *open system* consisting not only of branching tubes but of large spaces, the *lymph sinuses,* which occur beneath the skin (particularly well developed in such amphibians as the frog), between muscles, in the mesenteries, in the walls of the digestive system, and around the central nervous system. From the sinuses lymph is collected by lymph vessels, which open into the veins of the blood-vascular system. Small lymph vessels have an endothelial lining that is supported by connective tissue though larger vessels may have some muscle in the wall. Both lymph vessels and veins are well supplied with valves that permit flow in any one direction and the flow of lymph is largely brought about by muscular and respiratory movements. In fishes and amphibians the lymphatic system is supplied with *lymph hearts,* which are small, two-chambered, contractile structures lying at the point where lymph vessels enter veins and aid the flow of lymph into the general circulation. There are usually only a few such hearts, but in the long-bodied amphibians a large number of segmentally arranged lymph hearts may be found (fig. 11.3). In most cases, however, the flow of lymph is almost entirely dependent on extrinsic factors, such as rhythmic contractions of the intestine, changes in pressure in the thorax, and contraction of muscle groups in the limbs (fig. 11.4). The composition of lymph varies considerably according to the area from which it comes, but in general it has a protein content between that of blood and that of tissue fluid. Before entering the blood-vascular system the lymph passes through one or more *lymph glands (nodes).* The lymph is conducted to the nodes by *afferent vessels* whose entrance to the gland is guarded by *valves.* The

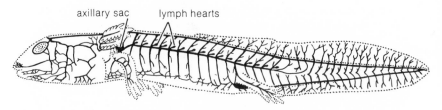

Fig. 11.3. Lymph hearts of the salamander. Dorsal, ventral, and lateral superficial lymph vessels are present and the lateral vessel carries a series of lymph hearts. The lymph hearts are contractile, supplied with valves, and ensure that lymph is kept in circulation. Lymph from each lateral vessel enters the venous system through an axillary sac. After Romer 1970 (*The vertebrate body*. Philadelphia: Saunders).

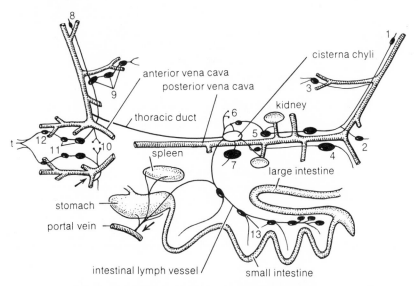

Fig. 11.4. Lymphatics of mammal (deep vessels of rat lymphatic system). The lymphatics are in solid black and a number of neighboring veins are shown. The lymph nodes are numbered: 1, knee; 2, tail; 3, inguinal; 4, lumbar; 5, kidney; 6, nodes in the region of the cistern chyli; 7, intestinal; 8, elbow; 9, axilla; 10, thoracic; 11, cervical; 12, submaxillary; 13, mesenteric; t, plexus of lymphatics around the tongue and lips. Arrows indicate the point of entrance of the lymphatics into the venous system at the junction of the subclavian and jugular veins and into the portal vein. After Romer 1970 (*The vertebrate body*. Philadelphia: Saunders).

nodes consist of a network of connective tissue embedded with masses of *lymphocytes* (a variety of white blood cell). The functions of the lymph glands are to destroy foreign particles and to add lymphocytes to the general circulation. Intestinal lymphatics are especially important since they absorb a large portion of digested fat. The absorbed carbohydrate and protein is carried in the hepatic portal vein to the liver. Lymph from the intestine traverses the mesenteric lymph vessels, which are often termed *lacteals* since they may assume a milky appearance when they are distended with fatty material. These vessels, together with those from the posterior part of the body, usually form a pair of longitudinal *tho-*

*racic ducts.* Paired thoracic ducts are common in fishes, reptiles, and birds, but in mammals they tend to fuse into a single structure, where a large swelling, the *cisterna chyli,* may develop. The lymphatic vessels frequently run parallel to the veins and, although they are thought by some to be derived from the embryonic veins, are probably of independent origin. Because of the delicate and diffuse nature of the lymph vessels, little of the lymphatic system can be seen in general dissection.

## C. Development of the Vascular System

This section describes the embryological development of the blood-vascular system from mesenchymal *angioblasts.* The nomenclatures of the heart chambers—that applied during the period of development and that of the adult stage—are discussed, in an attempt to clarify the ambiguities that arise in the literature. Such clarification facilitates meaningful comparisons of heart structure and functions in various vertebrate groups.

The blood and blood vessels arise in the embryonic mesoderm as a series of blood islands, which are masses of mesenchyme cells. These cells differentiate into blood-forming cells or angioblasts. The central angioblasts in each island form the blood cells and the peripheral ones flatten and form an endothelium. The developing blood islands fuse to form the blood vessels. Since blood cells survive only a relatively short time in the vascular system, animals must retain throughout life a tissue that is capable of producing blood cells. Such tissues are termed *hemopoetic,* and the location of such tissues varies considerably in different animals and even in the same animal at different stages in the life history. In many lower vertebrates such as sharks, amphibians, and reptiles the embryonic kidney and the thymus are important hemopoetic tissues; in embryonic mammals the liver and spleen are active in the formation of blood cells. In many teleosts the kidneys continue to produce blood cells throughout life, but in other teleosts, adult amphibians, and adult reptiles most of the hemopoetic tissue is located in the liver. In elasmobranchs and lungfishes the gonads are also important blood-cell-forming areas. In most vertebrates the lymphoid tissue which develops in the embryo, such as tonsils, lymph nodes, etc., continues to produce lymphocytes throughout life. In higher vertebrates the *bone marrow* and the spleen are important areas of cell production. In adult mammals the marrow of flat bones such as vertebrae, skull bones, and sternum become important hemopoetic areas if large numbers of cells are required.

The heart arises as a cluster of angiogenic cells which form a *heart rudiment* on each side of the embryo between the endoderm and the splanchnic mesoderm. Each heart rudiment acquires a lumen lined by cells which form the endothelium of an *endocardial tube.* Folding movements of the endoderm concerned with the formation of the foregut within the embryo bring together the two endocardial tubes which fuse to form a single *heart tube.* During this period the cells of the splanchnic mesoderm adjacent to each endocardial tube subdivide, and, as the two tubes fuse, the cells surround the resulting cardiac tube as an *epimyocardium (epimyocardial mantle).* Later this mantle of cells differentiates to form the muscular wall of the heart, the *myocardium,* and the outer covering of the heart, the *epicardium* or *visceral peritoneum.* During development the epimyocardium is usually separated from the endocardial tube by a gelatinous layer, the *cardiac jelly*. The cardiac jelly is soon invaded by mesenchyme cells, which are important in the formation of the heart valves. Initially the heart tube is connected to the wall of the pericardium by mesenteries, the *dorsal* and *ventral mesocardium*. The ventral mesocardium disappears almost as soon as it forms but the dorsal mesocardium persists for some time before it breaks down, leaving the heart freely suspended within the pericardium and attached only at its anterior and posterior ends. As the heart develops within the pericardium, it becomes "buckled" and eventually assumes an S shape (fig. 11.5). The anterior and midportions of the heart tube, since they are the first to be released from the dorsal mesocardium, bulge ventrally and posteriorly. The *cardiac loop* thus formed consists of an anterior limb, which eventually develops into the *bulbus cordis,* and a *posterior* or *descending limb,* which forms the ventricle. The portion of the cardiac tube that eventually forms the sinus venosus and the atrium remains temporarily embedded in the mesenchyme of the transverse septum. Later it becomes freed and assumes an intrapericardial position moving in a cranial and dorsal direction. It has frequently been stated that the folding of the heart tube is due to the fact that the elongated tube cannot be accommodated within the pericardium. It is now believed that formation of the cardiac loop is a basic property of the cardiac tube, since it will assume an S shape when removed from the constraints of the pericardium and grown in tissue culture. Fusion of the two heart rudiments to form the heart tube takes place from an anterior to posterior direction so that the truncoventricular part of the heart is formed first and the atria and sinus venosus later.

During development the first contractions of the heart tube occur in the ventricular myocardium, but when the atrium is "added," its rate of contraction is faster than that of the ventricle. The ventricular rate then accelerates to match that of the atrium. The sinus venosus has an even faster intrinsic rate of contraction, and the atrium and ventricle then speed up to match the rate set by the sinus. Thus, it is the sinus venosus that dictates the rate at which the heart beats and is therefore termed the *pace-maker*. It also appears that the early heartbeats are facilitated by the presence of the cardiac jelly. If the embryonic tubular heart is to expel an adequate volume of blood with each wave of contraction, the diameter of the vessel must be reduced to almost obliterate the lumen. In addition the heart must be able to relax sufficiently to accommodate enough blood for a substantial cardiac output. Experiments on the chick heart have shown that the relaxed myocardium can hold an adequate volume of blood but is unable to contract sufficiently to expel the required amount. This mechanical difficulty is resolved by the cardiac jelly, which provides an intervening medium through which the force produced by the relatively slight contractions of the myocardium can be transmitted radially to effect a closure of the lumen. Subsequently, the great buildup of cardiac muscle within the myocardium eliminates the necessity for the jelly. Twisting of the cardiac tube varies in different vertebrate groups. In elasmobranchs the ventricle bends only moderately to the right and most of the movement is ventral and caudal

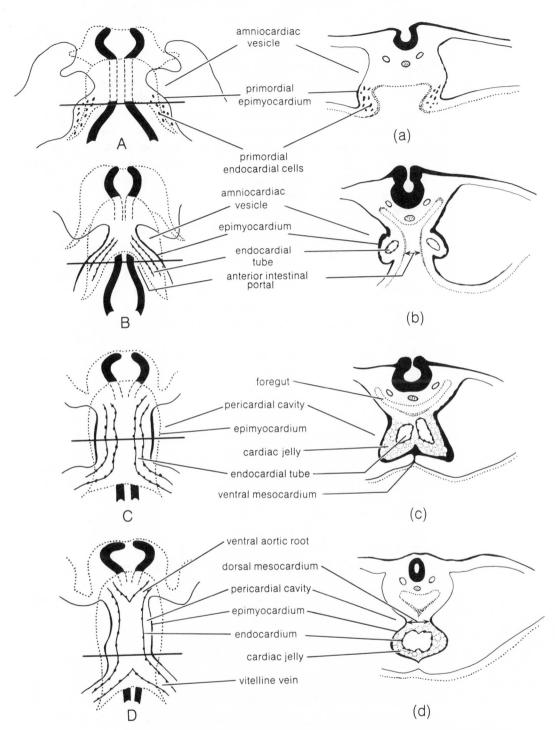

Fig. 11.5. Development of the heart in the chick (25–30 hours). Diagrams on the left (*A–D*) are ventral views and those on the right (*a–d*) are transverse sections at the levels indicated by the heavy lines across the ventral view figures. The heart develops from paired primordia, which form endocardial tubes surrounded by an epimyocardium. The endocardial rudiments fuse to form a single cardiac tube in the mid-ventral line. After Torrey 1967 (*Morphogenesis of the vertebrates*. New York: John Wiley).

and so it lies more or less directly below the atrium. In amphibians and amniotes bending is such that the ventricle is located nearer the posterior and the atria move forward to lie above the truncus arteriosus.

The vertebrate heart is therefore essentially a folded tube whose walls are composed of cardiac muscle (*myocardium*) lined internally by the *endocardium* and externally by the *epicardium*. The heart is divided into a number of *chambers* with valves which facilitate blood flow. In its simplest anatomical form (such as in the cartilaginous fishes) the heart consists of a posterior chamber, the *sinus venosus,* which collects all the blood returning from the body in the systemic veins. The sinus venosus opens into a single *atrium,* which has slightly thicker walls than the sinus venosus. The atrium leads into a thick-walled chamber, the *ventricle,* which provides the power to pump the blood through the vascular network. The blood leaves the heart via the *conus arteriosus* and then flows into the *ventral aorta.* The conus arteriosus has a number of valves that prevent the blood from flowing back into the heart. The ventral aorta gives rise to a series of *aortic arches,* which form the *afferent branchial arteries* carrying the blood to the gills for oxygenation. Within the gills the afferent branchial vessels break up into a network of capillaries from which blood is collected by a series of vessels which unite to form the *efferent branchial arteries*. Blood from these arteries eventually enters either the *carotid arteries,* which supply the head, or the *dorsal aorta,* the branches of which supply all areas of the body. This blood is then recollected and returned to the sinus venosus. This type of blood-vascular system is usually termed *single circulation,* since a red blood cell at any point within the vascular system has to pass through the heart only *once* in order to return to the same position. Modifications of the blood-vascular system in vertebrates are largely related to the incorporation of a *pulmonary circulation* and consequent separation of pulmonary blood and systemic venous blood within the heart. In general this involves the subdivision of the atrium and the ventricle by *septa* (*interatrial* and *interventricular septa*) into right and left sides. The left side receives blood from the lungs to be circulated through the body and the right side receives systemic blood from the body to be pumped to the lungs. This type of system is referred to as *double circulation,* since any cell must pass through the heart *twice* to return to any position within the system.

**1. Nomenclature of the Heart Chambers**

Considerable confusion exists about the nomenclature of the chambers of the vertebrate heart and the junction of the heart with the major arterial arches. The terms *conus arteriosus, truncus arteriosus, bulbus arteriosus,* and *bulbus cordis* have always been used ambiguously. In the descriptions given below the following terminology has been adopted:

*Conus arteriosus* is a narrow chamber leading from the ventricle into the ventral aorta. It is supplied with a number of *conal valves,* often arranged in rows, which prevent regress of blood into the ventricle. The walls of the conus are composed largely of cardiac muscle and the conus is regarded as a chamber of the heart.

*Bulbus arteriosus* lacks cardiac muscle and is regarded as the swollen, expanded base of the ventral aorta. *It is peculiar to teleosts.* The entrance to the bulbus from the ventricle is guarded by *bulbal valves,* and the walls of the bulbus contain large quantities of elastic tissue.

*Truncus arteriosus* is considered synonymous with ventral aorta.

*Bulbus cordis* should be used only to describe the *embryonic structure* formed from the ascending limb of the cardiac tube. It eventually produces a portion of the ventricular base, the exit from the ventricle, and the junction between the ventricle and the major arteries.

*Ventricular apex* is the narrow portion of the ventricle at the farthest distance from the atrioventricular junction.

*Ventricular base* is the anterior portion of the ventricle joining the atria and the arterial arches.

**2. Development of the Heart in Mammals**

In the early stages of development the sinus venosus has a small *transverse portion* and *right* and *left horns,* into which drain the *umbilical, vitelline,* and *common cardinal veins.* The left sinus horn gradually becomes less important as a *sinoatrial fold* develops. The right sinus horn becomes incorporated into the wall of the right atrium and forms a smooth-walled portion of the chamber, the *sinus venarium,* which is later marked off from the trabeculate portion of the atrium by the *crista terminalis.* The opening of the sinus venosus into the right atrium is at first bordered on each side by a *valvular fold,* the *right* and *left venous valves,* which fuse anteriorly to form a ridge, the *septum spurium.* When the right horn is incorporated into the atrium, the right venous valve disappears and the left forms the *valve of the inferior vena cava* or *eustachian valve* and the valve of the *coronary sinus* or *Thebesian valve* (fig. 11.6).

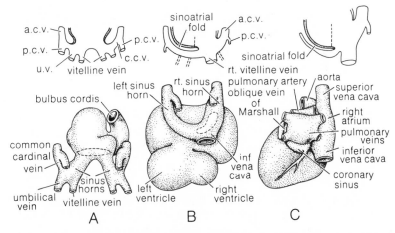

Fig. 11.6. Development of the sinus venosus in man. The sinus venosus and great veins are seen from the dorsal side. The diagrams show the formation of the coronary sinus and the incorporation of the right sinus horn into the wall of the right atrium. The broken line in *A* and *B* indicates the entrance of the sinus venosus into the atrial cavity. Above each drawing is a diagram of a transverse section showing the relation of the great veins to the atrial cavity. *A,* approx. 24 days; *B,* approx. 35 days; *C,* in the new born; a.c.v., anterior vena cava; p.c.v., posterior vena cava; c.c.v., common cardinal vein; u.v., umbilical vein; vit. v., vitelline vein. After Langman 1972 (*Medical embryology.* Baltimore: Williams and Wilkins).

The division of the atrial cavity begins with the formation of a sickle-shaped *septum primum,* which grows down from the dorsal roof of the cavity toward the *atrioventricular canal* connecting the atrium with the ventricle. At this stage the walls of the atrioventricular canal produce localized thickenings that form the *dorsal* and *ventral endocardial cushions.* The septum primum is separated from these by the *ostium primum.* The endocardial cushions soon fuse and thus divide the atrioventricular canal into right and left channels. This partition fuses with the base of the septum primum. Later the primary septum ruptures and forms a new opening (connecting right and left atria), the *ostium secundum.* At this stage a second interatrial septum appears in the roof of the right atrium and forms the *septum secundum,* part of which grows down to meet the primary septum. The result is that the atria communicate with each other via an elongate cleft (involving both septum primum and secundum), the *foramen ovale* (fig. 11.7). After fusion of the endocardial cushions, the area around each orifice forms localized proliferations which subsequently form valves covered by endocardium and connected to thickened *trabeculae* in the ventricle (*papillary muscles*) by the *chordae tendineae.* In this way two valve leaflets are formed on the left side and constitute the *bicuspid* or *mitral valve,* and three leaflets on the right form the *tricuspid valve.*

The interventricular septum develops largely from a prominent *endocardial ridge,* and initially the right and left chambers are in communication via an *interventricular foramen.* At this time a pair of *truncoconal ridges* appears in the conus arteriosus. These ridges fuse to form the *aorticopulmonary septum,* which descends toward the ventricles. As it does so, it follows a spiral course and hence the pulmonary and aortic arches twist around each other. In this manner, the aorta becomes associated with the left side of the ventricle and the pulmonary arch with the right. The truncoconal ridges extend toward the interventricular foramen and, together with some material from the dorsal atrioventricular cushion, complete the interventricular septum and obliterate the interventricular foramen. In addition to the two main truncoconal valves, two smaller ridges occur at the junction of the aortic and pulmonary canals with the ventricle. These ridges alternate with the main ridges and form the *semilunar valves* that prevent any regress of blood from the aortic and pulmonary arches into left and right ventricles.

## 3. Embryonic Circulation

A study of development yields fundamental information on vascular-system biology and allows specific comparisons to be made on the basic vascular plans in different groups. The most generalized pattern of embryonic development is found in the vascular system of the amphibians and lungfishes. The first blood vessels appear beneath the yolk-laden gut cells and run anteriorly as a pair of *vitelline veins,* which fuse anteriorly to form a *subintestinal vessel* that initially opens directly into the heart (fig. 11.8). The area below the gut soon becomes invaded by the developing liver, and the subintestinal vein breaks up in the substance of the liver. Anterior to the liver the subintestinal vein develops into an *hepatic vein* (liver to heart) and posterior to the liver into an *hepatic portal vein* (gut to liver). Anterior to the heart a *ventral aorta* gives rise

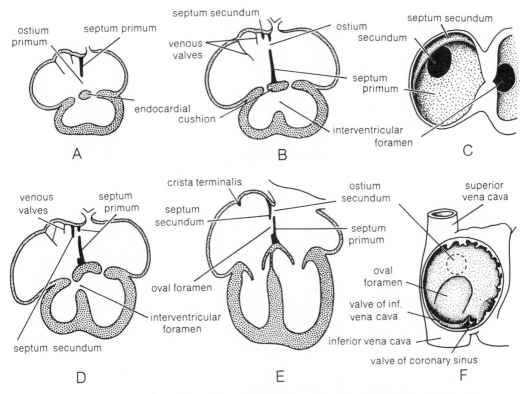

Fig. 11.7. Schematic representation of the interatrial septum in man. The subdivision of the atrial cavity into right and left sides begins with the formation of the septum primum, which later fuses with a partition formed in the atrioventricular canal. Before the two atria become completely separated by this fusion, the upper part of the septum primum ruptures, forming an ostium secundum. At about the seventh week of development a septum secundum develops in the roof of the right atrium, but this never completely divides the atrial cavity and has a persistent oval foramen. The oval foramen is obliterated at birth and the right and left atria are completely separate. *A*, 6 mm. approx. 30 days; *B*, 9 mm. approx. 33 days; *C*, same stage as in *B*, but seen from the right; *D*, 14 mm. approx. 37 days; *E*, newborn; *F*, view of atrial septum seen from the right, same stage as *E*. After Langman 1972 (*Medical embryology*. Baltimore: Williams and Wilkins).

to a vessel extending dorsally on either side of the foregut to a level just below the notochord. At this point each vessel turns posteriorly and then sends a number of *vitelline arteries* to the gut tube. The paired loops from the ventral aorta are the first of the *aortic arches* of the pharynx and, as development proceeds, other arches are added posteriorly. These arches may break up to form the capillary system of the gills, but the anterior arches either atrophy or become considerably modified as development proceeds. The aortic arches on each side lead into *lateral dorsal aortae,* which soon fuse to form a single *dorsal aorta.* The early branches of this vessel, such as the vitelline arteries, are paired, but, as the size of the gut decreases relative to the size of the animal, these vessels fuse. The single vessels descend in the mesentery and form the *mesenteric, coeliac arteries,* etc., of the adult. The initial circulation thus established

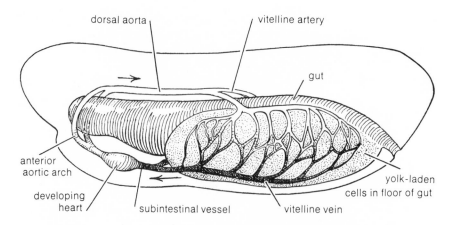

Fig. 11.8. Circulation of early tadpole. General circulation of young frog tadpole which has largely yolky cells forming the floor of the gut. The yolk provides the food supply for the developing organism and hence the vitelline circulation is well developed. After Romer 1970 (*The vertebrate body*. Philadelphia: Saunders).

in the embryo is concerned with the gut and is a *visceral circulation*. As the mesodermal structures of the body develop, it becomes necessary to supply the outer body tube with blood and the *somatic circulation* is formed. The afferent supply to the somatic circulation stems from the dorsal aorta as paired lateral and dorsal branches to the musculature and the nervous system. The *cardinal veins* from the tissue capillary networks, arise and they are essentially the efferent portion of the somatic circulation. The paired *anterior cardinal veins* drain the region anterior to the heart, and corresponding *posterior cardinals* drain the region posterior to the heart. The cardinals gain access to the heart by fusion to form the *common cardinals*. Other elements in the somatic circulation are the limb vascular systems and, in lower vertebrates, the veins draining the ventral part of the body wall, the *abdominal veins*.

In many vertebrates, such as elasmobranchs and amniotes, the early embryonic circulation is more elaborate because of the presence of an *extraembryonic network* that supplies the *yolk sac* and, in mammals, because of the placenta. In these vertebrates the gut is essentially without a floor; the paired vitelline veins arise on either side of the yolk sac, and there can be no formation of a single ventral vessel until relatively late in development. A highly developed system of vitelline arteries and veins persists for a considerable portion of embryonic life in reptiles and birds. Despite the absence of yolk from the mammalian yolk sac, vitelline vessels develop and may remain for some time. In mammals further series of vessels develop along the stalk of the allantois as the *allantoic* or *umbilical arteries,* and blood is returned by corresponding *allantoic* or *umbilical veins* that run forward in the body wall. These seem to be homologous with the lateral abdominal veins of the lower vertebrates, and initially they open directly into the heart. Later the right umbilical veins disappear, and the left enters the liver. In the adult the position of the left umbilical vein is marked by the *ligamentum teres hepatis*.

There are, however, numerous changes that take place in the vascular system during the ontogeny of mammals, and these are related to the existence of a placental circulation in this group (fig. 11.9). Before birth blood returns to the fetus by way of the umbilical vein. On reaching the vicinity of the liver the main portion of this blood bypasses the liver by flowing through a *ductus venosus* and into the *inferior vena cava*. A small portion of the blood from the umbilical vein enters the liver sinusoids to enrich the blood from the *hepatic portal vein*. The amount of blood entering the ductus venosus, and hence the liver, is controlled by a sphincter muscle in the umbilical vein. From the inferior vena cava blood enters the right atrium and a major portion of this blood is directed through the defect in the interatrial septum (*oval foramen*) into

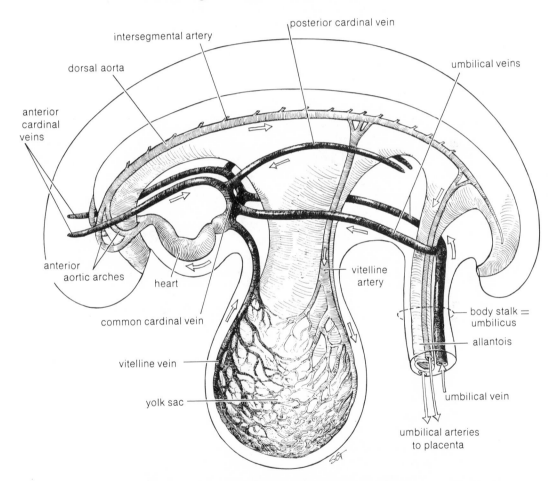

Fig. 11.9. Circulation of mammalian embryo. Amniotes possess a second series of embryonic vessels (in addition to the vitelline vessels), the allantoic or umbilical arteries and veins. Paired allantoic (umbilical) arteries arise from the dorsal aorta and run in the body stalk, or umbilicus, to the placenta. Gaseous exchange and exchange of nutrients takes place in the placenta, and blood is returned to the fetus in the umbilical vein. In the early embryo the umbilical vein drains directly into the heart but later enters the liver and forms part of the hepatic portal system. The allantoic vessels persist until birth, when they abruptly cease to function. After Romer 1970 (*The vertebrate body*. Philadelphia: Saunders).

the left atrium. The remainder of the blood diverted by the *crista dividens* (the lower edge of the *septum secundum*) remains in the right atrium and mixes with venous blood from the *superior vena cava*. Blood from the left atrium enters the left ventircle and then the left systemic arch from which the carotids arise. Venous blood from the right ventricle leaves via the pulmonary arch, which is connected to the aortic arch by a *ductus arteriosus*. Because of the high resistance to flow in the lungs most of the blood in the pulmonary arch is shunted through the ductus arteriosus and into the dorsal aorta. Blood is returned to the placenta by way of two *umbilical arteries*.

A number of significant changes in the vascular system occur at birth, when the placental flow is interrupted and the lungs become functional (fig. 11.10). At this time the walls of the ductus arteriosus contract, and

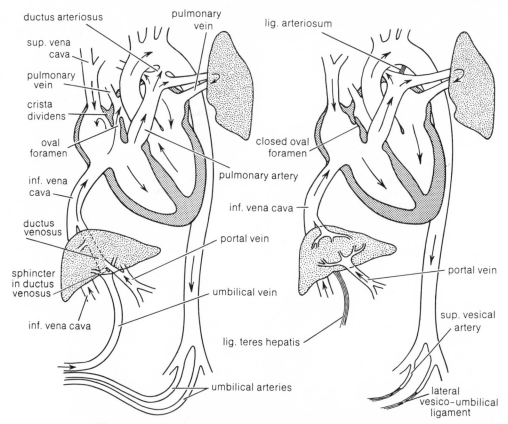

Fig. 11.10. Changes in mammalian circulation at birth. *A,* before birth; *B,* after birth. Before birth oxygenated blood returns to the fetus via the umbilical vein, and the main portion of this blood flows through the ductus venosus, into the inferior vena cava and thus bypasses the liver. A small portion of blood enters the liver sinusoids and the volume of blood doing so is regulated by a sphincter muscle in the ductus venosus. In the fetus the pulmonary and aortic vessels are widely connected by a ductus arteriosus. Since the resistance to flow in the pulmonary vessels is high, the majority of blood leaving the heart is directed into the aorta. After birth the ductus venosus is obliterated and forms the ligamentum venosum. The ductus arteriosus closes and its position is marked by the ligamentum arteriosum. After Langman 1972 (*Medical embryology*. Baltimore: Williams and Wilkins).

thus the blood flow to the lungs is increased. Consequently the blood return from the lungs to the left atrium rises and the pressure on the right side is lowered because of the interruption of the placental flow. The septum secundum therefore becomes opposed to the primary septum closing the foramen ovale and hence functionally separating the right and left atria. The umbilical arteries initially constrict by muscular contraction and later the distal portions become closed by fibrous tissue and form the lateral *vesicoumbilical ligaments,* while the proximal portions persist as the *superior vesical arteries.* The obliterated umbilical vein forms the *ligamentum teres hepatis* in the lower margin of the falciform ligament, and the ductus venosus is converted into the *ligamentum venosum,* which runs from the teres ligament to the inferior vena cava. The initial closure of the ductus arteriosus results from muscular action, but later the duct is obliterated by the proliferation of cells from its wall and forms the *ligamentum arteriosum.* During the early stages after birth, passage of blood through the ductus arteriosus and the foramen ovale is not uncommon and may persist until obliteration of the ductus and the fusion of the septa between the atria are complete.

**D. Structure and Evolution of the Vascular System**

A detailed study of the morphology of the arterial and venous system facilitates the description of the fundamental plan of vertebrate vascular systems. It also allows meaningful evolutionary relationships to be established.

**1. The Arterial System**

The structure and development of the anterior portion of the arterial system in modern vertebrates suggests that their ancestors had a series of vascular aortic arches in the wall of the pharynx. These arches arose from a ventral aorta and each one corresponded to a skeletal element in the wall of the pharynx. The aortic arches were probably developed primarily as a means of transferring blood from the ventral aorta to the dorsal side and hence also to the head and the body. They also supplied the large pharynx, which was essentially a food-collecting area, in much the same manner as they do in *Branchiostoma,* and the blood probably lost oxygen in this region, rather than collecting it. The blood picked up its oxygen by diffusion over the general body surface, again in much the same way as the modern cephalochordates. With the evolution of jaws and consequent reduction in the size of the pharynx, which no longer acted as a food collector, the aortic arches became associated with gills that were developed on the pharyngeal skeletal elements, the *gill arches.* Each branch from the ventral aorta (aortic arch) formed an afferent blood supply to the gills and was therefore an *afferent branchial artery* (fig. 11.11). After passing through a capillary network in the gill, the blood was collected into an *efferent branchial* or *epibranchial artery,* leaving the gill dorsally. The epibranchials flowed into paired *dorsal aortae.* Anterior extensions of the ventral aorta and the dorsal aortae supplied the head with blood and posteriorly the dorsal aortae united to form a single median *dorsal aorta.* This vessel extended posteriorly below the notochord and supplied the rest of the body with blood. The agnathans, which have 6–14 gill pouches, each supplied by an aortic arch, come closest to the supposed "ancestral" condition of the aortic

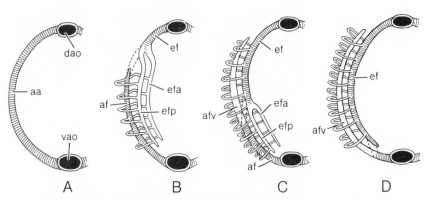

Fig. 11.11. Diagram of circulation in fish gill, from the left side and behind. *A*, embryonic condition with continuation of the aortic arch (aa) from the ventral aorta (vao) to the dorsal aorta (dao). *B*, condition in shark. Afferent gill vessel (af) formed from the aortic arch. Paired efferent vessels (efa, efp) are new formations. *C*, transitional condition seen in sturgeon. *D*, teleost condition. Embryonic arch gives rise to efferent vessel (ef). Afferent vessel (afv) is new formation. After Romer 1970 (*The vertebrate body*. Philadelphia: Saunders).

arches. However, the position of these arches differs from that of other vertebrates in that each arch runs to a single pouch, whereas in the gnathostomes each arch runs *between* successive pouches. The condition in agnathans must be regarded as a specialized feature of that group. Although the ancestors of vertebrates (like the agnathans) probably had a relatively large number of aortic arches (fig. 11.12), living vertebrates usually possess no more than six. In living gnathostomes the first and most anterior aortic arch to develop corresponds to the *mandibular gill arch*. Reference to the chapters on the skeleton and the respiratory system will show that the ancestral vertebrates may have had two (and

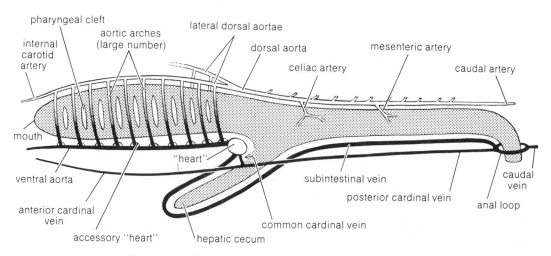

Fig. 11.12. Circulatory system of hypothetical "provertebrate" (lateral view). Heart was probably a simple pumping organ and accessory "hearts" may have been present to provide adequate pressure to force blood up the aortic arches. The pharynx, with a large number of gill slits, was a food-collecting organ.

possibly more) skeletal arches, and hence aortic arches, in front of the mandibular arch. However, since the mandibular gill arch and its aortic arch are common to all living gnathostomes, they are usually assigned the number I and more posterior arches are numbered II through VI. During development the mandibular gill arch forms the jaws, and a *complete* mandibular aortic arch is never present in the adult. The afferent portion of this arch is lost, and the gill on the anterior face of the hyoid gill slit (*spiracular slit*) is supplied, not by the first arch but by a branch from the efferent portion of arch II or from the dorsal aorta. Therefore, blood that reaches the spiracular gill is already oxygenated, and this gill is frequently referred to as a *pseudobranch*. The efferent portion of the first aortic arch in fishes persists and runs forward from the pseudobranch to the orbital region and then enters the brain case and joins a branch of the *internal carotid*. The second or *hyoidean aortic arch* is more persistent and exists as a typical aortic arch in the cartilaginous fishes and in some bony fishes. In most actinopterygians and the lungfish *Epiceratodus* it is lost. In the rest of the fishes arches III through VI are present in a more or less typical condition, but in the lungfish *Protopterus* arches III and IV pass directly through the gills without breaking up into capillaries and a branch from arch VI forms the *pulmonary artery*. In fishes the efferent branchial vessels usually leave the gills opposite the gill arch, but in many cases, for example, selachians, there is a shift so that the arteries emerge *opposite the gill slit*. Also communication between the efferent branchials is frequently established by dorsal connections; in some cases these junctions may be below the gill. The fourth efferent branchial in fishes often gives rise to a *hypobranchial artery,* which runs to the heart as a *coronary artery* and may also have an anterior branch to supply the mandibular region (fig. 11.13).

In selachians the anterior and posterior efferent branchial vessels develop individually in the embryonic tissue, separating the successive gill pouches parallel to a continuous aortic arch. The efferent vessels then become connected to the aortic arch by a series of capillary loops, and the primary arch becomes interrupted dorsally. Its long ventral portion becomes the efferent branchial. With the newly formed efferent vessels joined to the shorter dorsal portion of the arch, this vessel, the epibranchial artery, now functions as the efferent channel from the gill. In teleosts the primary aortic arch is interrupted ventrally and its greater part forms the efferent branchial. In this case the afferent vessels are newly formed to join the base of the aortic arch.

In the amphibians marked changes in the aortic system occur even in those groups that retain a more or less larval form. Arches I and II disappear early in development. The four remaining arches (III–VI) persist in adult urodeles as continuous tubes, since internal gills fail to develop and the external gills are supplied by accessory loops. In anuran larvae the development of the internal gill system results in the interruption of continuity of the vascular arches, but this is restored at metamorphosis. Therefore in adult amphibians arches III, IV, V, and VI may be present. All are found in urodeles, but anurans lack arch V and arch VI becomes modified in both groups to form a *pulmonocutaneous* arch.

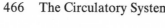

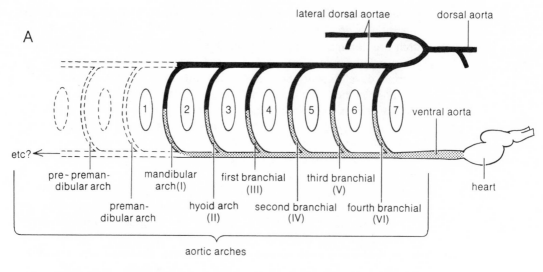

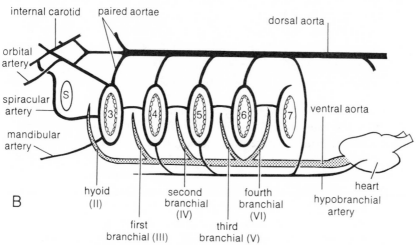

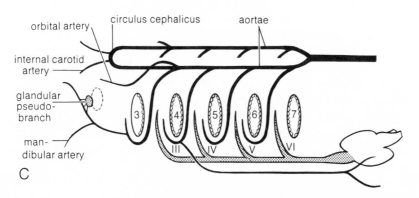

Fig. 11.13. Aortic arches in vertebrates. The ancestral vertebrates almost certainly had a number of vascular arches anterior to the mandibular arch. However, since the precise number of such arches is not known and the mandibular arch is the most anterior arch recognizable in all vertebrates, it is labelled 1. *A*, hypothetical ancestor; *B*, elasmobranch; *C*, actinopterygian; *D*, lungfish; *E*, larval urodele; *F*, adult urodele; *G*, anuran; *H*, lizard; *I*, mammal.

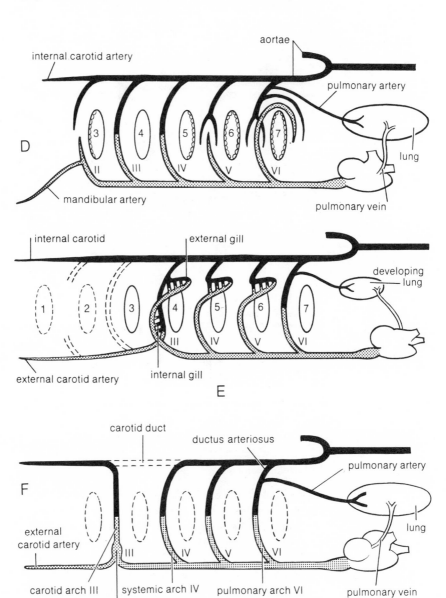

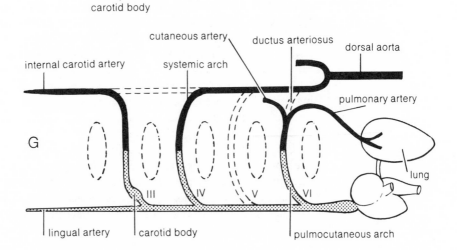

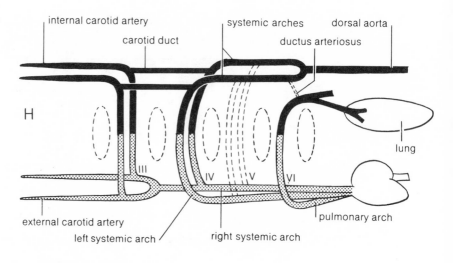

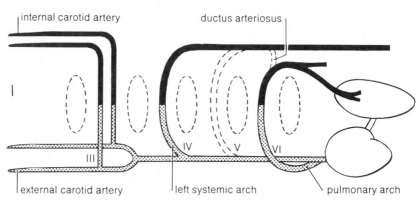

Experiments have shown that in many fishes the majority of blood in arch III is directed to the head, and in amphibians that this is the sole function of the arch, that is, the *carotid arch*. The connection between this arch and arch IV (corresponding to a portion of the lateral dorsal aorta) is usually lost, but in some amphibians and reptiles it persists as the *carotid duct (ductus caroticus)*. Arch IV is always a conspicuous arch in tetrapods since it forms the major channel for blood flow to the body and is the *systemic arch*. Arch V is lost in all tetrapods (except some urodeles), and in many cases it fails to develop even in the embryo. It is impossible to say what selective advantage there might be for arch IV rather than arch V to form the major systemic artery in vertebrates; the fact remains that is consistently used. As indicated above, arch VI forms the pulmonary (or pulmonocutaneous) arch and embryologically it is connected to the dorsal aorta. In many cases this connection, the *ductus arteriosus (ductus Botalli)*, carries large amounts of blood from pulmonary to systemic circuits before the lungs are fully developed. The ductus arteriosus frequently disappears at birth or at metamorphosis, and its position is often marked by a *ligamentum arteriosum*. It does, however, persist in some adult reptiles and urodeles. In such cases the function of the ductus arteriosus is uncertain, but it often acts as a pressure-regulating device that operates relative to changes in the resistance to flow of blood through the lungs. In living tetrapods and lungfishes the ventral aorta becomes much shortened and may be distinguished in some cases as the *truncus arteriosus*. A longitudinal splitting of the truncus occurs in a spiral fashion and isolates the pulmonary and systemic arches in such a way that the pulmonary arch becomes associated with the right side of the ventricle and the systemic arch with the left. In reptiles both the right and left systemic arches persist (joining posteriorly to form the dorsal aorta), whereas in birds and mammals a single systemic arch remains, the left in mammals and the right in birds. In birds and mammals the major variations in the aortic arches occur in the details of the arrangement of their associated arteries, such as the subclavians and the carotids.

### a. Blood Supply to the Head

In fishes the head is supplied with blood by vessels that are essentially continuations of the dorsal aorta. Each vessel gives rise to an extensive *orbital artery* (*stapedial artery* of higher forms), which supplies most of the face and the jaws before entering the brain case as the *internal carotid artery* immediately anterior to the pituitary. The internal carotid is the main vascular supply to the brain and the eyeball and is also connected to the *efferent pseudobranchial artery* within the cranium. A ventral artery to the floor of the mouth usually arises from the anterior aortic arch (arch II) corresponding roughly in position to the *lingual* or *external carotid* of other groups and is called that in many texts. However, this vessel is not homologous with the true external carotid (lingual), which arises from the base of arch III. The arrangement described above persists in the majority of fishes, but in amphibians it is modified in relation to the loss of the internal gills and aortic arch II, the carotid arch now being formed by aortic arch III. In typical amphibians and reptiles the external carotid or lingual artery arises near the base of aortic arch III and supplies the tissues of the tongue and throat. The main carotid artery enters the brain case after producing the *stapedial artery,* which supplies the outer part of the head and jaws. In primitive mammals and the embryos of all mammals the arterial system of the head is similar to that described for amphibians and reptiles. In most mammals, however, the external carotid grows anteriorly to tap the stapedial artery and the result is that most of the blood is then carried to the head region in the external carotid and the stapedial artery becomes reduced or absent.

### b. Blood Supply to the Body and Limbs

Blood from the heart in vertebrates is carried in the systemic arch (or arches) to a single dorsal aorta lying below the notochord and above the roof of the dorsal mesentery. In some cases the dorsal aorta is continued into the tail as the *caudal artery*. The arterial trunks that leave the dorsal aorta are of two types: single vessels that run ventrally in the mesentery and form a series of *visceral arteries;* and paired vessels, *somatic arteries,* to the body wall and limbs. The anterior portion of the gut is supplied by a *coeliac artery,* branches of which run to the stomach and the liver. One or more *mesenteric arteries* supply the intestine and in some cases the coeliac and anterior mesenteric arteries arise as a single vessel, the *coeliacomesenteric artery*. Paired lateral vessels to the kidney are the *renal arteries* and there are usually numbers of *intersegmental arteries* that supply the body wall. In primitive vertebrates these vessels leave the dorsal aorta and run in each myoseptum. In most cases, however, the regular arrangement is obscured by anastomoses between successive vessels and the formation of longitudinal channels. Each pectoral area is supplied by a *subclavian artery*, which may arise from the systemic arch or from the dorsal aorta. An *iliac artery* runs to each pelvic region. These vessels may arise as distinct branches from the dorsal aorta, but in many cases they appear as a bifurcation of the aorta. In the early stages of development the embryonic limbs or fins are supplied by a complex network of vessels. During ontogeny one of these assumes dominance and develops into the main artery to the limb in the adult. The artery thus formed is termed the *brachial* (forelimb) or *femoral* (hind limb), but it is not likely to be derived from the same vessel in every case.

## 2. The Venous System

The veins of vertebrates may be considered in three groups: (*a*) veins draining the gut tube, constituting the *visceral venous system,* derived largely from the *embryonic subintestinal vein;* these are the *hepatic portal vein* (from the gut to the liver) and the *hepatic veins* from the liver to the heart; (*b*) the *somatic venous system* draining the "outer body tube" (body wall, head, and often also the limbs) consisting essentially of the *cardinal* and *abdominal veins;* and (*c*) the *pulmonary veins*.

### c. The Visceral Venous System

The development of the hepatic portal vein has been described in the section on the embryonic circulation. The hepatic portal vein carries blood from the gut, pancreas, and spleen and breaks up into a complex network within the liver. In most fishes blood from the liver is carried by a large median vessel or pair of vessels, the hepatic veins, opening directly into the sinus venosus. In lungfishes and tetrapods a portion of this vein is involved in the formation of the *posterior vena cava* (*postcaval vein*), in which case the term hepatic vein describes the vessel from the liver joining the postcaval, which then also carries hepatic venous blood.

### b. The Somatic Venous System

The venous return from the body wall is largely in longitudinal vessels lying above the gut and mesenteries. In the lower vertebrates such vessels are the *cardinal veins* and in the more advanced forms are *venae cavae* or *caval veins*. The cardinal veins are found in embryos of all vertebrates. *Posterior cardinals* run in the dorsal body wall and corresponding *anterior cardinal veins* originate on either side of the head. At the level of the heart the anterior and posterior cardinals on each side unite to form a *common cardinal vein* or *Cuvierian duct,* which enters the sinus venosus. In the majority of vertebrates, except the mammals, most blood from the head and the eye drains into a longitudinal vein on either side of the brain case. This vessel is the *lateral head vein* or *vena capitis lateralis* and may be considered the forward continuation of the anterior cardinal (fig. 11.14). The lateral head vein (sometimes called the *head vein*) also collects blood from the brain via the *anterior, median,* and *posterior cerebral veins,* which issue through foramina in the wall of the brain case. The posterior cerebral vein (*internal jugular* of tetrapods) leaves the cranial cavity via the *jugular foramen* along with the vagus

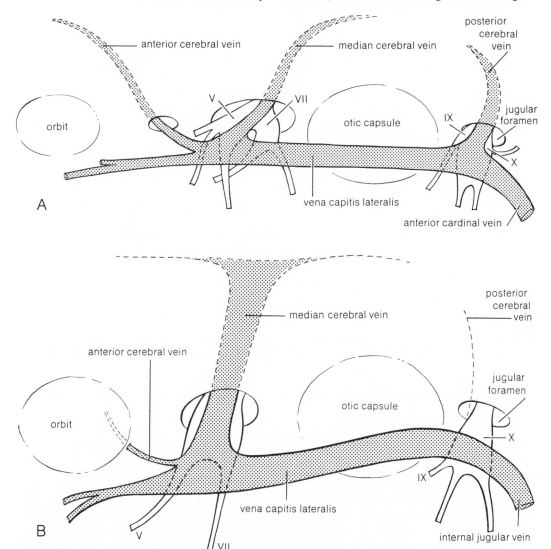

Fig. 11.14. Venous drainage of the head in vertebrates. *A,* elasmobranch; *B,* anuran; *C,* placental. V–XI, cranial nerves. In the elasmobranch the main drainage channel is the lateral head vein (vena capitis lateralis), which receives blood from the cranial cavity via the anterior, median, and posterior cerebral veins. In anurans the lateral head vein persists, and the main cranial venous channel is the median cerebral vein. The anterior cerebral vein is reduced in size as is the posterior cerebral vein, which in some cases may be absent. In mammals the lateral head vein has been abandoned and blood from the orbital region enters a large sinus system which drains the rest of the cranial cavity and leaves via the internal jugular vein.

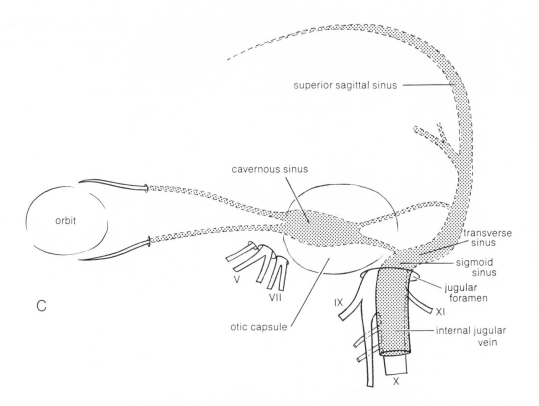

nerve (X); the portion of the anterior cardinal posterior to this junction is often referred to as the *jugular vein*. In primitive forms, as in elasmobranchs, the anterior cardinal runs posteriorly to join the posterior cardinal and the subclavian veins open directly into the common cardinal. In amphibians and reptiles the lateral head vein remains but becomes less important, and the blood from the more superficial head structures is collected by an *external jugular vein* that probably represents one or more of the finer anterior branches of the anterior cardinal. The external jugular vein joins the internal jugular issuing from the brain case, and the vessel thus formed is the *common jugular vein*. In these groups the common cardinals become incorporated into the jugular trunks (anterior cardinals), which open directly into the sinus venosus. Before entering the sinus, the jugular veins may be joined by the *subclavian veins*. The vessels thus formed are sometimes named *anterior venae cavae,* as in mammals. In mammals a major change in the head drainage pattern occurs and results in the disappearance of the lateral head vein. The blood from the structures that were previously drained by this vein, in the lower vertebrates, enters the brain case and is eventually collected by the internal jugular. The more superficial head tissues are drained (as in amphibians and reptiles) by the external jugular, which unites with the internal jugular to form the common jugular vein. Each common jugular is joined by a subclavian vein and thus forms an anterior vena cava. In many mammals, including man, the anterior drainage system is further modified. In the embryo a cross connection develops between the anterior cardinals a short distance in front of the heart (fig.

11.15) and the left common cardinal disappears. The blood from the left is thus carried to the right side and all the blood from the head enters the heart in a single anterior vena cava.

In the early embryos of vertebrates the posterior cardinal veins are paired vessels that receive blood from the caudal vein and then from the kidneys, gonads, and the dorsal parts of the body wall (from a series of *parietal veins* that roughly parallel the course of the intersegmental arteries) and finally join the anterior cardinals at the level of the heart. At this stage a new series of vessels appears ventromedial to the kidneys as the *subcardinal veins*. Next a pair of *lateral abdominal veins* develop in the ventrolateral part of the body wall and each receives an *iliac* and *subclavian vein* before joining the common cardinal. At this time the subcardinal veins develop numerous branches that drain the kidney. Hence blood from the caudal vein may pass through the kidney in the branches of the posterior cardinal and then be collected by the subcardinals. However, some of the blood may bypass the kidney in the posterior cardinal, and the portion of the posterior cardinal between the kidney and the subcardinal is soon lost; the posterior portion of the postcardinal routes all the caudal blood through the kidney and is the *renal portal vein*. This is roughly the pattern in elasmobranchs, where the vessel draining the posterior part of the body (which is usually called the postcardinal) is in reality formed both from the posterior cardinals and the subcardinals. In teleosts the lateral abdominal veins are absent and the iliac veins open into the renal portal (posterior cardinal) and the subclavian veins open directly into the common cardinals.

In lungfishes the lateral abdominal veins fuse to form a single *ventral abdominal vein* that opens directly into the sinus venosus. The subclavian veins enter the common cardinals and the iliacs the renal portal vein. But the iliac veins also communicate with the abdominal vein by newly formed *pelvic veins*. The caudal vein, instead of opening into the renal portal vein as in other fishes, drains into a large vessel between the kidneys. This median vessel formed by the fusion of the subcardinals is continued anteriorly as a pair to join the postcardinals. The right subcardinal channel is much larger than the left and joins the larger right postcardinal that forms a "postcaval" channel passing directly through the liver to enter the sinus venosus. The left posterior cardinal is small and enters the left common cardinal. Blood from the liver is still carried to the sinus venosus by a pair of hepatic veins. Little modification of the dipnoan pattern is found in amphibians, especially the urodeles. In amphibians however, the ventral abdominal vein breaks up within the substance of the liver and does not enter the sinus venosus directly. The development of the postcaval system differs from that of lungfishes, too. The posterior portion of the postcaval system (as in the lungfish) is derived from the right subcardinal, but the anterior portion is produced by a posterior extension from the right hepatic vein. The posterior cardinals are retained in adult urodeles but are lost in the anurans; in this case the sole channel of the blood from the kidney to the heart is the *postcaval vein* (fig. 11.16).

In reptiles the venous system is basically similar to that of amphib-

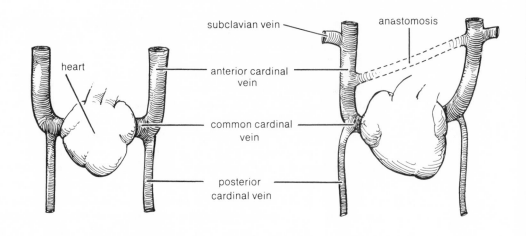

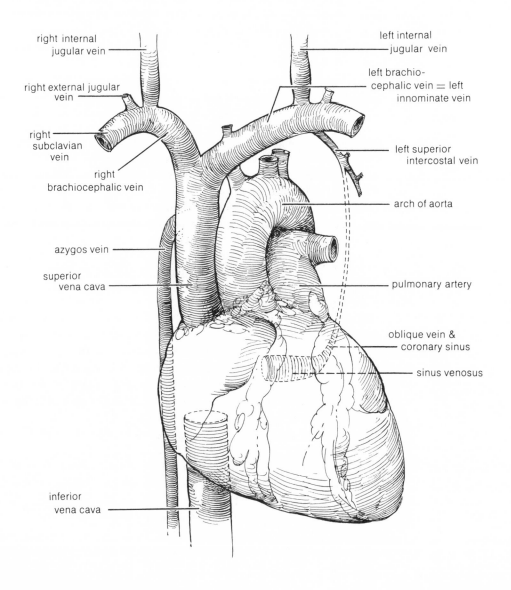

ians except that the renal portal system is reduced and some of the vessels pass directly through the kidney to enter the postcaval. This is also true of the birds but in this group by far the largest portion of blood bypasses the kidney, and so the term "portal" vein is hardly justified. Birds possess a vein that connects the caudal vein with the hepatic portal vein, and this is usually termed the *caudal-mesenteric vein*. There is some doubt about whether this represents the abdominal vein of the other vertebrates.

In mammals virtually all of the venous return from the posterior portion of the body is in the posterior vena cava, and the posterior cardinals almost disappear in the adult. The postcaval, while anatomically a fairly simple vessel, has a complex beginning. In the early stages of development the posterior cardinals and subcardinals bear much the same relations as they do in the lower vertebrates. At this stage the midportions of the subcardinals fuse to form an enlarged *subcardinal sinus,* which takes over the drainage of the kidney from the posterior cardinals as they begin to disappear. As a channel leading from the liver to the right subcardinal then develops, the bulk of the blood from the kidney flows into this vessel, and the posterior cardinals degenerate further. At this stage a pair of *supracardinal veins* develop from the extreme anterior ends of the postcardinals and extend backwards. At the same time a series of *sacrocardinal veins* develops posteriorly, and one of these vessels forms the extreme posterior end of the postcaval. By this time blood is carried anteriorly by a single channel, the postcaval or posterior vena cava, which is thus formed from a *sacrocardinal segment* derived from the right sacrocardinal vein, a *renal segment* from the subcardinal sinus, and an *hepatic segment* developed mainly from the embryonic right vitelline vein. The supracardinal veins form the main drainage of the dorsal body wall, and the left one, which forms the *hemiazygos vein,* loses its connection with the posterior cardinal (which forms the *coronary sinus*). The right supracardinal together with the proximal portion of the posterior cardinal forms the *azygos vein,* which develops a cross connection to receive blood from the hemiazygos.

### c. Pulmonary Veins

The pulmonary veins carry oxygenated blood from the lungs for recirculation by the heart. It has recently been reported that the walls of pulmonary veins in such mammals as the rodents are well supplied with cardiac muscle fibers. These fibers are essentially similar to those of the myocardium, and their contractions assist the return of pulmonary blood to the heart.

---

Fig. 11.15. Development of the anterior venous system in man. Ventral views of the anterior venous system showing successive stages in the development of the left innominate vein. The innominate vein develops as a result of an anastomosis between the left and right anterior cardinal veins so that all the blood from the anterior region enters the left atrium in a single great superior vena cava. An intercostal vein (left superior intercostal vein) and a small vein (oblique vein) are the persistent positions of the left anterior cardinal. After Romer 1970 (*The vertebrate body*. Philadelphia: Saunders).

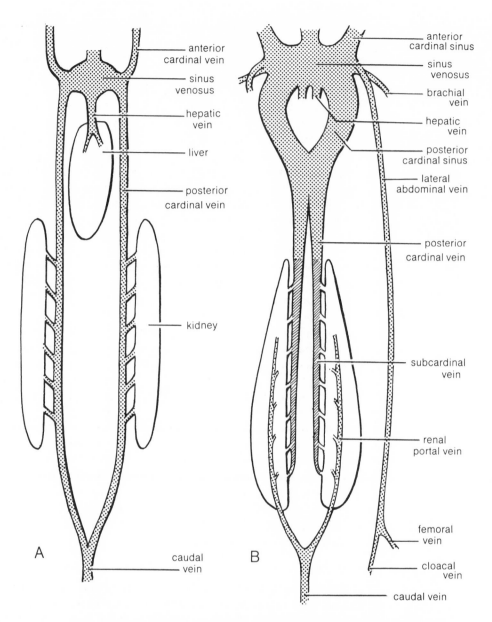

Fig. 11.16. Posterior venous system in vertebrates. Diagrams to show the structure and possible evolutionary history of the posterior venous system. *A*, larval lamprey: primitive venous system with unmodified posterior cardinal veins draining blood from the kidney and body wall by a number of vessels. *B*, elasmobranch: condition which is typical of fishes and in which a renal portal system has been introduced. The renal portal system carries blood from the caudal vein to the kidney and blood is then collected by the subcardinal veins (shaded). Some blood from the body wall is now returned in the lateral abdominal veins directly to the sinus venosus. *C*, lungfish: renal portal system persists, but some blood bypasses the kidney and returns directly to the sinus venosus in the ventral abdominal vein. The subcardinals unite as a single vessel. The right posterior cardinal is now much expanded as a postcaval vein. *D*, urodele: the ventral abdominal vein now opens into the liver and not the sinus venosus. The posterior drainage is now almost entirely in the postcaval vein

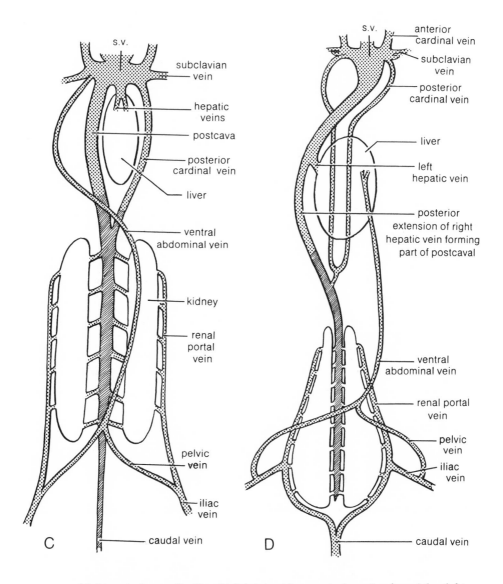

(right posterior cardinal), which is joined by a posterior extension of the right hepatic vein. *E,* mammal: renal portal system is eliminated and blood supply to the kidney is now entirely arterial. Venous return is via the massive posterior vena cava. This vein, although anatomically comparatively simple, has a complex developmental and evolutionary history. It is composed of segments derived from sacrocardinal, subcardinal, hepatic, and posterior cardinal veins. (For full description see text.)

**E. Functional Anatomy and Evolution of the Heart**

Although much information is available on the functional anatomy of the mammalian heart, such data on other vertebrates are rather scarce and scattered widely throughout the literature. An attempt is made to collate such information here in order to show how the heart may have changed during the course of vertebrate evolution.

**1. The Heart in Fishes**

In fishes the venous blood from the liver and the *ducts of Cuvier* (which collect all other systemic venous blood) drains into the sinus venosus.

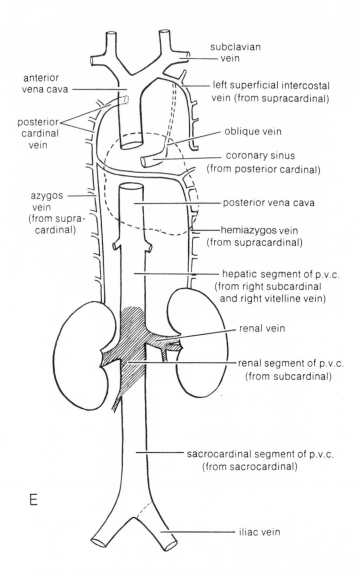

subclavian vein

anterior vena cava

left superficial intercostal vein (from supracardinal)

posterior cardinal vein

oblique vein

coronary sinus (from posterior cardinal)

azygos vein (from supracardinal)

posterior vena cava

hemiazygos vein (from supracardinal)

hepatic segment of p.v.c. (from right subcardinal and right vitelline vein)

renal vein

renal segment of p.v.c. (from subcardinal)

sacrocardinal segment of p.v.c. (from sacrocardinal)

E

iliac vein

There it enters a single atrium and then the ventricle and is finally pumped into the *ventral aorta* (*truncus arteriosus*) through the *conus arteriosus* in elasmobranchs or the *bulbus arteriosus* in teleosts. These chambers are separated by valves at the *sinoatrial* and *atrioventricular junctions*. The conus is a contractile chamber whose walls are composed largely of cardiac muscle, and it is therefore usually regarded as a chamber of the heart. The bulbus, however, is elastic and noncontractile. It is the swollen base of the ventral aorta and is peculiar to teleosts. The opening of the ventricle into the bulbus is guarded by a pair of *bulbal valves,* whereas the conus has a number of *conal valve*s arranged in rows; in some cases as many as seven rows may be present. In fishes, contractions of the atrium play an important part in filling the ventricle (fig. 11.17). The pressure produced by the contraction of the ventricle (*systolic pressure*) is generally higher in teleosts than in elasmobranchs. The conus and the bulbus act as distensible *elastic reservoirs* that empty *after* ventricular systole and thus maintain a flow of blood into the ven-

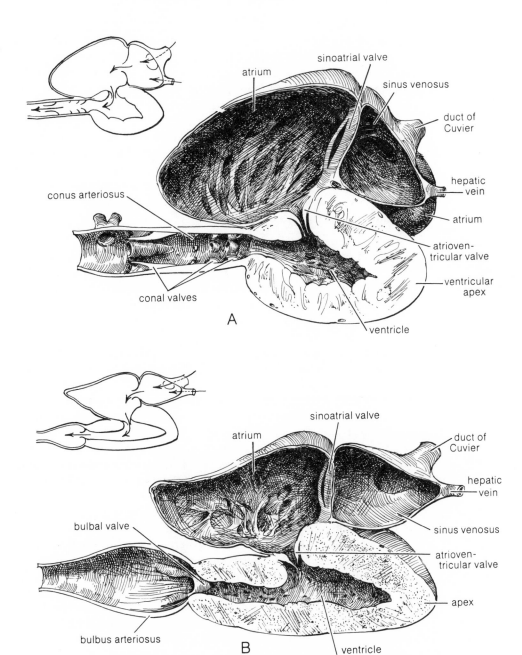

Fig. 11.17. Heart in fishes. *A*, shark; *B*, trout. In the shark blood leaves the muscular ventricle in the conus arteriosus, which is considered to be a chamber of the heart. In the trout the bulbus arteriosus receives blood from the ventricle and is essentially the swollen base of the ventral aorta. After Randall 1968.

tral aorta when the ventricle is relaxing (*diastole*). Thus, there is no reversal of blood flow within the ventral aorta after contraction of the ventricle such as there is in the major arterial arches in many higher vertebrates. It is believed that the presence of this mechanism in fishes protects the delicate gills from large fluctuations in pressure. In addition, contraction of the conus in elasmobranchs is necessary for the effective closure of its valves. In many bony fish the dorsal aorta is divided internally by a longitudinal ligament attached to its dorsal wall. Recent experimental work suggests that this ligament acts as an ancillary pumping mechanism. The output of the pump increases with the frequency of the tail beat and the pump facilitates the flow of blood to the muscles during swimming.

## 2. The Heart in Lungfish and Amphibians

The form and function of the heart in lungfish and amphibians is related to the presence of *accessory respiratory areas* such as the lungs and the skin. The presence of such areas in these groups is correlated, at least in part, with the subdivision of the heart chambers and the separation of blood returning from the accessory respiratory areas from the systemic venous blood.

Lungfish have a separate *pulmonary return* to the left side of the heart. Such a return is also present in the amphibians, but the blood returning from the skin in amphibians does so along with the general systemic venous blood. Among lungfishes there is some variation in the structure of the heart that is clearly related to the degree of air breathing. In *Protopterus* and *Lepidosiren,* which are predominantly air breathers, there is almost complete division of the heart into right and left sides. In *Neoceratodus,* which largely relies on gill respiration and uses its lungs only in hypoxic conditions, the division of the heart is much less complete, and the heart is much more "piscine" in nature. The structure of the gills is also closely correlated with the predominance, or otherwise, of lung breathing. In *Neoceratodus* the gill arches all bear well-developed functional gills, whereas in the other genera gills are absent from arches III and IV and the gill filaments on the other arches tend to be coarse and probably are not particularly effective areas of oxygen uptake. The gills in all genera, however, are important for unloading carbon dioxide.

The heart of *Protopterus* is surrounded by a thick, fairly rigid pericardium. The systemic veins open into the sinus venosus and the pulmonary vein transverses the pericardial space enclosed in a fold of fibrous tissue. The pulmonary vein enters the left side of the atrium; the fibrous fold is continued and forms a septum that partially divides the atrium into right and left portions posteriorly, but anteriorly the atrium is undivided (fig. 11.18). The ventricle is largely divided into right and left chambers by an *interventricular septum.* Toward the base of the ventricle there is communication between the right and left sides as the septum is incomplete and is attached to an *atrioventricular plug.* This is a structure unique to the Dipnoi and develops in the embryo at the junction of the atria and the ventricle, that is, the *atrioventricular canal.* In the mammal this canal becomes divided by dorsal and ventral endocardial cushions, which fuse to form a partition. In lungfish only the dorsal cushion develops and this becomes elaborated to form the atrioven-

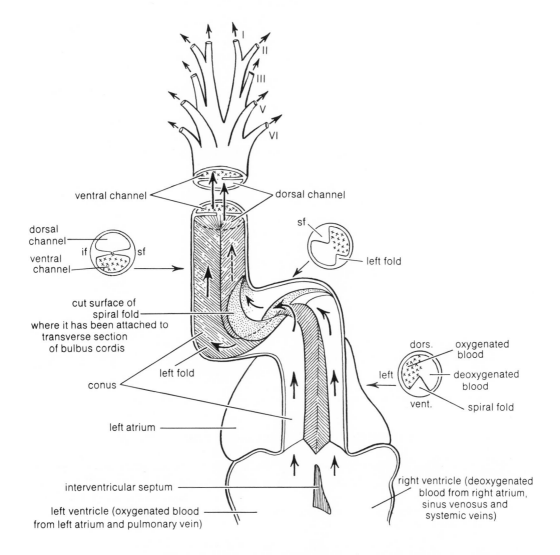

Fig. 11.18. Diagram of conus arteriosus in the lungfish, *Protopterus*. Heart viewed from dorsal side with dorsal wall of conus removed to show folds. The spiral fold and left fold unite to form the horizontal septum, which divides the conus into dorsal and ventral channels.

tricular plug. The lumen of the conus arteriosus is divided by a prominent *spiral fold* from the ventral wall and a small *left fold*. The conus and the folds within are much twisted, and the spiral fold rotates through an angle of 270° as it passes forward along the conus. Anteriorly the spiral fold and the left fold unite to form a *horizontal septum* that divides the cavity of the conus into *dorsal* and *ventral channels*. In *Protopterus* the arterial arches arise directly from the conus, and there is no recognizable ventral aorta. Arches II, III, and IV arise from the ventral channel and arches V and VI stem from the dorsal channel. Experimental work has shown that the blood entering the left atrium from the lungs is directed to the left side of the ventricular septum, then by means of the spiral fold into the ventral channel, and from there into arches II–IV. After passing through the right atrium, the venous blood from the sinus

venosus enters first the right side of the ventricle, then the dorsal channel of the conus, and thereafter arches V and VI. Clearly, within the
heart of *Protopterus* there is a good degree of separation of oxygenated
and deoxygenated blood and the conal folds are so arranged that the
blood rich in oxygen finds its way into the arches that pass directly
through the gills and then into the carotid arteries and thence to the
brain. The systemic blood, however, is directed through the gills that
possess filaments and also into the pulmonary arch (fig. 11.19). Varia-

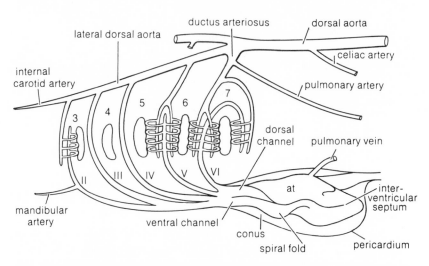

Fig. 11.19. Diagram of heart and arterial arches in *Protopterus*. Deoxygenated
blood from the dorsal channel of the conus enters arches V and VI and becomes
oxygenated in the gills. Blood from arch VI may be passed along the pulmonary
artery or may enter the dorsal aorta through the ductus arteriosus. Oxygenated
blood, from the ventral channel, passes directly through the gills (4 and 5)
which lack gill capillaries. After Johansen and Hanson 1968.

tions in the output of the heart in terms of the ratio of total pulmonary
flow have been recorded during respiration. A prominent increase in the
pulmonary blood flow occurs immediately after a long breath of air. It
seems likely that the *ductus arteriosus,* connecting the pulmonary arch
with the dorsal aorta, is important in regulating the amount of blood that
enters the lungs since the closure of the ductus would ensure that all
the blood in arch VI would pass to the lungs. A *vasomotor adjusting
mechanism* for the ductus has not yet been demonstrated, however.

The amphibian heart shows considerable structural variation that can
be correlated with the relative importance of nonpulmonary respiration.
In anurans the sinus venosus opens into the right atrium, which is separated from the left atrium (which receives pulmonary blood) by a complete interatrial septum. In many cases the right atrium is larger than the
left and may occupy a considerable space on the left side of the conus
arteriosus. The atria open into a single undivided ventricle whose walls
are composed of a spongy network of muscle that is drawn out into
*trabeculae* projecting into the cavity of the ventricle. The ventricle leads
into a conus arteriosus that is supplied with conal valves at its junction
with the ventricle and also around the entrance to the pulmonary arch.

In addition there is a spiral fold that arises from the left side of the
conus along the whole of its length and effectively subdivides the lumen
of the conus into dorsal and ventral chambers. The conus leads into a
short truncus arteriosus from which arise the carotid (IV), systemic (V)
and pulmonocutaneous (VI) arches (fig. 11.20). It is of great functional significance that the disposition of the arterial arches leaving the
truncus is *asymmetrical,* but this fact has very often been overlooked.
The pulmonocutaneous arches leave the base of the truncus at a single
aperture on the left side sheltered by the anterior limit of the spiral fold.
Early accounts of the anuran heart indicated that the blood returning
from the lungs was effectively separated within the ventricle from the
blood of the systemic venous return and that the pulmonary blood found
its way into the carotid arch, whereas the systemic blood was directed

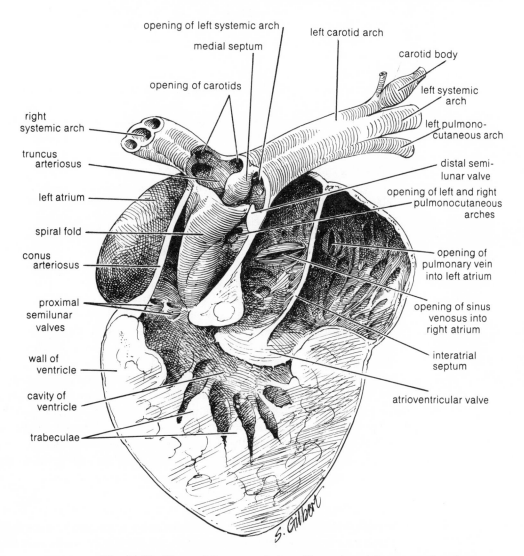

Fig. 11.20. Heart of *Rana catesbeiana.* Heart viewed from ventral surface.
Ventral wall of the conus and truncus have been removed to show spiral valve
and opening of arterial arches.

to the pulmonocutaneous arch. The systemic arches usually received a mixture of blood from these two sources. The separation of blood leaving the heart was assumed to result from pressure differences in the arterial arches. Blood under the lowest pressure at the beginning of the ventricular systole passed up the ventral side of the spiral fold into the pulmonocutaneous arch, which offered the least resistance to flow. As the pressure of expulsion increased, the blood forced its way dorsal to the spiral fold and into the systemic arches, and finally the blood under the highest pressure, which was oxygenated blood, entered the carotid arches. The high resistance to flow in the carotid arch was assumed to be due to the presence of the carotid body. Further, it was assumed that the systemic blood flowed up the dorsal side of the conus *before* the oxygenated blood flowed up the ventral side. Recent work indicates that there is a good separation of the oxygenated blood (from the lungs) from that returning by way of the sinus venosus and that the pulmonary blood almost exclusively is propelled into the carotid arches whereas that from the systemic return finds its way into the pulmonocutaneous arches. The right systemic arch carries pulmonary blood while the left systemic carries mixed blood. It is clear, however, that the separation of blood within the heart and arterial arches is *not* normally the result of large differences in resistance to flow within these arches. Various attempts have been made to account for the separation of blood within the undivided ventricle, and it has often been suggested that the trabeculae of the ventricle are responsible for such segregation. No satisfactory explanation for such a mechanism has ever been proposed. Correlated with this problem is the fact that as the blood leaves the ventricle, it does so in a clockwise spiral (fig. 11.21), and this phenomenon seems to be a basic property of the contractile mechanism of the amphibian heart (since it has also been described in urodeles). Further, it seems that *the fate of blood leaving the ventricle is determined by its position within the ventricle prior to ventricular systole.* It is also important to note that the relative position of blood entering the ventricle from the right and left atria will be greatly affected by the amount of blood returning to these chambers, which in turn will be related to the area being used for respiration (see below). Experiments show that the course of blood through the anuran heart may vary with the mode of life of the animal at any moment in time and with the related mode of respiration. For example, when the animal is aestivating and the skin alone is being used for respiration, the blood returning to the sinus venosus will be richer in oxygen than that returning from the lungs, for under these conditions the blood may actually lose oxygen in the lungs to sustain the lung tissues. It is clear that under such circumstances the blood within the ventricle will be more or less uniform in oxygen concentration and the absence of an interventricular septum allows this blood to be directed to the three arterial arches. Therefore the mechanism that causes the blood to pass through the anuran heart is of necessity a variable one since the blood return to the heart from the various respiratory surfaces is also variable. Further, the absence of a ventricular septum in this group allows varying volumes of blood to be sent to the respiratory surfaces

Fig. 11.21. Blood flow through heart in amphibians. *A, Rana catesbeiana; B, Xenopus laevis.* In *Rana* the right atrium is considerably larger than the left and cutaneous respiration is important. In *Xenopus* the lungs are much more important in respiration and the atria are more equal in size. In the two species the proportions of pulmonary and systemic blood within the heart varies and hence there are differences in the fate of the blood leaving the heart. (For full discussion see text.) After Foxon 1955.

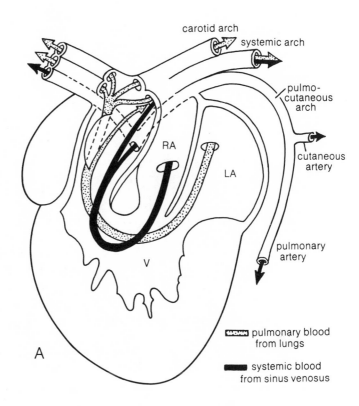

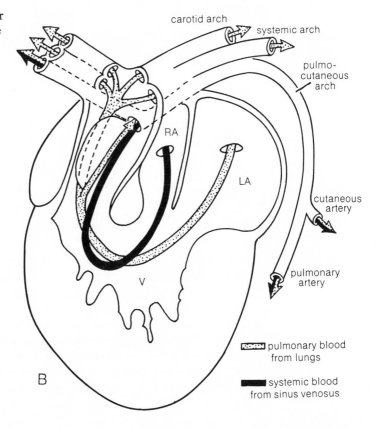

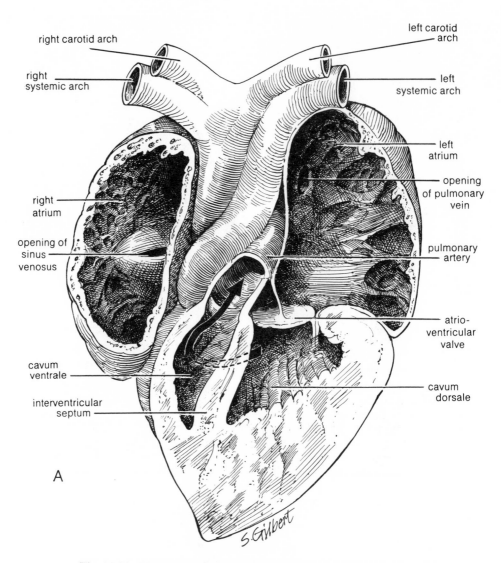

right carotid arch

left carotid arch

right systemic arch

left systemic arch

left atrium

opening of pulmonary vein

right atrium

opening of sinus venosus

pulmonary artery

atrio-ventricular valve

cavum ventrale

cavum dorsale

interventricular septum

A

S. Gilbert

Fig. 11.22. Heart and arterial arches of lizard, from the ventral side. In *A*, a portion of the ventral wall of the ventricle has been removed to show the interventricular septum. The course of blood, across this incomplete septum into the cavum ventrale and then into the pulmonary arch, is shown. In *B*, more of the ventricular wall has been removed to show the positional relationships of the pulmonary and systemic arches.

in a manner which would not be possible if a complete interventricular septum were present.

The importance of nonpulmonary respiration varies within the Anura and some species rely on cutaneous respiration far more than others. The prevalence (or otherwise) of cutaneous respiration can often be correlated with the structure of the heart and the way in which blood passes through the heart. For example, in *Xenopus* the lungs are more important areas for gaseous exchange than they are in *Rana,* and consequently the skin is less important than the lungs in *Xenopus*. In *Xenopus* the atria are more or less equal in size and the pulmonary artery is rela-

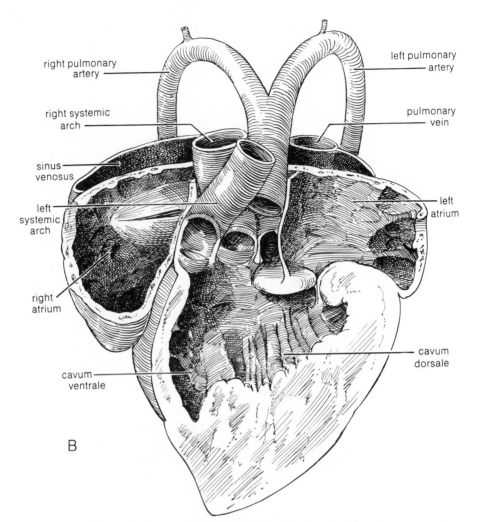

right pulmonary artery

left pulmonary artery

right systemic arch

pulmonary vein

sinus venosus

left atrium

left systemic arch

right atrium

cavum dorsale

cavum ventrale

B

tively larger than in *Rana* and the cutaneous artery relatively smaller. In *Xenopus* the pulmonary return is therefore larger than in *Rana* and the blood within the heart is therefore in different proportions. The carotid and systemic arches receive blood from the pulmonary circuit, and the pulmonocutaneous arch receives mixed blood (fig. 11.22).

In urodeles the skin generally assumes a more important role in respiration and smaller lungs are found in a number of species. In members of the family Plethodontidae the lungs are entirely absent, and cutaneous respiration is the only means of oxygenating the blood. In such cases there is a distinct correlation between the structure of the heart and the absence of the lungs. The disappearance of the lungs is related to the loss of the spiral fold within the conus and the breakdown of the interatrial septum; for example, in the lungless salamanders the heart assumes a "piscine" structure (each chamber contains blood of more or less uniform oxygen concentration). In such cases the pulmonary artery either fails to develop or supplies a small area of skin in the pectoral region and the anterior part of the stomach. The importance of the buccal cavity as a respiratory area has often been stressed in accounts of the gaseous exchange in amphibians, but recent accounts indicate that in most cases the movements of air into the buccal cavity are largely for the purpose of sampling the air and are "sniffing" movements. It has been calculated that in *Rana temporaria* the buccal cavity is responsible for only 0.9% of the total oxygen uptake and in *Xenopus laevis* only about 0.2%. The buccal cavity appears to be slightly more important in the urodeles, and in the newt *Triturus cristatus* it may be responsible for 3.0% of the respiratory exchange. The importance of the buccal cavity reaches its maximum in the plethodontids, and in *Plethodon glutinosus* the buccal cavity may account for up to 25% of the total oxygen uptake.

A point consistently overlooked in accounts of respiration in amphibians is that it involves the unloading of carbon dioxide as well as the uptake of oxygen. Regardless of the relative importance of the skin in amphibians as an area for accepting oxygen, it is always a major area for the release of carbon dioxide.

In many amphibian groups a spherical or oval swelling occurs at the point where the common carotid bifurcates into the external and internal carotid arteries. This structure is variously known as the *carotid body, carotid gland,* or the *carotid labyrinth* and is divided into an extensive ventral chamber and a smaller more dorsal sinusoidal plexus. The sinusoidal plexus has been shown to contain two types of cell ($\alpha$ and $\beta$ cells). The $\alpha$ cells are said to be associated with the termination of nerve fibers, while the $\beta$ cells form a complex network of intermingled cytoplasmic processes.

The function of the carotid body in amphibians is still a matter for speculation. Some accounts indicate that it detects pressure changes by virtue of stretch receptors in its walls, which are connected to the glossopharyngeal nerve (IX) and associated with the $\alpha$ cells. Other accounts insist that the carotid body is not innervated by the glossopharyngeal nerve and that it functions in the secretion of adrenalin, which helps to regulate the blood pressure in the internal carotid artery. It has also been suggested that the gland has a mechanical function in that it diverts blood into the external carotids, which arise at a rather awkward obtuse angle from the carotid arch.

**3. The Heart in Reptiles**

There are two major anatomical arrangements of the heart in modern reptiles. The first is found in the Squamata and the Chelonia. The interventricular septum is largely *horizontal* and divides the ventricle into dorsal and central cavities, rather than into right and left sides. The dorsal ventricular chamber is more extensive and is termed the *cavum dorsale* and the ventral chamber is the *cavum ventrale*. At the apex of the ventricle the septum is complete but toward the ventricular base the septum assumes a more oblique position and has a defect that allows communication between the dorsal and ventral chambers. The cavum ventrale, which is now located toward the right side, is closely associated with the pulmonary artery and is often termed the *cavum pulmonale*. A feature of the reptilian heart that is often overlooked, usually to facilitate description, is that *both atria open into the cavum dorsale*. The portion of the cavum dorsale that receives blood from the left atrium (i.e., from the pulmonary return) is the *cavum arteriosum* and the portion that receives blood from the right atrium the *cavum venosum*. The area connecting these is the *interventricular canal,* which is bounded anteriorly by the membranous right and left atrioventricular valves (figs. 11.23 and 11.24).

The second type of reptilian heart is that found in crocodiles where the heart has a complete interventricular septum dividing the ventricle into right and left sides. The *atria open separately into these chambers* and the right systemic arch, which gives rise to both carotid arches, stems from the left ventricle, whereas the left systemic and pulmonary arches arise from the right ventricle. At their bases the right and left systemic arches communicate via a *foramen of Panizza.*

While the blood passes through the heart in squamates and chelonians, there is good separation of oxygenated from deoxygenated blood: the pulmonary artery receives blood from the right atrium (systemic venous blood) and the aortic arches receive largely oxygenated blood from the left atrium. In lizards both systemic arches arise from the cavum dorsale and the right arch receives blood directly from the left atrium. The left systemic also acquires blood from this source, but in many cases this blood is mingled with blood from the cavum venosum. As the right and left arches leave the heart, they twist around each other and the right systemic gives rise to both carotids, which thus carry oxygenated blood. It is evident that blood is ejected into the pulmonary circuit before the systemic because of the low resistance to flow in the pulmonary bed. This also facilitates the movement of blood entering the ventricle from the right atrium across the septal defect into the cavum pulmonale during the early stages of ventricular systole. In the later stages of ventricular contraction the septum and the muscular ridge become opposed, thus effectively separating the pulmonary chamber (cavum pulmonale) from the cavum dorsale. In chelonians the left systemic arch arises far to the right side of the heart and there is no muscular ridge separating the cavum pulmonale. The systemic arches do not twist around each other as they do in lizards. The blood from the cavum venosum passes across the septal defect and enters the pulmonary artery as it does in lizards. A good deal of this blood also enters the left systemic arch because of its location. The arterial blood from the left atrium enters the right systemic mainly, although some of it may enrich the blood in the left arch. The volume of blood entering either the pulmonary or systemic arch has been shown to vary in the chelonians, especially in the diving forms, and blood within the ventricle may be moved (shunted) from the right to the left or vice versa. The direction of these *shunts* is related to the balance between the resistance to flow in the systemic and pulmonary circulations. When the animal is actively respiring, pulmonary resistance is relatively low, there is a distinct left-to-right shift so that some blood from the pulmonary return may be redirected to the lungs. When the animal is diving and pulmonary resistance is high, there is a right-to-left shunt so that blood may be rerouted to the systemic circulation, instead of entering the pulmonary circuit (fig. 11.25).

The anatomy of the crocodilian heart suggests that it is basically similar to that of birds and mammals except that it has two systemic arches. The left systemic arises from the right ventricle and the right systemic

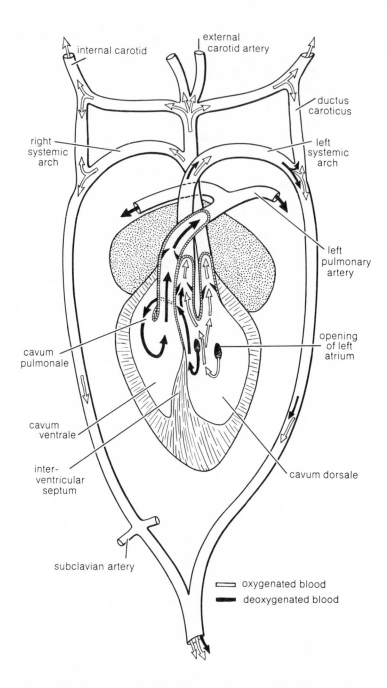

Fig. 11.23. Diagram of heart and arterial arches of lizard. *Both* of the atria open into the cavum dorsale of the ventricle. Most of the deoxygenated blood crosses the incomplete interventricular septum and enters the pulmonary arch. Some deoxygenated blood mixes with oxygenated blood and enters the left systemic arch. The right systemic, from which the carotid arteries arise, carry only oxygenated blood.

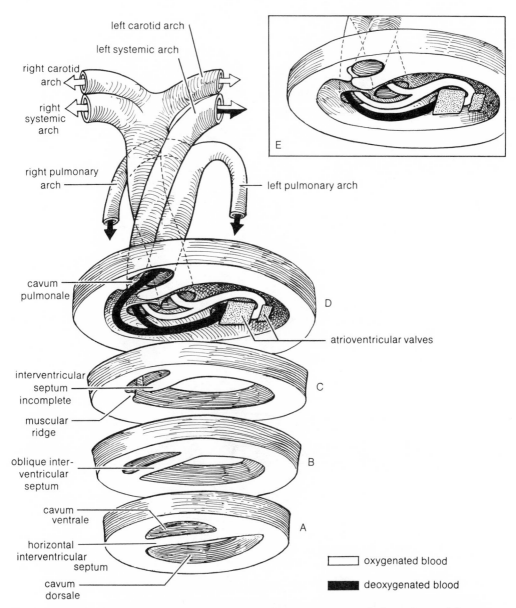

left carotid arch

left systemic arch

right carotid arch

right systemic arch

right pulmonary arch

left pulmonary arch

E

cavum pulmonale

D

atrioventricular valves

interventricular septum incomplete

C

muscular ridge

oblique inter-ventricular septum

B

cavum ventrale

A

horizontal interventricular septum

cavum dorsale

oxygenated blood

deoxygenated blood

Fig. 11.24. Heart of lizard. Series of sections (*A–D*) through ventricle starting toward apex (*A*) and moving toward base (*D*). Sections *A–B* reveal the horizontal nature of the interventricular septum and the relative proportions of the cavum dorsale and ventrale. *C* shows the incomplete nature of the interventricular septum and *D* the positions of the arches and pattern of blood flow. *E* shows the appositions of the interventricular septum and the muscular ridge during ventricular systole. After Hughes 1965.

from the left ventricle. It seems, therefore, that the right ventricle would would force blood into the left systemic and pulmonary arches while the left ventricle would supply the right systemic and hence the carotids. But, pressure recordings and measurements of oxygen levels in the systemic arches show that the oxygen levels in the right and left arches are *similar* and significantly higher than that of the pulmonary arch. The pressure

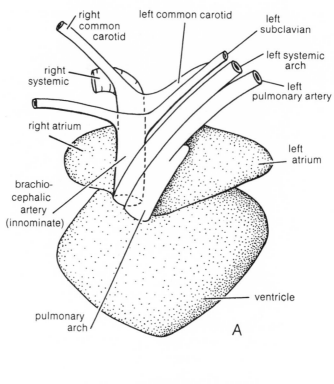

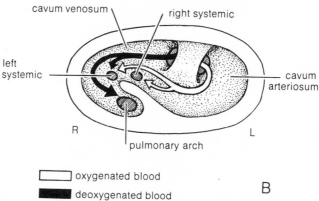

Fig. 11.25. Heart of chelonian, *Trionyx triunguis*. *A*, heart and arterial arches from the ventral side. *B*, section through base of ventricle indicating blood flow and how the orientation of the base of the arches is correlated with the types of blood they collect. After Girgis 1961. Permission of the Trustees of the British Museum (Natural History).

within the right ventricle, though it is sufficient to force blood into the lungs, which (as in other reptiles) have a low resistance to flow, is not normally high enough to overcome the resistance in the left systemic arch. Therefore the systemic arches are filled by the same "pump," that of the left ventricle, which forces blood directly into the right systemic and via the foramen of Panizza into the left systemic. Some evidence suggests that during diving a right-to-left shunt occurs because of an increase in the pulmonary resistance, which causes some blood to enter the left systemic from left ventricle and thus to bypass the lungs (fig. 11.26).

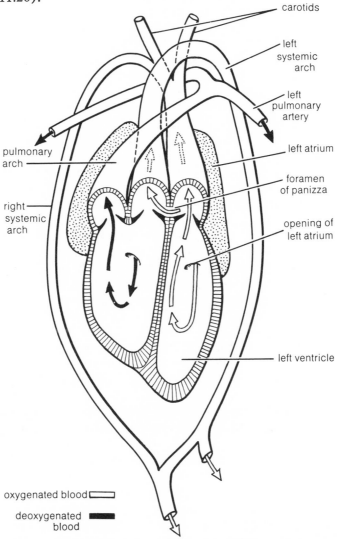

Fig. 11.26. Diagram of heart and arterial arches of crocodile. The interventricular septum is complete and the right and left atria open on either side of this septum. The right and left sides of the heart are completely separate except for the presence of the foramen of Panizza. This foramen has an important functional significance since it allows oxygenated blood to enter the left systemic arch. After Hughes 1965.

## 4. The Heart in Birds and Mammals

The heart of birds shows some resemblance to that of the crocodilian reptiles in that the carotids arise from the right systemic arch. In the case of birds, however, the left systemic arch and hence the foramen of Panizza is absent and there can be no communication between the right and left sides of the heart. Blood from the pulmonary vein enters the left atrium and is completely separate from the systemic venous blood that returns to the right atrium. Blood from the left atrium passes into the left ventricle and is forced into the right systemic and the carotids.

The pulmonary artery arises from the right ventricle and hence carries the systemic venous blood to the lungs. Thus a complete double circulation is established in this group.

A double circulation is also found in mammals, but in this group it is the left systemic arch that persists. Although there are minor differences in the shape and size of the heart in different birds and different mammals, in general there are no gross variations in structure in each group like those that occur in the lower vertebrates.

The beating of the heart is a basic property of cardiac muscle since the heart will continue to beat outside the body under appropriate conditions. During the development of the heart contraction rate of the various heart chambers depends on the sinus venosus, which acts as a pacemaker. In the adult, however, the heart beat may be modified by the nervous system, and in general stimulation of the vagus nerve, X (parasympathetic), slows down the heart rate. The cardiac vagus is often referred to as the *depressor nerve*. The beating of the heart may be accelerated by the stimulation of the sympathetic system whose fibers leave the central nervous system in one or more cervical nerves. In the lower vertebrates the heartbeat continues to be initiated by the sinus venosus at the point where it joins the atrium. This point is the *sinoatrial node,* which consists of muscle cells with branches at acute angles. The stimulus then spreads over the atria and from there over the other heart chambers. In the poikilotherms this stimulus is transmitted by the normal cardiac muscle fibers but in the homeothermic animals a specialized conducting system is developed to facilitate the spread of the stimulus from the sinoatrial node. The *conducting system* consists of a series of *muscle cells* specialized for the conduction of impulses rather than contraction. These fibers are usually especially rich in glycogen and striated only at the periphery. The stimulus from the pace-maker spreads over the atria and is transmitted to a second group of nodal fibers (at the base of the interatrial septum) forming the *atrioventricular node.* The impulse is then carried by the *atrioventricular bundle,* a special group of fibers that passes along the interventricular septum and then divides into two bundles that run down either side of the interventricular septum. In some forms though not in others the atrioventricular bundle is surrounded by a fibrous sheath. From the interventricular septum the conducting bundles break up into small branches and then into fine twigs that ramify among the cardiac muscles and form a reticulum below the endocardium (fig. 11.27).

Conduction through the atria, which lacks any special conducting system, is considerably slower than that in the ventricle. In the atrium of the dog the impulse travels at 1.6 meters per second whereas in the ventricle the speed is 4.0 meters per second. The impulse is slowed down considerably in the atrioventricular node and falls to about 0.2 meters per second. The retardation of the impulse ensures that the atria have an adequate time to empty before the ventricle commences to contract.

The way the heart contracts and the pressure produced are also related to the structure of the myocardium of the ventricle. In fishes and amphibians, where the blood pressure is relatively low, the wall of the ventricle consists of interlacing bundles of muscle forming a sponge-like

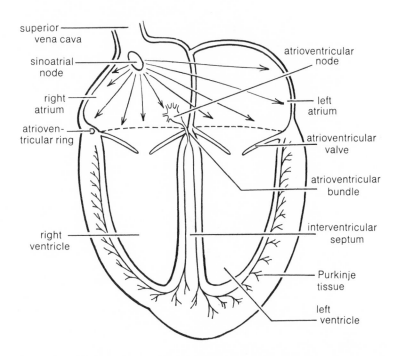

Fig. 11.27. Conducting system of mammalian heart. Contraction is initiated at the sinoatrial node and the stimulus spreads to the atrioventricular node and thence to the ventricle. Rapid conduction of stimuli is facilitated by the Purkinje tissue, which consists of highly modified muscle fibers.

arrangement. In reptiles the myocardium has a dense peripheral portion overlying a spongy interior. In birds and mammals the ventricular wall consists of a mass of firm muscle in which the fibers are arranged in the form of a double spiral. The *bulbospiral* starts at the mitral valve and the *sinospiral* at the tricuspid valve, and when the heart contracts, the blood is "wrung" out of the ventricle rather than squeezed, as it is in the lower vertebrates (fig. 11.28). The pumping action of the heart is also enhanced by the presence of an interventricular septum, which becomes firm as ventricular systole commences and forms a fixed fulcrum at the ends of the spirals.

Therefore the nature of the conducting system and the arrangement of the muscles and septum enable the higher vertebrates to have a high rate of contraction and also to produce a high pressure, and hence rapid circulation of the blood consistent with a high metabolic rate.

The increase in thickness of the myocardium and its more compact nature in the homeotherms are also associated with the presence of an extensive system of *coronary arteries* that supply the heart muscle with oxygenated blood. The coronary arteries arise from the base of the aortic arch and ramify throughout the cardiac muscle and finally blood is returned to the right atrium via the coronary veins. The heart in fishes also has a well-developed coronary system, which is associated not with the nature of the wall of the ventricle, but rather with the fact that the heart in most fishes contains only deoxygenated blood and therefore must have a source of blood that contains oxygen. The coronary artery

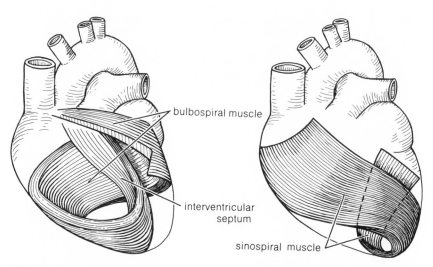

Fig. 11.28. Arrangement of muscle fibers in mammalian heart. Much of the cardiac muscle is arranged in two spirals, the bulbospiral and the sinospiral. During ventricular systole blood is "wrung" rather than squeezed out of the heart.

in most fishes is derived from a special *hypobranchial circulation* from one or more of the efferent branchial arteries. A subsidiary coronary artery may arise from the dorsal aorta. The heart in amphibians is generally supposed to be without a coronary system. This appears to be true of the anuran amphibians, at least as far as the ventricle is concerned, but a small coronary artery to the atria has been described in some forms. The ventricle has a spongy nature and it is clear that none of the cardiac muscle cells will be far away from the blood and this seems to be adequate for the anuran ventricle. An interesting modification is found in the caecilian amphibians: the walls of the ventricle are covered by a network of vessels that resemble the coronary arteries of amniotes. These vessels open into the sinus venosus, however, and are in fact *coronary veins*. They communicate with small spaces in the myocardium, which in turn are connected with the main ventricular cavity. During contraction of the ventricle, small quantities of blood are squeezed through the myocardium and into the coronary veins, thus ensuring a vascular supply to all the muscle cells (fig. 11.29).

## 5. Evolution of the Heart

Many accounts of the evolution of the vertebrate heart assume that there has been an increase in the complexity and the number of heart chambers thus:

Fish ⟶ Amphibian ⟶ Reptile ⟶ Birds and Mammals
1 atrium ⟶ 2  atria ⟶ 2  atria ⟶ 2  atria
1 ventricle ⟶ 1  ventricle ⟶ 2 ventricles ⟶ 2 ventricles
                      incomplete         complete
                      septum             septum

But examination of the heart in the squamate reptiles shows that completion of the interventricular septum (implicit in this scheme) would result in the isolation of the cavum pulmonale from the rest of the

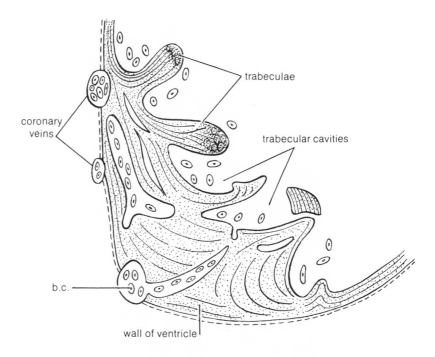

Fig. 11.29. Heart of apodan amphibian, *Hypogeophis*. Portion of ventricular wall. The ventricle is not supplied with coronary veins but some of the blood from the ventricular cavity is squeezed through the muscle of the ventricular wall and collected by the coronary veins; b.c., blood cells. From Lawson 1966.

ventricle. The result of this separation would be the total exclusion of the pulmonary artery from the circulation. Clearly the interventricular septum of crocodiles, birds, and mammals cannot have been evolved by the completion of the squamate septum. Hence the interventricular septum in the birds and mammals cannot be homologous with that of the lepidosaurian reptiles. The homologies and relationships of the septum within the amniotes can be established best by an examination of its embryology in these groups and also in the lungfishes. In the lungfishes the interventricular septum has a dual embryological origin. It is derived from an *endocardial ridge* that grows forward from the rear wall of the ventricle toward the conus and is the *primary septum*. While this ridge is developing, several muscle buds, covered by a thin cap of endocardium, arise laterally and coalesce to form a prominent network of trabeculae, the *secondary septum* that fuses with the primary septum to form the definitive interventricular septum. The horizontal septum in the Squamata and Chelonia is homologous with the primary septum of lungfishes (and is largely of endocardial origin). The secondary septa, which develop at right angles to this septum in some reptiles, probably correspond to the secondary septum of lungfishes. Developmentally the interventricular septum in birds corresponds to a secondary septum but that of mammals relates to a primary septum. It has been emphasized earlier that the evolution of the heart is closely associated with the mode of respiration, especially in amphibians. It is believed that the early amphibians evolved from the crossopterygian fishes, which inhabited water that was impermanent and hence was an environment similar to that of modern lungfishes. It is entirely probable that the ancestors of the first land vertebrates already had well-developed lungs and hence a heart which could in some measure separate oxygenated and deoxygenated blood. Since lungfishes as a group are, in the evolutionary sense, extremely conservative, it is likely that the heart in the crossopterygian ancestors of the tetrapods bore some resemblance to that of modern lungfishes. Such a heart was essentially composed of two atria and two ventricles and was capable of effecting a good degree of separation of the blood. An important event in the evolution of the Amphibia was that they became committed to the use of nonpulmonary respiration, mainly that of the skin. The existence of this type of respiration is correlated with the *breakdown of the interventricular septum* and the heart of modern amphibians has evolved from an essentially four-chambered structure by the loss of the interventricular septum. Therefore the absence of the interventricular septum in modern Amphibia is correlated with cutaneous respiration and clearly allows blood within the ventricle to be routed in a way that would not be possible if such a structure were present. Hence the heart of modern amphibians must be regarded as a specialized structure and the lack of an interventricular septum is indicative of this specialization and not of any intermediacy between the fish and reptiles. Further evidence for this hypothesis is that those groups of amphibians that rely heavily, and in some cases entirely, on nonpulmonary respiration lose not only the interventricular septum but also the spiral fold in the conus and even the interatrial septum. In the heart of caecilians and some urodeles the ventricle retains a prominent *central trabecula,* which may represent the *remnant* of an interventricular septum, and a recent description of the heart of the urodele *Siren intermedia* indicates the presence of a well-formed interventricular septum.

The early tetrapod stock that gave rise to lizards and snakes clearly relied on the primary septum for the separation of the ventricular chambers. This is also true for the ancestors of living tortoises and turtles. The reptiles from which the crocodiles and birds have evolved must, however, have utilized the secondary septum rather than the primary one. The similarities between the vascular system of the crocodiles and birds is emphasized by the association of the carotid arch with the right systemic arch. Further, during development the elongation of the third (carotid) arch, related to the development of the neck, occurs mainly after this arch has assumed a dorsal position. In lizards and snakes, however, this elongation occurs ventrally. Paleontological evidence indicates that crocodiles and birds have evolved from the archosaurian or "ruling reptiles," whereas lizards and snakes are derived from a lepidosaurian stock. The structure of the skeleton in chelonians suggests that they evolved directly from the "stem reptiles" or cotylosaurs.

The mammals arose from yet another reptilian stock, the *Synapsida,* which was one of the very first offshoots from the cotylosaurs. The ancestors of mammals retained a heart similar to that of the crossopterygian/tetrapod ancestors but relied primarily on an interventricular septum of the primary type and early in their history completely separated the pulmonary and systemic bloods.

Thus, it is apparent that in the evolution of vertebrates the secondary septum has been part of the hearts of birds and crocodiles, whereas squamates, chelonians, and mammals have adopted the primary septum to fulfill their particular needs (fig. 11.30).

In accounts of the evolution of the vascular system, the point is fre-

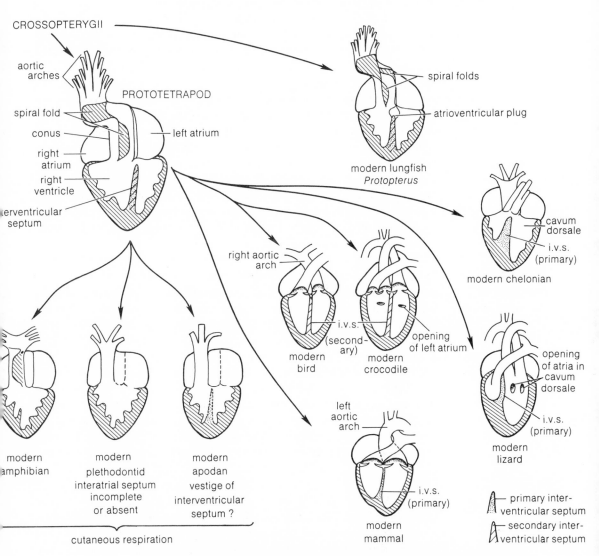

Fig. 11.30. Diagrammatic scheme of evolution of the heart in vertebrates. It is possible that the ancestors of the tetrapods had a four-chambered heart, not unlike that of a modern lungfish, in which the interventricular septum was of a dual origin (primary and secondary). In amphibians the acquisition of cutaneous respiration is associated with the disappearance of the interventricular septum. The primary interventricular septum is retained in the reptiles (except crocodiles) and mammals, whereas birds and crocodiles utilize the secondary interventricular septum.

quently made that the heart in amphibians and reptiles is inefficient since they are apparently "unable" to complete the separation of oxygenated blood effectively. Such statements must be treated with caution since they are obviously not applicable to modern amphibians, for the "mixing" of blood within the ventricle is clearly an adaptive feature correlated with cutaneous respiration. The situation within reptiles is less clear. Some of the mixing of blood within the reptilian heart and major arterial arches is of great functional significance and allows the blood to be shunted from one side to the other—often correlated with diving. In other cases it is impossible to establish the functional significance of mixing, but it may be that the exchange of blood within one part of the system and the other is necessary for the equilibration of pressures on either side of the system.

The evolution of the heart in the vertebrates also involves the decrease in the importance of the sinus venosus. In fishes this is a well-developed chamber that collects all the venous blood. In Amphibia the sinus venosus collects all the systemic venous blood and the pulmonary vein opens directly into the left atrium, much as in lungfishes. In reptiles and birds the sinus venosus becomes much reduced in size and incorporated largely into the wall of the right atrium. However, the sinus venosus is invariably marked off from the right atrium by a sulcus and internally by the presence of valves. In mammals the incorporation of the sinus venosus into the atrial wall is almost complete.

## F. Special Adaptations of the Vascular System

The prime function of the cardiovascular system is to ensure that all tissues are adequately supplied with blood. However, there are occasions when, because of changing demands, it is necessary that the established pattern of blood flow be changed.

### 1. Arteriovenous Anastomoses

Blood usually flows through the microcirculation from artery, arteriole, capillary, venule, and vein, but in some cases the accepted routing of blood may be varied so that the capillary circuit may be modified. In such instances blood is allowed to flow directly from an artery into a small vein. Such capillary bypasses are termed *arteriovenous anastomoses* and are of special importance in the skin of mammals, for they route blood directly from the artery to the vein during cold conditions, bypassing the capillaries and hence reducing heat loss (fig. 11.31).

### 2. Large-Scale Rerouting Pathways

A gross redistribution of blood supply may occur in order to concentrate blood (and hence oxygen) in the tissues that need it most. Such tissues are the brain and heart, which have a very low tolerance of oxygen deficiency, whereas other tissues, such as the skin and kidney, can survive for long periods in anoxic conditions.

#### a. Diving

Curiously enough, diving animals do not usually have unduly large lungs or the capacity to store large quantities of oxygen. It seems that the ability to remain submerged is related to the vascular system and its ability to redirect blood. During diving the rate of heart beat is lowered (bradycardia), often to one-tenth of that prior to diving. The

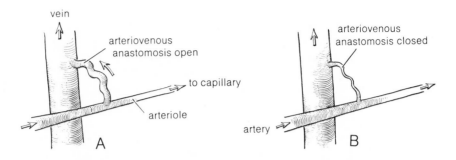

Fig. 11.31*A–B*. Arteriovenous anastomosis in skin. The arteriovenous anastomosis is an important physiological device for shunting blood directly from
an arteriole into a vein. Thus the blood effectively bypasses the capillaries and
this device is of great value in temperature regulation.

central blood pressure remains normal but there is a very slow drop in
the pressure between contractions (diastolic pressure), indicating that
there is constriction of peripheral vessels and some shutdown of the
vascular system in certain areas. There is also an overall slowing down
of the metabolic rate; such changes also occur in hibernating animals.
During diving there is a drastic reduction in the peripheral blood flow in
most muscles, skin, intestines, and kidneys. Blood flow through many
glands is also reduced, but this does not apply to the liver, thyroid, and
adrenals. The coronary and cerebral circulations are unimpaired, however, and may even be slightly increased.

In many diving animals there is also a marked enlargement of the
venous system associated with a relatively high blood volume, and enlarged hepatic sinuses or hepatic veins undoubtedly act as *blood reservoirs*. In some cases there is also a duplication of the posterior vena cava,
and the intrathoracic part of these vessels is enclosed in a thick layer of
striated muscle. This muscle can regulate the amount of blood returning
to the right atrium (and hence to the lungs); thus the venous system
provides a store of relatively well-oxygenated venous blood that can be
admitted to the system as required.

These modifications of the vascular system in diving forms ensures
that, during diving, there is a reduction in the peripheral circulation,
which ensures that the heart and brain are well supplied with blood.

### b. Retia Mirabilia

Many vertebrates incorporate into their vascular systems a complex
network of vessels that form retia mirabilia. Such structures occur in a
wide variety of organisms and in varying positions, but they are all designed on a countercurrent principle, so that the blood in one vessel is
flowing in the opposite direction from that in the vessel adjacent to it.
This countercurrent flow is used in nature for the exchange of gases, ions,
and heat. It is based on the principle that in two vessels lying side by side
the blood flowing in opposite directions is at different temperatures,
contains different concentrations of gases or ions, and so on. In such a
situation there is an exchange of heat, gas, etc., so that the blood in one
stream is potentially able to exchange up to the maximum concentra-

tion of material in the other stream, instead of equilibrating at a mean value as it would if the streams were flowing in the same direction.

Retia mirabilia are extensively used to maintain high pressure of gases in swim bladders and the countercurrent system is used in the oxygenation of blood in the gills of fishes. In addition retia are extensively used to conserve heat and are frequently found at the bases of the limb, fin, or tail, where they serve to trap the heat of the body before the blood enters the extremity. One extraordinary example of such a system is found in fast-swimming fishes such as the tuna and the mackerel shark, where it is used to trap the heat of the body in the muscles. The result is that the muscle is several degrees warmer than the rest of the fish or the sea water. In fishes blood is usually supplied to the muscles by a series of segmented arteries and returned to the heart in the posterior cardinal vein; the major portion of the heat, which the venous blood collects in the muscles, is lost as the blood passes from the heart through the gills. Therefore the body temperature of the fish is the same as that of the sea water. In the tuna and many other fishes, however, the blood to the muscles, particularly the red muscle, is supplied by an extensive series of cutaneous arteries and veins that break up into an enormous plexus of parallel vessels within the muscle. The net result is that the venous blood returning to the circulation passes its heat to the arterial blood (flowing into the muscle) and hence enables the muscle to be kept at a temperature above that of the environment.

The physiological advantage of this is that it increases the efficiency of the muscle; it is known that if two otherwise similar muscles have temperatures which differ by a factor of 10°C, the muscle with the higher temperature produces three times more power than the one with the lower temperature.

Retia mirabilia are also often found on the surface of the liver and function in a similar manner in keeping up the liver temperature and so increasing its activity while reducing the overall body heat loss.

General Comment on Dissection

The student should be aware that it may not be possible to locate all the structures described in a single dissection specimen (or even a small number of specimens). In many fresh specimens vessels may be difficult or impossible to locate because they have collapsed or are void of blood, and in injected specimens the injection medium may have failed to penetrate.

The student should also realize that considerable variation occurs in the detailed anatomy of dissection types. Such variations are usually limited and involve differences in the configuration of fine blood vessels and nerves, but, in some cases large-scale variations may involve major structures.

Thus the accounts given below are based on large numbers of dissections and are essentially those of "typical" specimens. It will be to your advantage to trace and identify blood vessels by following these accounts without reference to diagrams because of such variation. If illustrations are necessary, first consult the diagrams in the early part of this chapter, and then resort to accessory laboratory guides.

**G. The Circulatory System of Elasmobranchs**

In the circulatory system of elasmobranchs the first and second aortic arches have been modified as in all vertebrates. Changes have occurred in the posterior veins. The heart remains a one-way tube with the four typical chambers.

The account applies to the spiny dogfish and the skate.

**1. The Chambers of the Heart**

The heart of the elasmobranchs consists of four chambers. The pericardial cavity has already been exposed in the preceding dissection. Spread its walls apart. Identify the chambers of the heart as follows: on raising the heart a triangular chamber will be seen extending from the heart to the transverse septum, its two corners buried in the septum. This is the *sinus venosus,* the most posterior chamber of the heart. Each corner of the sinus venosus is continuous with a large vein, the *duct of Cuvier* or *common cardinal vein,* which is enclosed in the transverse septum and will be seen later. Anterior to the sinus venosus is the *atrium,* a large, thin-walled chamber expanded on each side of the heart and appearing as if paired. Between the two sides of the atrium rests the *ventricle,* a thick-walled chamber and the most conspicuous portion of the heart from the ventral side. The pointed posterior end of the ventricle is known as the apex; the broad anterior end, the base. From the base of the ventricle a thick-walled tube runs forward and penetrates the anterior wall of the pericardial cavity. This is the *conus arteriosus,* the fourth and most anterior chamber of the heart. The blood circulates through the chambers of the heart in the following order: sinus venosus, atrium, ventricle, conus arteriosus.

**2. The Systemic Veins**

Systemic veins have already been defined as those veins that enter the heart. All systemic veins in vertebrates open into the sinus venosus or its equivalent, but because of some differences the dogfish and skate will be described separately.

In dissecting the veins, one follows them away from the heart, and it is often convenient to speak of them as if they proceeded from the heart to body structures. But the student must always bear in mind the fact that they convey the blood from the parts of the body to the heart.

Dogfish: Insert one blade of a fine scissors in the sinus wall and slit the ventral wall of the sinus venosus open in a crosswise direction. The cavity of the sinus is thus exposed and should be washed out thoroughly. All of the systemic veins open into the cavity of the sinus, and the openings may now be identified if the cut edges of the sinus wall are spread well apart. Each lateral wing of the sinus that lies buried in the transverse septum has a very large opening, the entrance of the duct of Cuvier or common cardinal vein. The natural relations of this entrance are best observed on the intact right side. On the left side carry the slit in the sinus laterally to meet the incision previously made across the gill slits. The entrance of the common cardinal vein into the sinus is thus slit open. The following may then be noted. Just medial to the main opening of the common cardinal vein into the sinus are several small apertures, most of which appear to be subdivisions of the chief opening. The most anterior of these small apertures is, however, the opening of the *inferior jugular vein.* Probe into the opening and note that the vein stems from the floor of the mouth and pharyngeal cavity, where it runs alongside the ventral ends of the gill arches. By turning back the flap of the floor of the mouth and pharyngeal cavities (previously formed), you can more readily follow the course of the vein. Now look into the posterior part of the wall of the sinus, while pulling the wall taut. In the median line of the posterior wall is a white fold, and on each side is an opening. Probe into both openings and note that your probe passes internally to the coronary ligament and into the liver. Follow your probe into the liver by slitting the substance of the liver, and note the cavity in the right and left lobes of the liver thus revealed. These cavities, which extend nearly the entire length of the liver lobes, are hepatic sinuses. (Because many of the veins of elasmobranchs are not definite vessels but spaces in the tissues without definite walls, they are more correctly designated sinuses.) The two hepatic sinuses are the persistent proximal parts of the vitelline veins of the embryo.

Now probe posteriorly into one of the common cardinal veins. Raise the viscera in the anterior part of the pleuroperitoneal cavity and observe that your probe has entered a large bluish sac in the dorsolateral wall of the pleuroperitoneal cavity. Follow this sac and its fellow of the opposite side posteriorly. Each bends toward the median region, narrowing considerably; by pressing the viscera to one side you may trace each sac posteriorly as a narrow tube on each side of the attachment of the dorsal mesentery to the median dorsal line. These two vessels are the *posterior cardinal sinuses;* they are the chief somatic veins of the trunk. At the level of the anterior part of the liver the two posterior cardinal sinuses communicate with each other by a broad connection, which may be found by probing into the left vein and directing the probe toward the right one. In this same region each vein has on its ventral surface an extensive communication with a large blood sinus, the *genital sinus,* surrounding each gonad. Each posterior cardinal sinus also receives numerous segmentally arranged branches from the body wall (*parietal veins*) and from the kidneys (*renal veins*). The kidneys are the long, slender, flat, brownish organs that lie immediately lateral to the posterior cardinal sinuses and extend the entire length of the pleuroperitoneal cavity. The parietal and renal veins are readily identifiable if they happen to be filled with blood, but they are impossible to see when empty.

Now turn the animal dorsal side up and locate the lateral line. Make a longitudinal incision above the gill slits on the left side along the lateral line and deepen the incision until you break into a large cavity with a smooth lining, the *anterior cardinal sinus.* (If the study of the branchial musculature was carried out, the anterior cardinal sinus has already been exposed and this incision will be unnecessary.) Probe anteriorly into the anterior cardinal sinus and follow your probe by an incision. The cardinal sinus can thus be traced forward above the spiracle to the eye, where it connects with an *orbital sinus* surrounding the eyeball. At the level of the posterior end of the eyeball there is an opening in the ventral wall of the anterior cardinal sinus. On probing into this, you will find that it extends medially into the skull. It is the opening of the *interorbital sinus,* which connects the two orbital sinuses. Locate the hyoid arch. In the floor of the anterior cardinal sinus, between the hyoid and third gill arches, is an opening. On probing this, you will find that it leads into the *hyoidean sinus,* a vessel that extends ventrally along the outer surface of the hyoid arch and connects with the *inferior jugular vein.* Next trace the anterior cardinal sinus posteriorly. It turns abruptly ventrally and joins the posterior cardinal sinus. If pressed into the anterior cardinal sinus at the turn, the probe will emerge into the posterior cardinal sinus. The union of the two sinuses forms the common cardinal vein already described.

Along the lateral wall on each side of the pleuroperitoneal cavity, immediately external to the pleuroperitoneum, is a conspicuous vein, the *lateral abdominal vein.* Note the *parietal branches* that it receives intersegmentally from between the myotomes. Trace the right vein anteriorly and find its entrance into the duct of Cuvier, just in front of the expansion of the posterior cardinal sinus. Slit open the lateral abdominal veins at this entrance and find the opening of the *brachial vein,* which drains the pectoral fin. The very short common stem formed by the union of the lateral abdominal vein with the brachial is the *subclavian vein,* which opens into the duct of Cuvier. The brachial vein passes along the posterior surface of the pectoral girdle in contact with the cartilage and may be picked up easily on the left side, where the girdle has been cut across. The vein may also be found by cutting across the base of the fin, where it forms an opening dorsal to the fin rays in the posterior half of the fin. Next trace the lateral abdominal vein posteriorly. It passes along the inner surface of the pelvic girdle and then continues posteriorly along the lateral margin of the cloacal aperture as the *cloacal vein.* At about the middle of the base of the pelvic fin the lateral abdominal vein receives a *femoral vein* from the fin; the opening into the lateral abdominal vein will be found by slitting open the latter. The femoral vein is a short vessel situated just under the dorsal skin of the fin.

In the midventral line there is a *ventral cutaneous vein,* which anteriorly bifurcates and disappears into the musculature.

Skate: The sinus venosus consists of a tube on each side, and the central portion is smaller and is attached to the transverse septum by a sheet of connective tissue, which may be broken. Each side of the sinus is buried in the transverse septum. Follow the right side to the point where it disappears dorsal to the cartilage of the pectoral girdle. With a sharp scalpel carefully shave away the cartilage and surrounding tissues until the sinus venosus can be followed laterally. Insert one blade of a fine scissors into the ventral wall of the sinus and slit it open in a crosswise direction. The sinus is continuous on each side with a tube or chamber, the *common cardinal vein* or *duct of Cuvier,* which turns dorsally. All of the systemic veins open into the common cardinal veins, and their openings may now be identified. The junction of common cardinal vein with the sinus venosus is marked by a slight fold. Concealed by this fold in the anterior wall of the common cardinal vein is the opening of the *inferior jugular vein*. This opening is so small that the probe probably cannot pass into it. The inferior jugular vein drains the walls of the pericardial cavity and the floor of the mouth and pharyngeal cavities. In the posterior wall of the common cardinal vein, at its junction with the sinus venosus, is the opening of the *right hepatic sinus*. Probe posteriorly into this. It leads into the right hepatic sinus, a space situated between the anterior margin of the right part of the liver and the transverse septum; in the females the sinus lies dorsal to the beginning of the oviduct, which is enclosed in the falciform ligament. To locate the hepatic sinus press the liver caudad away from the transverse septum; the sinus forms a small sac between the liver and the septum ventral and somewhat to the side of the esophagus. A similar *left hepatic sinus* is present on the left side, and the two are connected in some specimens. Cut into the right hepatic sinus and look in its posterior wall for the small openings of the hepatic veins draining the liver.

Now probe into the main cavity of the common cardinal vein in a dorsal and posterior direction. The probe enters the *posterior cardinal sinus,* which is a broad, thin-walled tube lying against the dorsal wall of the pleuroperitoneal cavity. In females it is on the dorsal side of the oviduct; in males, dorsal to the testis. Follow the posterior cardinal sinus posteriorly. It turns toward the median line, where it soon meets its fellow of the opposite side to form a single large median sinus. This sinus communicates on its ventral surface with the large *genital sinus* within each gonad. More posteriorly the posterior cardinal sinus separates again into two veins that proceed caudad on the medial side of the kidneys. The kidneys are rounded lobes at the posterior end of the pleuroperitoneal cavity against the dorsal wall. To see them, remove the pleuroperitoneal membrane against the dorsal wall on each side of the cloaca. Do not injure the ducts from the kidneys. After exposure of the kidneys the posterior cardinal veins will be found on the medial side of the kidneys. They connect with each other between the kidneys.

Return to the common cardinal vein and probe into it in a dorsal direction. Turn the animal dorsal side up and locate the end of your probe. Make an incision into the spot indicated by the probe and extend the incision longitudinally forward to the eye. Deepen the incision carefully until an elongated cavity with a smooth wall, the anterior cardinal sinus, is exposed. It is situated just medial to the dorsal ends of the gill arches and visceral pouches. It may be followed forward with the aid of the probe. It turns laterally in front of the first visceral pouch, runs along the anterior border of this pouch, and then turns anteriorly again.

With the dorsal side of the animal still facing you, locate the chief anterior cartilage (*propterygium*) of the pectoral fin. It forms a crescentric ridge lateral to the gill region nearly halfway from the mid-dorsal line to the margin. Make an incision along the medial face of this cartilage on the left side. A vein will be exposed running along the cartilage here. It is one of the *brachial veins*. Follow it posteriorly. It will be found to enter the common cardinal vein.

Turn the specimen ventral side up again. Along the lateral wall of the pleuroperitoneal cavity runs the *lateral abdominal vein*. Note the *parietal branches* it receives from the body wall at each myoseptum. Trace the lateral abdominal vein anteriorly. It passes along the internal surface of the cartilages of the pectoral fin and pectoral girdle and enters the common cardinal vein. Immediately on the posterior side of the cartilage of the pectoral girdle a *brachial vein* will be found entering the lateral abdominal vein. Immediately posterior to this is another cartilage, and on the caudal side of that another *brachial vein* joins the lateral abdominal. A third *brachial vein* was mentioned in the preceding paragraph. The lateral vein, after it has received the brachial veins, is termed the *subclavian vein,* but it is extremely short and it enters the common cardinal vein almost immediately. Trace the lateral abdominal vein posteriorly and find its origin in a network of small vessels on the sides of the large intestine and cloaca. It passes on the inner surface of the cartilages of the pelvic girdle and pelvic fin. Slit the vein open along the surface of the cartilages; *femoral veins* will be found emerging from between the cartilages and entering the lateral vein. The largest of the femoral veins is located along the posterior side of the puboischiac bar. Probe into the femoral veins and note their distribution in the pelvic fin.

**3. The Hepatic Portal System**

The hepatic portal system consists of veins that collect the blood from the digestive tract and spleen and pour it into a network of capillaries in the liver. Locate the bile duct. Lying in the hepatoduodenal ligament alongside the bile duct is a large vein, the hepatic portal vein. Trace it posteriorly and identify the branches it receives from the digestive tract. These branches are generally filled with blood and are therefore easily traced. If they are empty, they may be readily injected through the hepatic portal vein, even in specimens which have been preserved for a long time.

Dogfish: Trace the hepatic portal vein posteriorly; first it receives a small *choledochal vein,* which runs along the bile duct. Posterior to this, the hepatic portal vein is formed by the union of three large branches. The left branch, the *gastric vein,* passes at once to the stomach, where it is formed by the union of the *dorsal* and *ventral gastric veins,* which branch on the dorsal and ventral surfaces of the stomach. The middle of the three branches of the hepatic portal is the *lienomesenteric vein,* which passes posteriorly, dorsal to the duodenum and is embedded in the substance of the dorsal lobe of the pancreas, from which it receives small *pancreatic veins*. At the posterior end of the pancreas the vein is seen to be formed by the union of two branches: the *posterior lienogastric vein,* from the spleen, and the *posterior intestinal vein,* from the left side of the valvular intestine. The posterior lienogastric vein collects branches from the spleen and the adjacent wall of the stomach. Note the numerous branches into the posterior intestinal vein from along the lines of attachment of the turns of the spiral valve. The right branch of the three that form the hepatic portal vein is the *gastrointestinal* or (*pancreaticomesenteric*). It passes dorsal to the pylorus, from which it receives a *pyloric vein,* as well as an *intraintestinal vein* from the interior of the spiral valve; then it lies embedded in the substance of the ventral lobe of the pancreas, where it receives the *anterior lieno-gastric vein* from the spleen and adjacent parts of the pyloric portion of the stomach. Its main trunk, the *anterior intestinal vein,* continues along the right side of the intestine, from which it receives branches along the lines of attachment of the spiral valve.

Skate: The hepatic portal vein is formed by the union of three tributaries: a *gastric vein* from the left, a *lienomesenteric vein* from the middle, and a *pancreaticomesenteric* vein from the right. The gastric vein passes to the right margin of the stomach and there receives the *dorsal and ventral gastric veins* from the dorsal and ventral surfaces of the stomach. The dorsal gastric vein receives tributaries from the spleen. The lienomesenteric vein receives a *lienogastric* branch from the spleen and adjacent stomach. Its main tributary, the *posterior intestinal vein,* runs along the left side of the intestine, beginning in the tip of the rectal gland; in its course along the intestine it receives branches along the lines of attachment of the spiral valve. It also collects blood from the pancreas. The *pancreaticomesenteric vein* collects from the pancreas and from the *anterior intestinal vein,* draining the duodenal region. It also receives a *posterior gastric vein* from the narrow portion of the stomach between the pylorus and the bend.

The *hepatic portal vein* reaches the dorsal surface of the liver and there divides into branches that penetrate the substance of the liver. In the liver the branches fork into smaller and smaller veins, which empty into the hepatic sinuses; we have already seen that the hepatic sinuses open into the sinus venosus (dogfish) or common cardinal vein (skate).

## 4. The Renal Portal System

In the renal portal system the venous blood passes into a network in the kidneys. Cut across the tail just posterior to the anal opening. In the cross section locate the caudal blood vessels, enclosed in the haemal arch. The *caudal artery* is dorsal, the *caudal vein,* immediately ventral to the artery.

Dogfish: Probe into the caudal vein. Observe that the probe can be passed either to the right or to the left and therefore that the vein forks at the anus. The two forks are the *renal portal veins.* Leaving your probe in one of the renal portal veins, locate the kidney in the pleuroperitoneal cavity. It is a long, brown organ situated against the dorsal body wall, and there is one on each side of the middorsal line external to the pleuroperitoneum. Slit the pleuroperitoneum along the lateral border near the posterior end of the kidney on the side where your probe is inserted and gently lift the kidney away from the body wall. A space will be found between the kidney and the body wall, where your probe has passed. This space is the *renal portal vein* or *sinus*. It branches into the kidney and also receives tributaries from the body wall.

Skate: The kidney has already been exposed. Look along the medial side of the posterior part of the kidney for a vein coming from the vertebral column. Do not injure any ducts on the ventral side of the kidney. In males the vein in question lies immediately to the dorsal side of the male duct, which will be seen passing along the ventral surface of the kidney to the cloaca. The duct may be lifted from the kidney surface and bent to one side. The vein is the *renal portal vein*. It lies along the medial border of the kidney and turns onto the ventral surface of the organ, giving off branches into its substance and receiving tributaries from the body wall lateral to the kidney. The renal portal veins are continuations of the *caudal vein;* the latter forks at the anus giving rise to the two renal portal veins. The forking is, however, difficult to trace in the skate.

**5. The Ventral Aorta and the Afferent Branchial Vessels**

Turn once more to the pericardial cavity of the specimen. The conus arteriosus runs forward and penetrates the anterior wall of the pericardial cavity. Carefully pick away muscles and connective tissue from the region extending from the anterior end of the pericardial cavity to the lower jaw. In the median ventral line you will find a large vessel, the *ventral aorta,* which continues forward from the conus arteriosus. By dissecting carefully to the left side, you can find the branches of the ventral aorta.

Dogfish: There are three main pairs of branches of the ventral aorta, two of which subdivide into two. The posterior pair of branches arises just at the point where the conus arteriosus passes into the ventral aorta. (At this point note the pair of *coronary arteries* passing along the conus arteriosus onto the surface of the ventricle and to the walls of the pericardial cavity. They should be preserved as far as possible.) Follow the most posterior branch of the ventral aorta. It divides in two, the posterior branch penetrating the intrabranchial septum of the sixth gill bar, the anterior branch the septum of the fifth. The middle branch of the ventral aorta rises shortly in front of the third branch and passes without division into the intrabranchial septum of the fourth gill bar. After giving off this branch the ventral aorta proceeds forward, without branching, to a point just posterior to the lower jaw. Here it forks to form its anterior pair of branches. Trace the left branch laterally. After some distance it forks and supplies the second and third gill bars. Trace any one of the branches of the ventral aorta out into the intrabranchial septum, slitting the septum. Note the small branches from the artery into the gill lamellae. The five pairs of branches of the ventral aorta are the *afferent branchial arteries*.

Skate: There are two main pairs of branches from the ventral aorta. The posterior pair arises where the conus arteriosus passes into the ventral aorta. (At this point note the paired *coronary artery,* passing to the conus arteriosus and to the branchial muscles. Preserve it as well as possible.) Follow the posterior branch of the ventral aorta. After some distance it subdivides into three branches, which pass to the fourth, fifth, and sixth gill bars, penetrating the intrabranchial septa. Trace them into the septum by slitting the septum, and note the branches from each to the gill lamellae of both hemibranchs of the septum. Follow the ventral aorta forward beyond the pair of posterior branches. It passes without branching for some distance and then forks into right and left branches. Follow the left branch. After a considerable distance it forks into two vessels, which penetrate the intrabranchial septa of the second and third gill bars. The five pairs of branches of the ventral aorta are the *afferent branchial arteries*.

In front of the anterior fork of the ventral aorta will be found some soft brownish diffuse material, the thyroid gland.

**6. The Efferent Branchial Arteries and the Dorsal Aorta**

Open the flap already formed on the floor of the mouth and pharyngeal cavities; turn it outward and fasten it in that position. The esophagus may be cut to make this procedure easier. With a forceps strip off the mucous membrane from the roof of the mouth and pharyngeal cavities. In the roof there are four pairs (dogfish) and three pairs (skate) of large blood vessels extending from the angles of the gill slits obliquely caudad. They are the *efferent branchial* or *epibranchial arteries* and represent the dorsal parts of the aortic arches. Clean away the connective tissues from these arteries so that they are clearly exposed and trace them to the left, noting that they disappear dorsal to the gill cartilages. Remove these cartilages carefully. This is best done by cutting across them, grasping the cut end, and loosening the cartilage toward the gills. Note that each artery is formed at the dorsal angle of the gill slit by the union of two vessels, a smaller *pretrematic branch* that comes from the hemibranch on the anterior face of the gill pouch and a much larger *posttrematic branch* from the hemibranch on the posterior wall of the gill pouch. Note the small vessel that runs from each gill lamella into the pre- and post-trematic branches. In the skate the first two epibranchial arteries unite and there are but three pairs of main vessels in the roof of the pharyngeal cavity.

Next, dissect on the right side, which has been kept intact, in order to see the full course of the epibranchial arteries. Remove the mucous membrane from the floor of the mouth and pharyngeal cavities, exposing the ventral portions of the gill arches. Remove these cartilages carefully without disturbing any of the arteries, and also remove the cartilage from the full length of the branchial bars. It will now be seen that they form a complete loop, the *efferent collector loop,* around each gill cleft. Note further that the posttrematic branch on the anterior wall of each branchial bar is connected with the pretrematic branch on the posterior face of the same bar by means of cross branches (there are between 3 and 5 in the dogfish and one in the skate). Thus all the efferent collector loops are interconnected.

Vessels springing from the ventral ends of the collector loops will now be described. These are the "external carotid artery" and the hypobranchial system. The "external carotid" (dogfish) from the

pretrematic branch of the first collector loop near its ventral end runs
forward along the lateral part of the lower jaw. The hypobranchial
system, which is very variable, stems from the ventral ends of the
second to fourth collector loops and supplies the wall of the heart,
the pericardial cavity, and hypobranchial musculature.

Dogfish: The main vessel of the hypobranchial system, the *com-
missural artery,* arises on each side from the ventral end of the second collector loop and may or may not receive tributaries from the
third and fourth collector loops (or these tributaries may first join a
longitudinal vessel, the *lateral hypobranchial artery*). The commis-
sural arteries then enter the anterior end of the pericardial cavity as
the *coronary arteries* and are here connected by an anastomosis along
the dorsal surface of the conus. From this anastomosis the coronary
artery of each side forks into one branch that passes along the side of
the conus arteriosus and spreads out on the ventricle and a more
dorsal branch that runs backward in the wall of the pericardial cavity,
then passes to the pharyngeal floor, and is finally distributed to the
esophagus. Connection occurs between the hypobranchial system and
the *subclavian artery.*

Skate: Contributions from the ventral ends of the collector loops,
especially the second, form a longitudinal vessel, the *lateral hypobranchial artery,* which anteriorly continues as the *external carotid
artery* into the lower jaw and posteriorly enters the pericardial cavity
as the *anterior coronary artery,* proceeding along the conus arteriosus
to the ventricle. In the posterior part of the pericardial cavity the
*posterior coronary arteries* (originating from the *subclavian artery*)
run along the sinus venosus.

Turn again to the roof of the oral cavity. From the dorsal end of
the collector loop around the second gill slit, a *hyoidean epibranchial
artery* runs forward and after a short distance bends toward the
median line. At this bend it gives off a branch, the *orbital* or *stapedial
artery,* which will be found by gently shaving away the cartilage at
this point. The orbital artery enters the orbit, where it supplies eye
muscles and other structures and then proceeds to the snout. Beyond
the orbital branch the main vessel, now known as the *internal carotid
artery,* passes to the median line and joins its fellows of the opposite
side; the main internal carotid artery so formed passes into the
cranial cavity, where it branches to the brain, eye, and internal ear.
In the dogfish a pair of slender vessels that represent the paired anterior parts of the dorsal aorta (radices) connect the hyoidean epibranchial artery with the first of the four epibranchial arteries.

Clear away all tissue from the pretrematic part of the first collector
loop. From the middle of the pretrematic part an *afferent* or
*spiracular artery* arises, passes forward, then turns sharply dorsally,
and disappears. Turn the animal dorsal side up and remove the skin
around and posterior to the spiracle. Pick up the spiracular artery
again below the spiracle and follow it to the inner side of a white
band, the hyomandibular nerve. Follow the spiracular artery to the
walls of the spiracle and note its branches to the rudimentary gill
(pseudobranch) in the spiracle. (In the skate there are numerous

branches to adjacent muscles and only very small branches to the spiracular walls.) Now turn the animal ventral side up, remove the mucous membrane between the spiracle and the upper jaw, and shave away the cartilage about halfway between the upper jaw and the hyoidean epibranchial artery. An artery of moderate size will be revealed and, when traced toward the spiracle, will be found to be the continuation of the afferent spiracular artery (already seen) formed by the reunion of the branches to the pseudobranch. It is the *efferent* or *epibranchial spiracular artery;* it passes into the skull ventral to the stapedial artery. When traced by scraping away the cartilage, it will be found to join the internal carotid artery.

The four epibranchial arteries (third to sixth) pass to the mid-dorsal line of the roof of the pharyngeal cavity. Here they join in pairs in the middorsal line to form a large trunk, the *dorsal aorta,* which passes into the pleuroperitoneal cavity. In the skate the first and second epibranchials unite, and shortly posterior to this junction a *vertebral artery* arises on each side and passes into the cartilage of the skull, where it is distributed to the brain and spinal cord.

## 7. The Dorsal Aorta and Its Branches

Separate the esophagus from the body wall on the left side and follow the dorsal aorta posteriorly. A *subclavian artery* is given off on each side from the dorsal aorta between the points where the third and fourth pairs of efferent branchial arteries unit with it. Trace the left one into the pectoral fin. It proceeds obliquely caudal and lateral, passing on the dorsal wall of the large posterior cardinal sinus. In the skate it gives off the *posterior coronary artery* and passes internal to a large white band, the nerve of the pectoral fin. (This may be cut through.) At the lateral boundary of the posterior cardinal sinus the subclavian artery gives rise to the small *lateral artery,* which branches into the body wall and usually proceeds posteriorly along the body wall in a position on a level with the lateral line. Further laterally, at the point where the lateral abdominal vein enters the duct of Cuvier, the subclavian artery gives rise to the *ventrolateral artery,* which proceeds caudad halfway between the lateral abdominal vein and the midventral line, giving off *intersegmental branches* into the body wall. At the posterior end of the pleuroperitoneal cavity it anastomoses with the vessels supplying the pelvic fins. After giving off these branches into the body wall the subclavian artery, now named the *brachial artery,* proceeds into the pectoral fin.

The dorsal aorta is a large vessel in the middorsal line of the pleuroperitoneal cavity. It has median, unpaired branches to the viscera and paired lateral branches to the urogenital system. Such branches form the *visceral arterial system.* The dorsal aorta also supplies paired branches to body wall and fins, forming the *somatic arterial system.* Work on the left side, turning the viscera to the right.

Dogfish: Just after it has penetrated the pleuroperitoneal cavity, the dorsal aorta gives rise to the large *coeliac artery,* which distributes blood to the gonads, stomach, and liver. Near its origin the coeliac artery gives off small branches into the adjacent gonads, esophagus, and cardiac end of the stomach. It runs posteriorly for a considerable distance without branching, until it enters the gastro-hepatic ligament and produces three branches: the *gastric, hepatic,* and *pancreaticomesenteric* (or *intestinopyloric*) *artery.* The *gastric artery* passes to the stomach and divides into *dorsal* and *ventral gastric arteries,* which branch on the surface of the stomach and penetrate its walls. The *hepatic artery* turns anteriorly, runs alongside the bile duct, and enters the substance of the liver. The *pancreatico-mesenteric artery* passes dorsal to the pylorus, gives off small branches into the pyloric portion of the stomach and the ventral lobe of the pancreas, a moderately large *duodenal artery* into the duo-denum, and a large *anterior intestinal artery* along the right side of the small intestine. The anterior intestinal artery gives off branches to the small intestines at the rings of attachment of the spiral valve. The dorsal aorta, after the origin of the coeliac artery, runs without further visceral branches to the free edge of the dorsal mesentery. Here it gives off two arteries, which course in the border of the mesentery. One of these, the *gastrosplenic* or *lienogastric artery,* passes to the spleen and bend of the stomach. The other, the *anterior mesenteric artery,* passes to the valvular intestine, where it runs caudad along the left side with branches at the turns of the spiral valve. Beyond the gap in the dorsal mesentery the dorsal aorta gives off the *posterior mesenteric artery* running along the free anterior border of the mesorectum to the rectal gland.

Skate: Shortly after entering the pleuroperitoneal cavity, the *dorsal aorta* gives off the *coeliac artery,* which supplies an *hepatic branch* to the liver; an *anterior gastric branch,* which divides into *dorsal* and *ventral arteries* to the stomach wall; *splenic branches* to the spleen; and a *gastroduodenal branch,* from which arise a *posterior gastric artery* to the posterior part of the stomach, *pancreatic branches* to the pancreas, and a *duodenal branch* to the pylorus and the duodenum. Shortly posterior to the origin of the coeliac artery, the dorsal aorta gives rise to the *anterior mesenteric artery,* which, after some small branches to the pancreas and spleen, proceeds posteriorly along the small intestine, to which it supplies branches at the lines of attachment of the turns of the spiral valve. Shortly caudad of the origin of the superior mesenteric artery, the *posterior mesenteric artery* branches from the dorsal aorta. It sends *genital arteries* to the gonads and their ducts and then passes in the mesen-tery to the rectal gland.

The lateral visceral and somatic branches of the dorsal aorta are similar in the dogfish and the skate. The former consist of the *genital arteries* already noted (but completely developed only in mature specimens) and the *renal arteries* into the kidneys. The latter can be seen by loosening the kidney from the dorsal body wall and looking on the dorsal surface of the organ. The somatic branches consist of paired *parietal* or *intersegmental arteries* to the body wall, which pass out along the myosepta. The *subclavian artery* to the pectoral fin has already been seen, and the paired *iliac arteries* to the pelvic fins arise from the dorsal aorta shortly in front of the cloaca. They course along the body wall and, after giving off a network of branches into the walls of the cloaca and anastomosing anteriorly with the posterior end of the ventrolateral artery, enter the pelvic fins. The dorsal aorta continues into the tail as the *caudal artery,* which is situated in the haemal canal immediately ventral to the centra of the vertebrae.

**8. The Structure of the Heart**

The heart of elasmobranchs is a tube bent into an S shape and differentiated into four chambers. These chambers have already been named. The sinus venosus has already been examined. It is a thin-walled chamber, triangular in form in the dogfish, tubular in the skate. Cut across the connections of the sinus venosus with the transverse septum and also across the base of the ventral aorta and remove the heart from the body. Look into the previously opened sinus venosus and find the large *sinoatrial aperture,* which leads into the *atrium.* It is guarded by a pair of valves formed from the smooth free edges of the sinus wall. Note the shape of the atrium. It is a broad, thin-walled chamber with large lateral expansions on each side of the ventricle. Slit open the atrium and wash out the blood clots it contains. Note the folds in its wall. Find the *atrioventricular* opening into the *ventricle,* which is guarded by two valves. Each of these is a pocket of thin tissue, the opening of the pocket being directed into the ventricle. Cut off the ventral half of the ventricle and also slit open the *conus arteriosus* by a longitudinal ventral incision. Note the small U-shaped cavity of the ventricle and its thick, spongy walls forming numerous cavities and crevices in which the blood is held. Examine the *atrioventricular valve* from the ventricular side and note its two pockets and the attachment of the pockets to the ventricular wall. On the wall of the conus arteriosus note the pocket-shaped *semilunar valves,* the pockets opening anteriorly. In the dogfishes there are nine valves in three circles of three each; two circles are near the posterior end of the conus, while the third circle, composed of larger and stronger valves, is near the junction of the conus with the ventral aorta. The *conus arteriosus* of the skate bears three longitudinal rows of valves with five valves in each row. To distinguish the valves, run the point of a probe along the conus wall from the anterior end backward, thus opening the pockets.

**9. The Pericardioperitoneal Canals**

Inspect the posterior wall of the pericardial cavity after the removal of the heart. In the dogfish a large opening will be found in the posterior wall dorsal to the previous attachment of the sinus venosus. The *pericardioperitoneal canal* is situated along the ventral wall of the esophagus internal to the visceral peritoneum of the esophagus. Eventually the canal opens into the pleuroperitoneal cavity by a small slit. In the skate there is an opening of moderate size in the center of the posterior wall of the sinus venosus. On probing into this, you will find that it leads into the *pericardioperitoneal canal,* which passes along the dorsal wall of the hepatic sinus. It forks into two canals lying on the ventral wall of the esophagus internal to its serosa. They open into the pleuroperitoneal cavity by minute slits.

The pericardioperitoneal canals serve to connect the pericardial and pleuroperitoneal cavities and arise through the failure of the transverse septum to close completely across the coelom.

## H. The Circulatory System of Necturus

In amphibians the heart has become more compact, although the same four chambers are present as in elasmobranchs. Although gills persist throughout life in some urodeles, such as *Necturus,* there is, nevertheless, a pair of lungs functional in respiration. The circulation through the gills is similar to that of fishes. Three or four aortic arches connecting with networks in the gills are present in those urodeles with persistent gills.

In the venous system changes have occurred, notably the formation of a new vein, the postcaval vein, which in time functionally replaces the posterior cardinals. The lateral abdominal veins anteriorly join the hepatic portal system and the renal portal system posteriorly.

## 1. The Chambers of the Heart

The pericardial cavity has already been exposed; if a new specimen is provided, the pericardial and pleuroperitoneal cavities are to be opened as before. The parts of the heart visible in ventral view are the *ventricle* and *atria* and the *conus arteriosus*. The ventricle is thick-walled and conical, and anterior to the ventricle on either side is a thin-walled atrium. Springing from the base of the ventricle and passing forward between the two atria is the tubular conus arteriosus. Anteriorly the conus passes into the base of the ventral aorta (*truncus arteriosus*), which in this case is within the pericardium. Lift the apex of the ventricle and note the *sinus venosus* situated dorsad to the ventricle, which receives two large venous channels lying in the dorsal wall of the pericardial cavity. These are formed by the union of the two *hepatic sinuses* with the two *common cardinal veins*. The hepatic sinuses are the two large veins that emerge from the transverse septum and pass into the sinus venosus. The common cardinal veins join the hepatic sinuses on their lateral surfaces.

## 2. The Hepatic Portal System and the Ventral Abdominal Vein

In the median ventral line of the body wall, posterior to the liver, enclosed in the falciform ligament, is the *ventral abdominal vein*. It is homologous to the lateral abdominal veins of the elasmobranchs and receives *parietal branches* from the body wall. At the level of the posterior end of the liver it leaves the body wall and passes into the dorsal surface of the liver. After a short course it joins the hepatic portal vein at the place of attachment of the hepatoduodenal ligament.

The *hepatic portal vein* is formed by the union of branches from the intestine, pancreas, spleen, and stomach. Stretch out the dorsal mesentery of the small intestine. In this mesentery, about halfway between the body wall and the intestine, runs a conspicuous *mesenteric vein*. Trace it posteriorly and note its beginning in the wall of the large intestine. As it passes forward in the mesentery, it receives numerous *intestinal veins* from the small intestine. It then passes into the substance of the pancreas, receiving small *pancreatic veins* from that organ. At the level of the pancreas the large *gastrosplenic vein* from the left also joins the mesenteric. The gastrosplenic vein is formed of *splenic branches* from the spleen and *gastric veins* from the stomach. The union of the mesenteric, gastrosplenic, and pancreatic veins produces a large vessel, the *hepatic portal vein,* which lies along the center of the dorsal face of the liver. It also receives the ventral abdominal vein, as already noted. Follow it along the surface of the liver. It branches into the liver substance and in its course also receives additional gastric veins from the stomach as well as veins from the ventral body wall, which pass into the liver by way of the falciform ligament. (These have probably been destroyed in the study of the digestive tract.) The hepatic portal vein subdivides in the substance of the liver.

## 3. The Renal Portal System

Trace the *ventral abdominal vein* posteriorly. It soon receives the *vesical veins* from the bladder. Shortly anterior to the hind limbs the abdominal vein is formed by the union of the two *pelvic veins* that run along the inner surface of the lateral body wall just in front of the pelvic girdle. Follow one of the pelvic veins. It is joined by the *femoral vein* from the hind limb. The vein formed by this union is the *renal portal vein,* which passes at once to the dorsal surface of the adjacent kidney. In male specimens the kidney is a brownish organ of considerable size situated at the side of the intestine. In female specimens the kidney is much smaller and more slender and is situated at the common point of attachment of the mesovarium and mesotubarium. It can be located by laying the ovary to one side and the oviduct to the other side. At the posterior end of the pleuroperitoneal cavity the kidney in females lies between the intestine and the oviduct. When you have located the kidney, identify the *renal portal vein* near the lateral margin of its dorsal surface. At the place where the renal portal vein passes from the body wall to the surface of the kidney, it receives the *caudal vein,* which ascends from the tail, forks, and passes to the surface of the posterior end of each kidney. Each renal portal vein runs forward along the surface of the kidney, sends numerous branches into the substance of the kidney, and also receives branches from the body wall.

## 4. The Systemic Veins

### a. The Anterior Systemic Veins

It has already been noted that the common cardinal vein joins the hepatic sinus on each side in the pericardial cavity. Turn to the pericardial cavity and locate the common cardinal veins. Trace one of them laterally, removing the muscles between the pericardial cavity and the base of the forelimb. Just outside of the pericardial cavity the common cardinal receives the *jugular* and *subclavian veins.* Trace the subclavian into the forelimb by removing the skin from the outer surface of the limb. The subclavian is formed at the shoulder by the union of a *cutaneous vein* from the skin and the *brachial vein* that runs along the surface of the limb muscles. The jugular vein is homologous to the anterior cardinal vein of the elasmobranchs. Follow it forward. It is formed by the union of the *external* and *internal jugular veins*. The external jugular vein then passes dorsally immediately behind the gills. It may be identified here by removing the skin behind the last gill. The external jugular may then be traced forward above the gills, where it enlarges, forming the *jugular sinus*. This sinus receives tributaries from the head and jaws. The internal jugular vein is a small vein that joins the external jugular posterior to the jugular sinus. It is difficult to find, but part of it will be seen later in the roof of the mouth.

The common cardinal vein also receives a *lateral vein* from the body wall. Remove the skin from the lateral line shortly posterior to the forelimb. Cut through the shoulder muscles to reveal the partition (horizontal skeletogenous septum) between the epaxial and hypaxial muscles. The lateral vein will be found along this partition and can be followed forward into the common cardinal vein.

### b. The Postcaval Vein

Turn to the pleuroperitoneal cavity. Examine the dorsal mesentery of the small intestine at its junction with the dorsal body wall (in female specimens spread the ovaries apart, laying one to each side). In the mesentery runs a large vein, the *postcaval vein,* which passes forward, receiving numerous *genital veins* from the adjacent gonads and *renal veins* from the kidneys. Trace it forward. At about the level of the spleen it turns ventrally and enters the dorsal surface of the right side of the liver. It can be seen best by laying the stomach to the left and the liver to the right. It passes forward, embedded in the liver substance, and should be followed by picking away the liver tissue. It receives several *hepatic veins* from the liver. At the anterior end of the liver the postcaval vein emerges as a large vessel situated in the coronary ligament. It pierces the transverse septum and forks into the two hepatic sinuses, which, after being joined by the common cardinal veins, enter the sinus venosus.

### c. The Posterior Cardinal Veins

At the place where the postcaval vein turns ventrally toward the liver, it is connected with a pair of veins, the *posterior cardinal veins*. Trace these anteriorly. They lie near the middorsal line of the anterior half of the pleuroperitoneal cavity, one to either side of the dorsal aorta. In females they are situated in the mesotubarium, along the line where this unites with the dorsal wall. Trace the posterior cardinals posteriorly and note connections between them and the renal portal veins. Note, also, the *parietal veins* that enter the posterior cardinals in their course along the body wall. The posterior cardinals may be traced anteriorly to the transverse septum. Shortly before reaching the transverse septum, the posterior cardinals diverge from the common cardinal vein and, penetrating the lateral portions of the septum, enter the common cardinal vein at almost the same point where the jugular and the subclavian enter.

**5. The Pulmonary Veins**

The *pulmonary vein* is a large vessel situated along the ventral side of each lung. The pulmonary veins run forward in the walls of the lungs, converge shortly caudad of the transverse septum, and at the septum become one vessel. This vessel passes through the transverse septum and, running forward in the dorsal wall of the left hepatic sinus, enters the left atrium.

**6. The Ventral Aorta and the Aortic Arches**

The conus arteriosus passes anteriorly into the ventral aorta. Trace the ventral aorta forward out of the pericardial chamber by dissecting away the anterior wall of the chamber. The ventral aorta soon forks into two vessels, which pass to the right and left. Trace the right one, since the gill bars have been left intact on that side. Follow it toward the gill arches. It divides into two vessels; and subsequently the posterior one again divides in two, making a total of three *afferent branchial arteries,* one to each of the gills. Trace each one into the gill, removing the skin from the gill. At the entrance into the gill the first afferent branchial artery gives off a *lingual* or *external carotid artery,* which turns medially, running beside the branchial artery, and then branches into the floor of the mouth. Within the gill each branchial artery sends up a loop, which branches among the filaments, from which other branches collect into a loop on the other side of the gill. This loop rejoins the branchial artery. In addition to the two loops, a short connecting branch runs through the base of each gill.

**7. The Branches of the Dorsal Aorta**

Trace the dorsal aorta posteriorly into the pleuroperitoneal cavity, where it runs in the middorsal line. Immediately beyond its origin it gives off a *subclavian artery* on each side. Trace one of them. The subclavian artery passes laterally, producing a conspicuous *cutaneous artery* lying on the inner surface of the pectoral girdle and branches to the skin and nearby muscles. The subclavian then emits an artery to the shoulder, and, as the brachial artery, passes into the forelimb, where it branches extensively.

In its course along the pleuroperitoneal cavity the dorsal aorta gives off both visceral and somatic branches. The first visceral branch of the dorsal aorta is the *gastric artery,* which passes to the stomach and forks into the *dorsal* and *ventral gastric arteries;* these supply the corresponding walls of the stomach. The ventral gastric artery also furnishes a few small branches to the spleen. Some distance posterior to the origin of the gastric artery, the *coeliacomesenteric artery* springs from the aorta. It passes ventrally in the mesentery, giving rise to some *mesenteric branches,* to the beginning of the small intestine and then proceeds to the region of the hepatoduodenal ligament, where it branches into a *splenic artery* to the spleen, a *pancreaticoduodenal artery* to the pancreas, duodenum, and pyloric region of the stomach, and a *hepatic artery,* which runs along the dorsal surface of the liver in contact with the hepatic portal vein and supplies numerous branches to the liver substance. Posterior to the point of origin of the coeliacomesenteric vessel, the dorsal aorta gives off a number of *mesenteric arteries* into the intestine.

The lateral visceral branches of the dorsal aorta consist of numerous *genital arteries* to the testes in the male and ovaries in the female and the *renal arteries* to the kidneys. The somatic branches of the aorta consist of the *parietal* or *intercostal arteries*. These arise from the dorsal side of the aorta and pass dorsally and divide in two, one branch to each side of the body. These branches pass laterally along the internal surface of the body wall and supply the body musculature.

Near the posterior end of the pleuroperitoneal cavity the aorta gives off an *iliac artery,* which passes laterally alongside the femoral vein toward each hind limb. Each iliac artery gives off an *epigastric artery,* which runs anteriorly along the body wall, and an *hypogastric artery* to the bladder and cloaca and then enters the hind limbs as the *femoral artery.* Each femoral artery runs along the medial side of the leg and at the knee gives rise to a number of branches. The dorsal aorta proceeds into the tail as the *caudal artery,* giving off a pair of cloacal arteries as it passes that region.

**8. The Structure of the Heart**

Remove the heart from the pericardial cavity by cutting across both ends. The sinus venosus has thin, delicate walls and receives the two large trunks formed by the union of the common cardinal vein and hepatic sinus on each side. Anteriorly the sinus passes into the atrium. Locate the *sinoatrial opening* guarded by a pair of *valves* by cutting open the sinus. Cut into one of the atria and wash out its contents. Look into the atrium and note the *interatrial septum,* which is perforated by a number of openings. Remove the ventral half of the *ventricle* and also slit open the *conus* and ventral aorta by a longitudinal incision. Note the thick, spongy walls of the ventricle and the numerous strands in the interior. Locate the single atrioventricular opening between the atria and ventricle; on the left side it is guarded by a pair of valves. Note the transverse row of three *semilunar valves* in the base of the opened conus arteriosus. In the truncus arteriosus is a longitudinal partition that divides the interior into right and left channels.

**I. The Circulatory System of the Turtle**

In reptiles the heart is still more compact than in fishes and amphibians. The sinus venosus remains and is attached to the atrium. The ventricle has become partially divided by an interventricular septum, which is complete in the Crocodilia. The conus arteriosus has become subdivided into three arterial trunks by the fusion of the ridges that in lower forms give rise to the conus valves. Here valves form only at the junction of the arterial trunks with the heart. In the venous system there are paired ventral abdominal veins in place of the single vein of amphibians, but these have the same relations to the two portal systems as the one in amphibians.

Specimens for the study of this system should have been doubly injected, i.e., into the arterial system, and in both directions into one ventral abdominal vein. Remove the plastron. With the bone scissors cut away the sides of the carapace between the fore and hind limbs.

**1. The Chambers of the Heart**

Examine the heart in the pericardial cavity, removing the ventral wall of the pericardial sac if this has not already been done. From the ventral view two parts of the heart are visible: the *ventricle,* the thick-walled, conical posterior part, and the *atria,* the thin-walled ones on each side anterior to the ventricle. The ventricle is attached to the posterior pericardial wall by a ligament, which is apparently a remnant of the ventral mesentery or ventral mesocardium of the heart. Cut through this ligament, raise the ventricle, and press it forward. A large chamber, the *sinus venosus,* is revealed dorsal to the atria and attached to the right atrium. The bases of the systemic veins will be seen entering the sinus. Put the ventricle back in place. Observe the large arteries that spring directly from the base of the ventricle. The *conus* becomes divided into three trunks in reptiles by lengthwise fusions along the longitudinal ridges from which in elasmobranchs the conus valves originate. Since this process is general throughout the amniotes, all of the latter lack a discrete conus arteriosus and ventral aorta.

**2. The Ventral Abdominal Veins and the Renal Portal System**

Running in the ventral pleuroperitoneum from the pelvic girdle up to the heart are two large veins, the *ventral abdominal veins.* They are homologous to the lateral abdominal veins of the elasmobranchs. The two veins are generally connected just anterior to the pelvic girdle by a cross branch. Trace the veins forward. They receive *pericardial veins* from the pericardial sac, and then turn dorsally to enter the liver. Just at this turn, each vein receives a *pectoral branch* from the pectoral muscles of that side. Trace the pectoral vein into the muscle. Slit the pleuroperitoneum alongside each abdominal vein and, by lifting the cut edges, find the places where the vein of each side penetrates the lobe of the liver.

Trace the ventral abdominal veins posteriorly. Make a longitudinal slit in the pleuroperitoneum midway between the two veins and, separating the cut edges, look within and locate the urinary bladder. Note the small *vesical vein* passing from the bladder into each abdominal vein. Continue to trace the abdominal veins posteriorly. Each passes to one side of the pointed anterior extremity of the pelvic girdle and at the same time gradually turns laterally. As it turns, it receives a *pelvic vein,* which runs over the ventral surface of the muscles of the pelvic girdle. The left pelvic vein is usually larger than the right one.

In their course between the heart and pelvic girdle each vein gives off laterally one or more small branches that pass to the borders of the carapace, where they join the *marginocostal vein* (to be described later).

Continue to trace the abdominal veins in a posterior direction. As both have identical branches, it is necessary to follow only one, selecting the one that has been most successfully injected. It passes along the dorsal surface of the pelvic girdle near the anterior margin of the latter; pull the girdle toward you in order to follow the vein. Grasp the hind leg on the side on which you are dissecting and work it back and forth until it is through the skin between the leg and the carapace back to the end of the tail. Remove the skin from leg and tail. Now trace the abdominal vein laterally along the base of the leg.

Just beyond the pelvic vein a small *crural vein* from thigh muscles
and a larger vein from the fat enter the abdominal vein. About an
inch and a half lateral to this the large *femoral vein* emerges from the
leg and joins the abdominal vein now termed the *external iliac vein*.
The femoral vein should be followed into the leg by separating the
muscles. The external iliac vein is now situated alongside a conspicuous artery, the *epigastric artery,* both being embedded in the abdominal wall, from which small veins pass into the external iliac vein.
The external iliac receives the epigastric vein, which accompanies the
artery of the same name anteriorly along the curve of the carapace.
It then turns abruptly posteriorly and runs between the base of the
leg and the carapace, deeply embedded in loose tissue, which should
be cleared away. The vein receives branches from the carapace and,
near the posterior part of the thigh, a well-marked *ischiadic vein*
from the thigh. Posterior to this point it receives several small
branches from the leg and, as the *caudal vein* passes along the side
of the tail, a *cloacal branch* from the anal region.

Return to the point where the epigastric vein enters the external
iliac vein. At this place a large vein continues forward from the anterior and dorsal surface of the external iliac. This vessel, the *renal
portal vein,* runs forward and dorsally, penetrating the pleuroperitoneum. Cut the pleuroperitoneum transversely halfway between the
heart and pelvic girdle, cutting across both abdominal veins. Cut also
into the pleuroperitoneum at the place where the renal portal vein
passes through it. A layer of muscle will be found outside the peritoneum at this point. Both muscles and membrane should be slit
ventrally to meet the transverse incision across the pleuroperitoneum.
In this way free access is gained to the pleuroperitoneal cavity. With
the left hand carefully press all of the viscera forward. It is usually
necessary to detach the lung from the dorsal wall and push it forward
also. With the right hand press the pelvic girdle caudad. A space
cleared of viscera is thus left dorsal to and in front of the pelvic
girdle. Look into this place near the median dorsal line for a somewhat flattened kidney, situated against the median dorsal wall. The
kidney is retroperitoneal, i.e., dorsal to the pleuroperitoneum. This
latter membrane should be stripped off from the ventral face of the
kidney. (In male specimens the rounded yellow testis and black
coiled epididymis will be noted attached to the ventral surface of the
kidney.) The renal portal vein may now be followed from the point
where it leaves the iliac through the pleuroperitoneum toward the
kidney. Before reaching the kidney it receives a vein from the carapace. At about the middle of the lateral border of the kidney is a
fissure; the renal portal vein enters this fissure and passes onto the
ventral face of the kidney, where it immediately forks. One of its
branches, the *vertebral vein,* runs forward and may be traced in well-
injected specimens by separating the lung from the carapace, raising
the lung, and stripping the pleuroperitoneum from the dorsal wall.
The vertebral vein passes anteriorly, dorsal to the arches of the ribs,
and receives laterally an *intercostal branch* at each suture between

the costal plates of the carapace. The *intercostal veins* anastomose with each other in the curve of the carapace by means of a longitudinal vessel, the *marginocostal vein,* which is an anterior continuation of the epigastric vein previously noted. The marginocostal vein also has connections with the abdominal veins. The posterior branch of the renal portal vein passes posteriorly over the ventral face of the kidney, and, as the *internal iliac* or *hypogastric vein,* receives branches from the male reproductive organs, bladder, and cloaca. The renal portal vein gives off branches into the kidney as it passes along the ventral face of that organ. Both the vertebral and the marginocostal veins are formed by the longitudinal anastomoses of transverse segmental branches of the posterior cardinal veins of the embryo.

3. The Hepatic Portal System

Lift up the lobes of the liver, separate them gently from the stomach and duodenum, and find a large *hepatic portal vein* on their dorsal surfaces at the place where the gastro-hepato-duodenal ligament is attached to the liver. The hepatic portal vein runs completely across the liver embedded in its wall, and at the right, at the point where the bile duct enters the duodenum, it turns abruptly posteriorly, penetrating the mesentery. Note that on the left numerous *gastric veins* enter the hepatic portal vein from the stomach. Just to the right of the bridge connecting the two lobes of the liver, two or three *anterior pancreatic veins* pass from the pancreas into the hepatic portal vein. Near the bile duct the hepatic portal vein receives *cystic veins* from the bile duct, *posterior pancreatic veins* from the right end of the pancreas, and a long *duodenal branch* from the first part of the small intestine. The hepatic portal vein should be followed posteriorly; it is embedded in the pancreas and at the bend of the duodenum penetrates the mesentery and emerges to the left of the duodenum. The liver and duodenum must be pressed forward to follow it. The vein next passes to the posterior side of the adjacent loop of the small intestine, which should also be pressed forward. The vein will then be found to pass on the left side of the spleen in contact with that organ from which it receives numerous *splenic tributaries.* Shortly posterior to the spleen, the hepatic portal vein reaches the central point of the mesentery, where the mesentery is thrown into a coil. At this place the numerous *mesenteric veins* accompanied by arteries will be seen passing in the mesentery from all parts of the intestine into the hepatic portal vein.

By cutting away the liver substance, trace the anterior portions of the ventral abdominal veins into the liver and find their union with the hepatic portal vein. Note how the hepatic portal vein breaks up into many branches in the liver substance. As in other vertebrates, the direction of flow in the hepatic portal vein is from the digestive tract into the liver.

4. The Systemic Veins

Four large systemic veins enter the sinus venosus. Turn the ventricle forward to obtain a clear view of the sinus. A large *anterior vena cava* or *precaval vein* enters the left wall of the sinus, passing around the border of the left atrium. The *left hepatic vein,* emerges from the bridge of the liver and enters the left angle of the posterior wall of the sinus. The large vein that passes into the right angle of the posterior wall of the sinus is the *posterior vena cava* or *postcaval vein,* which emerges from the right lobe of the liver. Just in front of the entrance of the postcaval vein is the right *precaval vein,* which passes into the right anterior angle of the sinus venosus; it can be seen best by pressing the heart to the left.

### a. The Branches of the Precavals

Each precaval enters the pericardial cavity by passing through the anterior wall of the pericardial sac. From this point it may be followed forward. Since both have identical branches, it is necessary to follow only one. In specimens that have been preserved for a long time the dissection of the branches of the precaval veins is generally unsatisfactory because the branches are often empty; but, of those named below, as many as the condition of the specimens permits should be identified. Be very careful not to injure the adjacent arteries springing from the ventricle. Trace the precaval forward out of the pericardial sac. Shortly anterior to the place where the precaval penetrates the pericardial sac the vein receives practically simultaneously four tributaries, three small and one large. The most medial is the small *thyroscapular vein,* which collects a branch from the thyroid gland (the gland is situated in the fork of the large arteries) and then passes to the inner surface of the shoulder, where it drains several muscles. Lateral to this vein is the slightly larger *internal jugular vein.* This runs anteriorly alongside the neck in contact with a white nerve (vagus or tenth cranial nerve). It receives medially an extensive network of *esophageal veins.* It may be traced anteriorly to the base of the skull from which it issues, making an anastomosis with the *external jugular vein.* The third tributary of the precaval vein is the large *subclavian vein* (by far the largest of the four branches), which passes along the side of the neck and, as the *axillary vein,* turns toward the shoulder. Here it is formed by the union of the two large branches, the *external jugular vein* from the neck and the *brachial vein* from the forelimb. The external jugular lies along the side of the neck, lateral and dorsal to the internal jugular. It collects blood from the head and, in its passage posteriorly along the neck, has at regular intervals *vertebral veins* passing into it from between the vertebrae. Near its junction with the brachial vein it receives the last of the vertebral veins, which descends from the junction between last cervical and first trunk vertebra, where it connects with the anterior end of the vertebral vein described with the renal portal system. The external jugular vein also receives branches from the skin and muscles of the shoulder region. The fourth and most lateral and dorsal of the tributaries of the precaval is a small *scapular vein* from the muscles covering the scapula.

Draw the branches of the precaval as far as you have found them.

### b. The Left Hepatic Vein

The left hepatic vein should be traced into the left lobe of the liver, from which it collects venous blood. To do this, clear away the intervening posterior wall of the pericardial sac and pleuroperitoneum.

### c. The Postcaval Vein

Trace the postcaval vein posteriorly into the right lobe of the liver. Its course may be followed by making a slight hole in the vein where it enters the sinus and probing posteriorly into the hole, cutting away the liver substance along the probe. Note the numerous hepatic veins that enter the postcaval during its passage through the liver. Find where the postcaval enters the liver from behind to the right of the hepatic portal vein. At this point the serosa of the liver is fused to the pleuroperitoneal membrane over the ventral face of the lung. This fusion should be broken and the postcaval vein freed. Posteriorly it curves toward the median line, where it runs alongside the dorsal aorta. The postcaval may be traced to the posterior end of the pleuroperitoneal cavity. Its relations there will be described later.

**5. The Pulmonary Veins**

A pulmonary vein is situated posterior to each bronchus, and passes from each lung to the left atrium of the heart. It passes dorsal to the precaval vein. The right pulmonary runs in the dorsal wall of the pericardial sac anterior to the sinus venosus and joins the left vein at the entrance of both into the left atrium. The point of entrance is near the left precaval vein.

**6. The Aortic Arches and Their Branches**

From the ventricle three large arterial trunks extend forward. Clean away the connective tissue from these arteries and separate them from each other. The trunk farthest to the left is the *pulmonary arch;* the vessel next to it is the *left systemic arch* or *left aorta;* the third is the *right systemic arch* or *right aorta,* but it is concealed from view by the *brachiocephalic (innominate) artery,* the large branch that the right aorta gives off immediately on leaving the heart. Note the small *coronary arteries* springing from the base of the brachiocephalic artery and branching over the surface of the heart. The brachiocephalic artery lies in the median line and forks at once into large branches. In the angle of the fork lies a reddish body, the *thyroid gland.*

### a. The Branches of the Brachiocephalic Artery

It divides at once into four trunks: the large medial ones are the *right* and *left subclavian arteries,* the smaller lateral ones are the *right* and *left carotid arteries.* Clean away the connective tissue from these vessels and follow their courses. The subclavians embrace the thyroid gland between their bases and supply small *thyroid arteries* into this gland. Each subclavian next gives off branches to the ventral side of the neck and to the trachea; of these the chief one is the *ventral cervical artery,* arising from the subclavian about one-half inch beyond the thryoid gland and branching profusely into the esophagus, trachea, muscles of the neck, and thymus gland. The thymus gland is a yellowish mass lateral to the ventral cervical artery and receiving branches from it. The subclavian artery, now named the *axillary artery,* turns laterally and passes to the inner surface of the pectoral girdle, where a large vessel arises and branches extensively into the pectoral and shoulder muscles. The axillary then turns abruptly posteriorly and about an inch beyond the turn gives off a small *dorsal cervical artery* into the neck, a *marginocostal artery* laterally, and a *vertebral artery* caudally. The marginocostal artery runs laterally and then turns posteriorly along the curve of the carapace. The vertebral artery passes backward along the vertebral column dorsal to the ribs alongside the vertebral vein. It gives off, at the sutures of the costal plates, the *intercostal arteries,* which run laterally into the marginocostal artery. At the point where the marginocostal and vertebral arteries arise from it, the axillary bends sharply laterally and, as the *brachial artery,* passes into the forelimb alongside the *scapular vein.*

Each *carotid artery* passes forward along the ventral side of the neck, soon crossing dorsal to the subclavian, and then lies medial to the subclavian. In specimens in which the neck is drawn into the shell the carotids usually make loops in the neck region. As the carotid artery passes the thymus gland, it gives branches into the gland. It then proceeds, without branching, the entire length of the neck in contact with the internal jugular vein and the vagus nerve and enters the skull by a foramen in front of the auditory region.

### b. The Pulmonary Arteries

The pulmonary arch divides immediately into right and left pulmonary arteries. To see this division, lift the pulmonary trunk and look to its dorsal side. The left pulmonary artery proceeds laterally posterior to the left systemic, to which it is more or less bound by connective tissues, marking the site of the *embryonic arterial ligament* or *ligament of Botallus.* The pulmonary artery proceeds directly to the left lung in company with the left bronchus and left pulmonary vein. Trace the right pulmonary in the same way; it is similarly bound to the right systemic.

### c. The Right and Left Systemics

Trace both of these arteries away from the heart. Each makes a curve as it leaves the heart and turns posteriorly, passing dorsal to the precaval vein, the bronchi, the pulmonary vessels, and disappearing above the lobes of the liver. Vessels already studied may be cut to follow the aortae posteriorly. Trace the left aorta first. Grasp the stomach and left lobe of the liver and press them to the right, separating the cardiac end of the stomach from the lung. The left aorta will be found passing to the left of the esophagus and dorsal to the stomach. It gives off three large branches simultaneously: One of these is the *gastric artery,* which passes to the stomach in the cardiac region and follows the curve of the stomacch along the length of this organ. After a short distance it forks into *anterior* and *posterior gastric arteries,* which supply the lesser and greater curvatures of the stomach, respectively. Another branch from the left aorta is the *coeliac artery,* which soon forks into *anterior* and *posterior pancreaticoduodenal arteries.* The anterior pancreaticoduodenal artery passes to the left end of the pancreas and supplies the pyloric end of the stomach and the liver, then turns to the right and runs along the pancreas, supplying the liver, pancreas, and duodenum with many small branches. The posterior *pancreaticoduodenal artery* enters the right end of the pancreas and, as it passes along the pancreas, supplies branches to the liver, pancreas, duodenum and gall bladder. The third branch of the left aorta is the *superior mesenteric artery,* which runs posteriorly in the mesentery; trace it, tearing the mesentery, to the center of the coils of the mesentery. At this point the artery breaks up in a fan-like manner into many radiating branches that traverse the mesentery to all parts of the small intestine. One branch, the *inferior mesenteric artery,* passes to the large intestine and accompanies it to the cloaca.

Now follow the left systemic posterior to the point of origin of the superior mesenteric artery. It becomes smaller and soon meets the right systemic. The two join in a V-shape and form the *dorsal aorta,* which continues posteriorly in the median dorsal line. Follow the vessel that meets the left aorta anteriorly. Separate the right lobe of the liver from the right lung, and turn the liver and duodenum to the left. The vessel in question can then be traced anteriorly dorsal to the right bronchus and pulmonary vessels to the heart. Immediately beyond its origin from the heart the right aorta gives rise to the large *brachiocephalic artery* whose branches were followed above. It has no other branches.

## 7. The Dorsal Aorta and the Postcaval Vein

The digestive tract may now be removed, except the large intestine, which is to be left in place. Follow the dorsal aorta posteriorly. It runs in the median line ventral to some long muscles accompanied by the postcaval vein, which courses at first to its right and later comes to lie ventral to the aorta. The postclaval vein is formed by two vessels running along the medial side of the kidneys. Each of these receives numerous *renal* and *genital veins* from the kidneys and reproductive organs, respectively. The postcaval vein thus originates between the kidneys.

The vein may be removed and the dorsal aorta studied. The aorta gives off a number of small branches into the muscles on which it rests and then passes between the two kidneys. Hold the large intestine backward and clear away the connective tissue between the two kidneys. The dorsal aorta gives rise to numerous *renal arteries* into the kidneys and *genital arteries* to the reproductive system. At the posterior end of the kidneys it forks into the *right* and *left common iliac arteries.*

Separate one kidney from the carapace and press it and the reproductive organs to the other side. Two large arteries will be seen emerging dorsal to the kidney. The anterior one is the *epigastric artery;* the posterior, the *common iliac* mentioned in the preceding paragraph. Follow the epigastric. (If it was injured in the dissection of the renal portal system, try the other side.) It runs laterally to the point where the renal portal vein enters the pleuroperitoneal cavity. At this point it divides. The anterior branch continues to the carapace and runs forward along the curve of the carapace, supplying the fat bodies and becoming continuous with the marginocostal artery described above. The posterior branch turns and passes medially parallel to the ventral abdominal vein. It supplies the base of the leg and the pelvic muscles and terminates on the ventral surface of the pelvis.

Next follow the common iliac artery of the same side. It divides at once before emerging from above the kidney into an *internal iliac* and an *external iliac artery*. The external iliac forks after a short distance. The ventral and larger branch supplies the muscles of the pelvis and enters the thigh as the *femoral artery*. The smaller dorsal branch passes to the point where the ilium articulates with the sacral ribs; here it passes dorsal to the nerve and then turns ventrally as the *sciatic artery,* into the hind leg. The internal iliac is best followed by replacing the kidney against the dorsal wall, pulling the large intestine backward, and locating the point of origin of the internal iliacs from the common iliac. The chief branch of the internal iliac is the *rectal (hemorrhoidal) artery,* which passes forward along the side of the large intestine; in addition there are branches to the accessory bladders, the lower ends of the oviducts, and the pelvic region in general.

## 8. The Structure of the Heart

Separate the heart by cutting across the great vessels and remove it from the body. The posterior chamber of the heart is the sinus venosus, which receives the four great systemic veins. Clean out the blood from the sinus. It is a thin-walled chamber attached to the right atrium, with access through the *sinoatrial opening,* which is guarded by a pair of thin *valves.* Open each atrium by making a slit in the margin and washing out the blood clots. The walls of the atria are somewhat spongy. Look into the left atrium and note the thin *interatrial septum,* which completely separates the cavity of the *left* and *right atria.* Find the opening of the pulmonary veins into the dorsal wall of the left atrium near the septum. Find on each side the large *atrioventricular opening* between each atrium and the ventricle. Make a cut all of the way around the margin of the ventricle, so as to make dorsal and ventral flaps of the ventricle. Spread the two flaps apart, cautiously extending your cut inward until the two flaps are attached only along the base of the ventricle. Note the thick walls of the ventricle and the muscular columns projecting into the interior. The cavity of the ventricle is a broad but flattened cavity usually containing a spongy network that may be cleaned out. As you spread the two flaps wide, note in the base of the ventricle a band passing across from one side to the other. On each side of this band is an atrioventricular opening. The band is a continuation of the interatrial septum and forms a fold or valve on each side, which partially occludes the atrioventricular opening. The right valve continues ventrally into a ridge on the ventral flap of the specimen. This ridge is the incomplete *interventricular septum.* On bringing the two flaps of the specimen together, it will be seen that the interventricular was connected with the muscular wall of the dorsal flap and that a space is left dorsal to the septum by which the right and left ventricles communicate with each other. The *right ventricle (cavum ventrale)* to the right and largely ventral to the septum is very small, while the *left ventricle (cavum dorsale)* is much larger and communicates with the cavity of both atria because of the incomplete character of the interventricular septum. Spread the flaps of the specimen again and pass a probe ventral to the interventricular septum. The probe emerges in the pulmonary artery. Probe into the other arterial trunks and find their openings into the ventricle. The opening of the left aorta is to the right of the interventricular septum, into the small right ventricle, while that of the right aorta is to the left of the septum, into the left ventricle. Slitting open the arterial trunks to find the pocket-like semilunar valves that guard their exits from the ventricle. Their presence shows that the bases of the arterial trunks represent the conus arteriosus.

**J. The Circulatory System of the Pigeon**

In birds the consolidation of the heart is completed. Through the incorporation of the greatly reduced sinus venosus into the right atrium and the splitting of the conus arteriosus to form the bases of the two great arterial trunks the heart comes to consist only of the two atria and the two ventricles (now completely separated). The entrances of the great veins, with the reduction of the sinus venosus, have moved anteriorly. The left (fourth) aortic arch is obliterated, and only the right arch remains. This means that only two great arterial trunks spring from the heart, the aorta and the pulmonary; the double circulation is now perfected. The general circulatory system is reptilian in character; the principal change is the direct connection of the postcaval vein with the renal portal system and the great reduction in the portal circulation through the kidneys.

In case a new specimen is provided for this work it should be opened, as before, by deflecting the pectoral muscles from either side of the keel of the sternum, then cutting through the sternum on each side of the keel and removing a median portion of the sternum including the keel. The peritoneal cavity is to be opened, as before, by a longitudinal incision. The specimen should have been injected through the pectoral artery.

**1. The Chambers of the Heart**

The heart is relatively large and more compact than in the forms previously studied. The chambers are more closely knit than in the lower vertebrates. The major portion of the heart is formed of the *right* and *left ventricles,* which together constitute a muscular, thick-walled cone, having a pointed *apex* directed posteriorly and a broad *base* directed anteriorly. The two ventricles are completely separated from each other, but the division between them is indistinct externally. The internal division between the ventricles passes obliquely from the left side of the base to about the middle of the right side of the heart; the left ventricle is therefore the larger of the two and includes the whole of the apex of the heart. Anterior to the ventricles are the two much smaller *atria,* thin-walled chambers. The external division between atria and ventricles is generally concealed by a line of fat, which should be removed. From the anterior end of the heart between the atria the great arteries spring without the intervention of a conus arteriosus. On raising the ventricles the dorsal portions of the atria become visible. There is no sinus venosus, and the great veins open directly into the right atrium. There are three such veins, two *anterior venae cavae* or *precavals* and a *posterior vena cava* or *postcaval.* The postcaval enters the right atrium from behind, emerging from the liver. The precavals travel from the anterior part of the body, one on each side, and, curving toward the heart at the level of the atria, enter the right atrium. The pulmonary veins open into the left atrium.

## 2. The Hepatic Portal System

Cut across the falciform ligament of the liver near the gizzard, noting the small vein passing from the ventral ligament of the gizzard in the falciform ligament to the liver. The lobes of the liver may now be turned forward. Running along the dorsal surface of the liver and branching into its substance is the large *hepatic portal vein.* The main part of the vein enters the right lobe of the liver, coursing between the two bile ducts. The remainder of it lies along the dorsal surface of the left lobe of the liver, sending branches into the liver, and at the left receives the *left* and *median gastric veins* from the margin and left side of the gizzard and from the proventriculus. Follow posteriorly the part of the hepatic portal that lies between the two bile ducts. It is formed by the union of three veins: a *superior mesenteric,* a *gastroduodenal,* and an *inferior mesenteric.* The superior mesenteric collects blood from the greater part of the small intestine. The gastroduodenal receives (1) the *right gastric vein* from the right side of the gizzard, (2) the *pancreaticoduodenal vein,* which runs along the duodenal loop draining the duodenum and pancreas, and (3) the *mesenteric vein* from the last loop of the small intestine. The inferior mesenteric vein runs along the large intestine from which it collects many branches. At its posterior end it turns dorsally and joins the renal portal system, where it will be followed later.

## 3. The Systemic Veins

### a. The Branches of the Precaval Veins

As both veins have identical branches, only one will need to be followed. Find the vein on each side and trace each into the right atrium by lifting the heart. The left precaval passes around the left atrium to enter the right atrium. The right precaval is much shorter and enters the right atrium directly.

Follow one precaval forward. It lies just posterior to a large artery and is formed by the union of three large veins, laterally the *pectoral vein,* slightly anterior and dorsal to this the *subclavian vein,* and anteriorly the *jugular vein.* Each of these veins should be followed. The pectoral vein at its union with the others receives the *internal thoracic (internal mammary) vein,* ascending from the inner surface of the ribs, and also has a tributary from the sternum and cora-coid. The main vein is formed laterally by the union of two veins emerging from the pectoral muscles. These may be followed into the muscles, where they are seen to collect many branches. The sub-clavian vein passes dorsally, ventral to a group of nerves (brachial plexus), and is somewhat concealed by arteries, which should not be injured. As the *brachial vein,* it emerges from the wing and then re-ceives a branch from the shoulder muscles. The jugular vein passes anteriorly on the dorsal side of the large arteries. On tracing the jugular forward it will be found to receive the following veins, named in order from the heart forward: on the medial side, some small and then a large branch from the crop (at the point of entrance of these into the jugular is situated a small reddish body, the *cervical lymph gland*); on the lateral side, a vein from the shoulder and, at the same level, the *vertebral vein* from the vertebral column; medially, another branch from the crop; laterally a large vein from a plexus of blood vessels in the skin of the neck; then small veins from the esophagus and trachea. On freeing the anterior end of the esophagus (also trachea) and cutting across it, you can follow the jugular vein to the soft palate, where it joins its fellow of the opposite side. Posterior to this union, each receives a plexus of veins from the skin of the face. On cutting away the soft palate from the anastomosis of the two jugular veins, you will find branches from the skull passing into the anastomosis.

### b. The Postcaval Vein

Raise the ventricles of the heart and note once more the large post-caval vein emerging from the liver and entering the right atrium between the two precaval veins. Note the large *hepatic veins* it re-ceives from the liver. The left hepatic vein receives the small vein of the falciform ligament mentioned previously. Follow the postcaval vein into the peritoneal cavity, turning all of the viscera to the left. The postcaval will be seen again at the posterior margin of the right lobe of the liver in contact with the dorsal body wall. The postcaval is here formed by the union of two *iliac veins.* In males the two oval testes will be noted at this point of junction. In females the single ovary and oviduct can be found on the left, concealing the left iliac vein. Each iliac runs along the ventral face of a three-lobed organ, the *kidney,* which is set close against the dorsal body wall. Follow the right iliac vein. From between the first and second lobes of the kidney it receives the large *femoral vein,* from the leg. The femoral vein receives a small branch from the body wall. Posterior to the entrance of the femoral vein the iliac vein corresponds to the *renal portal vein* of reptiles and amphibians and may be so named. From between the second and third lobes of the kidney it receives the *ischiadic vein,* which also comes from the thigh. At the posterior end of the kidneys the two renal portal veins anastomose. From this anastomosis arises the *inferior mesenteric vein,* already noted as a branch of the hepatic portal system. The anastomosis of the renal portals also receives a small *caudal vein* from the tail and on each side an *internal iliac vein* from the roof of the pelvic region. The left iliac and renal portal veins are the same as the right except that in the female the left veins receive the *genital veins* from the ovary and oviduct. These may be seen by turning the oviduct to the right. In their course over the kidneys the renal portal veins give off branches into the kidney, as in lower forms. There is probably some portal circulation in the kidneys, but most of the blood from the renal portals passes directly into the iliac veins. The iliac vein receives *renal veins* from the kidney; one of these is a major renal vein, which runs along the medial side of the kidney but it so embedded in the kidney substance that it is difficult to identify. The dorsal aorta runs between the two renal portals and iliac veins.

Turn the animal dorsal side up and remove the skin over the thigh. By separating the muscles identify the *femoral* and *ischiadic veins* and trace them into the leg. The ischiadic vein accompanies the large *sciatic nerve* and soon turns forward to run parallel to the femoral vein. Both veins are accompanied by arteries of the same name.

**4. The Pulmonary Veins**

The *pulmonary veins* emerge on each side from the lung and pass toward the heart immediately posterior to the precaval veins. There is usually one pulmonary vein from each lung, but there may be two. Note the branches collected by each vein from the lung. The veins pass to the dorsal side of the bases of the precavals and enter the left atrium.

**5. The Arterial System**

It has already been noted that the great arteries spring directly from the ventricle. Separate their bases from the atria. You will find that there are two arterial trunks. The larger, medial one is the *aorta* (*right systemic arch*). The smaller one, passing to the left and dorsal to the aorta, is the *pulmonary artery*.

*a. The Anterior Branches of the Aorta*

The aorta immediately gives rise in the median line to two large arteries, the *brachiocephalic* (*innominate*) *arteries,* and then turns to the right and disappears dorsally. It will be followed at a later time. Identify the branches of the brachiocephalic arteries; since both have identical branches, follow only one. Each proceeds laterally and slightly anteriorly and forks into two branches, an anterior *common carotid artery* and a lateral *subclavian artery*. The subclavian artery gives rise to a number of branches: the small *internal thoracic* (*internal mammary*) *artery* passing posteriorly along the inner surface of the ribs, two *pectoral arteries* to the pectoral muscles along with the veins of the same name, and an *axillary artery* to the wing. The axillary artery runs anteriorly and, after giving off a branch into the shoulder, enters the wing as the *brachial artery*. The two common carotid arteries pass forward and at the level of the cervical lymph gland each gives rise to a *vertebral artery,* which passes dorsally into the vertebroarterial canal of the vertebral column. The common carotid arteries then approach the median line and penetrate the muscles on the ventral surface of the vertebral column. On separating these muscles in the midventral line, the two arteries may be followed forward. They pass anteriorly side by side. Shortly before they reach the head, they diverge; and at the angle of the jaws each divides into an *external carotid artery* from which branches may be traced to the esophagus, palate, and head generally, and into a more deeply situated *internal carotid artery,* which passes through the skull to the brain.

### b. The Pulmonary Arteries

The pulmonary arch passes to the left side of the aorta and immediately forks into right and left pulmonary arteries. The left artery goes directly to the left lung. The right artery passes on the dorsal side of the brachiocephalic arteries and posterior to the turn of the aorta enters the right lung.

### c. The Aorta

The aorta turns to the right, forming the *arch* of the *aorta*. This may be followed by cutting across the right precaval vein and the right brachiocephalic artery. The arch of the aorta curves to the dorsal side of the right pulmonary artery, which can now be traced into the lung, and turns caudad. Follow it by dissecting away the tissue between the heart and the right lung and by breaking through the oblique septum. Turn the viscera to the left. Cut through the postcaval vein. The aorta, now the *dorsal aorta,* lies in the median dorsal line between the two lungs. It gives off small branches to the esophagus and body wall in its passage along the pleural cavities. At the entrance to the peritoneal cavity the large *coeliac artery* arises from the aorta. This runs posteriorly along the proventriculus, to which it branches. The coeliac artery then gives rise to the relatively small left *gastric artery* supplying the left side of the gizzard. The coeliac artery then passes to the spleen, to which it gives *small splenic arteries;* and just beyond the spleen it gives rise to the *hepatoduodenal branch*. This sends an *hepatic branch* into the liver and then, as the *anterior pancreaticoduodenal artery,* runs along the duodenal loop supplying duodenum and pancreas. The coeliac artery continues as the *right gastric artery,* which spreads out over the right surface of the gizzard. The right gastric sends a large *posterior pancreaticoduodenal branch* to the duodenal loop and pancreas and a *mesenteric branch* to the small intestine.

Immediately posterior to the origin of the coeliac artery, the *superior mesenteric artery* arises from the dorsal aorta and branches to the small intestine. One of these branches passes along the large intestine and anastomoses with the *inferior mesenteric artery* (described below).

The dorsal aorta now passes between the two kidneys. It gives off a *renolumbar artery* on each side, which supplies the anterior lobe of the kidney and then passes to the body wall and some muscles of the thigh. In female specimens *genital arteries* from the *renolumbar artery* and the *renofemoral artery* supply the ovary and oviduct. The renofemoral artery is a large vessel supplying the middle and posterior lobes of the kidney. It then proceeds to the lateral body wall, and, as the *femoral artery,* supplies the leg. It may be followed by turning the animal dorsal side up and looking between the muscles of the thigh along the course of the femoral vein previously identified. The femoral artery accompanies the large sciatic nerve and branches into the leg muscles. Returning to the peritoneal cavity, trace the dorsal aorta farther. As it passes between the kidneys, it gives off *lumbar arteries* into the dorsal body wall. At the posterior end of the kidneys it forks. At the point of forking arises the *inferior mesenteric artery* (which runs anteriorly in the mesorectum and anastomoses with a branch of the superior mesenteric artery) and the *caudal artery,* which proceeds posteriorly in the median line to the tail. The two forks of the dorsal aorta are the *internal iliac arteries.* They pass posteriorly along the roof of the pelvic region. The left one gives off branches into the oviduct.

## 6. The Structure of the Heart

Free the heart by cutting across the great vessels. Note that all veins enter the anterior end of the heart, having migrated forward out of the transverse septum. As already stated, there is neither sinus venosus nor conus arteriosus in the bird's heart. The atria are small, thin-walled chambers anterior to the ventricles. In the wall of the right atrium identify the openings of the systemic veins and in the left atrium the entrance of the pulmonary veins. Slit open the right atrium and note the thin *atrial septum* separating it from the left. A fold extends from this septum to the entrance of the postcaval vein, partly concealing the entrance. Note the deep cleft, the *atrioventricular opening,* through which the right atrium opens into the right ventricle. Open the left atrium similarly and find the left atrioventricular opening. Cut across the apex of the ventricles. Observe the crescentic form and relatively thin walls of the right ventricle and circular section and enormously thickened wall of the left ventricle. The *interventricular septum* completely separates the cavities of the two ventricles. Open the right ventricle by a slit extending from the previously cut apex to the base. Note the single *valve* that guards the right atrioventricular opening. In the left side of the anterior end of the right ventricle find the opening of the pulmonary artery, or probe into the base of the pulmonary artery and note the emergence of the probe into the right ventricle. At the base of the pulmonary artery are three pocket-like *semilunar valves.* Cut into the left ventricle and note the two thin *membranous valves* that guard the atrioventricular opening. They form the *mitral valve* and each flap is attached by delicate *chordae tendineae* to the wall of the ventricle. The wall of the ventricle has several muscular ridges that project into the cavity, the *columnae carnae.* On the medial side of the mitral valve, find the opening of the left ventricle into the aorta. Probe into this and satisfy yourself that it leads into the aorta. Note the three *semilunar valves* at the beginning of the aorta.

Removing the heart will permit you to trace the esophagus into the proventriculus and the bronchi into the lungs. You can also observe the form and extent of the lungs to advantage at this time.

**K. The Circulatory System of Mammals**

The mammalian heart is at an evolutionary level similar to that of birds, but is from a different reptilian stem. The sinus venosus and conus arteriosus as such are absent, and the compact heart consists of only two atria and two ventricles. These are completely separated and hence there is a perfect double circulation through the heart. The blood leaves the heart in the aortic and pulmonary trunks. The aortic trunk consists of the left half of the fourth aortic arch, whereas in birds it is the right half. In the circulatory system the chief change is the fusion of the renal portal system with the postcaval vein, with the complete elimination of the portal circulation through the kidneys. Mammals therefore have only a single portal circulation, that through the liver, and the hepatic portal vein is often simply termed the portal vein.

The specimen of rabbit or cat should have been injected in the arterial system.

**1. The Chambers of the Heart**

The heart is relatively large and compact, the chambers closely united with each other. The pericardial sac should be removed if this has not been done previously. The thymus gland may be well developed and in this case it will be necessary to dissect it away from the anterior part of the heart. The greater portion of the heart consists of the two *ventricles,* which constitute a firm thick-walled cone with a posterior pointed *apex* and a broad anterior *base.* The division between the right and left ventricles is marked externally by an indistinct line or groove extending from the left side of the base obliquely to the right and terminating to the right of the apex. The groove contains branches of the *coronary artery* and *vein,* which will be found ramifying over the surface of the ventricles. The *left ventricle* is much the larger of the two and includes the apex. Anterior to the base of each ventricle is a much smaller, thin-walled, generally dark chamber, the *atrium.* Each atrium in the contracted state presents a lobe, the *atrial appendage,* projecting medially and slightly posteriorly over the ventricle; in the cat (and man) this lobe has a scalloped margin. Extending anteriorly from the middle of the ventricular base forward between the two atria is a large *pulmonary artery.* Dorsal to the pulmonary is another arterial trunk, the *aorta.* These two trunks are generally embedded in fat, which should be removed. Grasp the apex of the heart, turn the heart forward and note the bases of the great veins (*pulmonary* and *systemic veins*) entering the atria. A sinus venosus is lacking as a distinct chamber, for it is greatly reduced and absorbed into the right atrium.

**2. The Hepatic Portal System**

Press the lobes of the liver forward and the other viscera to the left. Stretch the hepatoduodenal ligament by widely separating the stomach and liver, without, however, tearing the ligament. In the ligament lying dorsal to the common bile duct is the large *hepatic portal vein* (commonly called simply the *portal vein* in mammals). Free it by carefully cleaning connective tissue from its surface. Follow it anteriorly and note how it branches into the liver substance. Follow it posteriorly, dissecting away fat and connective tissue from its surface. Note the large branch it sends into the right lateral lobe of the liver. The branches received by the portal vein from the digestive tract are slightly different in the rabbit and cat. In preserved specimens the branches are not always easy to follow, but the student should identify as many as possible. The arteries accompanying the veins should not be injured.

Rabbit: Immediately posterior to the branch into the right lateral lobe of the liver, the portal vein receives on the right side the *gastroduodenal vein*. This vein is formed by the union of two veins, a larger *anterior pancreaticoduodenal vein,* which appears as a continuation of the main vein, and the smaller *right gastroepiploic vein*. The larger vessel runs in the tissue of the pancreas alongside the first part of the duodenal loop, collecting tributaries from both pancreas and duodenum. The right gastroepiploic vein stems from the pyloric region of the stomach and also receives branches from the great omentum. Shortly posterior to the entrance of the gastroduodenal vein into the portal vein, the portal receives, on the left side, the *gastrosplenic vein*. This vein is seen to be formed a short distance from the portal by the union of the *splenic* and *gastric veins*. The latter comes from the lesser curvature of the stomach, where it arises as numerous branches collecting from both surfaces of the stomach. The splenic vein is a large vessel running in the great omentum past the spleen and extending as far as the left end of the stomach. In its course it collects numerous splenic branches from the spleen and the left *gastroepiploic veins* from the stomach and the omentum. Some distance posterior to the entrance of the gastrosplenic vein, the portal vein receives the *posterior pancreaticoduodenal vein,* which runs in the mesentery of the duodenal loop, collecting blood from the pancreas and duodenum and anastomosing with the *anterior pancreaticoduodenal vein*. At the same level as the entrance of this vein the portal receives, on the opposite side, the *inferior mesentery vein*. This may be traced alongside the descending colon and rectum, from which it receives many branches, as well as some part of the transverse colon. The main trunk of the hepatic portal posterior to this point is now named the *superior mesenteric vein,* which is joined by the *intestinal* and *ileocaecocolic veins* draining all the parts of the intestine not already mentioned. In tracing its branches tear the mesenteries that bind together the coils of the intestine as far as necessary and also strip off the fat and lymph glands. The intestinal vein is the large vessel collecting from the greater part of the small intestine. It runs in the middle of the mesentery, receiving many tributaries in its course. The branches from the jejunum immediately beyond the duodenum, however, enter the posterior pancreaticoduodenal vein. The large ileocaecocolic vein drains the ileum, appendix, caecum, and ascending and transverse colons. Chief among its tributaries are: the *appendicular vein* from the appendix; the *anterior ileocaecal vein* from the sacculus rotundus, proximal part of the caecum, adjacent ileum, and ascending colon; and the *posterior ileocaecal vein* from the distal part of the caecum, adjacent ileum, and ascending colon.

Cat: On following the portal vein away from the liver, you will find three small veins that enter it: the *gastric,* the *anterior pancreatico-duodenal,* and the *right gastroepiploic.* These may enter separately or may unite in any combination; usually the last two unite before joining the portal vein. The gastric vein comes from the stomach and lies in the curve between the pylorus and stomach, formed at the lesser curvature by the union of many branches from both sides of the stomach. The anterior pancreaticoduodenal collects from the pancreas and duodenum. The right gastroepiploic vein stems from the pyloric region, greater curvature of the stomach, and adjacent greater omentum. Beyond the entrance of these three veins the hepatic portal receives a large tributary, the *gastrosplenic vein.* This passes to the left in the substance of the pancreas, receiving one or more small *middle gastroepiploic veins* from the stomach wall and omentum and a *pancreatic vein* from the pancreas. Beyond these tributaries the gastrosplenic is formed by the union of two main branches, the *right* and *left splenic veins.* The left splenic vein passes in the gastrosplenic ligament along the spleen, receiving branches from the spleen, the greater omentum, and several *left gastroepiploic veins* from the omentum and stomach. The right splenic vein comes from the right end of the spleen, receiving tributaries from the omentum and stomach wall as well. Beyond the entrance of the gastrosplenic vein, the portal is known as the *superior mesenteric vein.* This receives first a small *posterior pancreaticoduodenal vein* from the pancreas and distal part of the duodenum, and next the *inferior mesenteric vein* from the descending colon and rectum; it is then seen to be formed by numerous converging *intestinal branches* from the small intestine, caecum, and ascending colon. The lymph glands lying along the superior mesenteric vein, as well as the fat, should be removed in tracing the branches.

**3. The Systemic Veins**

There are three systemic veins in the rabbit, two precavals and one postcaval, and two in the cat, one precaval and one postcaval. The condition in the cat is due to the union of the two precavals anterior to the heart. Although the branches are similar in the two animals they will be described separately.

### a. The Branches of the Precaval Vein

This vein is also called the *anterior vena cava.*

Rabbit: Turn the apex of the heart forward and examine the great veins that enter the right atrium. The *left precaval vein* passes around the left atrium and enters the left side of the right atrium. It receives small *coronary veins* from the heart wall. The *right precaval* passes directly into the anterior part of the right atrium. Note additional cornary veins entering the right atrium directly.

Carefully trace the right precaval and clear away connective tissue and muscle from about its course and follow it away from the heart. At the point of entrance into the right atrium it receives, from behind, the *azygos vein.* Press the lungs to the left and follow the azygos posteriorly along the dorsal thoracic wall near the median line. Note the *intercostal veins* that enter it at regular intervals; they course along the posterior margin of each rib and are intersegmental. Entering the precaval immediately anterior to the entrance of the azygos is the *superior intercostal vein,* the first of the series of intercostal veins. Shortly anterior to this the *internal thoracic (internal mammary) vein* enters the precaval. This vein ascends on the internal surface of the chest very near the midventral line. Trace it posteriorly, noting branches from the intercostal muscles. It continues posteriorly on the abdominal wall as the *superior epigastric vein.* The next tributary of the precaval is the *vertebral vein.* It enters the medial side of the precaval at about the same level at which the internal mammary joins the lateral side. It may be traced dorsally to the cervical vertebrae, from which it emerges, receiving a *costocervical tributary* from the neck. Beyond this point the precaval receives the large *subclavian vein* from the forelimb. Follow this laterally. It passes between the first and second ribs into the axilla and is then known as the *axillary vein.* Expose the axilla by cutting down through the pectoral muscles near the midventral line and at their entrance on the humerus. The pectoral muscles should then be separated from the underlying serratus ventralis but should not be removed. The large, stout, white cords seen crossing the axilla are the nerves of the brachial plexus and should not be injured. The small rounded masses of lymph glands will also be noted in the axilla.

In the axilla the axillary vein receives the *long thoracic vein,* the *subscapular vein,* and the *cephalic vein.* The long thoracic vein runs caudad on the thoracic wall in the serratus muscle; it then passes to the inner surface of the skin and extends along the entire length of the abdominal wall; it is especially prominent in females, where as the *external thoracic (external mammary)* vein it collects blood from the mammary glands. (The greater part of this vein was probably removed with the skin.) The subscapular vein enters the axillary vein dorsal to the long thoracic vein. It collects a conspicuous branch (*thoracodorsal vein*) from the latissimus dorsi and cutaneous maximus muscles; it then passes through the teres major muscle to the external surface of the shoulder, where it collects from various muscles. The *cephalic vein* (so named because the corresponding vein in man was formerly thought to connect with the head) is the chief superficial vein of the arm. It can best be picked up on the outer surface of the upper arm; near the distal end of the upper arm it penetrates between muscles and, passing between the teres major and subscapularis muscles, emerges on the internal surface of the shoulder and enters the axillary vein at the same place as, or in common with, the subscapular vein. Immediately beyond the entrance of these tributaries, the axillary vein becomes the *brachial vein* of the arm. This proceeds along the inner surface of the upper arm in company with an artery and a nerve.

Return to the precaval vein. At the point of entrance of the subclavian vein the precaval vein receives the *external* and *internal jugular veins* from the neck. The external jugular vein is the large vein that extends forward in the depressor conchae posterior muscle (most superficial muscle of the ventral surface of the neck). The internal jugular vein is a small vein that runs alongside the trachea, passing the thyroid gland, and accompanying the carotid artery and the vagus nerve. The place where the internal jugular enters is highly variable, as are its general relations; it may enter the precaval after the latter has received the subclavian, but it usually enters the external jugular. The precaval vein may thus be said to be formed by the union of the subclavian, external jugular, and internal jugular veins. Follow the external jugular. Shortly anterior to its union with the subclavian it receives the *transverse scapular vein* from the ventral portion of the shoulder, and near the same level it has a cross connection (*transverse jugular vein*) with its fellow of the opposite side (this union was probably destroyed in the previous dissection). Along the neck it receives various small tributaries from muscles and about one inch posterior to the angle of the jaws is formed by the union of the *anterior* and *posterior facial veins*. The anterior facial vein proceeds to the angle of the jaws, where it is formed by the union of veins from the anterior part of the face and jaws. Its main tributaries are the *angular vein,* which passes over the ventral part of the masseter muscle and then turns to the region in front of the eye, and the *deep facial vein,* which emerges between the masseter and digastric muscles and passes along the surface of the masseter. Other tributaries of the anterior facial vein come from the nearby lymph and salivary glands. The posterior facial vein passes to the parotid gland, where it receives a superficial vein, the *posterior auricular vein,* from the back of the ear and head. The main vein beyond the entrance of this branch lies embedded in the parotid gland, which may be dissected from it. The vein is accompanied by the facial nerve. At the base of the ear it is formed by the *inferior ophthalmic vein* from the orbit, the *temporal veins* from the temporal region, and the *anterior auricular vein* from the region in front of the ear.

The internal jugular vein extends along the length of the neck, receiving a few small branches, of which the chief ones are those from the thyroid gland. It may be traced to the occipital region of the skull, from which it emerges via the jugular foramen; it collects blood from the brain. Its size and place of junction with the external jugular are highly variable.

The left precaval vein is identical in its tributaries with the right, except that there is no azygos vein on the left side.

Cat: Turn the apex of the heart to the left and note the large *right precaval vein,* which enters the anterior margin of the right atrium. Note that there is no such vein on the left. Instead there is a *coronary sinus,* which runs along the dorsal surface of the heart in the groove between the atria and ventricles and represents the proximal portion of the left precaval. The coronary sinus will be found by cleaning out the fat from this groove. Note the numerous *coronary veins* from the heart wall entering the sinus. The sinus itself opens into the left posterior corner of the right atrium. Again pressing the heart to the left, clean the base of the precaval and note the large vein that passes in front of the root of the right lung and joins the precaval as the latter enters the atrium. This tributary of the precaval is the *azygos vein.* Trace it posteriorly, pressing the right lung to the left. It passes along the dorsal thoracic wall near the middorsal line and receives the *intercostal veins* at regular intervals. These course along the posterior border of the ribs. The most anterior of the intercostal veins join into a common trunk that enters the azygos shortly caudad to the entrance of the latter into the precaval. The azygos also receives small branches from the esophagus and bronchi.

Trace the precaval anteriorly. It receives small branches from the thymus gland and then receives a tributary of moderate size, the *common stem of the internal thoracic veins ( = sternal vein),* which comes from the midventral wall of the chest. This stem is formed posteriorly by the union of two veins, the *internal thoracic veins,* which run posteriorly in the chest wall, one to each side of the midventral line, and are extended into the abdomen as the *superior epigastric veins.* In their course the internal thoracic (mammary) veins receive branches from the diaphragm, chest wall, and pericardium. The precaval vein next receives small branches from the thymus gland and adjacent muscles and at a level between the first and second ribs is formed by the union of two large veins. These are the *brachiocephalic* or *innominate veins.* They are the two precaval veins of embryonic stages that later unite to form the single precaval vein of adult anatomy by the crossing-over of the left vein to join the right one. The branches of the two brachiocephalic veins are identical and only one needs to be followed, preferably the right one, since this has not been touched in the previous dissection. The places of entrance of the various tributaries are, however, somewhat variable.

Immediately anterior to the junction of the two brachiocephalics, opposite the first rib each of them receives a large tributary on the dorsal side. This is located by dissecting on the dorsal side of the vein and lifting the vein. The main part of the tributary can be traced into the cervical vertebrae; it is the *vertebral vein* and courses in the vertebrarterial canal, draining the brain and spinal cord. Before it enters the brachiocephalic, the vertebral is joined by the *costocervical vein,* which comes from the muscles of the back and receives branches from the chest wall on the inner surface of the first two ribs. The costocervical vein may be traced by turning the animal dorsal side up and, on the side where the muscles were dissected, dissecting in the serratus ventralis and the epaxial muscles. The communication of the vertebral and costocervical veins with the brachiocephalic and with each other is variable and may not be exactly as described here.

The brachiocephalic, at the same place as the entrance of the veins just described, is formed by the union of a lateral *subclavian vein* and an anterior *external jugular vein.* The subclavian vein passes laterally in front of the first rib into the axilla, where it is known as the *axillary vein.* Expose the axilla by cutting through the pectoral muscles near the midventral line and at their insertion on the humerus. The pectoral muscles should then be separated from the underlying serratus ventralis but should not be removed. The stout white cords crossing the axilla are the nerves of the brachial plexus and are not to be injured. Lymph glands will also be noted in the axilla. The most medial tributary of the axillary vein is the large *subscapular vein,* which passes through the proximal part of the upper arm to the dorsal side of the humerus and collects from various muscles of the upper arm and shoulder, receiving also the *posterior circumflex vein* from the external surface of the upper arm. The beginnings of the subscapular vein will be found in the trapezius muscles. The axillary vein lateral to the entrance of the subscapular receives a small *ventral thoracic vein* from the medial portions of the pectoral muscles. Lateral to this it receives the *long thoracic vein,* which runs caudad along the inner surface of the pectoral muscles, and the *thoracodorsal vein,* which courses parallel to the long thoracic vein but dorsal to it and collects chiefly from the latissimus dorsi muscle. There is a broad connection between the thoracodorsal and subscapular veins. Lateral to these branches the axillary vein is known as the *brachial vein.* It runs along the inner surface of the upper arm in company with nerves and the brachial artery. These structures will be found by separating the muscles on this surface of the upper arm.

Return now to the jugular vein. It receives on its medial side the small *internal jugular vein* that passes forward in the neck alongside the trachea in company with the carotid artery and vagus nerve. The much longer external jugular vein assumes a more superficial position and in addition to small branches from adjacent muscles receives the large *transverse scapular vein* from the shoulder. This passes laterally in front of the shoulder and anastomoses with the cephalic vein of the arm. The *cephalic vein* is the superficial vein of the forelimb and will be found on the external or lateral surface of the upper arm. It also connects with the posterior circumflex vein described above. The external jugular anterior to the entrance of the transverse scapular vein is situated in the sternomastoid muscle. On following it forward, you will find that it arises at the angle of the jaw by union of the *anterior* and *posterior facial veins.* At their point of union they are connected across the ventral side of the throat by the *transverse vein,* which has probably been destroyed. The anterior facial vein collects from the face and jaws and submaxillary and lymph gland, its main tributary being the *angular vein* from the region of the eye. The posterior facial vein emerges from the parotid gland and receives the posterior *auricular vein* from the pinna and back of the head. The main vein then lies embedded in the parotid gland and may be followed by dissecting away the gland. It is then seen to be formed by the union of veins from the temporal region and region anterior to the ear.

### b. The Branches of the Postcaval

The following description applies to both the rabbit and the cat. Turn the apex of the heart forward and note the large vein that enters the right atrium from behind. This is the *postcaval vein* (also called *posterior vena cava*). It then passes posteriorly in the thorax lying slightly to the right of the median line, enclosed in the free dorsal border of the caval fold of the pleura. Follow it caudad. It passes through the diaphragm, from which it receives several *phrenic veins*. In the rabbit it then lies against the dorsal wall of the peritoneal cavity slightly to the right of the medial line, dorsal to the right median lobe of the liver, and in contact with the hepatic portal vein. It then passes into the right lateral lobe of the liver, from which it emerges near the right kidney. In the cat the postcaval vein passes into the right median lobe of the liver and, enclosed in the liver substance, traverses the length of the liver, emerging from the posterior lobule of the right lateral lobe. Note the large *hepatic veins* that flow from the liver into the postcaval vein. These are best seen by dissecting the substance from the liver. Follow the postcaval, posteriorly, carefully cleaning away the connective tissue and fat from it and its tributaries. It runs slightly to the right of the middorsal line of the peritoneal cavity alongside the dorsal aorta, which must not be injured. The first major tributary of the postcaval is the *right adrenolumbar vein,* which passes along the posterior surface of the adrenal gland, which lies anterior to the kidney. The adrenolumbar vein receives branches from the adrenal gland and also collects from the adjacent body wall. Immediately posterior to this vein the large *right renal vein* passes from the kidney into the postcaval. Next, by turning the viscera to the right, locate the left adrenal gland and kidney and find the *left adrenolumbar* and *renal veins*. They are situated posterior to the right ones. The left adrenolumbar and renal veins generally unite in a common stem before they enter the postcaval. The vein of the left gonad opens into the left renal vein. In male specimens this is the *left internal spermatic vein;* it may be traced posteriorly (in contact with the postcaval in the rabbit) to the scrotum. In female specimens it is the *left ovarian vein* from the ovary, a small oval body lying about the middle of the lateral wall of the peritoneal cavity. The *right internal spermatic* or *ovarian vein* enters the postcaval directly, in the cat shortly posterior to the right kidney, in the rabbit much farther caudad. The postcaval vein in its course along the body wall receives at regular intervals the paired *lumbar veins* from the wall; these can be seen by loosening the vein, raising it slightly, and looking on the dorsal surface. Near the posterior end of the peritoneal cavity the postcaval receives a pair of *iliolumbar veins*. Each of these, in company with an artery, extends laterally along the body wall and receives an anterior branch from the region of the kidney. Sometimes the left ovarian vein enters the left iliolumbar. Posterior to this point the dorsal aorta comes to lie ventral to the postcaval, concealing the latter. The dissection of the remainder of the postcaval will therefore be deferred until the aorta is studied.

Variations from the foregoing account of the postcaval and its branches are common and result from the persistence of parts of the complicated embryonic condition. An apparent splitting (really a failure of fusion) of the postcaval into two main trunks caudad of the kidneys is a common variation.

**4. The Pulmonary Veins**

Examine the roots of the lungs and note numerous veins, several on each side, entering the left atrium from the lungs. These are the *pulmonary veins*. They lie to either side of the postcaval vein; those of the right side pass dorsal to the postcaval, and in the rabbit those of the left side dorsad of the left precaval.

| 5. The Pulmonary Arteries | The pulmonary arch is a conspicuous vessel extending from the base of the right ventricle forward between the atria curving to the left. Its base is generally surrounded by fat, which should be cleaned away. It divides into *right* and *left pulmonary arteries*. Press the heart to the right and follow the left pulmonary artery into the left lung. In the rabbit it passes to the dorsal side of the left precaval vein, which may now be severed. The left pulmonary artery courses parallel to, and anterior to, the most anterior of the pulmonary veins. Now turn the heart to the left and similarly find the right pulmonary artery, proceeding to the right lung; to trace it, sever the precaval vein. It lies immediately next to the foremost pulmonary vein. The trachea lies dorsal to the right pulmonary artery. |

6. The Aorta and Its Branches

Springing from the base of the left ventricle to the left of, and dorsal to, the pulmonary artery is a large *aorta* or *aortic arch*. Right and left *coronary arteries* leave the base of the aorta. The left coronary artery lies between the pulmonary artery and the left atrium and ramifies over the ventral and left side of the heart. The right coronary artery lies along the groove between the right atrium and right ventricle and branches to the right and dorsal surface of the heart.

Follow the aorta forward, cleaning away tissue from its surface. It describes a curve to the left and from the arch of the aorta spring the large arteries of the neck, head, and forelimbs.

To trace the *brachiocephalic artery* forward, the precaval vein and its branches may be removed. The artery gives off small branches into the thymus gland and trachea lying dorsal to it and then divides into two branches in the rabbit and three in the cat. These are: *right subclavian* and *right common carotid* in the rabbit and *right subclavian* and *right* and *left common carotids* in the cat. Each of these will be traced separately.

*a. Subclavian Artery*

Trace the right subclavian; both right and left have identical branches.

Rabbit: From the posterior surface of the subclavian arises the *internal thoracic (internal mammary) artery,* which follows the vein previously described along the ventral chest wall and continues into the abdomen as the *superior epigastric artery.* At the same level the *supreme intercostal artery* arises from the posterior surface of the subclavian, practically in common with the superior epigastric artery. It runs posteriorly on the dorsal wall of the thorax and receives the first *intercostal arteries.* On its anterior surface at about the same level as these the subclavian artery gives rise (*a*) to the *vertebral artery,* which passes immediately dorsad toward the cervical vertebrae, where it enters the vertebroarterial canal, and (*b*) to the *superficial cervical artery,* which ascends in the lateral part of the neck supplying various muscles, its main branch (*ascending cervical*) accompanying the external jugular vein. The *transverse artery of the neck* leaves the subclavian at the same place or in common with the supreme intercostal artery. It passes dorsally in front of the first rib through a loop formed by two nerves and emerges on the medial side of the serratus ventralis muscle. It is best found by looking on this muscle and then tracing the artery toward the subclavian. After giving off the foregoing branches, the subclavian passes in front of the first rib into the axilla, where it is the *axillary artery*. This lies between two of the stout nerves belonging to the brachial plexus. Its branches are similar to those of the axillary vein and the accompanying veins. After giving rise to the small *thoracoacromial artery* to the pectoral and deltoid muscles, the axillary gives off the *long thoracic* and *subscapular arteries,* accompanying the veins previously described. The former runs posteriorly along the serratus muscle and then, as the *external thoracic* (*external mammary*) *artery,* passes to the under surface of the skin of the lateral abdominal wall, being especially conspicuous in females. (Most of this vessel was destroyed in removing the skin.) The subscapular artery has a conspicuous branch (*thoracodorsal artery*) passing caudad to the latissimus dorsi and cutaneous maximus muscles; it then turns dorsally and, perforating the teres major, emerges on the outer surface of the shoulder, supplying various muscles. Near the point of origin of the subscapular the *deep artery* of the arm arises and, after giving off branches into the subscapular muscles, passes between this muscle and the teres major to the dorsal part of the arm, where it runs in company with one branch of the cephalic vein and a nerve; these are situated internal to the lateral head of the triceps, which should be deflected. The axillary artery now passes to the upper arm, where, as the *brachial artery,* it courses along the inner surface of the limb in company with the brachial vein and nerves.

Cat: At the level of the first rib the subclavian has four branches: *internal thoracic* (*internal mammary*), *vertebral, costocervical axis,* and *thyrocervical axis.* The internal thoracic springs from the ventral surface of the subclavian, accompanies the corresponding vein along the chest wall, and passes on to the abdominal wall as the *superior epigastric artery.* The vertebral artery arises from the dorsal surface of the subclavian and passes dorsally into the vertebrarterial canal, giving off small branches into the neck muscles. The costocervical axis divides in two almost at once. One branch, the *superior intercostal artery,* passes posteriorly near the middorsal line of the thorax, giving off the *intercostal branches* and then supplying the deep muscles of the back. The other branch of the costocervical axis leaves the thoracic cavity, passing dorsally in front of the first rib, and divides into the *transverse artery of the neck,* supplying the serratus ventralis and rhomboideus muscles, and the *deep cervical artery* to the epaxial muscles of the neck. These branches are best found by looking among the muscles in question and tracing the vessels toward the subclavian. The *thyrocervical axis* generally arises anterior to the other branches. It passes forward near the carotid artery and, after branching to the muscles of the dorsal side of the neck, turns laterally in front of the shoulder, as the *transverse scapular artery;* it accompanies the external jugular vein for a short distance and supplies many muscles of the shoulder and neck.

The subclavian artery now runs in front of the first rib into the axilla, where it is named the *axillary artery.* This gives off several arteries: the *ventral thoracic artery,* which passes medially to the medial ends of the pectoral muscles; the *long thoracic artery,* which passes posteriorly along the middle region of the pectoral muscles and then to the latissimus dorsi; and the large *subscapular artery,* near the arm. The subscapular artery gives off the *thoracodorsal artery,* lying parallel but more dorsal to the long thoracic artery and supplying the latissimus dorsi; the subscapular then turns dorsally, passes through the proximal part of the upper arm dorsal to the humerus, and branches to the muscles of the upper arm and muscles of the back and shoulder.

The axillary artery then proceeds as the *brachial artery* to the medial surface of the forelimb, where it accompanies the brachial vein and some nerves, and branches into the limb.

### b. Common Carotid Artery

The two common carotid arteries arise in the cat from the brachiocephalic artery and immediately diverge; in the rabbit the right one arises in common with the right subclavian, while the left usually springs independently from the arch of the aorta. Trace the common carotids forward. Their branches are similar in the two animals. They pass anteriorly in the neck, one on each side of the trachea to which they give small branches. At the level of the anterior end of the thyroid gland each supplies a superior thyroid artery to the gland. At the level of the larynx there are branches into the larynx and adjacent parts (probably destroyed) and an *occipital branch* into the dorsal muscles of the neck. The common carotid at about this same level gives off the *internal carotid artery.* In the rabbit this artery arises at the place where the carotid passes to the dorsal side of the shining ligament of the digastric. In the cat it arises slightly caudad to the occipital artery. In both animals the internal carotid passes dorsally in company with nerves and enters the skull by a foramen in the tympanic bulla. However, in the adult cat, the proximal part of the internal carotid is vestigial and is represented by a small ligament, which may be revealed by careful dissection. The extension of the common carotid beyond this point is the *external carotid artery.* At the angle of the jaw it branches to all parts of the head. Its chief branches are: the *lingual artery* into the tongue and the *external maxillary artery* running along the ventral border of the masseter muscle and branching to the upper and lower lips and jaws. The main artery then passes along the posterior border of the masseter muscle. It receives *auricular* and *temporal branches* from the pinna and temporal regions and then, as the *internal maxillary artery,* turns internal to the masseter muscle.

### c. The Thoracic Aorta

After having given rise to the subclavians and the carotids, the aorta arches to the left. Note, where it passes the left pulmonary, the strong fibrous band which connects the two vessels. This is the *arterial ligament* or *ligament of Botallus* and is the remnant of the embryonic connection between the systemic and pulmonary arches. Follow the aorta posteriorly, pressing the left lung to the right. It descends posteriorly, lying against the dorsal wall of the thorax to the left of the median line. It is situated within the mediastinum; the mediastinal wall may be cleared away. The aorta courses along the thorax and its major branches are the paired *intercostal arteries,* which arise from the aorta at regular intervals and run along the thoracic wall along the posterior margin of the ribs. The aorta also has small *bronchial arteries* to the bronchi and *esophageal arteries* to the esophagus. Along the dorsal surface of the aorta on its left side runs a delicate tube resembling a streak of fat. This is the *thoracic duct,* the *main lymphatic channel* for the posterior part of the body. Trace it forward; its connection with the jugular vein, generally at the point of union with the subclavian, may be found.

The aorta penetrates the diaphragm, to which in the rabbit it gives *superior phrenic arteries* and passes into the peritoneal cavity, where it is termed the *obdominal aorta.*

### d. The Abdominal Aorta

Turn the digestive tract to the right and locate the dorsal aorta as it passes the diaphragm. It will be found against the dorsal wall in the median dorsal line. Clear away the mesogaster and clean the surface of the aorta. The branches consist of unpaired *median visceral branches* to the digestive tract, paired *lateral visceral branches* to the kidneys and reproductive organs, and paired *somatic branches* to the body wall.

Shortly posterior to the diaphragm the aorta gives rise to two large unpaired visceral arteries, the *coeliac* and the *superior mesenteric arteries.* In the cat the second is shortly posterior to the first, but in the rabbit the superior mesenteric artery lies one-half inch posterior to the coeliac. As the branches of these two vessels are different in the two animals, because of the differences in their digestive tracts, it will be necessary to describe them separately.

Rabbit: The coeliac artery near its origin from the aorta gives rise to the small *inferior phrenic arteries* to the diaphragm. Beyond this point the *splenic artery* arises from its posterior surface. This vessel passes in the mesogaster to the spleen, where it runs in the gastrosplenic ligament. In its course to the spleen it provides the short *gastric arteries* to the left limit of the stomach; along the spleen it supplies *splenic branches* to the spleen; beyond the spleen it branches into the omentum; at about the middle of the spleen a large branch, the *left gastroepiploic artery* arises from the splenic artery and passes to the greater curvature of the stomach. The coeliac artery beyond the splenic passes the lesser curvature of the stomach, where it may best be followed by turning the stomach forward.

Here it produces a group of vessels, the *left gastric* (or *coronary*) *arteries,* which radiate to the stomach wall on both sides of the lesser curvature and also send small branches to the esophagus. Shortly beyond this point the coeliac artery is known as the *hepatic artery,* which passes along the right end of the lesser curvature, very shortly giving rise to the *gastroduodenal artery.* This runs to the pyloric region and branches into the *anterior pancreaticoduodenal artery* to the pancreas and first part of the duodenum and the *right gastro-epiploic artery,* which returns to the stomach wall by way of the great omentum. The hepatic artery now passes to the dorsal side of the pylorus and enters the hepatoduodenal ligament. After giving off the small *right gastric artery* to the pylorus, it proceeds to the liver, lying to the right of the bile duct.

The superior mesenteric artery is the chief artery of the intestine and has many complicated branches in the rabbit; these branches, for the most part, follow the branches of the hepatic portal vein. Clean the surface of the superior mesenteric artery and follow it; it runs alongside the superior mesenteric vein. The first branch is the small middle *colic artery,* which arises from the ventral wall of the superior mesenteric and passes to the transverse colon and the beginning of the descending colon. At the same level but from the dorsal side, the *posterior pancreaticoduodenal artery,* which runs to the duodenal loop and pancreas, arises. The superior mesenteric artery then forks into the *intestinal artery,* which runs in the mesentery of the small intestine and gives off numerous branches ventrally into the intestine, and into the large *ileocaecocolic artery.* This has many branches to the ileum, the caecum, the appendix, and the ascending colon. Its branches are: the *anterior right colic artery,* which forks several times, supplying the greater part of that portion of the ascending colon that bears the haustra; the *posterior right colic artery,* arising near the preceding vessel and supplying the remainder of the haustra-bearing region of the ascending colon; the *appendicular artery,* arising with the posterior right colic and running along the appendix and that part of the ileum adjacent to the appendix; the large *posterior ileocaecal artery,* passing to the greater part of the caecum and to that portion of the ileum lying between the caecum and the ascending colon; the much smaller *anterior ileocaecal artery* to the more distal part of the caecum and adjacent ileum; and the *caecal artery* (or arteries) to that portion of the caecum that adjoins the appendix.

Cat: The coeliac artery passes toward the stomach and soon divides into three branches. The most cranial one is the *hepatic;* the second is the *left gastric;* and most caudal and the largest is the *splenic.* Trace the splenic artery as it courses in the great omentum toward the spleen and forks. One branch travels to the left of the spleen and also sends branches into the pancreas and *short gastric arteries* to the stomach; the other branch passes to the right of the spleen and also supplies branches to the pancreas and the omentum and the *left gastroepiploic arteries* to the greater curvature. The *left gastric artery* passes to the lesser curvature, where it splits into many branches, supplying both sides of the stomach. The hepatic artery passes along the border of the left end of the pancreas to the dorsal side of the lesser curvature and enters the hepatoduodenal ligament. It is best found by separating the stomach and the liver. It lies to the left side of the hepatic portal vein. As it passes the pylorus, it gives off the large *gastroduodenal branch.* This divides into the *anterior pancreaticoduodenal artery,* which descends along the beginning of the duodenum and supplies the pancreas as well; the *right gastroepiploic,* which passes from the pylorus along the greater curvature of the stomach to the left; and the small *pyloric artery,* which enters the pyloric region (this may also arise independently from the hepatic). The *hepatic artery* proceeds into the liver, sending a *cystic artery* to the gallbladder.

The *superior mesenteric artery* supplies the greater part of the intestine and initially gives rise to the *middle colic artery,* which passes to the transverse and descending parts of the colon. A little farther posteriorly the superior mesenteric gives rise simultaneously (*a*) to the *posterior pancreaticoduodenal artery,* which ascends along the duodenum, supplying it and the pancreas and anastomosing with the anterior pancreaticoduodenal, and (*b*) to the *ileocolic artery* to the caecum and terminal portion of the ileum and sending also a *right colic branch* to the ascending colon (this may arise independently from the superior mesenteric). The superior mesenteric then divides into numerous *intestinal branches* to the small intestine.

Return now to the dorsal aorta. Its next branches are the paired *adrenolumbar* and *renal arteries.* In the rabbit, the adrenolumbars are branches of the renals, but in the cat they arise independently. They pass close to the adrenal gland, to which they give an *adrenal branch,* and then course along the dorsal body wall. In the cat each sends a *phrenic artery* anteriorly to the diaphragm. The renal arteries are large vessels passing into the kidneys. The aorta posterior to the kidneys gives rise to paired arteries to the gonads (these may, however, branch from the renals). They are the *internal spermatic arteries* in the case of the male. In the female the corresponding *ovarian arteries* are larger, and in the cat are convoluted. In its passage along the middorsal line the aorta gives off paired *lumbar arteries* at intervals. These are found by loosening the aorta and looking on its dorsal surface. Posterior to the genital arteries the *inferior mesenteric artery* arises as an unpaired visceral branch and passes to the descending colon and rectum, running to the mesocolon. In the mesocolon it forks into the *left colic artery,* passing craniad along the descending colon, and the *superior rectal (hemorrhoidal) artery,* passing caudad to the posterior part of the descending colon and the rectum.

The digestive tract may now be removed, but the end of the large intestine should be left in place. Hold the stump of the colon, together with the urinary bladder and in female specimens the uterus (the forked coiled tube at the posterior end of the peritoneal cavity), back against the pubes and follow the aorta farther. Near the end of the peritoneal cavity it divides into the two *common iliac arteries* in the rabbit; in the cat it gives off a pair of *external iliac arteries,* followed by a pair of *internal iliac (hypogastric) arteries.* Anterior to this place in the cat, or in the rabbit at the level of the fork or from the common iliac arteries, a pair of *iliolumbar arteries* arises and passes laterally along the body wall. Each iliolumbar artery divides into an *anterior branch,* which passes forward toward the kidney, and a *posterior branch,* which extends to the thigh.

The two common iliac arteries in the rabbit fork into an anterior *external iliac* and a posterior *internal iliac.* In the cat the external and internal iliacs arise separately from the aorta, the latter immediately posterior to the former. After giving rise to the iliacs the aorta continues in the middorsal line as the small *median sacral* or *caudal artery,* lying halfway between the two internal iliacs. In the cat this vessel arises from the fork of the internal iliacs. In the rabbit it originates anterior to the forking of the aorta, its origin is concealed by the postcaval vein and will be seen later. The sacral artery supplies the sacral region and the tail.

Follow the external iliac. It passes laterocaudad out of the peritoneal cavity, in the rabbit to the dorsal side of the inguinal ligament. As it passes through the abdominal wall or shortly beyond the wall, it gives rise to the *deep femoral artery* (cat) or the *inferior epigastric* (rabbit). In the cat the deep femoral gives off branches into the thigh, but these are absent in the rabbit. In both animals the following branches are present: branches into the mass of fat between the thighs and into the external genital organs, of which one branch in male specimens constitutes the *external spermatic artery;* and the main vessel, which, as the *inferior epigastric artery,* turns craniad and ascends in the abdominal wall, running along the inner surface of the rectus abdominis muscle. It anastomoses with the superior epigastric artery. In the rabbit there arises, either from the inferior epigastric at the origin of the latter from the external iliac or from the external iliac itself nearby, the *superficial epigastric artery,* which extends forward on the inner surface of the skin of the abdominal wall and anastomoses with the external thoracic, a branch of the long thoracic. These vessels are particularly prominent in females, but the greater part of their course is destroyed in removing the skin. The external iliac, now named the *femoral artery,* proceeds along the center of the medial surface of the thigh, giving branches into the leg muscles.

Follow the internal iliacs, being careful not to injure the end of the postcaval vein lying in contact with them or any parts of the urogenital system (in males do not injure the male ducts curving around the base of the urinary bladder). The internal iliac arteries lie against the dorsal wall. At their origin from the common iliac (rabbit) or posterior to their origin from the dorsal aorta (cat) each gives rise to an *umbilical artery,* which passes to the bladder or in female rabbits to the uterus, first with a branch to the bladder. The internal iliacs then pass to the dorsal side of the postcaval vein. To follow them, dissect as deeply as possible between the rectum and the base of the thigh. The internal iliacs have branches to the pelvis; each gives off a *middle rectal artery* to the rectum, which accompanies the rectum to the anus but cannot be followed at this time. In female cats the *uterine artery* arises from the middle rectal (hemorrhoidal) and passes anteriorly again to the uterus.

**7. The Posterior Portion of the Postcaval Vein**

The postcaval vein may now be followed caudad from the point where it was previously left by removing the arteries that cover it. Its tributaries should be traced, as far as practicable, to the posterior end of the peritoneal cavity, dissecting deeply dorsally as before.

Rabbit: At the same level as the forking of the dorsal aorta, the postcaval receives the two large *external iliac veins*. It then continues in the middorsal line for a short distance caudad to this point; this portion is often termed the *common internal iliac vein* and is formed by the union of the two *internal ilias* or *hypogastric veins*. Trace the external iliac; its branches are similar to and accompany those of the artery of the same name. It receives the *vesical vein* from the bladder: this vein in females also drains the uterus. At the point where it passes through the abdominal wall, the external iliac receives the *inferior epigastric vein,* the main part of which runs forward along the internal surface of the rectus abdominis muscle and anastomoses anteriorly with the superior epigastric. The inferior epigastric near its entrance into the external iliac also receives tributaries from the fat between the bases of the thighs and the external genital region and sends a *superficial epigastric vein* along the inner surface of the skin of the lateral abdominal wall. This last vessel is particularly conspicuous in females but is destroyed in removing the skin; it anastomoses with the external thoracic, a tributary of the long thoracic. The external iliac passes to the dorsal side of the inguinal ligament and, as the *femoral vein,* continues along the medial side of the leg, in company with the femoral artery. Follow the internal iliacs. After a short distance the *sacral* or *caudal vein* enters one of them, usually the left one; it accompanies the caudal artery. Caudad of this, each internal iliac receives the *middle rectal vein,* which ascends from the anus and lies along the side of the rectum.

The iliacs and postcaval may then be removed and the origin of the caudal artery from the aorta traced.

Cat: The postcaval is formed dorsal to the bifurcation of the aorta by the union of the two large *common iliac veins*. One of them, usually the left one, receives the small *sacral* or *caudal vein,* which lies parallel to the artery of the same name. About one inch posterior to its junction with the postcaval, each common iliac arises by the union of the *internal iliac (hypogastric)* and the *external iliac veins*. The former receives branches from the gluteal region and the *middle rectal (hemorrhoidal) vein,* which runs along the sides of the rectum from the anus forward, and it also collects from the bladder. The external iliac passes out of the abdominal cavity. At its point of exit it receives the *deep femoral veins,* which collect from the thigh, from the fat between the thighs, and from the external genital region (receiving in males the *external spermatic vein* from the testes), and it also receives the *inferior epigastric vein* from the inner surface of the rectus abdominis muscle. The branches from the thigh may enter the external iliac separately. The external iliac, now known as the *femoral vein,* passes along the thigh, receiving branches from the leg muscles.

8. The Structure
of the Heart

Remove the heart from the body, cutting across the bases of the great vessels. Identify the systemic veins entering the right atrium (three in the rabbit, two in the cat) and the pulmonary veins entering the left atrium. Cut a transverse slit into the wall of each atrium and wash out the clotted blood, which generally fills the interior. Note the thick ridged walls of the atrial appendages and the thinner smoother walls of the remainder of the atrium. The *atrial septum* extends dorsally between the two atria and completely separates them. Find the large *atrioventricular openings.* In the cat, near the dorsal edge of the atrial septum, find the opening of the coronary sinus into the right atrium, noting the valve that guards the opening. Cut off the apex of the ventricle and note the thick walls and rounded form of the left ventricle and the smaller size, thinner walls, and crescentic form of the right ventricle. Cut open the right ventricle obliquely, beginning at the cut surface already made and extending out through the pulmonary artery, slitting open this artery. Wash out the right ventricle. Its cavity is rather small, and the walls are deeply cleft by muscular ridges, the *trabeculae carnae.* From the walls project a number of pointed finger-like muscles, the papillary muscles, which are connected by slender fibers, the *chordae tendineae,* to three flaps of the tricuspid valve. Two of the flaps can be stretched by pulling on the cut surfaces of the ventricle, while the third lies collapsed against the interventricular septum to which it is attached without the intervention of the papillary muscles. The tricuspid valve guards the right atrioventricular opening. In the base of the pulmonary arch note the three pocket-shaped *semilunar valves.* Similarly cut open the left ventricle by a longitudinal slit from apex to base. Wash out the interior. The cavity of the left ventricle is considerably larger than that of the right, and its walls thicker. The two ventricles are completely separated by the *interventricular septum,* which appears as the common internal wall of both ventricles. Note in the left ventricle the trabeculae carnae, the papillary muscles, and the chordae tendineae. The latter are attached to the membranous *bicuspid* or *mitral valve,* which is composed of two flaps guarding the left atrioventricular opening.

The removal of the heart permits a clearer view of some of the structures of the pleural cavity. The student should examine carefully the forking of the trachea into the bronchi, the form of the lungs and their relation to the pleural cavity, and the pulmonary arteries and veins.

L. Summary

1. The circulatory system consists of a blood-vascular (cardiovascular) system and a lymphatic system.

2. Arteries carry blood away from the heart and veins return blood to the heart.

3. Fluid passes out of the capillaries and forms the tissue fluid that is recollected by the lymph vessels and returned to the venous system.

4. Before its return to the blood or vascular system, lymph passes through one or more lymph glands, where it is filtered, foreign matter removed, and lymphocytes added.

5. Blood and blood vessels arise embryologically as masses of mesenchyme cells, the angioblasts.

6. Hemopoetic tissue, which produces blood cells, may occur in a variety of different organs, such as the kidney, thymus, and liver and not merely in the bone marrow.

7. The heart develops as a cardiac tube that forms a thick muscular myocardium covered internally by an endocardium and externally by an epicardium.

8. Initially the cardiac tube is connected to its coelomic chamber, the pericardium by mesenteries, but these soon disappear.

9. Differential growth converts the cardiac tube into an essentially S-shaped structure, which differentiates into the heart chambers.

10. In its simplest anatomical form the heart consists of a sinus venosus, atrium, ventricle, and conus arteriosus.

11. The conus leads into a ventral aorta where the aortic arches arise to form the afferent branchial arteries supplying the gills. Blood returns from the gills in a series of efferent branchial arteries (epibranchials).

12. A single circulation (as in fishes) is one in which a blood cell must pass through the heart once to return to its original position within the system. In a double circulation (for example, in mammals) a cell must pass through the heart twice to regain its position.

13. During development the first portion of the blood-vascular system to arise is that which supplies the gut and viscera.

14. The somatic circulation, established later, consists of a large number of arteries (intersegmental) from the dorsal aorta, and the blood is returned in the cardinal veins.

15. Many vertebrates have an extraembryonic circulation, and this is most highly developed in the placental mammals.

16. A number of ontogenic changes occur when placental flow is interruped at birth and the lungs become functional.

17. The ancestors of vertebrates probably had a large number of aortic arches, but in living vertebrates (with the exception of the highly specialized agnathans) the number does not exceed six.

18. In living gnathostomes the first gill arch to develop is the mandibular arch, the corresponding vascular (aortic) arch is hence labelled I, and the more posterior arches II through VI. Since the mandibular 'gill' arch becomes highly modified to form the jaws, a complete mandibular aortic arch is not present in any living vertebrate. Thus the gill on the anterior

face of the hyoid gill slit (spiracle), which lies between the mandibular 'gill' arch and the hyoid gill arch, is supplied by a branch from aortic arch II or even from the dorsal aorta. The gill that is termed a pseudobranch often loses its respiratory function and becomes glandular.

19. The anterior cardinal veins persist in all vertebrates as the internal jugular veins, which after collecting blood from other veins in the head region form large venous trunks, the anterior venae cavae (precavals). The anterior portions of the posterior cardinal veins lose their importance and generally disappear. The posterior portion of the postcardinals has a complex developmental and evolutionary history. In fishes, amphibians and reptiles they form the renal portal system that carries blood to the kidneys from the tail, legs, and pelvic regions. Within the kidneys the renal portal veins break up into capillaries and the blood is recollected by the subcardinal veins that lie between the kidneys. In some fishes, and in all tetrapods, the posterior venous return is largely taken over by a posterior vena cava (postcaval) that originates largely from the union of hepatic vein and subcardinals. In amphibians and reptiles the postcaval extends only as far as the posterior limit of the kidney, but in birds it establishes a direct connection with the renal portal veins, thus reducing the renal portal circulation. In mammals this connection is completed, and blood from the limbs and pelvic region passes directly into the postcaval. Thus, the renal portal system is lost.

20. In fishes the conus arteriosus is contractile and is regarded as a chamber of the heart. The bulbus arteriosus, peculiar to teleosts, is the swollen base of the ventral aorta. The bulbus and conus both act as distensible elastic reservoirs that empty at the end of each cardiac cycle. This prevents any reversal of blood flow in the ventral aorta and protects the delicate gills from any large pressure changes.

21. Lungfishes (Dipnoi) have a separate pulmonary return to the heart. The presence of an interventricular septum, atrioventricular plug, and complex spiral valve ensures an effective separation of oxygenated and deoxygenated blood. This separation ensures that blood rich in oxygen reaches the brain, while that depleted in oxygen is returned to the lungs.

22. The considerable structural variation in the amphibian heart is related to the relative importance of pulmonary and cutaneous respiration. While there may be an effective separation of oxygenated and deoxygenated blood within the heart the absence of an interventricular septum allows the maximum flexibility of blood flow pathways.

23. The interventricular septum in reptiles, other than crocodiles, is largely horizontal and divides the ventricle into dorsal and ventral rather than right and left chambers. Both atria open into the dorsal chamber (cavum dorsale). In crocodiles the interventricular septum is vertical and the atria open on either side of it.

24. The ancestors of the tetrapods probably possessed a functionally four-chambered heart (similar to that of lungfishes) that effectively separated oxygenated and deoxygenated blood. The interventricular septum in such forms probably had a dual nature and was developed from tissue of endocardial origin, which formed a primary septum, and ma-

terial derived from the myocardium, which formed the secondary septum. In amphibians the septum was lost in relation to the acquisition of cutaneous respiration. The interventricular septum of chelonians, lizards, snakes, and mammals is largely endocardial in nature (primary septum), whereas that of crocodiles and birds is myocardial (secondary septum).

25. Special adaptations of the vertebrate blood-vascular system enable blood to be redirected in response to varying demands placed on the system. Arteriovenous anastomoses allow blood to bypass the capillary network and thus to flow directly from a small artery to a small vein. Such bypass mechanisms are especially important as thermoregulatory devices in the skin of mammals.

26. Major changes in the blood vascular system occur in diving vertebrates. In such cases the heart is slowed down (bradycardia) and blood is concentrated in the heart and brain and curtailed in the viscera and muscles.

**References**

Adolph, E. F. 1967. The heart's pacemaker. *Sci. Amer.* 216(3):32–37.

Cox, C. B. 1967. Cutaneous respiration and the origin of the modern amphibia. *Proc. Linn. Soc. Lond.* 178:37–47.

Foxon, G. E. H. 1955. Problems on the double circulation in vertebrates. *Bio. Rev.* 30:196–228.

————. 1964. Blood and respiration. In *Physiology of the Amphibia*, ed. J. A. Moore, pp. 151–202. New York: Academic Press.

Foxon, G. E. H.; Griffiths, J.; and Price, M. 1956. The mode of action of the heart of the green lizard, *Lacerta viridis. Proc. Zool. Soc. Lond.* 126:145–57.

Francis, E. T. B. 1956. The vertebrate heart. *School. Sci. Rev.* 132:73–85 and 226–233.

Girgis, S. 1961. Observations of the heart in the family Trionychidae. *Bull. Brit. Mus. Nat. Hist.* 8: 73–107.

Gouder, B. Y. M., and Desai, R. N. 1966. Studies on the carotid body in the frog, *Rana tigrina. Daud. Naturwissenschaften.* 20:535–36.

Hughes, G. M. 1965. *Comparative physiology of vertebrate respiration.* London: Heinemann.

Johansen, K., and Ditadi, A. S. F. 1966. Double circulation in the giant toad, *Bufo paracnemis. Physiol. Zool.* 39:140–49.

Johansen, K., and Hanson, D. 1968. Functional anatomy of the hearts of lungfishes and amphibians. *Amer. Zool.* 8:191–210.

Kanno, T., and Matsuda, K. 1966. The effects of external sodium and potassium concentration on the membrane potential of atrioventricular fibres of the toad. *J. Gen. Physiol.* 50:243–53.

Kramer, A. W., Jr., and Marks, L. S. 1965. The occurrence of cardiac muscle in the pulmonary veins of Rodentia. *J. Morph.* 117:135–50.

Lawson, R. 1966. The anatomy of the heart of *Hypogeophis rostratus* (Amphibia, Apoda) and its possible mode of action. *J. Zool.* 149: 320–36.

Randall, D. J. 1968. Functional morphology of the heart in fishes. *Amer. Zool.* 8:179–89.

Satchen, G. H. 1971. *Circulation in fishes.* London: Cambridge University Press.

Simons, J. R. 1959. The distribution of the blood from the heart in some amphibians. *Proc. Zool. Soc. Lond.* 132:51–64.

——. 1965. The heart of the Tuatara, *Sphenodon punctatus. J. Zool.* 146:451–66.

Szidon, J. P.; Lahiti, S.; Lev, M.; and Fishman, A. P. 1969. Heart and circulation of the African lungfish. *Circulation Res.* 25:23–38.

Truex, R. C., and Smythe, M. Q. 1965. Comparative morphology of the cardiac conduction tissue in animals. *Ann. N.Y. Acad. Sci.* 127:19–33.

Turner, S. C. 1966. A comparative account of the development of the heart of a newt and a frog. *Acta Zool.* 48:43–57.

Wessels, N. K. 1974. *Vertebrate structures and functions.* Readings from Scientific American. San Francisco: Freeman.

White, F. N. 1968. Functional anatomy of the heart in reptiles. *Amer. Zool.* 8:211–19.

Whitford, W. G., and Hutchinson, V. H. 1965. Gas exchange in salamanders. *Physiol. Zool.* 38:228–42.

Wood, J. E. 1968. The venous system. *Sci. Amer.* 218:86–96.

# 12 The Comparative Anatomy of the Urogenital System
## Marvalee H. Wake

**A. Introduction**

The urogenital system of vertebrates is composed of functional units that (1) elaborate, transport, and store urinary wastes and water; (2) elaborate, transport, and occasionally store gametes; (3) provide for union of gametes; and (4) in many instances and many ways provide conditions facilitating development of embryos in or on a parent. In their evolution both urinary and reproductive components are closely tied morphologically and physiologically. Utilization of common transport passages and later separation of function with elaboration of accessory devices characterize the evolutionary history of the urogenital system. In adults there are two "primary" components of the system—the kidney and the gonad. These are the glands that manufacture the essential products of the system. "Secondary" components include the ducts that transport and sometimes store these products and the accessory structures of the ducts. The evolution of form and function of the "primary" components must be understood before the modification of form and function of the "secondary" components as associates of the "primary" can be evaluated. The evolution of these structures can be characterized in a phylogenetic order, and consideration of the embryonic development of each component contributes to an understanding of form and function in the urogenital system.

**1. Early Development of the Kidney and Gonads**

Differentiation of the somites results in establishment of a narrow longitudinal band of mesoderm along the lateral border of the somite. This is the *mesomere,* that is, the urogenital or nephrotomic mesoderm or plate. Like the epimere (the dorsal component of the somite), the mesomere becomes segmented anteriorly in all vertebrates and is segmented more extensively posteriorly in lower vertebrates. Each segment of the mass is called a nephrotome, and from each segment develop the functional kidney units, the *nephrons.* The term *holonephros* was introduced to designate a kidney derived from the entire nephrotomic plate in which a single nephron is derived from each segment, and this state is thought to be the primitive vertebrate condition. Early development in some hagfish, elasmobranchs, and caecilians simulates a holonephric condition. During total development, however, no postembryonic vertebrate retains a holonephros, but it utilizes instead a kidney produced in one of three general anteroposterior regions. Each kidney so derived is adapted to special developmental and functional conditions.

The site of origin of the gonads is the coelomic epithelium between the dorsal mesentery and the anterior portion of the middle kidney. A proliferation of cells under the surface of the coelomic epithelium produces a thickening. Primordial germ cells, which give rise to the haploid gametes, arise either near the floor of the archenteron or near the blastopore lip. These cells aggregate at the base of the dorsal mesentery and then migrate to the thickened areas of the coelomic epithelium. The mode of migration is controversial and four ways have been postulated, depending on the group of animals examined: (1) active migration, by amoeboid movement; (2) passive migration, by growth of surrounding tissues; (3) passive migration by circulatory transport; (4) chemotactic migration, by influence of gonadal inducers. It is significant that the origin of germ cells is extragonadal, and migration of the cells takes place. Experimental data are inconsistent and it cannot be determined whether primordial germ cells initiate development of the gonadal ridge, but their presence is necessary for normal development of gonadal components. The primordial germ cells and the proliferation of the coelomic epithelium constitute the epithelial nucleus whose growth forms the gonadal thickening beside the middle kidney. The thickening forms a ridge that elongates cranially and caudally.

With the elongation, the cells form cords. Contributions from nephrotomic tissue form (1) gonadal mesenchyme and (2) ducts that join the gonad to the kidney. These are distinct gonadal primordia that are identical in both sexes (indifferent, bipotential gonads). The gonads typically divide into a peripheral cortex and a central medulla (though some cyclostomes and teleosts lack a medulla), with peripheral germ cells. Finally, sexual differentiation occurs, characterized in the female by development of the cortex and involution of the medulla, and in the male by involution of the cortex, proliferation of the medulla, and migration of the germ cells in the medulla. Genetic sex and internal and external environment determine the direction of sexual differentiation.

## 2. Structure and Function of the Kidney

The functional unit of the kidney is the nephron. Typically it consists of a tubular structure with one end in association with a capillary network and the other end connected with a drainage duct. While each unit has as its function the removal of chemical waste and the maintenance of water balance, the units are modified according to developmental, physiological, and ecological constraints.

In its primitive state, the vertebrate kidney is thought to have consisted of single nephrons derived from each segment of nephrotome and retaining a connection to the coelom as well as to a draining duct. This archinephric or holonephric condition is approached in the early embryology of hagfish and caecilians but is not functional even in those forms at that stage.

### a. The Pronephros

In other vertebrates, the earliest nephric units develop only from the anterior nephrotomes. Of those that do develop, only a few are functional. The kidney thus derived is a *pronephros* and occurs during development in all vertebrates. The histological appearance of a functional pronephros is one of unorganized tubules in a roughly oval form covered by connective tissue. Each tubule has an opening into the coelom, the *nephrostome,* which indicates the segmental nature of the tubules (see fig. 12.1). Ciliated nephrostomial tubules extend from the coelomic

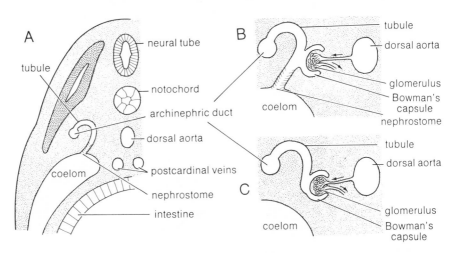

Fig. 12.1. Evolution and development of the nephron. *A,* the preglomerular condition is that of a nephrotomic tubule connecting the archinephric duct to the coelom. The coelomic part of the duct is ciliated. *B,* a functional pronephros is characterized by an opening to the coelom (the nephrostome), a glomerulus and capsule, and regional differentiation of the tubule into a partially ciliated proximal element and a nonciliated distal portion. The glomerulus and capsule form as an outpocketing of coelomic epithelium, which is invaded by blood vessels from the dorsal aorta; the blood vessels then organize into a capillary mass, the glomerulus. A portion of the wall of the nephron near the coelomic opening evaginates to form the capsule. The glomerulus is external to the tubule. *C,* a functional opisthonephric unit has lost its connection with the coelom and the glomerulus is internal to the capsule. The glomerulus is formed of capillaries from segmented renal arteries. The tubule is further differentiated regionally. The initially segmental origin of the nephric units is obscured as more tubules develop from the nephrotome.

openings and join to form a common proximal tubule. This continues as first an intermediate ciliated segment, then a nonciliated distal tubule, and finally the duct that drains all of the tubules. The presence of paired glomeruli that arise from the dorsal aorta is typical. An outpocketing of coelomic epithelium is invaded by dorsal aorta blood vessels, which organize into an elongate capillary mass, the *glomerulus.* A portion of the tubule wall near the nephrostome evaginates to approach the glomerulus, though the glomerulus remains external to the tubule.

### b. Phylogeny of the Pronephros

The trend from the presumed ancestral condition of one functional tubule per segment is one of progressive diminution in number and function of pronephric units. More pronephric rudiments develop than will become functional; the number is usually less than twelve but may be as high as fifty in caecilians. The functional pronephros is found in embryos and free-living larvae of cyclostomes, many fishes, and amphibians and has also been reported in adults of a few teleost genera. As the vertebrate series is ascended phylogenetically, the number of units that become functional decreases, and in birds and mammals the pronephros is vestigial and transitory. In dipnoans, the number of functional pronephric units is two, in anurans three. Urodeles show a predictable phylogenetic variation. The primitive salamanders, the cryptobranchids, have five functional units, the hynobiids four and three, and the more advanced salamanders two. The only caecilians that have been analyzed have eight to twelve functional units and most closely approximate the holonephros, since large numbers of nonfunctional units are present in series posterior to the functional pronephros.

### c. The Opisthonephros

The opisthonephros is evolved from the ancestral type of holonephros. It arises from nephrotomes posterior to that giving rise to the pronephros. The elimination of intermediate units, the change in time and development, and the elaboration of secondary and tertiary units and many glomeruli allows the opisthonephros to be recognized as a separate and distinct organ.

The development of opisthonephric tubules begins with a segmental solid nephrotomal outpocketing, which soon develops a lumen. One end joins the duct that had drained the pronephros, and the other end enlarges and becomes invaginated by capillaries from a segmental renal artery. This end becomes a double-walled capsule with an internal glomerulus (Bowman's capsule; the Malpighian body). Each tubule constricts behind the glomerulus to form a neck, and a midtubular constriction causes delimitation of a proximal and a distal secretory portion of the tubule. Although the tubule was initially segmental, additional tubules arise and so the metameric arrangement is obscured. Each tubule has its own glomerulus, but the collecting ends may unite to form small urinary ducts that join the nephric duct. The peritoneal funnels that empty into the coelom are usually lost in an opisthonephros.

### d. Phylogeny of the Opisthonephros

The cyclostome opisthonephros extends from the region behind the pronephros almost the length of the body and is the functional kidney of the adult. In hagfish, the pronephros persists but is not functional; it is absent in lampreys. The elasmobranch kidneys are long and strap-like (figs. 12.2C, 12.3A,B). In some forms peritoneal funnels are retained. Bony fish show more variety of gross aspect. Some primitive forms (Amia, sturgeons) retain peritoneal funnels. The kidneys of teleosts range from long, broad structures that may be fused in certain regions to short, compact, posteriorly positioned organs. Some marine fish lack glomeruli. The amphibian opisthonephros extends much of the length of the coelom. Some adult frogs retain ciliated nephrostomes, but these are associated with renal veins. Various modifications take place in the structure and function of anterior nephric units, especially of male amphibians. The anterior kidney tubules are utilized in sperm transport. In caecilians, the gross kidney shape and tubule structure and function are not affected by the acquisition of this function (fig. 12.2B). In most salamanders of both sexes the anterior kidney is thinner, and the urinary

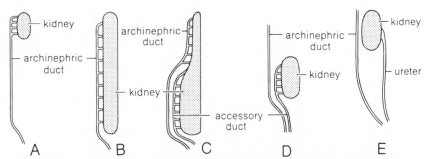

Fig. 12.2.  The evolution of the relationship of the kidney to the archinephric duct. *A*, the functional pronephric kidney of some adult teleosts is a compact anterior body connected to the archinephric duct by a number of small ducts. *B*, the opisthonephros of sturgeons, gars, and embyronic caecilians is an elongate body connected to the archinephric duct throughout its length. *C*, the opistho-nephros of some sharks and salamanders is constricted anteriorly. The anterior portion of the kidney is drained by the archinephric duct, the posterior (func-tional) part by an accessory duct. *D*, the opisthonephros of rays and frogs has lost the anterior portion. The mesonephric kidney is drained solely by the ac-cessory duct, and the kidney has no connection to the archinephric duct. *E*, the amniote metanephric kidney is drained by the ureter, which develops from a "ureteric bud" near the cloaca. See text for the evolutionary and functional implications of these changes.

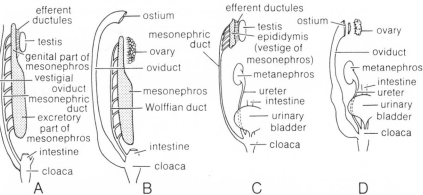

Fig. 12.3.  The urogenital systems of male and female anamniotes and amni-otes. *A*, male elasmobranch or amphibian; the mesonephros is differentiated into anterior genital and posterior excretory portions; the genital part is con-nected with the testis by means of the efferent ductules, which are outgrowths from the mesonephros; the mesonephric or archinephric duct serves as both genital and excretory duct; the oviduct or Mullerian duct is vestigial. *B*, female elasmobranch or amphibian; the ovary is not connected with the mesonephros; the mesonephros and mesonephric duct serve only excretory functions; the oviduct is well developed and opens into the coelom by the ostium near the ovary. *C*, male reptile or bird; the excretory part of the mesonephros has disappeared, but the genital part persists as the epididymis (in part), which is connected, as in anamniotes, with the testis by means of the efferent ductules; the mesonephric duct is purely genital and is renamed the deferent duct; the excretory function is served by metanephroi and ureters. *D*, the female reptile or bird; the mesonephros and mesonephric duct have entirely vanished; the condition of the ovary and oviduct is the same as in anamniotes; the excretory function is served by the metanephroi and ureters, exactly as in the male.

function of the tubules is reduced or lost, especially in males (fig. 12.2*C*). The entire opisthonephros of frogs is confined to the posterior part of the abdominal cavity, and there are no grossly marked anterior or posterior regions (fig. 12.2*D*). In males, certain anterior tubules function solely to transport sperm.

The amniote opisthonephros is functional during embryonic development. It may be designated a mesonephros because (1) it is the "middle phase" kidney of a pronephros, mesonephros, metanephros succession, and (2) it develops from the extensive region of the intermediate cell mass behind the pronephros but does *not* incorporate the most posterior region from which the metanephros develops, although adult opisthonephros does. The manner of development and differentiation of the amniote mesonephros is the same as that of the anamniote opisthonephros.

### e. The Metanephros

Reptiles, birds, and mammals are characterized by the development of a third, posteriormost kidney that is functional throughout adult life (figs. 12.2*E,* 12.3C,D). It consists of renal corpuscles and secretory tubules that develop from the continuous nephrogenous tissue. It lacks nephrostomes. One end of a developing tubule expands and incorporates a glomerulus from a renal artery branch. The other end contacts a collecting tubule that is developed from a diverticulum of the duct that drains the kidney. The pronephros and the mesonephros disintegrate as the metanephros becomes functional. In birds, and particularly in mammals, the tubule differentiates more elaborately than in the mesonephros so that a proximal convoluted tubule, an elongate central portion (the *loop of Henle*), and a distal convoluted tubule are formed. The tubules are numerous—between 1.3 and 4 million in the adult human. Arterial blood, then, circulates through the kidney so that nitrogenous wastes are removed, ionic balance effected, and water balance maintained.

### f. Phylogeny of the Metanephros

The kidneys of reptiles are small and compact (except in snakes and legless lizards, in which they are elongate) and usually have lobate surfaces. The posterior ends may be fused. Birds have highly lobed pelvic kidneys, and their posterior ends are frequently fused. Mammalian kidneys may be lobed superficially (as in seals, whales, artiodactyls, some carnivores, and some primates including human fetuses), or they may have the adult human compact, bean-shaped appearance. The human kidney may be used to illustrate the organization of the mammalian kidney. A connective tissue capsule covers the kidney. The cortex, containing the renal corpuscles and convoluted tubules, lies below the capsule and intimately above the medulla. One or more *pyramids,* comprised of loops of Henle and collecting ducts, are bounded by renal columns of connective tissue. The pyramid and its cortex cap constitute a kidney lobe. The tip of the pyramid extends into a minor calyx, which broadens and may fuse with others to form a major calyx. This structure opens into the cavity or pelvis of the kidney. The calyces and pelvis are derived from duct diverticula.

## g. The Environment and Tubule Morphology

Analysis of kidney tubule morphology has been used to support the
theory that vertebrates originated in fresh water and, conversely, to sub-
stantiate hypotheses of marine origin. The argument for fresh water
origins is as follows: cyclostomes and elasmobranchs have very
large renal corpuscles and rather short tubules; a large amount of
filtrate is produced. Amphibians and most teleosts have well-developed
corpuscles and tubules of variable length, but usually the tubules have
well-developed proximal convoluted segments. Again, a large amount of
filtrate is generated. Reptiles and marine teleosts, however, have reduced
or lost the corpuscle and drastically shortened the tubule (in addition,
marine teleosts have lost the distal convoluted tubule). Filtrate production is very low. Birds and mammals have large corpuscles but have
interjected the long, thin loop of Henle into the convoluted tubule. The
loop resorbs much water, so filtrate production is low (see fig. 12.4).

The primitive tubule type has been assumed by some to be a cyclo-
stome-freshwater fish type, since many of the primitive vertebrates still
inhabit fresh water. Since fresh water is more dilute than body fluids, the
osmotic gradient would tend to overdilute body fluids and eventually
cause death. The large corpuscle would filter great volumes of water
and maintain normal fluid concentrations, so it is suggested that the

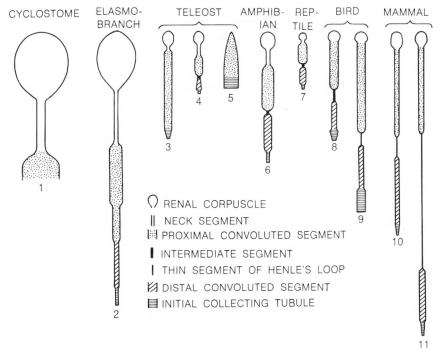

Fig. 12.4. Schematic representation of the nephron of various vertebrates:
cyclostome *Bdellostoma stouti* (1), elasmobranch (2), teleost *Myoxocephalus
octodecimspinosus* (3), teleost *Ameitrus nebulosus* (4), teleost *Opsanus tau*
(5), amphibian *Rana catesbeiana* (6), reptile *Chrysemys marginata* (7), bird
*Gallus domesticus* (8), bird *Gallus domesticus* (9), and mammal *Lepus cunicu-
lus* (rabbit, 10). From Prosser and Brown 1961 (*Comparative animal physiology*. Philadelphia: W.B. Saunders Co.)

glomerulus arose in fresh water for those purposes. In the marine environment, the high salt concentration would cause loss of body water to the environment, so the loss of the glomerulus by marine fish would prevent expression of filtrate. Thus water would be conserved, and excess salts would be secreted by the gills. The pattern of secondary loss of glomeruli, rather than secondary acquisition, is considered to be substantiated by the case of sharks. These marine animals have large glomeruli and manufacture much filtrate, but resolve the water loss problem by carrying urea in the blood to maintain a high internal concentration without affecting salt balance. Sharks and marine teleosts are thus concluded to have evolved two different means of dealing with the osmotic problem presented by the invasion of the marine environment.

There is evidence, however, that vertebrate origins were in a marine environment. The earliest vertebrate fossils are found in Ordovician deposits that may have been marine, and a number of early fossils of the same or very similar species are found in widely scattered localities, indicating an extensive migration possible only in the oceans. The kidney morphology can be interpreted as marine in origin. Other phyla have ducts opening to the coelom, as do embryonic vertebrates. These allow gametes to escape and may have served that function in prevertebrates. The development of a glomerulus is dependent only on blood pressure factors and has essentially no connection with osmoregulation. The mechanism and origin from coelomic ducts are the same as those of decapod crustaceans, whose marine origins are unquestioned. The kidney may have served initially to eliminate urea, rather than water. With a glomerulus present and with potential for water control, marine vertebrates were able to invade estuaries, and eventually freshwater rivers. Ostracoderms were apparently estuarine, and lampreys repeat this invasion at each breeding time. Both lampreys and hagfish have well-developed glomeruli that are compatible with marine life. Further, hagfish blood is isotonic with sea water, and there is no reason to think that hagfish ever had blood of lower osmotic pressure, or that their ancestors were freshwater forms.

Kidney morphology, then, offers no real clue to vertebrate origins and may be interpreted in the light of either theory. Fossil evidence may settle the question of the site of vertebrate origin, but with little or no likelihood that evidence will be found of the kidney morphology in the ancestral vertebrate, the nature of the kidney in that form remains hypothetical.

## 3. The Ovary

The female gonad, the ovary, has two essential functions: it produces ova, and it produces the female hormones estrogen and progesterone. The former function is, of course, vital to the maintenance of the species; the latter is vital to the maintenance of the individual  as well as the species. The hormones maintain ova production in the reproductive adult, and in many forms that retain young the hormones "condition" the female for that retention. Further, the hormones influence the behavior of the organism and the condition of bones, muscle, skin, viscera, and nerves throughout life.

New ova are formed through the mitotic divisions of primordial germ cells, which give rise to oogonia. The oogonia become surrounded by a layer of epithelial cells that is derived from the covering of the embryonic germinal ridge. The covering is the *primary follicle*. Many primary follicles degenerate; when one does develop further, the follicle cells divide and form several layers (the *secondary follicle*) and the oogonium enlarges and is termed a *primary oocyte*. The oocyte undergoes the first meiotic division, and a *secondary oocyte* and a small polar body are formed. In the majority of vertebrate species for which information is available, the secondary oocyte is ovulated and fertilization is required to initiate the completion (final division) of meiosis and to effect maturity of the ovum. The follicle ruptures to allow extrusion of the ovum (ovulation) into the coelomic cavity or the ovarian cavity of bony fish (see fig. 12.5 below). In some or all species in each vertebrate class

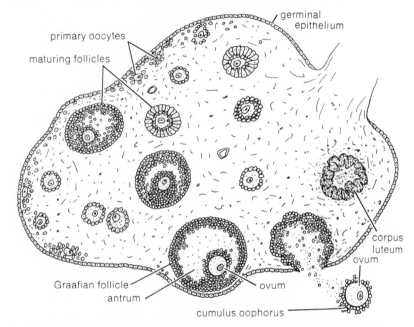

Fig. 12.5. The mammalian ovary. The section shows several stages of development of follicles and ova in the rat kidney. Peripherally are seen primary oocytes; a number of maturing follicles are also visible. One mature follicle is seen. This follicle, called a Graafian follicle, is characterized by the presence of a fluid-filled internal space, the antrum. The ovum is surrounded by follicle cells and these cells comprise the cumulus oophorus. At ovulation, the Graafian follicle and the adjacent ovarian wall break down, extruding the cumulus oophorus with the ovum into the coelom. Following ovulation, the remaining cells reorganize to form a corpus luteum. The preovulatory follicle produces the estrogenic hormones; the corpus luteum produces progesterones. This cycle is mediated by pituitary hormonal feedback. The pituitary secretes a follicle-stimulating hormone (FSH), which stimulates follicular maturation; as the estrogen titer increases, FSH production decreases and the pituitary secretes luteinizing hormone (LH). Ovulation follows follicular maturation, and LH stimulates corpus luteum formation. As the corpus luteum secretes more and more progesterones, LH secretion diminishes. FSH secretion resumes, the corpus luteum degenerates, another follicle begins to mature, and a new follicular cycle ensues.

except birds, the follicle cells reorganize or are invaded by other cells to form *corpora lutea*. The mammalian corpus luteum is comprised of follicular cells that modify to produce the hormone progesterone, which conditions the female for retention of a fertilized ovum. The "corpora lutea" of other forms have been demonstrated in some species to have that function, but in others the "corpora lutea" have been demonstrated *not* to have hormonal function, or evidence for such function is equivocal.

Ovaries are collections of follicles with a connective tissue cover, and range from having hollow centers to having masses of tissue stroma. The size of the ovary varies primarily with the size of the ovum produced, and, in most vertebrate species, with the season of the year and the correlated reproductive state.

*Phylogeny of the Ovary*

Cyclostomes have a middorsal single ovary (representing a fused pair, though with dominance of the left in hagfish) that extends the length of the body cavity. The ovary is a mass of follicles that are of uniform size. The caudal portion in lampreys may contain male gonadal remnants; hagfishes are hermaphrodites in which the anterior part of the gonad is ovarian, and the posterior portion is testicular. Only one region matures and is functional, however.

Among elasmobranchs, it is generally true that egg-laying species have paired large, functional ovaries, while viviparous sharks have a functional right ovary and viviparous rays a functional left one. Typically the ovary is a collection of small follicles, "corpora lutea," and blood vessels, with little stroma (fig. 12.6C). There is evidence for progesterone secretion from the "corpora lutea" of viviparous elasmobranchs. The medulla of the ovary is often replaced by hematopoetic tissue. Ova rupture through the ventrolateral surface of the somewhat elongate ovary or through an opening at the anterior pole.

In many teleosts, the ovaries are paired, though some have a single functional ovary. The ovaries are hollow, formed by folding of the germinal ridge (fig. 12.6A,B). Ovulation is into the ovarian cavity. In some live-bearing forms, fertilization occurs in the follicle, and development of the embryo takes place there or in the ovarian cavity. In some forms, extensions of the ovarian wall penetrate the mouth or gill cavities of the embryo and effect nutrition and gaseous exchange.

The amphibian ovaries are paired, hollow organs (fig. 12.6C). They are primarily cortex, though the internal lining is of medullary origin. Follicles project into the cavity as they mature, but ovulation occurs through the ventrolateral wall anteriorly to the coelom. Some amphibians, particularly caecilians, have postovulatory "corpora lutea." Hormonal function has been documented in the frog *Nectophrynoides occidentalis* and the salamander *Salamandra atra* and is associated with the development of oviducal young in these live-bearing species. The size of the ovaries is related to the general body form of the amphibian—frogs have massive ovaries that fill the compact coelomic region, caecilians have thin elongate ovaries, and salamanders have slightly elongate ovaries.

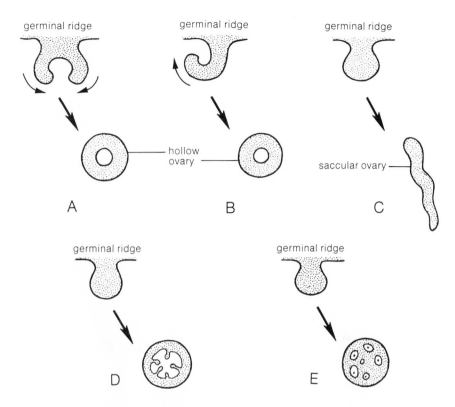

Fig. 12.6. The phylogeny of ovaries and their development. Development of the ovaries begins with a proliferation of the tissue of the germinal ridge. In teleost fish, the germinal ridge tissue forms paired outgrowths that join distally, as in *A*, or grows and folds back, as in *B*; in both types a hollow ovary results. *C*, in elasmobranchs and amphibians, the germinal ridge tissue proliferates as a mass. The mass loses its connection to the body wall and a space forms within the mass. The saccular ovary with oocytes embedded in its wall results. *D*, in reptiles and birds, a saccular ovary develops in a manner similar to amphibians, but the ovary is less distended as a sac and the ova protrude on pedicels into the ovarian lumen. *E*, the mammalian ovary also develops as a mass of germinal tissue but remains cellular throughout rather than becoming sac-like.

Reptiles have paired compact ovaries that are irregularly ovoid in shape. The medulla has many fluid-filled lacunae into which maturing, stalked follicles project (fig. 12.6*D*). The cortex has little stroma but contains many small follicles. "Corpora lutea" form after ovulation in viviparous forms, and there is considerable evidence for their role in maintaining internally developing young. As in some fish and amphibians, "corpora lutea" form in certain oviparous species as well. Their function is not well known.

The majority of bird species have only the left ovary developed and functional; exceptions are found among the Falconiformes. The cortex has little stroma but many pedunculate follicles. The ovary often looks like a bunch of grapes because of the presence of many follicles and ova at different developmental stages. The follicles are highly vascular, and ovulation is through a nonvascular strip, the stigma. Many ova are resorbed; few are ovulated.

Mammals have bilateral ovaries that are highly vascular and have a well-demarked cortex, which contains follicles, and a medulla. Mammals are distinguished by their highly differentiated follicles of several layers and by the presence in the follicle of a fluid-filled space into which the ovum projects (figs. 12.5A,B, 12.6E). Most follicles degenerate, but periodically a number are ovulated. In several species, copulation is the stimulus for ovulation. After ovulation, the follicle differentiates to form the corpus luteum. The hormonal function of the corpus luteum has been demonstrated in many mammals. If pregnancy does not occur, the corpus luteum degenerates; during pregnancy, however, the corpus luteum persists through part or all of the gestation period. Mammalian ovaries are usually small and compact and located near the dorsal wall of the pelvis. In monotremes, the left ovary is better developed than the right; in other groups the members of the pair are roughly equal in size.

4. The Testis

The vertebrate testis is a compact organ that consists primarily of numbers of *seminiferous tubules* (fig. 12.7). The tubules are comprised of numerous dividing sex cells, the *spermatogonia,* and fewer, large, supportive and nutrient cells, the *Sertoli cells.* The testis also produces the male hormones, or androgens. In lower vertebrates, specialized hormone-producing cells are variously reported to be obscure, absent, or present only at certain times. In mammals, however, masses of interstitial cells have been identified as the endocrine testis component.

The testes are usually located dorsally in the abdominal cavity, though in many mammals they descend outside of the peritoneal cavity into a special sac, the scrotum, carrying ducts, blood vessels, and nerves with them. The testes of monotremes, some insectivores, whales, hyrax, elephants, dugongs, sloths, armadillos, and some seals are permanently in the abdomen; the testes of some insectivores, all bats, most rodents, lagomorphs, some ungulates, and some carnivores are periodically withdrawn from the scrotum into the abdomen; the testes of most primates, most ungulates, most carnivores, and most marsupials are permanently scrotal. The scrotum, when present, serves as a temperature regulator, since temperatures below that of the core of the body are required for normal spermatogenesis. In homeotherms that lack a scrotum, there is still provision for spermatogenesis at reduced temperatures. In birds, active spermatogenesis takes place in the lower temperatures of night, or of a cool season of the year. Mammals that lack a scrotum often have the testes covered by a reduced skin layer, or they begin spermatogenesis during a cool time of the year.

Most vertebrates reproduce seasonally, and testis morphology reflects the seasonal changes. Testes generally are largest just before breeding, and after sperm are discharged, they are very reduced. Histologically, active spermatogenesis is indicated by well-developed seminiferous tubules with nests of sperm at a single or several developmental stages. The resting phase after breeding is represented by the presence of large spermatogonia on the tubule walls, without divisional stages or large masses of interstitial tissue.

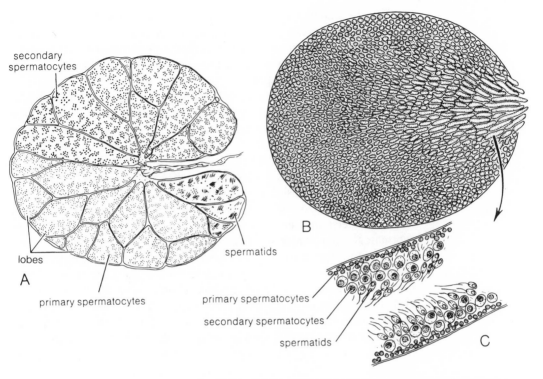

Fig. 12.7. Vertebrate testis morphology. *A*, in the amphibian testis spermatogenesis characteristically occurs in "locules" within the lobes of the testis. Each
locule incorporates the spermatocytes of a "cell nest," which is a clump of
spermatocytes that is the result of localized mitotic activity. The spermatocytes
in each locule are all at the same stage of development; as seen here, the locules
contain primary spermatocytes, secondary spermatocytes, or spermatids. Interstitial tissue is present and produces androgenic hormones. *B*, the mammalian testis is a collection of seminiferous tubules in interstitial tissue. Spermatogenesis occurs in the tubules. *C*, during active spermatogenesis, spermatocytes
and primary spermatogonia are found at the periphery of each tubule, secondary spermatocytes more medially, and maturing spermatids at the lumen of the
tubule. Amphibian spermatogenesis is characterized by synchronous maturation of the spermatocytes in a locule or tubule, while mammalian spermatogenesis shows sequential maturation with nearly all stages present in a tubule.

### Phylogeny of the Testes

Lampreys have a single, dorsal, lobular testis. Each lobule contains many
ampullae, which break down to discharge sperm to the coelom. The
testicular portion of the hagfish gonad is similar in morphology to that
of the lamprey. Elasmobranchs have paired dorsal testes, usually in close
association with paired lymphomyeloid epigonal organs (see chap. 11).
The testes contain several regions characterized by ampullae with spermatogenesis at several specific stages. Interstitial cells that produce androgens have been identified.

Bony fish usually have paired, dorsal testes. Two kinds of arrangements for spermatogenesis are found in the teleost testis. In the radial
arrangement, the testis is formed of tubules with spermatogonia at their
ends. The spermatocytes develop in clusters, and the clusters in a tubule

may be at different stages of development. In the acinous type of testis, spermatogonia develop in a small chamber bounded by a fibrous tissue membrane. All spermatogonia are at the same developmental stage in the chamber, and at maturation the connective tissue membrane disintegrates to release the sperm.

Amphibian testes are paired and are lobed to a greater or lesser degree. The lobed condition is most marked in some urodeles, in which the number of lobes increases with age. Conversely, numerous lobes seem to be a primitive condition in caecilians; an evolutionary trend toward fusion of lobes is evident. Lobes in frogs are not marked and are held together by a fibrous tissue sheath. The lobes are masses of seminiferous tubules, usually of considerable length. In salamanders, all of the developing sperm in a region of the testis are at the same maturation phase. In frogs and caecilians, cell nests or spermatocysts are formed as the primary spermatogonium divides mitotically to form numerous secondary spermatogonia. With this division, a membrane is formed around the secondary spermatogonia. The cells within a nest are therefore at the same maturation phase and nests in the same testis subunit are at the same phase in cyclically (seasonally) breeding frogs (fig. 12.7A). However, cell nests are in different maturation phases in tropical, acyclic frogs and in all caecilians, whether cyclic breeders or not. In cyclically breeding amphibians, the gonad size varies with season, that is, stage of spermatogenesis. The fat bodies, derived from the germinal ridge, also show variation in size according to season, as well as nutritional state, and may constitute a source of energy utilized during gametogenesis. Interstitial cells have been found in amphibian testes, and androgen production has been demonstrated for some forms.

Reptilian testes are paired, compact, oval structures composed of long seminiferous tubules and interstitial cells that secrete androgens. The pattern of spermatogenesis is similar to that of mammals. During maturation phases layers of cells form around the lumen of the tubule, with mature spermatozoa entering the lumen (fig. 12.7B). In reptiles that reproduce seasonally, testis size and rate of spermatogenesis are variable according to the time of the year in relation to breeding. Bird testes are round or oval, compact structures, and usually the left testis is larger than the right. Very long seminiferous tubules (to 250 m) and androgen-producing interstitial cells comprise the mass of the testis. In birds with a limited breeding period, testes enlarge greatly as spermatogenesis rate approaches maturation for the breeding season. Spermatogenesis has been demonstrated to be stimulated by an increase in the hours of daylight. However, no size variation occurs in forms with testes that are functional year round, such as in the domestic chicken. Bird testes are maintained at a temperature about equal to the core-body temperature and do not require a cooler temperature as do mammals. However, higher temperatures do impair spermatogenesis. Temperatures below core-body temperature in experimental situations do not reduce spermatogenesis and may, in fact, accelerate it.

Mammalian testes are internally lobulate, and each lobule contains several long, convoluted seminiferous tubules. Spermatogenesis follows the characteristic amniote pattern with each successive phase of sperm maturation apparent as a layer of spermatogonia around the tubule wall, progressing to mature sperm in the tubule lumen. Numerous interstitial cells are found among the tubules, and their hormonal function is well known. Testes vary little in size according to season in species with permanently scrotal testes, but the forms in which testes descend at breeding time show considerable enlargement just prior to breeding.

## 5. The Urogenital Ducts

The evolution of the ducts that transport the products of kidneys and gonads out of the body reveals the history of mechanisms for increasing the effectiveness of reproduction. Two kinds of ducts are involved: those that drain kidneys and those that drain the gonads. The development of each must be understood in order to appreciate the evolution of structure and function of these ducts among vertebrates.

### a. The Archinephric Duct

The pronephros, the embryonic head kidney, is drained by a duct that develops from fusion of the ends of the pronephric tubules. The duct grows posteriorly, utilizing nephrogenic tissue, and the ends of slightly more posterior tubules join the duct. The duct finally meets and opens into the cloaca. This duct has many names: because it is the duct that drains the primitive kidney condition, the archinephros or holonephros, it is called the *archinephric* duct; it is also called the Wolffian duct for its discoverer, and the pro-, meso-, or opisthonephric duct for its association with the embryonic or adult kidney type. For consistency, it will be called the archinephric duct here.

Cyclostomes demonstrate the primitive condition. The archinephric duct transports kidney products exclusively (fig. 12.8A). The gonads do not have ducts, but eggs and sperm are shed into the coelom and then into the aquatic environment via abdominal pores. This mechanism does not result in efficient fertilization. In males of all other vertebrate forms, the archinephric duct has been adapted for service in sperm transport. In fact, the duct has lost urinary function in many forms. The primitive gnathostome condition is one in which connections between sperm tubules of the testis and the tubules of the kidney have formed, and the archinephric duct transports both urine and sperm. This condition is seen in gar pikes, sturgeons, and caecilians (fig. 12.8B). The next step in this pattern results in the takeover by the testis of the anterior part of the kidney. This condition shows considerable variation in morphology. The primitive state is one in which testis tubules drain into kidney tubules and from there to the archinephric duct, as in gars, and others; but in *Necturus* the kidney tubules have reduced or lost glomeruli and do not function in urine production. In some sharks and urodeles, the anterior kidney is very reduced in size and has few tubules; they function only to carry sperm (fig. 12.8C). Finally, in many sharks, anurans, and urodeles, the anterior kidney is lost and testis ducts join the archinephric ducts directly (fig. 12.8D). In fact, in all sharks and skates, the archinephric duct transports sperm exclusively, and a network of tubules drains the posterior kidney. These tubules unite near the cloaca to form a common duct. This pattern is also seen in urodeles and anurans, but more often the archinephric duct transports sperm as well as urine from the remain-

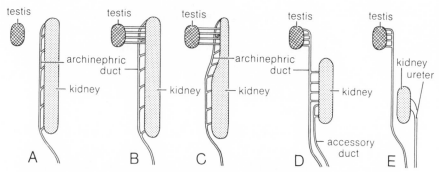

Fig. 12.8. The evolution of the relationship of the testis to the archinephric duct and the kidney. *A*, the testis of the cyclostomes emits spermatids into the coelomic cavity. The sperm are released from the coelom through abdominal pores to effect external fertilization. *B*, in gars, sturgeons, and caecilians, the testis is connected to the anterior part of the kidney by a system of ducts. That part of the kidney is partly functional in urinary production. Sperm are transported through kidney ducts to the archinephric duct, which then transports both sperm and kidney products to the cloaca. *C*, in certain salamanders, the testis is connected by ducts to the anterior part of the kidney, which is not functional, except to conduct sperm to the archinephric duct. The duct transports sperm and kidney products to the cloaca. *D*, in many sharks, frogs, and advanced salamanders, the anterior "sexual" part of the kidney has been lost, so the ducts from the testis join the archinephric duct directly. The duct carries both sperm and kidney products. The kidney is also drained by an accessory duct. *E*, in amniotes, the testis ducts join the archinephric duct directly and the duct transports only sperm. The ducts from the testis that joined the kidney or the archinephric duct are called vasa efferentia and the archinephric duct is the vas deferens. The kidney is evacuated by the ureter. The trend then is toward the assumption of sperm transport by the archinephric duct, with kidney transport assumed by other structures.

ing anterior or middle part of the opisthonephros, and new ducts are formed to drain the posterior kidney independently. These ducts are called ureters in both elasmobranchs and amphibians but are not homologous to amniote ureters, which have very different origins.

Dipnoans have concentrated their testis-kidney connections to the posterior part of the kidney, and the archinephric duct drains urine primarily. Teleosts advance that pattern, and the result is the acquisition of a new duct, the *vas deferens,* to transport sperm; the archinephric duct transports urine only.

In all amniotes, the archinephric duct is not utilized for urine transport, but is used exclusively for sperm conduction (fig. 12.8*E*). It does retain one important influence on the kidney early in development. As in all vertebrates, the archinephric duct is the inducer of the meso- or opisthonephric tubules and persists as the urine transport duct for those tubules. In amniotes the duct induces the metanephros from posterior nephrogenic tissue. This takes place before gonad differentiation and establishment of transport connections. The duct that drains the metanephros, the ureter, is the result of anterior growth of a bud that develops where the archinephric duct joins the cloaca. The anterior terminal tubules of the developing ureter form the calyces and collecting ducts of the metanephros.

The anterior end of the archinephric duct, which joins the testis tubules directly, is modified to form an *epididymus* in elasmobranchs and amniotes. It is a highly convoluted part of the duct that lies close to the testis and receives the connecting tubules that run from testis (the rete testis) to join the former kidney tubules (ducti efferenti). As it passes posteriorly to join the cloaca or urethra, the duct (now called the ductus deferens) may widen posteriorly to form a cavity for sperm storage before fertilization is attempted.

The modification of the archinephric duct is not so marked in female vertebrates. The gonads are not "competing" with the kidneys for use of the duct. In all anamniote forms, the duct persists for transport of urine, though in chondrichthyes, anurans, and urodeles, accessory "ureters" form to aid the draining of the posterior kidney. The duct degenerates in amniotes at the time of formation of the metanephric ureter, though some remnants are associated with the ovary.

### b. The Oviduct

A new duct developed to transport, and often maintain, ova (fig. 12.9) arises in different ways in different groups, except in cyclostomes, which lack such a duct. In elasmobranchs and salamanders, the archinephric duct splits and the oviduct (or Müllerian duct) is formed in parallel. Elasmobranch oviducts use one of the pronephric tubules to form the anterior end of the duct, which opens into the coelom. In other vertebrates, except teleosts, the peritoneum over the anterior end of the developing mid-kidney invaginates. The lips of the groove close and the tube then grows posteriorly until it joins the cloaca. The anterior end remains open to the coelom. The duct thus formed is present in both males and females, but in males it usually degenerates. It persists, however, as a thin longitudinal thread in many amphibians (Bidder's duct) and some fish, and in caecilians its posterior end is highly glandular and may secrete material for aid in sperm maintenance. In teleosts with saccular ovaries, the oviducts are extensions of the ovary and are not homologous with oviducts of other vertebrates. In other teleosts the duct is not continuous with the ovary, but its homologies too are questionable. Some teleosts conduct eggs via abdominal pores. Dipnoi, however, have oviducts that are homologues of typical vertebrate oviducts.

The elasmobranch oviduct (fig. 12.9B) shows a number of specializations. The oviducts are fused anteriorly so that a single opening to the coelom, the *ostium tubae,* is present. Each oviduct has an anterior enlargement, the *shell gland.* It is best developed in oviparous forms and secretes the horny egg case that protects the eggs. The posterior one-third to one-half of each oviduct is enlarged to form a uterus. The uteri are best developed in ovoviviparous and viviparous forms, and the walls of the duct may develop elongate folds to provide surface area for gaseous exchange when embryos are developing. The uteri join the cloaca.

Teleost oviducts are often fused for most of their lengths. The tube terminates in a genital pore or in a urogenital papilla, since teleosts lack cloacae. The papilla of some forms is elongate and acts as an ovipositor. In most ovoviviparous and viviparous forms, the development of

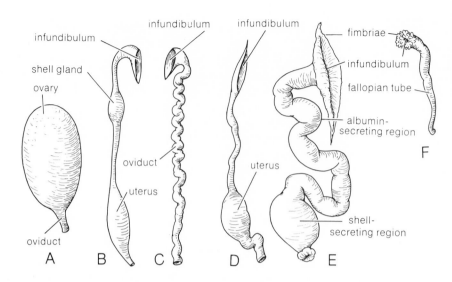

Fig. 12.9. The phylogeny of vertebrate oviducts. *A*, in many teleosts the oviducts are extensions of the ovaries and are not homologous with the oviducts
of other vertebrates. *B*, in sharks and skates the oviduct (Mullerian duct)
forms parallel to the archinephric duct. The duct has an open infundibulum,
which receives ova, and shows regional specialization for an anterior "shell
gland" and a posterior enlarged uterus. It opens into the cloaca. *C*, in amphibians and other vertebrates, the duct forms by the invagination of the peritoneum near the developing kidney. A groove forms, its lips close, and the tube
thus formed grows posteriorly to the cloaca. In some frogs and salamanders
the oviduct enlarges and becomes highly convoluted at the time of egg transport. *D*, the oviducts of reptiles show considerable regionalization. Anterior
regions are albumen-secreting glands and posterior regions shell-secreting
glands. As the ovum passes posteriorly through the duct, it is coated with these
materials. Some species retain the eggs in an enlarged posterior uterus for all
or part of the development of the young. *E*, in most birds, only the left oviduct
is functional. The duct is highly regionalized, and the ovum receives some five
or six different secretions of nutrient or fluid material and of enclosing membranes. *F*, the mammalian duct has an anterior fimbriated infundibulum and
distinct internal cellular regions. Its terminal portion is dilated to form the
uterus, and the terminal portions of the paired ducts are often fused to form
the uterus. The uterus opens externally via the vagina. There is no common
cloaca; urinary, intestinal, and genital openings are separate.

young takes place in the oviduct, rather than in the ovarian cavity,
though they are continuous. The oviduct lining, too, is often highly modified to accommodate the young.

The ducts in amphibians are paired and extend from near the anterior
end of the ovaries to the cloaca, which they usually join independently
(fig. 12.9*C*). Occasionally they reach the cloaca as a fused median unit,
as in some frogs. The ostium tubae of most frogs and urodeles has a
slightly expanded infundibulum. In members of those groups the oviducts become expanded and coiled at breeding season. The oviducts of
caecilians enlarge but do not become convoluted. The posterior ends of
the oviducts are often slightly enlarged, and in anurans and urodeles
act primarily to store ova before egg-laying. The ducts are also equipped

with glands that secrete a jelly-like coat over the descending eggs. In *Salamandra atra, S. salamandra,* the toad *Nectophrynoides,* the frog *Eleutherodactylus jasperi,* and possibly other frogs with direct development, fertilized eggs are retained in oviducts until miniature young are born. Many caecilians, perhaps the majority of species, retain young in the oviducts until they are fully metamorphosed. The oviduct linings are equipped with glands that produce a fatty secretion that is ingested by the young. The oviducts are highly vascularized and the gills of the embryos effect gaseous exchange when appressed to the oviduct wall.

The paired oviducts of reptiles have distinct regions. The ostium tubae are large and receive ova from the coelom (fig. 12.9*D*). The upper part of the oviduct of all reptiles except lizards and snakes has glands that secrete albumen around each ovum. Ova traverse the duct by ciliary action and pass the posteriorly located shell gland, which secretes a thin leathery shell. The oviducts independently join the cloaca. The oviducts enlarge and coil as breeding season approaches. Some snakes and lizards are ovoviviparous and viviparous. Developing embryos remain in the oviduct, and some forms derive nutriment from the maternal female. Among the viviparous forms, *yolk sac placentas* (in natricine snakes, for example) and *allantoic placentas* (in scincid, xantusiid, and gekkonid lizards) are known to occur (see chap. 4).

In all birds except raptors, the right ovary and oviduct are degenerate and only the left is functional. The regions of the long convoluted duct are more specialized than in reptiles (fig. 12.9*E*). A large infundibulum engulfs the ova. Anteriorly the glandular region of the duct secretes a thin layer of albumen, the chalaziferous layer. A lower part of that region secretes a thicker albumen layer around the chalaziferous layer. The ovum then enters the isthmus of the duct, where inner and outer membranes are deposited around the coated ovum. The posteriormost expansion of the duct is a "uterus" or shell gland, where the calcareous outer shell is deposited. Fluid albumen enters the egg in both the isthmus and the shell gland. The egg then passes through a short vagina, which secretes mucus over the egg, into the cloaca, from which it is laid. Fertilization takes place high in the oviduct, so that all layers can be deposited over a developing zygote that has a large yolk sac.

The oviducts of mammals are also differentiated into discrete regions (fig. 12.10*A*). Fertilization takes place in the anterior part of the ducts. In monotremes, paired ducts are present but only the left is functional. The anterior part is a thin Fallopian tube (named for an early describer), which joins an expanded uterus. Albumen is deposited in the Fallopian tube, and the shell in the uterus, where the egg more than doubles its size. The two uteri are separate and open into a urogenital sinus that leads to the cloaca. Marsupials have a similar morphology. Oviducts are paired, and paired vaginae open into the urogenital sinus. In some forms such as the kangaroo, the vaginae fuse and form a sinus that extends posteriorly as a blind sac. If the sac opens into the urogenital sinus, it is called a third vagina, and developing young pass through it into the urogenital sinus (see fig. 12.13*B*). All eutherian mammals have a median vagina, the result of fusion of the pair. The uterine portions of the oviducts are fused to various degrees (fig. 12.10*B–E*). The most primi-

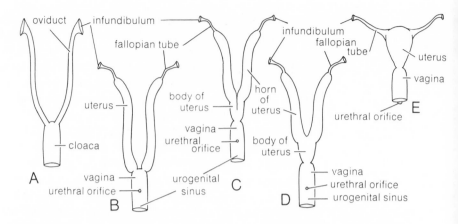

Fig. 12.10. Types of mammalian oviducts. *A*, condition found in the majority of female vertebrates; the two oviducts are completely separate and open independently into the cloaca. *B–E*, various conditions found in mammals, showing differentiation of the oviducts into uterine tube, uterus, and vagina, and progressive fusion of the lower parts of the oviducts: duplex type (*B*), found in rodents, in which the two vaginae are fused; bipartite type (*C*), occurring in carnivores, the vaginae are fused and the lower parts of the two uteri are fused to form a single body, divided in two by a partition that represents the fused walls of the two uteri; the upper parts of the two uteri remain separate as the horns; bicornuate type (*D*), found in many ungulates, similar to *C* except that the partition has disappeared; simplex type (*E*), occurring in man and the apes, in which both vaginae and uteri are fused along their entire lengths, leaving only the uterine tubes separate. Note that in *B–D* the urethra joins the vagina to form the urogenital sinus, which opens to the exterior while in *E* the urethra and vagina are wholly separate and open independently to the exterior. After Hyman, 1942.

tive condition is found in the aardvark, elephant, some bats, and many rodents; the two uteri have separate cervices (a slightly telescoped terminal uterine region) that enter the vagina. The two sides of the uterus fuse posteriorly and enter the vagina via a common cervix in the *bipartite* uterus of carnivores, lagomorphs, some bats and rodents, and some ungulates. A greater degree of fusion is found in the *bicornuate* uterus of most ungulates, some carnivores and bats, insectivores, and whales. In the *simplex* uterus, the fusion is complete and a single median uterus is found in armadillos, apes, and man. The oviducts or Fallopian tubes remain unfused and anteriorly terminate in an infundibulum. This usually opens to the coelom very close to the ovary, and in rats and mice it forms a closed capsule over the ovary. Thus it is only in these forms that ova do not pass briefly through the coelomic space. Posteriorly the vagina may join the urethra to form a urogenital sinus. In other forms the urethra and vagina have separate openings.

   Histologically the Fallopian tubes are composed of an outer serous layer, a middle muscular layer, and an inner epithelium that has ciliated and glandular cells. The uterus is similarly structured, but the inner layer, the *endometrium,* is particularly complex. It is highly vascularized and has tubular uterine glands. The condition of the endometrium varies according to the hormonal balance of the female, and in higher primates menstruation, a sloughing of the superficial endometrium, ensues in response to low hormone titer. The vaginal epithelium also changes according to the stage of the sexual cycle. The endometrium provides the conditions for implantation of the fertilized ovum, and, with the extraembryonic membranes, modifies to form the placenta, the organ of fetal nutrition and gaseous exchange.

In summary, the urogenital ducts have evolved as mechanisms for transport of gametes and of nitrogenous waste and water. Their phylogeny reflects the evolution of such aspects as water economy, efficiency of fertilization, egg protection, and protection and nurture of developing young.

**6. The Cloaca**    A cloaca (common chamber through which kidney, intestinal, and sometimes genital products exit) occurs in all vertebrates at some time during development. The intestine and urogenital ducts open into the cloaca (fig. 12.11A–D). It cannot be stated whether the ancestral vertebrate

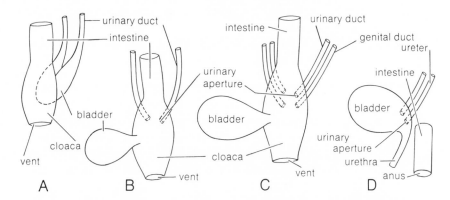

Fig. 12.11. The evolution of the cloaca and the bladder in vertebrates. The cloaca is a chamber that receives kidney and intestinal products and genital products in most groups. It is present only embryonically in cyclostomes, mammals, and many teleosts. A, a cloaca is small in embryonic teleosts. The cloaca in males receives the urinary ducts and the intestine, but is later lost. The bladder, when present, is a dilation of the terminal ends of the urinary ducts. B–C, in elasmobranchs, amphibians, reptiles, and birds the cloaca receives both urinary and genital ducts. The bladder, when present, is a ventral outgrowth of the cloaca. In many species the cloaca is partitioned into discrete regions, separating somewhat the orifices of the structures that open into the cloaca. (B represents a condition in which genital ducts fuse with urinary ducts, which then enter the cloaca.) D, separate partitioning of the duct orifices is achieved in mammals. The cloaca is absent. Ureters open into the bladder, and the genital ducts of males open into the urethra. The urinary, genital, and intestinal orifices of females are all separate and external.

had a cloaca or not, or whether its absence in adult cyclostomes is primitive or derived. It is often suggested that the kidney ducts originally opened directly to the exterior and that a cloaca is not an original vertebrate characteristic. In cyclostomes the urinary ducts open on a urogenital papilla behind the anus. Those who postulate the lack of a cloaca as a primitive feature may not have been aware that cyclostomes and teleosts have a cloaca early in embryonic development, but that it is shortly subdivided and lost. This supports the idea that the presence of a cloaca is indeed primitive.

In male sharks and amphibians the cloaca receives only the urinary ducts, but in the females both oviducts and urinary ducts enter it. However, the males often have rudimentary oviducts. Some female salamanders have a specialized portion of the cloaca, the *spermatheca,* in which sperm packets deposited by the male and picked up by the female are stored. Fertilization then takes place in the cloaca, when eggs pass through and meet the sperm.

In many vertebrates, especially reptiles and birds, the cloaca is more or less distinctly divided by folds into the *coprodaeum,* which receives the intestine; the *urodaeum,* or urogenital sinus, which receives the urinary and genital ducts; and the *proctodaeum* or terminal portion, formed of the ectodermal proctodaeal invagination and opening via the anus. In male reptiles and birds the cloaca receives the ducti deferenti from the testes and the ureters; in females, the oviducts and the ureters open into the cloaca (see fig. 12.9). In many fishes and in amphibians, reptiles, and the embryos of birds and mammals the urinary bladder (or the allantois) opens into the urodaeum.

The cloaca of mammals shows structural variation. A cloaca is present in monotremes, but a fold marks off the urodeal from the coprodeal region. The completion of such a fold in placental mammals separates the cloaca into two parts, each with an opening to the exterior. The dorsal part includes the rectum, which opens through the anus; the ventral part is the urogenital sinus or urogenital canal, which receives the stalk of the bladder, the urethra, and the urinary and genital ducts (fig. 12.11*C*). The division of the cloaca is incomplete in many marsupials and some other mammals; a shallow cloaca, one that corresponds to the proctodaeum, remains. The ureters open into the urodeal cloaca in the embryos of placental mammals but subsequently shift to open into the urinary bladder. Thus finally the ureters open into the bladder while the deferent ducts in males or the vagina in females unite with the urethra to form a common tube or chamber, the urogenital sinus or canal, which opens to the exterior in front of the anus by a urogenital aperture. In all placental mammals the male ducts join the urethra, and hence the sperm must, for a short distance, use a passage in common with the urine. Although this passage is very much shorter in amniotes than in anamniotes, the male amniote never achieves complete separation of urinary and genital paths. This has, however, been accomplished by the females of the "highest" placental mammals (order Primata, including man, apes, and monkeys), in which the urethra is separated from the vagina by another cloacal fold and the urinary and genital ducts have separate external openings.

## 7. The Urinary Bladder

Two distinct kinds of bladders occur among vertebrates: the mesodermal, essentially noncloacal bladder of elasmobranchs, holocephalans, and most teleost fish; and the endodermal, cloacal outpocket of amphibians, most reptiles, a few birds, and mammals. Cloacal bladders typically have an inner lining of transitional epithelium, a vascularized loose connective tissue, and layers of smooth muscle, bounded by a serous epithelium. Several vertebrates lack a bladder entirely, including hagfish, some teleosts, some reptiles, and nearly all birds.

Male elasmobranchs and holocephalans have ureters with enlarged cloacal ends that usually fuse to form a bilobed bladder. Females of these forms have bladders derived from expansions of the cloacal ends of the archinephric duct remnants. Teleost bladders are primarily swellings of the fused ends of the archinephric ducts, but in some forms there appears to be some contribution from the dorsal (urodeal) part of the cloaca. Such bladders are highly variable in size and shape, but all drain via a tube to an opening behind the anus.

The amphibian bladder is a simple midventral cloacal evagination (ig. 12.11$C$). It is usually a single anteriorly directed lobe, but in some forms, including most caecilians, it is bilobed. The kidney ducts open into the urodeal cloaca, and urine is forced into the bladder by gravity and the contraction of the cloacal wall.

In amniotes, the same cloacal evagination occurs. However, the outpocketing occurs before the embryo has formed a ventral body, so the pocket expands into the extraembryonic coelom. Urine is forced into the cavity from the cloaca. The fluid inflates the sac and presses it against the chorionic embryonic membrane. The evagination is the important allantoic sac (see chap. 4). As the body wall closes, the part of the allantois outside the wall is sloughed off. In some forms, including most turtles, lizards, and mammals, the part remaining inside the body wall enlarges to form the adult urinary bladder. Some aquatic turtles have additional paired cloacal outpocketings, called accessory bladders, which are apparently important respiratory structures.

In placental mammals, the degree of development of the allantois varies with the degree of efficiency of the placenta as an excretory structure. The allantois is sloughed with the placenta at birth. The allantoic stalk develops within the body wall as the urinary bladder. Some contribution of tissue from the cloacal wall is included. The enlargement of the bladder, the realignment of the ureters on the bladder, and the constriction of the embryonic cloaca to form the urethra give rise to the bladder associations of the adult. Urine flows directly from ureters into the bladder, a condition unique in mammals since a cloaca is not utilized.

In snakes, crocodiles, a few lizards and turtles, and most birds, none of the allantois survives the body-wall closure, and these forms lack bladders as adults. These animals conserve body fluids rather than eliminating them as a property of osmoregulation. Concentrated uric acid is the nitrogenous waste product in the absence of provision of water for dilution of ammonia or urea.

| | |
|---|---|
| 8. The Accessory Glands | A number of accessory glands in both males and females have been discussed in terms of their functional association with the genital ducts. The phylogeny of these glands provides further insight into the evolu- |

tion of vertebrate urogenital form and function. In addition, glands not yet discussed are associated with urogenital structures or are derived from presumptive urogenital tissue.

Most accessory glands in males are associated with provision of fluid for sperm transport or with mechanisms of sexual attraction. The *seminal vesicle* of mammals is a glandular outpocketing of the ductus deferens near the juncture with the urethra. The structure secretes seminal fluid. "Seminal vesicle" is a misnomer, since the vesicle of elasmobranchs, some teleosts, and anurans is a dilation of the end of the ductus deferens used for temporary sperm storage. It is therefore the homologue of the mammalian ampulla, rather than of the mammalian seminal vesicle. The "seminal vesicle" of caecilians is the dilated, glandular posterior end of the Müllerian duct. Its condition varies according to season, and it apparently secretes fluid during active spermatogenesis. It is not homologous to other "seminal vesicles."

In reptiles, some of the posterior kidney tubules are modified to produce an albuminous fluid for sperm transport. Birds lack accessory glands. Mammals have fluid-secreting glands not found in other vertebartes. An ampullary gland may join the lower end of the ductus deferens. Several glands contribute fluid to the urethra; the *prostate* gland (a lobular, multiducted gland) the *bulbourethral* (Cowper's) glands (which secrete an alkaline fluid), and the *urethral* (Littre's) glands (which secrete mucus) provide fluids that have specific functions in sperm transport and maintenance. Among non-fluid-producing glands in males are those associated with the cloaca in many animals. In many urodeles, special cloacal glands secrete the jelly that forms the base of the spermatophore. In reptiles, cloacal scent glands are used for attracting and marking females and possibly for marking the territory inhabited. In many mammals lacking a cloaca, anal musk glands serve those purposes.

Most of the accessory glands of females have been discussed previously. In fish and amphibians, the glands of the oviduct secrete fluid or jelly to protect the egg, and many fish secrete material that makes the egg adhesive. The shell gland of the lower oviduct in elasmobranchs, reptiles, birds, and monotremes secretes a horny material that may be impregnated with mineral salts. In viviparous forms in all classes in which they occur, the glandular oviducal wall provides for nutrition of young. The glands may secrete a "uterine milk" (elasmobranchs, teleosts, caecilians, some reptiles) or contribute to the formation of an allantoic, yolk sac, or chorionic placenta (elasmobranchs, teleosts, some reptiles, mammals). The modifications of the bird oviduct with its specialized glands that secrete albumen and membranes have been described. Female mammals have several glands that are homologous to the male fluid-secreting glands. These are reduced in size and produce a lubricant fluid during sexual excitation. They open into the vestibule and include *Bartholin's* glands (homologous to the male bulbourethral glands), the *paraurethral* glands (prostate homologues), and *vestibular* glands (homologous to the glands of Littre). The accessory glands of females, then, are associated with egg protection, nurture of young, and, in a few, with facilitation of copulation, and mate attraction.

## The Adrenals

Adrenal glands have special association with the urogenital system. Much of the tissue of these glands is derived from the cells that give rise to the kidney tubules; hence these specialized glands are considered at this point. The organization of the adrenal glands shows an interesting evolutionary pattern that is reflected phylogenetically among vertebrates. The mesoblast, or interrenal component, secretes the "cortical" hormones. The second kind of tissue forming the adrenal gland is derived from neural crest tissue. This latter kind of adrenal tissue, the chromaffin cells, secretes adrenalin. The adrenal, with an interrenal cortex and a chromaffin medulla, is a mammalian structure; it is the history of the association of these two parts that is of evolutionary interest.

Among cyclostomes, lampreys have lobulate masses of interrenal tissue that line the postcardinal veins and renal arteries. Chromaffin tissue exists as thin strips on the large arteries. Hagfish have a single strip of interrenal tissue on the kidney and have chromaffin tissue situated as it is in lampreys. Elasmobranchs have yellow, rodlike interrenal bodies on the posterior kidneys, and, in addition, skates have smaller interrenal bodies ascending the kidneys. Chromaffin tissue occurs as segmental bodies on aortal branches.

Interrenal bodies of teleosts are small and either free or embedded on the cranial kidney surface (forming the corpuscles of Stannius). Some may have chromaffin cells embedded in them, but most have cells in small clumps on blood vessels in the anterior part of the body.

Amphibians have the first intermingling of the two cell types. Typically, columns of interrenal tissue with chromaffin cells on their borders are scattered on the ventral surfaces of the kidneys and along blood vessel paths. Reptiles, too, show a transition in adrenal type. Boid snakes and land turtles are like the amphibians and sea turtles resemble birds. Other reptiles have elongate structures in which both tissue types intermingle. These are associated with the main blood vessels near the kidneys. Birds have paired or fused bodies in which cords of the two tissues intertwine. These structures are found near the anterior end of the kidney on each side of the vena cava.

Among mammals a different kind of association is seen. Adrenals are bean-shaped or elongate with chromaffin tissue (medulla) encapsulated by interrenal tissue (cortex). In monotremes, the two kinds of tissue intermingle at one pole of the structure. In marsupials, the adrenal is variable in position. It often has no contact with the kidney and is associated with the vena cava. In fact, the medulla of the Australian cat adrenal opens directly into the vena cava. Among eutherians, the adrenal may be far from the kidney, overlying the adrenal vein or near the vena cava, or it may form a cap on the kidney, as in primates. Some mammals have four or five large accessory adrenals that usually consist of interrenal tissue exclusively. The association of the gland with tissue of like derivation (the kidney) and with blood vessels (an advantage for an endocrine structure), as well as the degree and kind of contact of the two kinds of adrenal tissue shows an evolutionary trend in the phylogeny of living vertebrates.

**9. Modes of Reproduction**

It may be simplistic to state that reproduction is the key to maintenance of species, or indeed to the origin of species, but a discussion of the modes of reproduction in that context reveals much about the evolution of form and function in the urogenital system. The primitive condition of gamete production in order to effect fertilization involves the shedding of great quantities of ova and sperm in some proximity to each other in an aquatic environment and then abandonment of the gametes. Cyclostomes, some elasmobranchs, many teleosts, and some amphibians still practice this mode of reproduction. There are numerous dangers inherent in such a system—much energy is expended in gamete production, and, for this and other reasons, mating and gamete production occur only once in the lifetime of several species, and adults die shortly after mating. The large quantities of eggs may be only partly fertilized, since sperm are not conducted directly to them. The eggs are subject to death if their aquatic environment changes, and to predation by a variety of organisms. Typically, a larval form emerges if the egg survives. Often the larva can exploit aspects of the ecology that the adult cannot.

If this system is faulty, why has it persisted at all, and what has been derived to replace it? The huge numbers of gametes produced serve to insure that enough eggs will be fertilized to produce an adequate new generation. The fact that larvae can exploit other parts of the environment than adults can aid in their survival by reducing competition with adults of the same species. Often great numbers of larvae hatch and so the number that survives to adulthood is adequate to ensure perpetuation of the population.

A number of attempts have been made to reduce the hazards of such a system. Various means of protecting the eggs, once shed and fertilized, have evolved. Internal fertilization has been developed along many lines, often conmitant with formation of an intromittent organ. Parental care of eggs and/or young is seen. In addition, many forms have developed ways of retaining the fertilized eggs for some or all of their development, ultimately resulting in a means of maternal nutrition of developing embryos and elimination of the larval stage.

### a. Internal Fertilization and a Copulatory Mechanism

External fertilization often involves large numbers of gametes shed into water to effect gametic union. A reduction of the number of gametes required is concomitant with internal fertilization. The advantages of internal fertilization are fewer gametes are required, since sperm are conducted directly to ova, and fertilized ova can be protected within the female for a time. Cyclostomes have not effected means of internal fertilization. However, all elasmobranchs and holocephalans practice internal fertilization. The medial border of the pelvic fins of male elasmobranchs is modified to form a rolled tube, called the clasping organ (fig. 12.12A). Sperm emerge from the cloaca and flow into the tube, which is inserted into the cloaca of the female. Males of some species have the organ equipped with special glands or muscles to force materials through the tube. Sea water and sperm are then injected into the female. Holocephalans have similarly modified pelvic fins and also have anterior claspers in front of the pelvic fins and a frontal clasper on top of the head. It is not

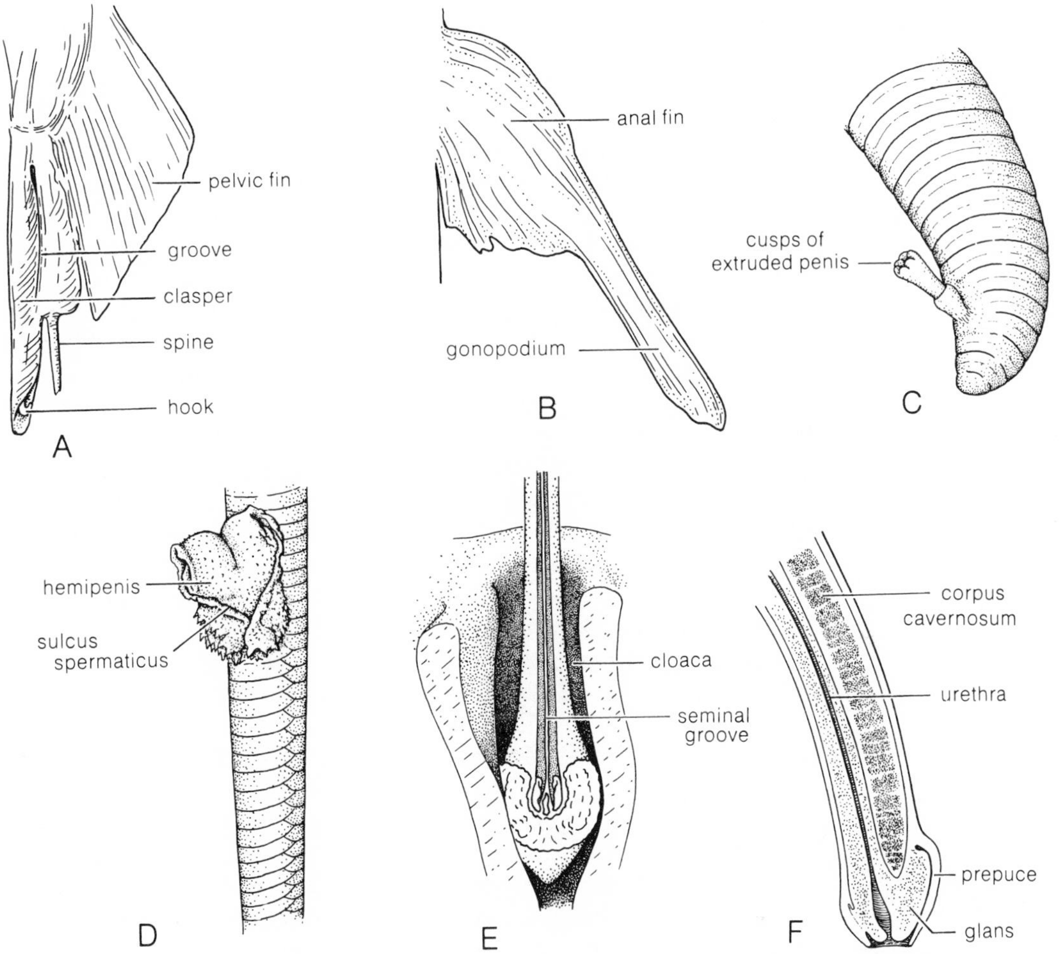

Fig. 12.12. Vertebrate intromittent organs. *A*, sharks have modified some of the rays in the pelvic fin to form an intromittent organ. The medial rays are elongated and bear a groove on the surface for sperm conduction. The tip of this "clasper" may bear a hook to facilitate positioning in the female. *B*, certain teleosts have modified the rays of the anal fin to form an elongate gonopodium. *C*, male caecilians evert the posterior part of the cloaca out through the vent. It is then inserted in the vent of the female. The shape of the everted organ is species-specific, and it "fits" the shape of the cloaca of the female. *D*, male snakes have paired hemipenes. One is inserted in the vent of the female. It is extruded by filling blood sinuses to make it erect. Hemipenis morphology is species-specific, and these structures may have ridges, spines, or knobs. An external groove conducts sperm. *E*, turtles have a single median penis derived from the cloaca. It is extruded by filling sinuses with blood. *F*, the placental penis is extracloacal in origin. It is comprised of the urethra, erectile tissues (corpus spongiosum and corpora cavernosa), nerves, blood vessels, a connective tissue sheath, and an outer layer of skin. The human penis (figured) has an aggregation of sensitive nerve endings at its tip in the glans, which is covered by a skin fold called the prepuce.

known how the latter function. Most teleosts have external fertilization, but some make use of copulatory organs formed from caudal hemal spines, anal outgrowths, or, most commonly, an elongation of the anterior border of the anal fin. Such an anal fin *gonopodium* may be inserted into or appressed to the cloaca of the female, and the sperm are directly transported to the ova (fig. 12.12*B*).

Among amphibians, several diverse means of effecting internal fertilization are utilized. *Ascaphus truei* is the only frog known to have internal fertilization, and male *Ascaphus* have a cloacal appendage that is used as an intromittent organ and appressed to the female cloaca. Nearly all other frogs practice external fertilization in water. However, a number of terrestrial frogs have direct development of young, and it is possible that some, if not all, practice internal fertilization, though the mechanism is not known. Most salamanders use a method of internal fertilization, but not that of an intromittent organ. After a courtship ritual, the male deposits a spermatophore, a jelly-like pedestal capped with a mass of sperm. The female then receives the spermatophores with the lips of her cloaca. Sperm are held in some form for varying lengths of time in a specialized cloacal chamber, the *spermatheca,* until fertilization takes place. All caecilians practice yet a third method of internal fertilization. In males the lower portion of the cloaca can be everted and extruded from the vent (fig. 12.12*C*). It is then inserted into the cloaca of the female and sperm are injected into the female. It is significant that as members of the three orders of amphibians explored the terrestrial environment in diverse ways, they developed three very different means of effecting internal fertilization—important for a totally terrestrial life.

All reptiles practice internal fertilization. The egg is fertilized high in the oviduct, then either retained during its development or laid after a leathery shell has been deposited. The significance of the development of the amniotic membranes (discussed in detail in chap. 4) cannot be overlooked. This acquisition enabled reptiles, birds, and mammals to become totally terrestrial, for they did not need to return to water during any reproductive or developmental stage of life. A mechanism had evolved that at once insured a fluid environment for the developing embryo and prevented desiccation of that environment. Internal fertilization is effected by copulation among reptiles. *Sphenodon* lacks an intromittent organ and uses cloacal apposition to transfer sperm. Two different kinds of intromittent organs are found in other reptiles. Male snakes and lizards have well-developed, paired *hemipenes;* typically one is inserted into the cloaca of the female during copulation (fig. 12.12*D*). Females, too, develop rudimentary, but nonfunctional, hemipenes. The hemipenes develop from body-wall outpocketings that lie lateral to the cloaca. Before young are born, the outpocketings are pulled inward so that the hemipenis is an inverted epidermis-lined structure that lies in highly vascularized loose connective tissue. Each male hemipenis has a pair of retractor muscles and a propulsor muscle. When the blood sinuses fill and the muscles are relaxed, the hemipenes are forcefully evaginated and pop out just above the cloaca. Each hemipenis has a groove or sulcus on its dorsal surface that is continuous with the cloaca and conducts sperm directly. The hemipenis may also have scales or spines on the distal end that stimulate the female, dilate her cloaca, and anchor the hemipenis. After copulation, the blood sinuses drain and the retractor muscles invert the hemipenes.

Turtles and crocodiles have a single median penis that develops from the tissue of the cloaca (fig. 12.12*E*). A central pad forms that is underlain by paired *corpora fibrosa,* which fuse to form a distal glans about a medial groove. Paired *corpora cavernosa* lie above the corpora fibrosa and, when filled with blood from the walls of a tube, extend the penis from the cloaca. Retractor muscles withdraw the penis when it resumes its flaccid state.

Birds, too, have internal fertilization and are the only class that lays eggs exclusively. Kinds of nests, nesting behavior, brooding, and care of young range from minimal to elaborately expressed mechanisms. Courtship behavior, as well, may be highly involved, culminating in cloacal apposition to effect copulation. Ratites, ducks, and a few other birds, however, have a well-developed penis. It has the same basic structure as that of the turtle but is more elaborate; it has an eversible blind sac that lengthens the erected organ and a well-formed spiral seminal groove.

All mammals utilize internal fertilization, and all copulate by means of introduction of the male penis into the female vagina. The monotreme penis is much like that of the turtle in development. It arises from ventral cloacal tissue and is surrounded by erectile tissue, the *corpus spongiosum.* The dorsal groove has become a closed tube, and the glans is surrounded by a skin fold, the *prepuce.* The platypus glans is split at the tip and covered with spines; the *Echidna* glans is two double-lobed knobs. The sperm canal carries only sperm (the urinary apparatus has a separate canal into the cloaca) and opens on the glans via several tubules. These forms lay eggs, which are cared for by the parents. Maintenance of hatched young will be discussed below.

In placental mammals, the rectum is partitioned from the urogenital apparatus. The penis and its female homologue, the *clitoris,* and their accessory structures, develop from tissue ridges beside the embryonic cloaca and are therefore extracloacal. This pattern resembles lizard-snake origin, rather than turtle-monotreme. The ridges, or urogenital folds, grow rapidly in the male, meet in midline, elongate, and acquire a lumen. Growth of the ridges is slower in the female and they meet only incompletely. The clitoris is the result of midline fusion; the *labia majora* and *minora* represent the unfused folds. In placental mammals the lumen of the penis, the urethra, transports both urine and sperm; muscular contraction blocks the urinary passage above the entrance of the sperm duct during copulation. The penis is comprised of the urethra, its coat of erectile corpus spongiosum, paired erectile corpora cavernosa, nerves, blood vessels, a connective tissue sheath, and an outer layer of skin (fig. 12.12*F*). In marsupials, the scrotum, containing the testes, is anterior to the penis (fig. 12.13*B*). The penis is anterior, and usually directed forward, in eutherian mammals that have a scrotum. In many mammals, a bone develops in the septum between the corpora cavernosa in both penis and clitoris. It increases the rigidity of the organ, especially the penis, during erection. The glans, the distally expanded corpus spongiosum, is highly variable in shape and structure and is often

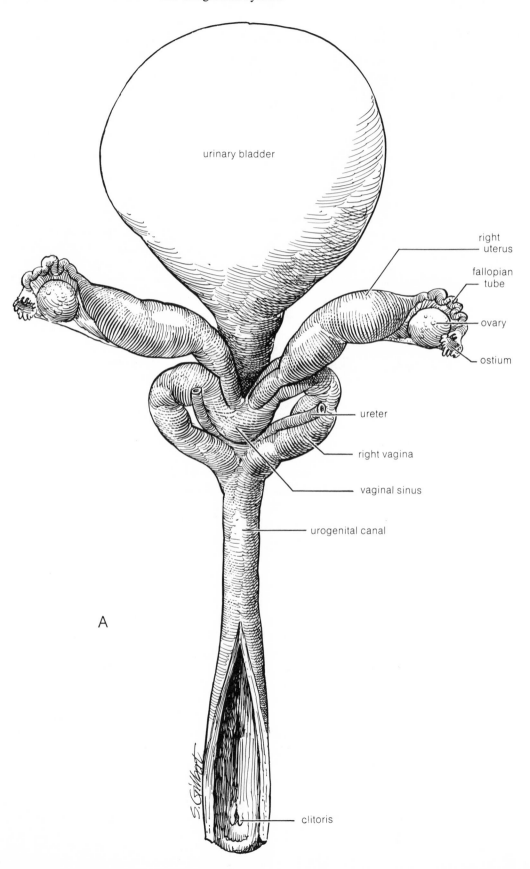

A

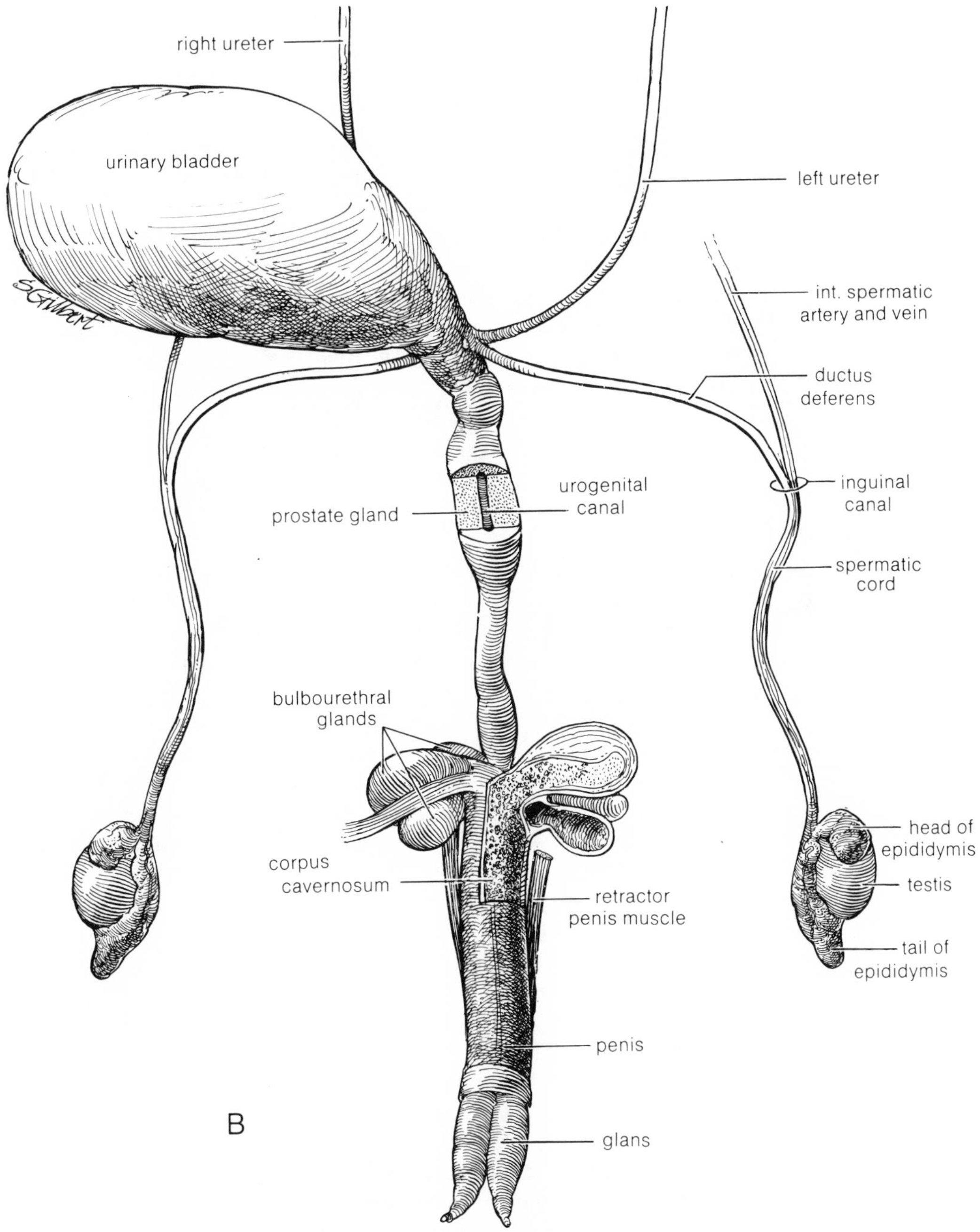

Fig. 12.13. Urogenital systems of marsupials. *A*, female opossum showing inward bend of vagina; *B*, male opossum, showing bulbourethral glands and forked penis.

equipped with spines, filaments, and the like. These are suspected to stimulate the female during copulation.

### b. The Evolution of Parental Care

Among forms (representing every class of vertebrates) that lay eggs, a great variety of ways of protecting the eggs is found. The most primitive kind of protection may be the act of seeking a sheltered place in which to lay the eggs, as some cyclostomes do. Oviparous elasmobranchs lay eggs that are enclosed in a horny egg case (one to four ova per case) secreted by the oviducal shell gland. A shell is transitory during the development of fertilized ova in ovoviviparous sharks, skates, and rays. Holocephalans also produce single ova that are encased in a horny structure. Such a case would protect the egg against predation and may aid in maintaining osmoregulation.

Teleosts protect their eggs in many ways. Some construct elaborate nests; in some species one of the parents guards and aerates the eggs until they hatch. In other forms, a parent picks up fertilized eggs and holds them in its mouth, not eating until hatchlings emerge. The eggs themselves may be floating (pelagic) or sunken (demersal) and may have adhesive areas. This would provide advantage in dispersal (or lack of it) and avoidance of predation in the adult habitat.

Amphibians, too, protect eggs. Frogs can build very elaborate nests such as the hylid foam nests from which tadpoles drop into water. Parental care exists in many forms; primitive pipid frogs seize eggs after they are fertilized and seal them to the parental back, where they are carried until hatching. Some hylids (*Gastrotheca*) and others carry the eggs in pouches or exposed on the back. The male *Rhinoderma darwini* carries eggs and developing young in its vocal sac. Certain female salamanders and caecilians protect their clutches by coiling around the egg masses.

The reptilian egg is also protected by a leathery outer layer that aids in withstanding desiccation. Birds all lay eggs that are protected by the calcareous outer shell. Further, in reptiles and particularly birds, the range of nest-building and parental care runs from none of either to very elaborate nests and brooding of the eggs. The trek of sea turtles to ancestral breeding sites is well known. The eggs are buried in sand which retains moisture and a constant temperature. Newly hatched young seek the sea and then return, years later, to the same breeding sites.

Monotremes are the only mammals that lay eggs. The embryo is fairly well developed by the time the calcareous shell is secreted and—uniquely —growth continues after the shell is formed. Hatching takes place two weeks after the eggs are laid. The mother does not leave the nest during that time.

An important trend in reduction of the number of eggs with increased protection of each egg is seen. Fish lay thousands of eggs and these eggs are essentially unprotected; monotremes usually lay two eggs that are well-protected structurally, physiologically, and physically by a parent.

### c. The Evolution of Viviparity

Parental protection of eggs in nests, internal incubation chambers, and so on, has been alluded to. Parental care of the young after hatching or birth is found in nearly all classes of vertebrates, and, further, some members of nearly all classes retain eggs in the female oviduct for varying lengths of time. Many forms lay eggs in which embryos are in advanced stages of development; other forms retain the eggs until "hatching" occurs, and larvae or miniature adults are "born." In some species, the hatched young are retained in the oviduct for a period of time. Still other forms provide oviducal secretions to nurture the developing young. Finally, species in several classes have a placenta that provides for nutrition and gaseous exchange to and from fetal and maternal bloodstreams. Those forms that lay eggs are called *oviparous*. Those that retain young but provide no nutrition in addition to the embryo's yolk are *ovoviviparous*. Some authors include those forms that provide nutriment but lack a placenta among ovoviviparous forms, while others consider any form of maternal nutrition, placental or not, to be a condition of viviparity. All species that provide any maternal nutrition will be considered *viviparous* here.

Cyclostomes do not care for the hatched young, but many elasmobranchs do provide parental care. The young often follow large objects, most frequently the mother. She can then protect and possibly guide her brood. The majority of sharks, skates, and rays retain eggs and developing young in the oviduct and miniature adults are born. These ovoviviparous forms have various rates of development. Forms of viviparity are found in some elasmobranchs. In some forms, specialized cells in the oviducal wall secrete a fatty "uterine milk," which is ingested by the developing young. The oviduct wall may have folds that have elongate villi that enter the spiracle of the embryo and directly nourish it via the esophagus (in the genus *Pteroplatea*, for example). In other forms, the highly vascularized yolk sac of the embryo effects a placental relationship with the maternal oviducal wall and exchange is effected. In *Mustelus* the oviduct wall has villi that lock with grooves on the yolk sac surface. Each embryo is in a separate oviducal compartment and has its own connection to the duct wall. In *Scoliodon* the yolk sac becomes obliterated, but the attachment remains, and so an exclusively vascular placenta and umbilical cord are effected.

Most teleosts lay eggs, but these are carried in brood pouches in some forms, such as seahorses. Other forms retain young in the oviduct (following the requisite internal fertilization) and in some cases the entire larval state is spent in the oviduct or is bypassed so that tiny adult replicas emerge from the female. Several families of cyprinodontoids are viviparous. In the Poeciliidae and Anablepidae the embryos are retained in modified ovarian follicles until birth. Unique transitory modifications develop to insure embryonic nutrition. In Poeciliidae the follicle becomes a syncytium, and the pericardial sac enlarges until it covers the head of the embryo. The follicle then develops highly vascular folds and, in contact with the pericardial sac, forms a close circulatory association. In *Anableps* the embryo hindgut expands and bulbs form on the portal system vessels, which effect the association with the follicle wall. In Goodeidae and Jenynsiidae the embryos are maintained in the cavity of the ovary. The ovarian epithelium becomes secretory and then desquamates and this material is absorbed by the embryos. Nonviable embryos autolyze and contribute to the material to be ingested. In jenynsiids ovarian vascular folds develop and overlie the embryonic gills, effecting a branchial placenta. Members of the Goodeidae lack ovarian flaps but embryos develop processes from the anus (trophotaeniae) that lie on the oviduct wall to absorb nourishment. In all of these forms the fluid surrounding the embryos is swallowed and some nutrients are ingested. Many fish show elementary forms of care of the young, such as school maintenance and defense of the area in which the young are located, for example.

Amphibians show many ways of retaining developing embryos. Young in the egg membranes are carried on the back in several forms: in *Pipa* the embryo's tail fin is highly vascularized late in development and it acts as an organ of gaseous exchange with the vascular epidermis of the back of the female. In *Gastrotheca,* embryos carried in the dorsal pouch effect a gill placenta with the interior skin ridges of the bearer. *Rhinoderma* retains developing young in the vocal sac of the male; *Nectophrynoides* carries the embryos in the female oviducts and may effect gaseous exchange via the tail fin of the embryo and the oviducal lining. *Rheobatrachus silus* broods eggs in its stomach, and fully metamorphosed froglets are "coughed up"! An interesting experiment in direct development is found in several families. Miniature froglets hatch from eggs laid on land, and the abandonment of the aquatic larval stage is an important step toward completely terrestrial adaptation.

Salamanders show less diversity in their means of protecting eggs and young. Members of the genus *Salamandra* are the only salamanders that retain developing young in their oviducts. Miniature adults are born, though often in *salamandra* and usually in *caucasica* eggs are laid or larvae of a late state emerge. Salamanders, too, are experimenting with direct development. Many species lay eggs on land, even in vegetation, and miniature adults that avoid an aquatic phase are hatched. Among caecilians that lay eggs, the female usually coils around the mass until larvae hatch and wriggle to a nearby stream. Many caecilians retain developing young in the oviducts. The larval stage is spent in the duct, and metamorphosed young are born, circumventing a free aquatic phase for larvae. Not only are the oviducts modified to produce a nutrient substance, but the developing young are equipped with a rasp-like fetal dentition that is used to scrape the oviducal wall. Gaseous exchange is effected by placement of the gills along the highly vascular duct wall. Several species of caecilians nourish young by this mechanism.

Turtles and crocodilians are oviparous, but they, too, have evolved ways of protecting eggs and young, as previously discussed. Lizards and snakes show a full range of parental care from depositing eggs and abandoning them to placental viviparity. Among ovoviviparous forms, for example some natricine snakes and anguid lizards, young are retained in the oviduct until miniature adults are born. They are dependent on their yolk for nutrition and are not thought to receive nourishment from the female. However, recent studies with labelled chemicals indicate that there may be nutrient absorption by embryos in the duct. The viviparous forms, as previously mentioned, have variously developed placentas. Those that have a yolk sac placenta show interlocking folds between chorion and endometrium, but the fetal tissue does not invade the maternal. Choriallantoic placentas, found in several lizards, involve either reduction of both maternal and fetal tissue, so that capillaries are apposed, or develop interlocking epithelial cells between fetus and parent. The young are usually able to fend for themselves when born.

Most birds have developed means of caring for eggs and hatched young that are exceeded only by higher mammals. Upon hatching, the young are only partially feathered and require a few days' feeding before they leave the nest (precocial), or they are naked and require long-term feeding and care before they can take care of themselves (altricial). Further, many parents give the young intensive training as they leave the nest, in flying, swimming, obtaining food, and defense, as necessary. In some species, birds return to a previous year's nest, and the pair bond is a stable, monogamous one, thought by some to be correlated with intensive care of the young.

All mammals except monotremes are placentate and, further, nourish the young after they are born. Marsupials have very short uterine pregnancies, but the young may be maintained for many months after they enter the pouch. Further, in some forms, a female lactating may become pregnant again, but the new embryo cannot implant in the uterus until the nursing pouch young is weaned. Marsupials usually have a yolk sac placenta. The yolk sac is large in comparison with the allantois. Only a few genera form an allantoic placenta. That placenta is an invasion of the maternal tissue by the fetal tissue with an interlocking circulation. The vitelline (yolk sac) vessels form the fetal contribution.

Eutherian mammals have placentas that are modified chorioallantoic structures. Their evolution is one of reduction of tissue layers separating maternal and fetal circulations until only the capillary endothelia maintain the discreteness of the two circulations. The morphological details of the placenta are not important here (see chap. 4); the nature of the maintenance of embryos and young is significant because of its highly derived nature. All eutherian embryos receive nutrition from the mother via the placenta; gaseous exchange is also effected. When the young are born, they are not able to fend for themselves, and, whereas birds procure food for their young, female mammals are equipped with mammary glands to provide a highly specific diet for newborns. Lactation usually continues for less than a year, and yearlings are able to find food for themselves. Mammals may spend considerable time instructing their young in getting food, in defense, and in a variety of other activities necessary for individual survival. The aggregations that mammals form are effective means of protecting and teaching young, and in highly organized social systems, characteristic of "advanced" mammals, the division of labor within the society provides a highly structured environment for long-term maintenance of the young. In fact, such a division of labor characterizes other highly evolved social systems, such as those of ants, bees, and termites. Parental care appears to be a marked correlate with complex social systems.

**B. The Urogenital System of Selachians**

The dissection directions apply to the spiny dogfishes and the skate. The dogfishes supplied for dissection are usually immature, and it is therefore difficult or impossible to locate in them all of the parts of the urogenital system. At least a few mature males and females should be on hand for demonstration. Skates are sexually mature while still relatively small.

Remove the digestive tract (if fully examined previously) except the cloaca and the liver.

1. The Female Urogenital System

*Dogfish:* The ovaries are a pair of soft, oval bodies situated dorsal to the liver, each with a mesentery, the mesovarium. In mature specimens the ovaries contain large eggs, consisting chiefly of yolk.

The kidneys are long, slender, brown bodies lying against the dorsal body wall, one to each side of the dorsal aorta; they are retroperitoneal. Free their lateral borders by slitting the pleuroperitoneum and note the thickness of the organ at various levels. The thinner anterior portion has lost its urinary function and is degenerate in females; the broader, thicker posterior part performs the work of excretion. Between the two kidneys is a tough, shiny ligament; it should not be mistaken for a duct. Both components of the adrenal complex are present in dogfishes but are not very easy to find. Notice the chromaffin masses, called *suprarenal bodies,* which are a longitudinal series of light spots near the medial border.

The oviducts in immature females are slender tubes running along the ventral face of the kidneys, without mesenteries. In mature females they are very large tubes that spring free from the kidneys by means of well-developed mesenteries, the *mesotubaria.* Trace the oviducts forward. They pass forward along the dorsal coelomic wall, curve around the anterior border of the liver, and enter the falciform ligament. Here the two oviducts are united to a common opening, the *ostium,* a wide funnel-like aperture lying in the falciform ligament, with the opening facing posteriorly into the pleuroperitoneal cavity. As already indicated, the ostium is formed by the fusion of peritoneal funnels of the pronephros. In order to find the opening it is usually necessary to separate the walls of the ostium, since they tend to adhere. Trace the oviducts posteriorly. They are narrow at first, but in mature specimens they soon present a slight enlargement, the shell gland or *nidamental gland.* In *Squalus* this gland secretes a thin membrane in which several eggs become enclosed. Posterior to this gland, the oviduct narrows again. In mature females it will enlarge greatly to form the uterus, which swings free by means of the mesotubarium; in immature females there is no uterine enlargement or any mesotubarium, but the oviducts widen slightly as they proceed posteriorly along the ventral faces of the kidneys.

Trace the oviducts to the cloaca. Cut open the cloaca by a median slit, which will open up the intestine. Note the opening of the intestine into the ventral part of the cloaca, the coprodaeum, and the slight fold that separates it from the dorsal urogenital region of the cloaca or urodaeum. Note the urinary papilla in the middorsal wall of the urodaeum. In mature specimens the large openings of the oviducts are readily seen on each side of the urinary papilla; in immature females they are in the same position but are quite small and can be found best by cutting into the posterior ends of the oviducts and probing toward the cloaca.

The kidney ducts are somewhat difficult to find in females. The duct lies along the ventral face of the kidney exactly dorsal to the oviduct in immature females, along the line of attachment of the mesotubarium in mature ones. Locate it in immature specimens by carefully stripping away the oviduct and also freeing the peritoneum from the ventral face of the kidney. The archinephric duct is a slender tube proceeding directly to the cloaca, where the two ducts join to open by the terminal pore of the urinary papilla (see fig. 12.14).

*Skate:* The ovaries, a pair of elongated, soft bodies containing large yellow eggs, are situated dorsally in the anterior half of the pleuroperitoneal cavity. The large oviducts pass dorsal to them. Follow one oviduct forward; its narrow anterior portion passes along the dorsal coelomic wall, curves around the anterior margin of the liver, and, after entering the falciform ligament, unites with its fellow at a single common opening, the ostium. This is a wide funnel-like aperture situated in the ligament and facing caudad. Trace the oviducts caudad. After a short distance they widen greatly to a conspicuous bilobed swelling, the shell or nidamental gland, which secretes the horny case in which the eggs are laid. The wide uterus continues from the shell gland and proceeds to the cloaca, supported by a thickened mesotubarium. Cut open the cloaca in the midventral line, also slitting the intestine. Note the opening of the intestine into the ventral part of the cloaca, the coprodaeum, and the conspicuous horizontal fold that separates it from the dorsal urogenital part, or urodaeum. Cut into the latter by cutting forward through this fold. The urodaeum is greatly extended and thickened craniad. Find the oviducal openings, one to each side of this thickened part of the cloaca, and the urogenital opening in the middorsal wall between them. The terminal part of the cloaca leading to the anus is the proctodaeum.

The kidney is an opisthonephros; its main part, the caudal opisthonephros, consists, in female skates, of a thick rounded lobe lying dorsad at each side of the cloaca, best revealed by stripping away the thick peritoneum that covers its ventral surface. The anterior or cranial part of the opisthonephros is nearly degenerate in females but can be found as diffuse brownish tissue extending forward ventral to the dorsal aorta. From the medial surface of the caudal opisthonephros several ducts, the accessory urinary ducts, pass anteriorly and medially in contact with the posterior cardinal vein and open into a small chamber, the urinary sinus, situated on the dorsal surface of the anterior end of the cloaca. The two urinary sinuses of the two sides unite into a common chamber, which is sometimes called the urinary bladder. It does not correspond to the bladder of higher forms, since it consists of the enlarged terminations of the mesoneph-

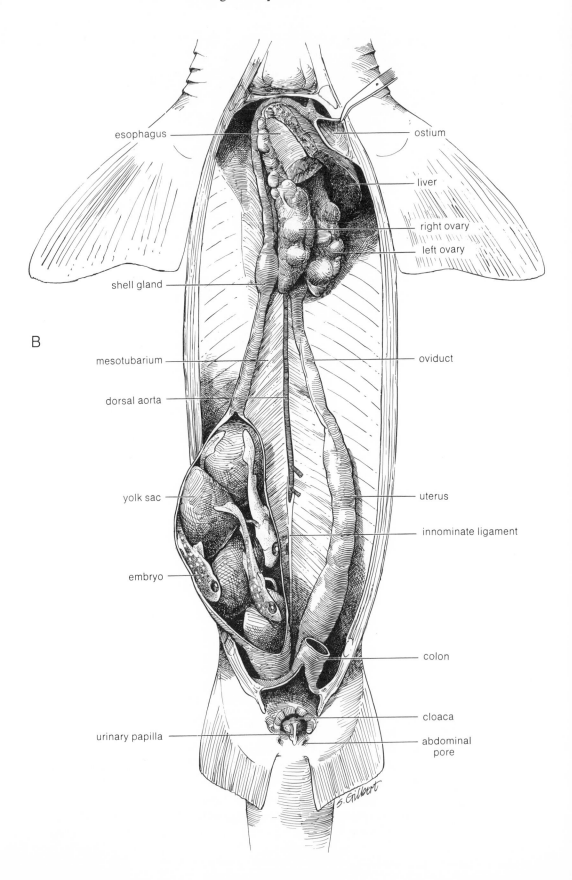

esophagus

ostium

liver

right ovary

left ovary

shell gland

B

mesotubarium

oviduct

dorsal aorta

yolk sac

uterus

innominate ligament

embryo

colon

urinary papilla

cloaca

abdominal pore

S. Gilbert

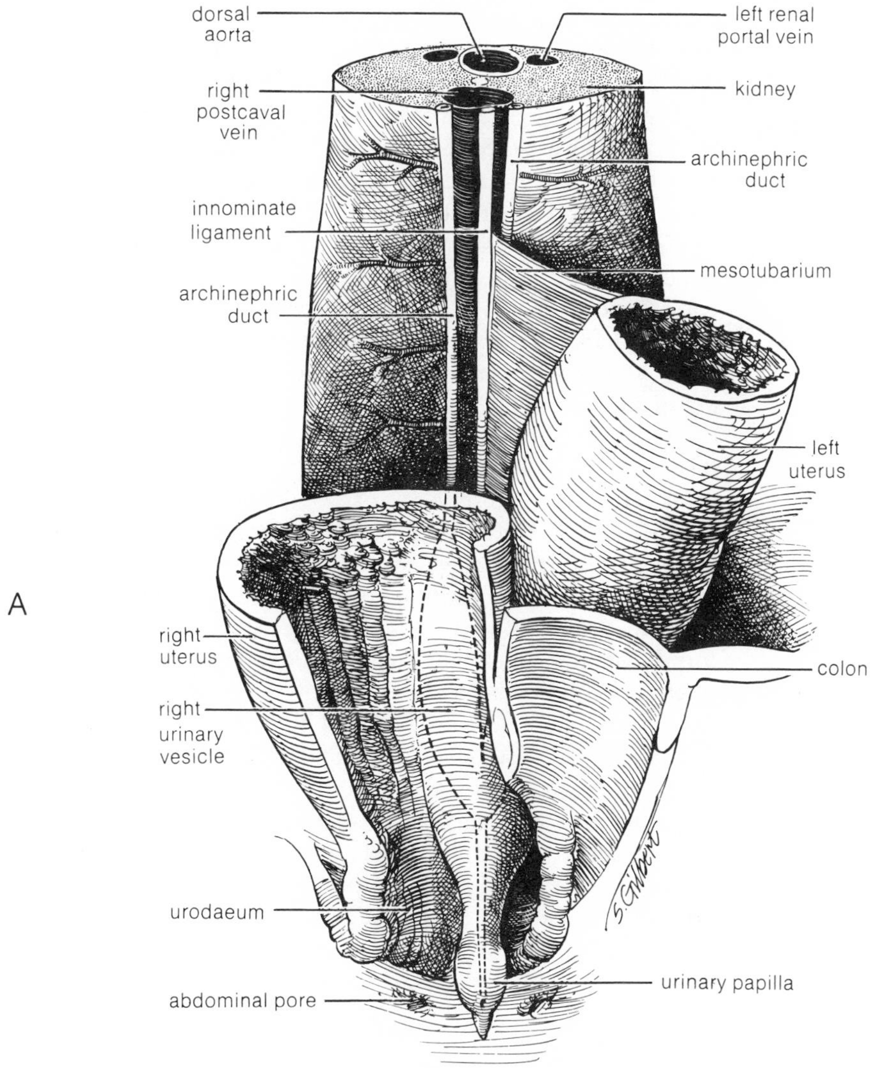

Fig. 12.14. Shark female urogenital system. Notice the relationship of ovaries to oviducts, and the ducts to the cloaca (*A*). The specimen shown was pregnant with three young with yolk-sac placentas (*B*).

ric ducts. Cut into this, note the entrance of the two urinary sinuses into it, and find in its middorsal wall the opening into the cloaca. The ducts of the cranial mesonephros are slender tubes extending anteriorly from the urinary bladder, lying on the dorsal surface of the strong white portions of the mesotubaria.

**2. The Male Urogenital System**

The testes are a pair of soft bodies dorsally situated, each with a mesorchium. In the dogfishes they are located in the anterior part of the pleuroperitoneal cavity, dorsal to the liver. In the skate they are broad, flat bodies against the dorsal wall.

In the dogfishes the kidneys are identical in both sexes and should be examined next according to the directions given for the females.

In the male skate, however, the cranial part of the kidney is very much better developed than in the female and extends forward as a firm cylindrical body on either side of the middorsal line.

As explained earlier in this chapter, the male ducts in the majority of vertebrates are the archinephric ducts. In mature males these ducts are consequently much larger than in females. The ducts run posteriorly along the ventral face of the kidneys. In immature specimens each is a slender, straight tube, similar to that of the female, but in mature males it is greatly coiled. Each testis is connected with the corresponding region of the opisthonephros by delicate ducts, the efferent ductules, which run in the mesorchium and can sometimes be seen by holding the mesorchium up to the light. They vary in number in different selachians (said to be from four to seven in *Squalus acanthias,* one in skates and rays). They connect with the tubules of the cranial part of the opisthonephros, which has practically lost its urinary function and serves as part of the male genital system, conveying the sperm into the kidney duct. Therefore in male selachians the part of the opisthonephros penetrated by the efferent ductules is an epididymis. From the tubes of the epididymis, the sperm pass into the kidney duct, now called male duct or ductus deferens, which forms a greatly coiled tube (epididymidal duct) on the ventral surface of the epididymis and on the ventral surface of *Leydig's gland.* Leydig's gland, believed to secrete a fluid beneficial to the sperm, is the part of the opisthonephros behind the epididymis. Leydig's gland has peritoneal funnels (lacking in the epididymis), although these do not connect with the ductus deferens, but lacks Bowman's capsules. As the ductus deferens approaches the caudal opisthonephros, it becomes less coiled and enlarges upon the surface of the caudal opisthonephros as a wide, straight tube, the *seminal vesicle.* Trace this to the cloaca by removing the peritoneum from the ventral face of the kidney. At its posterior end on the side of the cloaca the seminal vesicle terminates in a sac, the sperm sac, which projects craniad as a blind sac lying against the ventral surface of the seminal vesicle.

Cut open the cloaca and identify its parts as for the female. There is no difference between the male and female in the cloaca of the dogfish, but in the male skate the cloaca is very much smaller than in the female and is not divided into intestinal and urogenital parts. In the median dorsal line of the male skate there is a urogenital papilla.

The sperm sacs should now be cut open and the papillae identified where the seminal vesicles open into them. The two sperm sacs unite at their posterior ends to form a urogenital sinus, which opens at the tip of the urinary papilla.

The seminal vesicle of male *Squalus* does not receive any kidney tubules from the opisthonephros on which it lies. Instead, this part of the opisthonephros has an accessory urinary duct separate from the mesonephric duct (here seminal vesicle). This accessory duct will be found by lifting out the seminal vesicle (note the lack of tubules connecting this with the opisthonephros) and gently picking at the kidney tissue on the medial side of the groove where the vesicle lay.

The accessory urinary duct is a delicate, white tube, which, you will see, receives collecting tubes from the kidney substance at intervals. The accessory ducts are difficult to trace; they enter the sperm sacs where the minute openings occur to the medial side of the openings of the seminal vesicles. In male skates, conditions are the same as in the female (see fig. 12.15).

Look in male dogfishes for vestiges of the ostium and oviducts in the vicinity of the liver.

**C. The Urogenital System of *Necturus***

**1. The Female Urogenital System**

The ovaries have already been noted as elongated sac-like bodies bearing eggs of various sizes. Note the mesovarium. Lateral to each ovary, running along the dorsal body wall, is the oviduct, a thick, white, coiled tube supported by the mesotubarium. Follow it anteriorly. At the anterior end of the pleuroperitoneal cavity it becomes more delicate in texture and is fastened to the lateral wall. Here it has a funnel-shaped opening, the ostium; the dorsal rim of the oviduct is fastened to the body wall, but the ventral rim is free and can be lifted to expose the opening. Trace the oviducts posteriorly to the cloaca. They enter it, one on each side of the large intestine. Cut the cloaca open by a lateral slit extending up into the intestine. Note the papilla by which each oviduct opens into the cloaca. A transverse fold separates the coprodaeum from the urodaeum.

The kidneys, or opisthonephroi, are long, slender organs extending from the cloaca forward to the medial side of the oviducts and are enclosed in the mesotubarium. The kidneys of *Necturus* are thus covered on both sides by peritoneum, a rather unusual condition in vertebrates. The female kidney is divisible into a slender, anterior genital part and a thicker, caudal part. The adrenal glands consist of small bright flecks and patches along the sides of the postcaval vein. The archinephric duct lies along the lateral border of each kidney but is very delicate in females and difficult to locate. Seventy to eighty delicate collecting ducts cross from the kidney in the mesentery into the mesonephric duct. This duct proceeds to the cloaca, where it opens to the dorsal side of the oviduct, becoming imbedded in the medial wall of the oviduct. It opens into the cloaca by a minute pore on the dorsal side of the oviducal papilla. In tracing it, make a cut along one side of the cloaca, freeing the cloaca from the body wall. Note the urinary bladder extending from the midventral region of the cloaca and find its opening into the cloaca (see fig. 12.16).

**2. The Male Urogenital System**

The testes are a pair of elongated bodies situated to the sides of the small intestine, each supported by a mesorchium. Dorsal and lateral to each testis is the long, brown kidney, or opisthonephros, enclosed, as in the female, in the same mesentery that supports the mesonephric duct. The kidney, as in the female, is divisible into anterior genital and posterior urinary parts; but the former, although thin and flat, is much wider in the male. The genital part is an epididymis. Along the lateral border of the kidney runs the conspicuous coiled archinephric duct, which, as in male vertebrates in general, also acts

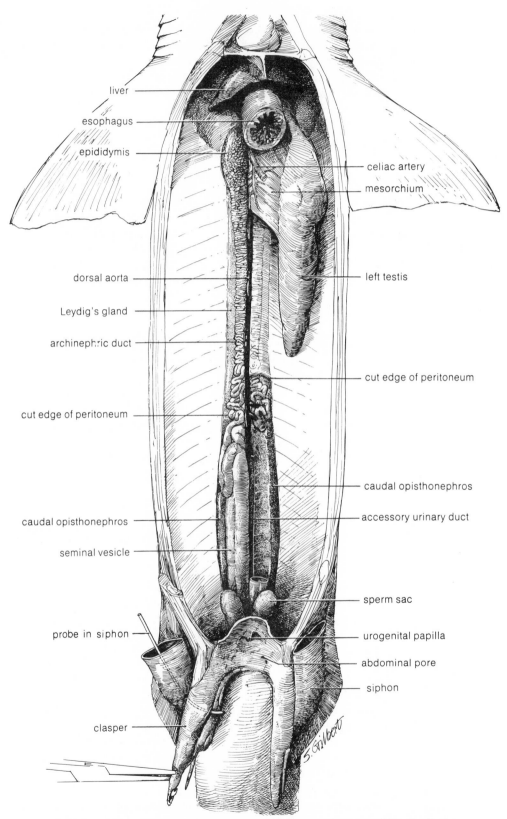

liver

esophagus

epididymis

celiac artery

mesorchium

dorsal aorta

left testis

Leydig's gland

archinephric duct

cut edge of peritoneum

cut edge of peritoneum

caudal opisthonephros

accessory urinary duct

caudal opisthonephros

seminal vesicle

sperm sac

probe in siphon

urogenital papilla

abdominal pore

siphon

clasper

S. Gilbert

Fig. 12.15. Male shark urogenital system. Note especially the ducts.

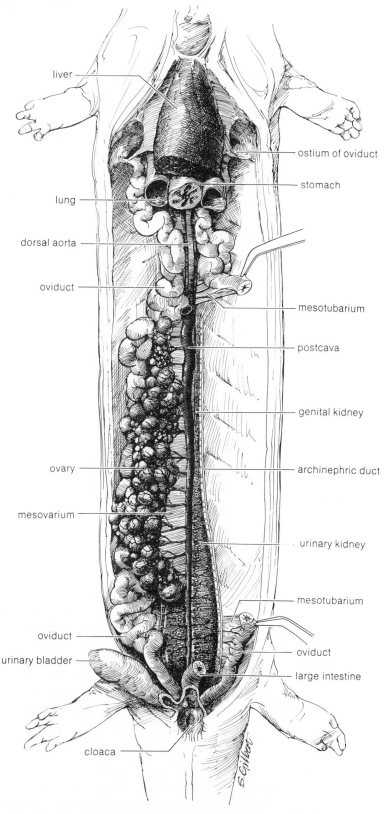

Fig. 12.16. Female *Necturus* urogenital system. Note the morphology of the ovary and oviduct, and especially the kidney and its duct.

as the male duct or ductus deferens. Two to four delicate efferent ductules cross the mesorchium from the anterior part of the testis into the genital kidney, from which about twenty-six collecting tubules run transversely into the archinephric duct. The genital kidney lacks peritoneal funnels but has renal corpuscles. Trace the mesonephric duct to the cloaca; along the caudal functional part of the kidney, the duct is narrower, loses its convolutions, and receives fifty to sixty collecting tubules from the kidney. Open the cloaca by a slit to one side of the midventral line, carrying the slit into the large intestine. Note the transverse fold dividing the cloaca into coprodaeum and urodaeum. The small openings of the archinephric ducts are difficult to find; they are located in the dorsolateral wall of the cloaca, just caudad of the fold. Note the urinary bladder and its opening into the ventral cloacal wall.

The adrenal glands are the same in the male as in the female (see fig. 12.17).

Remove the digestive tract, if this has not already been done, but leave the large intestine in place.

**D. The Urogenital System of the Turtle**

**1. The Female Urogenital System**

The female genital system consists as usual of a pair of ovaries and a pair of Müllerian ducts or oviducts. The ovaries have already been noted as large bag-like bodies in the posterior part of the pleuroperitoneal cavity. They usually contain yellow eggs in various states of development. Each ovary is supported by a mesentery, the mesovarium. Along the posterior border of each ovary runs the oviduct, a large white coiled tube, supported by the mesotubarium. Trace the oviduct forward and find the ostium, which lies in the mesentery and has wing-like borders that are generally closed together and should be spread apart to see the opening. Trace each oviduct to the cloaca. Each opens into the side of the anterior end of the cloaca, ventral to the opening of the intestine. The stalk of the large bilobed urinary bladder joins the cloaca midway between the two oviducts.

The cloaca has already been exposed. (If not, do so by cutting through the pelvic girdle on each side and removing the median portion of the girdle). Clear away connective tissue from around the cloaca. Attached to each side of the cloaca posterior to the oviducts are two elongated sacs, the accessory urinary bladders. Their function is uncertain, but in females, at least, they carry water, which she employs to soften the soil while digging a nest. A dark structure visible through the cloacal wall is the *clitoris,* homologous to the male penis and apparently without function in the female.

Now cut open the cloaca to one side of the clitoris, extending the cut in the median ventral line up to the stalk of the bladder. Look into the cloaca. Observe that the clitoris consists simply of thickenings in the ventral wall. Find the large openings of the accessory bladders. Next note the opening of the large intestine. This is the most dorsal of the openings and is somewhat separated by a fold from the urogenital openings, so that the cloaca is divisible into

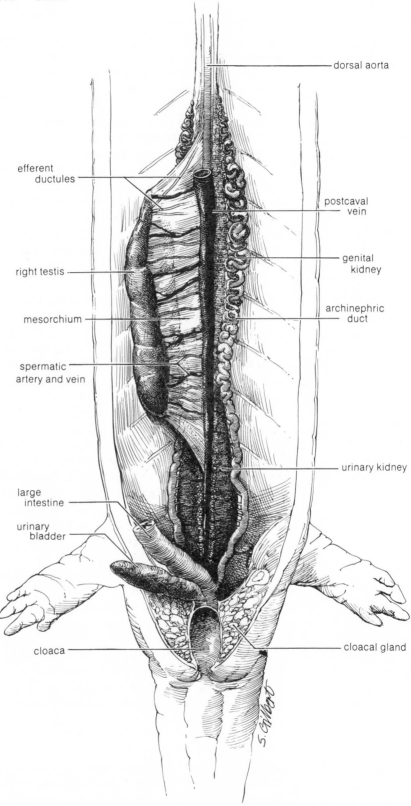

dorsal aorta

efferent
ductules

postcaval
vein

genital
kidney

right testis

archinephric
duct

mesorchium

spermatic
artery and vein

urinary kidney

large
intestine

urinary
bladder

cloaca

cloacal gland

Fig. 12.17.  Male *Necturus* urogenital system. Note the regionalization of the
kidney.

coprodaeum and urodaeum. Ventral to the opening of the intestine are the openings of the oviducts on thickened papillae. They are found best by cutting into the oviduct and probing posteriorly into the cloaca. Between and ventral to the oviducal openings is the opening of the urinary bladder.

The kidneys of the turtle are metanephroi. They have already been identified as flattened lobed organs fitting snugly against the posterior end of the pleuroperitoneal cavity. The renal portal vein and its tributary, the internal iliac, run along the ventral face of each kidney. Cut away this vein; directly dorsal to it is a tube, the metanephric duct or *ureter,* extending from the middle of the kidney to the cloaca (Fig. 12.18). It enters the cloaca at the base of the oviduct. By making a slit in it and passing a probe into it, its opening into the cloaca will be found just anterior to the thickening caused by the oviducal entrance.

## 2. The Male Urogenital System

The male urogenital system consists of the paired testes and their ducts; the ducts of the testes are mesonephric ducts, now termed *deferent ducts.*

Expose the cloaca as in the female and find the two accessory bladders attached to its lateral walls. Note the place of attachment of the rectum to the cloaca and ventral to this the attachment of the urinary bladder. The dark mass seen through the ventral wall of the cloaca is the penis or organ of copulation, which is inserted into the female cloaca at mating so that the sperm are injected directly into the female system. Note that a penis is first met with in reptiles. A pair of rounded masses projects from the anterior wall of the cloaca to either side of the stalk of the bladder; these are the bulbs of the corpora cavernosa, part of the penis. The muscles that retract the penis will be seen attached in the ventral wall of the cloaca.

The kidneys were previously identified as flattened, lobed bodies fitting against the posterior wall of the pleuroperitoneal cavity. Each testis is a yellow spherical body attached to the ventral face of the kidney by the mesorchium. Lateral and posterior to the testis is an elongated, dark coiled body, the epididymis. The testis is connected to the anterior part of the epididymis by the minute efferent ductules that run in the mesorchium. The efferent ductules and epididymis are remnants of the mesonephros. The male duct or deferent duct begins as a greatly coiled tube on the surface of the epididymis, here termed epididymidal duct. Remove the peritoneal covering of the epididymis, uncoil the deferent duct, and trace it to the cloaca, where it enters craniad, and at the base of the bulb of the corpora cavernosa.

Next cut open the cloaca, inserting the blade of the scissors into one corner of the anus and cutting far to one side to avoid injuring the penis. Spread apart the cloacal walls and study the penis. It consists of two spongy ridges, the corpora cavernosa or cavernous bodies, in the ventral wall of the cloaca. Between these folds in the midventral line is a deep groove, the urethral groove, which in its natural condition is practically converted into a tube by the approximation of the

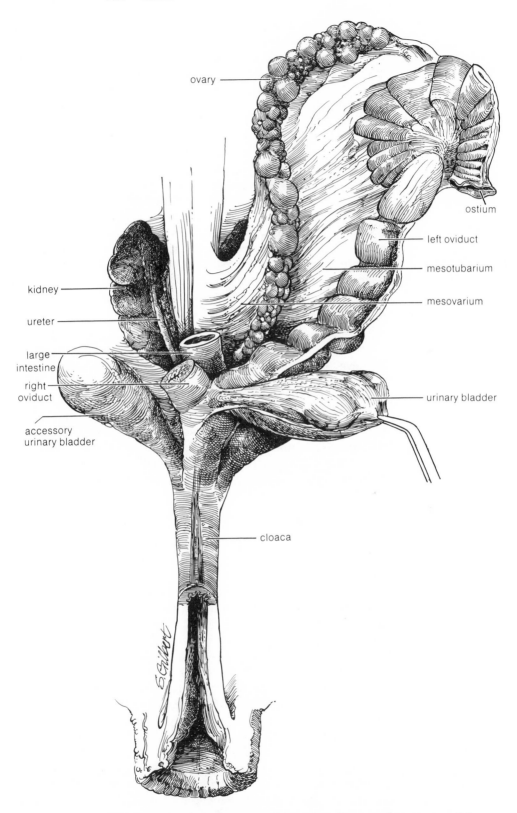

ovary

ostium

left oviduct

mesotubarium

mesovarium

kidney

ureter

large
intestine

right
oviduct

urinary bladder

accessory
urinary bladder

cloaca

Fig. 12.18. Female turtle urogenital system. Compare the structure of the
ovary and oviduct shown with that of females of other species illustrated.

cavernous bodies. The urethral groove terminates caudad at the base of a heart-shaped projection, the glans of the penis. The anterior ends of the cavernous bodies form bulbs, already noted, which project forward into the coelom at the sides of the stalk of the bladder. The bulbs are filled with blood, which they receive from the internal iliac vein. All parts of the penis are very spongy and vascular. During the sexual act the blood from the bulbs rushes into the spongy spaces of the cavernous bodies and the glans, erecting them and causing the cavernous bodies to come in contact above the urethral groove, converting the latter into a canal for the passage of sperm.

The kidneys are metanephroi, and their ducts the metanephric ducts or ureters. The ureter will be found immediately to the dorsal side of the epididymis, which should be removed. The ureter is a short, straight tube proceeding to the cloaca, where it opens just anterior to the opening of the ductus deferens. The two openings will be found at the sides of the anterior beginning of the urethral groove.

Find the openings of the accessory bladders, the urinary bladder, and the rectum into the cloaca. The latter is dorsal to the urogenital openings.

**E. The Urogenital System of the Pigeon**

**1. The Female Urogenital System**

Remove the digestive tract, leaving the large intestine in place. In adult birds there is a single ovary and oviduct on the left side. The right ovary and duct are present in the embryo but disappear almost entirely before hatching. The ovary is a mass containing eggs of various sizes, situated at the anterior end of the left kidney, attached by a short mesovarium. Posterior to the ovary the coiled left oviduct proceeds to the cloaca, being supported by the mesotubarium. The ostium is situated in the mesotubarium near the ovary; it is a wide opening with wing-like borders fastened to the mesotubarium. A small remnant of the right oviduct is attached to the right side of the cloaca.

The kidneys are metanephroi. Each is a flattened, three-lobed organ situated against the dorsal wall. The ureters or metanephric ducts are located just dorsal to the renal portal veins, which should be stripped from the face of the kidney. The ureter begins on each side at the groove between the anterior and middle lobes of the kidney and extends straight posteriorly to the cloaca. The left ureter is concealed by the oviduct.

The cloaca is an expanded chamber receiving the rectum on its median ventral surface, the left oviduct to the left, the very small right oviduct to the right, and the ureters dorsal to the oviducts. Cut into the cloaca to the right of the rectum. Note that the cavity of the cloaca is subdivided. There is a large ventral portion (coprodaeum) where the rectum opens. Dorsal to this and separated from it by a fold is the urodaeum, into which open the oviducts and ureters. The opening of the left oviduct is readily found here; the openings of the ureters are more medial and smaller. The most dorsal compartment of the cloaca is the proctodaeum, a small chamber with a raised rim, which opens to the anus. In the anterior wall of the proctodaeum dorsal to the rim an opening may be noted; it leads into a small pouch, the *bursa of Fabricius,* which seems to have some function in young birds but degenerates with maturity (see fig. 12.19).

**2. The Male Urogenital System**

The testes are a pair of oval organs at the anterior ends of the kidneys; their size varies considerably with the season. They lack definite mesorchia. The kidneys and the ureters should be studied according to the directions given for the female. The male ducts or deferent ducts spring from the medial surface of the testes near their posterior ends, with the intervention of an epididymis too small to be identified macroscopically. The deferent ducts are slender, convoluted tubes that pass caudad parallel to the ureters. Trace both ducts to the cloaca.

The cloaca is smaller in the male than in the female, and the lips of the anus more protruding. The rectum enters medially and ventrally, the urogenital ducts laterally. Cut into the cloaca as directed for the female and identify its chambers as described there. They are the same in the two sexes, except that the male urodaeum is smaller and receives the two deferent ducts instead of the oviducts. Ureters and deferent ducts open on small papillae in the lateral walls of the urodaeum (fig. 12.20).

**F. The Urogenital System of Mammals**

**1. The Kidneys and Their Ducts**

Remove the digestive tract, leaving the rectum in place.
The kidneys of mammals are metanephroi, and their ducts the metanephric ducts or ureters. The kidneys are large oval organs situated against the dorsal wall of the peritoneal cavity; they are retroperitoneal. The right kidney is usually considerably anterior to the left one. Clear away fat and connective tissue from around the kidneys and note their characteristic bean shape, convex laterally, concave on their medial faces. The concavity is termed the *hilus,* and from it a white tube, the ureter, passes out, turning posteriorly. Follow the ureters caudad, clearing away fat, and note their entrance into the urinary bladder. In females the ureters pass dorsal to the horns of the uterus; in males dorsal to a white cord, the male duct or ductus deferens, which loops over the ureter and disappears dorsal to the bladder.

With a cut remove the ventral half of a kidney. Within the hilus there is a cavity, the renal sinus, occupied by the renal artery and vein and by the expanded beginning of the ureter, termed the renal pelvis. Into this pelvis the substance of the kidney projects as the renal papilla, on which are situated the microscopic openings of the collecting tubules. Remember that the kidney substance is readily divisible into two areas, a peripheral cortex and a central medulla. The cortex contains the renal corpuscles and the convoluted and looped portions of the kidney tubules. The medulla is marked by lines that converge on the renal papilla; these lines are the collecting tubules. It will be recalled that the collecting tubules, the pelvis, and the ureter arise by outgrowth from the mesonephric duct. The collecting tubules and renal papilla together form a renal pyramid; there is only one in the rabbit and in the cat, but there are about twelve in the human kidney.

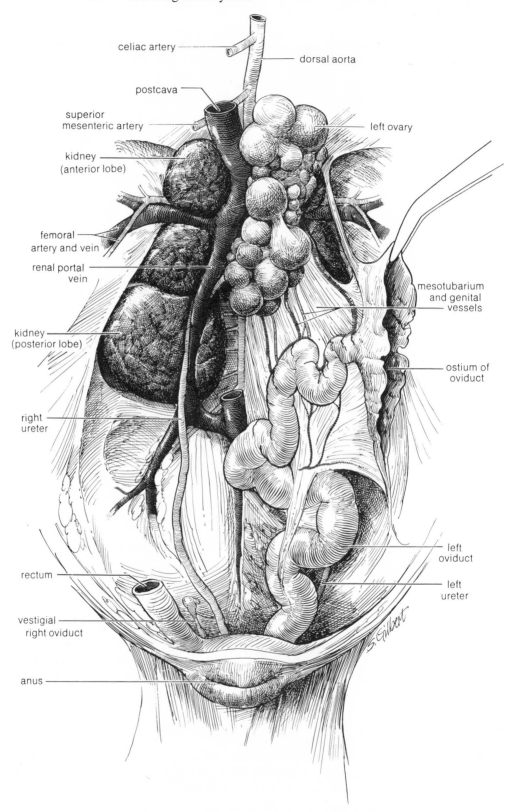

Fig. 12.19. Female pigeon urogenital system. Notice the structure of the ovary and the vestigial right oviduct.

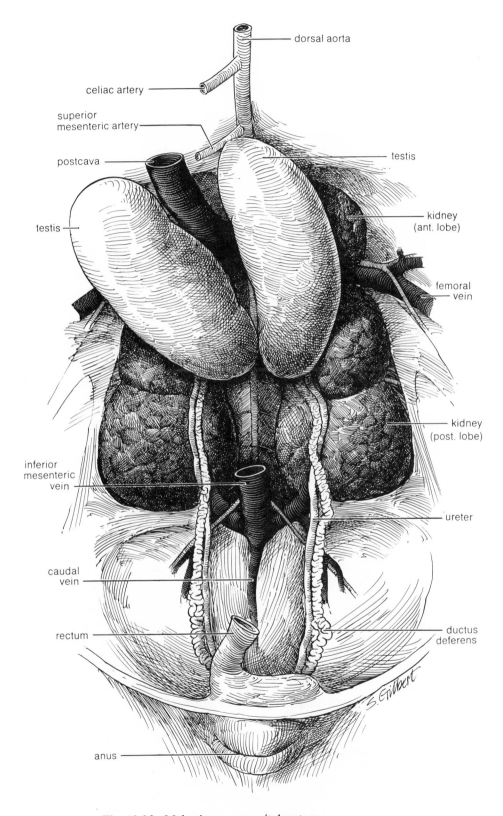

Fig. 12.20.  Male pigeon urogenital system.

The urinary bladder is a pear-shaped sac at the posterior end of the peritoneal cavity. It is ventral to the rectum in the male, ventral to both rectum and uterus in the female. The free anterior end of the bladder is named the apex or vertex, the posterior portion the fundus. The fundus continues posteriorly as a narrowed stalk, the *urethra* (also called the neck of the bladder). The bladder is covered by the peritoneum, which is connected with the peritoneum of the abdominal wall by means of the median and lateral ligaments previously noted. The pouch between the bladder and rectum (male) or bladder and uterus (female) is named the rectovesical or vesicouterine pouch, respectively.

## 2. The Female Reproductive System

The female reproductive system consists of a pair of ovaries and their ducts. The ovaries are very small oval bodies located at the sides of the peritoneal cavity at the anterior end of the coils of the uterus. Each will be seen to bear little clear vesicles, the *Graafian follicles,* each of which contains an egg or ovum; in pregnant females the ovary also bears small hard lumps, the *corpora lutea,* which represent follicles from which the eggs of the pregnancy were discharged. The ovary is suspended by the mesovarium, which extends forward to the kidney and is continuous posteriorly with the ligament of the uterus.

The ducts of the ovaries are, as in other vertebrates, the Müllerian ducts or oviducts, but they are differentiated into several distinct parts in mammals. The uppermost portion of the oviducts is a slender, convoluted tube that passes lateral to the ovary and curves over its anterior end; its mesentery, the *mesosalpinx,* forms a sort of hood, partly enclosing the ovary. This portion of the oviducts is the uterine or Fallopian tube. It opens in front of the ovary (rabbit) or to the lateral side of it (cat) by the ostium, which has fringed borders, the fimbriae. On tracing the uterine tube posteriorly, you will find it widens suddenly into a thick-walled tube, the uterus (rabbit) or horn of the uterus (cat). Its size depends on whether the animal is pregnant or not; in pregnant animals the uteri or horns are greatly enlarged and exhibit a series of swellings, each of which contains an embryo (these will be examined later). The strong fold of peritoneum supporting the uteri or horns is the mesometrium. Mesovarium, mesosalpinx, and mesometrium together are called the broad ligament of the uterus in mammals. The round ligament of the uterus is the fold extending from the beginning of the uterus or horn posteriorly to the body wall; it is continuous with, but at right angles to, the broad ligament. In the cat the two horns of the uterus unite in the median line, dorsal to the bladder, with a single tube, the body of the uterus. Body and horns together constitute the uterus or womb, but the young develop only in the horns. In the rabbit the two uteri are separate along their entire lengths, and consequently there is no division into body and horns. In the cat the body of the uterus continues posteriorly as the vagina; in the rabbit the two uteri join the vagina; the vagina is a tube situated in the median line between the bladder and the rectum. It exists through the ring formed by the pelvic girdle and vertebral column.

The external genital parts or external genitalia were described with the external anatomy. In the rabbit make an incision through the skin forward from the vulva. In the median line beneath the skin is a hardened body, the clitoris, homologous to the penis of the male. Its anterior end is attached by ligaments to the ischium and pubic symphysis. Cut across the clitoris and note the two cavernous bodies of which it is composed. In the cat the clitoris is minute.

Now cut through the pubic and ischial symphyses and spread the legs well apart. Trace the urethra, the vagina, and the rectum posteriorly. At first the urethra lies on the ventral face of the vagina, bound to it by tissue; it then unites with the vagina to form a common tube, the urogenital canal or urogenital sinus. Cut this free, lift it out, and follow it to the urogenital aperture. Cut open the urogenital aperture and note the free posterior end, or glans, of the clitoris projecting into the cavity in the rabbit. Free the rectum from the urogenital canal and follow it to the anus. Along its sides in the rabbit are a pair of elongated anal glands; in the cat, the rounded anal glands or sacs occur to either side of the rectum close to the anus. The secretions of the anal glands are strongly odoriferous and presumably sexual in nature.

Cut open the vagina. In the rabbit note the external uterine orifice with raised fringed lips by which each uterus opens into the vagina. The rabbit uterus is of the duplex type. In the cat the body of the uterus is divided into lateral halves by a median partition, and the horns open to either side of this partition. The cat uterus is of the bipartite type. The lower end of the uterus, called the cervix, projects into the vagina by a fold. The opening of the body of the uterus into the vagina is the external uterine orifice (see fig. 12.21).

## 3. The Male Reproductive System

The two testes are lodged in the scrotum, divided into two compartments by the internal partition. Cut through the skin ventral and in front of one testis, exposing the testis as an oval, white body. Clear away tissue anterior to the testis and find a white cord, the spermatic cord (cat), or ductus deferens (rabbit), passing forward and entering the peritoneal cavity through the inguinal canal. The two ends of this canal are the external and internal inguinal rings (cat); they are indefinite in the rabbit. The spermatic cord contains a white duct, the ductus deferens, and blood vessels and nerves. Trace the spermatic cord (cat) or ductus deferens (rabbit) by cutting open the inguinal canal. The two deferent ducts turn toward the median line, loop over the ventral surfaces of the ureters, and disappear on the dorsal surface of the urethra.

Now cut through the pubic and ischial symphyses and spread the legs apart. Trace the deferent ducts and the urethra caudad, separating them from the rectum. The ureters may be cut through and the bladder held caudad. The deferent ducts pass along the dorsal surface of the urethra. In the rabbit they enlarge and enter an expanded

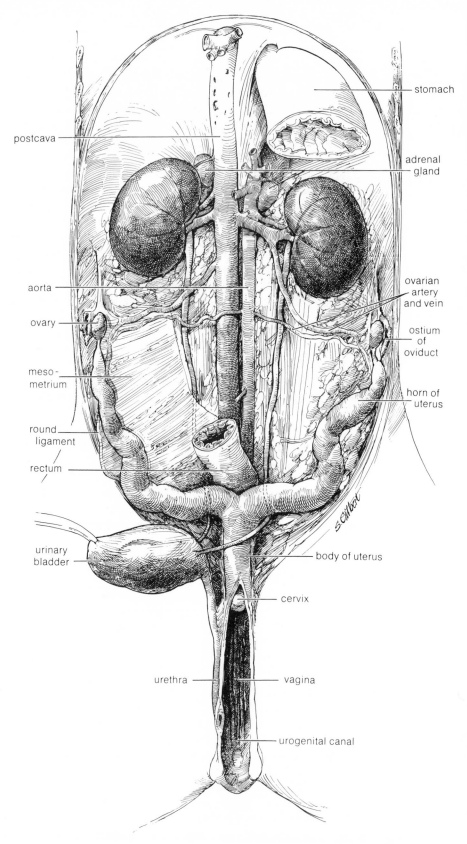

Fig. 12.21. Female cat urogenital system. Compare the relative size of the
ovary, and its vascularization, with that in females of other species illustrated.
Notice the proportions of the uteri.

sac, the seminal vesicle, which bulges forward between the rectum and the bladder. Cut into the seminal vesicle and note the openings of the two deferent ducts into its ventral wall, and the thickening in its posterior dorsal wall caused by the prostate gland. Find the union of the seminal vesicle and urethra that forms a common tube, the urogenital sinus. In the cat the two deferent ducts join the urethra without the formation of a seminal vesicle, the point of junction being surrounded by a slight enlargement, the *prostate* gland. The common tube thus formed is the urogenital canal or sinus. The *bulbo-urethral* glands or Cowper's glands are small swellings situated on the urogenital canal shortly posterior to the prostate gland in the rabbit, about an inch posterior in the cat. The terminal inch of the urogenital canal is enclosed in the penis. Cut into the prepuce and note the point projecting within it, called the *glans* of the penis. Note that the prepuce is simply a fold of skin around the glans. At the tip of the glans is the urogenital opening. The glans in the cat bears a number of minute spines. Dissect anteriorly from the glans, exposing the remainder of the penis, which is a hardened cylindrical structure. Find where the urogenital canal enters its anterior end. Note, also, the strong attachment of the penis to the pelvic region. Cut across the middle of the penis and note that it is composed of two cylindrical bodies, the corpora cavernosa or cavernous bodies, closely placed. The urogenital canal, here called the cavernous urethra, lies on the dorsal side of the penis, resting in a depression between the two cavernous bodies. At the anterior end of the penis the two cavernous bodies diverge, forming the crura of the penis, which are attached to the ischia. The cavernous bodies are spongy structures, and during the sexual act they become distended with blood, so that the penis is caused to project out of its sheath, the prepuce (see fig. 12.22).

Trace the rectum to the anus, following the directions given for the female.

The structure of the testis may now be investigated. Each testis is enclosed in a white fibrous sac, which is the peritoneal pouch made by the descent of the testis. Cut open this sac, exposing the cavity, or vaginal sac, in which the testis lies; this is a part of the peritoneal cavity. The tunica vaginalis, which lines this cavity, is reflected over the surface of the testis as a covering layer. This deflection lies along the middorsal line of the testis, and a mesorchium is thus formed between testis and the wall of the vaginal sac. The posterior end of the testis is attached to the posterior scrotal wall by a short, stout ligament, the *gubernaculum,* continuous with the mesorchium and homologous to the round ligament of the uterus. The ductus deferens lies along the dorsal surface of the testis and is quite coiled, forming the epididymidal duct. This begins at the anterior end of the testis as a coil, the head of the epididymis, which receives the invisible efferent ductules from the testis. The coiled epididymidal duct then passes along the dorsal surface of the testis as the body of the epididymis and finally, at the posterior end of the testis, makes another coiled mass, the tail of the epididymis, to which the gubernacu-

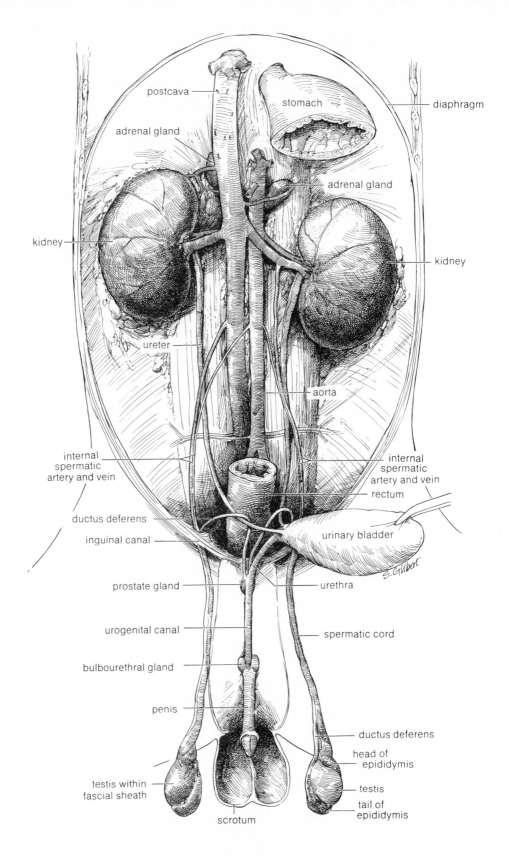

Fig. 12.22. Male cat urogenital system. The testes are removed from the scrotum. Notice the several accessory glands associated with the urethra.

lum is attached. From this the much convoluted ductus deferens proceeds anteriorly and passes into the inguinal canal, where it becomes a straight tube. The head of the epididymis corresponds to the epididymis of lower forms and is derived from the mesonephros.

**G. Summary**

1. The urogenital system is derived from the mesomere of the embryo.

2. The urinary or excretory system consists of the paired kidneys and their ducts. Primitively, each kidney extended the length of the pleuroperitoneal cavity and was composed of segmentally arranged tubules and a collecting duct. Such a kidney is termed the holonephros or archinephros, and its duct the archinephric duct.

3. The primitive kidney tubules consisted of the following parts: a ciliated peritoneal funnel opening into the general coelom; a tubule leading from this to the nephrocoel (coelomic cavity of each mesomere); an opening, the nephrostome, from this into the kidney tubule proper; the kidney tubule; and the opening of this into the archinephric duct. The archinephric duct was formed by the union of backward extensions of the kidney tubules.

4. A capillary network termed a glomerulus is primitively associated with each kidney tubule. It is either an external glomerulus, projecting into the coelom near the peritoneal funnel, or an internal glomerulus, pushing in the wall of the nephrocoel. The combination of glomerulus and nephrocoel wall is termed a renal corpuscle or Malpighian body.

5. A holonephros is approached only in the embryos of myxinoid cyclostomes and caecilians; in vertebrate embryos generally, the kidneys develop in two or three successive sections in an anteroposterior direction.

6. The most anterior vertebrate kidney is termed the pronephros; it is similar to the holonephros, composed of segmental tubules provided with peritoneal funnels. It appears in the embryos of all vertebrates but persists and remains functional throughout life only in the cyclostomes and a few teleost fishes. Its ducts, the pronephric ducts, grow backward and open into the cloaca.

7. As the pronephroi degenerate, there develop behind them the second kidneys, opisthonephroi or mesonephroi, which differ chiefly in having more than one tubule per segment. The pronephric ducts persist and are utilized by the mesonephroi as their ducts, being then termed the mesonephric ducts.

8. In amniotes the mesonephroi degenerate during development, and there appear behind them the third kidneys, or metanephroi. These develop in part from the remaining tissue of the mesomere and in part by outgrowth from the mesonephric ducts near their termination in the cloaca. The stalks of the outgrowth form the metanephric ducts or ureters. The metanephroi are the functional kidneys of adult amniotes; they have many tubules per segment and lack peritoneal funnels.

9. In anamniotes generally, the second kidney utilizes the tissue that furnishes both mesonephros and metanephros in amniotes. Hence, the kid-

ney does not correspond exactly to the mesonephros of amniote embryos, although it is often called that. It is here termed opisthonephros.

10. Pronephros, mesonephros, and metanephros appear to be successive alternations of the original holonephros.

11. A urinary bladder is generally present as an evagination of the ventral cloacal wall or as a chamber made by fusion of the cloacal ends of the kidney ducts (fish). The kidney ducts generally open directly into the cloaca, not into the bladder; but in mammals they shift so as to open into the bladder.

12. The genital system of vertebrates consists of the paired male sex glands or testes and the paired female sex glands or ovaries and their respective ducts. These ducts in males are kidney ducts; hence the close association of urinary and genital systems. Testes and ovaries arise as mesomere thickenings that project into the coelom near the kidneys.

13. The ducts of the ovaries are termed Müllerian ducts or oviducts; they have no direct connection with the ovaries but open nearby into the coelom by a funnel-like ostium believed to represent one or more pronephric peritoneal funnels. The oviducts of elasmobranchs originate by a longitudinal splitting from the pronephric ducts, and, although this method of origin does not obtain in other vertebrates, it may be regarded as the phylogenic source of the oviducts. The ducts in teleosts are part of the ovary and are not homologous to the ducts of other vertebrates.

14. Except in placental mammals the oviducts enter the cloaca separately. In placental mammals the oviducts are more or less fused and differentiated into uterine tube, uterus, and vagina. Fusion proceeds from the distal end proximally, involving first the vaginae, then uteri. Partial fusion of the two uteri results in a uterus with horns, complete fusion in the single uterus of the primates. The uterine (Fallopian) tubes always remain separate.

15. The testes are located inside the abdominal cavity except in mammals, where in many cases they descend temporarily or permanently into an inguinal pouch of the body wall, called the scrotum.

16. In all vertebrates except cyclostomes and teleosts the testes use as their ducts those of the kidney, which are then called deferent ducts. Hence, in males in which the opisthonephros is the functional adult kidney, the ducts convey both urine and sperm. In amniotes, where the metanephros is the adult kidney, the ducts have only genital functions.

17. The kidney duct is always connected to the corresponding testis by the intervention of a persistent part of the kidney, termed the epididymis, whose renal corpuscles connect with the seminiferous tubules of the testis by a varying number of tubules, called the efferent ductules.

18. The vertebrate embryo has all the structures of both sexes and is potentially hermaphroditic. In the development of a male the oviducts are suppressed; in the development of a female the kidney ducts are suppressed or, in anamniotes, limited to an excretory function.

19. A cloaca is characteristic of most vertebrates, but in placental mam-

mals it becomes subdivided by a fold into a dorsal rectum and a ventral urogenital canal or sinus in such a way that the latter receives the urogenital ducts and the bladder. In male placental mammals the deferent ducts (mesonephric ducts) join the stalk of the bladder (urethra) to form a urogenital canal, which pierces the penis and opens at its tip. In most female placental mammals the urethra and common vagina unite similarly to form a urogenital canal; but in female primates, urethra and vagina are completely separate and open separately to the exterior; thus a urogenital canal is absent and the path of the urine is wholly distinct from the genital passage.

20. In mammals the male urogenital canal is generally provided with a variety of glands, termed the accessory sex glands, whose secretions are of importance for the vitality of the sperm.

21. A definite organ of copulation, the penis, is first seen in reptiles, occurs in some birds, and is general throughout mammals. It is a differentiation of the ventral floor of the cloaca and consists primitively of a cylinder of spongy tissue, the cavernous body, on whose surface a groove conveys the sperm. In mammals the groove closes over, becoming the cavernous urethra; and the cavernous body subdivides into two bodies. Certain fish and amphibians modify various structures for use as copulatory organs.

22. Trends in parental care of eggs and young, and especially retention and nurture of developing embryos by the female, are correlated with morphological changes in the urogenital system.

23. The general direction of evolution in the urogenital system is toward the separation of excretory, genital, and intestinal functions.

**References**

Asdell, S. A. 1964. *Patterns of mammalian reproduction.* 2d ed. Ithaca: Cornell University Press.

Amoroso, E. C. 1952. Placentation. In F. H. A. Marshall, *Physiology of reproduction,* pp. 1–42. London: Longmans, Green.

Bauchot, R. 1965. La placentation chez les reptiles. *Ann. Bio.* 7:547–75.

Breder, C. M., and Rosen, D. E. 1966. *Modes of reproduction in fishes.* Garden City: Natural History Press.

Burgos, M. H., and Fawcett, D. W. 1955. Studies on the fine structure of the mammalian testis. I. Differentiation of the spermatids in the cat (*Felis domesticus*). *J. Biophys. Biochem. Cytol.* 1:287–300.

Cuellar, O. 1966. Oviducal anatomy and sperm storage structures in lizards. *J. Morph.* 119:7–20.

Dalton, A. J., and Hagueman, F., eds. 1957. *Ultrastructure of the kidney.* New York: Academic Press.

Eckstein, P., and Zuckerman, S. 1952. Morphology of the reproductive tract. In F. H. A. Marshall, *Physiology of reproduction,* pp. 43–155. London: Longmans, Green.

Denison, R. H. 1956. A review of the habitat of the earliest vertebrates. *Fieldiana: Geol.* 11:359–457.

Fox, H. 1963. The amphibian pronephros. *Quart. Rev. Biol.* 38:1–25.

———. 1977. The urinogenital system of reptiles. In *Biology of the Rep-*

*tilia,* ed. C. Gans and T. S. Parsons, 6:1–157. New York: Academic Press.

Franchi, L. L. 1962. The structure of the ovary-vertebrates. S. Zuckerman, ed. In *The ovary,* pp. 121–42. New York: Academic Press.

Fraser, E. A. 1950. The development of the vertebrate excretory system. *Biol. Rev.* 25:159–87.

Harrison, R. J. 1962. The structure of the ovary-mammals. In S. Zuckerman, ed., *The ovary,* pp. 143–88. New York: Academic Press.

Hoffman, L. H. 1970. Placentation in the garter snake *Thamnophis sirtalis. J. Morph.* 131:52–87.

Hureau, C.; Chevrel, J.; and Petal, J. 1966. Considerations sur le plan général d'organization du rein des mammifères. *C. r. Ass. Anat.* 50: 518–39.

Lofts, B. 1974. Reproduction. In *Physiology of the amphibia,* ed. B. Lofts, 2:107–218. New York: Academic Press.

Matthews, L. H., and Marshall, F. H. A. 1952. Cyclical change in the reproductive organs of the lower vertebrates. In F. H. A. Marshall, *Physiology of reproduction,* pp. 156–225. London: Longmans, Green.

Perry, J. S. 1972. *The ovarian cycle of mammals.* New York: Nafner Publishing.

Rouiller, C., and Muller, A. F., eds. 1969. *The kidney: Morphology, biochemistry, physiology.* 2 vols. New York: Academic Press.

Salthe, S. N., and Mecham, J. S. 1974. Reproductive and courtship patterns. In *Physiology of the amphibia,* ed. B. Lofts, 2:310–522. New York: Academic Press.

Sharman, G. B. 1976. Evolution of viviparity in mammals. In *Reproduction in mammals,* ed. C. R. Austin and R. V. Short, 6:32–70. New York: Cambridge University Press.

Smith, H. W. 1953. *From Fish to Philosopher.* Boston: Little, Brown. 264 pp.

Taylor, D. H., and Guttman, S. I., eds. 1977. *The reproductive biology of amphibians.* New York: Plenum Press.

# 13

## The Comparative Anatomy of the Nervous System and the Sense Organs
### R. Glenn Northcutt

**A. General Considerations**

The nervous system is anatomically and functionally the most complex system of the vertebrate body. Its organization is still not as fully understood as the other organ systems, although in the last decade, new and exciting experimental methods have become available that allow us to begin to understand its organization better. At present, the phylogeny of the vertebrate nervous system is known only in very vague outline. Tantalizing bits of information hint at its organization and history. This chapter will describe some of the variation that exists in vertebrate nervous systems and suggest the functional significance of that variation.

**1. The Parts of the Nervous System**

The structural unit of the nervous system is the *nerve cell,* or *neuron,* which consists of the cell body and one or more processes, or *neurites,* that may be very long. The most generalized neuron is the *neurosensory cell,* in which the usually elongated cell body is situated at the surface of the body and acts as a *sensory receptor* and also as a *conductor.* Its inner end continues as a nerve fiber, the *axon,* joining the central nervous system. Neurosensory cells are common in invertebrates, but among vertebrates they are limited chiefly to the olfactory epithelium and the retina of the eye (fig. 13.1). A more specialized neuron is the *unpolarized* nerve cell, usually with several neurites, along which impulses pass in either direction. This type of neuron is commonly found in invertebrate nervous systems. The most specialized neuron, the type found in vertebrates, is the *polarized* nerve cell, which receives impulses along one or more processes, the *afferent* processes, or *dendrites,* and sends them out along one process, the *efferent* process, or *axon.* Axons and dendrites differ histologically as well as functionally. Neurons are also classified by the number and arrangement of the neurites they possess. Neurons that possess a single dendrite and a single axon are called *bipolar* neurons. Bipolar neurons are usually neurosensory neurons. Neurons that possess a single neurite are called *pseudounipolar* neurons. These neurons develop from bipolar neuroblasts and secondarily fuse their two neurites. Pseudounipolar neurons make up the dorsal root ganglia of the spinal nerves and ganglia of the cranial nerves. Neurons that possess more than one dendrite are called *multipolar* neurons. Multipolar neurons constitute the largest population of the nervous system and are subdivided into groups according to whether or not they project outside of the cellular population of their origin and according to the shape of their cell bodies (fig. 13.1). While the evolution of neurons is not understood, recent work suggests that generalized neurons have long, radiate, rectilinear dendrites

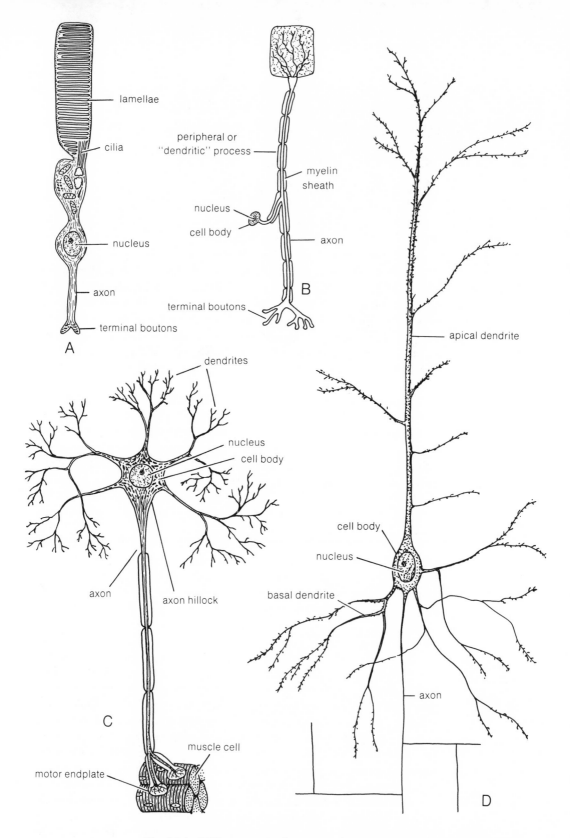

Fig. 13.1. Different types of neurons. *A*, neurosensory neuron (rod) from retina; *B*, pseudounipolar sensory neuron from dorsal root ganglion; *C*, multipolar motor neuron from ventral horn of spinal cord; *D*, pyramidal motor neuron from mammalian isocortex.

with a considerable degree of dendritic overlapping from adjacent neurons. These neurons appear to process afferent signals of very heterogeneous origin, and such populations are referred to as regions of *neuropil*. More specialized neurons have fewer dendrites with greater morphological complexity and little or no overlap from adjacent neurons. These more specialized neurons appear to process afferent signals of a homogeneous origin. Such cellular populations are most frequently found in multiple laminae called cortices.

An aggregation of nerve cell bodies outside the central nervous system is called a *ganglion*. A similar group inside the central nervous system is termed a *nucleus*. A *nerve* is a bundle of axons outside of the central nervous system. A similar group of axons inside the central nervous system is termed a *tract*. The axons of neurons are encased in a proteolipid sheath that begins just distal to the body and continues almost to the termination of the axon. This sheath lies external to the cellular membrane of the axon. Axons surrounded by such sheaths are said to be *myelinated* and are frequently referred to as *white* fibers. In the past, many axons were thought to lack such a sheath. They were said to be *unmyelinated* and were frequently referred to as *gray* fibers, but electron microscopy has shown that most gray fibers possess a thin sheath. The myelin sheath was once believed to be a nonliving structure secreted by neuroglial cells found near axons. It is now known, however, that the myelin sheath is a series of living plasma membranes of the neuroglial cells that are wrapped around the axon. The myelin sheath is not continuous along the length of an axon but is broken at intervals. In axons outside the central nervous system, these regions of discontinuity along the axon are termed the *nodes of Ranvier*. Electron microscopy has recently demonstrated that similar nodes are found around axons inside the central nervous system. Neuroglial cells and their myelin sheaths have been related to a number of functions: they electrically insulate axons; they are involved in the exchange of metabolites with neurons; they transport material differentially between blood and neurons; and they may have a trophic role in the regeneration of axons.

The distal ends of axons are finely branched and, except for those in specialized contacts on muscle, glands, and blood vessels, normally have small knobs (*boutons terminaux*) near cell bodies, dendrites, or the proximal portions of axons. In vertebrates there is not cytoplasmic continuity between cells, and the space between the boutons of one neuron and the cellular membrane of a second neuron is called a *synapse*. When a nerve impulse is generated in the dendrites and cell body of a neuron, it spreads down the axon to the boutons. There a chemical is released that may cause the second neuron to generate an impulse. Impulses reaching a bouton do not always cause a second neuron to generate an impulse. The second neuron may react only to a specific rhythm of boutonal transmissions, or frequently boutons may actually inhibit a second neuron from generating an impulse by their chemical transmitters.

The vertebrate nervous system consists of three parts: the *central* nervous system, the *peripheral* nervous system, and the *autonomic* nervous system. The central nervous system comprises the *brain,* situated inside the skull, and the *spinal cord,* situated inside the neural canal formed by the arches of the vertebrae. Brain and cord are composed of both nerve cell bodies, which form the *gray matter,* and nerve fibers, which form the *white matter,* so called because the thicker myelin sheaths lend a white color. The peripheral nervous system consists of the *cranial nerves,* arising from the brain, and the *spinal nerves,* arising from the spinal cord. Both kinds of nerves consist of the axons of nerve cell bodies found inside the brain or cord or in outside ganglia. The nerves, especially the spinal nerves, are markedly metameric in arrangement—typically there is a pair of nerves to each body segment—but this metamerism is probably imposed upon the nerves by the metamerism of the muscles and vertebrae. The autonomic nervous system controls and regulates in general the visceral activities of the body and the organs that subserve those functions, such as the heart and the digestive tract, the smooth musculature in several systems, secreting glands, blood vessels, and the respiratory and urogenital systems. It consists of a ganglionated cord lying to either side of the vertebral column along the dorsal coelomic wall, of ganglia in the head and among the viscera, and of connecting and distributing nerves.

**2. The Development of the Nervous System**

The central nervous system is formed by the enfolding of the ectoderm along the middorsal region of the embryo to form a tube (see chap. 4). That portion of the tube situated in the head becomes the brain, and that part posterior to the head becomes the spinal cord. The original single layer of enrolled ectoderm cells proliferates to form a thick zone of cells around the central cavity; most of these become nerve cells, but some give rise to supporting cells or neuroglia.

The location of this zone of nerve cell bodies, or gray matter, around the central cavity is believed to be the ancestral arrangement retained in a modified form throughout vertebrates in the spinal cord; but in the brain of all vertebrates there is considerable migration of the gray matter toward the periphery of the brain wall.

Neurites arise from the embryonic nerve cells and may remain in the central nervous system to form a peripheral zone of myelinated fibers, the white matter, or may grow out of it to form the nerves. The ganglia outside the brain and cord result from the migration of nerve cell bodies, mostly from the neural crests, a pair of longitudinal strands left outside the neural tube at the time of its closure.

In the development of the brain the original, simple tube becomes marked off by a transverse ventral fold into two regions, the primitive *forebrain* and *hindbrain* (fig. 13.2). This fold is situated just behind the future *infundibulum,* and it also marks approximately the anterior end of the notochord. The anterior part of the primitive forebrain, called the *telencephalon,* puts out paired lateral swellings, which become the *cerebral hemispheres.* The region behind this is the *diencephalon,* and it soon shows a dorsal evagination, the *pineal apparatus* (fig. 13.2C). The diencephalon becomes marked off by a constriction from the last part of the primitive forebrain, the *mesencephalon,* which develops a pair of pronounced dorsal swellings, the *optic tectum* (fig. 13.2C,D). A deep constriction forms between the mesencephalon and the primitive hind brain. The anterior part of the hindbrain, called the *metencephalon,* puts

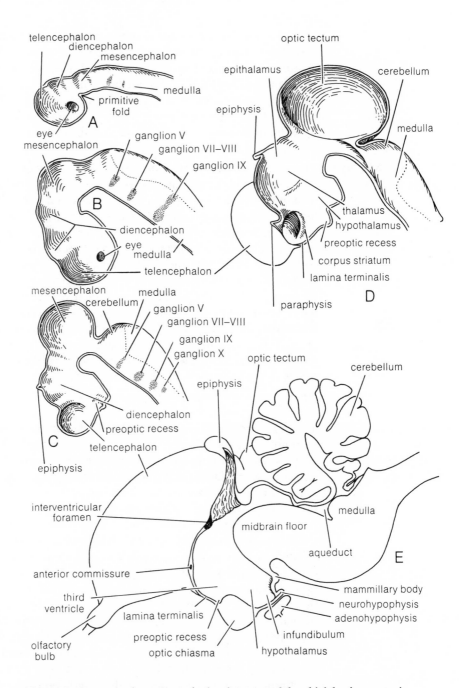

Fig. 13.2. Stages in the embryonic development of the chick brain as seen in sagittal view. *A*, primitive or ventral fold stage at which three major brain divisions can be recognized; *B*, main regions of brain have developed, as have the cranial nerve ganglia; *C*, epiphysis appears and cerebral hemispheres have expanded; *D*, paraphysis appears and optic tectum has expanded; *E*, olfactory bulbs evaginate and cerebellum is well developed.

out a dorsal enlargement, the *cerebellum* (fig. 13.2*E*), and the remainder and greater part of the hindbrain is the *myelencephalon* or *medulla oblongata,* characterized by its thin roof and the large nerve ganglia along its side walls (fig. 13.2*C*). The medulla oblongata is continuous with the spinal cord, from which it is not definitely delimited. These five main regions of the embryonic brain persist as such in the adult brain, developed by thickenings, folds, and outgrowths of their walls.

3. The Meninges    The central nervous system does not fill the bony canal in which it lies but is separated from it and protected by fluid and by connective tissue membranes, the meninges. In cyclostomes and fishes there is usually but one membrane, the *primitive meninx*. Amphibians, reptiles, and birds have two, an outer *dura mater* and an inner *secondary meninx*. In mammals there are three meninges: a relatively tough outer dura mater and two delicate inner ones, the *arachnoid* and the *pia mater,* both differentiations of the secondary meninx. The pia mater is closely associated with the brain and cord, into which it may send connective tissue partitions, and is separated by a considerable *subarachnoid space* from the web-like arachnoid. Outside the dura mater there is also a fat-vascular cushioning layer. The spaces around the brain and cord and between the meninges are in communication with the central canal of the cord and the ventricles of the brain, and all are filled with the *cerebrospinal fluid,* differing little from blood plasma.

4. Gross Features of the Adult Central Nervous System    The brain and spinal cord of vertebrates demonstrate the same kinds of morphological specialization as do other vertebrate body systems. That is to say, as vertebrates have adapted to different ways of life their brains reflect these specialized ways of life. The nuclei and tracts, which reflect much of this specialization, are described in the following section on the functional divisions of the nervous system.

All vertebrates possess the same number of brain divisions (fig. 13.3), but they vary in both absolute and relative volumes among species. These differences reflect both phylogenetic organization and physiological specialization for a particular ecological niche. Neurobiologists have often ranked living vertebrates and their brains in terms of their taxonomy and then spoken of their structure as though it had not changed in millions of years. The brain of a living frog may be a very poor substitute for the brain of an amphibian that lived 400 million years ago. We must always ask what structures in living vertebrate brains have been inherited from their ancient ancestors with little change, and what structures have been greatly modified in the course of their evolution. Does the brain of a living lamprey closely resemble the brain of an ancient ostracoderm, from which the lamprey sprang? How have nuclei and tracts changed from ancestral reptiles to modern reptiles, or to modern mammals? These questions and many more are only now beginning to be answered, and some of the answers will be explored in the following section on functional components. But first we must examine some of the gross variation that we see in the brains of living vertebrates.

The spinal cord in vertebrates is usually a slightly dorsoventrally flattened rod; this flattening is very pronounced in cyclostomes, where the band-like cord (fig. 13.4*B*) lies on the notochord. In tetrapods the cord presents *cervical* and *lumbar enlargements,* associated with the limbs; these are usually absent in fishes and limbless tetrapods such as snakes. The cord extends the full length of the vertebral canal in generalized

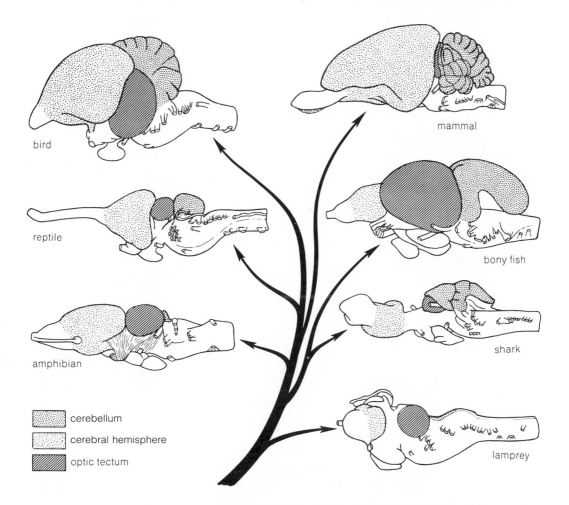

Fig. 13.3. Relative development of major brain divisions in a series of verte-brates. Brains are not drawn to scale. Modified after Northcutt 1977 (*McGraw-Hill encyclopedia of science and technology.* New York: McGraw-Hill).

fishes, reptiles, and birds; but in mammals and many teleosts it is much shorter than the vertebral column, tapering to a fine strand, the *terminal filament.* The cavity of the spinal cord is reduced to a very small *central canal* in the adult, and this is lined by a one-layered columnar epithelium, termed the *ependyma.* The gray matter of the cord, composed chiefly of nerve cell bodies and nonmyelinated axons, occupies its central region and in amniotes has in cross section the well-known H or butterfly-like shape, actually a four-fluted cylinder. The two dorsal limbs of the H are known as the *dorsal columns;* the two ventral limbs form the *ventral columns.* Between these there is in mammals a more or less definite *lat-eral column* on each side. The bar of the H encloses the central canal,

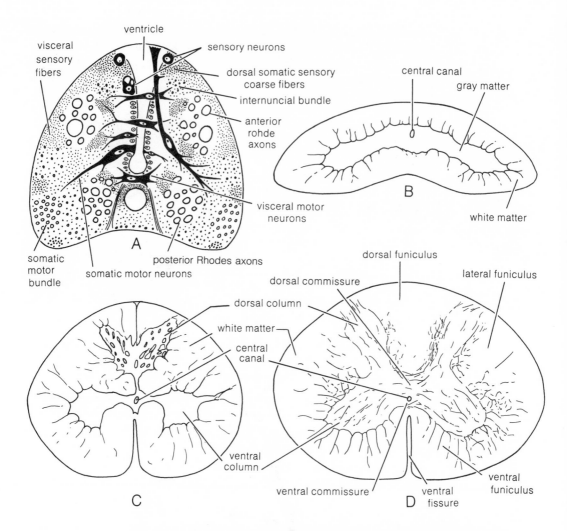

Fig. 13.4. Transverse section of the spinal cord showing the location and extent of neuronal cell bodies (gray matter) and more peripheral axonal pathways (white matter). *A, Branchiostoma; B,* larval lamprey; *C,* shark; *D,* mammal.

passing above and below it as the *dorsal* and *ventral commissures,* respectively. In anamniotes the gray matter varies in section from a rounded or quadrangular shape to a shape somewhat like an inverted T and Y (fig. 13.4*C*); in cyclostomes it forms a broad band (fig. 13.4*B*). The white matter, constituted of ascending and descending bundles or tracts of myelinated axons, forms a thick peripheral zone of the cord and is divided by the gray columns into *dorsal, lateral,* and *ventral funiculi* (fig. 13.4*D*). Columns and funiculi are, of course, less evident and defined in anamniotes, and they also differ in shape and proportions at different levels of the cord. The funiculi may be subdivided by longitudinal furrows, of which the deepest is the *median ventral fissure;* these also, except the last, are indistinct in anamniotes.

The vertebrate brain forms a conspicuous lobed enlargement continuous with the anterior end of the spinal cord. It occupies the cranial cavity of the skull, and the shape of this cavity conforms to the contours of the brain. The brain rests ventrally on the chain of cartilage bones ossified in the floor of the chondrocranium. Although the brain is bent at the mesencephalon in the embryos of all vertebrates, it straightens out in line with the longitudinal body axis in the later development of anamniotes. A permanent flexure is limited to amniotes and is especially notable in the brains of primates. The arrangement of the gray and white matter of the cord persists into the caudal part of the brain, but anteriorly the gray and white matter intermingle irregularly; in the cerebral hemispheres and cerebellum the gray matter forms a peripheral stratum over more central white matter. The white matter of the brain consists of ascending and descending tracts connecting various parts of the brain and cord; the gray matter consists mostly of "nuclei" (also called centers), that is, aggregates of cell bodies whose axons form the tracts of the white matter or issue in the cranial nerves.

The fifth, most posterior part of the brain, the myelencephalon or *medulla oblongata,* is continuous with the spinal cord through the foramen magnum and cannot be definitely delimited from the cord, although the boundary is usually placed just anterior to the level of the first spinal nerve. The rear part of the medulla is similar to the spinal cord, having a thick wall around the very small central canal. This canal soon widens out, however, into a large rhomboidal or triangular cavity, termed the *fourth ventricle;* its roof is greatly thinned out and consists only of the ependymal epithelium. To the outside of this, however, there adheres the richly vascularized pia mater, the two together forming the *tela choroidea.* In certain regions this is invaginated into the cavity of the fourth ventricle as vascularized tufts, known as *choroid plexi.* The walls of the medulla consist chiefly of the nuclei of origin and of termination of the fifth through the twelfth cranial nerves and their connections. These nuclei are arranged in generalized vertebrates according to their functions into four longitudinal columns (see below, under functional components), and this arrangement persists more or less throughout vertebrates.

The fourth division of the brain, or metencephalon, forms dorsally the *cerebellum.* This is of slight development in cyclostomes and amphibians, consisting of a ledge in front of the tela choroidea of the medulla, but is large in fishes of active habits, such as selachians and teleosts. It is of rather small size in most reptiles but increases in size and complexity in birds and mammals. Typically it consists of a median portion, the *body* or *corpus,* and a pair of lateral lobes, which are called *auricular lobes* after their shape in selachians, and which contain extensions of the fourth ventricle, the *auricular recesses.* In amniotes the auricular lobes are usually called the *floccular lobes.* The body of the cerebellum is generally marked by transverse fissures, which increase in number in birds and mammals, where the corpus seems more or less definitely divisible into anterior, middle, and posterior lobes. The floccular lobes are connected with the posterior lobes, and the marked lateral areas, termed cerebellar hemispheres, characteristic of mammals, appear to be developments of the anterior and middle lobes. The lateral areas (cerebellar hemispheres) have only been identified in tetrapods and appear to be concerned with muscles of the distal parts of the paired appendages. The cerebellum has a surface layer of gray matter known as the *cerebellar cortex,* and in birds and mammals this is underlain by a particularly thick stratum of white matter, so that sections of the cerebellum of these groups show the familiar "arbor vitae," lobulations of white matter covered by gray. It is characteristic of the teleost cerebellum that there is a thick invaginated lobe, termed the *valvula.* The cerebellum usually contains a cavity, the *cerebellar ventricle.*

The floor of the metencephalon shows little differentiation from the medulla in anamniotes, but, with increasing importance of the cerebellar hemispheres, its tracts and nuclei increase in volume and functional significance. Hence in mammals this floor, termed the *pons,* has a conspicuous fibrous structure, visible on the ventral surface in front of the medulla.

The functions of the cerebellum, insofar as they are known, are the coordination of muscular movements, the maintenance of muscular tone, and the equilibration of the body in space. Consequently, it has connections with many levels of the central nervous system, but particularly with sensory impulses from muscles and joints (*proprioceptive impulses*) and with the equilibratory mechanism of the inner ear.

The *mesencephalon* or *midbrain,* the third division of the brain, is also best developed dorsally, where its thick roof, known as the *optic tectum,* usually has two curved eminences (the *optic lobe*). In mammals there are four such externally apparent eminences, the *corpora quadrigemina;* the anterior pair are called the *superior colliculi,* the posterior pair the *inferior colliculi.* The inferior colliculi are present in fish but are not evident externally. Their exposure on the surface in mammals results from differences in embryonic development. The midbrain floor consists chiefly of tracts to and from other parts of the brain. The quantitative increase in tracts from the cerebral hemispheres of mammals results in a pair of conspicuous bundles, the *cerebral peduncles,* visible externally on the ventral surface of the midbrain. The midbrain is solid in mammals except for a narrow canal, the cerebral aqueduct (formerly the aqueduct of Sylvius), which passes through it from the fourth ventricle. In nonmammalian vertebrates the optic lobes are usually hollow, containing cavities, termed the *optic ventricles,* which open into the aqueduct.

The *diencephalon,* the second division of the brain, is chiefly a relay center for the cerebral hemispheres, and consequently its size and importance parallel those of the latter. It has a central cavity, the *third ventricle,* into which the aqueduct opens, and which in mammals is much compressed laterally. The roof of the third ventricle is thinned to an ependyma and forms a tela choroidea, which has choroid plexi projecting into the interior or, in certain anamniotes (petromyzonts, some ganoid fishes, selachians, some reptiles), *choroidal sacs,* often very large, bulging to the exterior. The term *parencephalon* is sometimes applied to the choroidal roof of the third ventricle. Ventrally the anterior boundary of the third ventricle is a thin membrane, the *lamina terminalis* (fig. 13.2*E*), believed to represent the original anterior end of the embryonic brain; more dorsally the diencephalon is bounded from the telencephalon by a transverse enfolding, the *transverse velum.* The diencephalon is divisible throughout vertebrates into a dorsal *epithalamus,* a middle thick *thalamus,* and a ventral *hypothalamus* (fig. 13.2*E*). The epithalamus consists chiefly of the *epiphyseal apparatus* and a pair of small masses, termed the *habenulae.* The epiphyseal apparatus apparently consisted primitively of two outgrowths of a sensory nature, and in lampreys, some ganoids, and many reptiles there are still two outgrowths—an anterior *parapineal* or *parietal body* and a posterior *pineal body* or *epiphysis.* In petromyzonts each develops a simplified eye. In *Sphenodon* and a number of other reptiles the anterior or parietal outgrowth forms a well-differentiated eye (usually called pineal eye, but more correctly termed parietal eye), and the posterior outgrowth is the pineal body proper. In other vertebrates only the pineal outgrowth appears in the embryo, and this develops into the pineal body, with no trace of eye formation. Except in lampreys, the pineal body contains glandular cells similar to the endocrine glands, and the pineal secretions are known to affect gonads and pigment cells. The parietal eye of reptiles occupies the parietal foramen of the skull, which is noticeable in the skull of primitive extinct tetrapods and their crossopterygian ancestors. These facts suggest that vertebrates originally had three eyes, but this question must be left undecided.

The thalamus comprises the lateral walls of the diencephalon; it is composed of many relay and integration centers for tracts passing to and from the cerebral hemispheres. In reptiles and mammals the dorsal parts of the thalami are frequently fused across the third ventricle by a connection of gray matter termed the *intermediate mass.*

The hypothalamus or ventral part of the diencephalon has the following main parts: *optic chiasma* (crossing of the optic nerves at their entrance into the brain), *tuber cinereum, infundibulum, hypophysis,* and *mammillary* region. The hypothalamus reaches its greatest development in fishes, where it is believed to be an important correlation center for olfactory, gustatory, and other sensory impulses. The tuber cinereum is an area of gray matter behind the optic chiasma. It continues ventrally as the infundibulum, a stalk in which the cavity is an extension of the third ventricle and from which the hypophysis or pituitary body extends. The hypophysis is composed partly of the oral epithelium of Rathke's pouch (*adenohypophysis*) and partly of nervous tissue from the infundibulum (*neurohypophysis*). The hypophysis is a very important endocrine gland that secretes a number of hormones. In fishes and amphibians the infundibulum is expanded posteriorly into a pair of swellings, the inferior lobes, and below them extends a soft-walled, very vascular structure, the *saccus vasculosus* or *vascular sac,* which is close to the pituitary. The mammillary region is the most posterior part of the hypothalamus and usually has a single or paired swelling, the mammillary bodies.

The telencephalon or forebrain, the most anterior division of the brain, consists of the paired *olfactory bulbs* and *cerebral hemispheres.* The olfactory bulbs are anterior outgrowths of the cerebral hemispheres (fig. 13.2*E*). They abut the rear wall of the nasal sacs; if the nasal sacs are considerably anterior to the brain, the bulbs remain in contact with them by way of a more or less elongated stalk, the *olfactory stalk* or *peduncle.* The olfactory bulbs are generally hollow but are secondarily solid in many teleosts and mammals. Their size is correlated with the development of the sense of smell, and hence they are very small in birds and relatively small in primates.

The cerebral hemispheres constitute the major part of the telencephalon. They form large lateral bulges to either side of the original median anterior brain wall, the lamina terminalis, which is thus left at the bottom of the cleft between the hemispheres (fig. 13.2E). Not infrequently, however, the olfactory bulbs or cerebral hemispheres or both may be secondarily fused together dorsally. The lamina terminalis extends forward and upward from the *preoptic recess,* a depression in front of the optic chiasma, to the *neuroporic recess,* the point of last closure of the neural folds of the embryo. Just below the neuroporic recess, the lamina terminalis contains two bundles, the *anterior* and the *pallial commissures,* which connect regions of the two hemispheres. Above the neuroporic recess the original telencephalic roof, extending backward to the velum transversum, is thinned to form a tela choroidea, from which choroid plexi project into the cavities of the cerebral hemispheres. Immediately in front of the velum transversum the telencephalic roof has a dorsal branched or folded evagination, the *paraphysis* (fig. 13.2E). This is evident in the embryos of all classes of vertebrates but does not persist in the adults of most tetrapods; its function is not known.

In most vertebrates the cerebral hemispheres are smooth, rounded, or elongated lobes not noticeably larger than the rest of the brain (fig. 13.3), but in amniotes they increase greatly in size.

Each cerebral hemisphere contains a cavity, known as the *lateral* ventricle, and these communicate with each other and with the third ventricle by a passage (or pair of passages) termed the *interventricular foramen* (formerly the *foramen of Monro*). In reptiles and birds these ventricles are somewhat reduced, but in mammals they are well developed though very irregular because of the bulgings of the hemisphere wall.

The cerebral hemispheres of vertebrates are formed embryonically in one of two ways (fig. 13.5). In cyclostomes, elasmobranchs, sarcopterygians, and tetrapods the cerebral hemispheres are said to be *inverted.* The unpaired prosencephalon embryonically forms a pair of cerebral hemispheres by evaginating the lateral walls of the early neural tube, and the cavity within each evagination is termed the lateral ventricle. In teleosts and other actinopterygians, the cerebral hemispheres are said to be *everted.* In these animals, the roof of the early neural tube proliferates and the thickened lateral walls bend outward on themselves so that the entire prosencephalon has opened out. These animals cannot be said to possess lateral ventricles in the strict sense. It appears that this divergence in telencephalic embryology arose with the ancient placoderms, but the functional significance of this divergence is not understood. It is difficult to recognize homologous regions between the everted and inverted cerebral hemispheres, and there is, at present, no general agreement on the subject.

The wall of the cerebral hemisphere may be divided into a dorsal

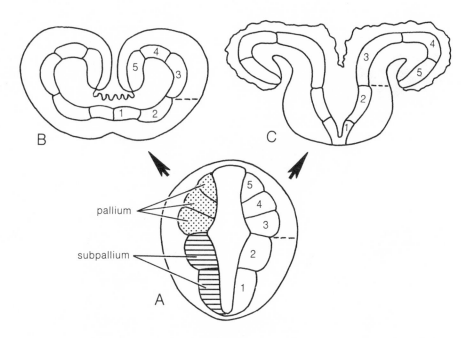

Fig. 13.5. Embryonic development of the telencephalon in vertebrates as seen
in transverse section. *A,* early stage in development prior to formation of
cerebral hemispheres. Neuroblasts are located next to central cavity (prosen-
coele). *B,* in all vertebrates except ray-finned fishes cerebral hemispheres form
by thickening and outpocketing of the lateral walls (evagination and inversion).
*C,* in ray-finned fishes, the hemispheres form by eversion of the roof (pallium)
resulting in a different topological organization. 1, septum; 2, striatum; 3,
lateral pallium; 4, dorsal pallium; 5, medial pallium.

half, the *roof* or *pallium,* and a ventral half, the *floor* or *subpallial region*
(fig. 13.5). In all tetrapods, the roof or pallium has three primary longitudinal zones (figs. 13.5, 13.6). The most medial zone (*PI* or *hippocampus*) is part of the limbic system and receives sensory information
from many brain regions. It and the medial half of the dorsal zone (*PIIa*
or *dorsomedial pallium*) are concerned with somatic and visceral integration. The lateral half of the dorsal zone (*PIIb* or *dorsolateral pallium*)
is concerned with processing visual information. This pallial zone along
with part of the lateral zone (*PIII* or *dorsal ventricular ridge*) forms a
laminated cellular gray area called the *isocortex* (*neocortex*) in mammals. This zone is concerned with auditory, visual, and somatic sensory
and motor processes in amniotic vertebrates. The other part of the
lateral zone (*PIII* or *pyriform pallium*) is concerned with olfactory
information.

The floor or subpallium is also divisible into three longitudinal zones
(fig. 13.6).The most medial zone (*BIII* or *septum*) is also part of the
limbic system and is concerned with visceral integration. The ventral
and lateral basal zones (*BI* and *BII*) fuse and form the *corpus striatum,*
which is concerned with somatic motor processes.

It is believed that in one group of ancestral reptiles the lateral zone
(PIII) proliferated into the lateral ventricle, while in a second group of

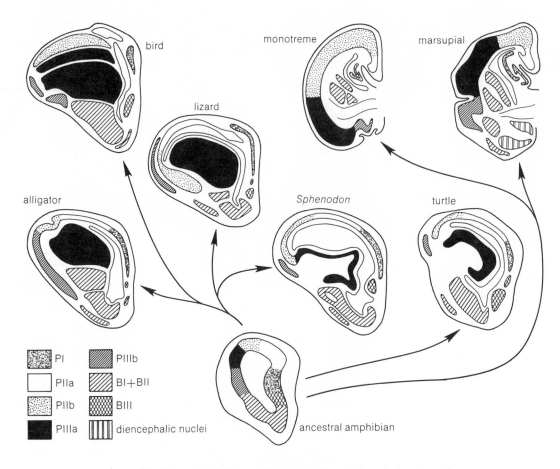

Fig. 13.6. Evolution of the cerebral hemisphere in land vertebrates as seen in transverse section of a single hemisphere. PI, hippocampus; PIIa, dorso-medial (cingulate) cortex; PIIb, dorsolateral cortex (isocortex); PIIIa, a second part of isocortex (dorsal ventricular ridge); PIIIb, lateral (pyriform) cortex; BI and II, corpus striatum; BIII, septal nuclei. Modified after Northcutt 1977 (*McGraw-Hill encyclopedia of science and technology*. New York: McGraw-Hill).

ancestral reptiles the lateral zone proliferated onto the surface to form part of the isocortex. In both cases the proliferation seems to have occurred in response to the increased functional significance of the processes being mediated by this lateral zone. While it is known that this region is involved in visual and auditory processes as well as motor control, the exact significance of the vast increase in volume is still not understood. It may, however, hold the key to an understanding of the evolution of the tetrapod telencephalon.

In eutherian mammals, a broad horizontal band of myelinated axons, the *corpus callosum,* connects the isocortices of the two hemispheres. The origin of the corpus callosum is not understood, but it allows parts of the isocortex of each hemisphere to specialize in the analysis of different types of information without losing communication with each other. Below the corpus callosum is another white fiber tract, the *fornix,* which runs from the hippocampus to the septum and mammillary bodies.

**5. The Functional Divisions of the Nervous System and the Peripheral Nervous System**

It is convenient in discussions of the nervous system to divide the body functions into two categories, *somatic* and *visceral*. The somatic functions are those mediated by the body wall, that is, the skin, the musculature, and the skeleton. The visceral functions are those carried on by the other systems, the digestive, respiratory, circulatory, urogenital, and endocrine systems. Each of these two categories has two further components, an *afferent* or *sensory,* and an *efferent* or *motor,* so that the nervous system in general is made up of four functional components. These are (1) the *somatic sensory component,* which handles the impulses from the sense organs of the skin, the special sense organs of the head (eye, ear, and nose), and the sensations from the deeper body-wall structures, such as muscles and joints (proprioceptive impulses); (2) the *somatic motor component,* which handles outgoing impulses to the striated musculature except the branchial musculature; (3) the *visceral sensory component,* which deals with sensations from the visceral systems, including the special sense of taste; and (4) the *visceral motor component,* which mediates impulses to smooth musculature of the viscera, to the branchial musculature, and to glands, and the like. The visceral components involve the autonomic as well as the central nervous system (see below).

The dorsal half of the central nervous system (gray matter) is sensory, the ventral half, motor. This division is well marked in the spinal cord, continues into the medulla, but is somewhat less clear in higher levels of the brain. In the spinal cord the order of arrangement of the four components is, from dorsal to ventral, somatic sensory, visceral sensory, visceral motor, and somatic motor (fig. 13.7*B*). The dorsal gray columns of the cord are somatic sensory; the ventral gray columns are somatic motor; the gray region between is visceral and in mammals

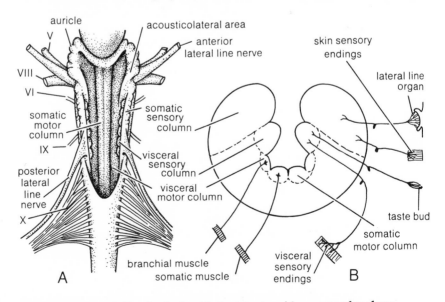

Fig. 13.7. *A*, medulla of a shark with the tela choroidea removed to show the longitudinal sensory and motor columns. *B*, transverse section through the medulla illustrating the positions of the sensory and motor columns and their respective organs and innervations.

forms a distinct lateral column connected with the autonomic system. The four components continue in the same order into the medulla, where they form four longitudinal areas, which are especially clear in fishes (fig. 13.7$A$). Large areas of the gray matter of brain and cord also serve for the association and coordination of impulses.

It has been shown by the brilliant work of a succession of American neurologists that each nerve of the peripheral nervous system is typically made up of fibers belonging to all four functional components. This concept has been of great value in understanding the cranial nerves and their central connections.

The spinal nerves issue from the spinal cord at segmental intervals, emerging through the intervertebral foramina. They are named after the vertebral regions, cervical, thoracic, lumbar, sacral, and caudal. In cases where the spinal cord does not reach to the caudal end of the neural canal, the last pairs of spinal nerves run inside the canal before issuing at the appropriate level and so form a bundle termed the *cauda equina*. The attachment of a nerve, often multiple, to the central nervous system is termed its *root*. Each spinal nerve has two roots, a *dorsal* or *sensory* root, which passes into the dorsal gray column, and a *ventral* or *motor* root, which issues from the ventral gray column (fig. 13.8). The dorsal root is composed chiefly of somatic sensory and visceral sensory fibers. It is a rule that the nerve cell bodies of sensory fibers are located not in the central nervous system but in outside ganglia. Each dorsal root bears a ganglion, the *dorsal* or *spinal* ganglion, which is made up of the

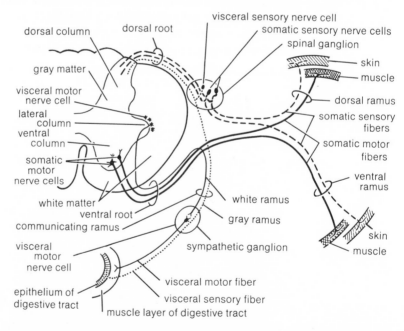

Fig. 13.8. Diagram of a transverse section through the mammalian spinal cord illustrating the peripheral spinal and sympathetic nerves. Sensory fibers enter spinal cord via the dorsal root while motor fibers leave the cord via the ventral root. Dashed lines are somatic sensory fibers; dotted lines are visceral sensory fibers; hatched areas receive visceral motor fibers; crosshatched areas receive somatic motor fibers.

cell bodies of the somatic sensory and visceral sensory fibers (fig. 13.8). The ventral root is composed of somatic motor fibers and in amniote vertebrates also contains most if not all of the visceral motor fibers, which in anamniote vertebrates exit, in part, by the dorsal root. The cell bodies of the somatic motor fibers are in the ventral gray column, and those of the visceral motor fibers are in the lateral column or the region of the gray matter corresponding to this position. It is a rule that the cell bodies of the somatic motor system are always inside the brain and cord.

Except in lampreys the dorsal and ventral roots unite distal to the spinal ganglion to form a *spinal nerve,* which then exits through the intervertebral foramen of its appropriate body segment and divides into three branches, the *dorsal ramus,* the *ventral ramus,* and the *communicating ramus*. Dorsal and ventral rami both contain somatic sensory and somatic motor fibers (also fibers of the autonomic system) and supply the skin and voluntary muscles of dorsal and ventral regions, respectively, of the appropriate body segment. The cutaneous (skin) areas supplied by the dorsal and ventral rami of each spinal nerve can be mapped, and such maps are useful in determining the level of injury to the spinal cord, since sensation is lost from skin areas below damage to dorsal parts of the cord. The somatic motor fibers in the dorsal ramus supply the epaxial muscles, those in the ventral ramus, the hypaxial muscles. The communicating ramus joins a ganglion of the autonomic system. For the original arrangement of the roots and branches of the spinal nerves see the nervous system of cyclostomes.

In connection with the paired appendages, the ventral rami of a variable number of successive spinal nerves of the region concerned branch and anastomose to form a network or *plexus,* from which the nerves going to the appendages emerge. These plexi are the *brachial plexus,* for the anterior appendage, and the *lumbar* or *lumbosacral* plexus, for the posterior appendage. In their passage through the plexus the nerve fibers intermingle and cross over, so that the emergent nerves have no correspondence in content with those that entered the plexus. The occurrence of these plexi is taken to indicate (1) that the limb muscles arise from the hypaxial parts of the myotomes, since only the ventral rami enter into these plexi; (2) that the limb muscles are derived from a number of myotomes, since each ramus supplies only the muscles derived from a single myotome and several rami are concerned in the plexi; and (3) that the appendage muscles must have undergone considerable torsion and displacement, resulting in a crisscross arrangement of their motor nerves, since each nerve retains connection with the muscle of its body segment. The plexi also carry sensory fibers. In the neck region the *cervical* nerves may form a *cervical plexus* for the neck muscles, which has connections with the cervical sympathetic and the spinooccipital nerves or their equivalents. It may be continuous with the brachial plexus, and a *cervicobrachial plexus* results.

In mammals the *phrenic* nerves, which innervate the diaphragm, are derived from the more anterior spinal nerves entering into the brachial plexus. This indicates that the myotomes that contribute to the diaphragm are of cervical origin and have migrated posteriorly.

In some fishes (selachians, Chondrostei, Dipnoi) there is associated with the lumbosacral plexus a longitudinal *collector* nerve that is joined by a number of spinal nerves anterior to the plexus and then enters the plexus.

In the transitional region between the medulla and the spinal cord, what were originally the most anterior spinal nerves undergo some modification, reducing their dorsal or sensory roots and becoming primarily somatic motor. These are termed the *spinooccipital* nerves. In cyclostomes these spinooccipital nerves are located outside the skull. The more anterior ones (generally two) are located within the skull in fishes and amphibians. When located within the skull, they are termed *occipital* nerves since they exit through the occipital region of the skull. In amniotes the more posterior ones, called *occipitospinal* nerves, are also in the skull. The number of head segments that are involved in skull formation, particularly the occipital region, vary in different vertebrates (see segmentation of the head), and no general agreement has been reached on the number of ancestral segments forming the head. The work of Fürbringer is usually cited in support of the view that the skull increases posteriorly in a phylogenetic scale so that cervical vertebrae have been incorporated into the skull as it has grown in a posterior direction. The spinal nerves associated with these cervical vertebrae are then said to become primarily somatic motor in function and are considered to become cranial nerves (*spinal accessory* and *hypoglossal nerves*). For this reason anamniotes are said to have ten cranial nerves whereas amniotes are said to have twelve cranial nerves. Even in mammals, the spinal accessory and hypoglossal nerves contain somatic sensory cells scattered along their roots and therefore these nerves cannot be considered to be pure somatic motor nerves. In all vertebrates these nerves, whether we call them occipital nerves or cranial nerves, innervate anterior trunk epimeric musculature and cranial hypobranchial musculature. It is less important to argue whether anamniotes possess only ten cranial nerves or twelve, as amniotes do, than to realize that the ancestral vertebrates probably possessed more than twelve cranial nerves.

The cranial nerves differ from the spinal nerves in that their dorsal and ventral roots do not unite and may appear to be quite separate nerves. As in the case of the spinal nerves, the visceral and somatic sensory fibers of the cranial nerves have their cells of origin in ganglia on the sensory roots close to the entrance of the roots into the brain. Because of the great changes involved in cephalization, the cranial nerves do not have the regular composition of the spinal nerves, and analysis of their components has been one of the important contributions of American neurologists.

### a. Nervous Terminalis

In 1878 Fritsch described a new cranial nerve in addition to the usual twelve in sharks, and this nerve has since been identified throughout the gnathostomes (except birds). It originates from the telencephalon near the neuroporic recess, runs along the inner side of the olfactory tracts, and terminates in the walls of the nasal sacs. It has one or more ganglia along its course. It is considered to be a visceral sensory and motor nerve to the blood vessels of the olfactory epithelium, but this has not been adequately demonstrated.

### b. Olfactory Nerve (I)

The olfactory nerves differ from all other vertebrate nerves in that their fibers originate from the neurosensory cells of the olfactory epithelium of the nasal sacs. The numerous fibers from these cells constitute the olfactory nerves, which run a very short course into the olfactory bulbs, in mammals passing through the pores of the cribiform plate of the ethmoid bone. Although the nasal sacs and the olfactory epithelium are of ectodermal origin, and hence the olfactory nerves should belong to the somatic sensory division, it appears from the central connections (i.e., brain pathways) of the olfactory tracts that the whole olfactory system is part of the visceral sensory division. The *organ of Jacobson* or *vomeronasal organ,* a sensory area of the interior of the nose, has a separate nerve, the *vomeronasal nerve,* considered a separate part of the olfactory nerve. This terminates in a distinct part of the olfactory bulb, called the *accessory olfactory bulb.*

### c. Optic Nerve (II)

The optic nerves are not really nerves at all but brain tracts. The sensory part of the eye, the retina, is embryologically part of the brain wall and consists of several layers of nerve cells. The fibers of the last layer form the optic nerve, which exits from the inner surface of the eyeball and runs a short course into the optic chiasma of the hypothalamus. Here the optic fibers cross wholly or, in many vertebrates, partly to the opposite side. The optic system belongs to the (special) somatic sensory division; although the fibers enter the ventral surface of the brain, they terminate in dorsal areas of thalamus and midbrain.

### d. Oculomotor Nerve (III)

The oculomotor nerves are motor nerves to certain muscles that move the eyeball (internal, inferior, and superior rectus and inferior oblique muscles). Embryology shows that these muscles arise from regular myotomes in series with the trunk myotomes; the oculomotor nerve is primarily a somatic motor nerve (it also carries autonomic fibers). The cells of origin lie in the midbrain floor in the somatic motor area, and the roots of the oculomotor nerves emerge on the ventral surface of the midbrain.

### e. Trochlear Nerve (IV)

The trochlear nerve supplies one of the eyeball muscles (superior oblique) and is therefore similar in content and connections to the oculomotor nerve. Although its roots emerge from the metencephalic roof just behind the optic tectum, the nuclei of origin are situated in the midbrain floor in the somatic motor area.

### f. Abducens Nerve (VI)

The abducens nerve is the third of the nerves of the eyeball muscles, innervating the external rectus muscle, and its nature and relations are similar to the others. The nuclei of origin lie in the floor of the medulla, and the roots emerge on the ventral surface of the anterior end of the medulla.

The reason that there are three cranial nerves to supply six small muscles is that these muscles come from three different myotomes (see discussions of head segmentation).

### g. Auditory or Acoustic Nerve (VIII)

The auditory nerve is the nerve of the inner ear, and, since this is an ectodermal sense organ, the auditory nerve belongs to the somatic sensory column of the medulla. As the inner ear has two functions—equilibration, vested in the semicircular canals and their ampullae, and hearing, vested in the cochlea or its phylogenetic forerunner, the lagena —so the auditory nerve carries two kinds of impulses and in mammals is accordingly divided into vestibular and cochlear components, each with a separate ganglion. In many nonmammalian vertebrates the fibers corresponding to the cochlear nerve do not form a separate nerve. The vestibular fibers of the auditory nerve have strong connections with the cerebellum, for this structure is concerned with equilibration.

Closely related functionally to the acoustic system is a second somatic sensory system peculiar to all aquatic anamniotes, the *lateralis system*. This system consists of sensory canals, their nerves, ganglia, and central pathways. The sensory canals form the lateral line canal along the side of the trunk and the canals distributed over the head. These canals contain sensory cells similar to those of the internal ear, and the entire lateralis system is so closely related to the acoustic system that it is customary to treat both together as the *acousticolateralis* system. The lateralis system is a skin sense system for the detection of water vibrations, and each of its nerves bears a sensory ganglion. However, these ganglia differ from other sensory ganglia in that their cells develop from ectodermal sensory patches (*dorsolateral placodes*), which form a series along the head in line with the *auditory placode* (the sensory patch that develops into the internal ear.) The lateralis ganglia lie close to the ganglia of the seventh, ninth, and tenth cranial nerves. The lateralis nerves accompany the branches of these three nerves, but the lateralis nerve roots enter the anterior part of the somatic sensory column of the medulla in the same region as the auditory nerve. This medullar region is well developed in aquatic vertebrates and is called the *acousticolateral area*. This area is continuous with the auricle of the cerebellum. The entire lateralis system degenerates in land vertebrates and in the anuran amphibians; it disappears when the aquatic tadpoles metamorphose into the adults.

### h. The Branchial Nerves

The following cranial nerves (i.e., fifth, seventh, ninth, and tenth) are closely related to the branchial bars and gill slits. Each of the four branchial nerves supplies a definite branchial bar and its musculature and continues to do so through all its phylogenetic changes. Typically, each branchial nerve forks around a particular gill slit, having a pretrematic branch in front of the gill slit and a posttrematic branch behind the gill slit (see chap. 10); there is, further, a pharyngeal branch supplying the pharyngeal lining. Since the branchial musculature is visceral, the branchial nerves lack somatic motor fibers. The pharyngeal branch consists exclusively of visceral sensory fibers; the visceral motor fibers to the branchial musculature are carried in the posttrematic branch, and both pre- and post-trematic branches have visceral sensory fibers. Somatic sensory fibers are well developed only in the fifth nerve. The sensory fibers of the branchial nerves have their cells of origin in ganglia close to or fused with lateralis ganglia; these branchial nerve ganglia also differ from ordinary sensory ganglia in that their cells are derived in part from the epibranchial placodes, sensory patches at the upper angle of the gill slits. The roots of the branchial nerves are attached to the sides of the medulla and relate to the appropriate functional columns of the medulla. Traditionally the lateralis system is considered to represent the special somatic sensory components of the branchial nerves. However, separate embryonic origins of the lateralis ganglia and the branchial nerve ganglia, their separation in the adult, and their termination in the acoustic area of the medulla, have led other researchers to consider them a series of separate cranial nerves, which also includes the eighth cranial nerve, rather than a component of the branchial nerves. It can be argued that the sensory ganglia of the branchial nerves have a multiple embryonic origin, from both the epibranchial placodes and the neural crest, and that their somatic and visceral sensory components are also segregated to different neural centers in the medulla. Furthermore, the somatic and visceral components of the different branchial nerves project to a single nucleus within the medulla. Thus the lateralis system may represent the special somatic sensory components of the branchial nerves and be separated from the other components because of the specialization and extreme development of this system in aquatic vertebrates, or it may be a completely separate series of cranial nerves. The question cannot be answered at present.

### i. Trigeminus Nerve (V)

The trigeminus nerve is the nerve of the first (mandibular) gill arch, that is, the upper and lower jaws; it forks around the mouth suggesting a gill-slit origin for the latter. It has a very large somatic sensory component to the eye region (ophthalmic branch), upper jaw (maxillary or pretrematic branch), and lower jaw (mandibular or posttrematic branch), supplying skin, teeth, and other areas. The mandibular branch contains the visceral motor fibers to the muscles of the mandibular arch (i.e., first constrictor and levator, adductor mandibulae, masseter, temporal, anterior belly of the digastric, the pterygoid muscles, etc.). The trigeminus is not accompanied by any lateral line component and lacks somatic motor and visceral sensory components. The somatic sensory part bears a large ganglion, the *Gasserian* or *semilunar ganglion,* attached to the somatic sensory column of the medulla; the visceral motor part leaves the medulla by a separate root, which has its nucleus of origin in the visceral motor column.

The *deep ophthalmic* or *profundus* nerve, apparently part of the trigeminus, is conspicuous in fishes but in mammals is not distinct from the ophthalmic branch of the trigeminus. This was originally a separate cranial nerve (see metamerism of the head).

Because of the very large spread of the somatic sensory part of the trigeminus, which must supply the skin of practically the whole head, the somatic sensory component of the remaining branchial nerves (excluding the lateralis system) is greatly reduced.

### j. Facial Nerve (VII)

The facial nerve is the nerve of the second or hyoid arch and its musculature; it forks around the first gill slit (spiracle). It has only a small somatic sensory component but has a large visceral sensory element from the taste buds and more anterior portions of the buccopharyngeal lining. The facial nerve is accompanied by large trunks of the lateralis system and has visceral motor fibers to the muscles of the hyoid arch (hyoid constrictor and levator, interhyoideus, constrictor colli, posterior belly of the digastric, depressor mandibulae, platysma and other muscles of facial expression). The sensory fibers have their cells of origin in the *geniculate* ganglion, attached to the side of the medulla. Commonly the ganglia of the fifth, seventh, and eighth cranial nerves are fused into one complicated mass, which in fishes also includes the lateralis ganglia.

### k. Glossopharyngeal Nerve (IX)

The glossopharyngeal nerve is the nerve of the third branchial bar; it forks around the second gill slit. It, too, consists chiefly of visceral sensory fibers from the more posterior taste buds and pharyngeal lining and of visceral motor fibers to the muscles derived from the third branchial arch. It is usually associated with a small lateralis accompaniment in fishes. The sensory ganglion, called *petrosal*, is attached to the side of the medulla in line with the facial ganglion.

### l. Vagus Nerve (X)

The vagus nerve is a large nerve with a wide distribution, supplying all the remaining branchial bars and slits and most of the viscera. It is accompanied by a very large lateralis nerve, which, after supplying a small part of the head canals of the lateralis system, runs as the large lateral line nerve along the side of the trunk under the lateral line. The vagus proper has a small somatic sensory part (with cells of origin in the *jugular* ganglion) from the skin around the ear, but its main or *visceral* trunk consists of visceral sensory and motor fibers. This runs along the upper ends of the remaining gill arches and gives off a ganglionated branch to each of these; from each ganglion there springs the usual pretrematic, posttrematic, and pharyngeal branches. With the loss of the branchial apparatus, the branchial part of the vagus is much reduced; but the visceral motor fibers in the posttrematic branches continue to supply the muscular derivatives of the more posterior branchial bars, that is, the striated muscles of the pharynx and larynx. After passing the branchial region, the visceral trunk of the vagus continues posteriorly and supplies visceral sensory fibers (cells of origin in the *nodosal* ganglion) to the esophagus, larynx, trachea, and viscera in general, except the extreme posterior ones. The visceral branches also carry visceral motor fibers (preganglionics) to the heart muscle and the smooth musculature of the same viscera, but these relay in ganglia of the autonomic system.

The only explanation for the supplying of several branchial bars by the vagus is that it must have appropriated at least parts of the nerves originally behind and separate from the ancestral vagus nerve.

### m. Spinal Accessory Nerve (XI)

The spinal accessory nerve is a visceral motor nerve whose roots are in line with the ventral or motor roots of the spinal nerves. It is compounded in amniotes of a part of the vagus and of contributions from the spino-occipital nerves. The spinal part is a visceral motor nerve to the cucullaris muscle of its derivatives; the vagal part goes to branchial musculature and also accompanies the visceral branches of the vagus, functioning by way of the autonomic system. In anamniotes the spinal accessory nerve is represented by a branch of the vagus and by the spino-occipital nerves. It is not well developed in reptiles.

### n. Hypoglossal Nerve (XII)

The hypoglossal nerve is primarily a somatic motor nerve innervating the muscles of the tongue. It will be recalled that the tongue muscles are hypobranchial muscles; that is, they come from regular myotomes behind the branchial region which curve around the rear of this region and grow forward in the midventral area. These myotomes are primitively innervated by the ventral roots of the spinal nerves.

## 6. The Segmentation of the Head

The question whether the vertebrate head was originally segmented, as the trunk is, has engaged the attention of leading anatomists for over a hundred years (see chaps. 6 and 10). From embryological research, the conclusion has been reached that the head is composed of a series of mesodermal segments which, through the process of cephalization, have become almost indistinguishably fused together. As in the case of the trunk, the segmentation is primarily myotomic, and that of the skeleton and the gill slits, for example, has been imposed upon these parts by the segmental arrangement of the epimeres. It is generally agreed that the otic capsule represents an important landmark and that there are three epimeres or somites in front of the capsule and a variable number behind it (fig. 13.9). The three prootic somites are termed the premandibular, mandibular, and hyoid somites. Each is hollow, containing a coelomic space, and except for the first (termed the anterior head cavity), they are continuous with the general coelomic space of the hypomere. The epimeres give rise only to somatic muscle, and in the head region the three prootic somites produce the extrinsic eye muscles. The premandibular somite gives rise to the four eye muscles innervated by the oculomotor nerve, the mandibular somite to the superior oblique, and the hyoid somite to the external rectus muscle. There are variations of this account of the origin of the eye muscles, and it must be regarded as only generally accurate.

Below and behind the ear the number of head somites (metaotic somites) varies from two or three to ten or twelve or more, according to

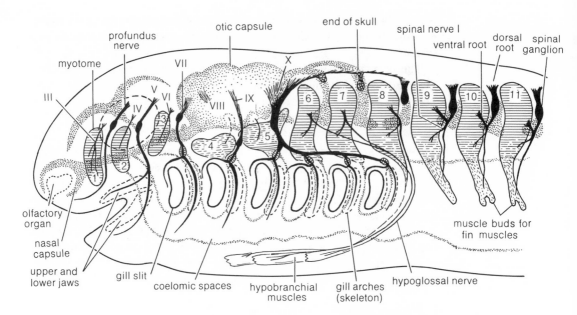

Fig. 13.9. Schematic lateral view of vertebrate head illustrating head segmentation. The somites (hatched areas) bear Arabic numerals and are innervated by somatic motor nerves. The branchiomeres (dashed outlines) located between gill slits and including the jaws are innervated by visceral nerves. The somatic and visceral head nerves bear Roman numerals. The heavily stippled areas indicate the chondrocranium, sensory capsules, and sclerotomes. The lightly stippled areas indicate the position of coelomic spaces. The myotomes and somatic motor nerves of segments 4 and 5 degenerate.

the species. In cyclostomes there is a complete set of metaotic epimeres, and all give rise to somatic muscle. In general, however, one or two metaotic epimeres fail to form muscle, presumably (at least in lampreys) because they are crowded out of existence by the enlarging otic capsule. The remaining ones contribute to the hypobranchial musculature.

The hypomeric portions of the head segments give rise to the branchial musculature. In embryonic stages a part of the hypomeric coelom runs in each visceral arch (branchial bar). The innervation of the head segments is of great interest. Each of the original head segments was innervated by a separate pair of dorsal sensory and ventral motor nerves. The oculomotor nerve is the somatic motor nerve of the premandibular segment; the profundus nerve is the somatic and visceral sensory nerve of this segment. The trochlear nerve is the somatic motor nerve of the mandibular segment; the trigeminus nerve is the somatic sensory, visceral sensory, and visceral motor nerve of this segment. The abducens nerve is the somatic motor nerve of the hyoid segment; the facial nerve is the sensory and motor nerve of the branchial musculature of this segment. The next two head segments (metaotic segments) normally do not form somatic musculature, and hence there are no somatic motor roots (ventral nerves) for these segments. The dorsal nerve bearing somatic sensory, visceral sensory, and visceral motor components is represented by the glossopharyngeal nerve for the fourth head segment (first metaotic). The vagus nerve represents the dorsal nerve of the fifth head segment (second metaotic). In addition to innervating the fifth head segment, the vagal-spinal accessory nerve complex innervates the branchial muscles of the remaining head segments. As has been pointed out earlier, the number of metaotic segments is variable among different vertebrate species, and thus the vagal-spinal accessory complex is variable since it represents a fusion of the remaining dorsal nerves of these head segments. The peripheral separation of this complex in land vertebrates is concomitant with the functional divergence of these remaining branchiomeres. The fourth and fifth branchiomeres form pharyngeal musculature while the sixth and seventh branchiomeres form axial and appendicular musculature. Thus, the spinal accessory nerve is not strictly a "new cranial nerve," but results as a separation of muscles having a common function into two sets of muscles having very different functions. The hypobranchial musculature is a product of the myotomes of these metaotic segments and is innervated by the hypoglossal nerve. This nerve also appears to be produced primarily by a fusion of the ventral somatic nerves of these head segments.

The gill slits are not thought to have been ancestrally segmental, but are believed to have had their segmental arrangement imposed on them by surrounding parts. The ancestral number of gill slits was greater than the number in existing jawed vertebrates.

Traces of additional prootic segments (besides the accepted three) have been reported from time to time. The labial cartilages of sharks have often been regarded as remnants of extra gill arches. When so regarded, it is claimed that the trigeminal nerve represents more than a fusion of the dorsal nerves of two head segments.

## 7. The Sense Organs

The sense organs are specialized cells or organized groups of cells and tissues that respond to environmental changes by initiating nerve impulses; these go to the central nervous system and there evoke appropriate responses. The sense organs are usually specific; that is, each kind responds only to a particular environmental agent such as light, temperature, and the like. The sensitive part may be either the nerve endings themselves or the neurosensory cells or the sensory cells with which nerve endings make contact.

### a. General Cutaneous Sense Organs

The cutaneous sense organs comprise the general sensory organs of the skin and body wall that mediate temperature, light and deep touch, pain, and joint sense. The morphology of these sensory organs varies from simple unmyelinated nerve endings (for pain) to highly structured encapsulated sense organs, called *tactile corpuscles,* in which the nerve endings are enclosed in layers of connective tissue. These include corpuscles of Ruffini (warmth), end bulbs of Krause (cold), Meissner's and Merkel's corpuscles (touch), and Paccinian corpuscles (pressure). The structural uniqueness of these corpuscles led to the notion that each corpuscle is the receptor of a specific sense, *the law of specific nerve endings.* However, free nerve endings transmit information on temperature changes, as the corpuscles of Ruffini and the end bulbs of Krause do.

In general, the anaminotes possess few encapsulated sense organs;

cyclostomes appear to possess none and rely primarily on free nerve-ending receptors. Most anamniotes inhabit an aquatic environment where there are only gradual changes in temperature and pressure on the body, while amniotes inhabit an environment that changes sharply and constantly. Encapsulated sense organs are probably a specialization to monitor these changes accurately. Two of the most specialized sensory receptors to evolve in vertebrates are the pit organ and labial organs of two families of snakes (the viperids and boids). These organs are heat receptors that can detect a change in temperature as slight as .01° C and are used by these snakes to detect warm-blooded prey, even in the dark.

### b. Proprioceptive Sense Organs

The proprioceptive sense organs are found in the muscles, tendons, and joints and play an important role in the maintenance of body posture. They also provide sensory information on limb and body position that the brain needs to coordinate movements of the limbs and body. The proprioceptive sense organ of muscles is the *neuromuscular* spindle of intrafusal muscle fibers (specialized small striated muscle cells located within the belly of skeletal muscles). These spindles receive sensory endings of the dorsal root spinal ganglion neurons. These endings are called annulospiral because the sensory nerve enters the connective tissue capsule of the intrafusal muscle spindles and arborizes by spiraling around the nuclear region of the muscle spindle. Information conveyed centrally from the muscle spindle is utilized as the input of the extensor or myotatic reflex described below. However, the annulospiral endings are not the source of information about limb position that reaches the cerebral cortex and thus consciousness. This information is thought to be provided by joint receptors and perhaps neurotendinous organs. These last receptors range from net-like ramifications of the nerve ending to end organs (Golgi organs) similar to Paccinian corpuscles and are located in tendons and the connective tissue of the joint capsule and within muscle sheaths. At present it is not known whether or not cyclostomes possess any of these sense organs, although sharks and all other vertebrates are known to possess them. They are particularly well developed in land vertebrates where the limbs become the major structures responsible for posture and locomotion.

### c. Special Cutaneous Sense Organs: Neuromast System

This system, limited to aquatic vertebrates, consists of sense organs in a linear arrangement. Primitively and still in cyclostomes, amphibians, and some other forms, the lines of sense organs are exposed on the body surface. More often they are sunk into canals, which may remain open as a groove (Holocephali, some sharks) but generally are closed over, with pores at intervals, as in most fishes. In teleosts the canals run through the scales. These canals are the canals of the lateralis system and consist of the main lateral line canal and four main branches on the head: a *hyomandibular* canal along the lower jaw, a *supraorbital* canal above the orbit, an *infraorbital* canal along the head below the orbit, and a *supratemporal* canal transversely across the rear part of the head. The head canals are supplied by branches of the anterior lateral line nerve accompanying the facial nerve, except that the supratemporal canal appears innervated by a branch of the posterior lateral line nerve that accompanies the vagus. A short section near the junction of supratemporal and main lateral line canals may be supplied by a lateral line component associated with the glossopharyngeal nerve. The lateralis system is found in cyclostomes and most fishes, in aquatic larvae of amphibians, and throughout life in those amphibians that lead a permanent aquatic existence.

The sense organs of the lateralis system are the *neuromasts*. Neuromasts (fig. 13.10) consist of ciliated neuroepithelial cells (hair cells) and supporting cells (sustentacular cells). The cilia of the hair cells extend into a gelatinous capsule called the *cupula*. The mechanical bending of these cilia results in a transduction that generates the nerve impulses from these hair cells. The cilia of each hair cell consists of many small stereocilia and one large kinocilium. A hair cell is directionally polarized toward the kinocilium. Thus, a fish swimming upstream will have only its hair cells with caudally directed kinocilia stimulated. The hair cells of neuromasts are receptors of hydrodynamic motion. They allow aquatic vertebrates to detect direction and rate of water current, as well as water displacement caused by stationary objects in the water or by predators and prey.

In addition to the neuromast system, there are several other related sense organs in fish. Among these are the *pit organs, ampullae of Lorenzini, vesicles of Savi,* and *mormyromasts.* Pit organs are isolated neuromasts sunk in pits on the head. The ampullae of Lorenzini are specialized neuromasts found on the heads of sharks. These neuromasts do not possess cilia and are located at the ends of tubes filled with a gelatinous substance. They appear to be electroreceptors monitoring the electric charges associated with the muscle contractions of the shark's prey. The vesicles of Savi appear to be similar receptors found in the electric skate, *Torpedo.* Mormyromasts are specialized neuromasts for the reception of electric pulses generated by mormyrid fish. These teleosts emit low voltage pulses and create a bipolar electric field around themselves. Their modified neuromasts detect distortions of this field produced by the entry of any solid object.

### d. Special Sense Organs

1) *The Inner Ear.* The sensory part of the ear is located in the inner ear, also called the *membranous labyrinth.* The middle and the external parts of the ear are accessory mechanisms for transmitting sound waves to the inner ear. The latter develops from an area of thickened ectoderm, the *otic placode,* which sinks below the ectoderm, forming a sac, the *otic vesicle* (fig. 13.11). This vesicle is connected to the surface by a canal, the *endolymphatic duct,* which in selachians remains open to the exterior but is closed in other vertebrates. Its blind end often expands to form a sac, the *endolymphatic sac,* which in amphibians and some reptiles is greatly expanded and branched. During embryonic development, the otic vesicle becomes divided into a dorsal chamber, the *utriculus,* and a ventral one, the *sacculus.* The endolymphatic duct enters at

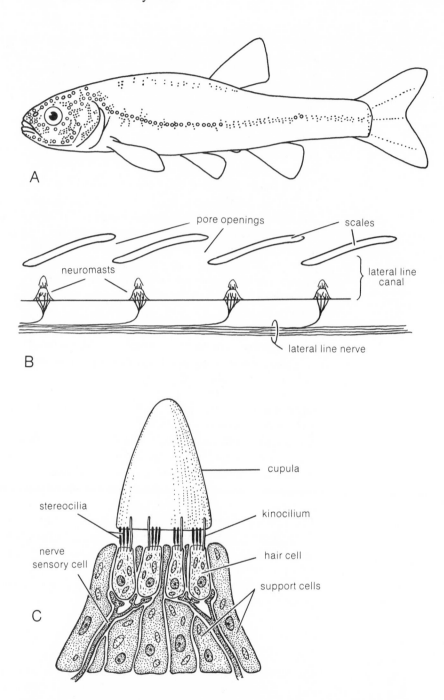

Fig. 13.10. *A*, the lateralis system of fishes consists of sensory organs located on the head and trunk as free isolated organs (pit organs) or arranged in lines. *B*, the line organs as seen in longitudinal view are usually located beneath the skin within a tube (canal) that communicates with the surface via pore openings. *C*, the sensory organs or neuromasts consist of support cells and sensory cells (hair cells) which are stimulated by mechanical displacement of a gelatinous cupula.

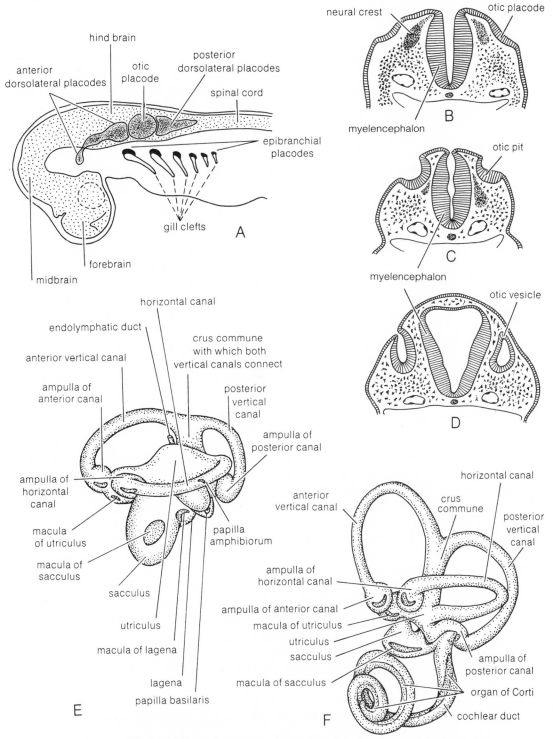

Fig. 13.11. Development and anatomy of the inner ear. *A*, the inner ear (otic organ) develops from an ectodermal thickening (placode) located lateral to the hindbrain. *B–D*, the otic placode invaginates, sinking beneath the surface, and forms a hollow otic vesicle. The wall of the vesicle forms the membranous labyrinth including the sensory organs of the inner ear, as well as the sensory neurons of the eighth cranial nerve that innervate the ear. *E–F*, external views of the left membranous labyrinth of an amphibian and a mammal illustrating major subdivisions and sensory epithelia.

the junction of these chambers or into the sacculus. Three canals, the *semicircular canals,* arise from the wall of the utriculus. These canals, two vertical and one horizontal, open into the utriculus at both ends. One end of each canal expands into a chamber, the ampulla, which contains a sensory patch of cells, the *crista ampullae.* There are one or more sensory patches, called *maculae,* in both the utriculus and the sacculus. Both of these sensory structures, crista and macula, possess hair cells similar to neuromasts. The cilia of the hair cells are embedded in a gelatinous membrane called the *cupula.* The cupulae of the saccular and utricular maculae are filled with inorganic crystals in which the tips of the hair cells' cilia are embedded. These inorganic bodies are called *otoliths.* The cupulae of the canal cristae do not form otoliths.

The internal ear is filled with a fluid, the *endolymph,* whose movements furnish the stimulus for equilibratory adjustments. When an animal's velocity changes, there is a short period before the movement of the endolymph also changes. This produces drag on the cupulae of the canal cristae, which bends the cilia of the hair cells. Thus, on the one hand, the semicircular canals are sensitive to changes in acceleration, particularly angular acceleration. On the other hand, the maculae are sensitive to changes in position, static equilibrium. When the head of an animal changes position, the otoliths slide over the maculae in response to gravity and produce a shearing force on the cilia of the hair cells.

In addition to functioning as organs of static equilibrium, the sacculus and utriculus (in a few teleosts) frequently possess sensory organs that are sensitive to propagated sound. Vertebrates that possess lateral line organs are sensitive to near-field sound, that is, focal disturbances of the water in which the vibrations are characterized by high amplitude and low frequency. Most vertebrates are also sensitive to propagated sound (pressure) in either water or air. Propagated sound sources are vibrations that are characterized by lower amplitudes and higher frequencies than near-field sound. These vibrations do not travel by gross movements of the media, but because of the elasticity and compressible nature of the molecules of either air or water.

Since most aquatic anamniotes possess nearly the same density as the water that surrounds them, they are essentially "transparent" to most propagated sound. However, many of these vertebrates possess morphological specializations that provide an organ whose density is significantly different from that of water to function as a more sensitive receptor of propagated sound.

In some teleosts (Ostariophysi) the perilymphatic spaces adjacent to the sacculus of the inner ear are in contact with a series of bones, the *Weberian ossicles,* derived from adjacent vertebrae that extend on each side from the anterior wall of the swim bladder (fig. 13.12). The Weberian ossicles convey pressure changes in the swim bladder to the saccular macula much as the middle ear ossicles transmit sound waves to the inner ear of tetrapods.

In sharks the skin over the endolymphatic fossa is especially sensitive to vibration, and this membrane may function in a way analogous to the tympanum of tetrapods.

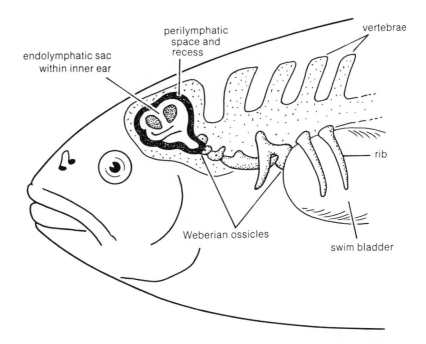

Fig. 13.12. Lateral view of the head of an ostariophysan teleost illustrating the chain of small bones (Weberian ossicles) linking the swim bladder with the inner ear.

Recently it has been shown that maculae located in the utriculus, in the sacculus, or in diverticuli of the sacculus may be responsive to propagated sound in different vertebrates. The differences in number and position of the maculae in different vertebrates strongly suggest that the ability to perceive propagated sound has arisen a number of times among living vertebrates and that the individual maculae are homologous only in the sense that they all arise embryonically from the otic vesicle.

In tetrapods the sensory cell patches that are sensitive to propagated sound are always associated with the sacculus. In amphibians, the sacculus forms two small diverticuli: the *amphibian papilla* and the *lagena*. Both the amphibian papilla and the lagena contain a macula that responds to sound. The lagena also contains a distal macula of unknown function in addition to the proximal macula, which responds to sound. All three maculae possess a thin tectorial membrane in which the cilia of the hair cells are embedded. The tectorial membrane is believed to be a modified cupula.

In reptiles and birds the lagena is enlarged by elongation, and the proximal macula (basilar papilla), as in reptiles, is the auditory sensitive epithelium.

In mammals the lagena is further enlarged and becomes a coiled structure, the *cochlear duct*. The homologue of the basilar papilla is termed the *organ of Corti* and is the transducing epithelium of propagated sound.

In all tetrapods airborne sound does not strike the body with sufficient force to compress the endolymphatic fluid and stimulate the hair cells within the lagena. For this reason, all tetrapods have developed such accessory structures as resonators, amplifiers, and conductors of air-borne sound to the lagena.

The membranous labyrinth is not only filled by the endolymph, but is surrounded by a space, the perilymphatic space, whose outer wall is the cartilaginous or bony otic capsule. The perilymphatic space is filled by a fluid, the *perilymph*. In tetrapods, the membranous labyrinth possesses two areas in which the bony otic capsule is reduced. These areas form membranous windows that face the inner wall of the middle ear cavity (fig. 13.13). The dorsal area, the *oval window,* vibrates as a

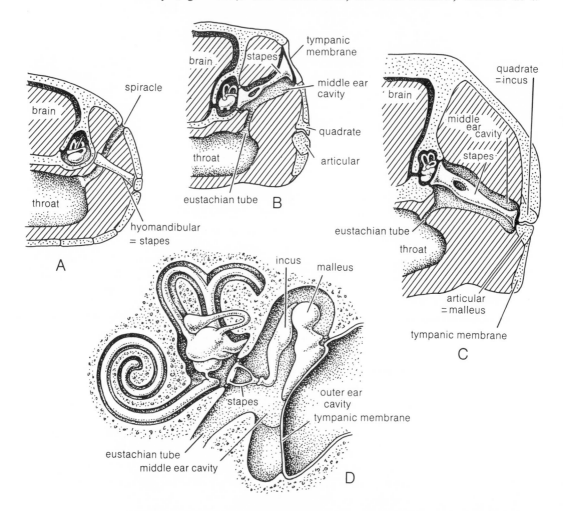

Fig. 13.13. Transverse sections through the otic region showing the evolution of the middle ear and auditory ossicles. *A*, fish; *B*, primitive amphibian; *C*, primitive reptile; *D*, mammalian level of organization showing only the ear region of the skull. Modified from Romer 1941 (*Man and the vertebrates.* Chicago: University of Chicago Press).

result of the insertion of the footplate of an ear ossicle on its surface. The outer end of the ear ossicle inserts on an eardrum, the *tympanum*. The tympanum forms the outer wall of the middle ear cavity and is free to vibrate against the low resistance of the air chamber on its inner side in response to a propagated sound source. Thus the tympanum forms a resonator, and one or more ear ossicles serve to amplify the resonance of the tympanum. Part of the amplification process is increased since the area of the oval window is smaller than the area of the tympanum. This reduction in area increases the force of vibration per unit area. Thus vibrations of the tympanum are transmitted to the perilymphatic fluid around the lagena by the ear ossicle via the oval window. In most tetrapods the perilymphatic space around the lagena turns ventrally and opens onto the middle ear cavity as a ventral window, the *round window*.

The lagena then is a membranous tube filled by a fluid, the endolymph, and surrounded by an outer fluid, the perilymph, enclosed in a membranous tube, which, in turn, is enclosed by bone except at two points, the oval and round windows. The lagenal macula (basilar papilla or organ of Corti) is formed by hair cells whose bases are attached to a thinned portion in the wall of the lagena, termed the basilar membrane. The cilia of these hair cells are embedded in a membrane termed the *tectorial* membrane, which stretches across the central lumen of the lagena. Vibrations at the round window create traveling waves in the perilymphatic fluid, which, in turn, move the basilar membrane of the lagena stimulating the hair cells. Both the endolymphatic and perilymphatic fluids can move because of the round window. Displacement changes induced by movements of the oval window are relieved by similar movements of the round window.

The hair cells of all portions of the membranous labyrinth are innervated by neurons that also form from the otic placode. These neurons form the ganglia of the eighth cranial nerve and convey both vestibular and auditory information into the brain.

2) *The Eyes.* Vertebrates possess a pair of lateral eyes and many also possess one or two dorsal eyes. The lateral eyes form the most complex sensory mechanism found in vertebrates, and that part of the eye that forms the sensory receptors, the *retina,* is a part of the brain itself. In vertebrate embryos, the lower region of the diencephalon bulges laterally as rounded eminences, the *optic vesicles.* The distal part of the vesicle invaginates, forming a two-walled cup, the *optic cup.* The wall that lines the cavity of the cup thickens and becomes the retina. The other wall, facing the brain, thins and becomes the *pigment layer* of the retina (fig. 13.14). Meanwhile, the surface ectoderm over the opening of the cup thickens, invaginates, and forms the lens vesicle, which differentiates to form the transparent *lens* of the eye. The distal rim of the optic cup does not form nervous tissue but, along with associated mesodermal tissue, forms the *ciliary body* and the *iris.* The ciliary body suspends the lens in front of the eyeball and is usually formed by connective and muscular tissues. Light enters the eye from various distances and must be focused onto the retina in order to form a sharp image. The process of changing the position of the lens, or the shape of the eyeball, in order to focus an image is termed *accommodation.* In amniotes, the ciliary body holds the lens in place and also contains muscle fibers that change the shape of the lens to facilitate focusing of a visual image. In most anamniotes, the ciliary body suspends the lens, but special muscles have evolved to move the lens forward or backward to facilitate focusing.

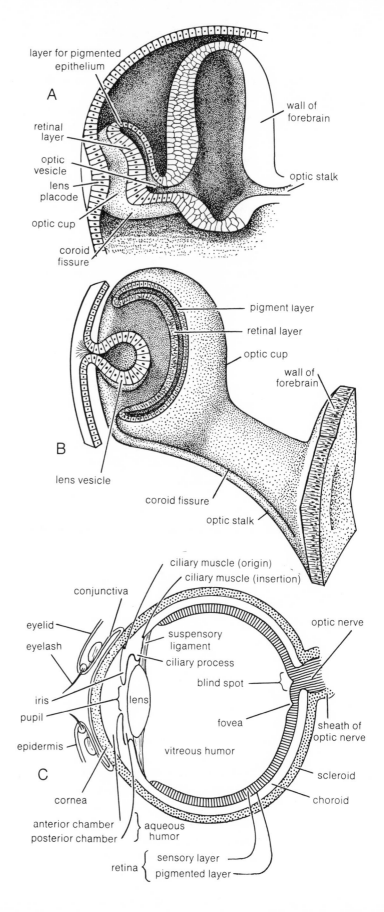

Fig. 13.14. Development and organization of the vertebrate eye. The optic vesicle forms by evagination of the diencephalon while the cornea and lens form from the surface ectoderm. *A*, optic vesicle; *B*, cornea and lens; *C*, diagrammatic vertical section through an adult eye.

**A**

layer for pigmented epithelium

retinal layer

optic vesicle

lens placode

optic cup

coroid fissure

wall of forebrain

optic stalk

**B**

pigment layer

retinal layer

optic cup

wall of forebrain

lens vesicle

coroid fissure

optic stalk

**C**

ciliary muscle (origin)

ciliary muscle (insertion)

conjunctiva

eyelid

eyelash

suspensory ligament

ciliary process

blind spot

optic nerve

iris

lens

pupil

fovea

sheath of optic nerve

epidermis

cornea

vitreous humor

anterior chamber
posterior chamber

} aqueous humor

sensory layer

pigmented layer

retina {

scleroid

choroid

The iris contains both radial and circular muscles and is perforated by a central hole, the *pupil,* which in tetrapods constricts in bright light and dilates in dim light. Variation in the diameter of the pupil allows the eye to form an image on the retina over a wide range of environmental conditions within a functional illumination range for the sensory receptors. The iris is formed by both mesoderm that surrounds the optic cup and by ectoderm from the edge of the optic cup.

In addition to the ectodermal parts of the eye, there develops from the surrounding mesoderm two supporting coats: an inner vascular layer, the *choroid* coat, which fuses to the pigment layer of the retina, and an outer tough protective layer, the *sclerotic* coat or *sclera.* The whole structure is termed the eyeball. Over the front of the eye, the sclera, plus the surface of the ectoderm, is transparent and is termed the *cornea.*

The cavity of the eye between the lens and the retina is filled by a viscous fluid, the *vitreous humor.* The cavity between the iris and the cornea is filled by a watery fluid, the *aqueous humor.*

As already mentioned, the eyeball is operated by a set of muscles so arranged as to move it in all directions; details of these muscles will be seen in the dissections.

The retina of the lateral eyes consists of a layer of sensory receptor cells and several layers of interneurons. The receptors are divided into two types: *rods* and *cones* (fig. 13.15). Both types of receptors contain

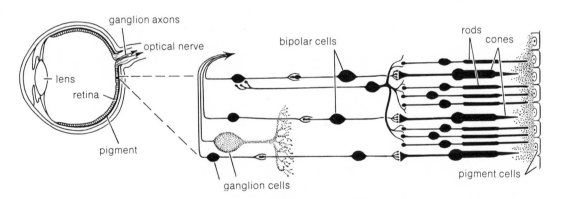

Fig. 13.15. Organization and neural connections of the retina. Note that the photoreceptors (rods and cones) are directed away from the entering light.

photosensitive pigments composed of protein and a pigment altered by the presence of light. The behavioral threshold for light perception by a single rod is a single photon (quantum). Thus one photon breaks down one molecule of visual pigment in one rod. The threshold for detection of light by a rod is, therefore, at its absolute limit. The breakdown of the photosensitive pigment triggers a series of events that results in a slow hyperpolarizing potential, a transient rather than an action potential as in most neurons. The effect of a sudden light stimulus then is to reduce the current and hyperpolarize the receptor.

Rods differ from cones in a number of ways. The visual pigment in rods is called rhodopsin and is sensitive to all wavelengths of light. Thus rods can only "sense" light or its absence, but cannot convey specific wavelength information (color) to the brain. Cones, however, are characterized by three different visual pigments, each one sensitive to different portions of the visual spectrum. A single cone is believed to possess only one of the three visual pigments, so that three cone populations exist. Cones have a much higher threshold of stimulation and only function in bright light. Thus only vertebrates that possess cones also possess color vision. These are usually diurnal vertebrates (most teleosts, frogs, most reptiles and birds, some mammals).

Increase in visual acuity is obtained by increasing the number of photoreceptors per unit area. Those vertebrates that rely heavily on vision for feeding or locomotion possess a region, *area centralis,* near the center of the retina that contains far greater numbers of photoreceptors than other parts of the retina. In some vertebrates this trend may continue by a thinning of the other retinal layers over the photoreceptors. This reduces the amount of tissue that light must pass through to stimulate the photoreceptors. Such an area in the retina is termed a *fovea.*

In addition to the layer of photoreceptors, the retina also contains a central layer of *bipolar* neurons and an inner layer of ganglion neurons (fig. 13.15). A single cone or several rods may synapse on a single bipolar cell. Thus the cones convey more precise pattern information than do the rods, even though cones have lower visual thresholds. The central bipolar layer also contains two additional neuronal populations: *horizontal* and *amacrine* neurons. Both horizontal and amacrine neurons perform complex integration functions between bipolar and ganglion neurons. Thus changes in signals occur not only in a vertical direction within the retina from receptor to ganglion neurons, but also horizontally between all three neuronal populations via the horizontal and amacrine neurons.

Bipolar neurons synapse on the cell bodies of the ganglion neurons. The axons of the ganglion neurons are the only ones that exit from the retina, and this collection of axons we term the *optic nerve.*

In addition to the lateral eyes, many vertebrates possess one or two dorsal eyes. Lampreys and many actinopterygian fish possess two dorsal eyes, the pineal and parietal eyes. Amphibians and reptiles develop only a single dorsal eye, the parietal eye, and many species of these two classes never form a dorsal eye at all.

Dorsal eyes develop as the lateral eyes in vertebrates do, by a dorsal evagination of the brain. The dorsal vesicle comes to lie just beneath the ectoderm, which forms a cornea (fig. 13.16). The roof of the vesicle forms a lens, and the floor differentiates into photoreceptors and a single layer of ganglion neurons whose axons project into the diencephalon.

While few physiological studies have been conducted on dorsal eyes, it is believed that they do not form visual images that are processed by brain centers, but only signal the presence or absence of light. Such a receptor may trigger or time circadian rhythms, activity, and reproductive cycles of vertebrates.

3) *The Olfactory Organ.* The olfactory organ develops from a pair of ectodermal placodes in all vertebrates. The placodes invaginate, forming the olfactory sac. The lining epithelium of the sac differentiates into

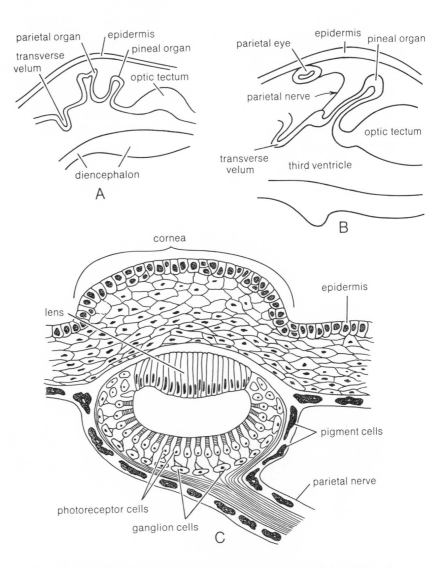

Fig. 13.16. Development and organization of the vertebrate dorsal (median) eye. *A–B,* the photoreceptors and lens evaginate from the roof of the diencephalon while the cornea arises from surface ectoderm. *C,* structure of the adult eye.

olfactory neurosensory cells and supporting cells (fig. 13.17). The neurosensory cells are columnar in shape and possess several nonmotile cilia on their free edges. These cilia increase the total surface area of the sensory cells, and it is believed that the receptor sites are located on the surface of the cilia. Each neurosensory cell possesses a single axon that projects to the olfactory bulb of the telencephalon. The axons of all the olfactory cells are collectively termed the *olfactory nerve.*

The second type of epithelial cell, supporting or *sustentacular* cells, possesses microvilli on their free surface, and these cells also secrete mucus, which moistens the entire olfactory surface.

In addition to these types, some cells, called basal cells, may possess the ability to differentiate continually into either sustentacular or neuro-

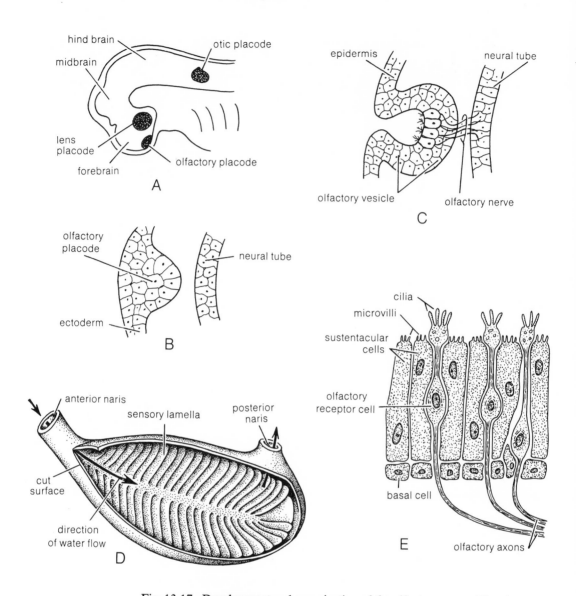

Fig. 13.17. Development and organization of the olfactory organ. The olfactory organ (epithelium) develops from an ectodermal placode (*A*) that invaginates toward the forebrain of the neural tube (*B–C*). The wall of the olfactory vesicle is thrown into complex lamellae in the adult (*D*), composed of sensory and support cells (*E*) as well as a basal layer that continues to give rise to new cellular populations.

sensory cells. Recent studies suggest that there is a continued turnover of neurosensory cells in the olfactory epithelium.

In fishes the olfactory surface is generally increased by thin folds that nearly fill the sac. In all gnathostomes the interior of the nasal cavities forms various folds; part of these folds bear olfactory epithelium while the other parts are nonolfactory mucus-secreting epithelium. In mammals the olfactory epithelium is primarily located on folds of the ethmoid turbinates, and these are more complicated and enfolded in those animals with a well-developed sense of smell.

In tetrapods a portion of the ventromedial wall of each nasal sac may secondarily evaginate to form a rounded or elongated blind sac, the *vomeronasal* or *Jacobson's organ*. It is absent in most turtles, crocodiles, birds, bats, primates, and aquatic mammals. It is well developed in lizards and snakes, where it opens directly into the oral cavity rather than into the nasal cavity as in many tetrapods. The neurosensory cells of the vomeronasal organ are similar to those of the main olfactory epithelium, but differ in that they do not possess cilia. The axons of the vomeronasal neurosensory cells project to a caudal portion of the olfactory bulbs of the telencephalon, where they frequently form a distinct structure termed the *accessory olfactory bulb*. Recent experimental studies have demonstrated that these two olfactory receptors, when present, project to very different brain centers; this fact suggests that there may really be two separate olfactory systems with very little functional overlap. The vomeronasal organ may be involved in social interactions, such as individual and sexual recognitions, and in some vertebrates it has been implicated in prey recognition.

In addition to the neurosensory receptors of the olfactory epithelium, there are also free nerve endings of the terminal and trigeminal nerves located among the cells of the olfactory epithelium. Recent neurophysiological studies have shown that the trigeminal nerve endings are also sensitive to the same molecules that excite the olfactory neurosensory cells. Thus it is possible that vertebrates are analyzing what we sense as smell by a number of different types of receptors projecting to very different parts of the brain.

### e. General Visceral Sense Organs

The sensory terminations in the viscera and mesenteries are similar to those of the skin. They consist of free nerve endings and encapsulated endings. These visceral sensory nerves proceed from the viscera through the trunks of the autonomic system to their cell bodies of origin in the spinal and cranial nerve ganglia.

### f. Special Visceral Sense Organs—Taste

The organs of taste are termed *taste buds*. These organs are similar to olfactory neurosensory cells in being chemoreceptors that possess a ciliated free surface. However, they differ from the olfactory receptors in several ways. The taste cells are of endodermal rather than ectodermal origin as are olfactory cells. The taste cells are located in barrel-shaped clusters rather than spread over a large uniform surface as are the olfactory cells. Taste cells do not project directly into the brain as do olfactory cells but form synapses with ganglion cells, which in turn project into the brain. Finally taste cells are less sensitive and are more restricted in the types of molecules to which they will respond than are olfactory cells.

The taste buds reach their widest distribution and greatest numbers in fish where they may spread over the external surface of the head and body. Some teleost fish (cyprinids and silurids) depend largely on external taste buds for finding food and may have as many as one hundred thousand taste buds located over the body as well as concentrated on special food-finding barbels located around the mouth. In such fish the taste centers in the visceral sensory area of the medulla are greatly enlarged and may form immense bulges, termed the *vagal lobes*.

In most vertebrates and in tetrapods, the taste organs are confined to the oral cavity and pharynx. Birds have few taste buds and, correspondingly, a poor sense of taste. In mammals the taste buds are well developed and they occur primarily on the papillae of the tongue. However, some also occur in the palate, pharynx, and epiglottis. Although taste buds may be located on tongue papillae throughout the vertebrates, this relation is more definite in mammals. The taste buds of mammals occur principally on the foliate, fungiform, and vallate papillae; they vary in numbers from several hundred to ten or more thousand.

The taste, or gustatory, fibers run in the seventh, ninth, and tenth cranial nerves. External taste buds, when present, and those of the anterior part of the oral cavity are supplied by the facial nerve; taste buds of the rear part of the oral cavity and of the pharynx are supplied by the glossopharyngeal and vagus nerves. The cells of origin of the gustatory fibers are in the ganglia of these three cranial nerves, and the fibers enter the side of the medulla by the dorsal roots of these nerves. In mammals the facial nerve innervates the fungiform papillae, the glossopharyngeal innervates the vallate and foliate papillae, and the vagus supplies taste buds on the epiglottis and adjacent regions.

## 8. The Autonomic Nervous System

The autonomic nervous system has been adequately investigated only in mammals; the description here chiefly concerns this class. The autonomic system is a complex of ganglia, nerves, and plexi through which the heart, lungs, digestive tract, other viscera, glands (including the sweat glands), walls of the blood vessels, and smooth musculature generally, including that of the feather and hair follicles, receive their motor innervation. For the autonomic system is, in effect, an elaboration of the *visceral motor division* of the nervous system. Although the visceral sensory fibers also run in the autonomic system, they have the same general relations as the somatic sensory fibers (having their cell bodies with the latter in the spinal and cranial ganglia) and hence require no special consideration. On the contrary, the arrangement of the visceral motor fibers differs from that of other parts of the nervous system in that their fibers relay in autonomic ganglia. These ganglia contain cell bodies whose fibers, known as *postganglionic* fibers, usually nonmyelinated, terminate in smooth musculature, heart muscle, or glands. These cell bodies receive their stimulation by way of myelinated *preganglionic* fibers that originate from cell bodies located in the visceral motor columns of the spinal cord or brain and pass out with the cranial or spinal nerves. It is another peculiarity of the autonomic system that it gives rise to a double innervation to most organs. In general one set of fibers is inhibitory and the other set is excitatory. The two sets also react differently to certain drugs. The elaboration of the autonomic system in vertebrates is readily understood if one considers that regulation of the heartbeat, respiratory movements, intestinal peristalsis, and many other important visceral functions are necessary to proper functioning, especially in relation to changing conditions.

The branchial musculature and its phylogenetic derivatives in vertebrates, although part of the visceral motor division, are not innervated by way of the autonomic system. There are no pre- and post-ganglionic fibers and no relays in outside ganglia, in connection with these muscles.

The autonomic system of mammals may be divided into the following parts: the *cranial outflow,* the *cervical component,* the *thoracolumbar chain of ganglia,* the *sacral outflow,* the *collateral ganglia,* and the *peripheral ganglionated plexi* (fig. 13.18).

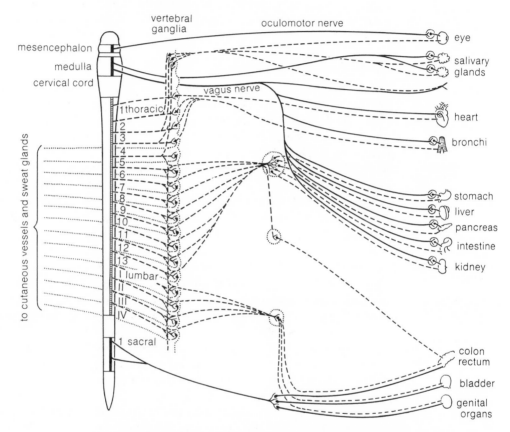

Fig. 13.18. Diagrammatic representation of the mammalian autonomic (visceral motor) system.

## a. The Cranial Outflow

The preganglionic fibers of the head autonomic system have their cells of origin in the visceral motor column of midbrain and medulla and pass out in the oculomotor, facial, glossopharyngeal, and vagus. The oculomotor fibers run to the *ciliary* ganglion in the orbit, where they relay; the postganglionic fibers supply the ciliary muscles of the lens and the iris muscles, which regulate the size of the pupil. The autonomic fibers of the facial nerve run (1) to the *pterygopalatine* ganglion, in the pterygoid fossa, from which postganglionic fibers supply the lachrymal or tear glands, and (2) to the *submandibular* ganglion near the submandibular gland, from which postganglionic fibers continue to the submandibular and sublingual salivary glands, causing them to secrete. The facial branch that carries these fibers to the submandibular ganglion is termed the *chorda tympani* because it traverses the middle ear along the inner surface of the eardrum and is quite famous because of the numerous experiments on salivary secretion performed by means of it. The preganglionic fibers in the glossopharyngeal proceed to the *otic* ganglion, near the foramen ovale; from this the postganglionic fibers go to the parotid salivary gland. The autonomic part of the vagus is very large and extensive; the visceral branches of the vagus consist chiefly of preganglionic visceral motor fibers, which relay in the peripheral ganglia and plexi in the thoracic and abdominal viscera as far posteriorly as the kidneys.

### b. The Cervical Component

This consists of three ganglia in the neck region: the *superior, middle,* and *inferior* cervical ganglia and the connections between them, composed of preganglionic fibers entering the cervical ganglia from the upper thoracic spinal nerves by way of the white communicating rami (see below). The inferior cervical ganglion may be fused with the first thoracolumbar ganglion, and the resulting ganglion is termed the *stellate* ganglion. After relaying in the cervical ganglia, the postganglionic fibers supply, in general, the same head parts as the cranial outflow (that is, the head glands and the intrinsic eye musculature, the heart and bronchi) and constitute the second autonomic innervation of these organs.

### c. Thoracolumbar Chain of Ganglia

This is a linear series of ganglia and connecting cords found on the dorsal coelomic wall to either side of the vertebral column. Together with the cervical ganglia, these are generally known as the *sympathetic trunks.* The number of ganglia only approximately corresponds to that of the vertebrae. The ganglia are connected to the spinal nerves by the communicating rami, each of which typically consists of two parts, a *white* and a *gray* ramus. The white ramus consists of myelinated fibers and conveys the preganglionic fibers from their cells of origin in the lateral gray columns of the spinal cord to the corresponding ganglion of the thoracolumbar chain; the fibers may relay there, but more often they relay in collateral or peripheral ganglia. The gray ramus carries postganglionic nonmyelinated fibers from cells of origin in the ganglia of the thoracolumbar chain to the spinal nerves. These fibers pass out with the spinal nerves to blood vessels and skin and terminate mostly in the smooth musculature of the walls of the blood vessels and of the hair and feather follicles, and in the skin glands. As already noted, there are no white rami, only gray rami, connecting the cervical ganglia with the cervical nerves; the same is true of the sacral part of the sympathetic trunks.

The preganglionic fibers of the white rami usually do not relay in the thoracolumbar ganglia but course in the sympathetic trunks and emerge in the cervical sympathetic or in special *splanchnic* nerves from the thoracolumbar trunks to collateral and peripheral ganglia, where the relay occurs. The postganglionic fibers supply the same viscera reached by the vagus nerve and constitute the second autonomic supply of these viscera;

the postrenal viscera, such as the colon, rectum, bladder, and genital organs, are reached from the lumbar part of the sympathetic trunks.

The sympathetic trunks continue into the tail in tailed mammals and have communicating rami with the caudal spinal nerves.

### d. The Sacral Outflow

Preganglionic fibers from the sacral spinal nerves run directly to collateral ganglia without passing through the sympathetic trunks; the postganglionic fibers supply the same postrenal viscera mentioned above, giving these the usual second autonomic innervation.

Because of their similar morphological and physiological relationships, it is possible to view the cranial and sacral autonomic pathways as a functional system, the *parasympathetic* system. The preganglionic fibers of this system pass directly to collateral or peripheral ganglia and plexi without passing through the sympathetic trunks. The neurotransmitter released at all synapses of the parasympathetic system is acetylcholine. The parasympathetic system stimulates activities of the body that are associated with conservation and restoration: decrease in rate of the heartbeat, decrease in blood pressure, increase in enzyme secretion and motility of the gut, and stimulation of voiding body wastes. In contrast, the thoracolumbar (including the cervical component) is termed the *sympathetic* system. The preganglionic fibers of this system do pass to the sympathetic trunks, where the cell bodies of the postganglionic fibers are frequently located. The cell bodies of postganglionic fibers projecting to organs of the thorax are always located in the sympathetic chain.

The cell bodies of the postganglionic fibers projecting to the abdominal cavity are found most frequently, however, in the collateral ganglia. Like the parasympathetic system, the synapses formed by the preganglionic fibers of the sympathetic system have acetylcholine as their neurotransmitter. However, the postganglionic fibers have noradrenalin (norepinephrine) as their neurotransmitter. The sympathetic system stimulates activities associated with stress such as aggression, fear, or flight. Thus the sympathetic system produces increase in rate of the heartbeat, increase in blood pressure, decrease in enzyme secretion and motility of the gut, and increase of blood flow to body wall musculature.

### e. The Collateral Ganglia

This includes the four head ganglia already mentioned—the ciliary, pterygopalatine, otic, and submandibular ganglia—and some abdominal ganglia closely associated with large arteries. These are the *coeliac* and *superior mesenteric* ganglia, located near the origin of the superior mesenteric artery (or coeliac axis) from the aorta, and the *inferior mesenteric* ganglion, alongside the inferior mesenteric artery. The first two are often fused together. The preganglionic fibers from the thoracolumbar trunks (by way of the splanchnic nerves) and from the sacral outflow terminate in the collateral ganglia, and the postganglionic fibers begin in the cells of these ganglia and proceed to their terminations in the viscera.

*ƒ. The Peripheral Ganglia and Plexi*

This comprises a vast and complicated system of sympathetic ganglia and networks upon or inside the viscera and blood vessels. Such ganglionated networks occur on the arch of the aorta in the heart wall, on the esophagus, in the lungs along the bronchi, and in the stomach, adrenal glands, spleen, gonads, intestine, along the arteries which supply them. There are also ganglionated plexi in the walls of the intestine, termed the plexi of Meissner and Auerbach. The large mass of plexi and ganglia associated with and including the coeliac and superior mesenteric ganglia is termed the *coeliac* or *solar* plexus. The preganglionic fibers of the parasympathetic system terminate in the peripheral ganglia and plexi, and from the cells of the latter the postganglionic fibers originate and pass into the musculature of the viscera. The plexi of Meissner and Auerbach have traditionally been considered to be composed only of the cell bodies of postganglionic parasympathetic fibers. It is now realized that in addition there are cell bodies of postganglionic sympathetic fibers as well as general visceral sensory neurons that form reflexes within the gut and with the motor neurons of the collateral ganglia. Thus there appear to be places in which the traditional view of the autonomic motor system —that it consists of only two motor neurons in a series—will not prove true. There is a *pelvic* or *hypogastric* plexus that receives the sacral outflow and from there postganglionic fibers reach the bladder, lower part of the intestine, external genitals, and so on.

Knowledge of the autonomic system in nonmammalian vertebrates is meager. Cyclostomes possess ganglia scattered throughout the body, but they are not divided into clear-cut sympathetic and parasympathetic systems. Selachians possess sympathetic trunks connected to the spinal nerves by communicating rami. Definite sympathetic trunks are found also in Dipnoi and teleosts, along with a highly developed cranial autonomic system. Anuran amphibians possess a sacral parasympathetic outflow, as do all other land vertebrates. In all amniotes the autonomic system appears very similar to that seen in mammals.

**9. Evolution and Organization of Neural Pathways**

The evolution of the central nervous system can be viewed as the evolution of functional systems or patterns whose specialization is reflected by increased complexity in certain brain regions. The individual neural regions associated with a particular pattern show a range of specialization. The most frequent types of specialization are increases in absolute volume of involved neural regions and a concomitant increase in the morphological complexity of the neurons forming these neural regions. Both sensory and motor pathways of the brain arise from, and terminate respectively upon, a limited set of neurons in the spinal cord. These spinal neurons form the basic reflex arcs and are the only elements other than the cranial nerves that transmit sensory information to the brain. In addition, the efferent pathways of the brain can only affect organs of the body via these spinal elements. These spinal patterns are first described and are then used to explain the action of brain patterns upon them. The mammalian patterns are emphasized because we know the most about them. After each mammalian pattern is described, the organization of these patterns in nonmammalian vertebrates is described if details of their organization are known.

### a. Spinal Patterns

These are the basic patterns used by all brain pathways to influence the organs of the body. These relationships are summarized in figure 13.19. Sensory receptors are responsive to a narrow range of energy (light, near-

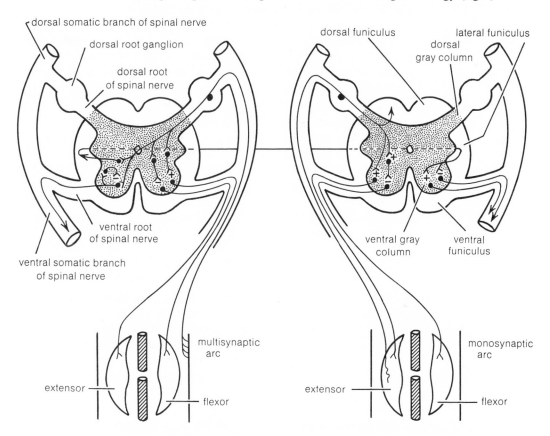

Fig. 13.19. Position and connections of spinal neurons forming multisynaptic and monosynaptic patterns responsible for postural and protective reflexes.

field sound, touch, pressure that damages tissues, i.e., pain, etc.). All receptors transduce fluctuations into electrical potentials that are transmitted into interneurons (internuncial neurons). Thus a single receptor, a sensory neuron, forms connections with several interneurons. These interneurons perform a number of different functions. Some convey receptor information to motor neurons, thus completing the circuits we call reflexes, which cause changes in muscle tonus or level of secretion by glands. Other interneurons convey receptor information to higher order neurons in the brain that may also monitor other kinds of information simultaneously. Finally, higher order neurons of various brain regions may alter motor neuron functions by a complex interplay of inhibition or excitation on the interneurons or motor neurons.

Two of the major spinal patterns are the monosynaptic and multi-synaptic reflex arcs. The *monosynaptic arc,* the anatomical basis of the myotatic reflex, subserves the maintenance of tonus and posture in all land vertebrates and consists of a two-neuronal chain (sensory and motor neuron). The cell body of the sensory neuron is located in the dorsal spinal ganglion (fig. 13.19), and its sensory ending (annulospiral ending) is located on intrafusal muscle fibers. When the muscle is stretched by the pull of gravity, the annulospiral ending is stimulated, and the sensory neuron is excited. The axon of the sensory neuron projects to the ventral gray column, where it is excitatory to somatic motor neurons that project to extensor muscles and by way of an internuncial neuron inhibitory to somatic motor neurons that project to flexor muscles. This mechanism functions to extend the limbs or stiffen the vertebral column and thus maintain the normal posture of the animal. This sensory neuron also sends an axon collateral to brain centers where sensory information is analyzed.

The *multisynaptic arc,* or flexor reflex, is the spinal pattern by which an animal withdraws a part of its body from a noxious stimulus. This spinal mechanism consists of at least three neurons: a sensory neuron, an internuncial neuron, and a motor neuron (fig. 13.19). The sensory neuron is located in the dorsal spinal ganglion, and its sensory ending is located in the skin and deep structures of the body wall. When skin or structures in the body wall are presented with a noxious stimulus, the sensory endings of these neurons are excited. The axon of the sensory neuron projects to the dorsal gray column, where it is excitatory to two different groups of internuncial neurons. The first group of internuncial neurons is excitatory to somatic motor neurons that project to flexor muscles, and the second group of internuncial neurons is inhibitory to somatic motor neurons that project to extensor muscles. This causes a limb or a body region so stimulated to be withdrawn from the noxious stimulus. The sensory neuron of the flexor arc, like that of the myotatic arc, also sends an axon collateral to brain centers for analysis of information.

The multisynaptic arc also gives rise to a mechanism that results in alternating extension of a forelimb and the diagonal hind limb. This mechanism is called reflex stepping. This reflex suggests that the spinal cord possesses the connections on which coordinated limb movements are based. The mechanisms of the spinal cord are similar among all vertebrates. Flexor and myotatic reflexes occur in all vertebrates. The myotatic reflex became increasingly important in tetrapods as they shifted from a sprawling gait to one in which the body is supported directly by the limbs. The major changes in the spinal cord have been increased segregation of cells and fibers of a common function from cells and fibers of other functions and increased elaboration of the lateral part of the ventral gray column. It is from this part of the spinal cord that the neurons that innervate the muscles of the limbs arise.

### b. Ascending Spinal Somatic Sensory Pathways

These pathways project to a number of brain centers and convey information on pain, temperature, touch, limb position (proprioception), and level of muscular tonus. Each of these types of information (modalities) are primarily conveyed by anatomically distinct and separate pathways.

In mammals information relating to pain, temperature, and gross touch (deformation of the skin) are transduced by free nerve endings and some encapsulated nerve endings whose cell bodies are located within dorsal root spinal ganglia. The axons of these cells terminate on interneurons located in the dorsal gray column. Some of these interneurons send their axons across the ventral white commissure of the spinal cord and into the lateral or ventral spinal funiculi, where they are collectively termed the *spinothalamic tracts* (fig. 13.20). The spinothalamic tracts

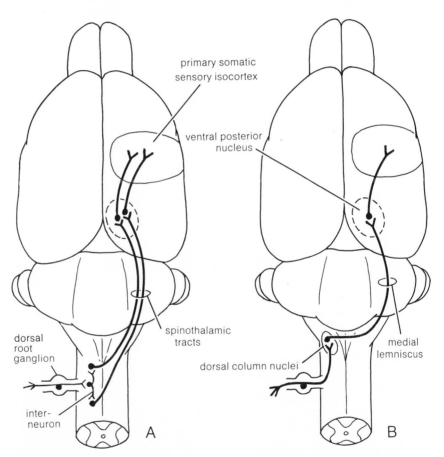

Fig. 13.20. Ascending somatic pathways in mammals. These pathways carry information on pain and temperature (*A*) and fine touch and limb position (*B*).

terminate in a posterior part of the thalamus termed the ventral posterior nuclear group. These thalamic neurons in turn project onto a part of the telencephalic roof termed primary somatic *sensory isocortex*.

In mammals information regarding fine touch and position of the limbs is transduced by encapsulated nerve endings of the skin, frequently associated with hair organs and connective tissue. As usual, the cell bodies of these receptors are located in the dorsal root ganglia of spinal nerves, and their axons enter the dorsal gray columns. In addition to these branches, each sensory neuron also possesses a long ascending branch that terminates in a collection of neurons, or dorsal column nuclei, that mark the dorsal boundary of the brain and spinal cord (fig. 13.20). In mammals the dorsal column nuclei are divided into a lateral *cuneate nucleus* and a medial *gracile nucleus*. The cuneate nucleus receives information only from thoracic and cervical spinal nerves, and the gracile nucleus receives information only from the lumbar, sacral, and caudal spinal nerves. The neurons of the dorsal column nuclei send their axons across the brain midline, where they turn rostrally and are termed the *medial lemniscus*. These axons also terminate in the ventral posterior nuclear group of the thalamus and, like the spinothalamic pathways, finally reach the primary somatic sensory isocortex of the telencephalon.

Information regarding muscle tone is conveyed to the cerebellum by the *spinocerebellar tracts* (fig. 13.21). Muscle tone or tension is transduced by annulospiral endings located on intrafusal muscle fibers and by Golgi tendon organs, sensory endings located in the muscle belly as well as in tendons. The cell bodies of these receptors are also located

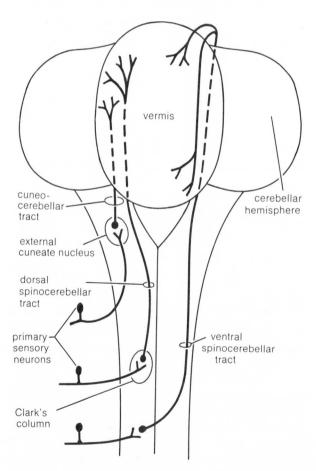

Fig. 13.21.  Ascending spinocerebellar pathways in mammals. These pathways carry information on muscle tone and limb position to the cerebellum.

in the dorsal ganglia of spinal nerves. Three very different pathways to the cerebellum exist, depending on where in the body the receptors are located and the amount of detail they transmit. The axons of sensory neurons whose receptors are located in the caudal, sacral, and lumbar portions of the trunk terminate on interneurons, *Clark's column,* located in the dorsal gray column of the spinal cord at cervical levels. These interneurons then project to the cerebellum, and their axons are collectively termed the *dorsal spinocerebellar tract.* This pathway primarily transmits information from the pelvic appendages and conveys information regarding the position and tonus of individual limb muscles.

The axons of sensory neurons whose receptors are located in the thoracic and cervical portions of the trunk terminate on interneurons located just lateral to the cuneate nucleus. These interneurons are termed the *external cuneate nucleus,* and their axons also project to the cerebellum. This pathway is termed the *cuneocerebellar tract,* and it transmits information primarily from the pectoral appendages and is the equivalent of the dorsal spinocerebellar tract.

The third spinocerebellar tract, the ventral spinocerebellar tract, also conveys information about muscle tone and position, but the detail it conveys is related to synergic muscle groups around joints rather than a single muscle. This type of information is probably used to guide posture and movements of an entire limb. The interneurons whose axons form the ventral spinocerebellar tract are located at the boundary of the dorsal and the ventral gray columns. This tract crosses in the ventral white commissure of the spinal cord and ascends rostral to the cerebellum and then turns caudally to enter the cerebellum.

Finally, *spinoreticular pathways* form ascending tracts to a number of brain centers (fig. 13.22). A large number of interneurons from all levels of the dorsal gray column send their axons into the lateral spinal funiculus where they ascend into the medulla to terminate among neurons that are collectively termed the *brain stem reticular formation.* The neurons of the reticular formation occupy the core of the medulla and

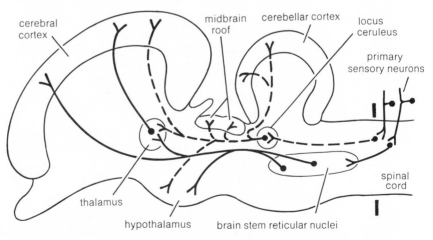

Fig. 13.22. Ascending spinoreticular pathways in mammals. Dorsal (dashed lines) and ventral (solid lines) pathways project to most regions of the brain.

mesencephalic tegmentum and are interposed between sensory and motor neurons of the cranial nerves in a position comparable to interneurons of the spinal cord. The neurons of the reticular formation are characterized as multimodal; they do not convey specific information regarding temperature, fine touch, and the like, but receive connections from different types of sensory neurons and receptors and mediate general arousal levels.

The brain stem reticular formation projects to the cerebellum, optic tectum, hypothalamus, and medial wall of the thalamus, as well as to most centers of the telencephalon. Recent experimental studies indicate that the reticular formation is not a diffuse network of neurons in which each cell projects everywhere in the brain, but that the reticular formation is composed of many localized groups with specific targets. Future studies will almost certainly reveal a large number of very specific pathways subserving much of the behavior that we now characterize as species specific.

Far less is known about somatic sensory pathways in nonmammalian vertebrates. In most anamniotes, except sharks, spinal neurons do not project directly to thalamic levels. Pathways do exist that occupy the same position in the lateral and ventral spinal funiculi as do the spinothalamic tracts of mammals. However, these pathways terminate in a brain area, the isthmus, located between the optic tectum and the cerebellum rather than in the thalamus.

In some sharks, sparse spinothalamic pathways exist, but they are better developed in some reptiles, such as crocodiles, and in birds. In these amniotes the spinothalamic tracts project to the posterior thalamus, which in turn projects to an area in the telencephalon called the dorsal ventricular ridge. The dorsal ventricular ridge of reptiles and birds is probably homologous to part of the isocortex of mammals. It is possible that the spinothalamic tracts of sharks arose independent of those in amniotes, but more studies are needed to resolve this question.

All vertebrates possess dorsal column nuclei, and all possess dorsal funicular fibers from the spinal cord that terminate on these neurons. However, the projections of the dorsal column nuclei have not been studied in any nonmammalian vertebrate, and at present it is not known whether or not such vertebrates possess a medial lemniscus and thus a fine sense of touch and detailed information on limb position.

Spinocerebellar tracts have been identified experimentally in all classes of vertebrates, even lampreys, which possess the most rudimentary cerebellum.

Spinoreticular pathways have been identified in all vertebrate classes and in most anamniotes. These constitute the major ascending somatic pathways. In actinopterygian fish and amphibians the reticular formation projects to the diencephalon, which, in turn, projects to the telencephalon. These pathways appear to constitute a second ascending system that may be more primitive than the spinothalamic and medial lemniscal tracts of amniotes. This hypothesis will only be confirmed, however, when additional information becomes available on the projection of the dorsal column nuclei in anamniotic vertebrates.

## c. Medullar Patterns

The functional components of the gray columns of the spinal cord continue into the medulla (fig. 13.23) and, like the spinal cord, are divided into dorsal sensory areas and ventral motor areas. Primitively these areas probably consisted of neurons scattered throughout the length of the medulla. Specialization of the vertebrate head resulted in the segregation of these columns into more or less distinct clusters of neurons that dorsally form the sensory nuclei and ventrally form the motor nuclei of the cranial nerves. Like the spinal cord, the medulla possesses interneurons sandwiched between sensory and motor neurons, and, like the interneurons of the spinal cord, these cells form pathways to and from higher brain centers and the spinal cord.

The dorsal medulla can be divided into a lateral somatic sensory zone and a medial visceral sensory zone. The lateral zone is further divided into three nuclei in a rostrocaudal direction: *mesencephalic, principal sensory,* and *spinal trigeminal* nuclei (fig. 13.23). These sensory nuclei convey all sensations of pain, temperature, touch, and muscle position from the skin and muscles of the head. Although these sensory nuclei are termed the trigeminal nuclei, they actually receive projections from the somatic sensory neurons located in the ganglia of cranial nerves VII, IX, and X as well. This is not surprising if we realize that each of these branchiomeric nerves innervates a portion of the skin over the head (see head segmentation). However, in all gnathostomes the skin and muscles associated with the first or mandibular arch hypertrophy in association with the formation of the jaws, and this specialization is reflected in the medulla with the hypertrophy of that portion of the somatic sensory column associated with the trigeminal nerve, which is the cranial nerve of the mandibular arch.

The information relating to pain and temperature in the head, like that of the body, is transduced primarily by free nerve endings in the skin and connective tissue. The cell bodies of these receptors are located in the sensory ganglia of the branchiomeric nerves and the axons of these cells terminate among the cells of the spinal trigeminal nucleus (fig. 13.23). Fine touch receptors are also located in the sensory ganglia of these branchiomeric nerves, but they project onto the principal sensory nucleus. The neurons of the principal and spinal nuclei form connections with interneurons of the reticular formation as well as with branchiomeric motor nuclei, thus completing reflex circuits similar to those seen in the spinal cord. The branchiomeric motor nuclei form part of a ventral medullar column that is closely associated embryonically with the visceral division of the nervous system. Again, this is not surprising since these nuclei innervate branchiomeric muscles that are the rostral continuation into the head of trunk lateral plate mesoderm. The branchiomeric motor column consists of the motor nuclei of V (the *masticator* nucleus), VII (the *facial* nucleus), and IX, X, and XI (the *nucleus ambiguus*). These motor neurons innervate the muscles derived from their respective head segments. At first it may seem surprising that somatic sensory nuclei should form reflex circuits with visceral (branchiomeric) muscles, but it must be remembered that in land vertebrates it is the

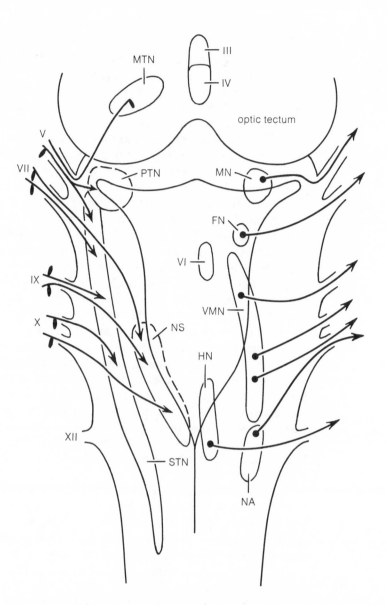

Fig. 13.23. Major sensory and motor nuclei of the vertebrate medulla oblongata. The sensory nuclei are illustrated on the left and the motor nuclei on the
right. FN, facial motor nucleus; HN, hypoglossal motor nucleus; MN, masticator (trigeminal motor) nucleus; MTN, mesencephalic trigeminal sensory
nucleus; NA, nucleus ambiguus (branchiomeric motor nucleus); NS, nucleus
solitarius (a visceral sensory nucleus); PTN, principal or main trigeminal
sensory nucleus; STN, spinal trigeminal sensory nucleus; VMN, visceral motor
nucleus of the glossopharyngeal and vagal nerves; III, oculomotor motor nucleus; IV, trochlear motor nucleus; VI, abducens motor nucleus; V, VII, IX,
X, XII, position of sensory and motor roots of the V–XIIth cranial nerves.

caudal branchiomeric muscles that are heavily involved in movements of the head on the trunk, particularly those movements that involve flexion and rotation. These reflexes are further aided by involvement of cervical somatic musculature, which is also under the influence of the spinal trigeminal nucleus via the reticular formation.

These trigeminal sensory nuclei and their closely associated interneurons, like their counterparts in the spinal cord, also give rise to ascending sensory pathways, *secondary trigeminal tracts,* that terminate in the ventral posterior thalamus. In mammals, these pathways convey pain, temperature, and touch information regarding the head to the primary sensory isocortex of the telencephalon. Again, similarities are seen with the ascending spinal sensory pathways; branchiomeric sensory receptors and interneurons project to a thalamic relay nucleus that projects to primary sensory isocortex.

The *mesencephalic trigeminal nucleus* is composed of large unipolar neurons that resemble dorsal root ganglion cells except that they are located within the brain. These giant neurons are believed to be primary sensory neurons that have been retained in the brain. Their single neurite bifurcates, and one process ends as a stretch receptor within the muscles of mastication or as pressure receptors within the connective tissue that surround the teeth and the central process terminates on the motor nucleus of V. Thus this pathway appears to be concerned with mechanisms that control the force of bite and tonus in the muscles of mastication.

Visceral sensory information from the head is conveyed to the medial, dorsal medulla. Here the visceral sensory column is termed the *nucleus solitarius* (fig. 13.23). This nucleus is divided into two parts; the rostral two-thirds is associated with special visceral sensory information from the pharynx (taste), and the posterior one-third with general visceral sensory information from the pharynx relating to pain, temperature, and touch. Taste buds are innervated by sensory neurons located in the sensory ganglia of cranial nerves VII, IX, and X. These sensory cells send their axons into the medulla, where they terminate in the rostral two-thirds of nucleus solitarius. The neurons of this nucleus form two different types of connections. Some neurons of nucleus solitarius project rostrally and end in a medullar nucleus, the *parabranchial nucleus.* The cells of this nucleus in turn project to the ventral posterior nuclei of the thalamus, which finally projects to a ventral part of the primary sensory isocortex in mammals.

The ascending medullar pathways that transmit visceral pain and temperature information from the head are not as well known as those that convey taste information from nucleus solitarius. Pathways are believed to exit to another part of the ventral posterior nuclei of the thalamus, which, in turn, projects to a visceral area of primary sensory isocortex in mammals.

The connections of nucleus solitarius are extremely complex and at present are only poorly understood even in mammals. The neurons of this nucleus not only form ascending sensory pathways to higher brain centers but also form complex connections with the medullar reticular formation and motor nuclei of other cranial nerves. Certain neurons located in the caudal part of nucleus solitarius constitute the cardiac and respiratory control centers much studied by physiologists.

The ventral medulla contains two other motor columns in addition to the lateral branchiomeric motor column already described as forming connections with the somatic sensory column of the dorsal medulla. These are the *somatic* and *visceral motor* columns of the ventral medulla. The somatic motor column (fig. 13.23) consists of four pairs of nuclei located near the midline of the medulla: the oculomotor, trochlear, abducens, and hypoglossal motor nuclei. These neurons innervate the head somites as described in the section on head segmentation. The neurons of the first three nuclei innervate the extraocular eye muscles, and the neurons of the hypoglossal nucleus innervate the hypobranchial muscles associated with the tongue. The somatic motor column forms extensive connections with certain elements of the acousticolateralis system as well as higher brain centers; these connections will be described later.

The remaining ventral medullar column to be discussed is the *visceral motor column*. This column consists of three motor nuclei (fig. 13.23): the *dorsal salivatory nucleus* of cranial nerve VII, the *ventral salivatory nucleus* of cranial nerve IX, and the *dorsal nucleus* of the *vagus* (cranial nerve X). These nuclei are composed of the preganglionic neurons of the cranial division parasympathetic nervous system. The dorsal salivatory nucleus innervates the submaxillary, sublingual, and lachrymal glands. The ventral salivatory nulceus innervates the parotid gland, and the dorsal vagal nucleus innervates the major thoracic and abdominal organs.

We know very little about the organization and variation of the medulla of nonmammalian vertebrates. Most of the nuclei that have been described in mammals have been identified in other vertebrates as well. Frequently there are spectacular enlargements of different parts of the medullar columns associated with particular adaptations. For example, the nucleus solitarius in many bottom-dwelling fish that rely heavily on taste has enlarged until it may form a lobe (usually termed vagal lobe) as large as the tectum or cerebellum in these fish. In animals with specialized tongue structure, such as the rasping lamprey or true chameleons, the hypoglossal nuclei increase their cell number and frequently their cell size. While specializations such as these have been noted, no systematic experiments have explored the medullae of vertebrates for major changes related to feeding and respiration, and our knowledge of this complex neural area so intimately related to head evolution in vertebrates remains fragmentary.

### d. Acousticolateralis Patterns

In all vertebrates a distinct swelling, the *acoustic tubercle,* occupies the dorsal medullar lip just caudal to the cerebellum (fig. 13.24). This medullar area is considered to be a specialized derivative of the somatic sensory column because of its dorsolateral position in the medulla and more importantly because the receptors (inner ear and neuromasts), which ultimately terminate in the tubercle, develop from ectodermal placodes. In anamniotic vertebrates the acoustic tubercle is divided into a dorsal component, the *lateral line lobe,* innervated by the anterior and

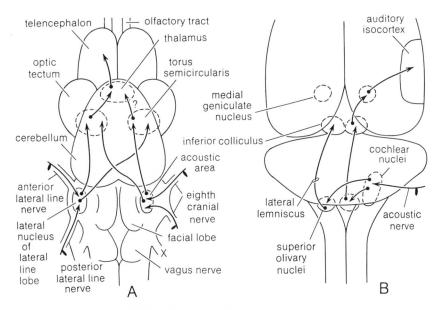

Fig. 13.24. Central auditory and lateralis pathways in a fish (*A*) and the central auditory pathways in a mammal (*B*).

posterior lateral line nerves, and a ventral component, the *acoustic area,* innervated by the eighth (acoustic) cranial nerve.

The anatomical bases of the central connections of the lateral-line lobe in vertebrates are not known. However, ascending pathways to the mesencephalon and telencephalon are known to exist from neurophysiological studies of sharks and rays. The lateral line lobe, its nerves, and receptors are lost in amniotic vertebrates.

The acoustic area in mammals is subdivided into dorsolateral cochlear nuclei and ventromedial vestibular nuclei (fig. 13.24). Axons of the neurons of the spiral ganglion, located within the cochlea of the inner ear, form the auditory branch of the acoustic nerve (VIII). These auditory fibers bifurcate as they enter the medulla and terminate on neurons of the dorsal and ventral cochlear nuclei. Most of the neurons of the cochlear nuclei project ventrally and end in a secondary auditory center, the *superior olivary complex.* These fibers terminate ipsilaterally as well as contralaterally after decussating in the medullar floor. Some neurons of the cochlear nuclei do not terminate in the superior olive, but decussate and turn rostrally just lateral to the superior olive. The axons of the superior olivary neurons, as well as those axons of the cochlear nuclei that have only decussated, are collectively termed the *lateral lemniscus.* This tract runs rostrally in the medulla and terminates in the caudal roof of the mesencephalon (the *inferior* or *posterior colliculus*). Physiological studies show that the neurons of the inferior colliculus are arranged in an orderly manner with respect to auditory frequencies (*tonotopic localization*). Also, many neurons are sensitive to interaural time differences. The neurons of the inferior colliculus form two major pathways; an ascending *brachium* of the inferior colliculus projects to the *medial geniculate* nucleus of the thalamus, and a descending medullar pathway forms complex connections with the cranial motor nuclei.

Observations suggest that the inferior colliculus plays an important role in localizing the source of sounds, that is, where sounds originate in space in relation to the animal perceiving the sound. The descending collicular pathway appears to be organized in such a way that ear and head movements are initiated to position the animal to maximize the information content of the perceived sound.

The thalamic medial geniculate nucleus forms a link in the ascending auditory pathway that terminates in the *temporal lobe* of the isocortex of the telencephalon. This pathway and its terminal auditory cortex appears to be concerned with what is the biological significance of the sound, that is, not where the sound is, but what the nature of the sound is.

In amphibians, reptiles, and birds, the primary auditory fibers project to the cochlear nuclei. Some cochlear fibers terminate in a superior olivary complex, but most continue rostrally, decussate, and terminate in the caudal mesencephalic roof. This mesencephalic target is termed the *torus semicircularis* and is homologous to the mammalian inferior colliculus. In reptiles and birds the torus projects bilaterally to a dorsal thalamic nucleus, which, in turn, projects to a part of the dorsal ventricular ridge of the telencephalon, which is the homologue of the mammalian auditory isocortex (fig. 13.24). The reptilian homologue of the medial geniculate nucleus of mammals is termed *nucleus reuniens,* and the avian homologue, *nucleus ovoidalis*.

The amphibian torus semicircularis is now known to project to the dorsal thalamus, but experimental studies have yet to solve the question of whether an auditory thalamotelencephalic projection exists. Nothing is known about the auditory projections in cartilaginous and bony fishes.

The most ventral component of the acoustic tubercle is composed of the vestibular nuclei. Axons of the neurons of the vestibular ganglion project to these nuclei, as well as to a caudal part of the cerebellum (floccular lobes). The vestibular nuclei form extensive connections with other parts of the brain stem and spinal cord forming circuits involved in orientational reflexes. A large number of the axons of the vestibular nuclear neurons collect near the midline forming a tract, the *medial longitudinal fasciculus*. This bundle forms extensive connections with the somatic motor nuclei of cranial nerves III, IV and VI. This circuitry serves the function of stabilizing eye position so that movements of an animal do not cause a visual image on the retina to wander. In addition to this pathway, the vestibular nuclei form connections with the cerebellum, the medullar formation, the medullar reticular formation, the branchiomeric motor nuclei and the somatic motor columns of the spinal cord to effect complex movements of head, trunk, and limbs that maintain equilibrium and orientation in three-dimensional space.

The vestibular system also appears to project to the isocortex via the thalamus, but the antomical details of this projection are not known in mammals. To date there are no experimental anatomical studies on the vestibular system in nonmammals.

### e. Cerebellar Patterns

In all vertebrates, the cerebellum forms the roof of the rostral medulla oblongata and is composed of a cerebellar cortex and, usually, deep

cerebellar nuclei. The cerebellar cortex in all vertebrates, except agnathans, possesses three distinct layers, the *molecular* layer, the *Purkinje*
cell layer, and the *granular* layer (fig. 13.25). These layers are underlain by a stratum of white matter of variable thickness composed of
fibers passing to and from the cerebellar cortex. The cerebellar nuclei,
when present, are embedded as cell masses in the white matter.

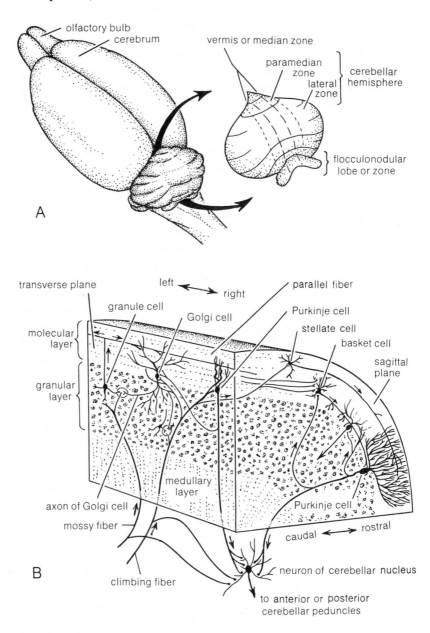

Fig. 13.25. *A*, simplified diagram of the mammalian cerebellar cortex, un-
folded, illustrating its major subdivisions. *B*, diagrammatic representation of
part of a cerebellar folium illustrating its neural elements and their interconnections. After Noback and Demarest 1975 (*The human nervous system.*
New York: McGraw-Hill).

The most conspicuous cells in the cortex are the Purkinje cells. The dendrites of these cells extend upward into the molecular layer and their axons pass into the deep white matter. Purkinje dendrites branch in one plane only, perpendicular to the longitudinal axis of the cerebellum, and each dendritic tree avoids overlap with other Purkinje cells. As the Purkinje axon projects toward the deep white matter, it gives off recurrent collaterals that form connections with *Golgi cells*. These cells are located in the granular layer, which is mainly composed of vast numbers of densely packed small cells with little cytoplasm called *granular* cells. These cells possess four or five dendrites, which end in claw-like extensions. These claw-like dendrites form synapses with one class of incoming fibers called mossy fibers. The axons of the granular cells ascend in the molecular layer, where they bifurcate forming two branches, called *parallel fibers,* which form synapses with the dendrites of Purkinje cells.

The other cell type found in the granular layer, the *Golgi* cells, possess dendrites that branch in the molecular layer and axons that end on the dendrites of granular cells.

The molecular layer is formed mainly by the dendrites of Purkinje cells, Golgi cells, and the parallel fibers of the granule cells. However, a few stellate neurons are also found in this layer. These cells are also called *basket cells* because their axons surround the Purkinje cell bodies branching in such a way as to form baskets.

The incoming or afferent fibers to the cerebellar cortex consist of two anatomical types, *climbing* and *mossy* fibers. The climbing fibers have smaller diameters than the mossy fibers, and they pass through the granular layer without forming collaterals. On nearing Purkinje cell bodies, each climbing fiber divides into several branches that climb and wind about the dendritic branches of a Purkinje cell. The majority of the climbing fibers arise as axons of cells of the inferior olivary complex. This complex is located in the caudal medulla and receives input from spinal neurons as well as from several different parts of the telencephalon.

The second type of afferent fibers, *mossy* fibers, branch repeatedly on entering the white matter of the cerebellum and form synapses with the granule cells. Most afferents to the cerebellar cortex, with the exception of the inferior olive, end as mossy fibers.

All parts of the cerebellar cortex are composed of these cell and fiber types, and this cortex is noted for its regularity and the distinct geometrical patterns among its different elements. This uniformity and orderliness has facilitated neurophysiological studies. The climbing fibers exert an excitatory action on the Purkinje cells. The parallel fibers of the granule cells have a similar effect. The parallel fibers also excite basket and Golgi cells. However, basket cells inhibit Purkinje cells, and Golgi cells inhibit granule cells. Thus, the inhibitory pathways to the Purkinje cells include one synapse more than the excitatory pathways. The effect is to limit the extent to which the cerebellar cortex can be excited by any single volley. Finally, the Purkinje cells have an inhibitory action on the cells of the deep cerebellar or vestibular nuclei with which they synapse. Since only Purkinje cell axons leave the cerebellar cortex, and since the arrangement of the cells is the same throughout all of the cerebellar cortex, functional specialization for different areas of the cerebellum is achieved by different kinds of sensory input and by differences in the targets to which the Purkinje axons project.

In mammals, the cerebellum receives information by pathways that enter the cerebellar cortex through one or more of its three pairs of fiber bundles, called *peduncles* (figs. 13.21, 13.26). The *anterior cerebel-*

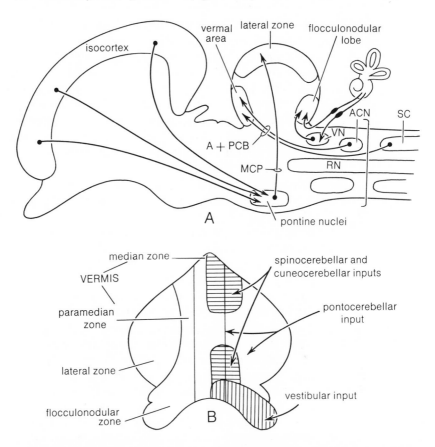

Fig. 13.26. *A,* sagittal diagram of a mammalian brain illustrating the course of the main afferent pathways to the cerebellum. *B,* sites of termination of the afferent pathways on the cerebellar cortex. A+PCB, anterior and posterior cerebellar peduncles; ACN, accessory or external cuneate nucleus; MCP, middle cerebellar peduncle; RN, reticular nuclei; SC, sensory column of spinal cord, VN, vestibular nuclei.

*lar peduncle* is composed in part of the ventral spinocerebellar tract (see Ascending Spinal Somatic Pathways, above), which conveys information regarding muscle tone and position of synergic muscle groups around an individual joint. This pathway conveys information from the hind limbs and caudal trunk. The *middle cerebellar peduncle* is composed entirely of axons of the cells of the *pontine nuclei (pons)*. The pons forms in the rostral floor of the medulla and receives the majority of its connections from the isocortex of the telencephalon. The pontine nuclei of each brain half project to the contralateral cerebellar cortex, but receive their input from the isocortex of the same side.

The *posterior cerebellar peduncle* is composed in part of several different afferent pathways, (1) the dorsal spinocerebellar tract, which conveys information regarding position and tonus of individual hind limb muscles, (2) the cuneocerebellar tract, which conveys information regarding position and tonus of individual forelimb muscles, (3) the olivocerebellar tract from the inferior olivary complex, and (4) the vestibulocerebellar tracts from the vestibular ganglion or vestibular nuclei, which convey information on changes in head position in space.

The differential distribution of these afferents to the cerebellar cortex occurs in the following manner (fig. 13.26): vestibulocerebellar pathways project to the flocculonodular lobe, the spinocerebellar and cuneocerebellar pathways project to the anterior and posterior medial part of the corpus (termed the *vermis* of the corpus cerebelli), and the middle cerebellar peduncle (pontocerebellar tract) projects to the lateral half and middle one-third of the corpus (termed the cerebellar hemisphere).

The axons of Purkinje cells in these different parts of the cerebellar cortex project to different deep cerebellar nuclei. In most mammals three pairs of deep cerebellar nuclei are located in the deep white matter of the cerebellum (fig. 13.27). The vermal cerebellar cortex receives spinal information, and its Purkinje axons project to a *medial* and middle (*nucleus interpositus*) cerebellar nuclei. The lateral corpus area (cerebellar hemisphere) receives information from the telencephalic isocortex via the pontine nuclei, and its Purkinje axons project to a *lateral* cerebellar nucleus (*dentate* nucleus). The flocculonodular lobe receives vestibular information, and its Purkinje axons project to the medial cerebellar nucleus as well as to the vestibular nuclei.

These three areas of cerebellar cortex and their cerebellar nuclei affect different parts of the medullar and spinal pathways and are involved in very different motor actions. The flocculonodular lobe projects to the medullar reticular formation via the medial cerebellar nuclei. The axons of the medial cerebellar nuclei neurons reach the medulla by passing through the posterior cerebellar peduncle and thus compose a part of this peduncle. Some of the Purkinje axons of the flocculonodular lobe also project directly to the vestibular nuclei. The medullar reticular formation gives rise to descending reticulospinal pathways that affect both alpha and gamma motor neurons in the spinal cord. The vestibular nuclei also give rise to a similar pathway called the *vestibulospinal tract*. Thus the flocculonodular lobe affects alpha and gamma motor neurons of head and body muscles in response to changes in vestibular information from the inner ear. This part of the cerebellum mediates changes in muscle tone and body position related to equilibrium.

The vermal area of the cerebellar cortex projects to medial and middle cerebellar nuclei. The course of the axons of the medial cerebellar nucleus has already been discussed. The axons of the middle cerebellar nucleus exit the cerebellum rostrally via the anterior cerebellar peduncle, and the majority of these fibers end in the red nucleus, a cell group located in the floor of the mesencephalon. The cells of the red nucleus form a descending tract (the *rubrospinal tract*) that again affects alpha and gamma motor neurons of the spinal cord. Gamma neurons only innervate intrafusal muscle fibers while alpha neurons innervate the bulk of mus-

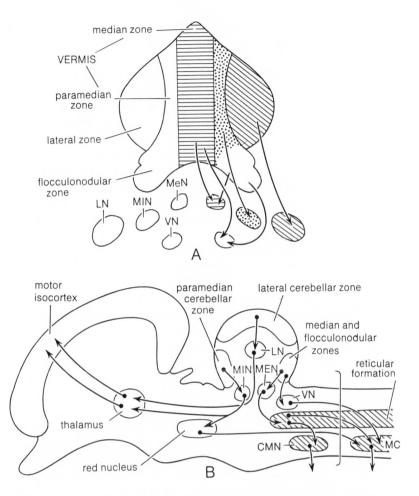

Fig. 13.27. *A,* dorsal view of a mammalian cerebellar cortex illustrating the major subdivisions and their efferent pathways to the cerebellar nuclei. *B,* sagittal view of a mammalian brain illustrating the major efferent pathways of the cerebellum. CMN, cranial nerve motor nuclei; MC, motor column of spinal cord; LN, lateral cerebellar nucleus; MEN, medial cerebellar nucleus; MIN, middle cerebellar nucleus; VN, vestibular nuclei.

cle cells termed extrafusal muscle fibers. Thus the vermal cerebellum receives information from the spinal cord regarding muscle tonus and position of limbs, and is involved in individual movements of limbs that relate to posture and locomotion. Some of the axons of the middle cerebellar nucleus do not end in the red nucleus of the mesencephalon, but continue rostrally and terminate in the thalamus (ventral lateral thalamic nucleus). From the thalamus vermal cerebellar information is conveyed to the motor area of the telencephalic isocortex. The importance of this connection will be discussed later.

The lateral area of the cerebellar cortex projects to the lateral cerebellar nucleus. This nucleus is the largest of the three cerebellar nuclei in mammals, and the axons of its cells leave the cerebellum in the anterior cerebellar peduncle and project to the ventral thalamic nucleus, as did a few axons of the middle cerebellar nucleus. Thus the lateral cerebellar cortex is clearly different from the other cerebellar cortical areas in that it´ does not influence motor neurons of the spinal cord via cell groups in the tegmentum or medulla but projects back to telencephalic isocortex from which its major afferent information originates. Lesions of the lateral cerebellar cortex in mammals result in loss of tone, interruption of the timing of muscle contraction, and alteration of length of muscle contraction of ipsilateral limb muscles involved in volitional movements. Physiological experiments indicate that the neurons of the telencephalic isocortex, which are also involved in volitional movements, cannot by themselves time the duration and sequence of muscle contractions that are involved in volitional movements. These functions now appear to be mediated by a partnership between telencephalic isocortex and lateral cerebellar cortex. Initial activation of the isocortex may occur, followed by excitation of the cerebellar cortex via cortico-ponto-cerebellar pathway. Feedback to the isocortex could then occur via the lateral cerebellar nucleus to the ventral lateral thalamic nucleus to the isocortex. This feedback could affect neurons of the isocortex so that their firing, modulated by the cerebellar feedback, would ultimately time and sequence the contraction of muscles involved in volitional limb movements.

The phylogeny of the cerebellum, like that of much of the brain, is poorly known and is only now beginning to be understood. The cerebellum of agnathans is poorly developed, and clear-cut molecular, Purkinje, and granular layers do not exist. But, since limbs are not found in these animals, rudimentary flocculonodular and vermal areas are the only cerebellar areas present. At present, nothing is known regarding the efferent connections of the cerebellum.

In cartilaginous and bony fishes, spinocerebellar and vestibulocerebellar connections have been experimentally demonstrated. Sharks are known to possess deep cerebellar nuclei, and an ascending projection to the thalamus from the deep cerebellar nuclei has been demonstrated. This projection is probably homologous to the middle cerebellar nuclear projection in mammals, rather than the lateral cerebellar nuclear projection, since sharks do not possess a well-developed pons nor well-developed appendicular muscles. Similar conditions appear in bony fish, except that these forms do not possess distinct deep cerebellar nuclei. The cells that are probably homologous to the cells of the deep cerebellar nuclei of other vertebrates are scattered among the Purkinje cell layer of the cerebellar cortex. A small ascending cerebellar tract to the thalamus has been reported in teleosts as well.

Amphibians possess a relatively poorly developed cerebellum in contrast to other vertebrates (except agnathans). However, they do possess deep cerebellar nuclei, and a small ascending cerebellar tract to the thalamus, as well as spinocerebellar and vestibulocerebellar tracts, are known.

Most reptiles have evolved more complex postural and locomotor mechanisms, and these trends are reflected in the size of the cerebellum. Cerebellar projections to the red nucleus and thalamus are known, and a rudimentary cerebellar hemispheric area is suspected to exist in crocodiles. At present there is no experimental anatomical evidence on the possible existence of connections between the cerebellum and the dorsal ventricular ridge of the telencephalon. Since the ridge is thought to be the homologue of much of the mammalian isocortex, evidence on projections to the pons from the ridge and projections to the ridge from the cerebellum via the thalamus would yield considerable insights into the organization of locomotion and behavior in reptiles.

Birds possess pontine nuclei that receive projections from the telencephalon, and physiological experiments indicate they possess a lateral cerebellar hemispheric area. This is not surprising when the motor behavior of many birds is considered. Parrots, for example, show control of individual digits when handling seeds and fruit with their feet, and the opening of mollusks by oyster catchers is another example. At present we do not know whether the lateral cerebellar area of birds is homologous to that of mammals or represents a case of parallelism. Only information on the cerebellar organization of reptiles will solve this question.

### f. Visual Patterns

Reception and the initial stages of information processing of visual stimuli occur in the retina of the eye. The retina of each eye receives light stimuli from an extensive area of space termed the *visual field,* which is usually measured by the number of degrees over which light can enter the eye and fall on the retina (fig. 13.28). The visual field of each retina in most vertebrates approximates 170° in diameter, but varies according to the curvature of the retina and the physical properties of the light-gathering and focusing mechanisms of the eye.

In most vertebrates, the eyes are positioned in the head so that overlap of the visual fields of the two eyes occurs. The area of overlap is termed a *binocular visual field.* Visual stimuli located in this field will project upon both retinas, which provides stereoscopic clues important to depth and size perception.

The axons of the ganglion cells of the retina are the only ones that leave the retina and project to the brain. These axons collect at the back of the eye and pierce the wall of the eye, where they form the optic nerve. In most mammals each optic nerve projects to both sides of the brain. The axons originating from ganglion cells in the anterior half of the retina cross to the other side, while the axons originating from ganglion cells in the posterior half of the retina project to the same side (fig. 13.28). This process of crossing, or decussation, occurs at the optic chiasma, which is located in the floor of the diencephalon. After decussating, the optic nerves continue and are called the *optic tracts.*

In mammals, the number of uncrossed optic fibers is correlated with the size of the binocular field; that is, as the binocular field becomes larger, the number of uncrossed, or ipsilateral, optic fibers increases. Until recently, mammals were thought to be the only vertebrates with ipsilateral optic fibers. However, experimental anatomical studies have revealed such projections in lampreys, chondrostean and holostean fishes, most amphibians, lizards and snakes. The optic nerves appear to be completely crossed in sharks, teleosts, lungfishes, turtles, crocodiles, birds, and a few mammals. At present there is no known correlation between the presence or absence of binocular visual fields in nonmammalian vertebrates and the development of ipsilateral optic fibers.

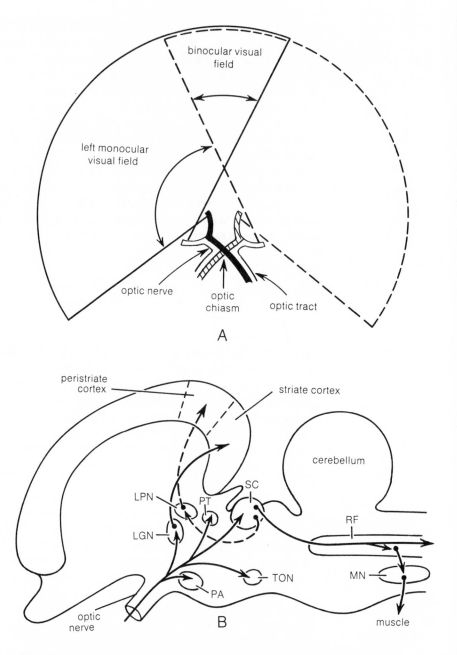

Fig. 13.28. *A*, diagram illustrating the relationships of the visual fields to partial decussation of the optic nerves. *B*, sagittal view of a mammalian brain illustrating the course of several major visual pathways.

The axons of retinal ganglion cells are not arranged irregularly as they project to different brain targets. Rather, they form highly organized point-to-point projections so that most retinal targets receive a retinotopic organized map of the visual field.

In most vertebrates each optic tract projects to five different neural areas: preoptic area of the hypothalamus, thalamus, pretectum, optic tectum, and tegmental optic nucleus (fig. 13.28). The second order projections of only the thalamus and the optic tectum are experimentally known even in mammals. In mammals many optic tract fibers end in dorsal and ventral components of the *lateral geniculate nucleus* of the thalamus. The cells of the dorsal lateral geniculate nucleus project to the ipsilateral visual isocortex (striate cortex) of the telencephalon. The cells of the ventral lateral geniculate nucleus project to the hypothalamus the pretectum, and the optic tectum (*superior colliculus*).

Other optic tract fibers continue caudal to the thalamus and end in the pretectum and the superior colliculus. The cells of the superior colliculus are arranged in sheets (laminae) sandwiched between fiber layers. The more dorsally located cells of the superior colliculus project rostrally to end in the caudal thalamus in a cell group termed the *lateral posterior nucleus* of the dorsal thalamus. The lateral posterior nucleus also projects to an area of the telencephalic isocortex called the *peristriate cortex* (fig. 13.28). Thus mammalian isocortex receives visual information by two different pathways, a retino-geniculo-striate pathway and retino-colliculo-thalamo-peristriate pathway.

The more ventrally located cells of the superior colliculus project caudally and terminate bilaterally in the medullar and spinal reticular formations.

Both reptiles and birds possess at least two visual pathways to the telencephalon. Retinal fibers terminate in a rostral thalamic complex, part of which is probably homologous to the lateral geniculate nucleus of mammals. In turtles the thalamus projects ipsilaterally to a part of the dorsal cortex of the telencephalon, which is probably homologous to the striate visual isocortex of mammals. The condition in birds is more complex. The rostral thalamic complex is composed of at least three separate nuclei, all of which project to a part of the dorsal telencephalon, termed the *wulst*. Unlike the condition in mammals, two of these nuclei project bilaterally to the dorsal telencephalon. At present very few reptiles have been examined experimentally, and future studies may reveal that bilateral thalamic projections are present in reptiles. If this does prove to be the case, then mammals may have secondarily lost the contralateral projection, and these retinothalamic pathways to the dorsal telencephalon in reptiles, birds, and mammals would be considered homologous.

An ascending pathway from the optic tectum to the thalamus occurs in reptiles and birds. This thalamic target is called *nucleus rotundus* in both classes. Rotundus projects ipsilaterally to a lateral area of the dorsal ventricular ridge in reptiles. The same region of the dorsal ventricular ridge in birds is termed the *ectostriatum*. These ridge areas are believed to be homologous to the peristriate visual cortex in mammals.

Physiological studies in agnathans, cartilaginous and bony fishes, and amphibians have revealed that visual information reached telencephalic levels in these forms. However, too little anatomical data is available to determine if both visual pathways to the telencephalon exist in these anamniotes.

In vertebrates, visual perception consists of many different components—detection of movement, perception of patterns, sense of direction, location of objects in visual space, perception of colors, perception of size and distance of objects, and the like. All of these aspects of vision, and their functional localization with respect to different visual pathways within vertebrate brains, have only begun to be studied.

In a number of vertebrates, physiological studies on the retinal ganglion cells and their axons suggest that at least two different trends exist in vertebrate retinal processing. Experiments on the retinal ganglion cells that project to the optic tectum in several amphibians reveal that these ganglion cells form a few distinct functional classes. At least four classes of ganglion cell axons terminate at different depths in the upper half of the optic tectum. Each class of ganglion cells transmits only a specific type of information to the tectum. The different cell types (rods, cones, bipolar and horizontal cells, etc.) located within the retina are arranged in such a manner that they perform operations or abstract certain features from a visual image. These operations are edge detection, convexity detection, and moving-edge and dimming detection. These different classes of cells also respond to differences in the size of visual images which range from 2° to 15° in angular diameter. Small moving objects elicit orienting and snapping responses in frogs while large moving objects elicit ducking, turning away, or jumping responses. Different classes of cells may thus mediate different visual behaviors. Ganglion cells that are activated by small moving objects project to the optic tectum, stimulation of which produces prey-catching behavior. While the anatomical pathways for avoidance behaviors are not known, pretectal cells are known to inhibit tectal cells that are selectively sensitive to prey-sized objects, and thus inhibit prey attack behaviors.

Similar retinal ganglion classes appear to exist in fish and reptiles. Thus, one trend in vertebrate visual processing involves the elaboration of different classes of retinal ganglion cells with peripheral abstraction of visual images. It is possible that different classes of ganglion axons project to each of the optic targets and that these targets compose complex pathways subserving different aspects of visual perception.

Physiological examination of the retinal ganglion cells in cats and monkeys has produced very different rules. These animals do not possess classes of ganglion cells that abstract visual images in the same manner that we have described for frogs. In cats, the receptive fields (the region of the retina over which a ganglion cell can be influenced to fire) of ganglion cells are organized in a manner to increase the contrast of an object against its background. Neurons of the lateral geniculate nucleus continue to enhance this contrast and thus appear to further "sharpen" the object against its background. However, the visual isocortex of the telencephalon contains a large number of functionally different cell classes. These cells will respond only to objects of certain shapes with a particular orientation and certain characteristic movements. Thus, mammals possess cortical neurons that abstract visual features, rather than classes of ganglion cells that operate to perform peripheral abstraction at the level of the retina.

The selective advantage of central abstractions over that of peripheral abstraction is that the isocortex is arranged in far more complex layers than the retina and contains far more neurons. This increases the total possible number and kinds of abstractions that can be performed on a visual image. As the types of abstraction increase, the kinds of visual information also increase; that is, only vertebrates with isocortex may be able to recognize the insect with the wistful expression.

### g. Thalamic Patterns

The thalamus constitutes the bulk of the diencephalon in vertebrates, forming the lateral walls of the unevaginated prosencephalon that we term the diencephalon. It is usually divided into dorsal and ventral regions. The *dorsal thalamus* is primarily composed of neurons that form ascending pathways for all sensory modalities to the telencephalon. The *ventral thalamus* is composed primarily of neurons that form descending pathways related to motor functions. Thus the thalamus is intimately linked to both sensory and motor pathways originating or terminating in the telencephalon.

The dorsal thalamus of mammals is traditionally divided into five major regions (fig. 13.29). These regions are characterized by different afferent and efferent connections. The *anterior* thalamic group receives projections from the caudal hypothalamus (*mammillary nuclei*) and projects to the dorsomedial roof of the telencephalon (*cingulate cortex*). This pathway forms part of the *limbic system,* which will be discussed in detail in a later section. The *ventral* thalamic group is divided into three major areas, ventral anterior nucleus, ventral lateral nucleus, and ventral posterior nucleus. The ventral anterior nucleus receives projections from the rostral ventrolateral wall of the telencephalon (*globus pallidus*) and from the cerebellar nuclei via the anterior cerebellar peduncle and projects to an area of isocortex related to motor functions (*premotor isocortex*). The *ventral lateral* thalamic nucleus receives projections from the lateral cerebellar nucleus and projects to the primary *motor area* of the isocortex. The *ventral posterior nucleus* is the primary target of the ascending medial lemniscus, spinothalamic, and secondary trigeminal tracts. Thus the ventral posterior nucleus is a primary target of all ascending somatic sensory information, and it in turn projects to the primary sensory isocortical area of the telencephalon.

The *lateral* thalamic group receives projections from the anterior colliculus (optic tectum) and pretectum and projects to peristriate isocortex.

The *dorsomedial* thalamic group receives projections from olfactory areas of the telencephalon and projects to an area of isocortex located under the orbital area of the skull, which is thus termed the orbital isocortex.

The fifth division of the dorsal thalamus is composed of the *intralaminar* nuclei. The neurons of this group are scattered along the margins of the other thalamic nuclei and constitute an extensive *arousal system* to the telencephalon. These neurons receive projections from all of the ascending sensory pathways we have discussed, and they project to the outer layer of the isocortex.

Such neurons appear to form an alerting or biasing mechanism for the isocortex. They probably do not convey information on where or what is happening in an animal's world, but rather convey that something new or different is happening that "needs" to be analyzed.

Finally, two other nuclei also form part of the thalamus, nuclei that are not assigned to any of the thalamic groups that we have described.

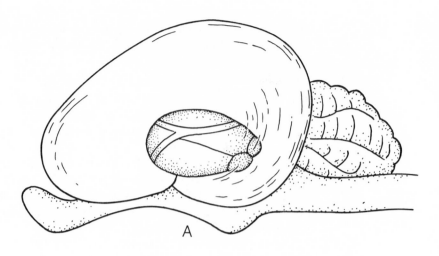

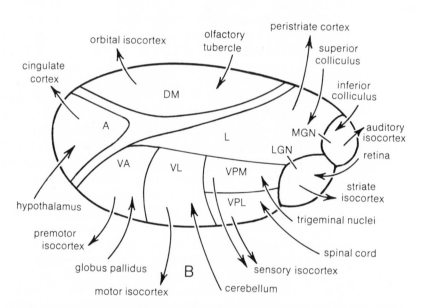

Fig. 13.29. *A,* diagram of a lateral view of a mammalian brain illustrating the relative position of the thalamus. *B,* schematic representation of the major subdivisions of the mammalian thalamus, and their connections, as seen in a lateral view. A, anterior thalamic group; DM, dorsomedial thalamic group; L, lateral thalamic group; LGN, lateral geniculate nucleus; MGN, medial geniculate nucleus; VA, ventral anterior nucleus of the ventral group; VL, ventral lateral nucleus of the ventral group; VPL, lateral division of the ventral posterior nucleus; VPM, medial division of the ventral posterior nucleus.

These are the *lateral* and *medial geniculate* nuclei. The lateral geniculate receives fibers from the eye and projects to the *primary visual isocortex* (striate cortex). The medial geniculate nucleus receives auditory projections from the posterior colliculus and projects to the primary auditory isocortex.

The dorsal thalamus clearly constitutes a major target of many ascending sensory pathways destined for the telencephalon. The dorsal thalamus has frequently been described as a relay area, and although this is true, we should not view the thalamic cells as merely passing information forward to the telencephalon as though the thalamus were a single link in a bucket brigade.

Information on the functions of retinal ganglion and tectal cells indicates that when neurons form complex chains or circuits, each level performs an operation (i.e., abstracts information) from the signal. The cells of the dorsal thalamus are clearly important in this regard. For example, neurons of the ventral posterior nucleus receive information regarding pain, temperature, touch, and limb position from all parts of the head and body, and in turn project to a single area of isocortex. We know that even when the thalamus and the isocortex are experimentally removed, the reflexes described in Spinal Patterns still occur. In fact, in most mammals locomotion still occurs in the absence of the thalamus and isocortex. Functions that are impaired relate to complex behavioral sequences such as food seeking, predator avoidance, establishment and defense of territories, and reproduction, for example. How thalamic and isocortical neurons are organized to abstract information critical to these behaviors and what sensory clues are common to many of these behaviors remains one of the major challenges to comparative neurobiology.

Information on the organization and connections of the thalamus in nonmammalian vertebrates is too fragmentary to characterize what major adaptative changes have occurred. Enough information is available, however, to suggest that the dorsal thalamus of all vertebrates probably possesses at least three major zones (fig. 13.30). All vertebrates exam-

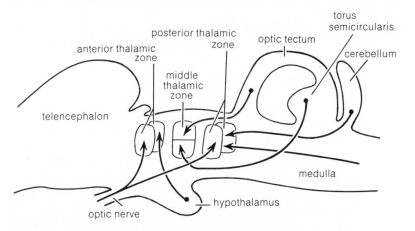

Fig. 13.30. Schematic representation of the major subdivisions of the dorsal thalamus in nonmammalian vertebrates.

ined to date possess an anterior thalamic zone divided into a medial area that receives information from the hypothalamus and a lateral area that receives direct projections from the eye. The lateral anterior area is probably homologous to the lateral geniculate nucleus of mammals, and the medial anterior area is probably homologous to the anterior thalamic group of mammals.

A central or middle thalamic zone can be recognized in all nonmammalian vertebrates. This zone receives projections from the optic tectum and, in all tetrapods, also receives auditory information from the torus semicircularis (inferior colliculus). This thalamic zone is probably homologous to the lateral thalamic group and medial geniculate nucleus of mammals.

Finally, a posterior thalamic zone in nonmammalian vertebrates can be divided into a lateral area that also receives direct projections from the eye and a medial area that receives ascending spinal projections and cerebellar nuclear efferents. The lateral posterior area is probably homologous to the mammalian pretectum, and the medial posterior area is probably homologous to the ventral thalamic group of the dorsal thalamus.

This pattern of thalamic organization suggests that each area of the dorsal thalamus is related to a specific type of sensory information and that different regions of the thalamus have hypertrophied or been reduced depending on the adaptations utilized by any group of vertebrates.

In most nonmammalian vertebrates visual information is heavily utilized in many species' typical behaviors. The optic tectum and pretectum are well developed in these forms. In most nonmammalian vertebrates the central and lateral posterior thalamic zones are also well developed since these areas are intimately related to the optic tectum (fig. 13.30).

In mammals the skin bears a heavy coat of hair that is also an important tactile sensory system, and the somatic sensory pathways are well developed. The posterior medial thalamic zone in mammals hypertrophied in relation to this tactile sense, and this zone becomes the single largest area of the dorsal thalamus (fig. 13.29).

### h. Striatal Patterns

The corpus striatum forms the rostral ventrolateral floor of the telencephalon. This nuclear area is divided into a dorsal striatum (*caudate* and *putamen* nuclei) and a *ventral* striatum (*globus pallidus*). The dorsal striatum has many of the characteristics of a sensory center, whereas the ventral striatum seems more like a motor center. The dorsal striatum in mammals receives afferents from a variety of different areas (fig. 13.31). Sensory and motor areas of isocortex project to the dorsal striatum as well as to the intralaminar thalamic nuclei. Neurophysiological studies indicate that the neurons of the dorsal striatum receive auditory, visual, and somatic sensory information. Tegmental and reticular areas of the midbrain are also known to project directly to the dorsal striatum. These pathways convey visceral as well as somatic sensory information. Thus a wide variety of sensory information converges in the dorsal striatum. The dorsal striatum projects to the ventral striatum and to a midbrain tegmental nucleus (substantia nigra). The strionigral pathway thus constitutes a feedback loop to one of the major nuclei that sends information to the dorsal striatum. Such feedback loops usually serve to modulate input to a higher center and are a common feature of nervous systems.

The ventral striatum forms a number of such feedback loops (fig. 13.31). It projects to the ventral anterior and ventral lateral thalamic nuclei, which in turn project to the motor isocortex. The motor isocortex projects to the dorsal striatum, which in turn projects to the ventral stri-

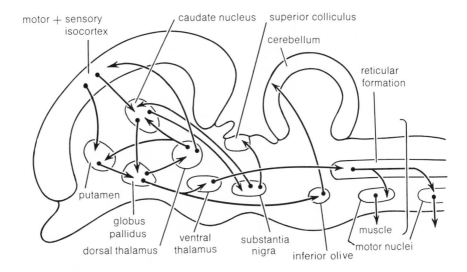

Fig. 13.31. Sagittal diagram of a mammalian brain illustrating some major pathways of the corpus striatum.

atum completing the loop. The ventral striatum also projects to the intralaminar thalamic nuclei, which in turn project to the dorsal striatum. Finally, the ventral striatum projects to a number of ventral thalamic and midbrain tegmental nuclei, which form a link in the descending reticulospinal pathways. Thus, the ventral striatum can affect motor behavior, but not directly at the spinal level.

The striatal system is further complicated by striotegmental pathways that end in the inferior olive of the medulla. You will remember that the inferior olive is a major source of afferents to the cerebellar cortex. This pathway introduces the possibility of more complex motor actions by way of the cerebellar cortex and its connections with the reticular formation, the thalamic nuclei and motor isocortex. Because of these complex interconnections it is not surprising that we do not understand the role of the striatum in motor movements. Damage to the striatal nuclei or their pathways results in involuntary tremor of the appendicular muscles. Such disease processes lead us to suspect that the striatum is involved in motor behavior but tells us little regarding its normal functions. Destruction of the striatum in many vertebrates results in the disruption or loss of species typical behaviors such as nest building or territorial displays.

In anamniotic vertebrates the isocortex is poorly developed and the thalamus projects primarily to the striatum. In these vertebrates the striatum may function in a comparable manner to isocortex in mammals. The striatum may analyze and integrate different types of sensory information and initiate motor sequences appropriate to the species. Such functions could account for its complex feedback loops and connections with the cerebellum. There are striking parallels between the striatum and isocortex in terms of their connections with the thalamus, cerebellum, and medulla. These similarities suggest that both of these telencephalic areas are involved in similar types of operations.

*i. Limbic Patterns*

The concept of the limbic system or limbic lobe was originally proposed for all those parts of the forebrain believed to be concerned with the sense of smell. While the pathways mediating smell form an integral part of the limbic system, accumulated evidence shows that the limbic system has a much broader role in that it mediates or regulates emotional behaviors and memory formation and controls general arousal states.

The word limbis means a fringe or border, and in a real sense most of the neural areas that form the limbic system border or surround the isocortex. In fact, the limbic system composes all of the telencephalon with the exception of the isocortex and striatum (fig. 13.6).

The limbic system receives many kinds of sensory information. Olfactory information will be described first. The primary olfactory epithelial receptors located in the olfactory sac or capsule possess axons that project to the *main olfactory bulb* of the telencephalon, where they ramify in the surface layers and form synapses with neurons in the olfactory bulb. Most land vertebrates possess both main olfactory bulbs as well as *accessory olfactory bulbs*. The main bulbs receive information from the major part of the olfactory epithelium and the accessory bulbs receive information via the vomeronasal nerve from a ventrolateral segment of the olfactory epithelium. This segment frequently evaginates and splits away from the parent epithelium during development, forming a separate organ termed the *vomeronasal organ*. Thus many land vertebrates actually posses two pairs of olfactory organs with very separate and distinct central pathways. Although the vomeronasal organ is located separately from the main olfactory organ, both possess similar cell types, and both have a chemoreceptor function. Recent studies suggest that the main olfactory organ may be sensitive primarily to airborne odors and thus is a distance receptor, while the vomeronasal organ is sensitive to odors that are liquid soluble and therefore is a proximal or near-field receptor. This suggests that behaviors that utilize olfactory clues, such as predation and species and sex recognition, may be mediated by two separate olfactory systems. Animals may first be alerted and track using airborne clues and subsequently use odors transferred to the vomeronasal organ by tongue or lips to initiate behaviors at close range.

The main olfactory bulb projects to a number of telencephalic areas. Its main targets are the lateral pallium, which in mammals is divided into an anterior *piriform cortex* and a posterior *entorhinal cortex* (fig. 13.32). Olfactory bulb fibers also terminate in the caudal ventrolateral wall of the telencephalon, termed the *amygdala*.

The accessory olfactory bulb also projects to the amygdala, but to a part that does not receive projections from the main olfactory bulb. Thus the piriform and entorhinal cortices and the amygdala constitute part of the limbic system that borders the isocortex ventrolaterally in the telencephalon (figs. 13.6, 13.32). Both the olfactory cortices and the amygdala project to the diencephalon, which in turn projects into the tegmentum and medullar reticular formation. Thus olfactory stimuli can facilitate complex hormonal and motor behaviors via the hypothalamus and motor centers of the medulla and spinal cord.

Olfactory information is conveyed to yet another major limbic struc-

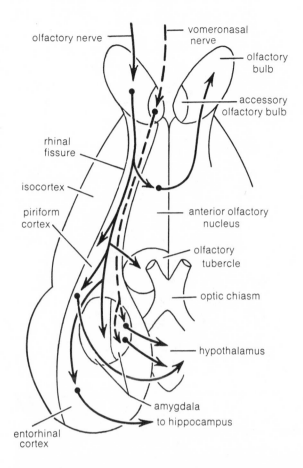

Fig. 13.32. Diagram of the right ventral surface of a mammalian forebrain showing the efferent pathways of the accessory olfactory and main olfactory systems. The efferents of the accessory olfactory bulb are indicated by dashed lines.

ture in the telencephalon via the entorhinal cortex. The entorhinal cortex projects medially to the *hippocampus*. The hippocampus, or medial pallium, along with the *cingulate* cortex forms the dorsomedial border of the isocortex (figs. 13.6, 13.33). The hippocampus not only receives olfactory information via the entorhinal cortex but also receives many other sensory modalities as well. Reticular nuclei located at the midline in the medulla receive various kinds of visceral sensory information and project directly to the hippocampus. The hippocampus also receives information from the hypothalamus by a complex route. The caudal hypothalamus projects to the anterior thalamic group of the dorsal thalamus, which in turn projects to the cingulate cortex. The cingulate cortex forms extensive connections with the hippocampus as well as with the dorsal striatum. As we have seen in the last section, the striatum is heavily involved in mediating motor activity that involves many species typical behaviors.

The hippocampus also receives auditory, visual, and somatosensory information, but the anatomical pathways are not understood. The axons of many hippocampal neurons collect on the ventromedial surface of this

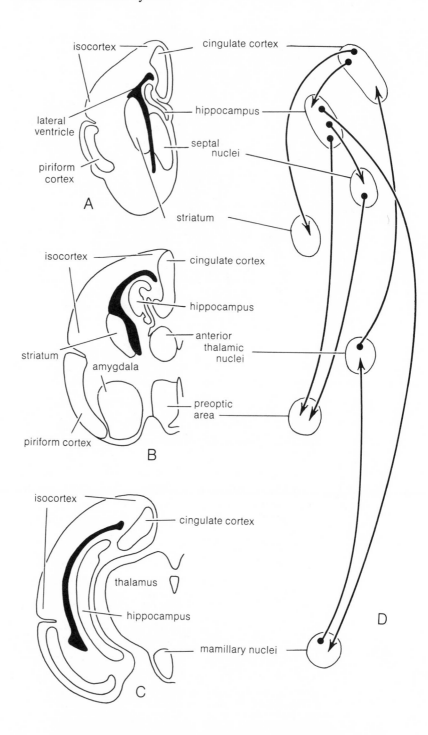

Fig. 13.33.  The mammalian limbic system. *A–C*, diagrams from rostral to caudal through the left cerebral hemisphere of a mammal illustrating the major limbic centers. *D*, summary of the interconnections of these centers.

cortex and form a tract, the *fornix,* which passes down the medial wall of the telencephalon. Many of the fornix fibers terminate in the ventromedial wall, termed the *septal nuclei,* while others continue ventrally and terminate in the rostral (*preoptic area*) and caudal (*mammillary nuclei*) parts of the hypothalamus (fig. 13.33). Thus the hippocampus possesses a feedback loop via the hypothalamic-thalamic-cingulate pathway as well as a major descending pathway to the medulla via the hypothalamic-tegmental pathways.

The septal nuclei also possess similar pathways. The septal nuclei receive information from the midline reticular nuclei and hypothalamus, and project to preoptic and caudal hypothalamic centers as well as directly to midbrain tegmental areas.

At this point you can realize that all of the telencephalic limbic centers possess three features in common. All receive multiple types of sensory information from a variety of other brain centers, all form complex feedback loops involving hypothalamus, and all possess descending pathways to the hypothalamus and/or tegmentum, which give these telencephalic limbic structures indirect access to medullar motor systems. These features also appear to characterize nonmammalian vertebrates as well, but details of these connections are not nearly as well understood as in mammals.

More behavioral experiments have been directed to an examination of the limbic system in mammals than any other single brain system. Yet our information is still meager on limbic functions and their neural mechanisms. The telencephalic limbic centers appear to guide behavior by regulating hypothalamic and tegmental pathways. This guidance of behavior seems to be primarily by inhibition of hypothalamic and tegmental centers. Artificial electrical stimulation of telencephalic limbic centers suppresses ongoing behaviors that are occurring at the time of the stimulation. Destruction of these same limbic centers seems to "release" various behavioral activities so that experimental animals frequently overreact to stimuli.

For example, lesions of the amygdala produce tame and placid animals that are insensitive to many types of environmental changes or to social signals from conspecifics. These animals do not lost their aggressive behavior, but the thresholds of stimuli that normally "release" aggressive behavior are radically changed. This suggests that the amygdala and its pathways do not mediate aggressive behaviors directly, but regulate or alter the probability of activity in tegmental pathways that directly mediate aggressive behaviors.

Similarly, destruction of the septal nuclei alters water and food intake, and mammals appear to overreact to stimuli. There is an increased tendency to flee from opponents. Thus the septal nuclei appear as part of a system that regulates ongoing behavioral sequences and motivational systems. Destruction of the cingulate cortex disrupts orderly sequencing of behavior. Cingulate damage in female rats disrupts maternal behavior. Such a female appears confused and may start to retrieve a pup that has strayed from the nest only to stop, drop the pup, pick up nest material, drop it, and return to the nest to continue nursing the other pups. Such behavior is best described as failure to complete a behavior sequence in a biologically successful fashion.

Animals with hippocampal damage demonstrate a similar syndrome. These animals show impairment in the ability to suppress behavioral sequences. They are prone to initiate and continue responses to appropriate stimuli, but with an intensity and duration that is not usually observed in normal animals.

### j. Isocortical Patterns

The isocortex (neocortex) of the telencephalon is a hallmark of mammalian brains and is clearly a major adaptive feature that is correlated with the mammalian radiation. In insectivores it constitutes only some 20 percent of the volume of the telencephalon and in primates it has hypertrophied to the degree that it constitutes some 80 percent of the telencephalic volume.

The mammalian isocortex is frequently divided into four lobes that are named for the bones that overlie these areas of the cortex: frontal, parietal, occipital, and temporal (fig. 13.34). All of these cortical areas can be characterized by a similarity in their internal histology. All isocortex possesses at least six cellular laminae at some developmental stage in the life of mammals. These laminae are formed by different types of neurons, and their afferent and efferent connections differ. The different isocortical lobes can be subdivided on the basis of differential development of the six laminae. In man, over 50 subdivisions of isocortex have been recognized. These subdivisions were originally recognized by a German anatomist, Korbinian Brodman, who assigned these subdivisions numbers as he examined different areas of the isocortex (fig. 13.34). His numbering system is still widely used, and subsequent work has demonstrated that most of his recognized anatomical subdivisions possess different functions referred to as *cortical localization*. These structural and functional differences are primarily related to the nature of the afferent thalamic projections and to the efferent connections of isocortex.

The occipital lobe can be subdivided into two areas (areas 17 and 18). *Area 17* (striate cortex) is the target of the lateral geniculate nucleus and is a primary visual area. *Area 18* (peristriate cortex) is the target of the lateral posterior thalamic nucleus, which receives its input from the superior colliculus. Thus area 18 is also a visual area, termed a secondary visual area because it does not receive projections from a thalamic nucleus that is primarily in receipt of sensory information.

The temporal lobe also contains primary sensory areas (*areas 41 and 42*). These areas receive projections from the medial geniculate nucleus and are termed the primary auditory isocortex. The parietal lobe contains *areas 1, 2, and 3,* which receive projections from the ventral posterior thalamic nuclei and thus constitute the primary sensory isocortex. The frontal lobe contains *areas 4 and 6,* which give rise to long descending pathways that terminate in the spinal cord. These pathways, termed the *corticospinal* or *pyramidal tracts,* end on interneurons related to the alpha and gamma motor neurons and thus constitute a fast velocity system that primarily controls distal appendicular muscles. Other long descending pathways also arise from areas 4 and 6 and terminate in the medulla,

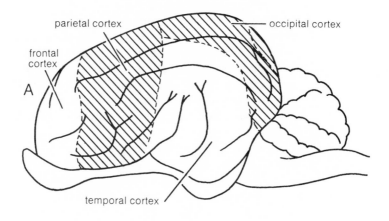

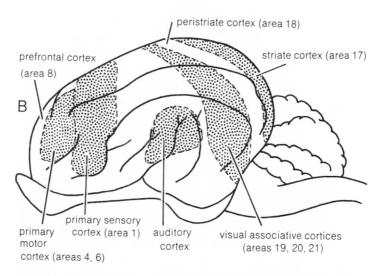

Fig. 13.34. Lateral views of the cat brain showing the position of the major lobes (*A*) of the isocortex and their functional significance (*B*).

influencing cranial motor nuclei. These pathways are termed the *cortico-bulbar tracts*. Area 4 is termed the primary motor isocortex and area 6 a supplementary motor area.

The isocortex rostral to area 8 in the frontal lobe is called the *prefrontal cortex*. This area does not receive primary sensory projections from the thalamus but does form extensive connections with most other isocortical areas. Prefrontal cortex is particularly well developed in primates, and damage or removal of prefrontal areas in man results in reduction of intellectual abilities, powers of concentration, and reduction in ability to solve abstract problems.

Large areas of the parietal and temporal lobes are termed *associative areas*. These areas consist of a number of secondary sensory areas. For example, areas 19, 20, and 21 are also concerned with visual processing. These areas receive projections from areas 17 and 18 as well as prefrontal cortex. Evidence is accumulating that each primary sensory area is bordered by rings of secondary cortices that continue the analysis and integration of sensory information related to their primary cortex. As the periphery of such a zone is approached the reciprocal connections with frontal cortex increases. Thus one of the primary functions of isocortex is to increase analysis of sensory information by mapping and remapping certain aspects of sensory signals. Such multiple representations or maps of each sensory modality also allows the integration of different sensory modalities, such as vision and somatic sensory information. These neural functions certainly underlie behavioral plasticity and complex learning.

Similarly, primary and secondary motor maps spread across the caudal frontal lobe increase the resolution and integration of motor control. Isocortex also gains access to the limbic system by projecting to the amygdala, cingulate cortex, and hypothalamus. Similar connections are also made with the caudate and putamen nuclei of the striatal system and with the cerebellum via the pontine nuclei.

Until recently isocortex was believed to be well developed only in mammals. Reptiles were believed to possess a small rudiment located in the dorsal pallium homologous to mammalian isocortex, as were birds. However, extensive experimental anatomical, embryological, and histochemical evidence now supports the hypothesis that the dorsal ventricular ridge of the telencephalon in reptiles and birds is homologous to all of the mammalian isocortex except area 17. The dorsal ventricular ridge (fig. 13.6) in these forms is now known to receive thalamic projections that appear homologous to the lateral geniculate (auditory) and lateral thalamic (secondary visual) and ventral nuclei of the dorsal thalamus (somatic sensory) of mammals. These studies support the hypothesis that the isocortex is not a single neural area, but in fact has a dual origin. Area 17, striate cortex, appears to originate as a dorsal pallial area, whereas the rest of the isocortex arose out of the lateral pallium (fig. 13.6). Whether or not anamniotic vertebrates such as sharks and bony fish may have independently evolved similar cortical tissues with complex integrative functions remains one of the many challenges of comparative neurobiology.

10. The Nervous System and Sense Organs of *Branchiostoma* (Amphioxus)

A brain, or cerebral vesicle, in amphioxus can be delimited from the rest of the neural tube by two criteria. The cephalic portion of the neural tube contains an enlarged lumen termed the cerebral ventricle, and its dorsocaudal boundary is marked by a series of large sensory neurons termed *Joseph's cells* (fig. 13.35). The rostral half of the brain wall is formed by a single layer of epithelial cells with extremely long apical cilium-like processes. Many of these cells may represent receptors, particularly photoreceptors. The midventral floor of the brain vesicle is characterized by a group of large neurons that possess a large number of granules and a well-developed endoplasmic reticulum. These neurons are believed to be neurosecretory and are collectively termed the *infundibular* organ.

The rostral pole of the brain is characterized by a group of pigmented epithelial cells that have been termed the pigment spot. These cells do not, however, possess an ultrastructure that is associated with photoreceptors, and they probably are not responsive to light. A pair of nerves arise from the rostral floor and are believed to be sensory since they terminate among the epithelial cells that form the neural wall at this point. These

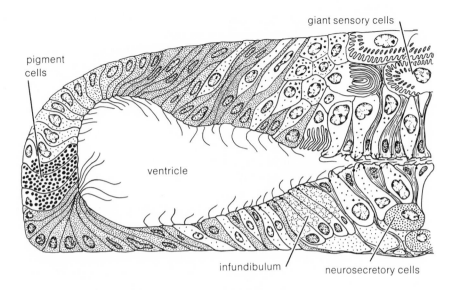

Fig. 13.35. Midsagittal schematic view of the brain of *Branchiostoma*. Modified from Meves 1973 (*Z. Zellforsch.*, 139:511–32).

nerves have frequently been homologized to the nervus terminalis of vertebrates. Like the vertebrate terminal nerve, the terminal nerve in amphioxus innervates a ciliated external pit that may be homologous to the vertebrate olfactory placode.

The absence of a nasal epithelium, lateral eyes, inner ears, and their associated nerves makes comparisons with vertebrate brains difficult, if not impossible. The first nerves to arise from the brain caudal to the terminal nerves are a pair of nerves located just caudal to the infundibular organ. These nerves are also sensory to the head and dorsal fin surfaces. They have been homologized to the trigeminal complex of vertebrates. The nerves caudal to this level are identical to those of the trunk except that a number of those immediately caudal to Joseph's cells innervate the pharynx. These nerves may correspond to the facial, glossopharyngeal, and vagus nerves of vertebrates.

Most workers have used this pattern of nerves and their origins to suggest that the rostral cerebral vesicle, including the infundibular organ, is homologous to the vertebrate forebrain, that the segment composed of Joseph's cells is homologous to the vertebrate midbrain, and that the more anterior spinal nerves should be homologous to the vertebrate hind brain.

The spinal cord of amphioxus resembles that of vertebrates. It possesses a central region formed by neurons, many of gigantic size, and a peripheral region of unmyelinated fibers. The central canal is continued dorsally by a narrow slit. The spinal nerves are divided into separate dorsal and ventral roots that do not unite, and the dorsal roots, or nerves, alternate with the ventral ones. The dorsal nerves carry all the visceral fibers and arise from the cord opposite the myosepta. Branches of the dorsal nerve penetrate the myosepta and innervate the skin, and also give off a visceral branch that innervates the wall of the gut. The dorsal roots lack ganglia and their sensory neurons of origin are located within the spinal cord.

The ventral spinal roots, or nerves, arise from the cord opposite the myotomes and enter these at once. Recent electron microscopic studies have revealed that the so-called ventral spinal nerves of amphioxus are really projections of the muscle fibers of the myotomes and that their neuromuscular junctions occur on the surface of the spinal cord. This is a peculiar form of muscular innervation that occurs in some echinoderms and nematodes as well. In addition, the spinal roots of the two sides of amphioxus alternate, but this is the result of a general right-left asymmetry that occurs at metamorphosis.

A number of similarities exist between the spinal cords of amphioxus and vertebrates. Both have a similar arrangement of functional columns with sensory neurons located dorsally and motor neurons located ventrally. Both have several types of interneurons that form connections at many levels within the spinal cord, and both possess giant interneurons that appear to be involved in conducting fast potentials that trigger escape movements.

Amphioxus also possesses photosensory cells that are located in the ventral wall of the spinal cord. These "eyes" consist of several neurosensory cells and a pigment cup, thus resembling the simple eyes of many invertebrates.

**11. The Nervous System and Main Sense Organs of Cyclostomes**

The brain of cyclostomes is much compressed in the anteroposterior direction of the great expansion of the buccal funnel and is correspondingly deep through the thalamic region, but it has the same general parts as other vertebrate brains (fig. 13.36). The compact telencephalon is divided into olfactory bulbs and cerebral hemispheres (also called olfactory lobes) of about equal size. The diencephalon contains a large third ventricle and is divisible into the usual three regions; the thalamus proper is cytologically as differentiated as in other anamniotes. In the epithalamus there are large habenulae, and the petromyzonts (but not the myxinoids) have both parts of the epiphyseal apparatus, a parietal and pineal body, each with a functional eye. There are large choroidal sacs near the epiphyseal apparatus. The hypothalamus is large and has a typical pituitary body, infundibulum (but no vascular sac), and mammillary region. The optic tectum (optic lobes) of the midbrain is evaginated and is well developed cytologically, forming at least five cellular laminae. The cerebellum, practically absent in myxinoids, is in petromyzonts a small shelf over the anterior end of the fourth ventricle. The rudimentary cerebellum may result from the absence of paired appendages. There are very large tela choroidea in the roof of the cyclostome brain.

The olfactory organ consists of one very large median unpaired olfactory sac, which was originally double, as there are two olfactory nerves. The peculiar relations of the olfactory sac to the hypophysis were explained previously. The eyes begin their development in the larva in vertebrate fashion but fail to differentiate very far and continue their development only after metamorphosis. Adult petromyzonts have well-developed eyes, which, however, appear to be generalized in some respects, especially in lacking an accommodatory mechanism for the lens.

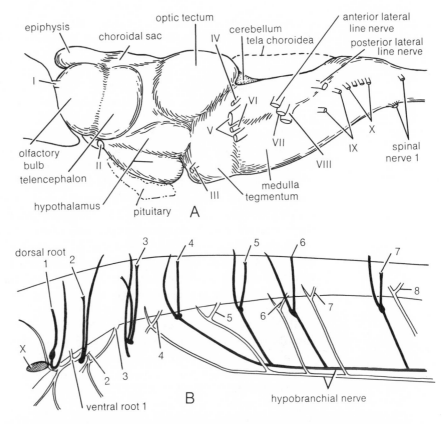

Fig. 13.36. *A*, lateral view of the brain of the marine lamprey, *Petromyzon marinus*. *B*, spinal cord of *Petromyzon* showing the sensory (black) and motor (white) roots of the spinal nerves. Note their separate and distinct courses, as well as the formation of the hypobranchial nerves.

In myxinoids the degenerate eyes are buried under the skin. The inner ears of cyclostomes are also peculiar, having but two semicircular ducts (petromyzonts) or only one (myxinoids), but this has two ampullae, one at each end, and hence would appear to be derived from the two-duct condition. The extinct anaspids, from which the cyclostomes evolved, also did not possess a horizontal semicircular canal. This suggests that these animals possessed little control over yaw during swimming. There is also no differentiation of the vestibule into sacculus and utriculus in cyclostomes. A lateralis system is present, better developed in petromyzonts than in myxinoids; it consists of groups of neuromasts in pits arranged linearly. There are external, as well as internal, taste buds.

The cranial nerves are typical of vertebrates in number and arrangement. There is no nervus terminalis. Myxinoids have no optic nerves, and the eye-muscle nerves are also degenerated. The profundus nerve is distinct, and the trigeminus proper is large and well developed, with two main branches, the ophthalmic and the maxillomandibular. The lateral line and auditory nerves are well developed, and they and their central connections show the same close relationship as in other vertebrates. The glossopharyngeal and vagus nerves have somatic sensory components (although the glossopharyngeal tends to be greatly reduced in myxi-

noids). The facial, glossopharyngeal, and vagus have the same relations to the gill slits as in gnathostomes; the vagus supplies all but the
first two.

The spinal nerves of petromyzonts possess generalized characters:
the dorsal and ventral roots alternate on the cord, and they do not unite
(fig. 13.36). There are, however, sensory ganglia on the dorsal roots.
In myxinoids the greater part of the two roots unite, although some
branches remain independent, so that this group is intermediate between
petromyzonts and gnathostomes. In the anterior part of the spinal cord
there is a complete set of dorsal and ventral roots (fig. 13.36), whereas in
gnathostomes the anterior sensory roots are generally missing. The dorsal
root of a given segment lies behind the ventral roots, and this appears to be
the primitive arrangement (fig. 13.36). In petromyzonts a number of the
more anterior sensory and motor nerves contribute a branch to a lengthwise nerve; and the two nerves so formed, one sensory and one motor,
run alongside each other posteriorly and then turn forward to innervate musculature corresponding to the hypobranchial musculature. Hence
the formation of a hypobranchial (or hypoglossal) nerve is already seen
in cyclostomes.

Besides the dorsal sensory ganglia, the spinal cord of lampreys contains dorsal cells located within the spinal cord and they are also sensory
in function. Some of their neurites enter dorsal nerve roots and are believed to be sensory from organs of the body. Their central processes
consist of a short posterior process and a long anterior process, which
frequently projects as far anteriorly as the gill region. The spinal cord in
the region of the dorsal fins possesses giant interneurons arranged in
an alternating pattern of six to eight pairs. These cells may be involved
in movement and in relaying information to the brain. In addition to these
giant neurons, a third class of giant neurons are found in the cord. Lampreys possess eight pairs of Muller neurons and two pairs of Mauth-
ner neurons. Their cell bodies are found in the tegmentum and medulla,
and their axons, 30 to 50 microns in diameter, are located along the
length of the spinal cord. Stimulation of these axons result in distinctive
movements such as flexion of the body and tail, movements of the fins,
rotations of the body, and undulations of the trunk. Thus, lampreys
possess a highly developed system within the medulla and spinal cord
for the control of swimming movements.

**B. The Nervous System and Sense Organs of Elasmobranchs**

A thorough knowledge of the nervous system of cartilaginous fishes
is indispensable for the understanding of the vertebrate nervous system. Not only is the nervous system of many of these animals generalized, but the cartilaginous nature of the skeleton and the relatively
large size of the cranial nerves make the nerve distribution easier to
follow than in other vertebrates.

**1. The Spinal Nerves and Fin Plexi**

Remove all of the viscera, including the kidneys and reproductive
organs, from the pleuroperitoneal cavity. Note against the dorsal
coelomic wall dorsal to the pleuroperitoneum the white nerves passing
out at segmental intervals. These are the *ventral rami* of the spinal
nerves. They lie along the myosepta, buried in the muscle, and can be

seen by cutting along the myosepta. Farther laterally they emerge to the internal surface. Trace the ventral rami into the hypaxial muscles.

In the regions of the paired fins the ventral rami supply the muscles of the fins and are more or less united with each other to form a plexus. The plexus for the posterior appendage is the *lumbosacral* plexus, for the anterior appendage, the *cervicobrachial* plexus.

Dogfish: The lumbosacral plexus to the pelvic fin is found by cutting through the skin on the dorsal side of the base of the fin. On separating the fin muscles carefully from those of the trunk, you will see the nerves of the plexus as white cords passing into the base of the fin. They are more or less embedded in connective tissue, which should be carefully cleaned away. There are ten nerves passing into the fin, of which, however, only the last ones are united by cross branches to form a plexus. The first of the ten is called the *collector nerve*. Trace it forward and note that it is formed by the union of branches from the ventral rami anterior to the fin.

The cervicobrachial plexus to the pectoral fin is located by cutting through the skin at the base of the fin on the ventral side. When the skin is separated from the muscles of the trunk, nerves will be seen passing in the connective tissue to the pectoral fin. Proceed carefully forward, carrying your cut into the coelom at the side of the esophagus. The plexus is then seen to consist of a number of nerves (eleven in *Squalus*) passing from the spinal cord into the fin. Only the first four or five of these, situated on the dorsal side of the bag formed by the posterior cardinal sinus, are united by cross branches to form a true plexus; the posterior ones pass directly into the fin.

Skate: A large number of ventral rami supply the pectoral fin; the anterior ones unite in a plexus. Strip off the pleuroperitoneum at the level of the subclavian artery and note there the enormous *nerve trunk* of the *brachial plexus*. It is formed by the union (within the neural canal) of a large number of ventral rami. This will be seen later. Follow out the nerve trunk to the pectoral fin. It lies along the posterior side of the curved cartilage (propterygium), which is situated in the pectoral fin about halfway from the middorsal line to the margin. Cut through skin and muscles on the dorsal side of the animal along the posterior and lateral side of this cartilage and expose the trunk. It supplies only the anterior part of the pectoral fin. The posterior part, as already noted, is supplied by direct ventral rami, not forming a plexus. The lumbosacral plexus for the pelvic fin is located as follows: Remove the skin from the base of the fin on the dorsal side. This exposes a fan-shaped layer of muscles. Cut through this, and just ventral to it you will find a number of nerves that diverge into the fin muscles.

**2. The Sense Organs**

*a. The Ampullae of Lorenzini*

It has been noted that the skin of the head is perforated by pores, from which mucus exudes under pressure. Note the distribution of the pores. Remove a piece of skin from a region bearing pores (in the skate from the ventral side of the head) and note that each pore leads into a canal of varying length lying beneath the skin. Each canal, named the canal of Lorenzini, terminates in a little bulb, the ampulla of Lorenzini, which is supplied by a nerve, a delicate white fiber easily seen attached to the ampulla. This system is known to respond with rhythmic discharges to mechanical and weak electrical stimuli. The ampullae are detectors of weak electrical fields used in the detection of prey.

### b. The Pit Organs

Pit organs occur throughout the selachians but are not easy to see or to distinguish from the pores of the canals of Lorenzini. In *Squalus* they occur in a row on the ventral side at the level of the bases of the pectoral fins, and there is also a row shortly in front of the level of the first gill slits; they also occur irregularly above the anterior part of the lateral line.

### c. The Lateralis System

In fishes and aquatic phases of amphibians there is present a system of sense organs, the *lateralis system,* related to the aquatic mode of life. This system, together with its nerves, is completely lost in land vertebrates. It consists of the *lateral line canals,* the sense organs of these canals, termed *neuromasts,* and the nerves supplying these. The following dissection will trace the course of the canals of the lateralis system; the neuromasts in the canals are microscopic.

Dogfish: Along the trunk the system consists of the lateral line, which marks the position of a canal. Find the lateral line on the head. Remove the skin at this place, noting the underlying canal and the pores connecting the canal with the surface. Trace the lateral line forward, removing the skin as you proceed. At the level of the spiracles the canals of the two lateral lines are connected by the *supratemporal* canal. Anterior to this, each forks into a *supraorbital* canal, passing forward above the eye, and an *infraorbital* canal, passing ventrally between the eye and the spiracle and then forward below the eye. Trace the supraorbital canal to the end of the rostrum; here it turns and proceeds posteriorly again, parallel to its former course, and becomes continuous with the infraorbital canal. The latter gives off a *hyomandibular* branch, running posteriorly along the sides of the jaws, and turns to the ventral surface of the rostrum, passing first posterior to the nostril and then turning forward between the two nostrils. There is also a short *mandibular* canal under the skin just behind the lower jaw; it is not connected with the other canals.

Skate: The lateral line system is more complex than in the dogfish and more difficult to follow. The lateral line canal runs on the dorsal surface just lateral to the middorsal spines. Remove the skin at this place and identify the canal. Trace it forward, removing the skin as you proceed. At the posterior end of the cartilage (propterygium) of the anterior part of the pectoral fin it gives off two canals, which proceed posteriorly over the surface of the fin. It then proceeds above the eye as the *supraorbital* canal, apparently connecting with its fellow by a cross union on the posterior part of the skull. The supraorbital canal passes in front of the eye and, as the *infraorbital* canal, below the eye.

In the region of the eye it gives off branches over the rostrum and a long branch that proceeds posteriorly along the lateral margin of the fin. On the ventral side of the skate there is a prominent canal passing just lateral to the gill slits. Trace this forward, noting branches behind and in front of the nostril and on the ventral surface of the rostrum. On the surface of the pectoral fins, after removal of the skin, the numerous, very long canals of Lorenzini are noticeable.

### d. The Olfactory Organs

These consist of a pair of olfactory sacs on the ventral side of the rostrum, opening externally by the nostril or external naris, with which various flaps of skin are associated, arranged so as to permit water circulation through the sac. Dissect the skin away from one olfactory sac and cut away the flaps so that you can look into the sac. Note the numerous plates or lamellae arranged in rows inside of the sac; these are covered with olfactory epithelium, for the sense of smell is well developed in fishes. Prove to yourself that the olfactory sac is closed internally, having no communication with the oral cavity.

### e. The Eye Muscles

Remove the tissue from around the eye on the same side of the animal as you did when studying the lateralis system, and completely expose the eyeball. In doing this, first cut away the upper eyelid (or in the skate the skin over the eye), noting that the inner lining of the eyelid is continuous with a thin layer (*conjunctiva*) that adheres closely to the external surface of the eyeball. Next cut away very carefully the cartilage between the eye and the brain (which is seen as a white, lobed structure in the median region) and also the cartilage in front of the eye. Do not injure the brain and do not cut into the elevation dorsal to the spiracle. The stout, white bands seen in this dissection are cranial nerves. In the skate very little cutting is required. The large, somewhat spherical body exposed is the *eyeball*. In the dogfishes it is embedded in a gelatinous material, which should be carefully cleaned out.

The eyeball reposes in a cavity, the *orbit,* to the walls of which it is attached by muscular bands, the extrinsic *eye muscles.* These are voluntary muscles derived from the myotomes of the first, second, and third segments of the head. There are six of these eye muscles, which should be identified as follows: From the dorsal view four of them will be seen. The one attached to the anterior wall of the orbit is the *superior oblique.* The other three originate from the posterolateral angle of the orbit and are named *recti* muscles. The most anterior one is the *internal* or *medial rectus;* its insertion on the eyeball is covered dorsally by the superior oblique. The next rectus muscle is the *superior rectus,* more dorsally situated than the others. The third, the *external* (or *lateral) rectus,* is inserted on the posterior surface of the eyeball. Next, raise the eyeball dorsally and note that the conjunctiva or most superficial coat over the external surface of the eyeball is continuous with the lining of the lower lid. Cut through this and free the eyeball ventrally, cleaning out the gelatinous and fibrous tissue there. On lifting the eyeball the remaining two eye muscles will be seen. The *inferior oblique* originates from the anteromedial corner of the orbit, the *inferior rectus* from the posteromedial angle of the orbit; both are inserted in contact with each other on the middle of the ventral surface of the eyeball. The white cords seen among the eye muscles are nerves.

The eye muscles originate from the orbit and are inserted on the eyeball. Their action is to turn the eyeball in various directions. As already stated, they are derived from the prootic head segments.

### f. The Structure of the Eyeball

Cut through the eye muscles at the insertions and remove the eyeball. The outermost coat covering the front of the eyeball is the conjunctiva, which is deflected onto the inner surface of the eyelids. Note the free edge of the conjunctiva clinging to the eyeball where the eyelids were cut. The conjunctiva is the epidermis of the skin and not one of the true coats of the eye. The outermost coat of the eyeball is the *sclera,* or *sclerotic* coat, a very tough membrane composed of connective tissue. The front part of the sclera is transparent and is named the *cornea;* the conjunctiva is inseparably fused to the outer surface of the cornea. Through the transparent cornea can be seen an opening, the *pupil.* Cut off the dorsal side of the eyeball so that you can look within the cavity. Place the larger piece under water. The large spherical body in the interior is the *crystalline lens.* Note that internal to the sclera is a black coat, the *chorioid* coat, and internal to this is a soft, often collapsed, greenish layer, the *retina.* Follow the choiroid coat to the front of the eye and note that there it is separated from the cornea, forming a black curtain, the *iris,* in the center of which is an opening, the *pupil.* The iris divides the cavity of the eyeball into an external cavity, the *anterior chamber* of the eye between the iris and the cornea, and an internal chamber, the *cavity of the vitreous humor,* between the lens and the retina. The anterior chamber contains a fluid, the *aqueous humor;* the cavity of the vitreous humor contains a gelatinous material, the *vitreous humor* or *vitreous body,* collapsed in the preserved specimen. The lens in life is attached to the junction of cornea and sclera by a circular membrane, the *ciliary body,* marked with radiating folds and applied to the inner surface of the iris; the attachment is lost in preserved specimens.

The retina is the nervous part of the eye containing the sensory cells (rods and cones) that are stimulated by light. The lens and the two humors focus the light upon the retina. The focus is changed in fishes by moving the lens back and forth. The pupil regulates the amount of light admitted. The coats of the eye serve for protection and to darken the interior.

After the eyeball has been removed, note in the orbit the origins of the six eye muscles, the *optic pedicel,* a cartilaginous stalk situated among the rectus muscles that helps to support the eyeball, and the *optic nerve,* a stout white stalk located in front of the rectus muscles. The stout white band in the floor of the orbit is the *infraorbital* nerve.

### g. The Internal Ear

The ear in fishes consists only of the *internal ear* or *membranous labyrinth*. This is embedded in the otic region of the skull, in a set of channels in the cartilage termed the *cartilaginous labyrinth*. In sharks and skates the internal ear is situated between the spiracle and the middorsal line in a pronounced elevation caused by the otic region of the chondrocranium. In the median line between these two elevations will be found a pair of small holes in the skin. Upon removing the skin bearing these holes, you will find the endolymphatic fossa of the chondrocranium beneath it. In this fossa are the two *endolymphatic ducts,* which open on the skin by the two holes just mentioned and connect the cavity of the internal ear with the surface. Shortly below the surface each endolymphatic duct has a slight enlargement, the *endolymphatic sac*. Very carefully shave off with a scalpel the cartilage of the elevation containing the ear, working on the same side as before. There will soon appear a canal in the cartilage containing a delicate curved tube; this is the *anterior semicircular duct*. Continue removing the cartilage without injuring this duct. The muscles posterior to the ear may also be removed. Another tube will soon be uncovered posterior to the first one; this is the *posterior semicircular duct*. There will next be revealed the thin-walled central chambers of the ear to which these ducts are attached. Continue picking away the cartilage in small bits, leaving all parts of the internal ear in place. A third duct, the *horizontal semicircular duct,* lying below and lateral to the others, will next be exposed. When the cartilage has been removed as far as possible, the parts of the internal ear may be identified. The central part, to which the ducts are attached, consists of three delicate chambers: the *anterior utriculus,* the *posterior utriculus,* and the *sacculus*. The semicircular ducts are slender tubes, curved in a semicircle and each terminating in a rounded sac, the *ampulla*. The ampullae of the anterior and horizontal ducts are in contact, and the chamber they enter is the anterior utriculus; these two ducts spring from the dorsal end of this chamber. The chamber to which both ends of the posterior duct are attached is the posterior utriculus. (This duct is considered by Retzius and others, however, to be part of the posterior duct; in such a case this duct is described as forming a circle and opening directly into the sacculus.) The larger chamber occupying a central position and receiving the endolymphatic duct (the connection is difficult to find) is the *sacculus,* fitting into a rounded depression in the cartilage. In each ampulla will be seen a white sensory patch or *crista,* to which a branch of the auditory nerve is attached. Larger sensory patches, termed maculae, occur in the sacculus and utricular chambers. Inside the sacculus is a white mass of sand grains, collectively termed the otolith; the movement of these grains may be concerned in equilibration. The lower posterior wall of the sacculus near the ampulla of the posterior canal has a slight bulge, the lagena, considered homologous to the cochlear duct of land vertebrates; it contains a macula. Compare your dissection with figure 13.37.

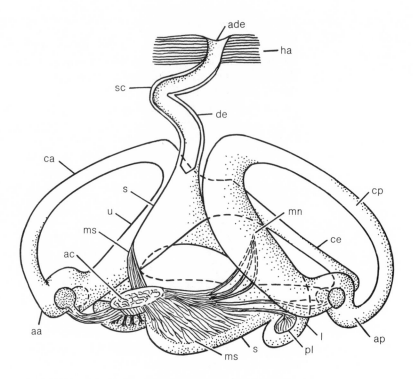

Fig. 13.37. Medial view of the right membranous labyrinth of *Squalus:* aa, anterior crista ampullaris; ade, opening of endolymphatic duct; ac, acoustic nerve; ap, posterior crista; ca, anterior canal; ce, horizontal canal; cp, posterior canal; de, endolymphatic duct; ha, chondrocranium; l, lagena; mn, macula neglecta; ms, saccula macula; pl, lagena macula; s, sacculus; sc, endolymphatic sac; u, utriculus. After Popper and Fay 1977 (*Amer. Zool.,* 17:443–52).

Open the sacculus and examine the otolith. The communications of the anterior and posterior utriculi with the sacculus may be sought.

The internal ear has two functions, that of hearing and that of equilibration. It is filled with a fluid, the endolymph, while the cartilaginous labyrinth is filled with perilymph. Sound waves impinging on the head or changes in the position of the head affect the endolymph, which in turn excites the sensory cells of the cristae and maculae, producing, in the first case, the sensation of hearing and, in the second, sensations of the animal's position in the water, enabling it to keep in the desired position.

**3. The Dorsal Aspect of the Brain**

The brain is now to be exposed by carefully picking away the cartilage in small pieces from its roof. The cranial nerves, white strands passing through the cartilage, must not be injured. One side of the head had thus far been left intact for the study of the cranial nerves. This side is now to be exposed along with the brain, as far as necessary. Remove the upper eyelid, as directed, under the eye, but leave all structures intact. In removing the cartilage between the brain and the eye the following nerves will be noted: the *superficial ophthalmic* nerve, running in the wall of the orbit near the dorsal surface; the small *trochlear* nerve, passing through the back wall of the orbit to the superior oblique eye muscle; in the skate the larger *oculomotor* nerve accompanying the trochlear. Dissect forward to the olfactory sacs, exposing them dorsally, leaving the ophthalmic nerve in place. Remove the skin behind the spiracle and note the *hyomandibular* nerve, passing posterior to the spiracle; this nerve is also to be preserved. To expose the posterior part of the brain the internal ears of both sides may be cut through and the mass of muscles posterior to the ear removed as much as necessary. Nerves will be seen passing through the cartilage in the ventral part of the ear but are not to be dissected out for the present. In short, the dorsal side of the brain is to be fully exposed, leaving all of the more superficial nerves intact. The dorsal aspect of the brain will then be studied first, and the cranial nerves afterward.

The brain is situated in a cavity in the chondrocranium, which it only partially fills (fig. 13.38). It is covered by a delicate membrane, the *primitive meninx,* in which the blood vessels of the brain are situated. The meninx is connected by strands with the membrane lining the cartilaginous walls of the cranial cavity. The space between brain and chondrocranium in live specimens is filled by a fluid.

The most anterior structures of the brain are the large *olfactory bulbs,* nervous masses situated in contact with the dorsal walls of the olfactory sacs. From the olfactory sac a number of very short fibers, which together constitute the *olfactory nerve,* pass into the olfactory bulb. The olfactory bulb is spherical in the dogfishes, elongated in the skate (fig. 13.39). Each olfactory bulb is connected with the next part of the brain by a stalk, the *olfactory stalk* or *peduncle,* also called the *olfactory tract.* The olfactory stalks enter the large rounded lobes, the *cerebral hemispheres,* which form the anterior (or lateral in the skate) part of the main mass of the brain. All of the parts mentioned so far belong to the *telencephalon.*

Posterior to the cerebral hemispheres is a depressed region, the *diencephalon,* which has a thin discolored roof, consisting of a *tela choroidea,* which is highly vascularized. From this, choroid plexi project into the interior, forming the choroid plexi of the third ventricle. The *optic nerves* pass from the orbit into the ventral surface of the diencephalon and are easily seen in the skate; in the dogfishes they can be seen by gently pressing the diencephalon to one side. Behind the diencephalon is the *midbrain* or *mesencephalon,* consisting dorsally of two rounded lobes, the *optic tectum.* A pair of nerves, the *trochlear* or *fourth* cranial nerves, arises from the posterior border of the optic tectum and passes forward to an eye muscle. By gently pressing the optic tectum to one side the *oculomotor* or *third* cranial nerve, will be seen emerging from the ventral surface of the midbrain and passing to the orbit, on each side.

Posterior to the optic tectum and somewhat overhanging it is the large main mass or *body* (corpus) of the *cerebellum,* slightly divided into four quadrants by faint longitudinal and transverse grooves; it belongs to the metencephalon. Behind this is the *medulla oblongata,* or *myelencephalon,* the elongated remaining section of the brain that

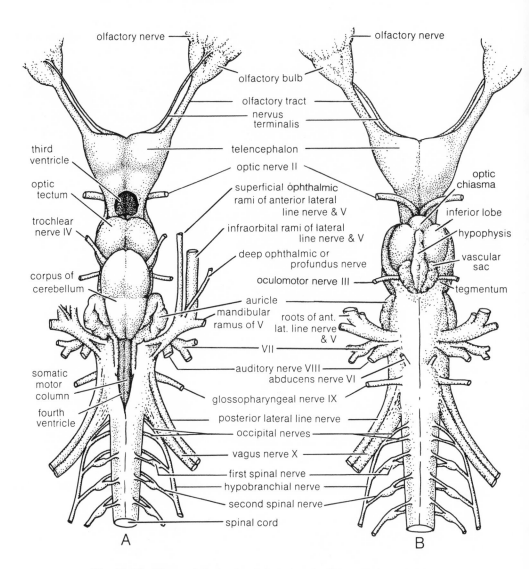

Fig. 13.38. Views of the brain of the spiny dogfish, *Squalus acanthias. A,* dorsal; *B,* ventral.

continues posteriorly with the spinal cord. The greater part of the roof of the medulla is also formed by a tela choroidea with choroid plexi, roofing the cavity of the fourth ventricle. The anterior end of the medulla is continuous with the *auricles* of the cerebellum, two ear-like projections at the sides of and below the body of the cerebellum. Remove the tela choroidea from the roof of the auricles and medulla, thus revealing the large cavity of the fourth ventricle. On lifting the posterior end of the body of the cerebellum, you will see that the auricles are continuous with this and with each other, thus providing the main pathway for the acousticolateral, auditory, and equilibratory impulses to reach the cerebellum. The entire dorsal rim of the medulla forms an elongated strip on each side, the *somatic sensory column;* the anterior part of this, continuous with the auricles, is the *acousticolateral area.* This somatic sensory column is the primary terminus for general cu-

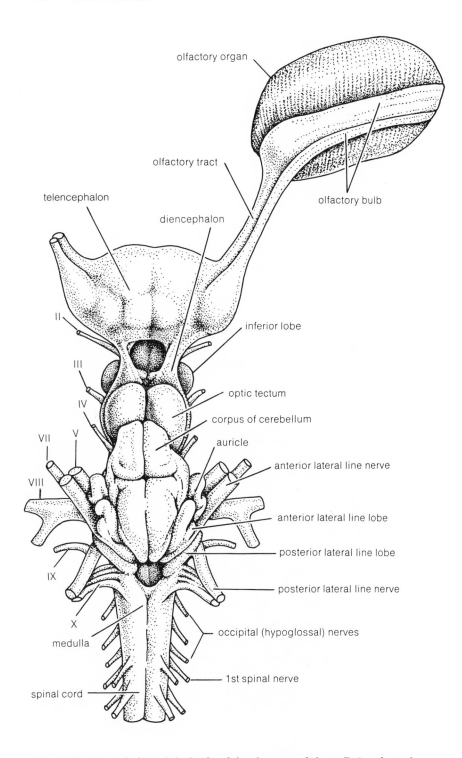

olfactory organ

olfactory tract

telencephalon

diencephalon

olfactory bulb

II

inferior lobe

III

optic tectum

IV

corpus of cerebellum

VII   V

auricle

VIII

anterior lateral line nerve

anterior lateral line lobe

posterior lateral line lobe

IX

posterior lateral line nerve

X

occipital (hypoglossal) nerves

medulla

1st spinal nerve

spinal cord

Fig. 13.39. Dorsal view of the brain of the clear-nosed skate, *Raja eglanteria.*

taneous and proprioceptive impulses coming in along the sensory nerves and spinal tracts. Near the anterior end of the somatic sensory column, at the middle of the acousticolateral area, will be seen, by pressing the latter toward the middle, the roots of a number of nerves. These are the roots of the *fifth, seventh,* and *eighth* cranial nerves, to be studied in more detail later. At the posterior end of the somatic sensory column, just anterior to the point where the walls of the medulla close together, the stout root of the *tenth* cranial nerve may be seen. In the lateral wall of the medulla, just ventral to the somatic sensory column, is another longitudinal area marked by a row of rounded elevations; this area is the *visceral sensory column.* As its name implies, it is associated with sensations from the viscera. In fishes the gills are important structures, so that a considerable portion of this column is connected with the gills; in fact, each of the little elevations is said to be a center for one branchial bar. Ventral to the visceral sensory column is the very slender visceral motor column, from which impulses go to the visceral muscles; in the head region these are the branchial muscles or their derivatives in advanced vertebrates. In the floor of the fourth ventricle are two conspicuous *somatic motor columns,* separated by a median groove. These are the places of origin of impulses to the somatic muscles, derived from the myotomes; in the head these consist of the six eye muscles and the hypobranchial musculature.

The fourth ventricle narrows posteriorly and is finally roofed over by the fusion of the walls of the medulla. Shortly beyond this point the medulla is continuous with the spinal cord. The posterior end of the medulla marks the posterior end of the brain but is not sharply defined.

## 4. The Cranial Nerves

The cranial nerves must be dissected with great care and their distribution noted. This distribution is, in general, similar in all vertebrates except that certain nerve trunks present in fishes disappear in the land vertebrates. One of the most striking examples of homology is found in this distribution of the cranial nerves, which in man still continue to supply the same parts as in fish.

### a. The First or Olfactory Nerve

This nerve has already been noted. It arises from the olfactory cells in the lamellae of the olfactory sac and passes by very short branches into the olfactory bulb. These branches are practically invisible. From the olfactory bulb, after a relay, the olfactory impulses pass along the olfactory stalks into the cerebral hemispheres. The olfactory nerve is a pure sensory nerve, belonging to, or at least closely related to, the visceral sensory division.

### b. Nervus Terminalis

The terminal nerve (not counted originally as one of the cranial nerves) is a fine white thread that will be found along the medial side of the olfactory stalk. Traced caudad, it will be found to separate from the stalk and come in contact with the middle of the anterior face of the cerebral hemisphere. It then runs along this face and into the fissure between the two cerebral hemispheres.

### c. The Second or Optic Nerve

The optic nerve arises in the retina of the eye and passes through the coats of the eye, emerging ventral to the internal rectus muscle. Find it there on the intact eye by pressing this muscle against the eyeball. The nerve is a stout, white trunk that pierces the cartilage of the orbit and passes to the ventral side of the diencephalon, where it may be seen by gently raising the diencephalon. As already explained, the optic nerve is not really a nerve but a brain tract; it also is a pure sensory tract, carrying the visual impulses and discharging them into the diencephalon and optic tectum.

### d. The Fourth or Trochlear Nerve

The trochlear nerve arises in the midbrain and emerges in the groove between the optic tectum and the cerebellum. Trace it on the side where the eye is still intact. It passes forward in the cranial cavity to about the level of the cerebral hemispheres; it then turns abruptly laterally, pierces the wall of the orbit, and is distributed to the superior oblique muscle of the eyeball. It is the motor nerve of this muscle and carries only somatic motor impulses. Although it appears externally to emerge from the roof of the midbrain, the motor cells from which it originates are, in fact, in the floor of the midbrain in a forward extension of the somatic motor column.

### e. The Third or Oculomotor Nerve

The oculomotor nerve arises from the floor of the midbrain. It is readily noticed in the skate, ascending to the orbit near the preceding nerve. In the dogfishes it is situated deep and is seen by pressing the cerebellum away from the wall of the orbit. Follow it into the orbit on both sides, getting its general relations first on the side where the eyeball was removed. It emerges into the orbit very near the insertion of the superior rectus muscle and is situated ventral to the superficial ophthalmic nerve, already noted. It should not be confused with the *deep ophthalmic* nerve, which is in contact with it as it enters the orbit; the deep ophthalmic nerve runs through the orbit in contact with the medial surface of the eyeball. This nerve will be better seen on the intact side. Observe the branches given by the oculomotor nerve, immediately after its entrance into the orbit, to the internal and superior recti muscles. On the intact side now loosen the eyeball and cut the insertions of the superior oblique and superior rectus close to the eyeball. Identify the deep ophthalmic nerve passing in the dogfishes dorsal to the internal rectus and lying against the eyeball; in the skate it passes ventral to the internal rectus. Free and preserve this nerve. Cut through the insertion of the inferior rectus and the optic nerve and, pressing the eyeball outward, note the branch of the oculomotor nerve that runs along the posterior side of the inferior rectus muscle, turns ventral to it, and then runs forward in the floor of the orbit to the inferior oblique. Note, also, that the branch to the inferior rectus also gives off a branch, one of the *ciliary nerves,* which enters the eyeball in company with an artery. Along this ciliary nerve small brown masses can be noted; they are part of a ganglionated *ciliary plexus* belonging to the autonomic system. The ciliary branches also enter the eyeball from the deep ophthalmic nerve. The function of the ciliary plexus and nerves is to control the smooth muscles of the iris, regulating the size of the pupil. Throughout vertebrates it also controls the accommodation mechanism of the lens, but this is poorly developed in selachians. The oculomotor nerve is a pure somatic motor nerve, except for the visceral motor fibers of the autonomic system that accompany it.

### f. The Sixth or Abducens Nerve

The abducens originates from the somatic motor column on the ventral surface of the anterior end of the medulla. Its origin will be seen later. It penetrates the orbit at the point of origin of the external rectus muscle and passes along the ventral surface of this muscle, to which its fibers are distributed. It will be seen as a white ridge on the ventral surface of the muscle. Like the other eyemuscle nerves, it is a somatic motor nerve.

As just seen, the third, fourth, and sixth cranial nerves are somatic motor nerves to the muscles of the eyeball. The reasons for this arrangement are explained above, where also the sensory parts of these nerves are indicated.

### g. The Fifth or Trigeminus Nerve

The trigeminus is a very large nerve with four main branches in elasmobranchs (three in land vertebrates). The trigeminus is attached to the medulla near the anterior end of the somatic sensory column, just behind the auricles of the cerebellum. Its roots here are inextricably mingled with the roots of the seventh and eighth nerves, and the three together form a conspicuous mass at the place stated. The trigeminus passes through the adjacent wall of the orbit and should be followed into the orbit by carefully picking away the cartilage around it. As soon as it penetrates the orbit, the trigeminus divides into four branches. The first of these, the *superficial ophthalmic* branch, is part of the *superficial ophthalmic* trunk, which has already been mentioned several times. This large trunk passes forward in the dorsal part of the cartilage of the medial wall of the orbit. Trace it forward. It passes out of the orbit through the ophthalmic foramen in the chondrocranium and above the olfactory bulb. Only a small part of this trunk is trigeminal; this is sensory to the skin dorsal to the orbit. The second branch of the trigeminus is the *deep ophthalmic* nerve. It passes through the orbit ventral to the preceding (giving off small ciliary nerves) and leaves the orbit by the orbitonasal canal. On tracing it forward through the canal, you will find that it comes in contact with the superficial ophthalmic trunk from which it again separates, supplying sensory fibers to the dorsal and lateral skin of the snout. Both the superficial and deep ophthalmic parts of the trigeminus are somatic sensory nerves.

The two remaining branches of the trigeminus lie in the floor of the orbit. To see them, remove the eyeball or study the side where the eyeball was previously removed. A broad white band, the *infraorbital* trunk, is seen in the floor of the orbit, passing obliquely laterally. In the dogfishes this trunk is composed of the mixed fibers of the *maxillary branch* of the trigeminus and the *buccal branch* of the seventh nerve. In the orbit the larger and more medial portion of the trunk is the maxillary branch, but farther out this becomes inextricably mingled with the buccal nerve. In the skate the infraorbital trunk is made up of three trunks; the outer one is the *maxillary* branch of the trigeminus, the middle one the *mandibular* branch of the trigeminus, and the inner one the *buccal* branch of the seventh nerve. As before, however, it should be remembered that there is an admixture of fibers of the fifth and seventh nerves in these trunks. Trace the maxillary branch of the trigeminus and buccal branch of the seventh out from the orbit, along the ventral surface of the rostrum. The branches pass to the region below and in front of the eye, to the medial side of the nostril (in the skate to the lateral side of the nostril also) and to the angle of the jaws. The maxillary branch is sensory to the skin of the rostrum, while the buccal nerve supplies the infraorbital lateral line canal and nearby ampullae of Lorenzini.

The fourth branch of the trigeminus is the *mandibular* branch. In the dogfishes it separates from the infraorbital trunk, where the latter enters the orbit from the brain and passes along the posterior wall of the orbit. The mandibular nerve is seen to branch to various muscles in the floor of the orbit (these are gill arch muscles) and, on following it out of the orbit, will be seen to be distributed to muscles of the lower jaw, this branch being situated just behind the teeth. In the skate the position of the mandibular nerve was described above as between the maxillary and the buccal nerves. Follow it forward. It curves around the angle of the jaw and supplies muscles of the lower jaw and the adjacent skin.

It will be observed that all of the branches of the fifth nerve are somatic sensory nerves coming from various sensory organs of the skin, except the mandibular nerve, which also contains some motor branches to muscles. As those muscles are branchial muscles, this part of the mandibular nerve belongs to the visceral motor system. As explained above, the deep ophthalmic or profundus nerve is not really part of the trigeminus nerve but an independent nerve with its own ganglion in cyclostomes and selachians. It is the sensory nerve of the same head segment of which the oculomotor nerve is the motor part. In advanced vertebrates the profundus nerve is incorporated into the ophthalmic branch of the trigeminus. The trigeminus nerve proper is the sensory nerve of the same head segment of which the trochlear nerve is the motor part. It is important to note further that the trigeminus is the *nerve of the upper and lower jaws,* i.e., that is, the *first or mandibular arch.* It is the sensory nerve of this arch and the motor nerve of its branchial muscles. The sensory ganglion (Gasserian ganglion) of the trigeminus nerve and the profundus ganglion are in the mass mentioned above, including the roots of the fifth, seventh, and eighth nerves.

### h. The Seventh or Facial Nerve

This nerve is intimately related to the trigeminus. It arises in common with the latter from the anterior end of the medulla and divides into three main branches. Two of these branches pass through the orbit in common with the trigeminus. The *superficial ophthalmic* branch of the facial nerve accompanies the same branch of the trigeminus and forms the greater part of the superficial ophthalmic trunk, supplying the supraorbital lateral line canal and adjacent ampullae of Lorenzini. The buccal branch of the facial nerve forms in the orbit the outer half of the infraorbital trunk and supplies the infraorbital lateral-line canal and nearby ampullae of Lorenzini. These two branches of the facial nerve are really parts of the lateralis system and together with the pretrematic portion of the hyomandibular trunk may constitute a separate cranial nerve termed the anterior lateral line nerve. The third branch of the facial nerve is the hyomandibular trunk, which has already been located posterior to the spiracle. Trace it inward toward the brain, cutting tissues in its path. It turns ventrally and runs through the anterior part of the ear capsule deep down. Follow it by removing the cartilage of the ear capsule in small pieces. The nerve passes ventral to some of the branches of the nerve of the ear and joins the anterior end of the medulla along with the trigeminus root. Near the brain it has an enlargement or ganglion (*geniculate* ganglion). From this ganglion is given off the *palatine* nerve. It will be found by dissecting carefully around and on the ventral surface of the ganglion. In the skate it is easily seen. It runs forward below the orbit along the roof of the mouth, where it supplies the taste buds and the lining epithelium in general. Now trace the hyomandibular outward, past the spiracle. It turns ventrally and breaks up into branches on the side of the head. These branches supply the hyomandibular and mandibular lateral-line canals, ampullae and similar sense organs, the muscles of the hyoid arch, and the lining of the floor of the mouth cavity and the tongue.

The hyomandibular is seen to be a mixed nerve with lateral-line (somatic sensory), visceral motor, and visceral sensory components. The sensory part of the facial nerve belongs to the same head segment as that of which the abducens constitutes the motor nerve. The facial nerve is the *nerve of the first gill slit (spiracle)* and of the *second or hyoid arch,* hence visceral motor to the musculature of that arch. With the loss of the lateralis parts, the facial nerve of land vertebrates consists only of the palatine ($=$ pharyngeal branch) and certain parts of the hyomandibular nerve ($=$ posttrematic branch).

### i. The Eighth or Auditory Nerve

The auditory nerve is a pure somatic sensory nerve extending from the internal ear to the brain. It enters the anterior end of the medulla and is there mingled with the roots of the fifth and seventh nerves. Follow it into the internal ear, on the side opposite that on which the hyomandibular was dissected. Note its branches to each ampulla and the fan-like arrangement of the branchlets to the crista of each ampulla. The auditory nerve also collects a number of branches from the walls of the sacculus and utriculus. It carries impulses for hearing and equilibration into the acousticolateral area of the medulla, where it is attached.

### j. The Ninth or Glossopharyngeal Nerve

This nerve passes through the floor of the middle of the ear capsule (where it is likely to be mistaken for a part of the auditory), parallel to the hyomandibular nerve. Pare away as much of the ear capsule as is necessary to reveal it. Find its attachment to the medulla posterior to the auditory nerve. Trace it out of the ear capsule. Just before it exits from the ear capsule, it bears a swelling, the *petrosal* ganglion. Insert a knife blade into the second (first typical) gill slit (in the skate into the dorsal wall of the corresponding visceral pouch) and slit the gill cleft open dorsally. The petrosal ganglion will now be seen to be located near the upper limits of the cleft. Dissect the nerve from the ganglion toward the gill slit. It very soon divides into three branches—two smaller anterior ones and a larger posterior one. The most anterior branch is the *pretrematic* branch; this passes to the anterior wall of the visceral pouch, to which it is a sensory nerve. The second branch is posterior to the pretrematic branch; it is named the *pharyngeal* branch and is a sensory nerve to the mouth cavity. (This branch appears to be lacking in the skate.) The third and largest is the *posttrematic* branch. It passes to the posterior wall of the visceral pouch and is both sensory and motor, its motor components supplying the muscles of the third branchial bar. The glossopharyngeal nerve is the nerve of the *second visceral pouch* and of the *third branchial bar*.

### k. The Tenth or Vagus Nerve

The vagus nerve is the very large trunk passing through the posterior border of the ear capsule. It is attached to the sides of the posterior part of the medulla. Dissect it out and follow its course. It passes medioventrally to the anterior cardinal sinus, the wall of which is formed of a tough membrane in the dogfishes. Open up the anterior cardinal sinus by a deep cut through the muscles above the upper ends of the gill slits. Follow the vagus nerve medioventrally to the anterior cardinal sinus. In the dogfishes it divides into two trunks at the point where it courses along the tough wall of the sinus. The medially situated trunk is the *lateral* branch of the vagus and passes posteriorly internal to the lateral line, whose canal it supplies. The lateral branch of the vagus possesses a separate ganglion and enters the medulla far more rostrally than the vagus proper. For these reasons the lateral branch may constitute a separate cranial nerve termed the posterior lateral line nerve. The lateral trunk is the *visceral* branch of the vagus, which continues along the anterior cardinal sinus. In the skate the vagus runs for a short distance in the sinus before dividing into a more dorsal *lateral* branch, which passes posteriorly internal to the lateral line canal to which it is distributed, and a more ventral *visceral* branch. In all three forms, open up the remaining visceral pouches as directed for the ninth nerve and determine the distribution of the visceral branch of the vagus. With the walls of the sinus well spread open, note the four branches crossing the floor of the sinus to the visceral pouches. Dissect out each of these and observe that each bears a ganglion, beyond which it divides into three branches: an anterior pretrematic branch, a middle pharyngeal branch, and a posterior posttrematic branch. The pharyngeal branch seems to be missing in the skate. As in the case of the ninth nerve, the pre- and post-trematic branches embrace the visceral pouch, which lies between them; all three branches have the same functions as described for the ninth nerve. We thus see that the vagus nerve supplies the remaining branchial bars, beginning with the fourth, and the remaining visceral pouches, beginning with the third. After supplying the gill apparatus the visceral branch of the vagus passes on into the pericardial and pleuroperitoneal cavities, supplying the heart, digestive tract, and other viscera (by way of the autonomic system).

## 5. The Occipital, Hypobranchial, and First Spinal Nerves

Very carefully expose the spinal cord posterior to the medulla by shaving away the cartilage of the neural arches in thin slices. On the dorsolateral surface of the cord note the little swellings, the *dorsal* or *spinal ganglia,* attached to the cord by the *sensory* or *dorsal root*. In the skate the spinal ganglia are elongated. Between the first spinal ganglion and the root of the vagus note two or three small roots springing from the side of the medulla. These are the *occipital* nerves, which represent the ventral roots of original spinal nerves of which the dorsal roots have disappeared. They innervate some muscles of this region, help to form the hypobranchial nerve described below, and contribute to the cervicobrachial plexus. On pressing the spinal cord to one side, you will see the *ventral* or *motor roots* of the spinal nerves arising from the ventrolateral region of the cord, at the same level as the occipital nerves. The ventral roots are formed by the union of several small rootlets coming from the cord. The most anterior ventral roots are situated anterior to the dorsal roots, which belong to the same segment.

In the dogfish the union of dorsal and ventral roots to form a spinal nerve is not easy to follow. It may usually be seen by carefully paring down the cartilage along the side of the spinal cord. In the skate the union is easily followed; the roots pass through the cartilage at the side of the cord and unite as they exit from the cartilage. A large number of the most anterior spinal nerves (really the ventral rami of the spinal nerves, the dorsal rami being very slender in the skate) are then seen to unite to form the very large nerve of the brachial plexus previously noted.

The *hypobranchial* nerve is a trunk formed by contributions from the occipital nerves and the first spinal nerves. In the dogfishes it may be located as follows: Insert one blade of the scissors in the angle of the jaw and cut back across the gill slits through to the side of the esophagus, as was done in an early stage of the dissection (if the same specimen is still being used, this cut will already have been made). Open the flap thus formed and expose the roof of the mouth. Make a longitudinal cut through the mucous membrane of the roof in the median dorsal line. Strip the membrane laterally, carrying with it the free dorsal ends of the gill cartilages, which will be readily located just lateral to the middorsal line. The visceral branch of the vagus nerve, which was already seen from the other side, is now exposed. It lies along the thin ventral wall of the anterior cardinal sinus. The ventral rami of the spinal nerves will be seen emerging from the muscle now exposed in the roof of the mouth. In the dogfishes two of these (they appear as one but will be found to consist of two on dissecting them toward the median line) pass obliquely toward the visceral branch of the vagus and enter its sheath, thus appearing to join it. The trunk they form, which is the hypobranchial nerve, can, however, be readily separated from the vagus. It lies just anterior to the nerves of the cervicobrachial plexus, to which it contributes branches. After passing dorsal to the dorsal side of the visceral branch of the vagus (which may be noted proceeding to the esophagus) and behind the last visceral pouch, the hypobranchial nerve turns ventrally and courses along the floor of the oral cavity supplying hypobranchial muscles.

In the skate the hypobranchial nerve leaves the spinal cord in common with the trunk of the brachial plexus. Locate this trunk again in the anterior wall of the pleuroperitoneal cavity, dorsal to the pleuroperitoneum and just behind the cartilage of the pectoral girdle. The entire trunk, with the exception of one branch, passes out dorsal to the cartilage to the pectoral fin. This one branch, the hypobranchial nerve, turns forward and is distributed to the muscles of the floor of the mouth.

The occipital and hypobranchial nerves are probably the homologues of the twelfth or hypoglossal nerve of amniotes, having been incorporated in the cranium in those forms. The muscles supplied by the hypobranchial nerve are derived from certain metaotic myotomes and in tetrapods furnish the extrinsic musculature of the tongue.

## 6. The Ventral Aspect of the Brain

Carefully free the brain from the chondrocranium; cut through the olfactory stalks and lift the anterior end of the brain. You will next see the two optic nerves entering the ventral surface of the diencephalon. Cut through them and lift the brain further. Next, pare away the wall of the orbit on one side. It will then be seen that certain structures attached to the ventral surface of the diencephalon extend ventrally into a deep pit(sella turcica) in the floor of the cranial cavity. Take special care to lift these out intact. Then cut through the remainder of the cranial nerves, cut across the spinal cord, and lift the brain out of the cranial cavity.

Examine the ventral surface of the brain. Note the forward continuation of the internal carotid artery on the midventral line of the brain. It forks around the ventral part of the diencephalon (farther posteriorly in the skate) and passes forward to the telencephalon, distributing many branches to all parts of the brain as well as to the orbit. The ventral surface of the brain presents nothing new except as regards the diencephalon, where several additional structures are visible. The two optic nerves are seen attached to the anterior end of the ventral surface of the diencephalon; as they enter the latter, they cross, the crossed region being named the *optic chiasma.* From the chiasma a broad band, the *optic tract,* extends dorsad and caudad into the dorsal part of the diencephalon and into the optic tectum. It is readily seen, especially in the spiny dogfish, by scraping off the primitive meninx at this place. It is thus evident that the visual impulses pass into these two portions of the brain. Posterior to the optic chiasma the floor of the diencephalon bulges ventrally and posteriorly as the *infundibulum,* consisting in large part of two rounded lobes, the *inferior lobes.* From between the two inferior lobes a stalk projects caudad and widens into a soft body. The dorsal part of this has thin discolored walls, being highly vascularized; this is termed the *vascular sac (saccus vasculosus)* and is a part of the infundibulum. The remainder of the structure hanging from the infundibulum is the *hypophysis* or *pituitary body,* generally more or less torn in removing the brain. This is an important gland of internal secretion. Dorsal to the vascular sac the roots of the oculomotor nerves will be found springing from the floor of the midbrain. The ventral surface of the remainder of the brain presents nothing new. The roots of the cranial nerves should be identified on the medulla. On the ventral surface of the medulla will be found the roots of the sixth nerves. If not identifiable on the brain, they will usually be found adhering to the floor of the cavity from which the brain was removed.

| 7. The Sagittal Section and the Ventricles of the Brain | The brain, like the remainder of the central nervous system, is hollow. Its cavities are known as *ventricles* and are continuous with each other by means of narrow passages. The *fourth* or last ventricle of the brain has already been identified as the cavity within the medulla oblongata. The ventral portion of this ventricle is named, from its shape, the *fossa rhomboidea.* Bisect the brain by a median sagittal cut. The cut surface can be examined best under water. From the fourth ventricle a narrow passage, the *aqueduct* of the brain, extends anteriorly. It communicates with the cavity of the cerebellum, the *cerebellar ventricle,* and the cavities of the optic tectum, the *optic ventricles.* Below the aqueduct is the thick floor of the midbrain. The aqueduct opens into the cavity of the diencephalon, the *third* ventricle. The thin roof of the diencephalon forms a tela choroidea, from which a choroid plexus is folded into the cavity. The anterior part of this roof in the dogfishes extends dorsally into a sac, the *paraphysis,* resting against the telencephalon, of which it is considered a part. Posterior to the paraphysis a thin transverse partition, the *velum transversum,* is seen in the roof of the diencephalon, particularly in |

the dogfishes. This marks the dorsal boundary between diencephalon and telencephalon. The small thickened region of the diencephalon just in front of the anterior end of the optic tectum is the *habenula*. From the habenula a slender process, the *pineal body,* may be seen in favorable specimens, extending dorsally, just back of the paraphysis. The entire roof of the diencephalon, including choroid plexus, habenula, and pineal body, is named the *epithalamus*. The lateral walls of the diencephalon constitute the *thalamus,* an important correlation center for various body senses. The ventral part of the diencephalon is named the *hypothalamus*. It consists of the infundibulum, including the inferior lobes, the hypophysis, and the *mammillary* bodies. The former should be identified on the section. The mammillary bodies are the thickened part of the ventral wall above the vascular sac. Note that the cavity of the third ventricle extends into all parts of the hypothalamus. The third ventricle connects by a passage, the *foramen of Monro,* or *interventricular foramen,* with the cavity in each half of the telencephalon. These two cavities are named the *first* and *second* or *lateral* ventricles. They extend out into the olfactory bulbs through the olfactory tracts. Cut into the telencephalon to see its ventricles.

**8. Autonomic System**

The autonomic system of selachians consists of a paired chain of ganglia along the middorsal region of the pleuroperitoneal cavity embedded anteriorly in the dorsal wall of the posterior cardinal sinuses and posteriorly in the kidney close to the suprarenal bodies and of ganglionated plexi in the viscera with which the ganglia of the chains connect. There is also a ciliary plexus in the orbit and ganglia along the gill arches, apparently for visceral sensory supply to the pharyngeal lining. There is no division into antagonistic sympathetic and parasympathetic systems. The communicating rami consist only of white rami.

**C. The Nervous System and Sense Organs of the Perch (Perca)**

Because of the small size of the spinal nerves and fin plexi, they will not be investigated.

**1. The Sense Organs of the Head**

*a. Lateral Line System*

The lateral line system is well developed in the perch. Along the trunk, this system consists of the lateral line, a longitudinal tube located in the dermis. This tube opens to the surface as it passes beneath each scale, and the neuromasts are located in the floor of the longitudinal tube between the serially repeating side tubes that reach the surface. A complex pattern is formed by the lateral line system on the head. It is subdivided into regions innervated by the facial and glossopharyngeal nerves. A *supraorbital line* passes rostrally from the lateral line system of the trunk over the upper border of the orbit and is supplied by the superior ophthalmic branch of the facial nerve. A *postorbital line* continues from the lateral line of the trunk and is supplied by the glossopharyngeal nerve. A third branch, the *infraorbital line,* arises from the postorbital line and passes just beneath the orbit. It is supplied by the buccal branch of the facial nerve. Finally, a *hyomandibular* and *mandibular* line arise from the postorbital line and run to the jaws. These last lines are supplied by the hyomandibular branch of the facial nerve.

### b. Olfactory Organs

These consist of a pair of olfactory sacs, which, as in sharks, do not open into the oral cavity. Dissect the skin away and note the lamellae arranged to form a bulb. The olfactory receptors are located on the internal wall of the sac, and their axons form the olfactory stalk, which is extremely long in the perch.

### c. The Eye Muscles and Eye

Remove the skull from above the eye by stripping it off in slivers with a knife. Take care not to cut so deeply as to damage the superior ophthalmic nerve, which lies in the soft tissue just medial to the eye and beneath the skull roof. The eye muscles are identical to those in the shark and after being identified should be transected next to the eye so that it can be removed from the orbit. Take care in doing this so that the nerves that run in the medial wall and floor of the orbit are not disturbed. The dorsal side of the eyeball can be cut off and the same structures identified as in the shark.

### d. Ear

The ear, as in sharks, consists of the internal ear only. The semicircular canals are extremely fine and can be seen only with difficulty. If the operculum is removed by cutting it free at its rostral attachment and the skull carefully shaved away, the internal ear may be seen. The sacculus is well developed in the perch, and a large, white, crystalline otolith can be carefully removed. Behind it will be seen the eighth cranial nerve arising from the medulla.

## 2. The Dorsal Aspect of the Brain

Remove the rest of the skull roof by stripping it away with a knife. The entire dorsal surface of the brain may be covered by a layer of white or yellow fat. This layer must be aspirated or gently washed away before the brain can be studied in detail. The brain of the perch is covered by a thin primitive meninx, as in sharks.

The most anterior portion of the brain (fig. 13.40) consists of the small *olfactory bulbs* that receive the extremely long *olfactory stalks*, which are the axons of the olfactory receptors located in the *olfactory sacs*. Just posterior to the olfactory bulbs are the cerebral hemispheres or *telencephalon proper*. The diencephalon cannot be seen in a dorsal view, since it is covered by the telencephalon. The *optic tectum* is seen just caudal to the telencephalon as the largest paired structure of the perch brain. Behind the optic tectum is the large unpaired *corpus* of the cerebellum. It is attached to the medulla by a second division of the cerebellum, the *eminentiae granulares,* which may be homologous to the auricle of sharks. The remainder of the brain is the medulla oblongata.

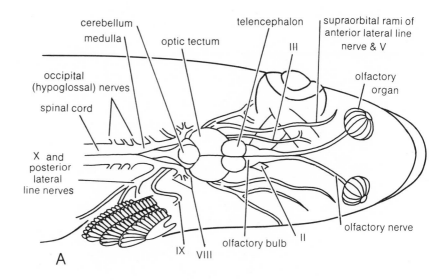

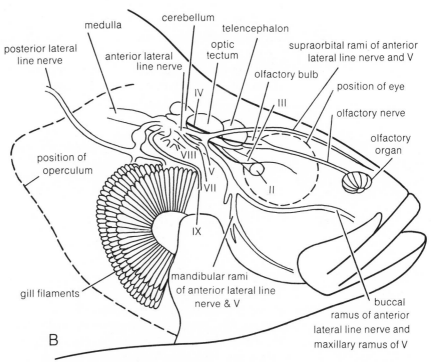

Fig. 13.40. Views of the brain and cranial nerves of the perch, *Perca flaves-cens*. *A*, dorsal; *B*, lateral.

3. The Cranial Nerves

*a. The First or Olfactory Nerve*

This nerve is often referred to as the olfactory stalk and is extremely long in the perch. It projects to the olfactory bulb.

*b. The Second or Optic Nerve*

This nerve is completely crossed in the perch and crosses the midline after arising from the eye to project to the diencephalon and optic tectum.

### c. The Eye Muscle Nerves

The third or oculomotor nerve arises from the ventral surface of the midbrain and can be seen passing through the orbit just lateral to the optic nerve (see fig. 13.40*B*). The *fourth* or *trochlear* nerve arises from the midbrain and passes over the lateral surface of the optic tectum. The *sixth* or *abducens* nerve arises from the ventral surface of the medulla and is so small as to be scarcely discernible in dissection (see fig. 13.40*A*).

### d. The Fifth or Trigeminus Nerve and the Seventh or Facial Nerve

These form a complex in the perch very similar to that seen in sharks. The trigeminus consists of four branches. The first of these, the *superior ophthalmic* branch, is part of the superior ophthalmic trunk, which also contains sensory fibers from the facial nerve. The second branch of the trigeminus, the *deep ophthalmic* nerve, runs just below the olfactory stalk and is frequently destroyed in removing the eye since it is a very fine nerve. The two remaining branches of the trigeminal complex form the *infraorbital trunk*. This trunk divides into a buccal branch and a hyomandibular branch, divisions of the facial nerve, and into an upper *maxillary* and lower *mandibular branch* of the trigeminus. As in sharks, the lateralis branches of the facial and vagal nerves may actually be separate cranial nerves secondarily associated with VII and X.

### e. The Eighth or Auditory Nerve

This nerve arises from the lateral wall of the medulla just ventral to the eminentia granularis of the cerebellum.

### f. The Ninth or Glossopharyngeal Nerve

This nerve arises from the lateral wall of the medulla just caudal to the auditory nerve. It turns caudally to run just below the vagus nerve and then turns rostrally to divide in the first gill pouch.

### g. The Tenth or Vagus Nerve

The vagus arises just below the caudal boundary of the eminentia granularis and divides almost immediately into a lateral and visceral branch. The lateral ramus supplies the lateral line of the trunk, and the visceral branch supplies the remaining gill apparatus and pericardial and pleuroperitoneal cavities.

**4. Ventral Aspect of the Brain**

Remove the brain by cutting across the spinal cord and the olfactory and other nerves. The ventral view reveals the optic nerves crossing to form the *optic chiasma*. Just caudal to the chiasma are a pair of large lobes, the *inferior lobes* of the *hypothalamus*. The *infundibulum* of the hypothalamus is revealed as an unpaired bulge located on the midline between the chiasma and the inferior lobes. The *hypophysis* is attached to the infundibulum but is usually torn free as the brain is removed. It is seen as a small lobe of tissue remaining in the floor of the skull just caudal to the point where the optic nerves were cut in removing the brain.

| | |
|---|---|
| **5. The Sagittal Section of the Brain** | Bisect the brain by a median sagittal cut. Examine the cut surface under water using a dissecting microscope. The *fourth ventricle* is the large cavity located within the medulla. This ventricle continues rostrally under the corpus of the cerebellum and then expands as the ventricle of the optic tectum. The roof between the corpus of the cerebellum and the optic tectum is pushed into an S-shaped curve. This portion of the roof is termed the *valvula* of the cerebellum. It is believed to be a division of the corpus cerebelli concerned with sensory information from the lateral-line system. The continuation of the central cavity is termed the *third ventricle*. It is bordered dorsally by the *habenula* and ventrally by the optic nerve and *hypothalamus*. Its lateral walls are the *thalamus*. The third ventricle continues as the ventricles of the telencephalon. |

The *fourth ventricle* is the large cavity located within the medulla. This ventricle continues rostrally under the corpus of the cerebellum and then expands as the ventricle of the optic tectum. The roof between the corpus of the cerebellum and the optic tectum is pushed into an S-shaped curve. This portion of the roof is termed the *valvula* of the cerebellum. It is believed to be a division of the corpus cerebelli concerned with sensory information from the lateral-line system. The continuation of the central cavity is termed the *third ventricle*. It is bordered dorsally by the *habenula* and ventrally by the optic nerve and *hypothalamus*. Its lateral walls are the *thalamus*. The third ventricle continues as the ventricles of the telencephalon.

**6. The Autonomic System**

Teleosts possess segmentally arranged ganglia forming a sympathetic chain that reaches as far rostrally as the trigeminus nerve. The preganglionic fibers passing to organs of the head from the sympathetic ganglia arise in the spinal cord. Sympathetic connections (gray rami) are present. The parasympathetic system is almost totally represented by branches of the vagus nerve. No sympathetic cardiac-accelerating system appears to be present.

**D. The Nervous System and Sense Organs of Necturus**

**1. The Spinal Nerves**

The spinal nerves are best found as follows: Make a longitudinal cut along the side of the body below the lateral line. Cut through the external and internal oblique muscles and separate this mass of muscle from the thin layer of the transverse muscles lying next to the coelom. The *ventral rami* of the spinal nerves will now be seen running between the oblique and transverse muscles, along the myosepta, and supplying the hypaxial muscles. Trace one of these toward the vertebral column, cutting away muscles from its course. It lies just behind the rib, which may be cut away. The nerve may be traced up to the vertebra, where it is embedded in an orange-colored material. If this is carefully cleared away, the *dorsal ganglion* of the spinal nerve will be found imbedded in it. It is a rounded, brownish body from which spring two nerves: the ventral ramus, just followed, and the smaller *dorsal ramus,* which supplies the epaxial muscles.

**2. The Limb Plexi**

The ventral rami of the spinal nerves form a plexus for each limb, the motor nerves for the limb muscles arising from this plexus. The *brachial* plexus is located as follows: Make a cut in the ventral body wall just medial to the base of the forelimb. Separate the pectoral and shoulder muscles from the sternohyoid muscle. The brachial plexus is then easily seen running posterior to the scapula. It consists of the ventral rami of three spinal nerves (3, 4 and 5) which have cross connections with each other. Beyond the plexus, nerves proceed into the forelimb.

The *lumbosacral* plexus is located by cutting through the skin and both layers of oblique muscles longitudinally just dorsal to the base of the hind limb. When the oblique is separated from the transverse muscle layers, the plexus is exposed. It consists of the ventral rami of three (sometimes four) spinal nerves. The anterior one of the three gives off a slight branch to the nerve next posterior to it, sometimes receives a contribution from the ramus anterior to it, and, as the *crural* nerve, passes into the limb. The middle of the three nerves is much the largest and, after receiving contributions from the nerve next posterior to it, enters the limb as the ischiadic (sciatic) nerve.

**3. The Sense Organs of the Head**

Because of the small size of the eye and its similarity to that of other vertebrates, it will not be investigated.

*a. Lateral Line System*

Lateral line sense organs are present in *Necturus* but are impractical to find in gross dissection. They are situated along lines similar to those of fishes.

*b. Nose*

Probe into the external naris and follow your probe with the cut, opening the entire nasal passage to the internal naris. Note the folds or *olfactory lamellae* in the interior of the passage. Unlike those in fishes, the olfactory sac opens into the oral cavity and has both olfactory and respiratory functions.

*c. Ear*

The ear, as in fishes, consists of the internal ear only, embedded in the otic region of the skull. Expose the dorsal surface of the skull by cleaning away the muscles. Locate the otic capsules, one at either side of the posterior end of the skull. Cautiously shave away the cartilage here and locate the three semicircular ducts and vestibule, as done in the dogfish. There are three of the former—*anterior vertical, posterior vertical,* and *horizontal*—each with an *ampulla,* all arranged exactly as in the dogfish. The vestibule is divided indistinctly into a dorsal *utriculus,* from which the ducts spring, and a ventral *sacculus.* The latter contains a crystalline mass, the *otolith.*

**4. The Dorsal Aspect of the Brain**

Remove the roof of the skull. This is best done by stripping it off in slivers with a knife. After the brain is revealed, study its dorsal surface (fig. 13.41). The brain is covered by a membrane, the *primitive meninx,* which is more or less divisible into the *pia mater,* a delicate pigmented membrane adhering to the brain, and an outer *dura mater,* which is separated from the skull by the *peridural* space.

The most anterior portion of the brain consists of the two elongated cerebral hemispheres. The olfactory bulbs are not distinct externally, being included in the cerebral hemispheres. Between and behind the posterior ends of the cerebral hemispheres is a thin roof constituting a tela choroidea, the anterior part of which forms a choroidal sac (parencephalon), projecting forward in the groove between the two hemispheres. Immediately behind this is another dorsally projecting process, the pineal body or epiphysis. Behind the diencephalon is the optic tectum, which is the dorsal part of the

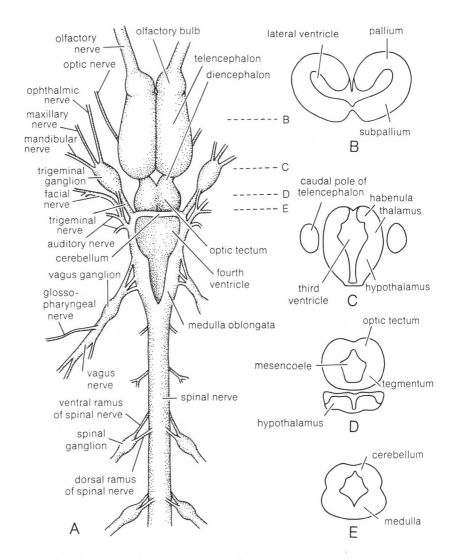

Fig. 13.41.  *A,* dorsal view of the brain of the mudpuppy, *Necturus maculosus.*
*B–E,* diagrams of transverse sections through the different major divisions of
the brain.

mesencephalon or midbrain. Behind this is another dark-pigmented
tela choroidea, which should be removed, revealing the triangular
cavity of the fourth ventricle. The anterior end of this cavity is overhung by a narrow shelf just back of the optic tectum; this shelf is
the cerebellum, very undeveloped in urodeles. The remainder of the
brain, enclosing the cavity of the fourth ventricle, is the medulla oblongata. Anteriorly the medulla is continuous below the optic tectum,
with projections that form the auricles of the cerebellum. The walls
of the medulla are divided into dorsal or *sensory* portions and ventral
or *motor* portions, the latter forming a broad band on either side of
the median ventral groove. The further subdivisions of these are not
evident in *Necturus.*

5. The Cranial
Nerves

*a. The First or Olfactory Nerve*

This is a stout band passing from the olfactory sac to the anterior
end of the telencephalon.

*b. The Second or Optic Nerve*

This small nerve may be seen in the floor of the cranial cavity by
pressing the telencephalon to one side. It passes obliquely caudad to
the ventral surface of the diencephalon.

*c. The Eye Muscle Nerves*

The *third,* or *oculomotor, fourth,* or *trochlear,* and *sixth,* or *abducens,*
nerves are so small in *Necturus* as to be scarcely discernible in
gross dissection. They originate from the same regions of the brain
and supply the same eye muscles as in elasmobranchs.

*d. The Fifth or Trigeminus Nerve*

The fifth nerve is the large trunk arising from the side of the anterior
end of the medulla. Trace it out through the skull. It passes in front
of the otic capsule and immediately enters a large ganglion, the
semilunar or Gasserian ganglion of the trigeminus. From the ganglion
three large nerves are given off—the *ophthalmic, maxillary,* and
*mandibular* branches. The former passes forward through the quadrate cartilage and runs anteriorly alongside the frontal bones. The
maxillary nerve proceeds along the margin of the upper jaw. The
mandibular nerve passes laterally to the angle of the jaws and then
turns forward along the lower jaw.

*e. The Seventh or Facial Nerve*

This arises just behind the trigeminus and sends a branch forward
to join the latter at the semilunar ganglion. This branch of the facial
nerve passes out with the ophthalmic nerve as the *superficial ophthalmic* branch of the facial and with the maxillary nerve as the *buccal*
branch. Both of these go to lateral-line organs. Again the lateralis
branches of VII and X may actually be separate cranial nerves
termed the anterior and posterior lateral line nerves. The greater
part of the facial nerve arises from the medulla in common with the
auditory nerve ventral to the above-named branch. This common
*acousticofacial* trunk of facial and auditory passes into the anterior
part of the otic capsule. From here the main trunk of the facial or
hyomandibular may be followed laterally. It branches to muscles,
lateral-line organs, and so on.

*f. The Eighth or Auditory Nerve*

This arises in common with the facial nerve and is distributed to
the internal ear. Its branches are readily noted in the otic capsule.

*g. The Ninth or Glossopharyngeal and Tenth or Vagus Nerves*

These arise together from the medulla posterior to the acousticofacial
trunk, by three roots. The common trunk passes along the posterior
margin of the ear capsule and enters a large ganglion. From this arise
several nerves that may be traced into the external gills (these nerves

are not homologous to the pre- and post-trematic branches of elasmobranchs) and to the branchial bars. The most posterior branch of the vagus gives off a *lateral* branch that passes to the lateral line, which it accompanies. The vagus also supplies the viscera.

| | |
|---|---|
| 6. Ventral Aspect of the Brain | Remove the brain by cutting across the spinal cord, the olfactory, and other nerves. The ventral view reveals some additional parts of the diencephalon. At the anterior end of the ventral surface of the diencephalon is the small *optic chiasma,* formed by the optic nerves. Posterior to this is the large *infundibulum,* from the posterior end of which projects the hypophysis. |
| 7. The Autonomic System | The urodele amphibians have a pair of sympathetic trunks alongside the dorsal aorta, extending from the vagus region to the end of the tail; and there are also the usual plexi along the large blood vessels and in the viscera. It appears that a cephalic part of the autonomic system is usually absent. |
| **E. The Nervous System and Sense Organs of the Painted Turtle (Chrysemys)** | The brain of reptiles is more specialized than that of fishes and amphibians in the increased size of the cerebral hemispheres. The cerebellum is also more specialized than that of amphibians and is particularly well developed in crocodilians. The entire lateralis system has vanished. |
| 1. The Spinal Nerves, the Autonomic System, and the Spinal Cord | Remove all of the viscera from the pleuroperitoneal cavity, leaving the large neck muscles in the middorsal region undisturbed. Identify the spinal nerves as the white cords passing along the sutural lines of the costal plates. The sympathetic chain should be identified as a white cord or cords located on the sides of the mass of neck muscles. |

*a. Spinal Nerves and Limb Plexi*

Carefully expose the spinal nerves of the trunk, avoiding injury to the sympathetic system. These nerves, called the *dorsal* spinal nerves, run along the sutures between the costal plates of the carapace. In most cases each consists of two branches, a smaller *dorsal ramus* and a larger *ventral ramus.* Toward the vertebral column they come from a large ganglion, the *dorsal* or *spinal* ganglion, situated in contact with the center of the centrum.

Expose the *brachial plexus* in the depression between the neck and the dorsal end of the scapula. It is generally formed by the cross unions between the ventral rami of the last four *cervical* spinal nerves and the first dorsal spinal nerve, which may be identified as the one in front of the first typical rib (really the second rib). The four cervical nerves form a complex network on the surface of the shoulder muscles. From this network the large *median* nerve proceeds along the anterior surface of these muscles and the smaller *ulnar* and *radial* nerves along the posterior surface; the radial is the most dorsally situated one. The first dorsal nerve sends a branch near its ganglion to the brachial plexus and is then distributed to the carapace just posterior to the forelimb.

The next six dorsal nerves are similar to the first description given. The ventral rami of the eighth, ninth, and tenth dorsal nerves, together with the two *sacral* nerves, form the *lumbosacral plexus* for the hind limb. This lies on the medial surface of the muscles covering the ilium and is found by separating the ilium, with its muscles, carefully from the median region. The branches from the three dorsal nerves unite to a sort of knot, from which several nerves proceed to the anterior part of the leg. The two sacral nerves, receiving also a contribution from the tenth dorsal nerve, unite to form a large trunk, the sciatic nerve, situated among the muscles on the posterior side of the leg.

There is a pair of *caudal* spinal nerves corresponding to each caudal vertebra, but these need not be looked for. There are nine pairs of *cervical* nerves, which will be found by looking in the neck at the same level as the level of emergence of the nerves of the brachial plexus.

### b. The Sympathetic System

Locate the vagus nerve in the neck. It is the conspicuous white cord running along the side of the neck. Trace it posteriorly. The sympathetic trunk is bound with it but at about the level of the first nerve of the brachial plexus separates from the vagus and enters a swelling or ganglion, the *middle cervical* ganglion. The sympathetic cord proceeds dorsally from this ganglion and lies on the ventral surface of the brachial plexus, where it presents two successive swellings, which together constitute the *inferior cervical* ganglion. Observe branches from the ganglia. The sympathetic cord passes to the ganglion of the first dorsal spinal nerve, to which its own ganglion is fused. It then proceeds as a delicate white cord across the second rib and again forms a ganglion, which is fused to the ganglion of the second dorsal nerve. The sympathetic cord then passes more ventrally, lying on the side of the long neck muscles. Follow it here and note the ganglia,which it bears at intervals and the branches from these ganglia. Note particularly the branches between the sympathetic ganglia and the adjacent spinal ganglia. These branches constitute the *ramus communicans* and consist of the visceral motor and visceral sensory fibers passing between the sympathetic and central nervous systems. The ganglia and branches of the sympathetic are particularly noticeable in the urogenital region.

## 2. The Sense Organs of the Head

### a. The Nasal Cavities

The external nares lead into wide chambers, the nasal cavities. Cut off the external nares and the roof of the skull posterior to them, thus revealing the nasal cavities. They are separated by a median *septum,* partly bony. From the ventral region of the septum a conspicuous fold projects into the nasal cavity. On the posterior wall of the nasal cavity is a slight projection, a *concha* or *turbinal.* Posterior to this the nasal cavity connects by a passage with the roof of the mouth cavity, the nasal cavities thus serving as respiratory passages.

### b. Eye

Although the eye is small, it can be dissected with a little care. Its parts are very similar to those of the elasmobranch eye. Make an incision through the skin around one eye and with the bone scissors remove the skull dorsal to and between the eyes. The two eyes are seen to be close together, separated by a median membranous *interorbital septum*. Near this septum on each side runs an artery. On the anterior dorsal surface of the eyeball is a gland, the *Harderian* gland. Over the posterior and ventral surface of the eyeball extends the much larger *lacrimal* or *tear* gland. Remove these glands, thus exposing the surface of the eyeball and the eye muscles. Extending from the interorbital septum to the dorsal surface of the eyeball is the *superior oblique* muscle. Posterior to this and inserted on the eyeball near it is the *superior rectus*. Between and ventral to these two is the *internal rectus*. Passing above the internal rectus are two nerves, the *trochlear* to the superior oblique muscle and the *ophthalmic* branch of the trigeminus. Loosen the eyeball ventrally and, raising it as far as possible, examine the ventral surface. The anterior part of this surface is covered by a flat muscle, the *pyramidalis,* which originates on the eyeball and passes to the eyelids and nictitating membrane. Remove this and clean the ventral surface of the eyeball. The *inferior oblique* and *inferior rectus* muscles are then seen converging to their insertions on the ventral surface of the eyeball. The *external rectus* is posterior to them.

Remove the eyeball and open it by cutting off its dorsal side. Identify the coats of the eyeball, the lens, the cavities of the eye, and the two humors, as in the elasmobranch eye, where the structure is practically identical. Note, however, the difference in the shape of the lens of the turtle eye.

### c. The Ear

The ear consists of two parts, a *middle* ear and an *internal* ear. The former is located posterior to the angle of the jaws internal to a circular area of skin. Remove this piece of skin and find beneath it a smaller cartilaginous circular plate, the *tympanic membrane* or *eardrum*. Make a cut around the margin of this and carefully raise it. Attached to its internal surface, posterior to the center, is a rod-shaped bone, the *columella,* whose inner end is fastened to the wall of a large cavity. This cavity is the *tympanic cavity* or *cavity of the middle ear*. It is an evagination from the first visceral pouch. Ventral to the inner end of the columella is a slit bounded by raised lips. This slit is the opening of the auditory tube connecting the pharyngeal cavity with the cavity of the middle ear and representing the stalk of the evagination by which the latter was formed. Considerably internal to the point of attachment of the inner end of the columella lies the internal ear. It will be more definitely located later. It is similar in structure to the internal ear of elasmobranchs.

**3. Dorsal Aspect of the Brain**

Remove the roof of the skull and expose the brain. The brain is coverel by a tough membrane, the *dura mater*. On cutting carefully through this, you will find a more delicate membrane, the *secondary meninx,* adhering to the brain; this is more or less pigmented and vascularized. The space between the two membranes, crossed by strands, is the *subdural space,* and that between the dura mater and the skull is the *peridural space.* Remove the dura mater from the dorsal surface of the brain.

The brain (fig. 13.42) has the same divisions as in the preceding

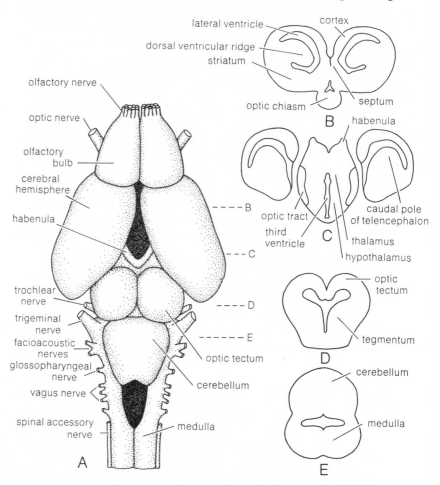

Fig. 13.42. *A,* dorsal view of the brain of the painted turtle, *Chrysemys picta. B–E,* diagrams of transverse sections through the different major divisions of the brain.

animals, but the proportions of the parts are somewhat altered. The anterior half of the brain is composed of the conspicuous enlarged *cerebral hemispheres,* relatively much larger than in fish and amphibians; at their anterior ends, separated from them by a slight groove, are the smaller *olfactory bulbs.* Between the posterior ends of the cerebral hemispheres lies the choroid roof of the diencephalon, projecting dorsally as a choroidal sac that adheres to the dura mater and is generally torn off in removing the latter. The pineal body or epiphysis lies on this sac. With the removal of the choroidal sac the diencephalon is seen as a depressed area posterior and ventral to the cerebral hemispheres. Behind it is the rounded optic tectum and behind that the cerebellum, smaller than in selachians but larger than in amphibians. The cerebellum overhangs the medulla, whose roof is made of the usual tela choroidea; when this is removed, the cavity of the fourth ventricle is revealed. The dorsal rims of the medulla are, as usual, the somatic sensory columns, and the anterior ends of these are continuous with the cerebellum by way of the small auricular lobes of the latter. In the floor of the medulla the two somatic motor columns are conspicuous, one to either side of the midventral groove.

## 4. The Cranial Nerves

The dissection of the cranial nerves is difficult, and consequently the following description is not complete.

### a. The Olfactory Nerves

These are the two stout nerves extending from the dorsal portions of the olfactory sacs to the anterior end of the olfactory bulbs of the brain.

### b. The Optic Nerves

On cutting through the olfactory nerves and raising the anterior end of the brain, you will see the optic nerves as two stout trunks situated below the cerebral hemispheres and passing out of the orbit.

### c. The Trochlear Nerve

This small nerve arises on each side of the brain in the dorsolateral angle between the optic tectum and the cerebellum. It passes ventrally and forward and can be seen by pressing the cerebral hemisphere away from the skull. It lies behind the larger oculomotor nerve. To find the course of the trochlear nerve in the orbit, expose the undissected eye as before, clearing away the glands. Cut through the superior oblique muscle at its point of insertion on the eyeball, and find below it the trochlear nerve terminating in this muscle. Medial to the trochlear is the ophthalmic branch of the trigeminus.

### d. The Oculomotor Nerve

This nerve originates from the floor of the midbrain immediately in front of the trochlear nerve and can be seen by pressing the cerebral hemisphere away from the skull. Loosen the ventral side of the eyeball and raise it. Among the loose tissues between the eyeball and the floor of the orbit, generally adhering to the eyeball, is a nerve, the *maxillary* branch of the trigeminus. Free this from the eyeball. Cut away the pyramidal muscle, and, raising the eyeball and pressing it medially as far as possible, separate the inferior and external rectus muscles and find between and above them the stout white trunk of the optic nerve. The oculomotor nerve is in contact with the ventral surface of the optic nerve and branches to the same four eye muscles as in the dogfish. These branches are not easy to follow.

### e. The Trigeminus Nerve

This is a stout trunk whose origin from the anterior end of the medulla will be seen by pressing the cerebellar region of the brain away from the skull. The trunk passes laterally and enters its ganglion, the semilunar ganglion, which lies in a depression in the medial wall of the skull. The trigeminus has three branches, the *ophthalmic*, the *maxillary*, and the *mandibular*, distributed to the orbit and nose, the upper jaw, and the lower jaw, respectively. Remove the eyeball on the side where it is still present, leaving the ophthalmic and maxillary nerves intact. Cut through the roots of the nerves anterior to the trigeminus in order to raise the brain, and bend it away from the side being dissected. Follow the ophthalmic nerve forward and note its distribution to the nasal sacs. Follow it posteriorly toward the root of the trigeminus. It enters the skull, runs with the trochlear nerve between the dura mater and the skull, and finally joins the semilunar ganglion. Follow the maxillary nerve posteriorly. Besides the branch below the eyeball, already noted, there is a branch in the floor of the orbit, running obliquely forward. These two branches unite to form the main trunk of the maxillary nerve at the posterior end of the orbit. Trace this nerve posteriorly among the muscles to where it pierces the skull. At this point it is joined by the *mandibular* branch of the trigeminus. Trace this laterally. After branching into adjacent muscles, the mandibular nerve proceeds ventrally and enters the lower jaw. Mandibular and maxillary branches pass together through a foramen in the skull and connect with the semilunar ganglion.

### f. The Facial and Auditory Nerves

These arise together from the side of the medulla just behind the roof of the trigeminus and immediately separate into an anterior facial nerve and a posterior auditory. The latter is distributed to the internal ear. This is situated in the skull opposite the acousticofacial root. This part of the skull may be broken open with the bone forceps. The semicircular ducts, ampullae, and vestibule of the internal ear will be noted. The auditory nerve will be seen branching among these structures. The facial nerve passes through the anterior part of the ear capsule and will be seen again later.

### g. The Glossopharyngeal Nerve

This arises by a small root from the medulla immediately posterior to the acousticofacial root and passes out through the posterior part of the ear capsule.

### h. The Vagus, the Spinal Accessory, and the Hypoglossal Nerves

The vagus and spinal accessory (eleventh) nerves arise together by a number of roots from the side of the medulla posterior to the preceding nerve, the more anterior roots belonging to the vagus and the posterior ones to the accessory. On cutting through these roots the more ventrally situated roots of the hypoglossal (twelfth) nerve will be seen. The three nerves pass out from the skull close together.

### i. The Abducens Nerve

Cut through all of the nerve roots on one side of the brain and tilt the brain toward the opposite side. The abducens nerves will be seen springing from the ventral surface of the medulla at about the same level as the acousticofacial root.

The seventh, ninth, tenth, and twelfth nerves may be traced farther, as follows: Turn the head ventral side up and remove the skin and superficial muscles from the hyoid apparatus. Locate the anterior and posterior horns of the hyoid. On the side of the neck, near the dorsal end of the anterior horn and posterior to it, the hypoglossal nerve will be seen emerging. It branches into the muscles over the anterior horn and sends a branch forward into the tongue muscles. Very near the point of emergence of the hypoglossal, but situated more deeply and nearer to the cartilage of the anterior horn, will be found the glosso-pharyngeal nerve. It runs between the two horns toward the median ventral line and supplies adjacent muscles and lining of the mouth cavity. Lateral to these nerves, a branch of the facial will be found crossing the anterior horn near its dorsal end and passing into the muscles lying along the posterior border of the mandible.

Make a median longitudinal incision through the whole floor of the mouth and pharyngeal cavities and open the two flaps so that the roof of these cavities is revealed. Locate the vagus (properly, the vagosympathetic) trunk in the neck and trace it anteriorly to its point of exit from the skull, removing the mucous membrane from the roof of the pharyngeal cavity. The vagosympathetic trunk passes to the dorsal side of the hypoglossal nerve, seen above, and there enters a ganglion, the *superior cervical* ganglion of the sympathetic. From this ganglion numerous branches pass out. Internal to the hypo-glossal nerve, locate the glossopharyngeal nerve; the carotid artery is situated between the two. Slightly anterior to these will be found the facial nerve as it exits from the skull. Its branches pass to the muscles between the anterior horn of the hyoid and the lower jaw, one of them curving over the ventral surface of the horn. The vagus nerve proceeds posteriorly and supplies the heart and other viscera. It will be noted that the facial, glossopharyngeal, and vagus nerves are much reduced, because of the loss of the lateral-line system and the gill apparatus. Note, however, that these nerves continue to supply the remains of the branchial bars and their muscles.

**5. Ventral Aspect of the Brain**

Remove the brain from the skull and examine the ventral surface. On the ventral surface of the diencephalon note the optic chiasma, the infundibulum just behind this, with the hypophysis projecting ventrally from the latter. Note the roots of the abducens and hypoglossal nerves arising from the ventral surface of the medulla.

**6. Median Sagittal Section**

Make a median sagittal section and study the cut surface. Identify the *fourth ventricle* in the medulla, the aqueduct or passage below the cerebellum, the *optic ventricle* in the optic tectum, and the *third ventricle* in the diencephalon. Note the larger size of the diencephalon, as compared with that of elasmobranchs, and the backward extension of the cerebral hemisphere over the diencephalon. The diencephalon is divided into *epithalamus, thalamus,* and *hypothalamus,* as in the dogfish, each including the parts previously enumerated. Note that the cerebral hemisphere presents a solid medial wall, called the *septum.* Cut into the roof or pallium of the hemisphere. Note its cavity, the *lateral ventricle,* and the large mass protruding from the floor into the ventricle; this mass is the *dorsal ventricular ridge.*

**F. The Nervous System and Sense Organs of the Pigeon**

The brain of the pigeon and other birds contrasts with that of the animals discussed earlier in the greatly enlarged cerebral hemispheres and cerebellum and the pronounced curvature and in the greatly reduced olfactory bulbs and other parts of the olfactory apparatus.

1. The Spinal Nerves and the Autonomic System

Carefully remove the remaining viscera from half of the trunk. Note the ventral rami of the spinal nerves passing laterally along the dorsal body wall between the ribs in the trunk region. Trace them toward the vertebral column and note, at the points where they emerge from the vertebrae, the ganglia of the autonomic system lying on the spinal nerves and the delicate white cords connecting the ganglia, forming the sympathetic trunks.

*a. Spinal Nerves and Limb Plexi*

In the neck the *cervical* spinal nerves can be seen by separating the vertebral column from the skin. They pass out at segmental intervals. The vagus nerve is the white cord that passes ventral to the proximal portions of the cervical nerves.

The ventral rami of the last cervical nerves, together with that of the first of the trunk, form the *brachial* plexus to the wing. This network, formed by the union of branches of four stout nerves, receives a small branch from the succeeding nerve.

The next five ventral rami pass out between the ribs. Following them is the *lumbosacral* plexus, divisible into three parts: the *lumbar,* the *sacral,* and the *pudendal* plexus. The lumbar plexus is formed by three nerves; nerves pass from it into the thigh. The sacral plexus arises from the union of five nerves, the first of which is the same as the third nerve contributing to the lumbar plexus. These five unite to produce a large trunk, the *sciatic* nerve, which passes along the dorsal side of the thigh between the muscles and proceeds down the leg. It will be found by separating the muscles along the middle of the dorsal surface of the thigh. It courses alongside the femoral artery and vein.

The remaining spinal nerves posterior to the sacral plexus form the pudendal plexus and pass obliquely posteriorly to the tail and cloacal region.

*b. The Autonomic System*

This has already been identified on the sides of the vertebral column. It consists, on each side, of a chain of two cords and segmental ganglia. One of the cords passes ventral to the head of the rib, the other dorsal to it. A sympathetic ganglion lies fused to each spinal nerve in the trunk region as the latter emerges from the vertebral column. On scraping off one of these sympathetic ganglia, you will find the *spinal ganglion* belonging to the spinal nerve dorsal to it. At about the middle of the rib-bearing region a plexus of nerves and ganglia will be seen extending ventrally from the main sympathetic cords and surrounding the dorsal aorta and its main branches to the digestive tract. This is the *coeliac plexus*. Posterior to this region the sympathetic cords are reduced and consist of a single trunk on each side. A sympathetic cord accompanies the pudendal plexus and has a ganglion in the middle of this plexus. Anteriorly the sympathetic cords pass across the ventral side of the brachial plexus, with its ganglionic enlargements, and then enter the vertebrarterial canals.

## 2. The Sense Organs of the Head

### a. The Nasal Cavities

Open one nasal cavity by a longitudinal slit just above the margin of the upper jaw from the external naris to the head. Note the median septum between the two nasal cavities and the swellings, the *turbinals* or *conchae,* projecting from the septum into the nasal cavity. There are three turbinals in a row: the first two large and conspicuous, the third and most posterior one consisting only of a small rounded swelling on the roof of the cavity in close contact with the posterior end of the second concha. Only this third concha is provided with olfactory epithelium. Beyond the conchae the nasal passages connect with the pharyngeal cavity.

### b. The Eye

Cut through the skin around the eyeball and also remove the roof of the skull between the two eyes. Note the relatively large size of the eyeballs and the *interorbital septum* between them. Along the dorsal margin of the septum course the two *olfactory nerves.* Press the eyeball outwardly away from the skull. Two thin, flat muscles will be seen extending to the eyeball from the orbit; the anterior one is the *superior oblique,* the posterior one the *superior rectus.* Cut through the superior oblique at its insertion on the eyeball and press it against the orbit. The *internal rectus* will now be seen extending to the eyeball ventral to the superior oblique. The white nerve crossing the orbit against the internal surface of the superior oblique is the *ophthalmic* branch of the trigeminus. Dorsal to it the smaller *trochlear* nerve is seen terminating on the superior oblique. The thin sheet of muscles on the surface of the eyeball is the *quadrate,* a muscle of the eyelids. On the anterior surface of the eyeball ventral to the superior oblique is a white, fat-like mass, the *Harderian gland.* Press the eyeball posteriorly, and find anterior to this gland, against the anterior wall of the orbit, the *inferior oblique* muscle. With the eyeball pulled forward the *external rectus* is seen extending to the posterior surface of the eyeball. Free the ventral margin of the eyeball. In the posterior ventral region, on raising the eyeball, you will see a small gland, the *lacrimal gland.* Two muscles will be seen on the ventral surface of the eyeball. The anterior one is the *inferior rectus,* the posterior one the *external rectus.* By cutting through the inferior rectus, you will find the *pyramid,* a muscle of the eyelids, internal to it. Cut through all of the rectus muscles and the inferior oblique at their insertions on the eyeball and remove the eyeball, severing the optic nerve. The pyramid and quadrate muscles are now more readily seen extending on the surface of the eyeball to the optic nerve; the quadrate muscle is broad and dorsally situated, the pyramid narrow and ventral. They are concerned in operating the nictitating membrane. In the orbit note the extent of the Harderian gland.

Cut off the dorsal part of the eyeball and identify the structures of the eye. Note the *sclerotic* coat, continuing as the transparent cornea over the exposed part of the eye; the *conjunctiva,* passing over the external surface of the cornea and continuing onto the eyelids; the black *choroid* coat internal to the sclerotic and forming the *iris* in front; and the soft *retina.* Note the peculiar ridged structure, the *pecten,* projecting from the choroid coat through the retina in the medial wall of the eyeball and extending to the lens. The pecten is a structure found in the eyes of birds but its function is uncertain; according to one researcher, it throws a shadow on the retina, hence making birds very sensitive to any movement in their visual field. Loosen the lens and observe that it is encircled by a structure continuous with the choroid coat and marked by radiating ridges, the *ciliary processes.* The whole structure, which is called the *ciliary body* and contains the *ciliary* muscles, holds the lens in place and can change the curvature and position of the lens. Note the shape of the lens—flat externally, more complex internally. The chambers of the eye and the two humors are the same as in the dogfish. Peel the iris from the cornea and note the stiff bony ring, composed of the small *sclerotic* bones, encircling the cornea.

### c. The Ear

The ear of birds consists of three parts, the *external* ear, the *middle* ear, and the *internal* ear. The external ear comprises the external auditory meatus, a passage situated below and behind the eye. Cut into this on the same side of the head on which the eye was dissected, and, at its internal end, find a circular transparent membrane, the *tympanic membrane.* Through the membrane the *columella* can be seen extending from its internal surface inwardly. Remove the tympanic membrane, noting the columella adhering to its internal surface. The cavity of the middle ear is now exposed; medially and ventrally it is connected to the pharyngeal cavity by the auditory tube; posterior and slightly dorsal to it is situated the internal ear. The inner end of the columella adjoins a tiny bone, the *stapes,* which fits into an opening, the *fenestra ovalis* or *vestibuli,* which leads into the internal ear. Look for these at the inner end of the columella. Next, carefully break away in small pieces the spongy bone behind the middle ear. Three bony *semicircular canals* are revealed. Each of them contains a membranous *semicircular duct,* as will be seen by breaking open one of them. The three ducts are situated in the same planes and have the

same names as in elasmobranchs. The remaining structures of the
internal ear, consisting of two small chambers, the *utriculus* and the
*sacculus,* are difficult to dissect because they are soft and deeply
embedded.

**3. Dorsal Aspect
of the Brain**

Expose the brain, removing the roof of the skull and the side of the
skull where the sense organs were dissected, including the medial
wall of the orbit. Note the *dura mater* enclosing the brain; on removing this, note the very delicate *secondary meninx* next to the brain
substance. Unlike the preceding forms, the brain closely fills the
cranial cavity.

   The brain (fig. 13.43) is short and broad and is strongly curved,
in correlation with the biped gait. The curvature results from flexures
of the brain in three regions. The chief or *primary flexure* occurs in

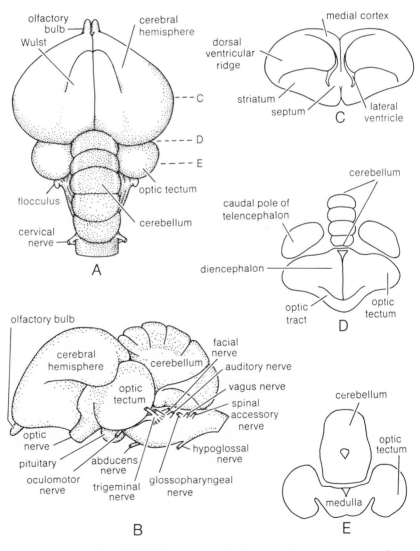

Fig. 13.43.  *A–B,* dorsal and lateral views of the brain of the pigeon, *Columba
livia. C–E,* diagrams of transverse sections through major regions of the brain.

the region of the midbrain, with the result that the posterior part of the brain is bent nearly at right angles to the anterior part. The second or *nuchal flexure* takes place in the medulla, bending the medulla at an angle to the spinal cord. The *pontal flexure* in the region ventral to the cerebellum bends the brain in the opposite direction from the other two flexures, with the result that this region of the brain is depressed.

The anterior end of the brain consists of the two very small *olfactory bulbs.* Posterior to them are the large *cerebral hemispheres,* separated by a deep *sagittal fissure.* These are so enlarged posteriorly that they completely conceal the diencephalon from dorsal view, with the exception of the delicate *pineal body,* which is seen in the posterior end of the sagittal fissure. The large *optic tectum* of the midbrain is ventral to the posterior ends of the cerebral hemispheres. Posterior to the hemispheres is the curved *cerebellum,* marked by transverse grooves. Posterior and ventral to this is the *medulla oblongata,* its anterior end depressed beneath the cerebellum. The roof of the medulla is composed, as usual, of a tela choroidea.

## 4. The Cranial Nerves

These are somewhat difficult to follow in detail. Work on the side left intact. There are twelve pairs of cranial nerves.

### a. The Olfactory Nerves

These are two stout and elongated nerves passing from the nasal sacs along the dorsal margin of the interorbital septum to the olfactory lobes.

### b. The Optic Nerves

On the side where the wall of the orbit was removed note the stout, white *optic tract* in front of the optic tectum. Follow this toward the orbit and find the optic nerve connected with its anterior end.

### c. The Trochlear Nerve

The cranial origin of this nerve is difficult to see at the present stage of the dissection. It arises in the deep groove between the optic tectum and the cerebellum and passes ventrally between the optic tectum and medulla. It runs forward in the floor of the cranial cavity to the orbit. To find it in the orbit, expose the intact eye as before. Cut through the superior oblique at its insertion on the eyeball and, lifting it, note the trochlear nerve passing to it and spreading out on its ventral surface. The ophthalmic branch of the trigeminus runs close to the trochlear nerve.

### d. The Oculomotor Nerve

The cranial origin of this nerve will be seen later. It branches to the inferior oblique and the superior, inferior, and internal rectus muscles. Remove the eye that is still in place, cutting the eye muscles as near the eyeball as possible and preserving the ophthalmic nerve intact. Look for the branches of the oculomotor nerve among the eye muscles in question. The branch to the inferior oblique in the floor of the orbit is the most conspicuous of them.

### e. The Abducens Nerve

This will be found on examining the posterior surface of the external rectus muscle. This nerve also supplies the pyramid and quadrate muscles.

### f. The Trigeminus Nerve

This has three branches, the *ophthalmic,* the *maxillary,* and the *mandibular.* The ophthalmic nerve has already been noted in the dorsal part of the orbit. Follow it forward, noting its distribution to the walls of the nasal cavities. In the floor of the orbit, near its outer margin, locate the maxillary nerve. Trace it forward, noting its branches to the orbit and upper jaw. Trace the maxillary nerve posteriorly, carefully cutting away tissues in its path. Caudad of the orbit it is joined by the mandibular nerve. Trace this, noting branches to muscles and main trunk passing into the lower jaw. Trace the common trunk of the maxillary and mandibular nerve toward the skull and note that they are joined in the skull by the ophthalmic nerve. At the point of union is the *semilunar ganglion,* lying in the skull. From the ganglion the trigeminus nerve may be traced to its origin from the side of the medulla below the optic lobe.

### g. The Facial Nerve

This arises from the medulla just back of the root of the trigeminus and passes through the anterior part of the ear capsule, where it will be found when the latter has been scraped away.

### h. The Auditory Nerve

This arises close to the facial nerve and passes out with it into the ear capsule, to the various parts of which it is distributed.

### i. The Glossopharyngeal and the Vagus

These nerves arise close together just behind the ear capsule and will be found there by carefully dissecting in the muscles. The glossopharyngeal is the smaller of the two and is anterior in position. It enters a ganglion, the *petrosal* ganglion, beyond which it is distributed to the palate, pharynx, and larynx. The vagus nerve is considerably larger than the glossopharyngeal. It passes laterally parallel and posterior to the glossopharyngeal and enters its ganglion, the *jugular* ganglion, which is united with the petrosal ganglion, the two forming a mass. Beyond this the vagus turns posteriorly and passes down the neck, supplying respiratory system, heart, and other viscera. Portions of the sympathetic system are intermingled with the ninth and tenth nerves.

### j. The Spinal Accessory and the Hypoglossal

The former passes out with the vagus and is distributed to certain muscles. The hypoglossal is found just posterior to the vagus. It is distributed to certain neck muscles and sends a branch forward to the tongue.

**5. Ventral Aspect of the Brain**

Remove the brain from the cranial cavity, preserving the roots of the cranial nerves as far as possible. Those not kept attached to the brain will be found in the cranial cavity.

Note the form of the olfactory lobes and cerebral hemispheres from the ventral aspect. Between the optic tecta is the diencephalon. In the center of this is the *optic chiasma,* marked by cross lines. From the optic chiasma the strong white *optic tracts* pass laterad and dorsad to the optic tecta and dorsal part of the diencephalon. Behind the chiasma is a depressed area, the *infundibulum,* from which the *hypophysis* extends ventrally. The hypophysis is usually left behind in removing the brain and will be found in a deep pit, the *sella turcica,* in the floor of the cranial cavity. The infundibulum bears a central cleft where the hypophysis was torn from it. At the sides of the infundibulum are the roots of the oculomotor nerves. Posterior to the diencephalon is the depressed medulla. Between the medulla and the optic tectum is the slender root of the trochlear nerve. On the ventral surface of the medulla are the roots of the abducens nerves; they should also be sought in the floor of the cranial cavity. On the sides of the medulla look for the roots of the fifth, seventh, eighth, ninth, and tenth nerves, situated in a row. The twelfth nerve arises from the ventral surface of the medulla about on the same level as the ninth and tenth roots. The eleventh nerve arises from the spinal cord by several roots and ascends to a position immediately behind the tenth root.

**6. Sagittal Section**

Make a median sagittal section of the brain and study the cut surface. In the medulla note the fourth ventricle overhung by the cerebellum. Note the thick ventral wall of the medulla and the pontal flexure causing a ventrally directed bend in the medulla. In the cerebellum observe the small *cerebellar ventricle* and the arrangement of the gray and white matter resulting in section in a tree-like appearance, called the *arbor vitae.* Each fold of the cerebellum consists of a central plate of white matter surrounded by a thick covering of gray matter. Anterior to the cerebellum is a region consisting dorsally of the mesencephalon and ventrally of the diencephalon. The optic tectum does not appear in the section, but the median part of the midbrain forms the dorsal part of the section. A narrow cavity, the *third ventricle,* is present in the diencephalon and extends into the infundibulum. In front of the latter appears the optic chiasma. Note that the cerebral hemisphere arches back over the diencephalon and midbrain, and note the strong connection of the diencephalon with the hemisphere. The cavity of the cerebral hemisphere is not visible in the median section. The medial wall of the hemisphere is called the septum, its dorsal wall the pallium. Cut into the latter, noting its thinness, and inside note the cavity or *lateral ventricle* of the hemisphere and the great mass, the *dorsal ventricular ridge,* bulging from the lateral wall.

**G. The Nervous System and Sense Organs of Mammals**

The mammalian brain is notable for the large size of the cerebral hemispheres, whose surface is still further increased in many mammals by convolutions, and for the high development of the cerebellum. Although it is not evident on the surface, the thalamus is also greatly enlarged and differentiated and has extensive connections with the cerebral hemispheres. There are four lobes visible in the midbrain, instead of the usual two. The olfactory apparatus is usually well developed and presents the same tracts and connections as in other vertebrates. A pronounced curvature is characteristic of mammalian brains.

For the complete dissection of the nervous system a new specimen is necessary, but the greater part of this system can be worked out on the same specimen as used for preceding systems. If a new animal is provided, open it by a longitudinal cut from the perineum through the anterior end of the sternum. If the old specimen is used, it will not be possible to see the branches of the autonomic system and the vagus to the viscera or the peripheral distribution of some of the cranial nerves. In the work on the nerves all structures other than nerves may be removed in order to expose them. In following one nerve, be careful not to destroy adjacent nerves.

**1. The Spinal Nerves, the Autonomic System, and the Vagus Nerve**

*a. Cervical Portion of the Sympathetic and the Vagus Nerve*

Locate the vagus nerve at a point near the larynx. It lies alongside the carotid artery. The nerve crossing the vagus near the larynx and giving off branches into the sternohyoid, sternothyroid, and related muscles is the *descending branch* of the twelfth or *hypoglossal* nerve.

Rabbit: The vagus nerve and the cervical part of the sympathetic trunk lie together on the dorsal surface of the carotid artery. The vagus is larger and more lateral. Toward the medial side of the sympathetic trunk, posterior to the larynx, may be separated a delicate nerve, the *cardiac branch* of the *vagus* (*depressor nerve* of the heart). Trace the sympathetic posteriorly. Just in front of the subclavian artery it enters a ganglion, the *inferior cervical* ganglion. From this ganglion cords pass to either side of the subclavian artery, forming the *ansa subclavia,* and unite again to another ganglion, the *first thoracic ganglion,* situated posterior and dorsal to the artery.

Cat: The sympathetic trunk is inseparably bound with the vagus, the two forming a large *vagosympathetic* trunk coursing lateral to the carotid artery and bound with it by a common sheath. Trace it caudad. Just in front of the first rib, branches of sympathetic origin arise from the trunk and proceed toward the esophagus. Shortly posterior to this point the sympathetic separates from the vagus and generally enters a ganglion, the *middle cervical ganglion,* which lies in contact with the vagus. From this ganglion cords pass on either side of the subclavian artery to form the *ansa subclavia* and, proceeding directly dorsally, unite to form a large ganglion, the *inferior cervical ganglion,* which lies against the neck muscles between the heads of the first and second ribs.

From the inferior cervical ganglion in both animals *cardiac* branches pass to the heart. The conspicuous nerve lying lateral to the vagus is the *phrenic* nerve or nerve of the diaphragm. The right vagus, just after it passes ventrally to the subclavian artery, gives off the *recurrent* or *posterior laryngeal* nerve, which runs anteriorly along the side of the trachea to the larynx. The left recurrent nerve arises much farther posteriorly from the left vagus.

### b. The Anterior Cervical Spinal Nerves

The spinal nerves emerge from the spinal cord in pairs between successive vertebrae, passing out through the intervertebral foramina. Those of the cervical region are called the *cervical* nerves; there are eight pairs of them. The ventral rami of the first four cervical nerves are loosely united with each other to form the *cervical* plexus; the last four, together with the first thoracic, form the *brachial* plexus. Since the first two are small and more or less mingled with the posterior cranial nerves, they will not be studied at this stage of the dissection.

To expose the cervical nerves, pull the muscles that are inserted on the anterior end of the sternum (sternomastoid, sternohyoid, sternothyroid) laterally or cut across them where necessary, thus exposing the musculature of the vertebral column. Look along the side of this, dorsal to the carotid artery, and note the ventral rami of the spinal nerves emerging at intervals. At about the level of the posterior end of the larynx lies the third cervical nerve in the rabbit, fourth in the cat. The nerves thus exposed are the ventral rami only; the dorsal rami are exposed only by more radical dissection, which will not be attempted here. The dorsal rami supply the epaxial musculature. Note the branches of the exposed ventral rami to the muscles of the side of the neck.

From the ventral ramus of the fourth cervical nerve (rabbit) and fifth cervical nerve (cat) arises the *phrenic* nerve. It passes posteriorly parallel to the vagus, in the rabbit close to the vertebral musculature. It receives a branch from the fifth (rabbit) or sixth (cat) cervical nerve and then continues posteriorly into the thorax. As it passes the sympathetic ganglia, it receives contributions from them. In the thorax it lies at the side of the pericardial sac, just ventral to the root of the lung. Trace it posteriorly and note how it spreads on the surface of the diaphragm. The phrenic nerves are the motor nerves of the diaphragm; their origin from the cervical nerves shows that the muscles of the diaphragm are derived from the cervical myotomes.

### c. The Brachial Plexus

The ventral rami of the fourth to eighth (rabbit) or fifth to eighth (cat) cervical nerves, together with the ventral ramus of the first thoracic nerve, are united by intercommunicating branches, called *ansae,* to form the brachial plexus, which innervates the muscles of the shoulder, breast, forelimb, and diaphragm. The fourth cervical (rabbit) or fifth (cat) takes part in the plexus only through its contribution to the phrenic nerve.

To expose the brachial plexus, cut through the pectoral muscles near the midventral line and separate the pectoral muscles from the underlying serratus muscle. The plexus lies in the axilla along with the axillary artery and vein. Then cut through the pectoral muscles as near as possible to their insertion on the humerus and separate them from the muscles of the upper arm. In this way the course of the nerves into the forelimb is exposed.

The connections of the nerves of the plexus are so intricate that it is impossible to describe them, but certain points may be noted. In the rabbit the fifth cervical immediately sends a branch to the sixth cervical and then proceeds laterally into the neck muscles. The sixth cervical is a broad nerve that passes to the shoulder muscles after communicating with the seventh nerve. The seventh is smaller and, after contributing to the eighth, likewise innervates the shoulder muscles. The eighth cervical and first thoracic unite in one trunk as they emerge from the vertebral column. From this trunk arise the nerves of the limb. In the cat the sixth cervical has a broad connection with the seventh and then proceeds to the shoulder. The seventh and eighth cervicals and the first thoracic are very stout, intricately connected trunks from which the nerves of the forelimb proceed.

The chief nerves from the brachial plexus are the following:

1. *The phrenic nerve* arises from the fourth (rabbit) or fifth (cat) cervical nerve, receives contributions from more posterior nerves, and innervates the diaphragm.

2. *The suprascapular nerve* is the most anterior nerve arising from the sixth cervical. The main part of this nerve passes between the supraspinatus and subscapular muscles to supply the supraspinatus and infraspinatus. In the cat a branch of this nerve passes over the shoulder to more superficial parts.

3. *The ventral thoracic nerves* supply the pectoral muscles and will be found entering the inner surface of these muscles between the two incisions made above. They are the most ventral of the nerves of the plexus. There are two of these nerves, one arising from the seventh cervical, the other from the eighth cervical and first thoracic. The former is small in the rabbit.

4. *The subscapular nerves,* of which there are three, are dorsally situated and pass into the inner surface of the shoulder. The first arises from the sixth cervical and passes to the subscapular muscle; the second arises from the seventh cervical and supplies chiefly the teres major; the third comes from the seventh and eighth cervicals and runs posteriorly along the internal surface of the latissimus dorsi muscle.

5. *The axillary nerve* originates chiefly from the seventh cervical. It passes through the upper part of the upper arm, ventral to the triceps, and, emerging on the lateral surface of the upper arm, supplies chiefly the deltoid muscles.

6. *The dorsal or long thoracic nerve* is best located by examining the outer surface of the serratus ventralis muscle. Traced anteriorly, the nerve will be found to pass internal to the scalenes and to spring from the seventh cervical nerve close to the vertebral column.

7. *The musculocutaneous nerve* (*cat only*) arises from the ventral surface of the sixth and seventh cervicals. It passes to the biceps

muscle, forking as it approaches the muscle. The posterior branch continues along the surface of the muscle and at the elbow passes to the lateral surface of the arm and supplies the skin of the forearm.

8. *The radial nerve* is the largest nerve arising from the plexus. Seventh and eighth cervicals and first thoracic nerves contribute to its formation. It passes to the upper arm and, coursing between the humerus and the triceps, turns distally. It supplies many muscles of the forelimb.

9. *The median nerve* lies posterior to the radial. It arises in the cat by branches from the last three nerves of the plexus and in the rabbit chiefly from the first thoracic. It passes to the upper arm and then turns distally running along with the brachial artery.

10. *The ulnar nerve* lies just posterior and parallel to the median nerve, originating chiefly from the first thoracic nerve. The ulnar and median nerves supply the limb distal to the elbow, although in the rabbit the median nerve innervates the biceps.

11. *The medial cutaneous* is the small nerve that runs in contact with the ulnar nerve. It turns superficially just above the elbow and is distributed to the skin of the forearm.

### d. The Thoracic Portions of the Vagus and the Sympathetic

Trace the two vagi toward the heart. They pass dorsal to the roots of the lungs. The left vagus, just caudad of the aortic arch, gives off the left recurrent laryngeal nerve, which turns cephalad, passes on the dorsal side of the aorta, and ascends along the side of the trachea. At the roots of the lungs the vagi give rise to the *pulmonary* plexus to the lungs. This plexus also extends to the heart as the *cardiac plexus*. The cardiac branches of the sympathetic system, noted above, join the cardiac plexus. The cardiac plexus is situated at the bases of the aorta and pulmonary arteries. In the rabbit the cardiac branches of the vagus may be traced into this plexus.

Caudad of the pulmonary plexus the two vagi in the rabbit continue posteriorly along the sides of the esophagus, to which they furnish small branches, and penetrate the diaphragm. In the cat each vagus divides just posterior to the root of the lungs into dorsal and ventral branches. The ventral branches of the two sides immediately unite into one trunk, which proceeds posteriorly, lying on the left ventrolateral surface of the esophagus. The two dorsal branches continue posteriorly, lying along the right and left sides of the esophagus; near the diaphragm on the dorsal side of the esophagus they unite in one trunk. In this manner the *dorsal* and *ventral* divisions of the vagi are formed; they pass through the diaphragm. In their course along the esophagus they furnish branches to it.

Locate again the inferior cervical ganglion. Note the communicating branches from this ganglion to the brachial plexus. In the cat a particularly stout branch extends anteriorly ventral to the bases of the sixth to eighth cervical nerves, giving branches to them. Trace the sympathetic trunk posteriorly from the inferior cervical ganglion. The contents of the pleural cavities may now be cleaned out. The sympathetic trunk is a white cord lying to each side of the vertebral column, passing ventral to the heads of the ribs. At segmental intervals, generally in the places between the ribs, it presents a ganglionic enlargement.

### e. The Thoracic Spinal Nerves

The first thoracic nerve contributes to the brachial plexus. The ventral rami of the remaining thoracic nerves pass laterally as the *intercostal* nerves, lying along the posterior side of each rib. These nerves are readily exposed by running the point of an instrument along the posterior side of each rib, slitting open the fascia of the intercostal muscles. As each nerve emerges from the intervertebral foramen, it receives one or two *communicating branches* (*rami communicantes*) from the adjacent sympathetic ganglion. These branches are rather delicate, and it may not be possible to see them. The dorsal rami of the thoracic spinal nerves supply the epaxial muscles. To see them, turn the animal dorsal side up and carefully cut down through the mass of epaxial muscles close to the vertebrae. The dorsal rami will then be seen emerging from the vertebral column and penetrating the epaxial mass, accompanied by blood vessels. There are twelve or thirteen pairs of thoracic nerves.

### f. The Abdominal Portions of the Vagus and the Sympathetic

Trace the vagi into the peritoneal cavity, removing the liver if this has not already been done. In the *rabbit* the left vagus crosses the ventral surface of the esophagus obliquely to the right and is distributed to the lesser curvature and ventral surface of the stomach. The right vagus crosses the dorsal surface of the esophagus obliquely to the left and is distributed to the dorsal surface of the stomach. In the *cat* the ventral division of the vagus passes to the lesser curvature, the dorsal division to the greater curvature. In both cases the vagi form plexi on the stomach, called the *ventral and dorsal gastric plexi,* which also connect with the nearby sympathetic plexi.

Locate again the posterior part of the thoracic portion of the sympathetic trunk. Expose it and note the nerve, the *greater splanchnic nerve,* which arises from the sympathetic trunk on each side and passes obliquely ventrally toward the diaphragm. In the cat this nerve is accompanied by additional smaller nerves, the lesser splanchnic nerves, arising from the sympathetic shortly posterior to the origin of the greater splanchnic nerve. The splanchnic nerves pass to either side of the crura of the diaphragm into the peritoneal cavity. (The crura of the diaphragm are the muscular cords that fasten the diaphragm to the lumbar vertebrae.) Turn the abdominal viscera to the right and look on the left surface of the superior mesenteric artery near its origin from the aorta. Two prominent sympathetic ganglia will be found lying on the superior mesenteric artery. These are the *coeliac* and *superior mesenteric* ganglia; the former lies in front of, or on the left surface of, the artery, and the latter behind, or on the ventral surface of, the vessel. The two ganglia are bound together by a strong connection. The splanchnic nerves of both sides may be traced into the coeliac ganglion. From this ganglion a prominent *coeliac plexus* will be seen extending toward the stomach, where it connects with the gastric plexi of the vagi. This great sympathetic plexus, formed around and dorsal to the stomach, is often called the *solar plexus*. From the coeliac and superior mesenteric ganglia and adjacent plexi also arise plexi for the liver, spleen, adrenal glands, gonads, and the great blood vessels. Some of these will probably be seen. The *inferior mesenteric* ganglion of the sympathetic system lies in the mesocolon alongside the inferior mesenteric artery. It is situated in the *inferior mesenteric plexus,* from which networks extend to adjacent structures.

The main sympathetic trunk of the abdominal region should now be traced caudad from the place of origin of the splanchnic nerves. The two trunks descend deep dorsally, lying in the groove between two muscle masses. At segmental intervals they have ganglionic enlargements from which nerves pass to the ganglia and plexi, already noted. At the posterior end of the peritoneal cavity the sympathetic trunks gradually diminish and disappear.

### g. The Lumbar and Sacral Spinal Nerves and the Lumbosacral Plexus

There are seven pairs of *lumbar* nerves and four (rabbit) or three (cat) pairs of *sacral* nerves. The ventral rami of the last four lumbar nerves form a *lumbar* plexus, those of the sacral nerves a *sacral* plexus, but since the two plexi are united with each other, they may be considered together as the lumbosacral plexus.

Remove all viscera from the peritoneal cavity, including the postcaval vein and aorta. In the dorsal wall is a muscular mass extending from the vertebrae to the pelvic girdle. This consists of a lateral larger muscle, the *iliopsoas,* and a smaller medial one, the *psoas minor.* In the *rabbit* the psoas minor is a slender muscle that occupies only the posterior part of the middorsal region; its stout shining tendon passes to the dorsal side of the inguinal ligament. In the *cat* the psoas minor extends nearly as far anteriorly as the iliopsoas; it narrows posteriorly to a tendon that passes obliquely laterally on the ventral surface of the iliopsoas, which is thus exposed both medially and laterally to the tendon of the psoas minor. The psoas minor covers a part of the iliopsoas in both animals, and the greater part of the lumbar plexus is situated between the two muscles. Note the abdominal parts of the sympathetic cords between the posterior portions of these muscles.

Locate the last thoracic spinal nerve. It lies about one-half inch posterior to the last rib. The first nerve posterior to this on the dorsal wall is the ventral ramus of the first lumbar nerve. Shortly posterior to this is the second lumbar nerve. These two nerves pass to the muscles and skin of the abdominal wall; in the cat each divides into two branches. The third lumbar nerve emerges dorsal to the iliopsoas muscle and divides into a larger lateral branch to the abdominal wall and a more slender medial branch, which passes obliquely caudad, reaching and following the course of the iliolumbar artery and vein. The fourth lumbar nerve is the first of the lumbar plexus. It has two main branches, the *lateral cutaneous* nerve and the *genito-femoral* nerve. The former is the stout trunk that emerges between the iliopsoas and psoas minor muscles and accompanies the course of the iliolumbar artery and vein, passing to the thigh. The genito-femoral nerve is a long slender nerve that runs along the medial border of the psoas minor muscle, lateral to the sympathetic cords. In the posterior part of its course it accompanies the external iliac artery. It supplies the thigh and the abdominal wall and adjacent areas of the inguinal region. After locating these two branches of the fourth lumbar, trace them toward the vertebral column, removing the psoas minor as far as necessary. Find the point of emergence of the fourth lumbar from the vertebral column and note the connection, very stout in the cat, between the fourth lumbar and the fifth.

The fifth lumbar contributes by means of its connection with the fourth lumbar to the lateral cutaneous branch named above and also forms a strong union with the sixth lumbar. To expose these, remove the rest of the psoas minor. The common trunk, formed by the union of branches from the fifth and sixth lumbar nerves, passes laterally as the large *femoral* nerve. Trace this to the thigh. It courses along the center of the medial surface of the thigh in company with the femoral artery and vein. It innervates adjacent muscles of the thigh and then continues down the shank and foot as the *saphenous* nerve.

The *obturator* nerve arises from the connecting band between the sixth and seventh lumbar nerves and passes obliquely caudad, dorsal to the pubis, through the obturator foramen and into the gracilis and other muscles.

The seventh lumbar and the first sacral unite to form a very large trunk, the *sciatic* nerve. The sixth lumbar and second sacral also contribute small branches to this nerve. Follow the sciatic nerve. It turns dorsally, passing between the ilium and the vertebral column. Thrust an instrument through the place where it turns and dissect where the instrument emerges on the dorsal side of the animal. At the separation of the muscles there, is the sciatic nerve. Expose it as near to the vertebral column as possible. The *gluteal* nerves will be seen separating from the anterior side of the main trunk and passing into the gluteus muscles. (The nerve on the posterior side of the sciatic trunk is the posterior cutaneous, described below.) Follow the sciatic nerve down the leg. After giving off branches to the thigh muscles, it divides shortly above the knee into a lateral branch, the *peroneal* nerve, which passes between the insertions of the biceps femoris and the gastrocnemius, and a more medial branch, the *tibial* nerve, passing between the two heads of the gastrocnemius.

The sacral nerves are united by ansae to form the sacral plexus. The first sacral also takes part in the formation of the sciatic nerve. The chief nerves arising from the sacral plexus are the *pudendal* nerve and the *inferior hemorrhoidal*. The latter arises in the cat from the point of union of the three sacral nerves and passes to the bladder and rectum. The pudendal nerve arises from the large trunk formed by the union of the second and third sacral nerves and in the cat may also receive a branch from the sciatic. This trunk passes laterally parallel and posterior to the sciatic. From it arise the pudendal nerve, which turns toward the rectum and urogenital organs, and the *posterior cutaneous* nerve, which continues laterally into the biceps femoris muscle. It will be found by turning the animal dorsal side up and looking where the sciatic nerve was exposed. The nerve in question lies immediately posterior to the sciatic nerve and enters the biceps femoris. The fourth sacral nerve in the rabbit is moderate in size; it passes laterally, then turns, and innervates, along with the pudendal nerve, the sides of the rectum.

The foregoing nerves are all the ventral rami of the lumbar and sacral nerves. To see the small dorsal rami of the lumbar nerves, proceed as directed for the dorsal rami of the thoracic nerves. The *caudal* spinal nerves will not be considered.

<table>
<tr><td>

2. The Spinal Cord and the Roots of the Spinal Nerves

</td><td>

With bone scissors cut out a piece of the vertebral column two or three inches long from the posterior thoracic and anterior lumbar region. Remove the epaxial muscles from this piece to expose the vertebrae, and with the bone scissors cut off the neural arches of the vertebrae, thus exposing the neural canal. The spinal cord lies in this canal without completely filling it. Note that the spinal cord is loosely enclosed in a tough membrane, the *dura mater,* from which strands pass to the walls of the neural canal. The space between the dura mater and spinal cord is the *subdural space.* Slit open the dura mater. The spinal cord is closely invested by a membrane the *pia mater,* which cannot be separated from its surface. Between these two is a delicate membrane, the *arachnoid,* which is almost impossible to identify in gross dissection. The arachnoid and pia mater of mammals together correspond to the secondary meninx of lower vertebrates. In living animals the spaces around and between these membranes are filled with the *cerebrospinal* fluid, which is a modified lymph.

</td></tr>
</table>

Observe the roots of the spinal nerves arising in pairs at segmental intervals from the sides of the spinal cord. They are ensheathed in the dura mater, which follows them to their exit from the intervertebral foramina and is continuous with their sheaths outside of the vertebral column. Examine one of the roots in detail. Although it appears at first glance to be single, a little gentle picking in the center of the root with the point of a probe will reveal that it is composed of two parts. One of these, the *dorsal* root, is attached to the dorsolateral region of the cord and near the intervertebral foramen bears a large oval swelling, the *dorsal* or *spinal ganglion.* The dorsal root in mammals carries sensory fibers only, and the nerve cells from which the sensory fibers originate are located in the *spinal ganglion.* The other, ventral root arises from the ventrolateral region of the cord by several branches, which unite in one trunk. The *ventral* root carries motor fibers only, arising from motor cells in the cord. The dorsal and ventral roots unite beyond the ganglion to form the spinal nerve, which then exits through the intervertebral foramen and divides into the dorsal ramus to the epaxial muscles and adjacent skin, the ventral ramus to the hypaxial muscles and adjacent skin, and the communicating rami to the sympathetic system. These rami were already seen.

Cut through the roots of the spinal nerves and remove a small section of the spinal cord for examination. In the median dorsal line identify a groove, the *dorsal median sulcus;* in the median ventral line, another groove, the *ventral median fissure.* Lateral to the dorsal median sulcus is the dorsolateral sulcus, along which the dorsal roots enter the cord. The region between the dorsal median and dorsolateral sulci is called the *dorsal funiculus.* The lateral region of the cord between the dorsolateral sulcus and the line along which the ventral roots emerge is the *lateral funiculus.* Between this and the ventral median fissure is the *ventral funiculus.*

Make a clean cut across the cord and examine the cut surface. In the section you will discern a central darker material, the *gray matter,* shaped like a butterfly, in which the nerve cells of the cord are located, and a much thicker, white material, the *white matter,* surrounding the gray matter and composed of nerve fibers. The white matter can be divided into the funiculi named above. Each funiculus consists of a number of tracts or bundles of fibers, whose functions are known, but these tracts are not visibly differentiated from each other.

| | |
|---|---|
| 3. The Peripheral Distribution of the Posterior Cranial Nerves | In this section will be described the peripheral course of the fifth, seventh, and ninth to twelfth cranial nerves. For the complete dissection of these it is necessary to have a specimen of which the head is intact, but most of them can be found, in part at least, on the same specimen on which the previous dissections were made. |

### a. The Eleventh or Spinal Accessory Nerve

This nerve supplies the sternomastoid, cleidomastoid, levator scapulae ventralis, and trapezius muscles. It is a pure motor nerve and is apparently derived from the vagus.

Rabbit: Separate the sternomastoid and cleidomastoid, on the one hand, from the basioclavicularis and levator scapulae ventralis, on the other. Running near the dorsal border of the levator scapulae ventralis and parallel to it is the spinal accessory nerve. Branches of the second to fourth spinal nerves pass ventral to it and unite with it by branches. Trace it posteriorly and note its branches on the inner surface of the trapezius. Trace it anteriorly and note the branches to the levator scapulae ventralis, sternomastoid, and cleidomastoid.

Cat: Cut through the clavotrapezius near its origin and deflect it ventrally, thus exposing the levator scapulae ventralis. On the inner surface of the clavotrapezius along the dorsal border of the levator scapulae ventralis runs the main part of the spinal accessory nerve. Trace it posteriorly, noting the branches into the trapezius muscles and the levator scapulae ventralis. Trace it anteriorly. It passes dorsal to the second cervical nerve, to which it is connected by a network, and near this region it gives branches to the sternomastoid and clei-domastoid muscles. It then passes through the cleidomastoid muscle.

### b. The Vagus, the Sympathetic, and the Hypoglossal Nerves

Follow the vagus and sympathetic anteriorly. Stretch the head forward by cutting across the lateral muscles of the neck. At about the level of the posterior end of the larynx the vagus and carotid artery are crossed ventrally by the *descending branch* of the *hypoglossal* or twelfth cranial nerve. This passes obliquely caudal toward the median line and supplies the sternohyoid, sternothyroid, and thyrohyoid muscles. Continue forward. At about the place where the common carotid artery divides into external and internal carotids, a conspicuous nerve is seen crossing the ventral surface of the vagus and carotid artery and curving anteriorly. This is the main part of the *hypoglossal* nerve. Follow it forward. It passes to the dorsal side of the mylohyoid muscle (which may be cut) and innervates the muscles of the tongue.

About halfway between the descending branch and main part of the hypoglossal nerve, but deeper dorsally and passing to the dorsal side of the carotid artery, is situated the *superior laryngeal* branch of the vagus nerve. It runs obliquely caudad to the larynx, which it penetrates, passing through the fibers of the thyrohyoid muscle.

Follow the vagus and sympathetic once more. At the place where the descending branch of the hypoglossal crosses them, the two separate in the cat. Shortly anterior to this the vagus in both animals presents an elongated swelling, the *nodosal* ganglion. At about the same level, but more medial in position, the sympathetic trunk enters an elongated pinkish body, the *superior cervical ganglion* of the sympathetic. The two ganglia lie just posterior to the hypoglossal as it curves forward into the tongue.

The hypoglossal, the accessory, the vagus, and the sympathetic are all involved in a plexus in which the first cervical nerves also take part.

### c. The Ninth or Glossopharyngeal Nerve

This lies very close to the main part of the hypoglossal nerve but more deeply dorsal. Dissect directly internal to the hypoglossal, where it curves anteriorly to the tongue. The *glossopharyngeal* is a smaller nerve lying dorsal to the hypoglossal along the sides of the pharynx anterior to the larynx. It is situated between the two horns of the hyoid. It divides into two branches; a smaller *pharyngeal* branch passing medially into the pharynx and a main lingual branch that enters the tongue. The former is a motor nerve to muscles of the pharynx, and the *lingual* branch is a nerve of taste.

Follow the nerves thus far described toward the point where they emerge from the skull. They are found to converge on a point to the medial side of the tympanic bulla. Here the ninth, tenth, and eleventh nerves emerge from the brain through the jugular foramen, located on the medial side of the bulla. The twelfth nerve emerges near the others through the hypoglossal foramen (consisting of several open-ings in the rabbit).

### d. The Seventh or Facial Nerve

The main part of this nerve is very superficial in position. It emerges at the posterior end of the masseter muscle at the base of the ear, in a sort of depression. On carefully searching in this region, you will find a stout white band in contact with the main part of the external carotid artery. At this place the facial nerve gives off a branch to the posterior part of the digastric muscle, and the *posterior auricular* nerve to the pinna. (The large nerve to the pinna, which may be noticed dorsal to this branch of the facial, is the *great auricular* nerve originating in the cervical plexus.) The facial nerve then pro-ceeds forward, branching over the external surface of the masseter muscle, and passes to the lips and region of the eye. It supplies the various parts of the platysma muscle, which, it may be recalled, is a dermal muscle of the head and neck, serving to move the ears, lips, eyelids, whiskers, and the like. The platysma muscle is a branchial muscle originally belonging to the hyoid arch, hence its innervation by the facial nerve.

### e. The Fifth or Trigeminus Nerve

This nerve has three main branches, the *ophthalmic,* the *maxillary,* and the *mandibular.* The former is best studied with the eye, since it passes into the orbit.

To locate the mandibular branch of the trigeminus, proceed as follows, freeing half of the mandible. Cut through the attachment of the digastric to the mandible and deflect the digastric backward. Cut through the attachments of all of the muscles along the medial sur-face of the mandible, keeping the knife against the bone. Next, free the lateral or outer surface of the body of the mandible from muscle attachments, chiefly the masseter. Cut through the symphysis of the mandible (place of junction of the two halves of the mandible at their anterior tips). Carefully bend the free half of the mandible outward to expose the side of the muscular mass that forms the floor of the mouth and pharyngeal cavities. The main part of the mandibular branch of the trigeminus, the *inferior alveolar* nerve, will now be seen passing into the mandibular foramen, situated on the medial sur-face of the mandible. In the rabbit the *mylohyoid* nerve, another branch of the mandibular, will be noted to the medial side of the in-ferior alveolar proceeding ventrally to the muscles of the floor of the mouth cavity. The inferior alveolar nerve runs in the interior of the mandible supplying the teeth and then emerges through the mental foramen on the lateral surface of the mandible at the level of the diastema. There the nerve, now named the *mental* nerve, may be found and followed into the lower lip.

Trace the inferior alveolar nerve posteriorly. It converges toward another branch of the mandibular nerve, the *lingual* nerve, which should then be followed forward. It passes into the tongue, lying close to the hypoglossal. The lingual branch of the trigeminus innervates the mucous membrane of the tongue but is not a nerve of taste.

Follow both lingual and inferior alveolar nerves centrally again. In front of the tympanic bulla, behind the point where the body of the mandible bends dorsally into the ramus of the mandible, you will see the *auriculotemporal* branch of the mandibular nerve, joining the other two. Traced peripherally, it is found to pass to the skin of the cranial side of the pinna, and in the cat it also sends branches along the side of the face, in company with the branches of the facial nerve.

The tympanic bulla may now be exposed. Emerging from the bulla will be found a slender nerve that very soon joins the lingual branch of the mandibular. This is the *chorda tympani* (so called because it runs in the tympanic membrane), a branch of the facial nerve. Its fibers pass out with the lingual nerve and supply the taste buds on the anterior part of the tongue; it also innervates the sublingual and submaxillary salivary glands, by way of the submaxillary ganglion of the autonomic system.

Besides the branches of the mandibular nerve named here, there are branches to the muscles of mastication, namely, the temporal, the masseter, the anterior part of the digastric, and the pterygoids.

Remove the half of the mandible. This will reveal additional branches of the mandibular nerve. One of these, the *buccinator,* will probably be noticed extending to the angle of the mouth, where it supplies the masseter muscle and the lips. The main trunk of the *maxillary* nerve, the second branch of the trigeminus, may now be sought. It is a very stout trunk lying at the sides of the palate in front of, and more deeply situated than, the main trunk of the mandibular nerve. It is somewhat concealed by an artery (internal maxillary) that runs along its ventral surface and should be removed. The maxillary nerve is then revealed as a large trunk that passes forward along the side of the hard palate and disappears dorsal to the teeth. Cut away the zygomatic arch on the same side on which the half of the mandible was removed; in the rabbit cut away the ridge that holds the molar and premolar teeth also. By this operation the contents of the orbit are revealed. Note in the cat the small reddish *infraorbital* salivary gland lying close to the maxillary nerve. In the rabbit the very large reddish mass of the *Harderian* gland and the smaller yellowish mass of the *infraorbital* salivary gland anterior to it are quite noticeable. The maxillary nerve should now be investigated. It divides into a large main trunk, the *infraorbital* nerves, and a small medial branch, the *sphenopalatine* nerve, which passes into the hard palate. The infraorbital nerves pass forward above the teeth, which they supply, and emerge through the infraorbital foramen, situated internal to the root of the zygomatic arch. If the upper lip is separated from the teeth, the foramen is easily found, and the nerve can be seen emerging from it to supply the upper lip and the side of the nose. Follow the sphenopalatine nerve toward the palate, cutting away the bone. It connects with a ganglion, the *sphenopalatine* ganglion of the sympathetic system. This ganglion lies near the sphenopalatine foramen. The chief branch of the sphenopalatine nerve is the *palatine* branch, which passes into the hard palate by a foramen. In the cat this nerve arises before the ganglion is reached, but in the rabbit, beyond the ganglion. Other branches of the sphenopalatine nerve pass from the ganglion into the nasal cavity.

<table>
<tr><td>

4. The Sense
Organs of the
Head

</td><td>

*a. The Eye, the Eye Muscles, and the Nerves of the Orbit*

Dissect on the other side from that on which the cranial nerves were worked out. Identify the upper and lower eyelids and the nictitating membrane, a fold projecting from the anterior corner of the eye. Make a slit through the junction of the upper and lower lids at the posterior corner of the eye so that the eyelids can be pulled away from the eyeball. Note that the skin passes onto the inner surface of the eyelids and continues over the exposed surface of the eyeball, thus forming the outermost covering membrane, the *conjunctiva,* for this part of the eyeball. Make an incision through the skin above the eye and deflect the skin downward toward the eye on which you are working, stretching the skin away from the head. On the skin of the inner surface of the upper eyelid note a thin sheet of muscle fibers, proceeding in a somewhat circular direction. This is the *orbicularis oculi,* a part of the platysma, and has the function of closing the eyelids.

Rabbit: Stretch the upper eyelid away from the head and clean away the connective tissue between it and the eyeball. A thin sheet of muscle will be found extending from beneath the supraorbital arch to the upper eyelid. This is the *levator palpebrae superioris,* which raises the eyelid. Repeat the foregoing directions on the lower eyelid, stretching the skin away from the eyeball. On the inner surface of the lower eyelid note the rest of the orbicularis oculi. The *depressor palpebrae inferioris* extend from the surrounding skin and eyelids; cut them away from the eyeball. With the bone clippers cut away the supraorbital arch and clean away tissue between the dorsal surface of the eyeball and the orbit. A slender but strong muscle will now be seen extending from about the middle of the wall of the orbit to the dorsal surface of the eyeball; this is the *superior oblique* muscle. It separates the thin sheet of the levator palpebrae superioris into two parts, which pass on either side of it. Trace the superior oblique to the wall of the orbit. Here there will be found a tendinous cord, the *trochlea,* over which the muscle passes. Next, remove the levator palpebrae superioris and find underneath its posterior portion the thin, flat *superior rectus* muscle. The insertion of the superior oblique on the eyeball is concealed under the margin of the superior rectus.

Remove the half of the mandible and the zygomatic arch on the side on which you are working. This fully exposes the ventral side of the eyeball. Along the ventral surface of the outer part of the eyeball extends the yellowish *infraorbital* salivary gland. Medial to this, extending beneath the eyeball, is the larger *Harderian* gland, which pours its secretion onto the nictitating membrane. Remove these glands; note the white part of the Harderian gland extending far medially. The *inferior oblique* muscle is now seen extending to the

</td></tr>
</table>

eyeball from the anteroventral region of the orbit. Posterior to it is the *inferior rectus muscle,* originating from the posteroventral region of the orbit. Note the branch of the *oculomotor* nerve running along the anterior border of the inferior rectus and supplying both muscles. The nerve that runs along the posterior border of the inferior rectus and innervates the lower eyelid is the *zygomatic* branch of the maxillary nerve. Immediately behind the inferior rectus is the *external* or *lateral rectus.* The nerve passing along the posterior margin of the external rectus is the lacrimal branch of the maxillary. It passes to the lacrimal gland and to the skin between the eye and base of the pinna. The lacrimal gland, a small, reddish body, can be found by pressing the eyeball forward and searching against the posterodorsal wall of the orbit. Two nerves pass the point of origin of the external rectus from the orbit. The larger is the *oculomotor,* the smaller the *abducens.* Cut through the insertions of the inferior oblique and inferior and external recti at the eyeball and deflect them ventrally. Above the inferior rectus the *internal* or *medial rectus* will be seen inserted on the eyeball. Look on the inner surface of the external rectus and find the abducens nerve curving around the posterior border of the origin of this muscle and passing onto its surface. Return to the dorsal surface of the eyeball, cut through the insertion of the superior oblique at the eyeball, and press the eyeball ventrally. Two nerves will be seen on the medial wall of the orbit. The lower one is the *trochlear* nerve. Trace it to the medial surface of the superior oblique. The upper nerve is the *frontal* nerve, one of the main branches of the *ophthalmic* branch of the trigeminus. It passes to the dorsal part of the orbit and exits through the anterior supraorbital foramen, to be distributed to the upper eyelid and skin in front of the orbit. It may have been cut in removing the supraorbital arch. The white part of the Harderian gland will be noted in the anterior part of the orbit. Cut through all of the insertions of the eye muscles at the eyeball and through the optic nerve, removing the eyeball. The *optic nerve* is the stout white trunk near the superior rectus. The muscles around the optic nerve, exclusive of those already identified, belong to the *retractor bulbi.* Find the main trunk of the oculomotor nerve and trace its branches to the retractor bulbi and superior and internal recti. The main nerve curves below the optic nerve. The *nasociliary* branch of the ophthalmic nerve may be noted passing between the superior oblique and the retractor bulbi. Its main portion, the *ethmoidal* nerve, leaves the orbit by a small foramen in front of the superior oblique muscle. When this nerve is traced posteriorly, fine branches to the orbit may be seen.

Trace the nerves of the orbit to their exits from the skull. The third, fourth, and sixth nerves and the ophthalmic and maxillary branches of the trigeminus pass through the orbital fissure. The mandibular branch of the trigeminus passes through the foramen lacerum.

Cat: Remove the eyelids and the surrounding skin, cutting them away from the eyeball. Remove the half of the mandible and the zygomatic arch from the side on which you are working. Press the eyeball ventrally away from the supraorbital arch. In the anterodorsal angle of the orbit a strong fibrous connection will be found between the wall of the orbit and the eyeball. On investigating this, you will find that it consists of two fibrous bands that form a pulley; this is known as the *trochlea*. The tendon of the *superior oblique* muscle passes over the trochlea and is inserted on the eyeball. Its insertion is much expanded and extends caudad from the trochlea. Posterior to the insertion of the superior oblique is a thin, flat muscle, the *levator palpebrae superioris,* or elevator of the upper eyelid. This passes to the dorsoposterior surface of the eyeball. Cut through this at its insertion. Posterior to this muscle in the dorsoposterior angle of the orbit is the flattened lacrimal gland. Cut out the nictitating membrane and examine its internal surface. It is roughened because of the presence of the *Harderian* gland in its wall.

Turn to the ventral surface of the eyeball, exposed by removal of the mandible, the zygomatic arch, and part of the hard palate. Identify again the small reddish *infraorbital* salivary gland, situated back of the last tooth. With connective tissue and fat cleared away, the *inferior oblique* eye muscle will be seen extending from the anterior part of the orbit to the ventral surface of the eyeball. Ventral and at right angles to the inferior oblique is the *inferior rectus*. The branch of the oculomotor that innervates the inferior oblique runs along the posterior border of the inferior rectus. The *zygomatic* branch of the maxillary nerve passes along the posterior border of the inferior rectus to the lower eyelid but may have been destroyed. Posterior to the inferior rectus is the *external rectus;* between and internal to them appears one of the four parts of the *retractor bulbi* muscle. Along the posterior border of the external rectus runs the *lacrimal* branch of the maxillary nerve, supplying the lacrimal gland and adjacent skin. On detaching the eyeball from the posterior wall of the orbit, you will see another part of the *retractor bulbi,* next posterior to the external rectus; dorsal to this is the *superior rectus*. Cut through both obliques at their insertions and press the eyeball posteriorly. Note the *internal rectus* on the anterior surface of the eye and above it the remainder of the retractor bulbi.

Cut through all of the eye muscles and the optic nerve at their insertion on the eyeball and remove the latter. Note the four parts of the retractor bulbi around the optic nerve. Deflect the external rectus ventrally and note the *abducens nerve* ascending on its inner surface. Running along the ventral surface of the optic nerve is a slender nerve, the *long ciliary* branch of the *ophthalmic* branch of the trigeminus; it accompanies the optic nerve into the eyeball. Look on the inner surface of the inferior rectus for the branch of the *oculomotor* to this muscle. Note the *ciliary ganglion* of the autonomic system near this branch and observe the branches between this ganglion and the oculomotor and long ciliary nerve, as well as the short ciliary nerves passing from the ganglion along the optic nerve. The ciliary ganglion and ciliary nerves belong to the cephalic part of the autonomic system and carry visceral motor fibers to the smooth muscles of the iris and the ciliary muscle. Find the main trunk of the oculomotor ventral to the optic nerve at the place of passage of both through the wall of the orbit and note the branches of the oculomotor to the retractor bulbi and superior rectus muscles. Bend all eye muscles except the superior oblique ventrally, leaving the superior oblique against the medial wall of the orbit. Crossing the inner surface of the superior oblique obliquely forward are two nerves. They are parts of the ophthalmic branch of the trigeminus. The lower one is the *ethmoidal* nerve, which passes through a foramen into the nasal cavity, and the upper one is the *infratrochlear* nerve, which goes to the anterior part of the upper eyelid. Posterior and parallel to the posterior margin of the superior oblique is the *frontal* branch of the ophthalmic. It innervates the upper eyelid and integument anterior to the eyelid. The *trochlear* nerve lies slightly dorsal and medial to the proximal portions of the ethmoidal and infratrochlear nerves. It runs obliquely dorsad and anteriorly and enters the superior oblique muscle at about the middle of its posterior margin.

The structure of the eyeball may now be investigated. It is very similar to that of all vertebrates. The outer, tough *sclerotic* coat, or *sclera,* is continuous with the transparent *cornea* covering the exposed surface of the eye. As found above, the cornea is covered externally by the *conjunctiva.* Cut off the top or dorsal side of the eyeball and look within. The large *lens* will be observed. Internal to the sclera is the black *choroid* coat of the eye, and internal to that the greenish-gray *retina.* Between the lens and the retina is a large chamber, the *cavity of the vitreous humor,* containing a gelatinous mass, the *vitreous humor* or *vitreous body.* Remove the lens. The choroid coat terminates behind the cornea as a black curtain, the *iris,* bearing in its center a round hole, the *pupil.* The space between the cornea and the iris is called the *anterior chamber* of the eye, and in live animals this is filled with a fluid, the *aqueous humor.* The boundary between the iris and the rest of the choroid coat constitutes a ring known as the *ciliary body.* It consists of two parts: a ring of thickened processes, the *ciliary processes,* next to the iris, and a ring of radially arranged ridges, the *orbiculus ciliaris,* extending to the main part of the choroid coat. Both parts of the ciliary body contain the ciliary muscle; this is a smooth muscle having both meridional and circular fibers. It originates on the sclera, is inserted on the walls of the ciliary body, and has the function of changing the shape of the lens. By making a new cut parallel to the first around the equator of the eyeball, you will be able to observe the relations of cornea, iris, and ciliary body more clearly. Note the marked thickening that results from the ciliary body. Examine the lens. Note its biconvex form, as compared with the spherical form of the lens of the fish eye. Around the equator of the lens will be found the torn attachment of a membrane. This membrane holds the vitreous body. Where it is attached to the lens, it exhibits parallel ridges, the *zonular fibers,* which in living animals fit into the hollows between the ciliary processes. The zonular fibers constitute the *suspensory ligament* of the lens, which passes from the lens to the ciliary processes. By means of this ligament traction can be exerted on the lens and its shape altered to some extent. The small space between the suspensory ligament and the iris is the *posterior chamber* of the eye. Peel the lens and note that it is composed of concentric coats or *lamellae,* like coats of an onion, each lamella being composed of lens fibers.

### b. The Nasal Cavities

Detach the head of the animal at the joint between the occipital condyles and the atlas, and discard the body. Cut off the pinnae. Clear the dorsal surface of the skull down to the bone. Saw completely through the head slightly to one side of the median sagittal plane. Use the saw only for the bony parts. After having sawed through the roof of the skull, cut down through the brain with a single sliding stroke of a blunt knife. The brain and skull should thus be cut in two, one part being slightly larger than the other. Wash the cut surfaces gently under the tap, and study the nasal cavities.

The nasal cavities are very long in the rabbit and shorter in the cat. They are divided into *right* and *left* cavities or *fossae* by a perpendicular plate, the *septum* of the nose, which is present on the larger section of the head. The septum consists of cartilage anteriorly and of thin bone posteriorly, the latter being the *perpendicular plate* of the *ethmoid* bone. On the smaller section the lateral and posterior walls of the nasal fossa are seen to contain delicate scrolled and folded bones, the *turbinated bones* or *conchae.* In the rabbit these are easily separated into an anterior concha, the *inferior concha* or *maxilloturbinal,* with many folds and located on a separate small bone of the skull; a *middle concha* or *nasoturbinal,* a long single fold dorsal to the inferior and dependent from the nasal bone; and the *superior concha* or *ethmoturbinal,* part of the ethmoid bone. In the cat the turbinals are closely crowded together, but by prying them apart gently, you can distinguish a small anterior *maxilloturbinal* on the maxilla; above this a single fold, the *nasoturbinal,* dependent from the nasal bone; and a great mass of folds, the *ethmoturbinal,* filling the greater part of the nasal fossa. The ethmoturbinals are also called the *ethmoid labyrinths,* and the spaces enclosed by the bony folds are called the *ethmoid cells.* Definite passages known as the meatuses of the nose run between the conchae and conduct air to the nasopharynx. They connect with the nasopharynx below the ethmoturbinals.

The posterior dorsal part of the nasal fossa is closed by the *cribriform* plate of the ethmoid, which unites with the perpendicular plate of the ethmoid bone medially and with the parts of the ethmoid that bear the labyrinths laterally. The anterior end of the brain (olfactory bulbs) is readily seen to abut against the cribriform plate, and through this plate the fibers of the olfactory nerve pass from the olfactory membrane covering the ethmoid labyrinths to the olfactory bulbs.

### c. The Structure of the Ear

Carefully remove the brain from the two halves of the skull, preserving the latter. In doing this the roof of the skull may be cut away.

Loosen the brain on all sides by passing a blunt instrument between the brain and the skull. The tough membrane, the *dura mater,* which covers the brain should be retained with the brain. Carefully cut the cranial nerves where they pass through the foramina of the skull, leaving their roots attached to the brain. Note the small, round, reddish pituitary body attached to the ventral surface of the brain and set into a depression in the floor of the skull; keep this body attached to the brain if possible. Preserve the two halves of the brain in a vessel of water or, if they are to be kept for some time, in weak formaldehyde.

After removing the brain, examine the cavities of the skull on the larger piece. Anteriorly behind the cribriform plate is the small *olfactory fossa,* where the olfactory bulbs are situated. Posterior to this is the large *middle* or *cerebral fossa* that holds the cerebrum. Behind this is the smaller *posterior* or *cerebellar fossa* for the cerebellum. The cerebral and cerebellar fossae are partly separated by a bony ledge, the *tentorium,* which is continued in live specimens by the dura mater. In the floor of the cerebral fossa in the basisphenoid bone is the *sella turcica,* lodging the pituitary body. In front of this note also the *optic foramen* and behind this, near the ventral end of the tentorium, the foramina for the passage of the third to sixth cranial nerves. In the wall of the cerebellar fossa observe an area of hard white bone; this is the *petrous* portion of the temporal. In the center of this is a foramen for the passage of the auditory nerve into the internal ear. Above this in the rabbit is a depression, the *floccular fossa,* which lodges a part of the cerebellum called the *flocculus.* In removing the rabbit brain, leave the flocculus behind in the fossa. In front of the ventral part of the petrous bone, just behind the tentorium, is the internal opening of the *facial canal* for the passage of the facial nerve. Behind the middle of the petrous bone is the jugular foramen for the passage of the ninth, tenth, and eleventh nerves. Behind this the twelfth nerve passes through one or more foramina.

The ear of mammals consists of three parts, the *external,* the *middle,* and the *internal* ear. The external ear includes the pinna or auricle and the *external auditory meatus* leading into the interior of the bulla; these have already been noted. The middle ear is situated in the tympanic bulla, and the internal ear in the petrous portion of the temporal bone. Consequently both are in the wall of the cerebellar fossa. With the bone clippers remove this wall in one piece and discard the remainder of the skull. Clean away the muscles from its external surface, exposing the tympanic bulla.

Rabbit: With the bone clippers cut away the ventral wall of the tympanic bulla. The large cavity of the middle ear, or *tympanic cavity,* is revealed. In the lateral wall of this cavity is a ring-like elevation of bone across which is stretched the delicate *tympanic membrane,* or *eardrum.* By probing into the external auditory meatus, determine that the meatus terminates at the eardrum, which closes its internal opening. The tympanic membrane has a nearly vertical position. Extending toward the tympanic membrane from the medial wall is a short calcareous process that supports the *chorda tympani* branch of the facial nerve as it courses from the facial to the tympanic membrane. Anterior to the tympanic membrane is a depression in which the three small ear bones are lodged. These bones are so small and so deeply lodged in the depression that they cannot be seen distinctly, but it is usually possible to extract one or more of them by picking in the depression with a forceps. (Compare the malleus, incus, and stapes with illustrations in chap. 8.)

Cat: Remove the fleshy part of the external auditory meatus down to the tympanic bulla. The meatus will be found to terminate at an oval opening with a slightly elevated rim. Across the rim is stretched the delicate *tympanic membrane,* or *eardrum.* The *handle* of the *malleus,* or *hammer,* is visible through the eardrum attached at its internal surface. Next, remove the ventral wall of the bulla with the bone clippers. The interior is the *tympanic cavity of the middle ear.* Note that it is divided by a bony plate into a larger, posteroventral chamber and a smaller, anterodorsal chamber. The latter is the one covered above by the tympanic membrane. Break open the plate of bone, exposing this cavity, which is called the tympanum proper and which contains the ear bones. Note the membrane that lines it and the eardrum forming its anterodorsal wall. From the posterodorsal region of the cavity a calcareous process projects toward the eardrum and carries the *chorda tympani nerve,* a branch of the facial, to the eardrum. From the internal surface of the eardrum the three small ear bones are plainly seen extending into a depression in the internal wall of the tympanum. These may be extracted and examined. (Compare the malleus, incus, and stapes with illustrations in chap. 8.)

There now remains between the bulla and the cerebellar fossa the hard white mass of the petrous bone. This contains the internal ear. Because of the complexity and small size of the internal ear, a dissection of it is impractical, but its main parts can be seen by breaking away the petrous bone in small fragments. The tiny, spirally coiled chamber in the bone is the *cochlea;* it contains a spiral tube, the *cochlear duct,* in which the organ of sound perception (organ of Corti) is located. In the thicker, harder part of the petrous bone are the *semicircular canals,* enclosing the *semicircular ducts.*

The internal ear is thus seen to be enclosed in channels in the petrous bone, consisting of the cochlea, the semicircular canals, and the vestibule or connecting chamber; together these constitute the *bony labyrinth.* The internal ear proper or *membranous labyrinth* is contained in the bony labyrinth. Its parts are the *sacculus* and *utriculus* enclosed in the vestibule; the *semicircular ducts* arising from the utriculus and situated inside of the semicircular canals, and the *cochlear duct* arising from the sacculus and enclosed in the cochlea. The cochlear duct is a mammalian feature developed from the lagena, present at that time in fishes; the lengthening and spiral coiling of the lagena also occurs in birds.

5. The Structure of the Brain

*a. The Membranes or Meninges of the Brain*

With the two halves of the brain previously removed before you, study the membranes of the brain. The brain is covered by a tough membrane, the *dura mater*. This consists of the dura mater of lower forms fused to the internal lining (periosteum) of the skull. A considerable space, the *subdural space,* is present between the dura mater and the other membranes of the brain. The dura mater dips down between the larger divisions of the brain. The surface of the brain is covered by the delicate *pia mater,* in which the blood vessels run. The pia mater follows closely all of the folds of the brain surface. Between the pia mater and the dura mater is another membrane, the *arachnoid,* which is very delicate and difficult to see. It can be found covering the depressions on the surface of the brain, for the pia mater dips down into these depressions, whereas the arachnoid passes over them. Between the arachnoid and the pia mater is the *subarachnoid* space crossed by a delicate web of tissue. In live specimens all of the spaces between the meninges of the brain are filled with the *cerebrospinal fluid*.

### b. The Dorsal Aspect of the Brain

Remove the dura mater. Fit the two halves of the brain together and study the dorsal surface. At the anterior end of the brain (figs. 13.44 and 13.45) are the two *olfactory bulbs,* relatively small, rounded masses into whose anterior surfaces the fibers of the olfactory nerve enter. Posterior to them are the enlarged pear-shaped *cerebral hemispheres*. Their surfaces are quite convoluted in the cat, consisting of folds, the *gyri,* with grooves, the *sulci,* between the gyri. The two hemispheres are separated from each other by a deep median sagittal fissure, the *longitudinal cerebral fissure* (which is on the larger piece of the brain). Gently spread open the fissure and note at its bottom a thick, white mass connecting the two hemispheres. This is the *corpus callosum,* a structure that is characteristic of the mammalian brain but lacking in monotremes and marsupials. It is composed of nerve fibers passing between the hemispheres. At the posterior end of the longitudinal fissure is a small, reddish mass of folded tissue, which is part of the *choroid plexus* of the roof of the diencephalon. The diencephalon or region of the brain posterior to the cerebral hemispheres is completely concealed in most mammals from dorsal view by the posterior extension of the hemispheres above it. The posterior ends of the cerebral hemispheres are in contact with the *cerebellum,* a large mass with a very convoluted surface. Between the cerebellum and the cerebral hemispheres is the *midbrain,* also concealed from dorsal view by the hemispheres. It is readily revealed by bending the hemispheres and the cerebellum apart. It consists of four rounded lobes or hillocks, known as the *corpora quadrigemina* or *colliculi*. The two anterior ones are named the *anterior* colliculi, the two posterior ones the *posterior* colliculi. The cerebellum consists of a median lobe, the *vermis,* and a pair of lateral lobes, the hemispheres. From each hemisphere in the rabbit there arises by a narrow stalk another lobe, the *flocculus,* which, as already seen, is left behind in the floccular fossa of the petrous bone when the brain is removed from the skull. Identify on the hemispheres the cut surfaces where the flocculi were attached. In the cat the floccular lobes are not definitely sep-

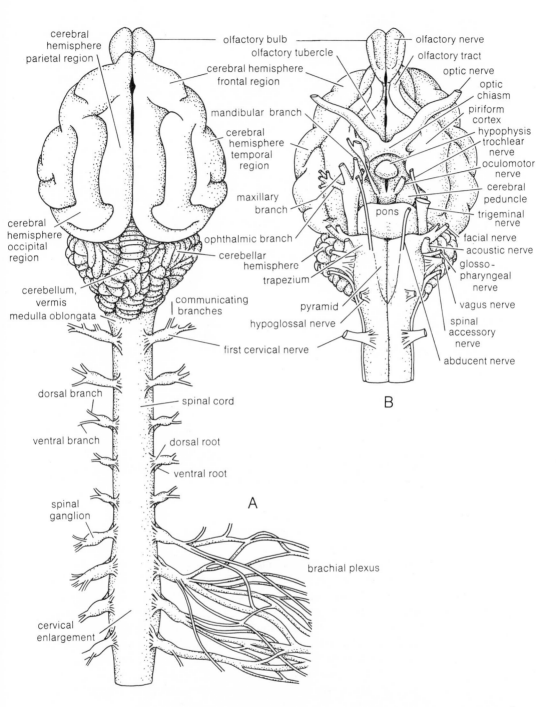

Fig. 13.44. Views of the brain and spinal cord of the domestic cat. *A*, dorsal; *B*, ventral. After Field and Taylor 1950 (*An atlas of cat anatomy*. Chicago: University of Chicago Press).

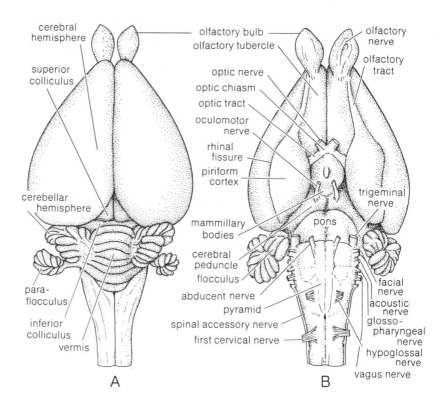

Fig. 13.45. Views of the brain of the domestic rabbit. *A*, dorsal; *B*, ventral.

arated from the main mass of the cerebellum. The floccular lobes are homologous to the auricular lobes of sharks. Posterior to the cerebellum and partly overlapped by the vermis is the *medulla oblongata;* lift the vermis and note beneath it the cavity of the *fourth ventricle* in the medulla. This is normally roofed over by a membrane, the *medullary velum,* which contains a choroid plexus, probably removed in sectioning the brain. In the cat the choroid plexus projects on each side between cerebellum and medulla as a little tuft of vascular tissue. At each side of the posterior pointed end of the fourth ventricle is a tract terminating in a club-shaped area, the *clava.* Lateral to each clava is another elongated area, the *tuberculum cuneatum.* These two belong to the *somatic sensory column.* A white bundle passes anterior to them toward the cerebellum, disappearing ventral to an elevation that lies just ventral to the hemisphere of the cerebellum. The bundle is the *restiform body* or *posterior peduncle* of the cerebellum, which conveys impulses from the medulla and spinal cord to the cerebellum. The elevation over the restiform body is the *area acoustica* or *primary auditory center.* The general resemblance of these structures to those found in the dogfish should be evident. Compare your dissection with figures 13.44 and 13.45.

### c. The Ventral Aspect of the Brain

Note the *basilar* artery (continuation of the two vertebral arteries) running in the midventral line and forming a circle around some structures in the center of the ventral surface. This circle, the *circle of Willis,* is joined on each side by the *internal carotid* artery. Note the arteries arising from the basilar and circle of Willis and distributed over the brain, coursing in the pia mater. The arteries should be removed.

At the anterior end of the ventral surface are the two olfactory bulbs (figs. 13.44 and 13.45). From each one a definite white tract, the *olfactory tract,* extends obliquely caudad and terminates posteriorly in a lobe, the *pyriform lobe,* which forms the posteroventral part of the cerebral hemispheres. The fissure, or sulcus, that separates the pyriform lobe from the rest of the cerebral hemisphere is called the *rhinal fissure.* Enclosed between the two pyriform lobes is the ventral side of the *diencephalon.* At the anterior end of this is the *optic chiasma,* from which the *optic nerves* project. The region between the optic chiasma and the olfactory tracts is called the *anterior perforated substance.* Behind the optic chiasma is a slight, rounded elevation, the *tuber cinereum,* from which the *pituitary body* or *hypophysis* hangs by a stalk. In case the pituitary body was torn off in removing the brain, a slit-like aperture will be noticed in the center of the tuber cinereum marking the place of attachment to the pituitary body. Immediately posterior to the attachment of the pituitary body is the *mammillary body,* which is not distinctly marked off from the tuber cinereum. Posterior to this is a depressed area, the *posterior perforated substance,* from which arise the two third, or *oculomotor,* nerves. From beneath (dorsal to) the pyriform lobes a thick white bundle will be seen passing obliquely backward on each side of the posterior perforated substance. These bundles are the *cerebral peduncles,* belonging to the midbrain. The fourth or *trochlear* nerves arise on the side of the brain between the cerebellum and the posterior colliculi and pass ventrally over the outer surface of the peduncles.

The remainder of the ventral surface of the brain belongs to the hind brain and consists of the *pons* and the medulla oblongata. The pons is the heavy band of fibers that crosses the ventral surface of the hind brain immediately behind the posterior perforated substance. By following it around to the sides of the brain, you will see that it narrows to a white cord, the *brachium pontis,* or *middle peduncle,* of the cerebellum, which passes into the substance of the cerebellum. Immediately posterior to the brachium pontis and partly concealing it is the thick root of the *trigeminus* nerve. On close examination this will be seen to consist of a large dorsal portion, the *sensory* root (*portio major*), which consists of the somatic sensory fibers of the trigeminus, and a very small ventral portion, the *motor* root (*portio minor*), which contains the visceral motor fibers for the muscles of mastication (masseter, temporal, digastric, etc.). Posterior to the pons is another bundle of transverse fibers, the *trapezoid* body, which is about half the width of the pons. Close inspection will show that the trapezoid body originates from the area acoustica or auditory center; it passes toward the median line but, before reaching it, turns forward and disappears dorsal to the pons. The trapezoid body is the main tract that carries the auditory impulses to the more anterior portions of the brain. Attached to the side of the area acoustica is the root of the eighth or auditory nerve. Just ventral to this and behind the root of the trigeminus, the root of the *facial* nerve emerges through the trapezoid body. In the median ventral line of the medulla is a groove, the *median ventral fissure*. Along each side of this runs a narrow bundle of fibers; each emerges dorsal to the posterior margin of the pons and proceeds straight posteriorly. These two tracts are the *pyramids* or *somatic motor tracts;* they convey impulses from the cerebral hemispheres to the voluntary muscles. At the place where the pyramids emerge from above the pons are the roots of the sixth or *abducens* nerves. The small root of the ninth or *glossopharyngeal* nerve will be found at the posterior boundary of the acoustic area, at the point where the restiform body passes dorsal to it and about on a line with the root of the eighth nerve. The equally small root of the tenth or *vagus* nerve lies immediately posterior to and on a line with the root of the ninth nerve. Posterior to the vagus are the numerous roots of the eleventh or *spinal accessory* nerve, arising in a line. The main root of the accessory ascends from the spinal cord but is probably missing in the specimen. The roots of the twelfth or *hypoglossal* nerve emerge along the lateral border of the pyramid, posterior to the preceding roots.

### d. The Median Sagittal Section

Now cut the larger half of the brain along the longitudinal cerebral fissure to obtain an exact median sagittal section (figs. 13.46 and 13.47). In making such a cut, use a knife and pass it through the brain with one sliding stroke. Examine the cut surface. The cerebral hemisphere forms a thick roof, which arches posteriorly above the diencephalon and midbrain. In the cerebral hemisphere identify the section of the *corpus callosum*. This is an obliquely placed longitudinal band of white material. Both anterior and posterior ends are enlarged, the former being named the *genu,* the latter the *splenium*. From about the middle of the corpus callosum a band of fibers, the *fornix,* curves ventrally. Between the fornix and the anterior half of the corpus callosum stretches a thin membrane, the *septum pellucidum,* consisting of two leaves. If the brain is sectioned exactly in the median sagittal plane, the section will pass between the two leaves of the septum pellucidum; but often the whole septum is left on one half; in this case a slit-like opening into a cavity, the *lateral ventricle,* will appear on the other half between the fornix and the corpus callosum. The fornix passes downward and soon turns (as the *column* of the fornix) into the interior of the brain, where it is lost to view. Immediately in front of the point where it disappears is the section of a small round bundle, the *anterior commissure*. From the anterior commissure a delicate membrane, the *lamina terminalis,* extends ventrally to the optic chiasma. The fornix, the anterior comissure, and the lamina terminalis form the anterior boundary of a deep but narrow chamber, the *third ventricle,* which lies in the middle of the diencephalon. The cavity of the third ventricle extends ventrally into the tuber cinereum and the pituitary body.

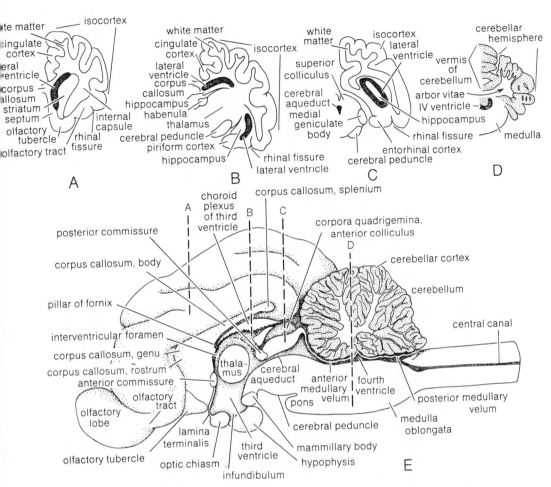

Fig. 13.46. *A–D*, diagrams of transverse sections through the left half of the brain of the domestic cat. *E*, midsagittal view of the brain of the domestic cat. After Field and Taylor 1950 (*An atlas of cat anatomy*. Chicago: University of Chicago Press).

The diencephalon is the massive region extending between the fornix and lamina terminalis and midbrain. It consists of three parts: a dorsal region, the *epithalamus;* a central and lateral region, the *thalamus;* and a ventral region, the *hypothalamus.* The hypothalamus includes the optic chiasma, the tuber cinereum, the mamillary body, and the hypophysis or pituitary body, all of which should be identified in the section. The epithalamus includes the structures in the roof of the diencephalon. These are the *tela choroidea,* a thin folded vascular membrane between the cerebral hemisphere and the diencephalon; the *pineal body,* a stalked body lying in the *tela choroidea,* the *habenula,* a small mass just in front of the attachment of the pineal body to the diencephalon; and the *posterior commissure,* a small circular area just posterior to the habenula. The *thalamus* constitutes the greater part of the *diencephalon.* On the cut surface it presents a large, round mass, the *intermediate mass* or *middle commissure;* this is not really a commissure but merely the cut median mass of the thalamus.

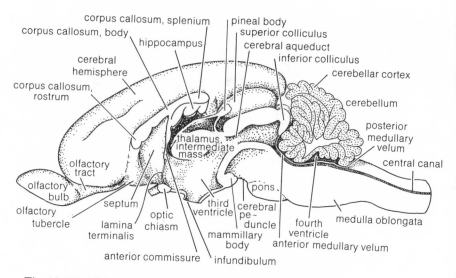

corpus callosum, splenium
corpus callosum, body
hippocampus
cerebral hemisphere
corpus callosum, rostrum

pineal body
superior colliculus
cerebral aqueduct
inferior colliculus
cerebellar cortex
cerebellum
posterior medullary velum
central canal

thalamus, intermediate mass

olfactory tract
olfactory bulb
olfactory tubercle
septum
lamina terminalis
optic chiasm
third ventricle
mammillary body
cerebral pe- duncle
pons
fourth ventricle
anterior medullary velum
medulla oblongata

anterior commissure
infundibulum

Fig. 13.47. Midsagittal view of the brain of the domestic rabbit.

The greater part of the thalamus is concealed by the overhanging cerebral hemisphere. On the smaller piece of the brain remove the cerebral hemisphere and then examine the dorsal and lateral regions of the thalamus. Three low elevations are present. The most dorsal and medial one is the *pulvinar.* Lateral to this and whiter in color is the *lateral geniculate body.* A white band, the *optic tract,* is plainly seen ascending from the optic chiasma and terminating on the lateral geniculate body. Posterior and ventral to the lateral geniculate body is a smaller swelling, the *medial geniculate body.* Behind the geniculate bodies will be recognized the corpora quadrigemina as two low hillocks. Ventral to them runs the stout cerebral peduncle, the anterior part of which is crossed externally by the optic tract.

Returning to the medial sagittal section, identify in the roof of the midbrain the two hillocks formed by the anterior and posterior colliculi or corpora quadrigemina. Below them is a narrow passage, the *aqueduct* of the brain, which connects the third ventricle in the diencephalon with the fourth ventricle in the medulla. Below the aqueduct is the thick floor of the midbrain, the *tegmentum.* At the sides of this are the cerebral peduncles, not exposed in the section. In the section of the cerebellum note the curious branching tree-like arrangement of the white matter, forming the arbor vitae or tree of life. This appearance is brought about by the fact that each fold of the cerebellar surface consists of gray matter or nerve cells, with a central plate of white matter or nerve fibers. The cerebellum fits into the fourth ventricle, from which, however, it is separated in the normal condition by a membrane, the *medullary velum.* Part of this velum will probably be found below the anterior part of the cerebellum. Identify in the section the mass formed by the pons. The section of the medulla has nothing of additional interest.

*e. Further Structure of the Cerebral Hemispheres*

On the intact half of the brain begin to cut away the roof of the cerebral hemisphere in thin slices. Note that the superficial substance of the roof is gray, the interior white—a reversal of the original condition. This gray outer coat of the mammalian cerebral hemispheres is called the cortex and is composed of a characteristic stratified arrangement of nerve cell bodies. The white matter under the cortex consists of fibers carrying impulses to and from the cortex. In the cat the cortex is much convoluted, and each convolution consists of a central core of white matter covered peripherally by a thick coat of gray matter. Continue to shave the brain ventrally until the corpus callosum is exposed as a narrow band of fibers conveying impulses between the two hemispheres. Remove the corpus callosum and the cortex lateral to it. This exposes the cavity of the cerebral hemisphere, called the lateral ventricle. It is filled by two conspicuous elevations, an anterior, smaller darker one, the *corpus striatum,* and a posterior larger one, the *hippocampus.* Remove the side of the hemisphere to expose the hippocampus. It is a curved body with an anterior free margin, the *fimbria.* Cut through the medial attachment of the hippocampus, raise the anterior border, and roll the hippocampus back. Observe that the part of the hippocampus still attached is continuous with the pyriform lobe. Turning back the hippocampus reveals the thalamus. Note the thick stalk extending from the thalamus into the cerebrum, immediately in front of the pulvinar and lateral geniculate body. Scrape the surface of this and note that it consists of a great mass of fibers radiating from the thalamus into the cerebral hemisphere. This is more evident in the rabbit than in the cat, since in the cat the fibers turn dorsally. This radiating mass is called the *corona radiata.*

---

**H. Summary**

1. The nervous system and the nervous parts of the sense organs are derived from the ectoderm.

2. The nervous system is subdivided into the central, peripheral, and autonomic nervous system. The first includes the brain and spinal cord; the second is made up of the cranial and spinal nerves; and the third is a system of ganglia, cords, and plexi controlling visceral motor functions.

3. The brain and cord are protected by membranes termed meninges, of which there is one in fishes, two in amphibians, birds, and reptiles, and three in mammals.

4. Functionally the nervous system can be divided into four components: the somatic sensory, which handles sensory impulses from the skin and other layers of the body wall; the visceral sensory, which handles sensations from the viscera; the visceral motor, which conveys motor impulses to the smooth musculature and the branchial muscles; and the somatic motor, which deals with impulses to the voluntary musculature (except the branchial muscles). These components are arranged in the spinal cord and medulla from dorsal to ventral in the order named. The autonomic system is primarily an elaboration of the visceral motor component.

5. The spinal cord consists of a central gray region and a peripheral

white region. The gray matter is of a sensory nature dorsally, receiving the somatic and visceral sensory impulses. The ventral half of the gray matter is of a motor nature, containing the cells of origin of the somatic and visceral motor impulses. The cells of origin of sensory impulses are always outside the brain and cord in ganglia.

6. The spinal nerves arise from the spinal cord at segmental intervals by way of two attachments or roots, a dorsal or sensory root and a ventral or motor root. The sensory root bears a ganglion, the dorsal or spinal ganglion, which contains the cell bodies of the sensory fibers.

7. In primitive groups (*Branchiostoma,* petromyzonts) the dorsal and ventral roots remain separate, but in gnathostome vertebrates they unite just beyond the spinal ganglion to form a spinal nerve. This soon divides into a dorsal ramus, which passes to the epaxial musculature and adjacent skin, a ventral ramus, to the hypaxial musculature and adjacent skin, and a communicating ramus, which connects with the autonomic system.

8. In the region of the appendages the ventral rami of the spinal nerves are intricately united by cross connections to form plexi from which the nerves to the muscles of the appendages arise. The chief plexi are the brachial plexus to the anterior appendages and the lumbar or lumbo-sacral plexus to the posterior appendages. These plexi indicate that the limb muscles come from the hypaxial parts of the myotomes, that several myotomes contribute to the limb musculature and that the limb muscles must have undergone torsion and change of position.

9. Fifteen different pairs of cranial nerves are known to occur in verte-brates. No single living vertebrate, however, possesses all fifteen pairs. Fishes and amphibians possess anterior and posterior lateral line nerves that are lost in amniotes. Amniotes subdivide the vagal complex of fishes into a vagus nerve proper and a spinal accessory nerve and thus are frequently said to possess one more pair of cranial nerves than fishes. In the cranial nerves, sensory and motor roots do not unite, and the cranial nerves are irregular in their functional components.

10. Most vertebrates possess a pair of terminal cranial nerves (num-bered 0 because they were  discovered after the other cranial nerves) that innervate the olfactory region. The function of these nerves is not known at present.

11. The olfactory nerve (1) extends from the olfactory epithelium in the nose to the olfactory bulbs. It is a pure sensory nerve.

12. The optic nerve (II) extends from the retina to the diencephalon. It is not a true nerve but a tract of the brain.

13. The oculomotor nerve (III) is a somatic motor nerve to the inferior oblique, superior, inferior, and internal recti and some accessory muscles of the eyeball. It originates in the midbrain.

14. The trochlear nerve (IV) is a somatic motor nerve to the superior oblique muscle of the eyeball. It arises from the midbrain.

15. The abducens nerve (VI), originating in the floor of the medulla, is a somatic motor nerve to the external rectus muscle of the eyeball.

16. The auditory nerve (VIII) is a somatic sensory nerve with its cells of origin in a ganglion or ganglia in the internal ear. It supplies the cristae and maculae of the internal ear and carries equilibratory and auditory impulses into the acousticolateral area of the medulla.

17. Trigeminal (V), facial (VII), glossopharyngeal (IX), and vagus (X) nerves are known as the branchial nerves, since they are definitely related to the branchial bars and gill slits. Typically, each one forks around a gill slit, sending a pretrematic branch in front of the slit and a posttrematic branch behind; there is, further, a visceral sensory pharyngeal branch to the pharyngeal lining. The posttrematic branch carries visceral motor fibers to the branchial muscles belonging to that particular branchial bar.

18. The trigeminus nerve (V) is the chief somatic sensory nerve of the head and the nerve of the first branchial (mandibular) arch. In all vertebrates it has three branches: the ophthalmic branch, to the orbit and nasal region (composed in fishes of two parts, the superficial and deep ophthalmic); the maxillary, to the upper jaw and roof of the oral and pharyngeal cavities; and the mandibular, to the lower jaw and floor of these cavities. The first two are pure somatic sensory nerves; the mandibular also carries visceral motor fibers to the branchial musculature of the mandibular arch. The trigeminus is attached to the side of the medulla, and its sensory part has a ganglion.

19. The facial nerve (VII) is the nerve of the second or hyoid arch. It possesses visceral sensory branches to taste buds and the pharyngeal lining, and visceral motor fibers to the musculature of the hyoid arch. The facial nerve arises from the side of the medulla, and its root is provided with a sensory ganglion.

20. The glossopharyngeal nerve (IX) is attached to the medulla and has a sensory ganglion; it is the nerve of the third branchial bar and, like the other branchial nerves, has visceral sensory fibers to taste buds and the pharyngeal lining and visceral motor fibers to the muscles of its particular bar.

21. The vagus nerve (X) is the nerve of the remaining branchial bars. It appears to have appropriated the visceral components of several originally distinct spinal nerves posterior to it. With the shift from a branchial mode of breathing, the vagus is somewhat reduced but continues to supply the corresponding region of the pharynx with visceral sensory fibers and to be the visceral motor nerve of the pharyngeal and laryngeal musculature belonging to the appropriate branchial bars. In addition, the vagus extensively innervates the heart, lungs, stomach, and other viscera with preganglionic visceral motor fibers belonging to the autonomic system. The vagus is attached to the medulla and has two or more ganglia.

22. The spinal accessory nerve (XI) of amniotes is compounded of vagal and spinal nerves; it arises from the medulla and upper spinal cord and is a visceral motor nerve to muscles of branchial origin.

23. The hypoglossal nerve (XII) is composed of a variable number of nerves of spinal origin that become included in the cranial cavity. They are somatic motor nerves to hypobranchial musculature.

24. Embryological studies suggest that the vertebrate head was originally composed of a number of segments each with a sensory and a motor nerve. In general there are three segments anterior to the otic capsule and a variable number behind it. The eye muscles arise from the three prootic segments, the hypobranchial musculature from metaotic segments; two or three of the latter nearest the otic capsule degenerate in most vertebrates without developing any musculature. Of the three prootic segments, the profundus branch of the trigeminus, the trigeminus proper, and the facial are the sensory nerves; the oculomotor, trochlear, and abducens, the motor nerves. The glossopharyngeal and vagus are sensory nerves of metaotic segments, but the somatic motor parts are reduced or lost through the degeneration of the myotomes in these segments. Each head segment except the first also possesses a branchiomeric muscle (lateral plate or visceral muscle) innervated by the branchial nerves.

25. The chief sense organs of the head are the nose, eyes, ears, and lateral line organs.

26. The sensory part of the nose consists primitively of a pair of olfactory sacs invaginated from the surface epithelium. These at first are blind sacs, but in land vertebrates they establish communication with the oral cavity for respiratory purposes. Thereafter the nasal cavities have both respiratory and olfactory functions; the latter is limited to certain areas of the wall of the cavities. The nose itself is formed by the fusion of certain projections around the mouth and is supported by bones and cartilages of the skull.

27. The eyes are compound structures. The nervous part of the eye is formed by an evagination from the brain. The lens of the eye is an invagination from the adjacent ectoderm. The coats of the eye, sclera and choroid, are formed in the surrounding mesenchyme. The eye is moved by muscles that are quite constant in arrangement in the different vertebrate classes.

28. The ear consists of internal, middle, and external portions. The internal ear is an invagination from the ectoderm. It differentiates into the three semicircular canals, the sacculus, the utriculus, and the endolymphatic duct. Fishes and many urodeles possess only an internal ear. The internal ear of cyclostomes possesses only one or two semicircular canals. The internal ear of mammals is more complicated than that of other vertebrates because of the development of a spiral outgrowth, the cochlear duct, from the sacculus. The cochlear duct functions in hearing, and the semicircular canals and the utriculus are concerned with equilibration.

29. Amphibians possess both a middle ear and an internal ear. The middle ear consists of a chamber developed by outgrowth from the first gill pouch. The outer wall of this chamber comes in contact with the skin, producing a double-walled membrane, the tympanic membrane or eardrum. Within the middle ear is a chain of small bones, derived from the gill arches but differing in nature in different tetrapods.

30. In reptiles the tympanic membrane sinks into the skull, leaving a

passage, the external auditory meatus, extending from the tympanic membrane to the exterior. This passage is deepened in birds and mammals, and in the latter a fold of skin, the pinna, develops around the external rim of the meatus. Pinna and meatus constitute the external ear.

31. The lateralis system, which is limited to aquatic vertebrates, is a system of skin sense organs arranged in lines over the head and along the side of the trunk. The sense organs consist of neuromasts and serve to detect vibrations in the water. The neuromasts may occur naked on the surface but are usually embedded in canals; they are innervated by a pair of anterior and posterior lateral line nerves (closely but secondarily associated with the facial and vagal nerves) that disappear in land vertebrates.

32. The autonomic system consists of ganglia in the head region; a pair of ganglionated cords in the cervical, thoracic, and lumbar regions; sacral nerves; and a complicated set of ganglia and ganglionated plexi among and in the viscera. The visceral sensory fibers have relations similar to those of the somatic sensory fibers; their cells of origin are in the spinal ganglia. But the visceral motor fibers of the autonomic system are peculiar in that a chain of two neurons is involved in the innervation of the smooth musculature. The first neuron in the brain or spinal cord sends a preganglionic fiber to a collateral or peripheral ganglion, and from this the second or postganglionic fiber proceeds to the musculature or to the glands.

33. The vertebrate brain consists of five main parts, named rostral to caudal: telencephalon, diencephalon, mesencephalon or midbrain, metencephalon, and myelencephalon or medulla oblongata.

34. The brain is hollow, containing cavities termed ventricles. The first two or lateral ventricles are located in the telencephalic hemispheres; the third ventricle is in the diencephalon; the fourth in the medulla. They are all connected with each other. In many vertebrates the olfactory bulbs, optic tectum, and cerebellum are also hollow. The roofs of the ventricles are thinned to a vascular membrane termed the tela choroidea, from which vascular tufts or choroid plexi project into the ventricles, producing cerebrospinal fluid.

35. The medulla oblongata undergoes little evolutionary change in vertebrates. It contains most of the nuclei of origin and the terminations of the fifth to twelfth cranial nerves. The medulla possesses the same functional columns as the spinal cord and forms important circuits underlying neural control of head musculature and glands. The medulla is also the site of termination of many ascending and descending pathways and thus is a major integrating and modulating region of the brain.

36. The metencephalon forms the cerebellum dorsally and the tegmentum ventrally. The cerebellum is divided into a corpus and flocculonodular regions in all vertebrates. The corpus is concerned with coordination of muscular movements and maintenance of muscle tone. The flocculonodular regions are concerned with equilibrium of the body in space. The lateral regions of the corpus in mammals and birds expand and are concerned with coordination of the distal muscles of the limbs.

37. The mesencephalon or midbrain consists of a roof or optic tectum (anterior colliculus in mammals) and torus semicircularis (posterior colliculus in mammals) and a floor or mesencephalic tegmentum. The optic tectum is a major visual center mediating pattern vision and visual orientation in space. The torus is concerned with similar auditory functions. The tegmentum contains the motor nuclei of the oculomotor, trochlear, and abducens cranial nerves.

38. The diencephalon differentiates into the hypothalamus, the thalamus, and the epithalamus. The hypothalamus includes the optic chiasma, infundibulum, tuber cinereum, and mammillary bodies and is particularly well developed in fishes. The ventral part of the infundibulum unites with Rathke's pouch, growing in from the roof of the oral cavity, to form the hypophysis or pituitary. In fishes the infundibulum also includes the inferior lobes and vascular sac. The thalamus forms the central region of the diencephalon and primarily consists of a number of nuclei (neuronal populations) that complete sensory circuits to most areas of the telencephalon. Particularly prominent are the lateral geniculate nucleus (vision), the medial geniculate nucleus (audition), and the ventral posterior nuclei (somatic sensory). These thalamic nuclei project upon telencephalic isocortex and are particularly well developed in amniotic vertebrates. The epithalamus or dorsal part of the diencephalon includes the epiphyseal apparatus and the habenula. The former originally comprised an anterior parapineal and a posterior pineal outgrowth, both of which originally bore an eye; but only the pineal body or epiphysis (without an eye) persists in most vertebrates.

39. The telencephalon differentiates into olfactory bulbs and cerebral hemispheres. The olfactory telencephalon is composed of the olfactory bulbs, olfactory tracts and the olfactory cortex (termed piriform cortex in mammals). All vertebrates also possess extensive nonolfactory sensory areas located in the telencephalon. The dorsal roof or dorsal pallium is concerned with vision; the lateral wall (striatum) and, in amniotic vertebrates, the dorsal ventricular ridge or its homologue, isocortex in mammals, is concerned with audition, somesthesis, and vision. In all vertebrates the telencephalon also modulates the activity of lower brain centers by extensive descending "motor" pathways.

**References**

Boudreau, J. C., and Tsuchitani, C. 1973. *Sensory neurophysiology.* New York: Van Nostrand Reinhold.

Brodal, A. 1969. *Neurological anatomy.* London: Oxford University Press.

Cahn, P. H. 1967. *Lateral line detectors.* Bloomington: Indiana University Press.

Eakin, R. M. 1973. *The third eye.* Berkeley: University of California Press.

Heimer, L. 1971. Pathways in the brain. *Sci. Amer.* 225 (July): 48–60.

Herrick, C. J. 1948. *The brain of the tiger salamander.* Chicago: University of Chicago Press.

Hubbard, J. I. 1974. *The peripheral nervous system.* New York: Plenum Press.

Jacobson, M. 1970. *Developmental neurobiology*. New York: Holt, Rinehart and Winston.

Jerison, H. J. 1973. *Evolution of the brain and intelligence*. New York: Academic Press.

Johnson, J. B. 1906. *The nervous system of vertebrates*. Philadelphia: P. Blakiston's Son.

Kappers, C. U. A.; Huber, G. C.; and Crosby, E. 1936. *The comparative anatomy of the nervous system of vertebrates, including man*. 2 vols. New York: Macmillan.

Kuhlenbeck, H. 1967–. *The central nervous system of vertebrates*. 4 vols. to date. New York: Academic Press; Basel: S. Karger.

Ochs, S. 1965. Elements of neurophysiology. New York: Wiley.

Papez, J. W. 1929. *Comparative neurology*. New York: Thomas Y. Crowell.

Petras, J. M., and Noback, C. R. 1969. Comparative and evolutionary aspects of the vertebrate central nervous system. *Ann. N.Y. Acad. Sci.* 167:1–513.

Prince, J. H. 1956. *Comparative anatomy of the eye*. Springfield: Charles C Thomas.

Sarnat, H. B., and Netsky, M. G. 1974. *Evolution of the nervous system*. London: Oxford University Press.

Shepherd, G. M. 1974. *The synaptic organization of the brain: An introduction*. London: Oxford University Press.

Truex, R. C., and Carpenter, M. B. 1969. *Human neuroanatomy*. Baltimore: Williams and Wilkins.

Van Bergeijk, W. A. 1967. The evolution of vertebrate hearing. In *Contributions to sensory physiology* ed. W. D. Neff. 2:1–40. New York: Academic Press.

# List of Contributors

Herbert R. Barghusen
Departments of Anatomy and Oral Anatomy
University of Illinois School of Medicine/Dentistry

E. J. W. Barrington
Department of Zoology
University of Nottingham

James A. Hopson
Department of Anatomy
University of Chicago

Richard J. Krejsa
Department of Biological Sciences
California Polytechnic State University, San Luis Obispo

Ronald Lawson
Department of Biology
University of Salford

R. Glenn Northcutt
Division of Biological Sciences
University of Michigan

Dennis R. Paulson
Burke Museum
University of Washington

Leonard Radinsky
Department of Anatomy
University of Chicago

David B. Wake
Museum of Vertebrate Zoology
University of California, Berkeley

Marvalee H. Wake
Departments of Zoology and Biology
University of California, Berkeley

George Zug
Division of Reptiles and Amphibians
National Museum of Natural History

# Index